ANNUAL REVIEW OF MATERIALS SCIENCE

ANNUAL REVIEW OF MATERIALS SCIENCE

KEYNOTE TOPIC: ELECTRONIC MATERIALS

VOLUME 25, 1995

BRUCE W. WESSELS, *Guest Editor*
Northwestern University

ELTON N. KAUFMANN, *Editor*
Argonne National Laboratory

JOSEPH A. GIORDMAINE, *Associate Editor*
NEC Research Institute

JOHN B. WACHTMAN, JR., *Associate Editor*
Rutgers University

ANNUAL REVIEWS INC 4139 EL CAMINO WAY P.O. BOX 10139 PALO ALTO, CALIFORNIA 94303-0139

ANNUAL REVIEWS INC.
Palo Alto, California, USA

International Standard Serial Number: 0084-6600
International Standard Book Number: 0-8243-1725-4
Library of Congress Catalog Card Number: 75-172108

The paper used in this publication meets the minimum requirements of American National Standard for Information Sciences—Permanence of Paper for Printed Library Materials, ANSI Z39.48-1984.

TYPESET BY BPC DIGITAL DATA LTD., SCOTLAND
PRINTED AND BOUND IN THE UNITED STATES OF AMERICA

PREFACE

This is the 25th volume of the *Annual Review of Materials Science*, one of the 26 series published by Annual Reviews, Inc., a not-for-profit scientific publisher dedicated to promoting the advancement of the sciences.

This series publishes significant, timely, and critical appraisals of research across the materials science and engineering fields. The *Review* emphasizes areas where substantial recent progress is seen and where new subfields may be anticipated. Materials science encompasses a broad spectrum of contributing fields and disciplines. This series seeks to provide an integrated picture of materials research by reviewing a wide variety of scientifically interesting and technologically important materials studies that often rely on the applications of common techniques and interpretive tools.

The importance of materials and their technology to the health and well-being of people and nations is now well recognized. Materials research runs the gamut from fundamental studies, bordering on condensed matter physics, to applied studies directed toward improving and exploiting specific properties for the marketplace. The broad spectrum of topics in the *Annual Review of Materials Science* reveals the ubiquity of advanced materials in technological innovation.

The current volume is the third in which a group of chapters is focused on a functional class of materials, in this case on electronic materials. This focused set views materials from a single perspective of current and potential application and also emphasizes the interrelationships between the keynote topic chapters and the more general topics treated in the volume. The Editorial Committee intends such an approach to serve not only the specialist in the keynote topic, but also the educator who endeavors to provide a broad perspective for the study of materials science. A context-setting introduction to the electronic materials chapters precedes the prefatory chapter.

We express our sincere appreciation to Bruce Wessels who, as guest editor, organized the electronic materials keynote chapters of this volume. We also thank the authors of all chapters appearing in this volume for contributing their expertise and diligence. We are pleased to give particular mention to the prefatory chapter by Arthur Freeman, which speaks eloquently to the impressive advances of recent years in computational methods for the solid state. With co-author Erich Wimmer, a detailed exposition is presented in a later chapter.

In these times of intense scrutiny of the costs and benefits of research, promised long-term genetic fruits of research are less compelling than immediate and tangible technological advances. Thus the diversity of materials science is both a blessing and a curse. On the one hand, great effort is required to demonstrate that a continuity and focus tie together the disciplinary and topical extremes of our field. On the other hand, this demonstration, to which we hope our series contributes, confirms the special bridging role played by materials science between science and technology.

A volume such as this requires a great deal of support in the production process as well, and as always we thank Sandra Cooperman, the series production editor, for her help in assembling and producing this volume.

ELTON N. KAUFMANN
FOR THE EDITORS AND THE EDITORIAL COMMITTEE

Annual Review of Materials Science
Volume 25, 1995

CONTENTS

▷ Keynote Topic: Electronic Materials

SOME RELATED ARTICLES IN OTHER *ANNUAL REVIEWS*

From the *Annual Review of Biophysics and Biomolecular Structure*, Volume 24 (1995):

Mass Spectrometry of Nucleic Acids: The Promise of Matrix-Assisted Laser Desorption-Ionization (MALDI) Mass Spectrometry, Michael C. Fitzgerald and Lloyd M. Smith

NMR Spectroscopic Studies of Paramagnetic Proteins: Iron-Sulfur Proteins, Hong Cheng and John L. Markley

Membrane Structure Studies Using X-Ray Standing Waves, Martin Caffrey and Jin Wang

From the *Annual Review of Physical Chemistry*, Volume 46 (1995):

Energy Transfer in Condensed Phases, Raoul Kopelman and Joseph Klafter

Imaging Techniques for the Study of Chemical Reaction Dynamics, Albert J. R. Heck and David W. Chandler

Nucleation: Measurements, Theory, and Atmospheric Application, David W. Oxtoby

Heteronuclear NMR Pulse Sequences Applied to Biomolecules, Jeffrey G. Felton and David Wemmer

Algebraic Methods in Spectroscopy, Michael E. Kellman

Density of Functional Theory, R. G. Parr and W. Yang

Effective Hamiltonians Theory and Molecular Dynamics, Georges Jolicard

Theory and Application of Photoionization Processes, Vincent McKoy and Kwanghsi Wang

Wave Dynamics, Bret E. Jackson

Coherent Molecular Processes in Solids, Keith A. Nelson

High-Pressure Structural Transformations in Semiconductor Nanocrystals, Sarah H. Tolbert and A. P. Alivisatos

4-Wave Mixing Spectroscopy, D. J. Rakestraw

Dynamics and Photochemistry of Neutral van der Waals Clusters, Elliot R. Bernstein

External Infrared Reflection Absorption Spectrometry of Monolayer Films at the Air-Water Interface, Richard Mendelsohn, Joseph W. Brauner, and Arne Gericke

Molecular Dynamics of Polymer Systems: Theory and Experiment, Julian Clarke

Potential Energy Surfaces for Chemical Reactions at Solid Surfaces, Barbara J. Garrison

ANNUAL REVIEWS INC. is a nonprofit scientific publisher established to promote the advancement of the sciences. Beginning in 1932 with the *Annual Review of Biochemistry*, the Company has pursued as its principal function the publication of high-quality, reasonably priced *Annual Review* volumes. The volumes are organized by Editors and Editorial Committees who invite qualified authors to contribute critical articles reviewing significant developments within each major discipline. The Editor-in-Chief invites those interested in serving as future Editorial Committee members to communicate directly with him. Annual Reviews Inc. is administered by a Board of Directors, whose members serve without compensation.

ELECTRONIC MATERIALS

Materials have been the cornerstone of the modern information age marked by the invention and development of the integrated circuit in the early 1960s. From then on, the push has been to make circuits that are faster, denser, and cheaper. They have indeed become ever more complex, with circuits having more than a billion elements now being planned and designed. To realize these goals, innovations in new materials and processes have to be developed. Although silicon remains the workhorse of integrated circuit technology, many other materials play a central role as dielectrics and interconnects. With increasing circuit complexity, advanced interconnect technology based on optoelectronics may be required, and optically active materials that are compatible with silicon technology are needed.

Fundamental to continued future development is processing. Of particular importance are thin film deposition techniques. The integration of a wide range of dissimilar materials is essential and may be accomplished through techniques such as molecular beam epitaxy and metal-organic chemical vapor deposition. These methods are central to the development of device quality materials for nano-scale electronic and optoelectronic devices.

The goal of this volume is to give a picture of some of the important advances in electronic materials and processing, particularly in areas that have seen rapid and significant progress over the last several years.

I wish to thank *Annual Reviews of Materials Science* editor Elton Kaufmann for the opportunity to serve as special editor of this volume. I would also like to thank Sandra Cooperman and the Annual Reviews editorial staff. A special thanks goes to the authors for taking the time to write the reviews as a service to our community.

Bruce M. Wessels
Guest Editor

Arthur J. Freeman

Annu. Rev. Mater. Sci. 1995. 25: 1–5

ELECTRONIC STRUCTURE THEORY IN THE NEW AGE OF COMPUTATIONAL MATERIALS SCIENCE

Arthur J. Freeman

Department of Physics and Astronomy, Northwestern University, Evanston, Illinois 60208–3112

KEY WORDS: electronic structure, density functional theory

Materials science is now in a golden age—as is clear, for example, from the breadth and depth of the chapters in this volume. Because it is contributing so much to this age, it is timely to describe a recently developed new branch of materials science, now recognized as Computational Materials Science. This development has followed rapidly from the application of recent advances in computational physics and from an enormous increase in the power (speed and memory) of computers and their easy availability. In the dawn of this new age, it is now recognized that Computational Materials Science has reached a new stage of development—the growing ability to design materials by computer. In all fields of computational materials research, a high level of reliability concerning predictions of physical and chemical properties and phenomena has been achieved—thanks in large part to the continued rapid development and availability of high-performance computers and their easy accessibility (via networks and workstations) and to the new computational methods/algorithms these computers make possible. A prime area of interest continues to be electronic structure theory, where new theoretical/computational methods—based on the density functional theory of Hohenberg, Kohn, and Sham—have been developed (by us and others) that permit highly precise solutions of complex many-body systems at the level of the local density approximation.

The present state has been slow in coming and has taken many years.

0084–6600/95/0801–0001$05.00

The way was marked by pitfalls and frustrations in (*a*) attempting to satisfy the ever-increasing demands of experimentalists seeking to understand results obtained with ever more sophisticated equipment, and (*b*) the struggle with performing increasingly complex calculations without having adequate computer power. Looking back, these decades were also marked by triumphs and the growing acceptance of the value and importance of these complex numerical investigations. To illustrate this on a personal note, when I was a graduate student at the Massachusetts Institute of Technology 40 years ago, computational physics had not even been named, nor was it recognised, as it is today, as a separate and distinct branch of physics. Pioneering work in computational electronic structure on a computer was being carried out for the first time in Prof. John C. Slater's newly formed Solid State and Molecular Theory Group. Slater himself was even then a long time leading proponent of the computational physics approach to studying electronic structure. The computer we used, called Whirlwind I, a really impressive name for a supercomputer, was one of the first supercomputers of its time. With its five thousand vacuum tubes, it occupied many large rooms in the building. Paper punch tapes read in our program and data through a photoelectric reader. We were allowed to run on the graveyard shift, i.e. from midnight to 8:00 a.m. because the other two shifts were devoted to Air Force–sponsored research carried out mostly at the Lincoln Laboratories. Still, we considered ourselves lucky in being able to harness the power of this machine, even though for my own work it could only do Simpson's rule integrations of atomic radial integrals. All the rest of the computations, including diagonalization of matrices, still had to be done by hand. Thus the determination of the electronic structure of such a simple molecule as OH by the then newly emerging Hartree-Fock configuration interaction method "only" took one and one-half years, thanks to the help of two "computresses," who helped by doing independent numerical calculations!

As the power of computers developed, so did our methods and thus the ability to carry out increasingly complex calculations. Thus in the 1960s, we were able to perform Hartree-Fock calculations on atoms and to use these results to understand a wide range of phenomena as diverse as X-ray atomic scattering factors and neutron form factors, crystal field splittings, and hyperfine interactions. The latter was particularly exciting because with our results and the rules we developed, we were able to explain a variety of experimental data obtained by NMR and the newly discovered Mössbauer effect, for which Rudolph Mössbauer was awarded a Nobel Prize soon thereafter. Pursuing a new direction in studying solids, we were able to calculate theoretically, by means of Slater's augmented plane wave (APW) method in the so called muffin-tin approximation, the non-self-consistent energy band structure of a variety of heavy metals. These included, for the first time—and much to Slater's surprise—the calculation of the energy band structure and properties of the rare-earth metals. Significantly, the results provided a completely new and very different understanding of their electronic, magnetic, and optical properties.

The increase of computing power that became available with IBM and especially the CDC computers (the precursors to the Cray machines) in the 1970s, provided the means for my post-docs and students to use expanded methodology and hence to solve more complex problems, including the fully relativistic APW energy band structure of the actinide metals and their intermetallics; the determination of the electronic wave functions, which permitted us for the first time to understand charge and spin density phenomena such as X-ray and neutron form factors; Fermi surfaces and Fermi surface driven phenomena; generalized wave vector-dependent susceptibilities (for understanding phonon anomalies); the development of a numerical (LCAO)-linear combination of augmented orbitals based method for the self-consistent solutions of the Kohn-Sham equations for bulk solids and thin films; and the development and application of the linearized APW (i.e. LAPW) method to obtain more precise solutions of electronic structure and properties, notably magnetism, of surfaces and overlayers.

The advent of Cray supercomputers and the easy access to them in the 1980s led to an explosion of activity and applications to complex problems of real materials that dramatically and decisively changed the nature and impact of computational electronic structure theory. Thus, for example, freed from the limitations of insufficient speed and memory, we developed the total energy full potential LAPW method (called FLAPW) for treating complex solids, surfaces, and interfaces. This approach (described in our chapter Density Functional Theory as a Major Tool in Computational Materials Science) went beyond the limitations of earlier approaches in solving the Kohn-Sham equations without shape approximations to the wave functions, charge density, and potential. With the power of the Cray, we could determine the potential from the charge density by solving a three-dimensional Poisson equation whereas previously, in order to simplify the computations, the one-dimensional Poisson equation resulting from the muffin-tin approximation to the potential was used.

In the following contribution to this volume, Erich Wimmer and I discuss some aspects of the recent state of density functional theory. It will help make clear that electronic structure theory has entered a new phase of predictability and reliability—due in large part to the impact of the continuing introduction and ready availability of ever more powerful supercomputers and high-performance workstations. In addition to their providing the raw computing power, these computers have had the important synergistic effect of permitting sophisticated, powerful new mathematical/computational algorithms and methods to be implemented with which to carry out complex quantum mechanical solutions on what are, after all, complex many-body problems.

Today, computational science is at a threshold: Advanced theoretical/computational techniques (already in hand and to be developed), combined with existing and new computational power, will provide an understanding of matter at the atomic scale with an unprecedented level of detail and accuracy. With the computational materials science approach we will be able to not only simulate experiment but also ultimately to design new molecules and materials and to predict their properties before actually synthesizing them. In contrast to an analytical/theoretical approach, which isolates and idealizes real systems to unravel fundamental relations and laws, the computational approach is synthetic: Its goal is to simulate more and more complex systems including as much of the environment as possible.

Powerful, first principles computational approaches now allow theorists to study real, rather than model, materials and to give precise answers and interpretations to the experimental data. Very often, they have even managed to go farther by proposing new structures and new materials, to realize and provide data, on the atomic scale, that are often impossible or hard to obtain experimentally. The approach followed is pragmatic: It simulates real systems in detail including as much as possible of the environment and without idealizing or simplifying any fundamental physical phenomenon. Of course, all this has been possible thanks to the great advances made in computer technology and fast algorithms that allow the numerical solution of complex equations in a short time.

Many of the theoretical efforts are focused on the atomic scale because the theoretical concepts and methodologies being developed are bringing us closer to the goal of a unified approach. The theoretical approach, i.e. density functional theory, that underlies most of electronic structure theory is appealing because it promises to provide such a unified approach, which is particularly suited for bridging the gap between our understanding of atomic scale properties and the macroscopic properties essential for producing useful materials applications.

The kinds of answers that this approach can give are broad and include quantities that are experimentally accessible (such as stable crystal structures, bulk moduli, phonon frequencies, densities of states, and energy level transitions) as well as some that are not (such as, for example, in-layer bond-length relaxations) but that are still important in affecting the overall properties of a given material. Of course, when direct comparisons are available, they provide powerful methods to test theoretical calculations and eventually to improve them.

Designing new materials with specific mechanical, thermal, chemical, electronic, magnetic, and optical properties hinges on one basic assumption: the properties of the macroscopic ensemble are related to and can be derived from the properties of individual molecules and atomic building blocks (such as crystallographic unit cells in solids). But which atomic-scale physical quantities and observables need to be looked at? This is the important question today and the key to successful materials-by-design in the future. The answer will come, in any given situation, from the close interaction between experimentalists who know and understand the material and its properties and the computational theorists. For this collaboration to succeed, it is therefore just as essential that experimentalists understand the basics of the tools used by the theorists, and especially their limitations, as it has traditionally been the case of the theorists understanding in detail the experiment. In an effort to bridge the gap, we focus in the next contribution on the present state of computational electronic structure theory and describe the what, how, and why of recent progress in theoretical/computational methods, notably local density functional theory, for determining the electronic structure and properties of materials. Its purpose is to provide insight and understanding of what has become such a powerful tool in computational materials science today.

Finally, it is a pleasure to acknowledge and thank my former (and present) post-docs, students and other collaborators, both here and abroad, for the pleasure of doing computational physics together and for the fun of discovery that we have shared over the years. They have been, in many instances, friends as well as colleagues. While I cannot thank each of them by name here, it is fortunate that their work shines through the literature of what has been, and continues to be, an exciting journey in computational physics. Over these years, the work was made possible through support from the National Science Foundation, the Air Force Office of Scientific Research, the Office of Naval Research, and the Department of Energy.

Annu. Rev. Mater. Sci. 1995. 25: 7–36

DENSITY FUNCTIONAL THEORY AS A MAJOR TOOL IN COMPUTATIONAL MATERIALS SCIENCE

Arthur J. Freeman

Department of Physics and Astronomy, Northwestern University, Evanston, Illinois 60208–3112

Erich Wimmer

European Centre for Computational Science and Technology, Biosym Technologies, 91893 Orsay, France

KEY WORDS: electronic structure, Kohn-Sham equations, atomistic simulations

---

INTRODUCTION

It is now widely recognized that atomistic simulations—especially those employing density functional theory—are becoming an increasingly important part of materials science today. However, a great variety of approaches and implementations have been developed that make the situation confusing not only to the non-expert, but perhaps even to the specialist. For this reason, we describe here the local density formulation of density functional theory in enough detail to provide some understanding of this now fairly well-developed, important, and most successful approach for studying electronic structure phenomena in materials.

It should be borne in mind that the fundamental atomistic principles underlying the structural and functional behavior of materials are astonishingly simple: (*a*) For most purposes, atomic nuclei can be treated as classical particles with a given mass and positive charge; (*b*) electrons are particles of spin one half, thus obeying the Pauli exclusion principle; their kinetic behavior is described by quantum mechanics; and (*c*) the only relevant interactions are of an electrodynamic nature, in particular, attrac-

0084–6600/95/0801–0007$05.00

tions and repulsions governed by Coulomb's law. Based on these fundamental principles it is conceptually possible to explain and predict the wonderful richness of most physical and all chemical properties of matter such as the structure and stability of crystalline phases, the mechanical properties of alloys, the optoelectronic properties of semiconductors, the magnetic properties of transition metals, etc.

Further, it needs to be understood that the density functional approach is quite general; it is applicable to all atoms of the periodic table, provided relativistic effects are taken into account for heavier elements such as third-row transition metals, rare-earths, and actinides. The range and variety of materials under investigation is enormous. The approach can be used for metallic, covalent, and ionic bonds. The kinds of answers that this approach can give are broad and include quantities that are experimentally accessible (such as densities of states, magnetic moments, and energy level transitions), as well as some that are not (such as, for example, in-layer bond-length relaxations), and yet are still important in affecting the overall properties of a given material. Of course, when direct comparisons are available, they provide powerful methods to test theoretical calculations and eventually to improve them. Its greatest strength lies in calculations for metallic condensed systems, yet its range also includes organic molecules.

With the inclusion of gradient corrections for the exchange-correlation term, even weaker interactions such as hydrogen bonds can be reasonably well described, although they are significantly overestimated in the local density approximation (LDA). Furthermore, so-called difficult molecules such as ozone seem to be treated by density functional methods with the same level of accuracy as other molecules. Within molecular applications, the approach is particularly useful for organometallic systems as well as large molecules (of the order of 100 atoms), which have remained out of reach for traditional ab initio quantum chemistry methods. Local density functional calculations do encounter problems for narrow-gap insulators and certain oxides. The LDA tends to overemphasize the metallic character, and one needs to be careful in the interpretation of the density functional one-electron energies.

Finally, our aim is to provide a general overview of the complexity of the different issues involved and at the same time to illustrate the kind of answers that the theoretical-computational approach can offer to the understanding of these issues by yielding well-defined results and making precise predictions. It has now been demonstrated that it is possible not only to make quantitative predictions for real systems, but more importantly, to gain insights into the underlying physics of these materials that are essential not only for the pure science but also for potential device applications.

BASIC CONCEPTS OF DENSITY FUNCTIONAL THEORY

The Kohn-Sham Equations

Hohenberg & Kohn (1) and Kohn & Sham (2) formulated a rather remarkable theorem that states that the total energy of a system such as a bulk solid or a surface depends only on the electron density of its ground state. In other words, one can express the total energy of an atomistic system as a functional of its electron density $E = E[\rho]$. The idea of using the electron density as the fundamental entity of a quantum mechanical theory of matter originated in the early days of quantum mechanics (3, 4). However, in subsequent decades, the Hartree-Fock approach (5, 6) developed and was applied to small molecular aystems. Calculations on realistic solid-state systems were then not possible, but Slater (7) used ideas from the electron gas with the intention to simplify Hartree-Fock theory to a point where electronic structure calculations on solids became feasible. Slater's work, which led to the so-called $X\alpha$ method (8), has contributed tremendously to the development of electronic structure calculations. Today, Slater's $X\alpha$ method can be seen as an early, simplified form of density functional theory. However, the $X\alpha$ method is hardly used in present electronic structure calculations and therefore will not be discussed further.

The electron density is a scalar function defined at each point $\mathbf{r}$ in real space, $\rho = \rho(\mathbf{r})$. The electron density and the total energy depend on the type and arrangements of the atomic nuclei. Therefore, one can write

$$E = E[\rho(\mathbf{r}), \{\mathbf{R}_\alpha\}]. \qquad 1.$$

The set $\{\mathbf{R}_\alpha\}$ denotes the positions of all atoms α in the system under consideration. Equation 1 is the key to the atomic-scale understanding of electronic, structural, and dynamic properties of matter. If one has a way of evaluating this equation, one can, for example, predict surface reconstructions, the equilibrium geometry of molecules adsorbed on surfaces, and the cohesive energies of solids.

Furthermore, the derivative of the total energy (Equation 1) with respect to the nuclear position of an atom gives the force acting on that atom. This enables the efficient search for stable structures and, perhaps more importantly, the study of dynamical processes such as diffusion or the reaction of molecules on surfaces. Most of the considerations discussed are based on the Born-Oppenheimer approximation in which it is assumed that the motions of the electrons are infinitely faster than those of the nuclei. In practice this means that the electronic structure is calculated for a fixed atomic arrangement, and the atoms are then moved according to classical mechanics. This is a fairly good approximation for heavy atoms like W, but can cause significant errors for light atoms like H or Li.

In density functional theory, the total energy is decomposed into three parts: a kinetic energy, an electrostatic or Coulomb energy, and a so-called exchange-correlation energy,

$$E = T_o + U + E_{xc}. \qquad 2.$$

The most straightforward term is the Coulomb energy U. It is purely classical and contains the electrostatic energy arising from the Coulombic attraction between electrons and nuclei, the repulsion between all electronic charges, and the repulsion between nuclei

$$U = U_{en} + U_{ee} + U_{nn}, \qquad 3.$$

with

$$U_{en} = -e^2 \sum_{\alpha} Z_{\alpha} \int \frac{\rho(\mathbf{r})}{|\mathbf{r} - \mathbf{R}_{\alpha}|} d\mathbf{r}, \qquad 4.$$

$$U_{ee} = e^2 \iint \frac{\rho(\mathbf{r})\rho(\mathbf{r}')}{|\mathbf{r} - \mathbf{r}'|} d\mathbf{r}\, d\mathbf{r}', \qquad 5.$$

and

$$U_{nn} = e^2 \sum_{\alpha\alpha'} \frac{Z_{\alpha} Z_{\alpha'}}{|\mathbf{R}_{\alpha} - \mathbf{R}_{\alpha'}|}, \qquad 6.$$

where e is the elementary charge of a proton and Z_{α} is the atomic number of atom α. The summations extend over all atoms and the integrations over all space. Once the electron density and the atomic numbers and positions of all atoms are known, Equations 4–6 can be evaluated by using the techniques of classical electrostatics.

The kinetic energy term T_o is more subtle. In density functional theory, the "real" electrons of a system are replaced by "effective" electrons with the same charge, mass, and density distribution. However, effective electrons move as independent particles in an effective potential, whereas the motion of a real electron is correlated with those of all other electrons. T_o is the sum of the kinetic energies of all effective electrons moving as independent particles. Often, one does not explicitly make this distinction between real and effective electrons.

If each effective electron is described by a single particle wave function ψ_i, then the kinetic energy of all effective electrons in the system is given by

$$T_o = \sum_i n_i \int \psi_i^*(\mathbf{r}) \left[-\frac{\hbar^2}{2m} \nabla^2 \right] \psi_i(\mathbf{r}) \, d\mathbf{r}. \qquad 7.$$

Equation 7 is the sum of the expectation values of one-particle kinetic energies; n_i denotes the number of electrons in state i. By construction, dynamical correlations between the electrons are excluded from T_o.

The third term of Equation 2 includes all remaining complicated electronic contributions to the total energy and is called exchange-correlation energy, E_{xc}. The most important of these contributions is the exchange term. Electrons are fermions and obey the Pauli exclusion principle. In real space, the Pauli principle implies that, around each electron with a given spin, all other electrons with the same spin tend to avoid that electron. As a consequence, the average Coulombic repulsion acting on that electron is reduced. This energy gain is called exchange energy. Effectively, each electron is surrounded by a positive exchange hole (7, 8). It can be shown (8) that the total charge integrated over the entire exchange hole equals $+e$. By definition, the additional many-body interaction terms between electrons of opposite spin are called correlation energy.

As an illustration, the total energy of a single C atom is approximately -1019 eV, that of a Si atom -7859 eV and that of a W atom $-439{,}634$ eV. The kinetic energy and the Coulomb energy terms are of similar magnitude, but of opposite sign, whereas the exchange-correlation term is about 10% of the Coulomb term and attractive for electrons (because the exchange-hole is positive). The correlation energy is smaller than the exchange energy but plays an important role in determining the details in the length and strength of interatomic bonds. In fact, compared with the total energy, the binding energy of an atom in a solid or on a surface is quite small and lies in the range of about 1 to 8 eV. Energies involved in changes of the position of atoms on a surface can be even smaller. For example, only about 0.03 eV are required to flip an asymmetric Si-dimer on a reconstructed Si(001) surface from one conformation into another where the role of the upper and lower Si atom are reversed. It is a tremendous challenge for any theory to cope with such a range of energies. Density functional theory, as it turns out, comes amazingly close to this goal.

The Hohenberg-Kohn-Sham theorem, which is a central part of density functional theory, states that the total energy is at its minimum value for the ground state density and that the total energy is stationary with respect to first-order variations in the density, i.e. $(\partial E[\rho]/\partial \rho|_{\rho=\rho_o} = 0)$.

In conjunction with the kinetic energy, we have introduced one-particle wave functions $\psi_i(\mathbf{r})$ that generate the electron density

$$\rho(\mathbf{r}) = \sum_i n_i |\psi_i(\mathbf{r})|^2. \tag{8}$$

As in Equation 7 for the kinetic energy, n_i denotes the occupation number of the eigenstate that is represented by the one-particle wave function ψ_i. So far, one has a formally exact theory in the sense that no approximations have been made yet to the many-electron interactions. By construction, $\rho(\mathbf{r})$ in Equation 8 is the *exact* many-electron density.

The goal of the next step is the derivation of equations that can be used for practical density functional calculations. Through Equations 7 and 8 we have introduced one-particle wave functions. A change of these wave functions corresponds to a variation in the electron density. Therefore, the variational condition can be used to derive the conditions for the one-particle wave functions that lead to the ground state density. To this end, one varies the total energy with respect to each wave function. This procedure leads to the following equations:

$$\left[-\frac{\hbar^2}{2m}\nabla^2 + V_{eff}(\mathbf{r})\right]\psi_i(\mathbf{r}) = \varepsilon_i\psi_i(\mathbf{r}), \tag{9}$$

with

$$V_{eff}(\mathbf{r}) = V_C(\mathbf{r}) + \mu_{xc}[\rho(\mathbf{r})]. \tag{10}$$

Equations 9 and 10 is called the Kohn-Sham equations. The electron density that corresponds to these wave functions is the ground state density that minimizes the total energy. The solutions of the Kohn-Sham equations form an orthonormal set, i.e. $\int \psi_i^*(\mathbf{r})\psi_j(\mathbf{r})\,d\mathbf{r} = \delta_{ij}$. This additional constraint is achieved by introducing Lagrange multipliers, ε_i in Equation 9. The Lagrange multipliers are effective one-electron eigenvalues, and their interpretation is discussed below. These eigenvalues are used to determine the occupation numbers n_i by applying the Aufbau principle. The eigenstates are ordered according to increasing eigenvalues. For systems non-spin-polarized, each state is occupied by two electrons until all electrons are accommodated. In spin-polarized systems, each state is occupied by at most one electron.

As a consequence of the partitioning of the total energy (Equation 2), the Hamiltonian operator in the Kohn-Sham Equations 9 and 10 contains three terms: one for the kinetic energy, the second for the Coulomb potential, and the third for the exchange-correlation potential. The kinetic energy term is the standard second-order differential operator of one-particle Schrödinger equations, and its construction does not require specific knowledge of a system. In contrast, the Coulomb potential operator

$V_C(\mathbf{r})$, and the exchange-correlation potential operator μ_{xc}, depend on the specific electron distribution in the system under consideration.

The Coulomb or electrostatic potential $V_C(\mathbf{r})$ at point $\mathbf{r}$ is generated from the electric charges of all nuclei and electrons in the system. It can be evaluated directly in real space,

$$V_C(\mathbf{r}) = -e^2 \sum_{\alpha} \frac{Z_\alpha}{|\mathbf{r}-\mathbf{R}_\alpha|} + e^2 \int \frac{\rho(\mathbf{r}')}{|\mathbf{r}-\mathbf{r}'|} d\mathbf{r}'. \qquad 11.$$

In condensed systems it is more convenient to use Poisson's equation $\nabla^2 V_C(\mathbf{r}) = 4\pi e^2 q(\mathbf{r})$ to calculate the electrostatic potential. Here, $q(\mathbf{r})$ denotes both the electronic charge distribution $\rho(\mathbf{r})$ and the positive point charges of the nuclei at positions $\mathbf{R}_\alpha$.

The exchange-correlation potential is related to the exchange-correlation energy by $\mu_{xc} = (\partial E_{xc}[\rho]/\partial \rho)$, which is formally exact in the sense that it does not contain any approximation to the complete many-body interactions. In practice, however, the exchange-correlation energy (and thus the exchange-correlation potential) is not known, and one has to make approximations.

As a consequence of the Kohn-Sham theorem, the exchange-correlation energy depends only on the electron density. As a simple and, as it turns out, surprisingly good approximation the exchange-correlation energy is assumed to depend only on the local electron density around each volume element $d\mathbf{r}$. This is called the local density approximation (LDA),

$$E_{xc}[\rho] \approx \int \rho(\mathbf{r}) \varepsilon_{xc}[\rho(\mathbf{r})]\, d\mathbf{r}. \qquad 12.$$

The basic idea is that in any atomic arrangement such as a crystal, a surface, or a molecule, there is a certain electron density $\rho(\mathbf{r})$ at each point $\mathbf{r}$ in space. The LDA then rests on two basic assumptions: (*a*) The exchange and correlation effects come predominantly from the immediate vicinity of a point $\mathbf{r}$, and (*b*) these exchange and correlation effects do not depend strongly on the variations of the electron density in the vicinity of $\mathbf{r}$. If conditions (*a*) and (*b*) are reasonably well fulfilled, then the contribution from volume element $d\mathbf{r}$ would be the same as if this volume element were surrounded by a constant electron density of the same value as within $d\mathbf{r}$.

A system of interacting electrons with a constant density is called a homogeneous electron gas. Substantial theoretical efforts have been made to understand and characterize such an idealized system. In particular, the exchange-correlation energy per electron of a homogeneous electron gas, $\varepsilon_{xc}[\rho]$, has been calculated by several approaches such as many-body perturbation theory (9) and quantum Monte-Carlo methods (10). As a result,

$\varepsilon_{xc}[\rho]$ is quite accurately known for a range of densities. For practical calculations, $\varepsilon_{xc}[\rho]$ is expressed as an analytical function of the electron density. These are different analytical forms with different coefficients in their representation of the exchange-correlation terms. These coefficients are not adjustable parameters, but rather they are determined through first-principles theory. Hence, the LDA is a first-principles approach in the sense that the quantum mechanical problem is solved without any adjustable, arbitrary, or system dependent parameters. Table 1 shows an example of such local exchange-correlation terms. Note that there are two types of exchange-correlation terms, one for the energy and one for the potential. The energy ε_{xc} is needed to evaluate the total energy, and the potential term μ_{xc} is required for the Kohn-Sham equations. The two terms are related by $\mu_{xc} = \{\partial[\rho\varepsilon_{xc}(\rho)]/\partial\rho\}$.

Using the explicit formulas given in Table 1, one can evaluate the exchange-correlation potential for any electron density $\rho(\mathbf{r})$. Thus all terms

Table 1 Explicit forms for the local density exchange (Gáspár 1954, Kohn & Sham 1965) and correlation terms (Hedin & Lundqvist 1972) for a non-spin-polarized system. Exchange and correlation energies per electron are denoted by ε and the corresponding potentials by μ. Both quantities are given in Hartree atomic units (1 Hartree = 2 Rydeberg = 27.21165 eV). The units for the electron density are number of electrons/(Bohr radius)^3.

	Energy	Potential
	$\varepsilon_{xc} = \varepsilon_x + \varepsilon_c$	$\mu = \dfrac{\partial(\rho\varepsilon)}{\partial\rho}$
Exchange	$\varepsilon_x = -\dfrac{3}{2}\left(\dfrac{3}{\pi}\rho\right)^{1/3}$	$\mu_x = -2\left(\dfrac{3}{\pi}\rho\right)^{1/3}$
Correlation	$\varepsilon_c = -c\left[(1+x^3)\ln\left(1+\dfrac{1}{x}\right)+\dfrac{x}{2}-x^3-\dfrac{1}{3}\right]$	$\mu_c = -c\ln\left(1+\dfrac{1}{x}\right)$

$$c = 0.0225, \quad x = \frac{r_s}{21}, \quad r_s = \left(\frac{3}{4\pi\rho}\right)^{1/3}$$

of the effective one-particle operator in the Kohn-Sham equations are defined, and one can proceed with a computational implementation.

Spin-Polarized Systems

So far, the discussion of density functional theory has been restricted to non-spin-polarized cases. However, many systems such as magnetic materials or the dissociation of molecules on surfaces involve unpaired electrons or molecular radicals and thus require a spin-polarized method. In such systems, the number of electrons with "spin-up" can be different from that with "spin-down." Density functional theory has been generalized to accommodate spin-polarized systems. That resulted in spin density functional theory with the local spin density (LSD) approximation (11, 12).

In the local spin density functional (LSDF) theory, the fundamental quantities are the electron density $\rho(\mathbf{r})$ and the spin density $\sigma(\mathbf{r})$. The spin density is defined as the difference between the density of the spin-up electrons and the density of the spin-down electrons, with the total electron density $\rho(\mathbf{r}) = \rho_{\uparrow}(\mathbf{r}) + \rho_{\downarrow}(\mathbf{r})$. In LSDF theory, the exchange-correlation potential for spin-up electrons is in general, different from that for spin-down electrons. Consequently, the effective potential Equation 10 becomes dependent on the spin. Thus the Kohn-Sham Equations 9 and 10 in their spin-polarized form are

$$\left[-\frac{\hbar^2}{2m}\nabla^2 + V^{\sigma}_{eff}(\mathbf{r})\right]\psi^{\sigma}_i(\mathbf{r}) = \varepsilon^{\sigma}_i\psi^{\sigma}_i(\mathbf{r}), \quad \sigma = \uparrow \text{ or } \downarrow, \qquad 13.$$

with

$$V^{\sigma}_{eff}(\mathbf{r}) = V_C + \mu^{\sigma}_{xc}[\rho(\mathbf{r}), \sigma(\mathbf{r})]. \qquad 14.$$

The exchange-correlation potential in LSDF theory depends on both the electron density and the spin density, as written in Equation 14. There are two sets of single-particle wave functions, one for spin-up electrons and one for spin-down electrons, each with its corresponding one-electron eigenvalues. For the case of equal spin-up and spin-down densities, the spin density is zero throughout space, and LSDF theory becomes identical with the LDF approach. Notice that in spin-polarized calculations, the occupation of single-particle states is 1 or 0, but there is still only one Fermi energy. In magnetic systems, the spin-up and spin-down electrons are often referred to as majority and minority spin systems.

GEOMETRICAL MODELS AND COMPUTATIONAL APPROACHES

Geometric Models

One of the most crucial steps in computational science is the creation of relevant geometric models. Many, but by no means all, phenomena are relatively short-range in nature. This makes it possible to choose geometric models that are small enough to be tractable by today's electronic structure methods, yet still large enough to be physically meaningful. Systems containing of the order of 100 atoms per repeat unit can be treated on a first-principles level with today's programs and computer hardware. A particular choice depends on the physical or chemical question to be answered, and each geometric model has its strengths and limitations. In the following, we discuss the most common geometric models for electronic structure calculations of surfaces (cf Figure 1) and outline their range of applicability.

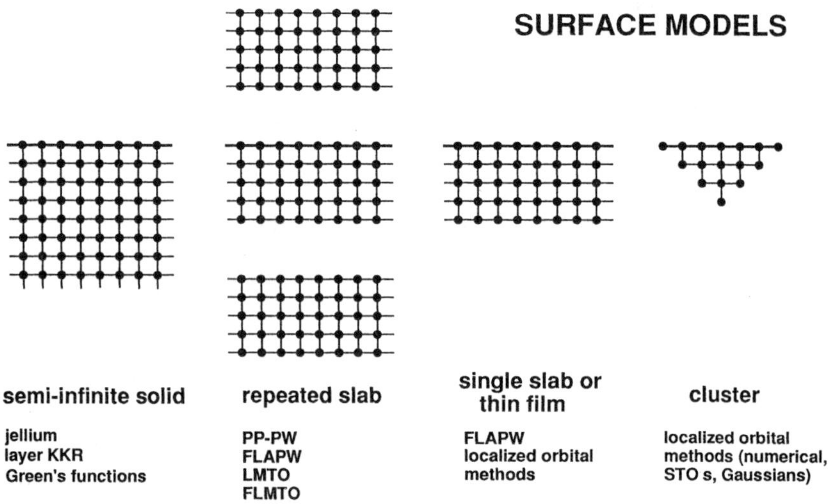

Figure 1 Geometric models for practical surface calculations. The geometry of a semi-infinite solid is mainly used in jellium calculations and methods employing Green's functions and matching techniques. Standard band structure methods using three-dimensional periodicity can be directly applied to a repeated slab geometry. These standard methods include the pseudopotential plane wave (PP-PW) methods, the full-potential, linearized, augmented plane-wave (FLAPW) method, the linearized muffin-tin orbital (LMTO) method, and the full potential LMTO method. The FLAPW method has been implemented for the single-slab geometry, which also can be used with localized orbital methods. The cluster geometry is amenable for localized orbital methods with numerical functions, Slater-type orbitals (STOs) or Gaussians as basis set, as used for molecular quantum chemical calculations.

Conceptually, the most satisfying surface geometry is that of a semi-infinite solid. This geometry can be used for the simple jellium model of surfaces, where the positive charge of the atomic nuclei is simply represented by a uniform, constant-positive background inside the solid and by zero outside an appropriately chosen surface plane. Effectively, the system is thus reduced to a one-dimensional problem, and the distribution of the electrons are then calculated using density functional theory.

The use of a semi-infinite solid is much more difficult if a full three-dimensional solution of the density functional problem is attempted. However, it is reasonable to assume that any material becomes bulk-like at a certain distance away from the surface. A priori one does not know that distance, but experience shows that typically of the order of 10 atomic layers are sufficient. Hence, one can use the electronic structure of the bulk system, calculated with the three-dimensional periodicity, up to a plane about 10 atomic layers underneath the surface. In the top layers or surface region, the electronic wave functions are then chosen to match the bulk states inside the solid and satisfy the vacuum boundary conditions above the surface. Green's functions techniques as used, for example, in the layer Korringa-Kohn-Rostoker method (13–16) provide the necessary mathematical apparatus to accomplish this matching procedure.

A simple but effective geometric surface model is the repeated slab geometry (cf Figure 1). Thin films consisting of about 5 to 20 layers are repeated in the direction perpendicular to the surface. The layers are made thick enough to approach bulk-like behavior near the center of each layer, and the spacing between is made large enough so that any artificial interactions across the vacuum region between the slabs are minimized. About 10–20 Å are usually sufficient to fulfill this requirement. For such a geometry, any three-dimensional band structure method can be used. The most common approaches for three-dimensional band structure calculations are the pseudopotential plane wave method, the full-potential linearized augmented plane wave (FLAPW) method, and the linearized muffin-tin orbital (LMTO) method with its generalization to full potentials. Practical applications of these approaches are limited by the number of atoms in the three-dimensional supercell. Thus a compromise needs to be found between slab thickness, spacing between the slabs, and computational effort.

One way to overcome at least one of these limitations is the use of a single-slab geometry (cf Figure 1). The slab still has to be thick enough to achieve bulk-like behavior in its interior, but correct vacuum boundary conditions are fulfilled above and below the slab. Transition-metal films as thin as give layers already exhibit the major characteristics of transition metal surfaces, such as surface states and reasonable values for work functions, while approaching bulk-like behavior near the center of the film. Simple metals such as Al require a larger number of layers, preferably more than about 10, to approach bulk-like behavior in the interior.

Finally, solids and surfaces can be modeled by finite clusters. This approach is being widely used for the investigation of chemisorption because it allows the application of standard quantum chemistry programs. Although reasonable structural information such as adsorption geometries can be obtained with relatively small clusters of 10 or 20 atoms, much larger clusters of preferably well over 100 atoms are required to achieve reliable results for sensitive quantities such as adsorption energies or the distinction between different adsorption sites with similar energy. Large clusters pose serious computational challenges, especially for transition metals where the convergence of the self-consistency procedure (see below) can be rather slow. Finally, even for large clusters, termination effects can have unpredictable side effects.

Overview of Computational Methods

Figure 2 provides an overview of the major choices between different surface electronic structure methods. If the system contains elements with atomic numbers greater than about 40, a scalar-relativistic or fully relativistic treatment is appropriate. Most electronic structure calculations are done either non-relativistically or on a scalar-relativistic level. The neglect

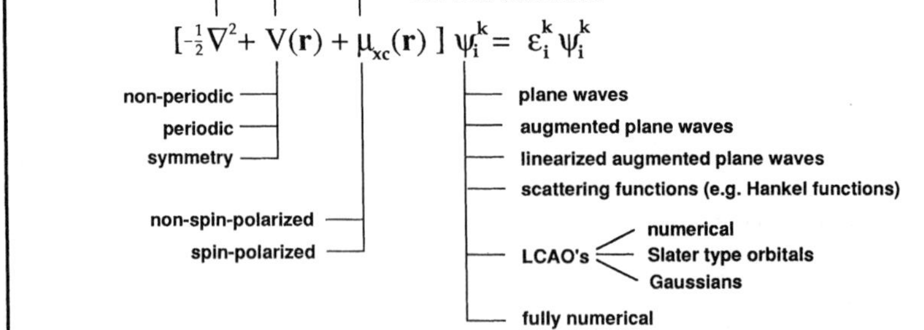

Figure 2 Overview of theoretical and computational choices for the solution of the density functional one-particle equations.

of relativistic effects can be quite severe, leading to an overestimation of bond distances of perhaps 0.1 Å and incorrect ordering of energy bands.

Choices of specific geometric models and their space group symmetry are reflected in the effective potential operator. Symmetry such as an inversion center and mirror planes can greatly reduce the computational effort. In practical density functional implementations, the computational effort scales approximately with a third power or less in the number of atoms. If the number of inequivalent atoms of a system is reduced, for example, by a factor of 4, the computational time can thus be reduced by about 50 times.

The next major choice is between a non-spin-polarized and a spin-polarized calculation. In most cases, structural information such as equilibrium geometries can be quite reliably obtained from the non-spin-polarized calculations. Also, the characteristic features of the charge density distribution usually differ little between non-spin-polarized and spin-polarized treatments. Therefore, one often performs non-spin-polarized calculations first followed by spin-polarization.

Characteristic features of specific computational methods are the various simplifications to the form of the effective potential. In the most general case, the potential is constructed from all electrons and no additional approximation is made to its shape. In the late 1930s, the so-called muffin-tin approximation was introduced in which the effective potential within a sphere around each atom is taken to be spherically averaged, thus resembling a free atom. In the remaining interstitial region of a solid, the potential is simply taken to be constant. This approximation can be generalized to surface calculations by adding in the region above the surface a potential that depends only on the distance from the surface, obtained by averaging the potential in planes parallel to the surface. Muffin-tin potentials are reasonable for densely packed metallic systems, but lead to uncontrolled errors if applied to more open structures. Both the full-potential and the muffin-tin potential approach are used in all-electron calculations.

Pseudopotentials are frequently employed for calculations on semiconductors. In this approximation, the core electrons and the Coulomb singularity from the atomic nuclei are replaced by a smooth pseudopotential around each atom. At a certain cut-off radius away from each atomic position, the pseudopotential becomes the actual effective potential. There are three major reasons for using pseudopotentials: (*a*) Simple basis functions such as plane waves work well with pseudopotentials, (*b*) there are fewer electrons thus simplifying the calculation, and (*c*) relativistic effects, which are mainly due to core electrons, can be included in the pseudopotential. Together, these features lead to computationally efficient and relatively simple computational implementations. The major drawbacks of pseudopotentials are (*a*) an additional approximation is introduced; (*b*) pseudopotentials for transition metals, rare earths, and actinides are possible in principle, but their effectiveness and accuracy have yet to be demonstrated; and (*c*) properties and effects involving the core electrons such as core level shifts, hyperfine fields, and spin polarization of the core electrons are not directly accessible from pseudopotential calculations. For these reasons, both all-electron and pseudopotential methods have found their place in electronic structure theory.

A central part of any electronic structure calculation is the treatment of the exchange and correlation effects. Within density functional theory, the appropriate choices have to be made in the selection of the exchange-correlation potential operator. We have discussed the local density approximation, its generalization to the spin-polarized case, and the fact that the local density approximation is uniquely defined (see above). However, different ways, e.g. many-body perturbation theory and quantum Monte Carlo methods, to evaluate the exchange-correlation energy in the homogeneous interacting electron gas have led to slightly different explicit forms for the exchange-correlation potentials. However, the deviations between these forms are typically smaller than the discrepancies between calculated and experimental values. Thus one can consider these potentials as practically equivalent.

As discussed above, there are several ways to include effects beyond the local density approximation. Although improvements such as the density-gradient corrections to exchange and correlation have become popular in quantum chemistry calculations, their advantages and benefits for solid state and surface calculations are still a matter of discussion and ongoing research.

Self-Consistency Cycle and Geometry Optimizations

The effective potential in the Kohn-Sham Equations 9 and 10 depends on the charge density, which is constructed from the one-particle wave functions. These wave functions are solutions of the Kohn-Sham equations. In other words, in order to set up the Kohn-Sham equations, one needs to know their solutions. This problem is solved by an iterative, self-consistent procedure, as shown in Figure 3. Start geometries for density functional calculations are constructed either by using experimental data such as bulk lattice constants, from simpler computational approaches such as force field or by semi-empirical methods, or from intuition. The start densities are then constructed from a superposition of atomic densities (calculated for individual free atoms or ions), or results from previous density functional computations are reused.

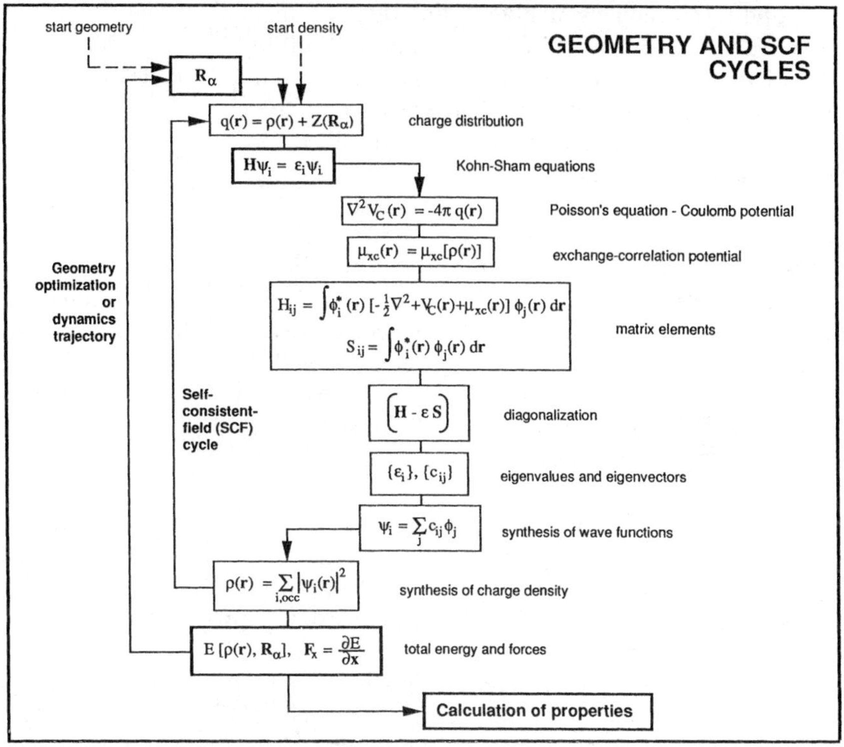

Figure 3 Scheme of typical electronic structure calculations. The outer cycle represents the geometry optimization or other manipulation of the geometry such as dynamics trajectories or Monte Carlo procedures. The inner cycle is the self-consistency procedure to solve the Kohn-Sham equation. In methods such as the Car-Parrinello (45) approach, both cycles are carried out simultaneously. The matrix diagonalization can be done either through explicit diagonalization or through iterative procedures. For simplicity, atomic units are used in the formulas.

The fundamental quantities of density functional theory are the electron density and, in the case of spin-polarized systems, also the spin density. With these quantities, and a given geometry, the Kohn-Sham equations are completely defined. Through the solution of Poisson's equation, the electrostatic Coulomb potential is obtained. Using an explicit form for the exchange-correlation potential, the exchange-correlation potential operator is constructed. For a given variational basis set $\{\phi_j\}$, the Hamiltonian and overlap matrix elements can then be computed as indicated in Figure 3. Subsequently, the matrix **H**-ε**S** is diagonalized, resulting in a set of one-particle eigenvalues, with the variational expansion coefficients corresponding to each eigenvalue. Using these coefficients, the single-particle wave functions are synthesized and a new ("output") electron density can be calculated. To this end, the occupation of each state is determined by using Fermi-Dirac statistics. This means that the eigenvalues are sorted with increasing energy, and all states are filled until the total number of electrons of the system is exhausted. In the case of systems with two-dimensional periodicity such as ordered surfaces, the synthesis of the charge density from one-particle wave functions involves the integration of the two-dimensional Brillouin zone in reciprocal space.

The new electron density is often referred to as output density, indicating that each cycle in the self-consistency procedure essentially maps an input density into an output density. By definition, self-consistency is achieved when the output density equals the input density. In practice, one iterates sufficiently long until the residual difference between the input and output densities does not cause any significant errors in the total energy or other properties of interest.

Although this self-consistency procedure is straightforward in principle, practical calculations, especially for magnetic systems, are often plagued by aggravating convergence problems. If one would feed the full output density back as input density, the self-consistency process would quickly diverge. Therefore, many computational schemes have been devised to enable and accelerate convergence. The simplest method consists in mixing the output density with the input density, e.g. using 10% of the new output density, in order to construct the input density for the next iteration. In magnetic systems, feedback of more than 3% can sometimes already lead to divergence. More sophisticated convergence and extrapolation schemes have thus been developed such as that suggested by Andersen (17). In many non-magnetic systems, sufficient self-consistency can be reached within 10–20 cycles, but 50 steps or more may be necessary in more difficult cases.

Once self-consistency is achieved, the total energy and the forces on each atom can be calculated. Using this information, it is possible to optimize geometries, i.e. searching for minima or saddle points on the energy hypersurface. Geometries can be optimized by using only the total energy for various atomic arrangements. In practice, this is an extremely tedious process, and only a few geometric parameters such as the lattice constant of a highly symmetric solid or the distance between an adsorbate atom and a surface can be optimized in this way. Knowledge of the forces on each atom greatly facilitates such geometry optimizations and makes it possible to relax many degrees of freedom simultaneously.

An exciting and promising use of these forces is the computation of dynamic trajectories. This enables the search for structural minima by simulated annealing techniques and, in principle, the computation of diffusion, phase transitions, and entire chemical reactions. However, the time-scales necessary to tackle these problems are in many cases far beyond the computational speed of present density functional calculations, but very encouraging first results have already been obtained. The final geometries are then used for the calculation of properties that can be derived from the charge density, the potential, and the one-electron wave functions. These include the structural, energetic, electronic, optical, and magnetic properties accessible through density functional calculations.

Specific Computational Implementations

In this section, we discuss two characteristic computational implementations for electronic structure calculations on surfaces. The dominant feature of a computational method is the choice of the variational basis set, as illustrated in Figure 2. We discuss first the FLAPW method, which is perhaps formally the most complex and computationally also the most demanding. The FLAPW method has been extensively used for surface electronic structure calculations, it is often taken as benchmark because of its accuracy, and it shows all major features of surface electronic structure calculations. The second approach discussed here employs numerically generated atomic orbitals. This approach is closer to quantum chemical calculations and thus is playing an increasingly important role in the investigation of localized properties in solids and in chemical reactions on surfaces.

FULL-POTENTIAL, LINEARIZED, AUGMENTED PLANE-WAVE (FLAPW) METHOD

The augmented plane wave (APW) method was introduced by Slater (18). The basic idea is the partitioning of real space into spherical regions around atoms, so-called muffin-tin or atomic spheres, and an interstitial region (cf Figure 4). The dominant part of the effective potential inside the spheres has spherical symmetry and is constant in the interstitial region. Thus one chooses as basis functions the solutions of Schrödinger's equations for these types of potentials, namely products of radial functions and spherical harmonics inside the atomic spheres and plane waves for the interstitial region. It is important to note that the muffin-tin potential is used here to construct the basis functions. Once such a basis set is defined, the solutions to any potential can be represented. A major step forward in the development of the APW method was the linearization of the eigenvalue problem (19, 20) by introducing two radial wave functions per angular momentum quantum number, which are the solutions of the radial Schrödinger equation, solved at fixed trial energies E_l

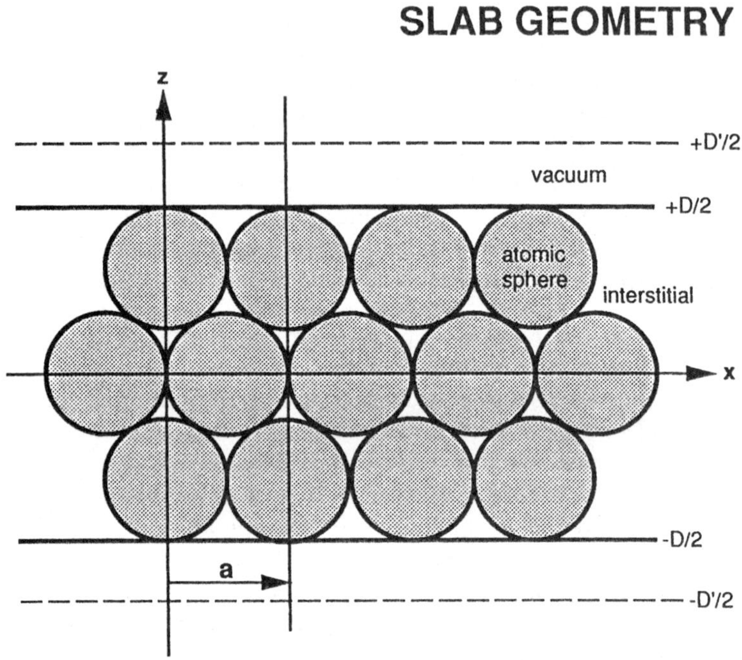

Figure 4 Single-slab or thin-film geometry used in FLAPW calculations. The interstitial region extends between the two vacuum boundaries at $z = \pm D/2$. The vacuum regions above and below the film extend to $\pm\infty$. The arbitrary boundaries at $z = \pm D'/2$ are used to define a three-dimensional plane wave basis set within the interstitial region. These plane wave basis functions are not used in the region between $\pm(D'/2$ and $D/2)$. They are also not used inside the atomic spheres. The vector **a** denotes the two-dimensional translational basis vector parallel to the film surface.

$$\frac{1}{2r}\frac{d^2}{dr^2}[ru_{nl}(r)] - \frac{l(l+1)}{2r^2}u_{nl}(r) + [E_l - V_{eff}(r)]u_{nl}(r) + h_r u_{nl}(r) = 0, \qquad 15.$$

and the energy derivative of these radial solutions (Hartree atomic units are used here with $h = \mathrm{m} = \mathrm{e} = 1$). The term h_r is a semi-relativistic correction following Koelling & Harmon (21).

Now, in the more complex case of a single-slab or thin-film FLAPW approach (22–25), one defines vacuum regions above and below the slab as shown in Figure 4. For the construction of the wave functions in these vacuum regions, one uses only the z-dependent part of the potential, starting at $z = \pm D/2$. The explicit form of the FLAPW one-particle wave functions in the thin-film geometry is thus

$$\psi_i(\mathbf{r}, \mathbf{k}^{\|}) = \sum_j c_{ij}\phi(\mathbf{r}, \mathbf{K}_j); \quad \mathbf{K}_j = \mathbf{k}^{\|} + \mathbf{G}_j. \qquad 16.$$

Here $\mathbf{k}^{\parallel}$ is an arbitrary vector of the two-dimensional Brillouin zone, and $\mathbf{G}_j$ is a three-dimensional lattice vector of the reciprocal space, with its k_x and k_y components as integer multiples of the basis vectors $\mathbf{b}_1$ and $\mathbf{b}_2$. In the z-direction, an artificial periodicity between the boundaries at $z = \pm D'/2$ is used to define the third component of $\mathbf{G}_j$. With these definitions, the FLAPW basis functions $\phi(\mathbf{K}_j)$ are

$$\phi(\mathbf{r}, \mathbf{K}_j) = \begin{cases} \Omega^{-1/2} e^{i\mathbf{K}_j \cdot \mathbf{r}} & \text{for } \mathbf{r} \in \text{interstitial} \\ \sum_{lm} [A_{lm}^{\alpha}(\mathbf{K}_j) u_l(E_l^{\alpha}, r_\alpha) + B_{lm}^{\alpha}(\mathbf{K}_j) \dot{u}_l(E_l^{\alpha}, r_\alpha)] Y_{lm}(\hat{r}_\alpha) & \text{for } \mathbf{r} \in \text{sphere } \alpha \\ \sum_{q} [A_q(\mathbf{K}_j) u_{\mathbf{k}q}(E_v, z) + B_q(\mathbf{K}_j) \dot{u}_{\mathbf{k}q}(E_v, z)] e^{i(\mathbf{k}^{\parallel} + \mathbf{K}_q^{\parallel}) \cdot \mathbf{r}} & \text{for } \mathbf{r} \in \text{vacuum.} \end{cases} \qquad 17.$$

Ω is the volume per unit cell between the boundaries at $z = \pm D/2$. $u_l(E_l, r)$ are solutions of the radial Schrödinger equations inside each sphere, calculated at a trial energy E_l and classified according to the angular momentum quantum number l. The functions $\dot{u}_l(E_l, r)$ are the energy derivatives of these functions at $E = E_l$. The coefficients A_{lm} and B_{lm} are determined by the requirement that each plane wave, $\exp(i\mathbf{K}_j \cdot \mathbf{r})$, is continuous in value and radial derivative at all atomic sphere boundaries. Hence the matching coefficients A_{lm} and B_{lm} for each sphere α depend on $\mathbf{K}_j$. The functions Y_{lm} are spherical harmonics centered at each atomic position.

The basis functions in the vacuum involve functions that depend only on the distance from the surface, multiplied by two-dimensional plane waves. Similar to the functions inside the spheres, the z-dependent functions are obtained from one-dimensional Schrödinger equations of the form

$$\left[-\frac{1}{2} \frac{\partial^2}{\partial z^2} + V_{eff}(z) - (E_v - K_q^2) \right] u_{\mathbf{k}q}(E_v, z) = 0. \qquad 18.$$

$V(z)$ is the planar average of the effective potential perpendicular to the surface, E_v are fixed energy parameters, and $\mathbf{K}_q^{\parallel}$ denote reciprocal lattice vectors of the two-dimensional lattice parallel to the surface. The matching coefficients $A_q(\mathbf{K}_j)$ and $B_q(\mathbf{K}_j)$ are chosen such that each basis function in the vacuum region is smooth in value and derivative (perpendicular to the surface) across the vacuum boundaries at $z = \pm D/2$.

Another way of looking at the FLAPW basis functions is as follows: In the FLAPW method, one uses numerical atomic-like functions in regions around the atoms and numerical functions for the vacuum region. These functions are joined together by plane waves. Continuity in value and derivatives throughout space, as required for any acceptable solution of Schrödinger's equation, is ensured by appropriate matching conditions. The radial functions in each atomic sphere depend on the actual spherical part of the effective potential, which varies from iteration to iteration in the self-consistency procedure. In standard FLAPW calculations, the radial functions are recalculated in each iteration. Typically, functions up to $l = 8$ are used with two functions, $u(r)$ and $\dot{u}(r)$, for each l value. Similarly, the z-dependent part of the vacuum potential is used to generate the basis functions in the vacuum region. Also, these functions are recalculated as the form of the potential changes with the iterations. FLAPW basis sets can be systematically improved by increasing the cut-off for the maximum length of the reciprocal lattice vectors $\mathbf{G}_j$ and by increasing the highest l component inside the atomic spheres. For densely packed systems such as transition metal surfaces, about 50 basis functions per atom are needed.

In contrast, standard basis sets in quantum chemistry are not recalculated in each iteration (in fact, they are defined just once for each atom type), and usually functions up to $l = 2$ are used. This shows the high variational quality of FLAPW basis functions compared with other methods using fixed functions.

Unlike earlier implementations of the APW and LAPW method, no shape approximations to the charge density or the effective potential are made in the FLAPW method (23). Both the charge density and the effective potential are represented by the same analytical expansions, i.e. a Fourier representation in the interstitial region and an expansion in spherical harmonics inside the spheres. In the vacuum region, two-dimensional Fourier expansions in planes parallel to the surface are used. Thus the charge density $\rho(\mathbf{r})$ is given by

$$\rho(\mathbf{r}) = \begin{cases} \sum_j \rho_j e^{i\mathbf{G}_j\cdot\mathbf{r}} & \text{for } \mathbf{r} \in \text{interstitial} \\ \sum_{lm} \rho^{\alpha}_{lm}(r_\alpha) Y_{lm}(\hat{r}_\alpha) & \text{for } \mathbf{r} \in \text{sphere } \alpha \\ \sum_q \rho_q(z) e^{i\mathbf{K}^{\parallel}_q\cdot\mathbf{r}} & \text{for } \mathbf{r} \in \text{vacuum} \end{cases}, \qquad 19.$$

with an analogous representation of the effective potential. The coefficients $\{\rho_j, \rho_{lm}(r), \rho_q(z)\}$ are used in each iteration to create the Coulomb potential via the solution of Poisson's equation and to synthesize the exchange-correlation potential. These coefficients of the input and output densities are also used to enable and accelerate the self-consistency procedure via mixing of the input and output density in each iteration.

In the FLAPW method, Poisson's equation is solved by the following technique (23, 26). The actual electron density inside the atomic spheres is replaced by a smooth pseudo-density, which is constructed in such a way that the multi-pole moments outside the atomic spheres are identical to those of the actual density. The smooth pseudo-density is Fourier transformed to a form that is analogous to the expansion of the interstitial charge density (Equation 19). In this representation, the solution of Poisson's equation is very simple, yielding the correct electrostatic potential everywhere outside and on the surface of the atomic spheres. Using the correct potential on the surface of the atomic spheres and the original electron density inside each sphere, the correct potential can subsequently be calculated also inside the spheres by a Green's function method. The potential in the vacuum region is also obtained by a Green's function method involving one-dimensional integrations of the z-dependent part of the vacuum potential.

The exchange-correlation potential is evaluated on a real-space grid and then fitted to analytic representations in each of the regions corresponding to the form of the wave functions in the various regions. Thus the evaluation of the FLAPW matrix elements (cf Figure 2) can be done analytically. This is straightforward but tedious both in programming effort and in computational demand. The difficulty arises from the fact that each function has to be integrated over certain regions of space.

The FLAPW method is an all-electron approach, but only the valence electrons are expanded in a variational basis set as described above. The core electron states, which are localized within the atomic spheres, are recalculated numerically in each iteration using a scheme equivalent to free atoms. Only the spherical part of the current FLAPW potential is used for these calculations. In the implementation mentioned above (23), the core electrons are treated fully relativistically, and a scalar-relativistic approach (21) is adopted for the valence states.

One of the most powerful capabilities of present density functional methods is the ability to predict the total energy as a function of nuclear positions. As discussed above, this allows the determination of important structural properties such as surface reconstructions, adsorbate geometries and, at least in principle, the simulation of diffusion and chemical reactions. Once the total electron density, the effective potential, and the one-particle wave functions have been calculated, all terms necessary for the evaluation of the total energy are given, as can be seen from Equations 2–7. Measured on the scale of total energies originating from all electrons, only very small and delicate energy differences determine the structural and chemical behavior of solids, surfaces, and molecules. For example, the energy change involved in the reconstruction of a surface can be eight orders of magnitude smaller than the absolute value of the total energy. Therefore, care has to be taken to ensure algorithmic and numerical stability.

The total energy capability has been implemented in the thin-film FLAPW method (24) and applied to a large number of systems. These calculations as well as many other systematic studies demonstrate that the local density functional approach provides a surprisingly accurate tool for structural predictions of solids, surfaces, and molecules with a typical agreement of about ± 0.02 Å between measured and calculated interatomic bond distances. However, the local density approximation overestimates the binding energy and underestimates the bond distance between weakly bound atoms such as a Mg dimer or in hydrogen bonds. Gradient corrections to the exchange and correlation terms seem to overcome some of these limitations.

Now it is known that a well-optimized atomic structure—generally unknown for most low-dimensional systems explored in the laboratory or modeled and studied theoretically for the first time—is a prerequisite for obtaining reliable predictions of their structural, electronic, and magnetic properties. This is especially the case for transition metal systems with strong interfacial hybridization and exchange interactions. Of course, geometry optimizations are greatly facilitated if the forces on each atom are known. To this end, one has to differentiate the total energy with respect to the nuclear coordinates of each atom. In principle, this is straightforward, but the complexity of the LAPW basis functions makes this a tedious task, and only recently has this feature been implemented for bulk solids by Yu et al (27) and Soler & Williams (28). Following their work, Wu & Freeman (29) recently developed a highly precise and efficient procedure to determine the atomic forces based on the FLAPW thin-film method. They have already applied this force approach for the determination of multilayer relaxation and reconstruction for a number of systems such as W(001), Fe(111), Mn/Fe(001), MgO(001), and 4*d* overlayers on MgO(001). The results on single layer relaxation show very good agreement between force and total energy results. In addition, Wu & Freeman (29) found that the forces are even numerically more stable with respect to charge self-consistency, a number of sample *k*-points, energy cutoff, etc and thus are reliable to more digits.

Finally, there are other full-potential and all-electron methods that can be applied to the study of electronic structures. In particular, we mention the recently developed full-potential, linearized muffin-tin-orbital (FP-LMTO) methods of Methfessel et al (30). Similar to the FLAPW method, the space is partitioned into spheres around atoms and the interstitial space. Inside the spheres, both the FLAPW and FP-LMTO methods use atomic-like functions. The methods differ conceptually in the treatment of the interstitial region. Whereas plane waves are used in the FLAPW approach, localized functions derived from scattering theory such as Hankel functions are employed in the FP-LMTO method. Such an approach can be computationally more efficient for systems with open structures, but these localized functions do not offer a systematic convergence.

LOCALIZED ORBITAL METHODS The FLAPW method, like the pseudopotential plane wave approach, is particularly well suited for the calculation of fairly densely packed solids as well as clean surfaces or surfaces with high-coverage overlayers. Because these methods require periodicity, the study of low coverages of atoms or molecules interacting with surfaces, for example, would require very large unit cells, and the computational effort would be high. The use of a cluster approach or, preferably, a cluster embedded in a semi-infinite crystal are better suited for the investigation of chemically localized phenomena on surfaces such as chemisorption and surface reactions. Linear combinations of atomic orbitals have emerged as a successful approach to such localized systems, and one can build on the experience gained in quantum chemistry during the past three decades.

Gaussian-type orbitals have become the standard in Hartree-Fock-based quantum chemical calculations. Gaussians have convenient mathematical properties that make it possible to calculate efficiently the four-center integrals occurring in Hartree-Fock theory. However, Gaussian-type orbitals do not have the correct shape required by wave functions, especially near the nucleus and also in the tails of the wave functions. Therefore, a tremendous effort has been made by quantum chemists to construct appropriate combinations of Gaussians that mimic true wave functions. In density functional methods, these four-center integrals can be avoided; therefore, there is no need for Gaussians, and a range of alternatives have been explored. Among the possible alternatives, the use of numerical functions is particularly elegant and powerful.

The concept of a localized orbital method is based on the fact that in condensed systems, the electron wave functions maintain their atomic-like character to a large extent. Therefore, one can express the one-electron wave functions of a molecule, a surface, or a bulk solid as a linear combination of atomic orbitals $\phi_j(\mathbf{r})$, i.e. $\psi(\mathbf{r}) = \sum_j c_j \phi_j(\mathbf{r})$ provided that enough variational freedom is given to the radial wave functions for the contraction or expansion caused by neighboring atoms. Each atomic orbital is constructed from a radial part and a spherical harmonic, $\phi_j(\mathbf{r}) = R_{nl}(r) Y_{lm}(\theta, \varphi)$ with $R(r)$ being the radial part of the wave functions depending on the principle quantum number n and the angular momentum quantum number l. The functions $Y(\theta, \varphi)$ denote spherical harmonics labeled by the quantum numbers l and m.

Within such an atomic-orbital basis, the radial contraction or expansion is made possible by the right choice of the radial functions $R_{lm}(r)$. For each state such as those derived from O-2s or O-2p levels, one should use at least two radial functions such as

$$\phi_{nl}^{(1)} = R_{nl}^{(1)}(r)\,Y(\theta, \varphi), \quad \phi_{nl}^{(2)} = R_{nl}^{(2)}(r)\,Y(\theta, \varphi). \qquad 20.$$

The function $R^{(1)}(r)$ is obtained, for example, from a neutral atom and the function $R^{(2)}(r)$ from a positive ion. For a given pair of quantum numbers n and l, the function of the positive ion is more contracted. Figure 5 gives an illustration of such a pair of wave functions.

In fact, this construction of basis functions from pairs of radial functions with the same quantum numbers also underlies the basis functions of the FLAPW method and the LMTO method mentioned above. In FLAPW, for each angular momentum-quantum number, one uses an atomic-like radial wave functions calculated at a trial energy E and a second radial function that is the energy derivative of the first function. In the LMTO method, one employs two radial wave functions per angular momentum-quantum number corresponding to two different trial energies.

The density functional method for molecules (DMol) (31, 32) uses

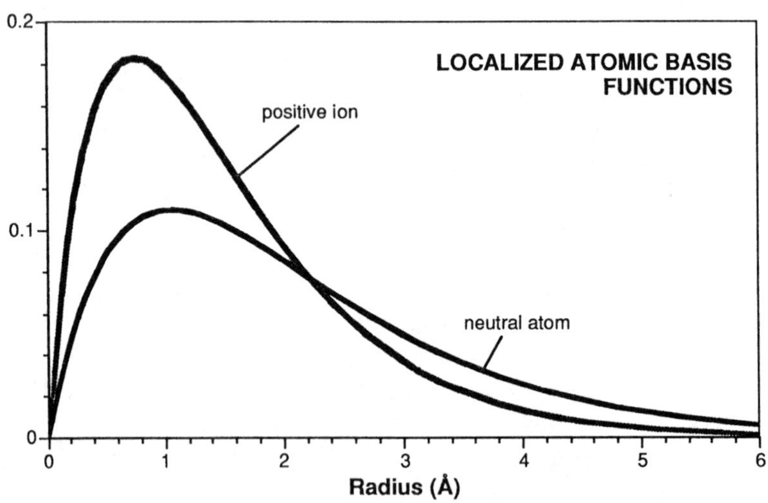

Figure 5 Atomic wave functions as used to construct localized basis sets. Functions from a neutral atom and positive ions provide radial variational freedom. The functions shown above are hydrogen-like $2p$ wave functions.

numerically generated atomic orbitals as basis functions and a self-consistent solution of the Kohn-Sham equations with a conventional matrix diagonalization. As described above, the self-consistency procedure requires the sequence: $\rho^i \rightarrow V^i \rightarrow \psi \rightarrow \rho^{i+1} \rightarrow V^{i+1} \rightarrow \ldots$. The potential consists of an electrostatic part and the exchange-correlation part. Because of its nature, the exchange-correlation potential has to be evaluated on each grid point of a numerical grid. In the DMol approach, this grid for the exchange-correlation terms is the same as that for the numerical integration. Even in the methods using analytic basis functions, such as Slater-type orbitals or Gaussians, the exchange-correlation terms involve grid-based numerical integrations, whereas the electrostatic potential terms can be evaluated analytically.

The electrostatic potential is obtained either by solving Poisson's equation and then calculating the matrix elements or by expanding the charge density in a set of analytical functions $g(\mathbf{r})$, namely $\rho(\mathbf{r}) = \Sigma_q \rho_q g_q(\mathbf{r})$. The Coulomb integrals originating from electron-electron interactions are then evaluated explicitly as

$$U_{ij} = \sum_q \rho_q \iint \frac{\phi_i^*(\mathbf{r}) g_q(\mathbf{r}') \phi_j(\mathbf{r})}{|\mathbf{r}-\mathbf{r}'|} d\mathbf{r}\, d\mathbf{r}'. \qquad 21.$$

In the DMol method, the former approach is taken. The charge density is projected onto a multi-center, multi-polar expansion that enables the solution of Poisson's equation and hence the evaluation of the electrostatic potential on the numerical grid.

In localized orbital methods using Slater-type orbitals or Gaussians, the charge density is fitted to an auxiliary basis set and then integrals (Equation 21) are calculated. Therefore, analytic-localized orbital methods are actually hybrids between analytic- and grid-based numerical methods. Contrary to earlier expectations, analytic-localized orbital methods still require numerical grids almost as dense as the grid for the exchange-correlation part of purely numerical orbital methods.

Localized orbital methods allow the calculation of forces, i.e. the derivatives of the total energy with respect to nuclear displacements. Contrary to the pseudopotential plane wave method, these forces are quite complicated, involving terms that originate from the incompleteness of the localized basis sets. Atomic force determinations and attendant automatic geometry optimizations are now a standard, and powerful, feature of the DMol code.

We have discussed the concept of pseudopotentials to eliminate the core electrons from the calculations. In principle, this approach can also be taken in methods using localized basis sets. Alternatively, one can freeze the wave functions of the core electrons and include only the valence states in the variational procedure. This reduces the size of the Hamiltonian matrix and thus saves computer time. However, the core and valence states need to remain orthogonal, which requires explicit orthogonalization steps.

The power of localized basis sets comes from the fact that this approach works not only for clusters describing a surface, but from ample evidence (32, 33) that such approaches give accurate geometries, vibrational frequencies, and binding energies for a wide range of organic, inorganic, and organometallic molecules, especially if nonlocal gradient corrections are included. Therefore, this approach lends itself particularly well to the study of chemical reactions of molecules on surfaces.

CAPABILITIES AND FUTURE PERSPECTIVE

The quantitative prediction of structural, electronic, optical, and magnetic properties from first principles is perhaps one of the greatest successes of computational materials science. The key for this success is the capability of density functional methods to allow the prediction of the total energy of a system as a functional of nuclear position. No empirical or other system-specific parameters are necessary to predict surface geometries to within a few hundredths of an Ångstrom, thus opening the possibility to predict surface relaxations, surface reconstructions, adsorption geometries and, in principle, the entire energy-hypersurface of atoms and molecules on surfaces. Impressive successes have already been achieved: for example, the quantitative explanation of surface diffusion, the reconstructions of the Si(111) and Si(001) surfaces, and a detailed, quantitative picture of the energetics of the reconstruction of the W(001) surface.

The same approach enables the calculation of elastic constants such as bulk and shear moduli and the behavior of solids under pressure. The accuracy and predictive capability is impressive. For systems of vastly differing chemical bonding character—metallic, covalent, ionic—and for any combination of atoms, the interatomic distances between chemically bonded atoms are usually predicted within a few hundredths of an Ångstrom and bond angles within a few degrees of experiment. It thus comes as good news that the theoretical and computational approaches used in these studies are maturing and are becoming increasingly available to a wider community of scientists and engineers as sophisticated, integrated software systems with graphical-user interfaces for easy use, accessibility, and visualization of the results.

While it is possible to compare these results with crystallographic data for perfectly ordered solids, the computational approach can be extended into domains where experiments become very difficult or even impossible. One can answer, for example, the questions of what happens with the atomic arrangements when an impurity or vacancy is introduced. An intriguing illustration of this capability is a recent study of the nature of phosphorus embrittlement of a grain boundary in iron (34, 34a). The first-principles calculations provide quantitative information on the segregation energy and the phase transition of the phase boundary, and also provide a detailed understanding of the chemical interactions of a phosphorus impurity with the surrounding iron atoms.

Surface and interface phenomena play a critical role in materials science. This is particularly evident in microelectronics, where the continuing reduction in feature size implies a dramatic change in the surface-to-volume ratio. The surfaces and interfaces even of well-known materials such as Cu and Fe show novel effects that represent difficulties to traditional approaches, but at the same time offer opportunities for novel systems. Heterogeneous catalysis represents another area where surface effects are of critical importance.

The progress in experimental surface techniques with the use of the scanning tunneling microscope, the atomic force microscope, and with photoemission experiments using synchrotron radiation is providing a wealth of new data on the atomistic scale. Computational methods are indispensable to interpret and understand these experimental data, and major successes have already been achieved. For example, first-principles total-energy calculations give a detailed understanding of the surface reconstruction of the Si(001) surface (which is the most technologically important Si surface). In a recent study, Tang et al (35) calculated the geometric parameters of the Si-dimers on the reconstructed Si(001) surface. Furthermore, the calculations revealed a very small energy difference of about 0.02 eV between the symmetric and buckled arrangement of the surface dimer. In contrast, computational approaches on the same level of theory on the Ge(001) surface performed by Spiess et al (36) reveal that for Ge, the buckled dimer is significantly more stable than the symmetric form. This detailed understanding of clean semiconductor surfaces sets the stage for the quantitative investigation of adsorbate atoms and molecules interacting with semiconductor surfaces. For example, Tang & Freeman (37) investigated the passivation of the Si(001) surface by Sb, showing the importance of Sb-dimer formation in this process.

Density functional theory has proven to be a unified theoretical approach, applicable to solids and surfaces as well as to isolated molecules (33, 38). Therefore, it is ideally suited to investigate the interactions of molecules with surfaces, as required, for example, to obtain the thermochemical data for chemical vapor deposition processes.

The understanding and the creation of new magnetic materials continues to be an extremely exciting field of both scientific interest and technological importance, especially in the field of magnetic recording. Computational materials science has made major contributions to the advancements in this field and continues to inspire the development of novel materials. This contribution has been most timely because during the course of the last decade, low-dimensional magnetism of surfaces, interfaces, and thin films has matured into a major branch of modern condensed matter physics and is likely to open vast vistas for practical applications. The abrupt termination of the lattice or composition in these systems leads to a variety of new phenomena such as a localized electronic surface states, magnetic moment enhancement, perpendicular magneto-crystalline anisotropy, complex magnetic ordering, etc. Fortunately, today it is possible to synthesize and study thin films with either stable or metastable phases, which has dramatically and importantly increased the range of materials that are magnetic and hence the challenges for understanding magnetism in low-dimensional systems (39).

It is known that ab initio energy band methods, mainly the FLAPW method described above, have played an important role in the understanding and development of low-dimensional magnetism. As discussed recently (40), theoretical calculations predicted the large enhancement of the magnetic moment for $3d$ transition metal (TM) surfaces or TM ultra-thin films deposited on inert substrates and possible magnetism in some normally non-magnetic materials; some of these predictions have already been verified experimentally. Stable magnetic structures, especially some antiferromagnetic (AFM) configurations, can now be predicted by comparing total energies and, in the same way, equilibrium atomic geometries and lattice relaxation (including surface and interface) can also be determined. For example, early experiments on surface magnetism of Fe and Ni surfaces seemed to indicate magnetically dead surface layers. These experimental findings were in bold contradiction to computational studies, which indicated enhanced surface magnetism (e.g. 41). The computational results proved to be correct because early experiments were clouded by surfaces covered with adsorbed atoms. In fact, theory is also capable of describing the quenching of magnetism by chemisorbed species.

Given this remarkable reliability in the calculation of surface magnetism, computational approaches can be used to predict the relative stability of ferromagnetic vs antiferromagnetic ordering in artificially layered Fe/Cr superlattices as a function of the thickness of the Cr layers (42). Even subtle quantities such as Kerr rotations—a fundamental quantity in the development of magneto-optic recording devices—can be calculated successfully from first-principles theory (43). Even extremely sensitive properties such as magnetocrystalline anisotropy energies can now be obtained from accurate first-principles calculations (44).

These few highlights demonstrate only a small fraction of the impressive predictive capabilities of the present state-of-the-art in computational physics, especially in the area of first-principles calculations discussed here. There remain, however, major obstacles. Perhaps the biggest is the simulation of dynamic processes that involve many orders of magnitude in time and length scales, such as crack formation and propagation, materials fatigue, and corrosion. In many cases, it is the processing of materials that gives rise to microstructures that are critical in determining the mechanical properties. Clearly, great efforts still have to be made to capture the full potential of computational materials science. Even today, however, with all their limitations and shortcomings, computational methods have proven to be of real value to the experimentalist. With the continuing developments of computer hardware and software, the scope and importance of computational materials science can only increase. Compared with the history and tradition of experimental materials science, the computational branch is quite young. Given its present successes and the rapid evolution of this field, exciting and most promising opportunities are ahead. It is hoped that these new capabilities will enable the creation of novel materials and processes to meet the changing requirements of our society in the twenty-first century.

ACKNOWLEDGMENTS

This work was supported by the MRL Program of the National Science Foundation at the Materials Research Center of Northwestern University under Award No. DMR-9120521.

Literature Cited

1. Hohenberg P, Kohn W. 1964. *Phys. Rev. B* 136: 864
2. Kohn W, Sham LJ. 1965. *Phys. Rev. A* 140: 1133
3. Thomas LH. 1926. *Proc. Cambridge Philos. Soc.* 23: 542
4. Fermi E. 1928. *Z. Phys.* 48: 73
5. Hartree DR. 1928. *Proc. Cambridge Philos. Soc.* 24: 89
6. Fock V. 1930. *Z. Phys.* 61: 126; 62: 795
7. Slater JC. 1951. *Phys. Rev.* 81: 385
8. Slater JC. 1974. In *Quantum Theory of Molecules and Solids*, Vol. 4. New York: McGraw-Hill
9. Hedin L, Lundqvist SJ. 1971. *J. Phys.* C4: 2064
10. Ceperley DM, Alder BJ. 1980. *Phys. Rev. Lett.* 45: 566
11. von Barth J, Hedin L. 1972. *J. Phys. C* 5: 1629
12. Gunnarsson O, Lundqvist BI, Lundqvist S. 1972. *Solid State Commun.* 11: 149

13. Korringa J. 1947. *Physica* 13: 392
14. Kohn W, Rostoker N. 1954. *Phys. Rev.* 94: 1111
15. Cooper BR. 1973. *Phys. Rev. Lett.* 30: 1316
16. Kar N, Soven P. 1975. *Phys. Rev. B* 11: 3761
17. Andersen DG. 1965. *J. Assoc. Comput. Mech.* 12: 547
18. Slater JC. 1937. *Phys. Rev.* 51: 846
19. Koelling DD, Arbman GO. 1975. *J. Phys. F* 5: 2041
20. Andersen OK. 1975. *Phys. Rev. B* 12: 3060
21. Koelling DD, Harmon BN. 1977. *J. Phys. C* 10: 3107
22. Krakauer H, Posternak M, Freeman AJ. 1979. *Phys. Rev. B* 19: 1706
23. Wimmer E, Krakauer H, Weinert M, Freeman AJ. 1981. *Phys. Rev. B* 24: 864
24. Weinert M, Wimmer E, Freeman AJ. 1982. *Phys. Rev. B* 26: 4571
25. Wimmer E, Krakauer H, Freeman AJ. 1985. *Advances in Electronics and Electron Physics*, ed. PW Hawkes, 65: 357. Orlando, FL: Academic
26. Weinert M. 1981. *J. Math. Phys.* 22: 2433
27. Yu R, Krakauer H, Singh D. 1992. *Phys. Rev. B* 45: 8671
28. Soler JM, Williams AR. 1989. *Phys. Rev. B* 40: 1560
29. Wu R, Freeman AJ. 1995. *Phys. Rev. B* In press
30. Methfessel M, Rodriquez CO, Andersen OK. 1989. *Phys. Rev. B* 40: 2009
31. Delley B. 1990. *J. Chem. Phys.* 92: 508
32. Delley B. 1991. *J. Chem. Phys.* 94: 7245
33. Andzelm J, Wimmer E. 1992. *J. Chem. Phys.* 96: 1280
34. Wu R, Freeman AJ, Olson GB. 1994. *Phys. Rev. B* 50: 75
34a. Wu R, Freeman AJ, Olson GB. 1994. *Science* 265: 376
35. Tang SP, Freeman AJ, Delley B. 1992. *Phy. Rev. B* 45: 1776
36. Spiess L, Freeman AJ, Soukiassian P. 1994. *Phys. Rev. B* 50: 2249
37. Tang SP, Freeman AJ. 1993. *Phys. Rev. B* 47: 1460
38. Wimmer E, Freeman AJ, Fu C-L, Cao P-L, Chou S-H, Delley B. 1987. In *Supercomputer Research in Chemistry and Chemical Engineering*, ed. KF Jensen, DG Truhlar, Ser. 353. p. 49. New York: Am. Chem. Soc.
39. Falicov, LM, Pierce TD, Bader SD, Gronsky R, Hathaway KB, et al. 1990. *J. Mater. Res.* 5: 1299
40. Wu R, Freeman AJ. 1991. *J. Magn. Magn. Mater.* 100: 497
41. Ohnishi S, Freeman AJ, Weinert M. 1983. *J. Magn. Magn. Mater.* 31–34: 889
42. Herman F, Sticht J, Van Schilfgaarde M. 1991. *J. Appl. Phys.* 69: 4784
43. Oppeneer, PM, Sticht J, Herman F. 1991. *J. Magn. Soc. Jpn.* 15: 73
44. Wang D-S, Wu R, Freeman AJ. 1993. *Phys. Rev. Lett.* 70: 869
45. Car R, Parinello M. 1985. *Phys. Rev. Lett.* 55: 2471

Annu. Rev. Mater. Sci. 1995. 25: 37–68

MOLECULAR ORBITAL MODELS OF SILICA

L. L. Hench and J. K. West

Advanced Materials Research Center, University of Florida, Alachua, Florida 32615

KEY WORDS: quantum, silica, SiO_2, fracture, hydrolysis, optics, infrared

---

ABSTRACT

Amorphous silica (a-silica) has a distribution of various ring sizes. It is generally known that these rings vary in size from 2-member rings through 6-member (or possibly higher) rings of silica tetrahedra. These silica rings, as well as chains, are modeled by the following semi-empirical quantum molecular orbital (MO) methods: AM1, PM3, and INDO.[1] The calculations predict a range of Si-O bond lengths to be from 1.65 to 1.74 Å and a range of Si-O-Si bond angles to be from 102 to 174°. These values compare well with previous ab-initio results. The predicted electronic and infrared spectra also agree well with previous models and experimental results. These MO models yield significant insight into the understanding of structure-property relationships for bulk amorphous silica, as well as for porous type VI sol-gel silica. The same MO methods are applied to a variety of topics: quantum clusters of $Si\text{-}SiO_2$, silica densification, silica-water interactions, silica pore models, Cr^{3+}-doped silica, silica fracture, and silica-amino acid interactions.

[1] Abbreviations used in text: AM1, Austin Method 1 parameterization based on MNDO; CNDO, Complete Neglect of Differential Overlaps; Extended Huckel, postulated Hamiltonian with Slater-type orbitals; G, Gaussian-type orbital of the form e^{-ar^2}; INDO, Intermediate Neglect of Differential Overlap; MNDO, Modified Neglect of Diatomic Overlap (has more correlation terms than INDO); MP2, Multibody Perturbation ab-initio theory with two centers; MP4, Multibody Perturbation ab-initio theory with four centers; PM3, Parametric Method 3 based on MNDO theory; RHF, Restricted Hartree Fock theory; STO, Slater-type orbital of the form e^{-ar}; UHF, Unrestricted Hartree Fock theory; ZINDO, Zerner's version of INDO.

0084–6600/95/0801–0037$05.00

Introduction

Understanding the relationship between the molecular structure and the physical and chemical properties of silica and silicates is one of the most significant topics in materials science because of the ubiquitous nature of silicon-oxygen containing compounds in the geosphere and biosphere (1). Silicon-based chemistry is a well-established field, as is structural analysis of silicon-based compounds using a variety of diffraction and spectroscopic methods (2). However, until recently there have been only a few attempts to correlate the molecular structure of silica and silicates with physical properties at a quantum mechanical level (3–13). These previous studies have been restricted in complexity due to the limitations in the various quantum mechanical models used to represent the electronic interactions in the SiO_2-containing clusters. Consequently, many of the structures calculated have been very small and involve only a few tetrahedra and siloxane bonds. Prior molecular orbital (MO) methods have also been restricted in their ability to model simultaneous changes in bond angles and bond distances. Calculations of transition states of activated intermediate complexes were also not possible for larger clusters until recently.

During the last three years, the ease of modeling silicon-based compounds has improved dramatically. Quantum chemists have produced new semi-empirical MO methods that are capable of calculating silica clusters with up to six tetrahedra per cluster in very reasonable time periods (9). The MO models are sufficiently rapid that the transition states associated with environmental effects can be calculated. Thus it is possible to achieve a quantitative theory of fracture of silica and compute the interaction of biological molecules with silica surfaces. This chapter reviews the calculational methods used in these studies and the results obtained for a range of SiO_2 structure property-environment interactions.

Molecular Orbital Methods

Calculational chemistry (9) is the study of the interaction of atoms by solving Schrödinger's equation $H\psi = E\psi$ for the electrons in the system of atoms being studied. Figure 1 shows the field divided into three main areas: statistical mechanics, quantum chemistry, and molecular dynamics (14). Quantum chemistry is generally split into ab-initio (rigorous) and semi-empirical (simplified) methods.

Ab-initio methods treat all of the electrons in an assembly of atoms, whereas semi-empirical methods consider only the valence electrons. Excellent reviews of these methods and their differences are given by Boyd (9) and Zerner (15).

Figure 2 is a schematic depicting some of the calculational methods in

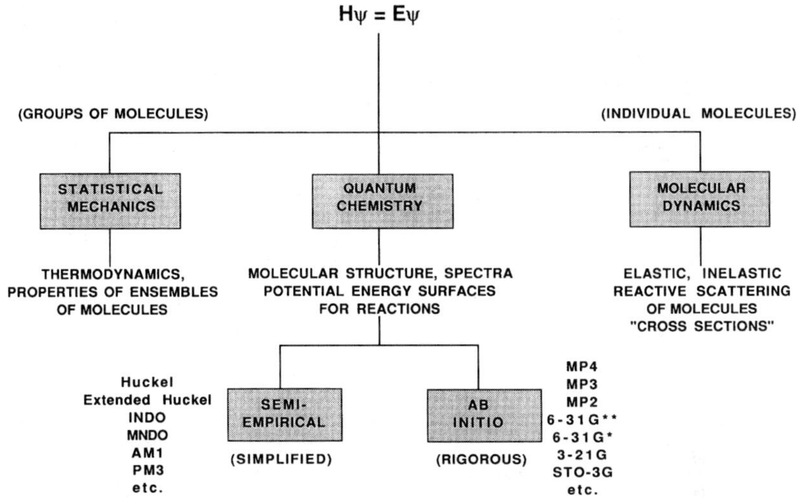

Figure 1 The relationship for different areas of calculation chemistry (Schrödinger equation).

relationship to one another. The computational (CPU) time increases for ab-initio methods (STO-3G, 3–21G, 6–31G*, 6–31G**, MPZ, MP4) as well as for cluster size, i.e. number of atoms in a molecule or molecular cluster. The semi-empirical methods AM1 and PM3 give good results with only a modest increase in computational time. The errors increase dramatically for the Huckel methods, but very large collections of atoms may be modeled.

In this review we discuss silica-based calculational models using a CaChe Worksystem®. The models provide an interface to ZINDO (Zerner's Intermediate Neglect of Differential Overlap program; 16), through a Macintosh environment, as well as to MOPAC (17), which provides an interface to MNDO, AM1, and PM3 semi-empirical methods.

Davis and colleagues have shown that the modified neglect of diatomic overlap (MNDO) semi-empirical MO method can be used to model the reactions of Si-O-containing molecules (8, 10). However, there are two situations in which MNDO produces large errors. MNDO overestimates core-core repulsions between atoms when they are separated by approximately van der Waals distances. This means that MNDO cannot reproduce H-bonds or accurately calculate the heat of formation, H_f, for strained or crowded molecules, such as those involved in bond fracture, or many transition state structures. The MNDO method also overestimates tran-

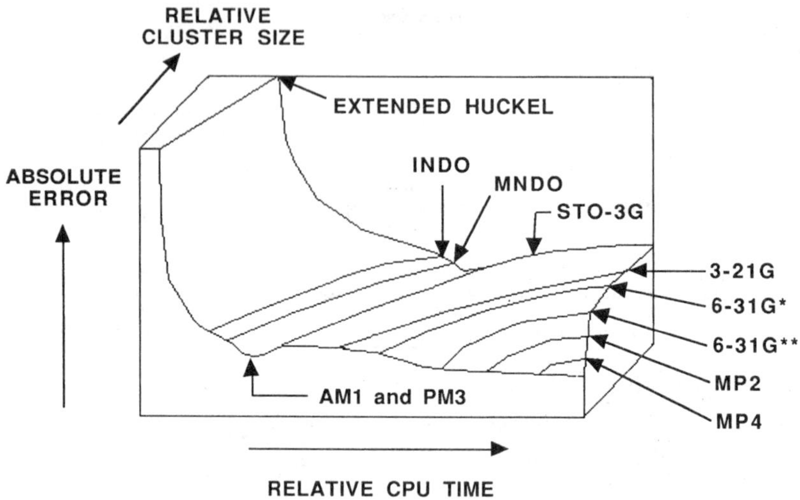

Figure 2 The relationship for different quantum mechanical methods. This study uses AM1 and INDO to interpret experimental data and predict material-property relationships. These methods are installed on a CaChe Scientific, Inc. worksystem.

sition state energies and distorts transition state geometries, again fatal flaws for calculating bond fracture mechanisms or the interaction of organic-inorganic clusters. The semi-empirical Austin Method (AM1) (17–19) reduces these problems by modifying the core repulsion function used in MNDO with additional Gaussian terms, and the calculated structures and heats of formation match well with experimental values (9).

ZINDO was developed at the University of Florida Quantum Theory Project (15). ZINDO provides two valence-electron-only semi-empirical procedures: a method for computing spectroscopic properties (electronic spectra), and a method for computing molecular geometries (conformations and structures).

CNDO (Complete Neglect of Differential Overlap) and INDO (Intermediate Neglect of Differential Overlap) methods were developed by Pople (20). Interested readers are encouraged to read Pople's book for a valuable introduction to the fundamental theory behind semi-empirical methods (20). A more recent review by Zerner is also excellent (15).

ZINDO uses the theoretically based INDO parameterization and therefore contains parameters for more elements than does the experimentally based MOPAC. Thus ZINDO can be used to calculate geometries for molecules that cannot be calculated with MOPAC. However, because MOPAC has been specifically parameterized for ground-state geometries and heats of formation, MOPAC (with the AM1 or PM3 parameters) is the better method for computing molecular structures and energetics (17).

ZINDO can use *d* orbitals and therefore can be used for calculations involving most transition series metals. By contrast, MOPAC uses only *s* and *p* orbitals and cannot be used for transition metals. ZINDO has an alternative parameterization for electronic (UV/visible) spectra called INDO/S, which produces good results. Since MOPAC is parameterized for ground-state heats of formation, excitation energies determined by MOPAC will probably not be as good as those from ZINDO. Although, the ZINDO geometry optimization is faster than MOPAC, the geometries obtained are not as reliable as those from most MOPAC models (MNDO, AM1, and PM3). ZINDO geometries are nearly as accurate as the MOPAC MINDO/3 geometries (15).

A comparison of various sizes of silica rings modeled using AM1 and PM3 shows that both models represent the geometries and infrared (IR) spectra better than previous MNDO IR spectra and geometries (21, 22). AM1 appears to represent the bulk distribution of rings slightly better than PM3 (23).

As a calculational strategy, the geometries for silicate rings and chains reviewed herein were optimized using AM1. These optimized geometries were then submitted to ZINDO configuration interaction (CI) calculations to predict the electronic spectra. The same AM1 geometries were also submitted to a FORCE calculation within MOPAC to calculate the IR spectra.

Silica Rings

The relationship between the molecular structure of amorphous silica (a-silica) and its physical properties has been examined by many investigators (1). Various ab-initio and semi-empirical quantum MO methods have been used to calculate the theoretical geometries of silica rings of different sizes and the effect of ring size on bond energy and IR and Raman vibrational modes. However, the early ab-initio calculations of Revesz & Gibbs (3, 4) and Hess et al (5) were limited to very small clusters containing only a few Si and O atoms. Small clusters of only one or two silica tetrahedra are insufficient to model a-silica because various structural analyses show that there is a distribution of Si-O-Si bond angles and distances (24) and a distribution of 3-, 4-, 5-, and 6-member silica rings (25). The semi-empirical MO method of modified neglect of differential overlays (MNDO) of atomic orbitals (MOPAC 3.0) used by Uchida & Shinmei (7) was capable of computing ring and chain clusters with as many as six silica tetrahedra. However, the theoretical approximations used in the MNDO method restricted the bond angles to $\sim 180°$, and therefore "the theoretical vibrational spectra contain some systematic errors" (7). Similar limitations and errors exist in Galeener's analysis of silica rings based on ab-initio calculations (6). Consequently, earlier theoretical analyses do not yield calculated vibrational spectra that match very closely with experiment.

We (21–23) have studied silica rings using both AM1 and PM3 (26) methods. Figures 3 and 4 show the optimized structures for 2-, 3-, 4-, 5-, and 6-member silica rings. The two models produce similar lowest energy geometries for small rings (two to four tetrahedra) but larger rings modeled

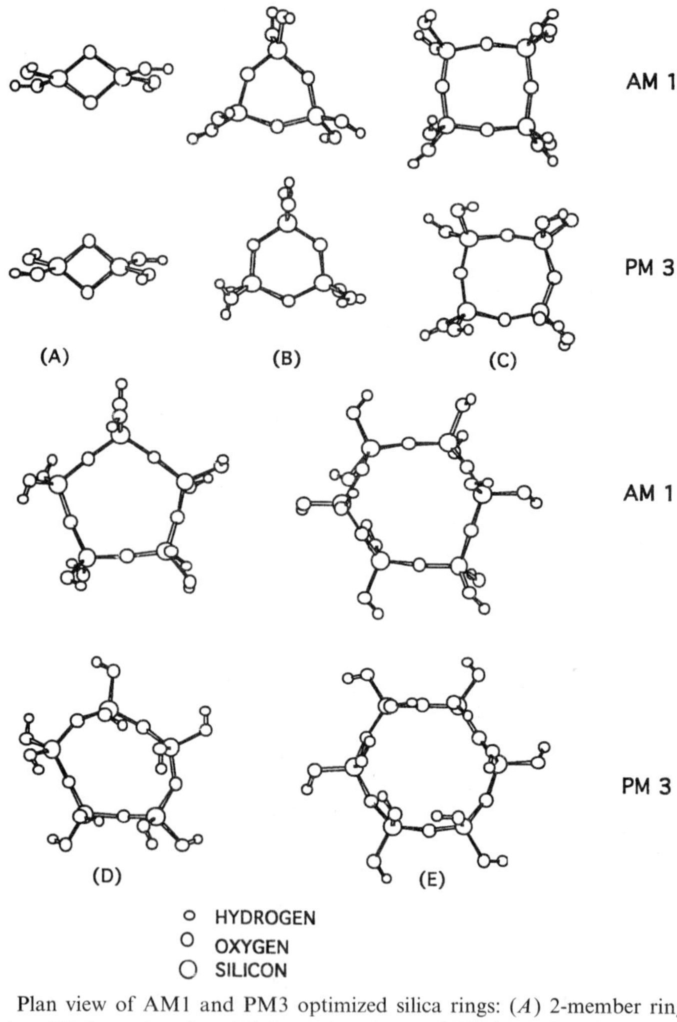

Figure 3 Plan view of AM1 and PM3 optimized silica rings: (*A*) 2-member ring, (*B*) 3-member ring, (*C*) 4-member ring, (*D*) 5-member ring, and (*E*) 6-member ring.

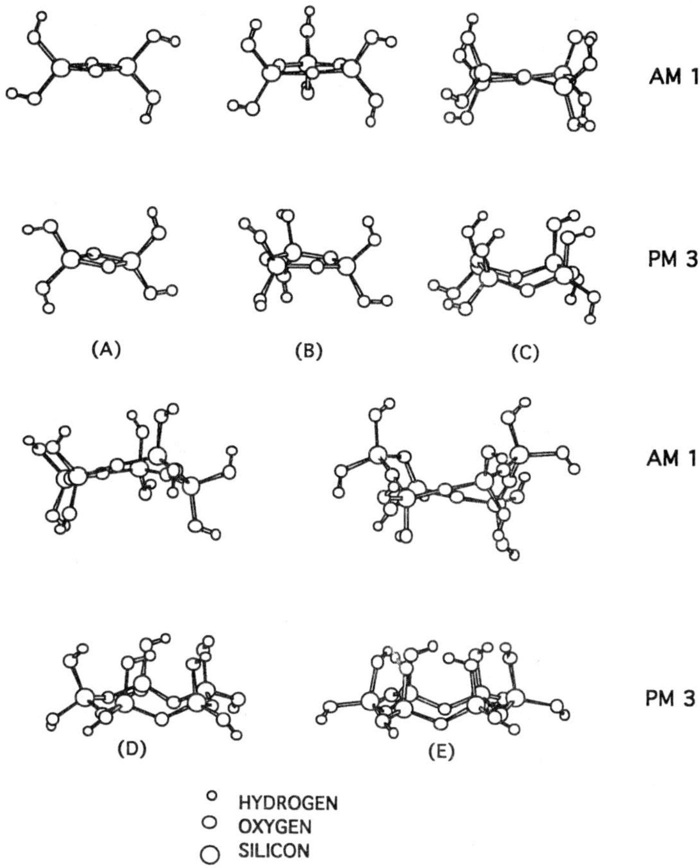

Figure 4 On-edge view of AM1 and PM3 optimized silica rings: (*A*) 2-member ring, (*B*) 3-member ring, (*C*) 4-member ring, (*D*) 5-member ring, and (*E*) 6-member ring.

by PM3 have different symmetry than AM1 rings. Figure 5 shows that AM1 predicts that the 5-member ring is the most stable while PM3 predicts that the 6-member ring is more stable. The Galeener ab-initio study (6) predicts the 4-member ring to be most stable. These three studies are also compared in Figure 6 showing the ring size dependence of bridging Si-O-Si bond distances and angles. Both Galeener's study and AM1 match the range of angles found experimentally (24). PM3 predicts bond lengths that match XRD pair function analyses very well (24).

Both AM1 and PM3 methods can be used to calculate force constants between all neighboring atoms that are converted to a vibrational spectrum using a FORCE, normal mode analysis, calculation with MOPAC; for

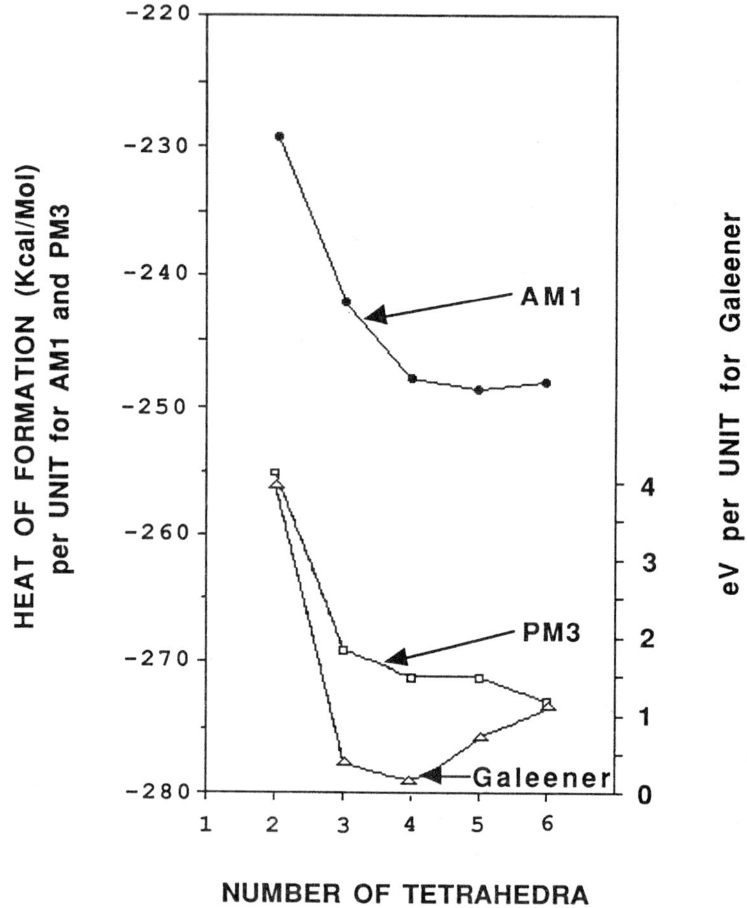

Figure 5 The distribution of theoretical energies per tetrahedron in AM1, PM3, and Galeener's models of silica rings.

details, see Stewart (26, 27). Thus the frequency of vibrational transitions are predicted by the calculations that compute the force necessary to distort atom pairs in the molecular cluster from their equilibrium geometries. The intensities of the IR-active transitions are plotted. The vibrational spectra obtained in this manner are idealized gas phase spectra and therefore are only an approximation to a-silica properties. Also, molecular transitions that are forbidden by symmetry in the MO models are present in experimental spectra of the condensed phase spectra. Anharmonicity effects, overtones, or combination transitions are also absent from the theoretical IR spectra. Thus the calculated IR spectra are sharp discrete lines, whereas

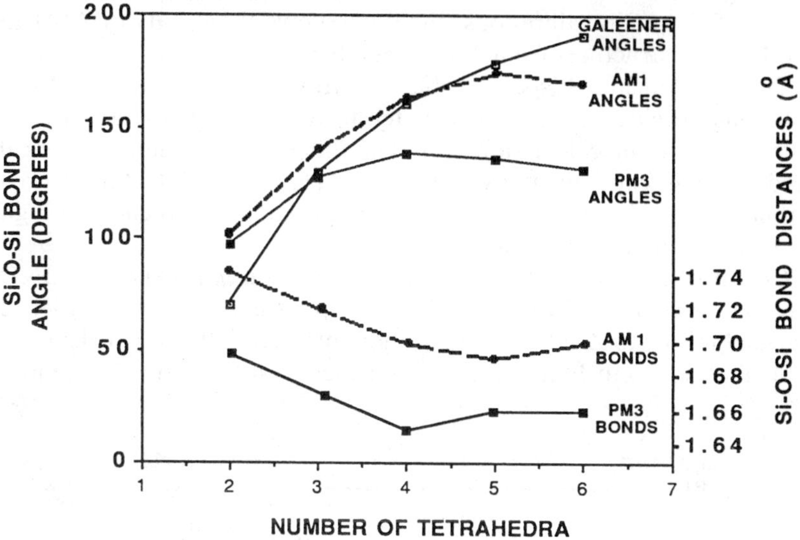

Figure 6 Comparison of AM1, PM3, and Galeener's bond lengths and bond angles for models of silica rings.

experimental a-silica modes are broadened by rotational levels, instrument optics, molecular motions, and other solid-state effects. These effects are approximated in the calculated spectra by introducing a Gaussian-like distribution for each vibrational mode with a half width of 50 cm^{-1}.

Figure 7 shows the experimentals IR spectra for a-silica with the assign-

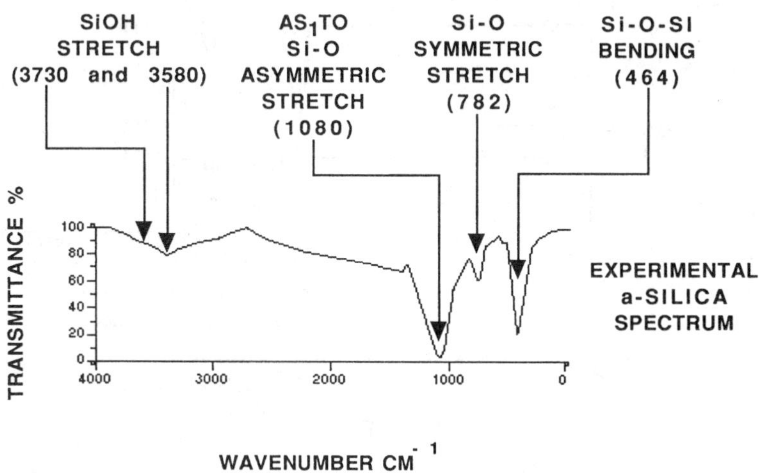

Figure 7 FTIR peak assignments for a-silica.

ments of the major vibrations. Figure 8 shows the results of AM1 and PM3 for the structures from 2- to 6-member silica rings.

The theoretical IR spectra obtained from the AM1 and PM3 calculations provide a much closer match to the experimental a-silica spectrum than earlier calculations (7). The difference between theoretically predicted and experimental values for the 1120 cm^{-1} Si-O ASTO stretch is small. The AM1 spectra are shifted approximately 30 cm^{-1} to higher frequencies compared to PM3.

Chains of silica tetrahedra were also modeled using AM1 and ZINDO (28). These chain structures and the ring structures presented earlier (23) were submitted to ZINDO configuration interaction (CI = 6) calculations (16). The results in Figure 9 show that there is little or no change in the

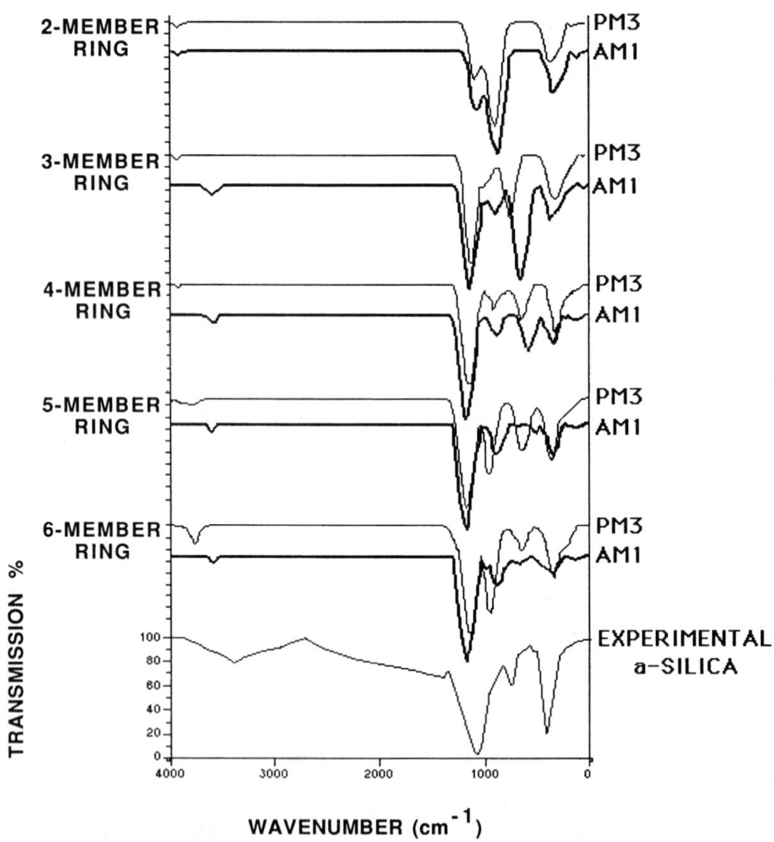

Figure 8 Comparison of AM1, PM3 calculated IR spectra of silica rings, and experimental IR spectra of a-silica.

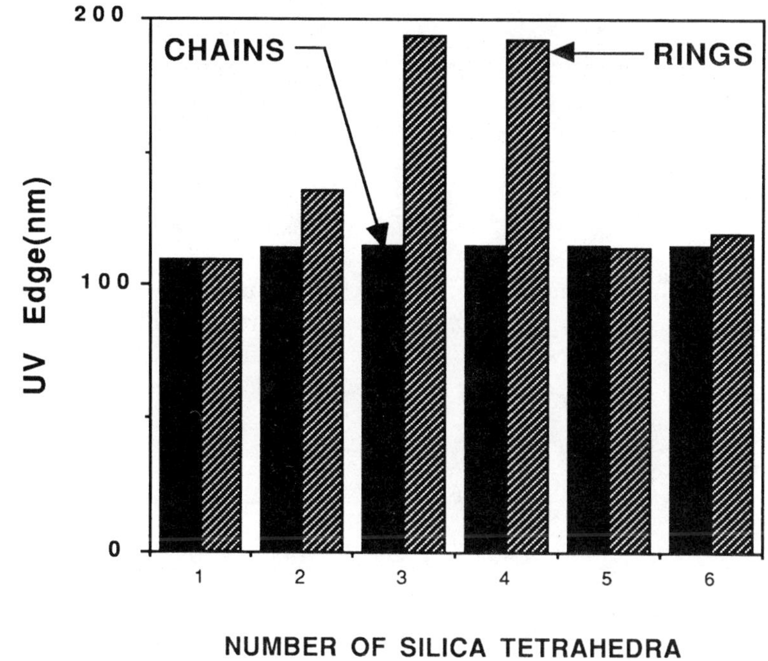

Figure 9 Comparison of the theoretical electronic UV edge between silica rings and chains.

UV absorption edge of the various silica chain structures. The silica rings, however, show a dramatic change of nearly 80 nm in the UV edge, and the calculations predict that a maximum (longer wavelength) absorption should occur for materials with a large concentration of 3- and 4-member rings (28). Wallace, using RAMAN spectroscopy, has shown that the concentration of 3- and 4-member rings in porous Type VI gel-silica changes with processing temperature (29) and with water content (30). The gel-silica network density also increases as processing temperature is increased (29, 31). Figure 10 shows that the 50% UV cutoff goes through a maximum (longer wavelength) as the processing temperature increases because of structural changes. This effect is probably not seen in small-pore (1.2 nm) gels because the [OH] is much higher than for larger pore gels (28, 32).

We propose the following equation as a first approximation to predict the wavelength, λ_{cutoff} of the UV cutoff in porous Type VI optical silica:

$$\lambda_{\text{cutoff}} = \alpha[\text{OH}] + \beta[\text{D}_2] + \gamma[\text{D}_1] + \delta(\Delta)^{+3/2}, \qquad 1.$$

where [OH] is the concentration of OH^-, $[\text{D}_2]$ is the concentration of 3-

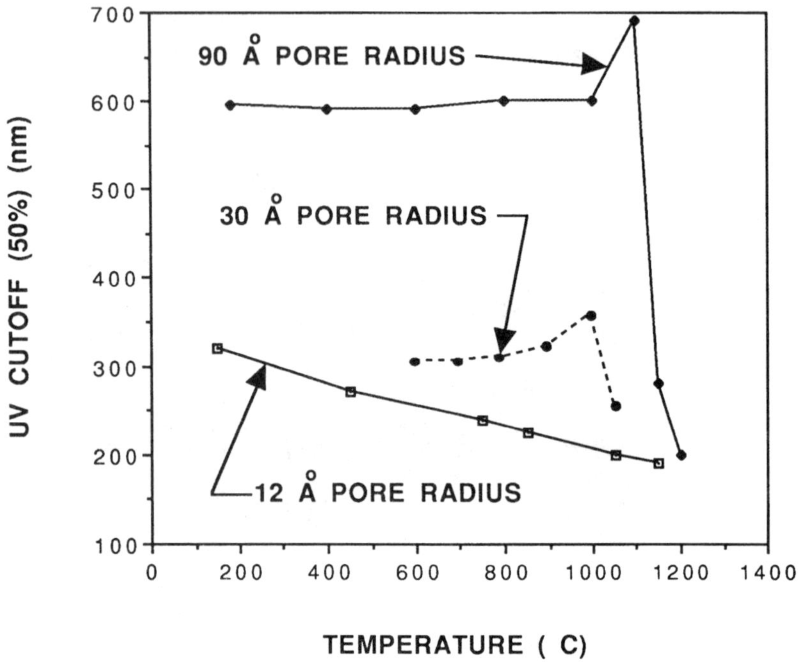

Figure 10 Experimental UV cutoff for porous sol-gel silica with different pore size as a function of processing temperature.

member silica rings, $[D_1]$ is the concentration of 4-member silica rings, Δ is the network wall thickness (Raleigh scattering component), and α, β, γ, and δ are linear coefficients.

The effect of the 3- and 4-member rings on UV absorption is not seen in bulk silica because of the very low concentrations of these structural units as compared to the porous materials.

Silicon-Silica (Si-SiO_2) Clusters

The investigation of spark-processed silicon-silica nanocrystalline structures has been reported by Hummel (33–36). It has been speculated that the ongoing research on "porous silicon" may eventually lead to efficient silicon-based light-emitting diodes, advanced display systems, optical memories, or lasers.

Figure 11 shows the results of spark processing performed in dry-air at room temperature and at liquid nitrogen temperature. In some cases, the experiments are conducted in a dried, high purity nitrogen atmosphere in order to exclude possible hydrogen or oxygen involvement during sample preparation, or in a carbon dioxide atmosphere.

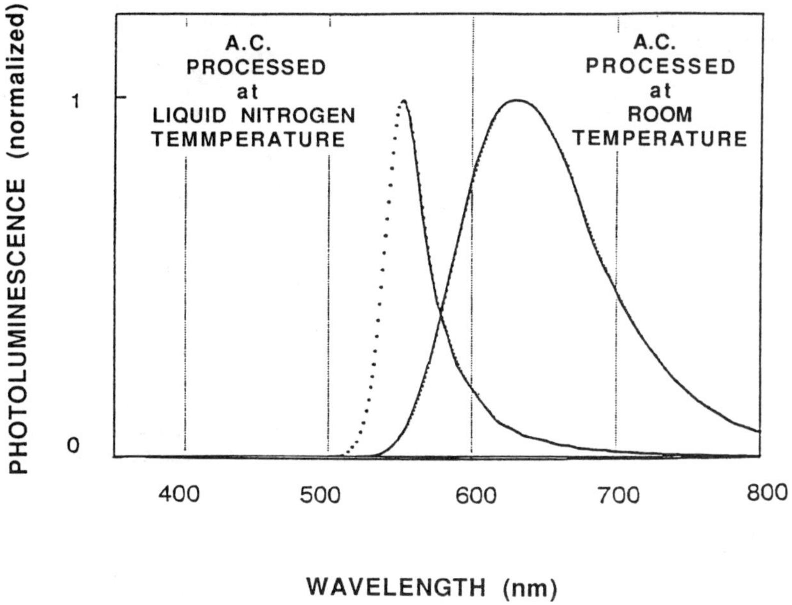

Figure 11 Normalized photoluminescence spectra of A.C. processed porous silicon wafers prepared at liquid nitrogen and room temperature and laser excited at 488 nm. (After Reference 36.)

These two normalized photoluminescence spectra were measured by excitation with the 488 nm beam of an argon laser. A photoluminescent peak at 630 nm (1.97 eV) was observed, as previously reported for air and nitrogen atmospheres (34, 35). In another experiment, the silicon wafer was spark-processed in an environment that exists slightly above liquid nitrogen temperature near 77 K. A blue shift was observed for both the ac and dc processing conditions (36).

AM1 and ZINDO semi-empirical MO methods were used to model small silica rings with nitrogen defects proposed to explain the photoluminescent shifts. Figure 12 compares the theoretical electronic spectra for an isolated 3-member silica ring with one that has a NH_2 substitution of OH. The spectra show that the nitrogen defect in the silica cluster produces absorption peaks in the 250 and 300 nm regions. The predicted blue absorption is very near the photoluminescence results observed for liquid nitrogen and air spark–processed porous silicon wafers. Thus even

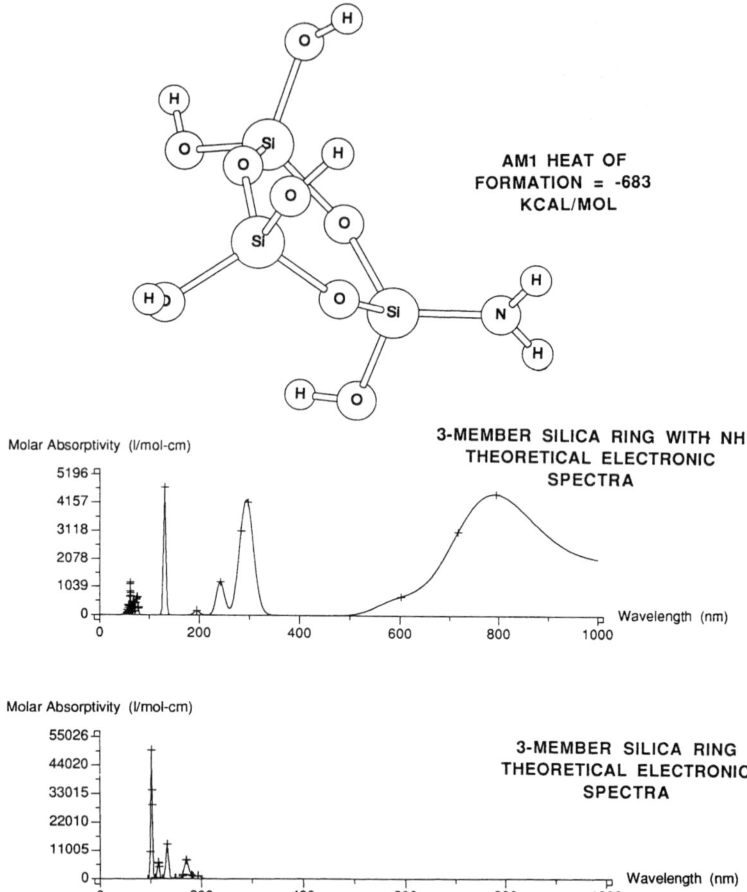

Figure 12 AM1 optimized structure of a 3-member silica ring bonded to NH_2 and the resulting change in the theoretical (ZINDO) electronic spectra.

MO calculations of small clusters reveal electronic spectra similar to experiment.

Silica Densification

Brinker & Scherer (2) divide the shrinkage of sol-gel–derived silica during sintering into three regions. Region 1 is defined as the temperature range in which weight loss occurs, and there is very little gel shrinkage. Region 2 is where both shrinkage and weight loss occur. In this temperature range, the shrinkage and weight loss are large. Region 3 is defined as the final stage where shrinkage occurs through viscous flow with very little weight loss. Wallace (29), Vasconcelos (31), and others have shown that the structure of the network also changes through a relaxation mechanism. In porous gels, the structural density actually decreases in Region 3 to that of a-silica, 2.2 gm/cc, as illustrated in Figure 13.

This relaxation mechanism has been modeled using AM1 (37, 38). If ring contraction is a model for slow crack growth, then the reverse action represents structural relaxation through ring expansion as shown in Figure 14 (39).

This mechanism of ring expansion reduces the network (structural) density as the smaller (3- and 4-member) rings expand. This is also consistent with RAMAN spectra (29) that show that the concentration of larger (D_1, 4-member silica) rings increases at the expense of smaller (D_2, 3-member silica) defects.

Silica-Water Interactions

During densification, silicate rings form on the pore surface of silica gels (2, 29, 32, 40, 41) by condensation of adjacent Si_sOH. These rings and the residual Si_sOH influence the behavior and properties of guest molecules and their interactions with the silica matrix. [Si_sOH] is determined by the sintering temperature, T_p, and the degree of rehydroxylation caused by subsequent adsorption of water. [Si_sOH] of a gel cannot be determined by

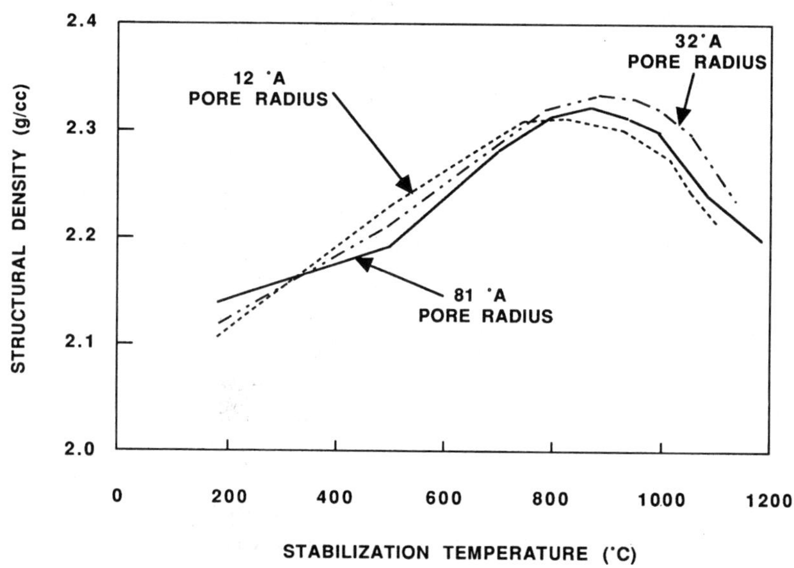

Figure 13 Variation of structural (network) density as a function of densification temperature for silica gel samples with 12, 32, and 81 Å pore radii (after Reference 31).

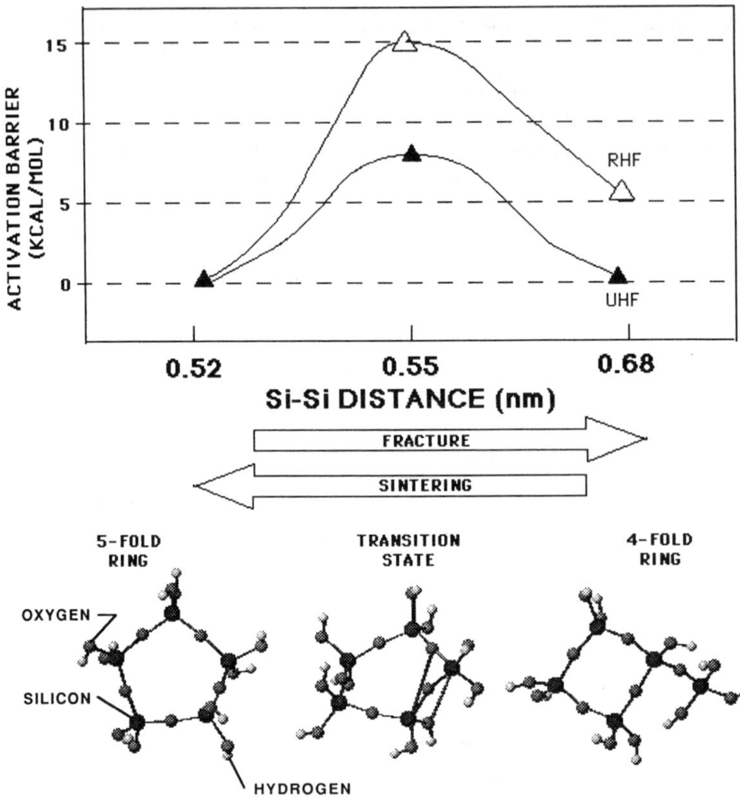

Figure 14 Five-member ring contraction sequence leading to the formation of a 4-member ring while extracting an a-silica tetrahedron.

IR spectroscopy because the IR SiOH peak saturates. Raman spectroscopy shows the interrelated changes in intensity in silica gels during densification and water adsorption. The 980 cm^{-1} Si_sOH peak, the 605 cm^{-1} D_2 trisiloxane ring peak, and the 2700–3800 cm^{-1} adsorbed water peak can all be measured as a function of processing and environmental conditions (2, 42–45). Raman spectroscopy has been used to ascertain (*a*) the rate determining step during D_2 ring hydrolysis, (*b*) quantitative values for $[Si_sOH]$ and $[D_2]$, and (*c*) the ratio of $[H_2O$ molecules$]/[D_2]$ rings during D_2 hydrolysis in silica gel monoliths (30).

As T_p increases during silica gel densification, the condensation of Si_sOH on the pore surface produces H_2O molecules and 2-, 3- and 4-member silicate rings known as cyclodisiloxane, D_2 cyclotrisiloxane, and D_1 cyclotetrasiloxane rings, respectively. The less strained, puckered D_1 rings form in the initial sol stage, whereas the planar D_2 rings, which are more strained than the D_1 rings, only form when $T_p \geqslant 180°C$. Some highly strained cyclodisiloxane rings also form during condensation of adjacent Si_sOH, but only at $T_p \geqslant 600°C$, giving a maximum concentration of ~0.1 disiloxane ring/nm^2 (2).

The pore surface of a stabilized silica gel rehydroxylates on exposure to water (41) due to the presence of the strained silicate rings. Eventually [Si_sOH] increases to a value of ~5 Si_sOH/nm^2 of pore surface (41). The adsorption of H_2O molecules onto acidic Si atoms in these strained rings is the rate determining step during both disiloxane and D_2 ring opening via hydrolysis:

$$\equiv\text{Si-O-Si}\equiv + H_2O \rightleftarrows \equiv\text{Si-OH} + \text{HO-Si}\equiv. \qquad 2.$$

Figure 15 shows a Raman spectrum of the hydrolysis of a D_2 ring (30). Because the rate of hydrolysis of silicate rings increases as the strain in their siloxane bonds increases, disiloxane rings hydrolyze most rapidly. D_1 rings are puckered, so they are less strained, and their rate of hydrolysis is much slower than D_2 rings (30). A plot in Figure 16 of [Si_sOH] against exposure to water shows an increase in [Si_sOH] and a decrease in [D_2] during D_2 ring hydrolysis.

The concentration of the surface disiloxane rings is an order of magnitude smaller than [D_2], D_2 rings hydrolyze much more quickly than D_1 rings (hours vs weeks), and the surface concentration of [D_2] is much larger than [D_1]. Thus when a stabilized silica gel adsorbs water, the initial rehydrolysis of the pore surface is dominated by the behavior of the large concentration of D_2 rings.

To find the reaction path for D_2 ring opening via hydrolysis, we chose the atomic distances associated with forming Si-O bonds. This distance was then held fixed at a number of selected values while allowing the rest of the molecule to optimize completely within MOPAC. The geometry along this reaction path that had the highest enthalpy was then used to begin an optimization of the transition state geometry. All the stationary points on the potential surfaces calculated in this study were fully optimized using AM1 PRECISE in the MOPAC program. The transition state was verified by showing that the first derivative of the enthalpy went to zero, and only one negative force constant was found in a normal mode (FORCE) calculation. The transition state involves the formation of a pentavalent silicon structure (46).

Figure 17 shows the reaction path enthalpy for the hydrolysis of the 3-member (D_2) ring into a chain with an activation barrier of 2.8 Kcal/mol. The reverse reaction or condensation of 2 Si-OH (silanols) has an activation barrier of 5.6 Kcal/mol.

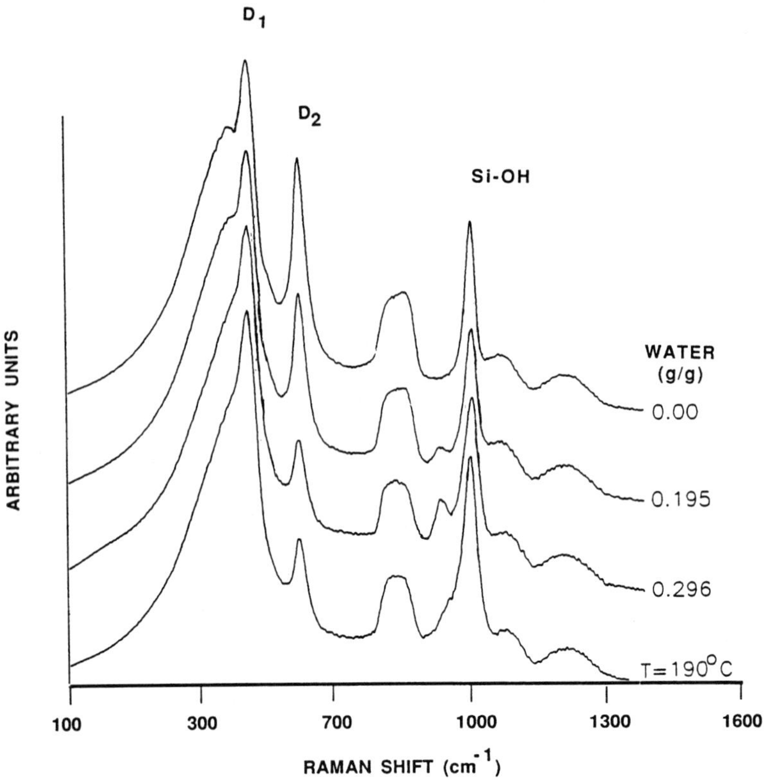

Figure 15 Evolution of the Raman spectrum (100–1350 cm^{-1}) for a sample of porous Type VI sol-gel silica stabilized at 650°C as the water content (W) increases from 0.08 H_2O g/g to 0.329 H_2O g/g (after Reference 30).

Silica Pore Model

Understanding the stability of porous Type VI silica optical matrices during diffusion of water is important because this new material is being used as a host for optically active composites (40). Aqueous solutions are being used during processing as carriers of optically active phases for filters, graded index lenses, laser dyes, sensors, and scintillators. A pore wall model has been developed to calculate the subnanometer wall thickness of Type VI porous silica-gel matrices. The AM1 MO calculations model silica structures with dimensions similar to those assumed for pore walls based upon X-ray small-angle scattering and NMR results. A baseline value for the mean pore wall thickness was determined from comparisons between the model and experimental results. This value separates

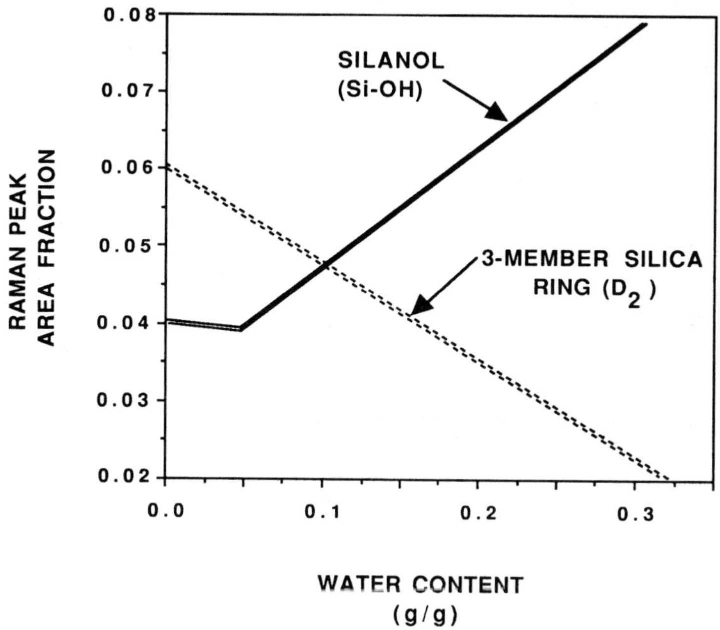

Figure 16 Dependence of the fractional area of the 605 cm^{-1} D_2 ring and the 980 cm^{-1} Si-OH Raman peaks on the water content, W.

environmentally stable matrices from unstable gels during water diffusion (or percolation). A ring-opening hydrolysis mechanism is proposed to explain the role of water in fracture of the porous sol-gel silica matrices (47).

The model treats the porosity as an array of cylindrical pores open to the atmosphere that are surrounded by the silica skeleton, as shown in Figure 18. The parameters of the model are r_p, the pore radius; $\mu+\delta$, half of the maximum dimension for the pore wall thickness; and L, the average pore length.

The mean wall thickness, Δ_M, can then be written as

$$\Delta_M = \delta + \mu \qquad 3.$$

and may be evaluated using (47):

$$\Delta_M = \left[\left(\frac{1+\sqrt{2}}{2}\right)\sqrt{\frac{\pi}{V_v}} - 2\right] r_p. \qquad 4.$$

Therefore, for a given V_v (volume fraction of porosity), the dependence of

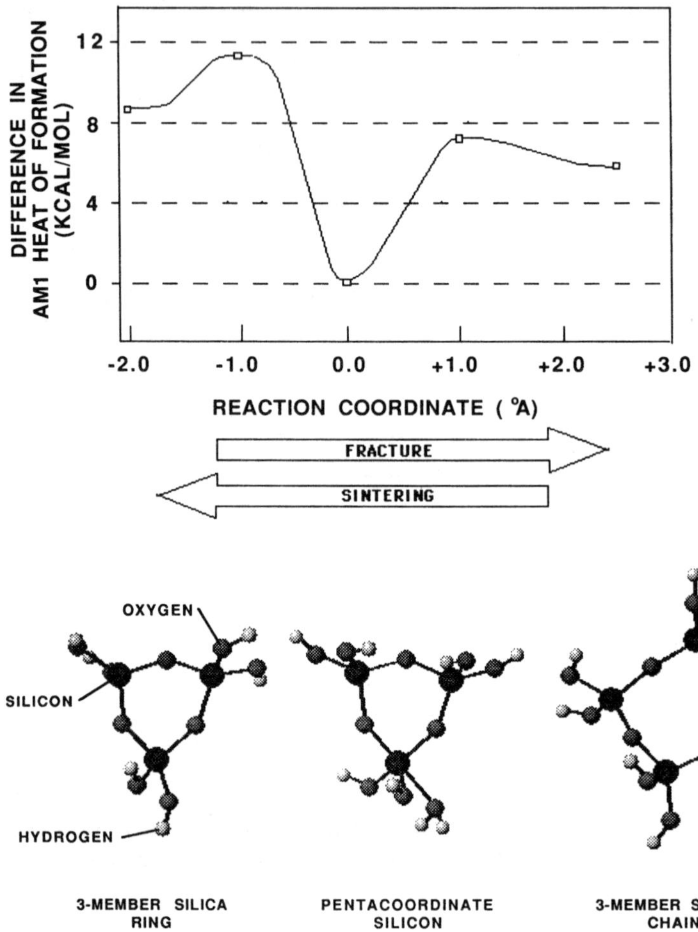

Figure 17 Reaction path for the formation of a trisiloxane chain via hydrolysis of a 3-member (D_2) ring. (After Reference 46.)

wall thickness on pore size becomes a linear relationship as shown in Figure 19.

The heavy black line in Figure 19 corresponds to the minimum conditions for stability and coincides with a mean pore wall thickness of ~4.5 nm. The stability boundary provides a method to design a silica-gel with a given pore size that will be stable under water adsorption and desorption.

The AM1-optimized structure of two 3-member or D_2 rings bonded together by one bridging oxygen bond is shown in Figure 20. This model

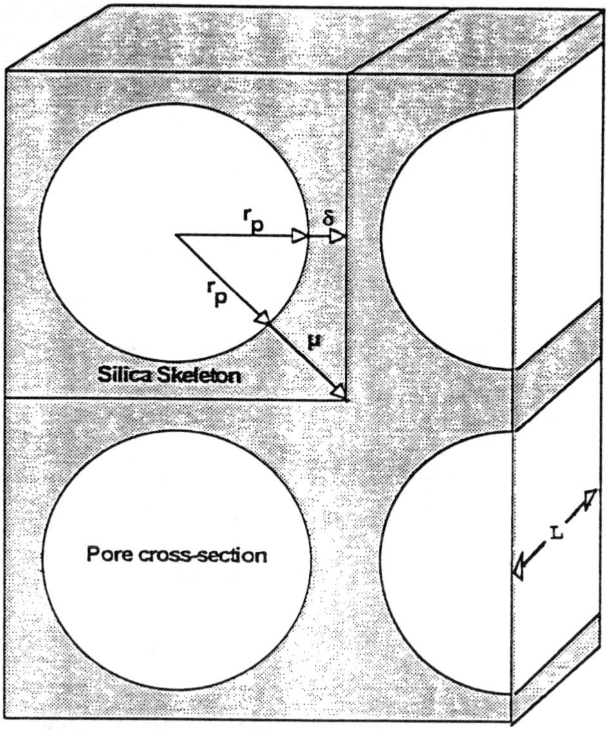

Figure 18 Schematic model of cylindrical pores in Type VI silica.

of dicyclotrisiloxane shows one possible structure that could bridge between two pores for low density porous silica. The length of this structural bridge is 1.1 nm between 0(12) and 0(27). This model is consistent with the experiment because the hydrolysis of the D_2 ring in porous silica is rapid (29) and would most likely lead to gel failure.

Cr^{3+}-Doped Silica

The effect of transition metal ions within a silica pore network has been modeled using ZINDO (48). The electronic spectra for the Cr^{3+} hexaaqua species was accurately reproduced with an improved geometry over previous calculations. The calculations based on geometrical arguments and electronic states indicate that hydrogen bonding, the deprotonation of a coordinated water ligand, and/or the substitution of a charged hydroxyl for a ligated water molecule are plausible mechanisms to explain the red-shifting of the visible absorptions seen experimentally. Thus semiempirical modeling methods are suitable to study interactions occurring

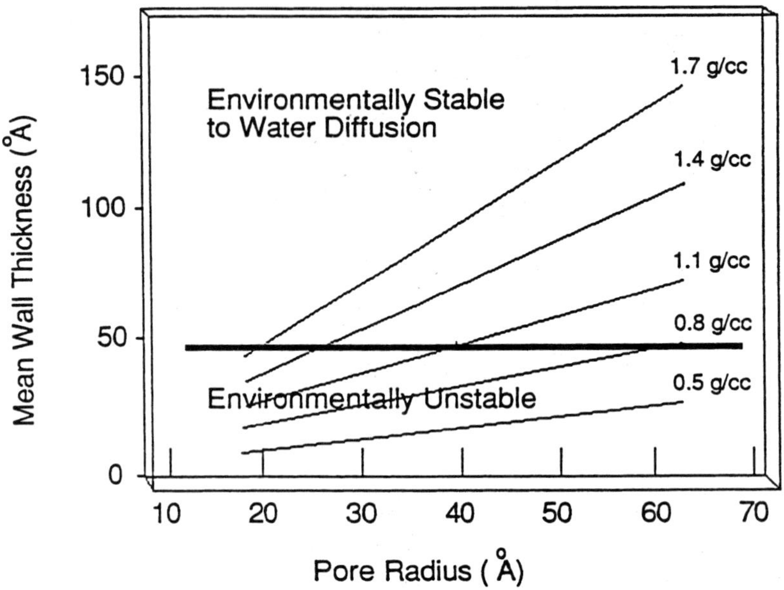

Figure 19 Calculated pore wall thickness for various pore radii for a specific bulk density.

between the surfaces of porous silica matrices and impregnated species. However, additional calculations and experiments are required before the local environment of the Cr^{3+} within the silica-gel can be fully comprehended (48).

Silica Fracture

It has been recognized since the time of Inglis (49) and Griffith (50) that fracture occurs on an atomistic level when the bonds between the atoms are strained past the breaking point. When a tensile stress at the tip of a crack exceeds a critical value, crack growth will occur (11). This fundamental concept underlying fracture is typically modeled as a continuous stress-displacement curve, approximated as a sine curve. However, because of structural constraints and the finite size of molecular bonds, a crack tip must advance in increments with the size of the step dictated by the localized structure containing the crack tip. A continuum of strain applied to a material must be translated mechanistically into quantized steps of bond breaking (38).

Molecular models have been proposed for the environmental effects on fracture (51, 52), but they are not quantitative and do not describe reaction pathways for the fracture process. A quantitative ring contraction model

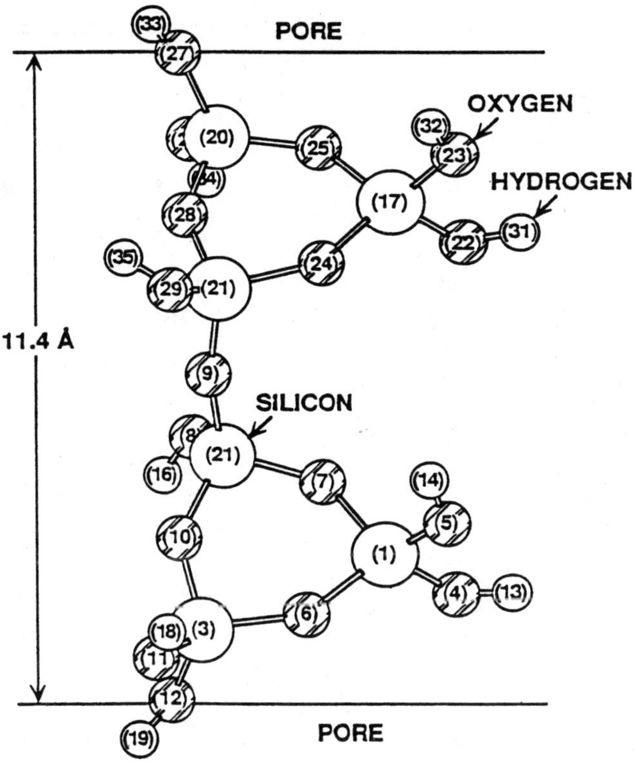

Figure 20 AM1 optimized model of dicyclotrisiloxane with an AM1 heat of formation of −1405 Kcal/mol.

for the fracture of a-silica is described based on AM1 semi-empirical MO calculations of strained 3- and 4-member silica rings and a 5-member ring-chain structure. The fracture barrier for 5-member ring-chain structures is 103 Kcal/mol. The barrier for fracture of a 3-member ring is +96 Kcal/mol. Fracture by contraction of 4-member rings has a lower energy barrier of +77 Kcal/mol due to the formation of pentacoordinate silicon transition states that produce trisiloxane rings and a broken siloxane bond. Thus the ring contraction model predicts that a crack will follow a path that depends on the distribution of 4-member (or larger) rings in a vacuum or fast fracture (38).

Using the same AM1 method, the transition state has been found for the contraction of a 5-member ring into a 4-member ring. The activation barrier to fracture for this contraction is only $E_f = +14.82$ Kcal/mol. This confirms our initial hypothesis that ring contraction can lead to much lower fracture energies than expected from simple Si-O bond breaking. Several different schemes of ring contractions are possible for both 5- and 6-member ring structures. All contraction paths have different intermediate structures that lead to the same end point of crack growth. The various barriers to fracture range from +8 to +52 Kcal/mol for the 5-member ring contractions, and from +9 to +33 Kcal/mol for 6-member ring contractions. This step-wise contraction mechanism for crack propagation is schematically shown in Figure 21 (39).

Figures 22–24 show the sequence of structural evolution at a crack tip as a 5-member ring is strained and the ring contraction occurs. A silica

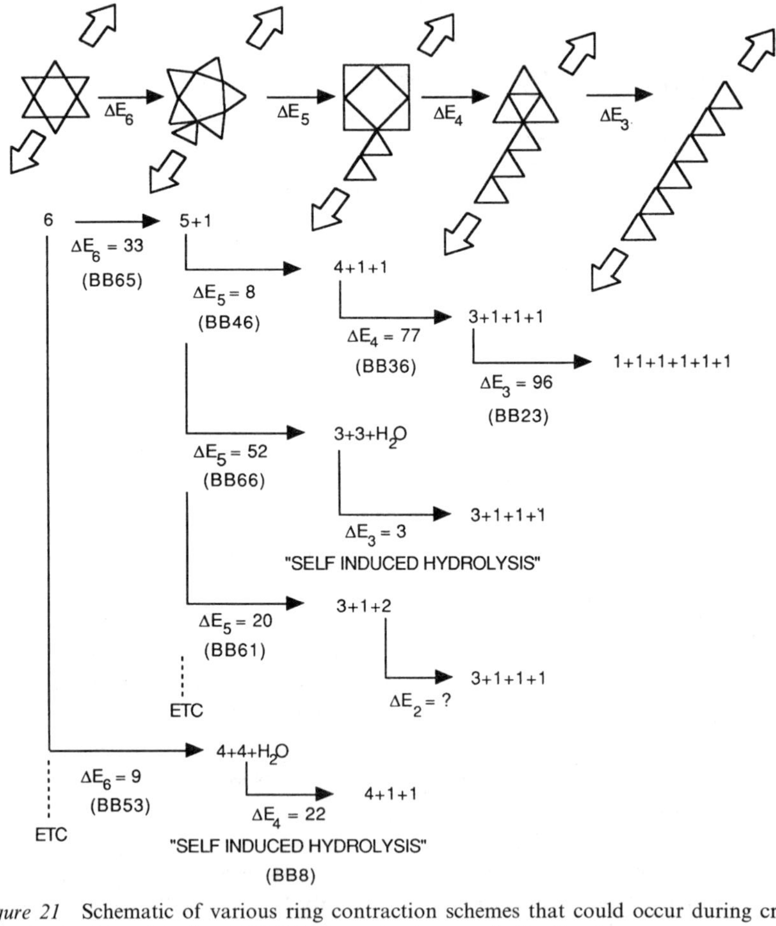

Figure 21 Schematic of various ring contraction schemes that could occur during crack growth of a-silica, showing AM1 UHF reaction barriers in Kcal/mol.

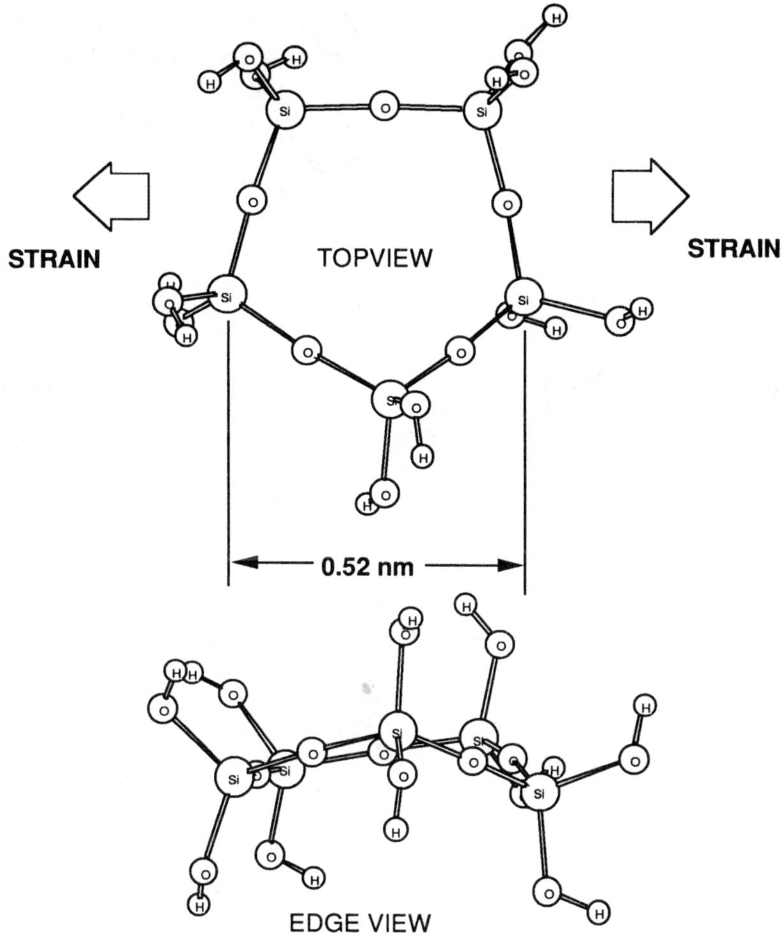

Figure 22 AM1 model of cyclopentasiloxane: a 5-member silica ring with an AM1 heat of formation of −1243 Kcal/mol (Model R5).

tetrahedron is extracted from a 5-member ring as shown in Figure 24. This reaction pathway was summarized in Figure 14.

Hydrolysis of the ring structures can also occur if a polar molecule similar to a water molecule attacks a strained silicon-oxygen bond (37).

A quantitative model of environmentally sensitive crack growth of a-silica is based on semi-empirical MO calculations (AM1 method) of a water molecule interacting with strained 3- and 4-member silica rings and a 5-member ring-chain structure. The energy barrier for hydrolysis of strained 3-member rings is only +7 Kcal/mol, the energy barrier for

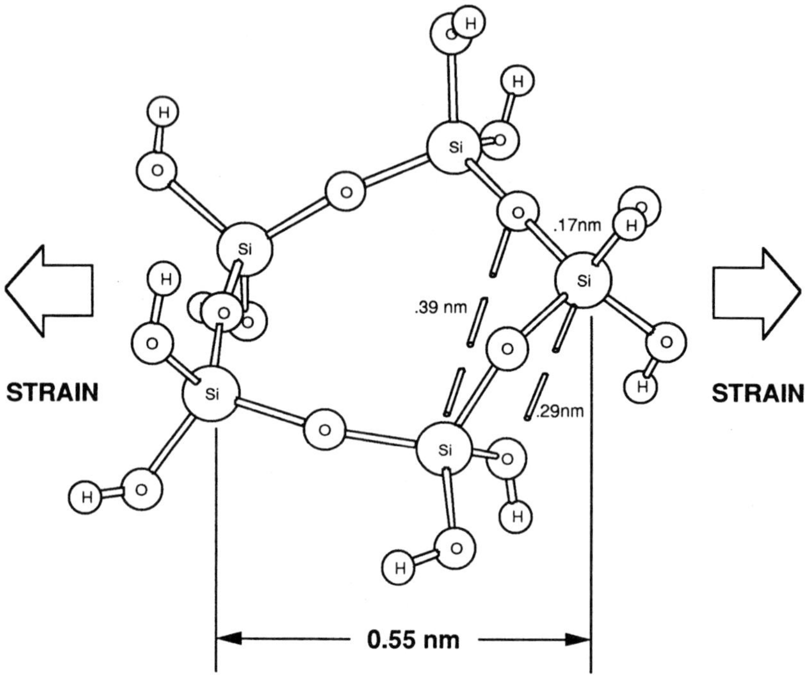

Figure 23 AM1 UHF transition state (saddle) for a 5-member ring contraction into a 4-member ring with a heat of formation of −1235 Kcal/mol, which yields a fracture barrier of +8 and a sintering barrier of +7 Kcal/mol (Model BB46).

hydrolysis of strained 4-member rings is +29 Kcal/mol, and for a 5-member ring-chain, it is +39 Kcal/mol. Thus the MO model predicts that the energetics of Region 1 crack growth (see Figure 25) slows crack growth (53) and is controlled primarily by the distribution and hydrolysis of 3-member silica rings in the a-silica structure. Region III crack growth is controlled by the distribution and energy of contraction of 4-, 5-, and 5+-member rings.

Silica-Amino Acid Interactions

Specific compositions of silicate glasses and glass-ceramics form a mechanically strong bond with bone (54). These implant materials, used clinically to replace or repair parts of the skeleton, are called bioactive materials (55).

There are eleven reaction stages that occur on the surface of bioactive glasses as shown in Figure 26. The results, wherein 22 glass compositions are compared, are summarized by Hench & LaTorre (56). All melt-derived

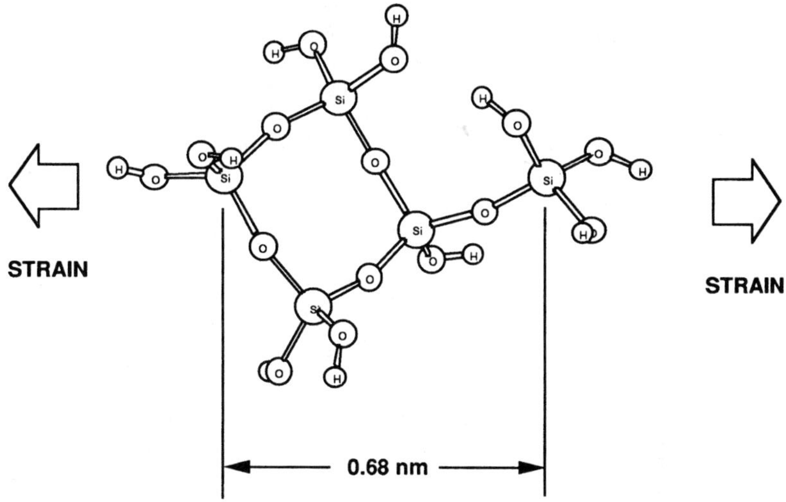

Figure 24 End point structure for a 5- to 4-member ring contraction with a AM1 UHF heat of formation of −1242 Kcal/mol (Model BB49).

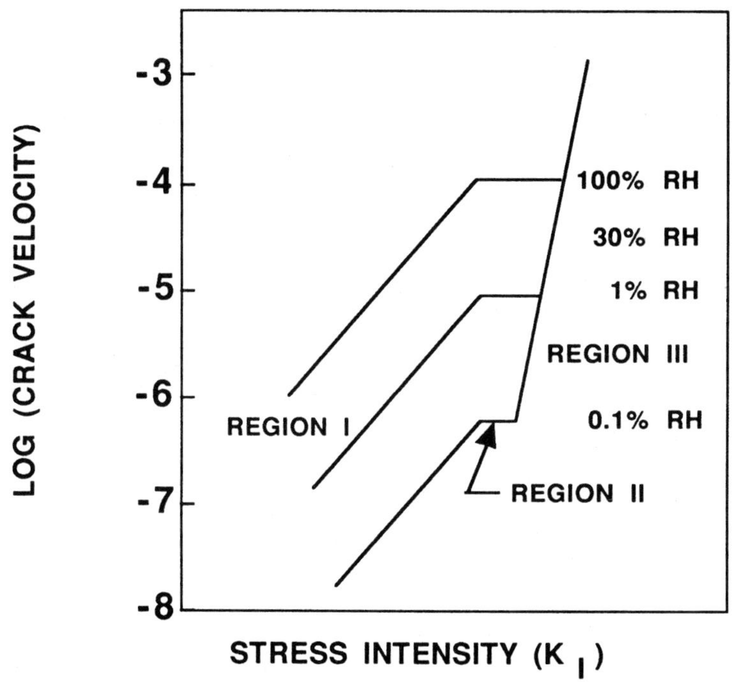

Figure 25 Velocity vs stress intensity (V-K) curve for slow crack growth in silica: Region I, hydrolysis controlled; Region II, diffusion controlled; Region III, ring contraction controlled.

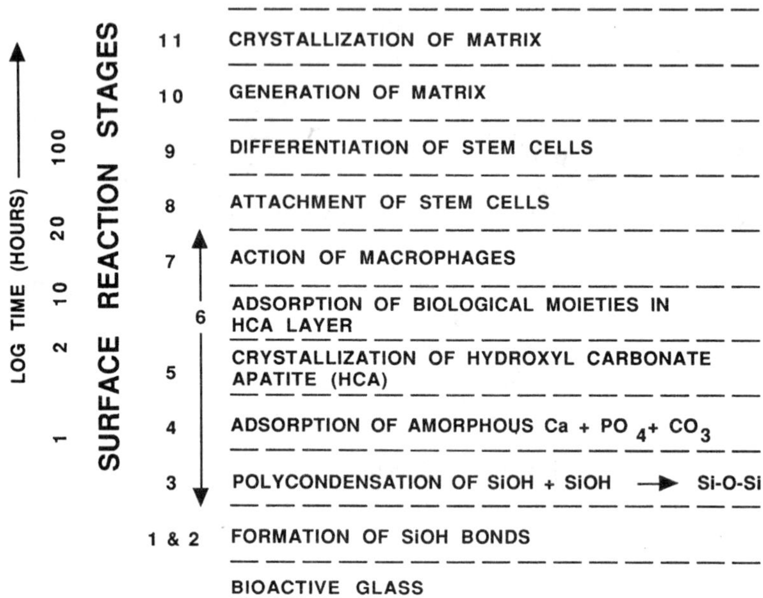

Figure 26 Sequence of interfacial reactions involved in forming a bond between tissue and bioactive ceramics.

SiO_2-CaO-Na_2O-P_2O_5 glasses of <60 mol% SiO_2 rapidly develop a crystalline hydroxy-carbonate apatite (HCA) layer on the glass surface regardless of the type of simulated body fluid (SBF) or glass composition. Glasses with <53 mol% SiO_2 form an HCA layer within 3 h; these glasses bond to both hard and soft tissues. Glasses between 53 and 56 mol% SiO_2 take 2 to 3 days to form an HCA layer and bond to bone but not soft tissues. The nucleation and crystallization kinetics of HCA (Stage 5) are controlled by silanol sites and porosity of the silica-gel layer formed on the glass by ion exchange and polycondensation reactions (Stages 1–3). Gel-derived glasses develop a crystalline HCA phase without requiring ion exchange reactions or formation of an intermediate amorphous calcium-phosphate layer (Stage 4). AM1 semi-empirical MO calculations have been used to study the reaction pathways on bioactive glass surfaces (57, 58). The calculations show favorable energetics for the nucleation of calcium-phosphate chains extending from a hydrated silica cluster. This indicates that HCA crystals may nucleate on a silanol-rich surface and grow by means of a screw dislocation mechanism.

Selective adherence on bioactive glass surfaces of various biological molecules associated with osteogenesis (Stage 6) appears to be responsible for the correlation of in vitro reaction kinetics of HCA formation with the rate of bone bonding. Chemical interaction of organic molecules with hydrated silica is also important in osteogenesis and bone mineralization and affects, in as yet an unknown manner, extracellular bone and cartilage matrix formation (59) and osteoblast proliferation (60, 61). Other biological processes including silicosis, enzyme immobilization within silica gels, diatom frustule development, silicification of plants, atherosclerosis, and aging also involve silica-protein interactions (57). However, the nature of chemical bonding reaction pathways, energetics, mechanisms, and structure of the silica-organic interactions is poorly understood.

AM1 was used to calculate the reaction pathways for bond formation between alanine and 3-member silica rings (57). The alanine molecule [H_2N-$CH(CH_3)$-COOH] was chosen for the study because it is one of the simplest amino acids and because experimental data also exist for the interaction of alanine with hydrated silica surfaces as shown in Figure 27*a*

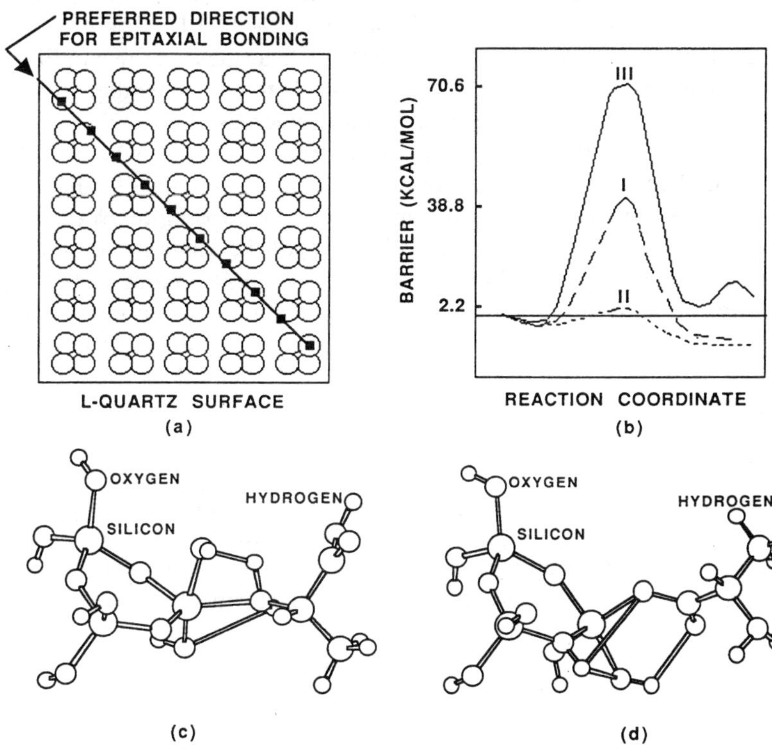

Figure 27 Silica alanine interactions: (*a*) epitaxial bonding to quartz surface; (*b*) reaction barriers for (I) amine-cyclotrisiloxane, (II) carboxy-cyclotrisiloxane, and (III) alanine-glycine; (*c*) alanine (NH_2)-silica transition state; (*d*) alanine (COOH)-silica transition state.

(62). This makes it possible to compare experimental results with the MO results. Epitaxial bonding of poly-l-alanine has been observed between specific crystallographic orientations of alpha-quartz crystals where the surface silanols match the bond distances of the alanine molecule (62).

The MO results show that there are low-energy reaction pathways for organic-inorganic reactions between alanine molecules and hydrated silica clusters. The energy barrier is very much lower than the barrier that exists for formation of a dipeptide bond between alanine and glycine, calculated using Figure 27*b* (57). The large energy barrier for peptide bond formation is circumvented in biological systems by use of enzymes, which are specific for each type of dipeptide formed. However, the MO calculations suggest that inorganic surfaces can provide a catalytic-like function in the binding of organic molecules similar to that of enzymes. Comparable results were obtained for the interaction of alanine with tetrasiloxane rings (63). However, the 4-member silica rings have a substantially higher activation barrier for the alanine -COOH interaction, i.e. +24 Kcal/mol. The barrier to the -NHH interaction is about the same, i.e. +49.4 Kcal/mol.

The interaction of alanine with 3-member silica rings is shown in Figures 27*b,c,d*. The amine group of the alanine bonds to the 3-member ring with an activation barrier of +38.8 Kcal/mol, whereas the carboxy groups bonds with an activation barrier of only +2.2 Kcal/mol, which is easily surmounted at body temperatures. The various MO results show that different types of interfacial bonds are formed depending upon the structure and orientation (stereology and chirality) of the organic and inorganic molecules. The differences in reaction pathways and binding energies for formation of -Si-O-C- linkages with amino acids vs formation of -Si-N- bonds may account for the variety of biochemical effects observed for silicon in nature and the influence of a hydrated silica gel layer on the rates of tissue bonding to bioactive implants.

Future Directions

The goal of future theoretical calculations is to simulate surface chemistry at the atomic level where the structure and influence of the bulk (or media) is included. There are several schemes where high-level quantum calculations could be carried out on interesting clusters that are embedded in much larger assemblies modeled by molecular mechanics (64, 65). There are also special approaches to solving Schrödinger's equation (variously known as the X-α method, electron density functional theory, propagator methods, generalized valence bond theory, etc) (66). These methods should certainly be used to investigate the structure-property relationships for silica and silica-environment systems because the structural details of rings and chains appear to be important in understanding physical, electronic, and optical properties of silica. An increased understanding of the relationship between molecular structure and properties is expected as we explore progressively larger MO clusters and eventually develop the means to correlate MO and MD calculations.

ACKNOWLEDGMENTS

The authors gratefully acknowledge financial support of the Air Force Office of Scientific Research Grant No. F49620-92-J-0351. We also acknowledge the technical advice of CaChe Scientific, Inc., Beaverton, OR, especially Sam Cole and George Pervis.

Literature Cited

1. Iler RK. 1955. *The Chemistry of Silica*, New York: Wiley
2. Brinker CJ, Scherer GW. 1990. *Sol-Gel Science*, San Diego: Academic
3. Revesz AG, Gibbs GV. 1980. In *The Physics of MOS Insulators*, ed. G Lucovsky, ST Pantelides, FL Galeener, pp. 92–96. New York: Pergamon
4. Gibbs GV. 1982. *Am. Mineral.* 67: 421–50
5. Hess AC, McMillan PF, O'Keeffe M. 1986. *J. Chem. Phys.* 90: 5661
6. Galeener FL. 1982. *J. Non-Cryst. Solids* 49: 53
7. Uchida N, Shinmei M. 1990. *J. Non-Cryst. Solids* 122: 276–84
8. Davis LP, Burggraf LW. 1986. In *Science of Ceramic Chemical Processing*, ed. LL Hench, DR Ulrich, pp. 400–11. New York: Wiley
9. Boyd DB. 1990. In *Reviews in Computational Chemistry*, ed. KB Lipkowitz, DB Boyd, pp. 321–54. New York: VCH
10. Davis LP, Burggraf LW. 1988. In *Ultrastructure Processing of Advanced Ceramics*, ed. JD Mackenzie, DR Ulrich, p. 367. New York: Wiley
11. Freiman SW. 1980. In *Glass: Science and Technology*, ed. DR Uhlmann, NJ Kreidl, 3: 21–78. New York: Academic
12. Wong-Ng W, White GS, Freiman SW. 1992. *J. Am. Ceram. Soc.* 75(11): 3097–102
13. Davis LP, Burggraf LW, Gordon MS. 1988. *J. Am. Ceram. Soc.* 110: 3056–62
14. Bartlett RJ. 1993. *Florida School on Applied Molecular Orbital Theory*, pp. 1–3. Gainesville: Univ. Fla. Press
15. Zerner MC. 1991. In *Review of Computational Chemistry II*, ed. KB Lipkowitz, DB Boyd, pp. 313–66. New York: VCH
16. ZINDO Version 3.0, CaChe Scientific, Inc., Beaverton, OR
17. MOPAC Version 6.1, Tektronix, Inc., CaChe Scientific, Beaverton, OR
18. Dewar MJS, Zoebisch EG, Healy EF, Stewart JP. 1985. *J. Am. Chem. Soc.* 107: 3902–9
19. Dewar MJS, Jie C. 1987. *Organometalics* 6: 1486–90
20. Pople JA, Beaveridge DL. 1970. *Approximate Molecular Orbital Theory*, New York: McGraw-Hill
21. Chia T, West JK, Hench LL. 1994. In *Better Ceramics Through Chemistry VI*, ed. AK Cheetham, CJ Brinker, ML Mecartney, C Sanchez, 346: 727–32. Pittsburgh, PA: Mater. Res. Soc.
22. West JK, Hench LL. 1994. *J. Non-Cryst. Solids*. 180: 11–16
23. West JK, Hench LL. 1995. *J. Am. Ceram. Soc.* In press
24. Bell RJ, Dean P. 1972. *Philos. Mag.* 25: 1381–98
25. Mozzi RL, Warren BE. 1969. *J. Appl. Cryst.* 2: 164
26. Stewart JJP. 1989. *J. Comput. Chem.* 10(2): 209–21
27. Stewart JJP. 1990. *J. Comput. Chem.* 11(4): 543
28. West JK, LaTorre GP, Hench LL. 1995. *J. Non-Cryst. Solids*. In press

29. Wallace S. 1991. *Porous silica gel monoliths: structural evolution and interaction with water*. PhD thesis. Univ. Fla., Gainesville. 290 pp.
30. Wallace S, West JK, Hench LL. 1993. *J. Non-Cryst. Solids* 152: 101–8
31. Vasconcelos WL. 1989. *Topological evolution and properties of sol-gel silica monoliths*. PhD thesis. Univ. Fla., Gainesville. 261 pp.
32. Hench LL, West JK. 1990. *Chem. Rev.* 90: 33–72
33. Hummel RE, Chang S-S. 1992. *Appl. Phys. Lett.* 61: 1965
34. Hummel RE, Chang S-S, Ludwig M, Morrone A. 1993. In *Microcrystalline Semiconductors—Materials Science and Devices*, 283: 45. Philadelphia: Mater. Res. Soc.
35. Hummel RE, Chang S-S, Ludwig M, Morrone A. 1993. Presented at *Intl. Workshop on Light Emission and Electronic Properties of Nanoscale Silicon*, Charlotte, NC
36. Hummel RE, Morrone A, Ludwig M, Chang S-S. 1993. *Appl. Phys. Lett.* 63: 2771
37. West JK, Hench LL. 1994. *J. Mater. Sci.* 29: 5808–16
38. West JK, Hench LL. 1994. *J. Mater. Sci.* 29: 3601–6
39. West JK, Hench LL. 1995. *J. Mater. Sci.* In press
40. Hench, LL, Wang SH, Nogues JL. 1988. In *Multifunctional Materials*, ed. RL Gunshor, 878: 76–85. Bellingham, WA: SPIE
41. Bunker BC, Haaland DM, Ward KJ, Michalske TA, Smith WL, et al. 1989. *Surf. Sci.* 210: 406–28
42. Brinker CJ, Bunker BC, Tallant DR, Ward KJ. 1986. *J. Chimie Phys.-Chim. Biol.* 83(11–12): 851–58
43. Gottardi V, Guglielmi M, Bertoluzza A, Morelli MA. 1982. *J. Non-Cryst. Solids* 18: 117
44. Krol DM, van Lierop JG. 1984. *J. Non-Cryst. Solids* 63: 131
45. Krol DM, van Lierop JG. 1986. *J. Non-Cryst. Solids* 86: 241
46. West JK, Wallace S. 1993. 1993. *J. Non-Cryst. Solids* 152: 109–17
47. West JK, Kunetz JM, Araujo FG, Hench LL. 1994. In *Sol-Gel Optics III*, ed. JD Mackenzie, Bellingham, WA: SPIE. 2288: 724–32
48. Kunetz J, Bendale R, Hench LL. 1994. See Ref. 47, pp. 700–8
49. Inglis CE. 1913. *Trans. Inst. Nav. Arch.* 55: 219–29
50. Griffith AA. 1921. *Trans. R. Soc. London* 221: 163–98
51. Michalske TA, Freiman SW. 1983. *J. Am. Ceram. Soc.* 66: 284
52. Michalske TA, Bunker BC. 1984. *J. Appl. Phys.* 56(10): 2686–93
53. Wiederhorn SM. 1967. *J. Am. Ceram. Soc.* 50: 407–14
54. Hench LL. 1991. *J. Am. Ceram. Soc.* 74(7): 1487–1510
55. Hench LL, Wilson J, eds. 1994. *Introduction to Bioceramics*, London: World Scientific
56. Hench LL, LaTorre GP. 1992. In *Bioceramics* 5, ed. T Yamamuro, T Kokubo, T Nakamura, pp. 67–74. Kyoto, Japan: Kobonshi Kankoki
57. Hench LL, West JK. 1994. *J. Vac. Soc. Technol. A* 12(5): 2962–65
58. West JK, Hench LL. 1992. See Ref. 56, pp. 75–86
59. Carlisle EM. 1986. In *Silicon Biochemistry*, ed. D Evered, M O'Conner. New York: Wiley
60. Vrouwenvelder WCA, Groot CG, de Groot K. 1993. *J. Biomed. Mater. Res.* 27: 465–75
61. Keeting PE, Oursler MJ, Wiegand KE, Bonds SK, Spelsberg TC, Riggs BL. 1992. *J. Bone Mineral Res.* 7(11): 1281–89
62. Hartwig B, Hench LL. 1972. *J. Biomed. Mater. Res.* 6(5): 413–24
63. Hench LL, West JK. 1993. In *Bioceramics* 6, ed. P Ducheyne, D Christiansen, pp. 35–40. Oxford: Butterworth-Heinemann
64. Allinger NL. 1977. *J. Am. Chem. Soc.* 99: 8127
65. Burkert U, Allinger NL. 1982. *Molecular Mechanics*, Washington, DC: Am. Chem. Soc.
66. Lowdin P-O. 1986. In *Science of Ceramic Chemical Processing*, ed. LL Hench, DR Ulrich, pp. 577–82. New York: Wiley

Annu. Rev. Mater. Sci. 1995. 25: 69–96

MODELING THE DEVELOPMENT AND RELAXATION OF STRESSES IN FILMS

M. D. Thouless

IBM Research Division, T. J. Watson Research Center, Yorktown Heights, New York 10598[1]

KEY WORDS: cracks, deformation, diffusion, dislocations, substrate

ABSTRACT

The mechanics for the development and relaxation of stresses in films is briefly reviewed. The discussion initially focuses on how stresses are distributed within a film and substrate, and on the resultant deformations that these stresses induce. Although the particular case of an elastically homogeneous system with a uniform stress within the film is addressed in detail, the modifications that would result from a nonhomogeneous system, or a nonuniform stress, are also outlined.

A brief discussion of stress relaxation in epitaxial systems and the glide of dislocations in thin films then follows. In polycrystalline films, diffusion of atoms as well as the motion of dislocations can contribute to the relaxation of stresses. It is possible to use constitutive models for the different relaxation processes to predict the development of stresses during thermal cycling. A comparison of these predictions with experimental observations allows general conclusions about the dominant relaxation mechanisms that operate in particular thin-film systems to be made.

An additional mechanism of stress relief is that of cracking and delamination. In this paper, just one mechanism is addressed—that where a brittle film cracks. An energy-balance calculation leads to results for a critical film thickness below which cracking will not occur and for the crack spacing as a function of both film thickness and residual stress.

[1] Current address: Department of Mechanical Engineering and Applied Mechanics, University of Michigan, Ann Arbor, Michigan 48109.

0084–6600/95/0801–0069$05.00

INTRODUCTION

Concerns of the electronics industry have been a major driving force in recent years behind the study of the properties of thin films (1, 2). Although thin films for microelectronic applications are the primary concern of this paper, many of the concepts discussed are also applicable to a wide range of other technologically important systems. The common element between these systems is that they all involve configurations in which one or more relatively thin layers of material are bonded to a thicker substrate. Often, the film may be the active component imparting an important property to the system, with the substrate merely providing structural support. Alternatively, the film may serve in a passive role providing protection or decorative covering for the substrate. Despite the diverse nature of the different applications, there is much that is common in the mechanics that describes their response to stress.

There are several different scales at which electronic components can be considered to be essentially multilayer composites. The smallest scale, where the films have a thickness in the order of tens to hundreds of nanometers, involves electrically active regions of semiconductors in which the band-gap can be controlled by layers of different compositions, or by doping with impurity atoms (2). These active areas are connected by successive levels of conducting films, or lines, consisting of metal interconnects electrically isolated from one another by layers of insulators (3). In addition to these metallization and passivation layers, a chip contains other films used for a variety of purposes such as etch stops and diffusion barriers. The same scheme of alternating layers of conducting and insulating material is continued at the next scale of packaging, where individual chips are connected together (4). Even though the mechanical properties of these components are of secondary importance to the electrical properties, the layers must maintain structural integrity. The stresses inevitably introduced into the components during processing and service can cause problems such as fracture of the passivation, which may result in electrical shorts, and void growth, which may lead to open circuits in the metal lines. Also, the introduction of dislocations and cracks into active regions of the semiconductor can seriously affect the electrical performance. Storage devices that may be connected to these electronic devices consist of thin films chosen for their magnetic or optical properties and supported by a substrate (5). Similar reliability issues are involved, and an understanding of the mechanical properties of these films is required to avoid failures caused by fracture or delamination.

Another important technological use of films is in protecting and/or decorating a substrate. Traditionally, films used for these applications have been known as coatings, a term reflecting their essentially passive role. Paint is perhaps the most ubiquitous form of coating, serving both an aesthetic purpose and protecting underlying structures from corrosion. Coatings are also used in the electronics industry, where the top layer on a chip is often a polymer designed to keep moisture out of the chip and to protect it against abrasion. Because of their inertness and hardness, ceramics are frequently used as protective coatings such as in thermal barrier coatings for engine components (6). Indeed, what are essentially ceramic coatings are often grown in situ as a result of oxidation of a metal (7, 8). The oxide scale, once formed, can provide an important barrier to further environmental attack. However, whether these or any other coatings remain useful barriers depends on their response to the stresses that may develop during growth or fabrication. If cracking or spalling occurs, the host substrate is left vulnerable to further environmental attack (8–10).

Finally, it is worth noting that the scale at which some of the mechanics of films are important also extends into other engineering disciplines. For example, many of these general concepts are applicable for understanding cracking of stucco and plaster on buildings, spalling of ice sheets (11), the delamination of ice from aircraft and other structures, and the buckling and cracking of road surfaces under extreme temperatures.

Despite the different scales and materials, a common feature to all the examples described above is that stresses may develop within a surface layer. The relaxation of these stresses can lead to failure; therefore it is important to develop an understanding of the relaxation processes so they can be controlled in a benign fashion. However, it should be appreciated that the definition of what constitutes failure depends on the engineering application. In many cases, the obvious definition is a loss of structural integrity, either by cracking, void growth, or delamination; a plastic or elastic response to the stress may be quite acceptable. There are, however, more sensitive applications in which a plastic response is not acceptable. The generation and movement of dislocations in semiconductor devices is one example, but rumpling of a decorative film by plastic deformation (or the introduction of a crack that has no structural significance) may also be considered a failure if the design criterion is based on aesthetic considerations. It should also be remembered that even a purely elastic response may be unacceptable in situations that call for extremely tight control of dimensional stability or surface flatness.

ORIGIN OF STRESSES

A stress may be induced in a film by the application of external forces or moments to the film-substrate system. For coherency to be maintained between the film and substrate, the strains in the two must match at the interface. However, any modulus mismatch between the film and substrate will result in a discontinuity in the stresses parallel to the interface. In addition to stresses induced by external loads, residual strains and, hence, stresses, can be introduced during and after fabrication of the film. These are established when the constraint of the substrate forces the atoms in the film to maintain a spacing different from their equilibrium positions under the ambient conditions.

Residual stresses that develop after processing are often referred to as extrinsic stresses. One common cause of an extrinsic stress is a thermal-expansion mismatch. If the thermal-expansion coefficients of the film and substrate are α_f and α_s, respectively, then an incremental change in temperature ΔT would lead to a displacement mismatch of

$$\varepsilon_0 = (\alpha_s - \alpha_f)\Delta T = \Delta\alpha\Delta T \tag{1.}$$

per unit length of the two components, if they were unbonded. Coherent bonding between the two enforces continuity of displacements across the interface, and the atoms are constrained away from their equilibrium positions. When the substrate is much less compliant than the film—either because it is much thicker or has a much higher modulus—the overall dimensions of the system are governed by the substrate, and the film dimensions are forced to match it. For example, if the thermal-expansion mismatch and moduli are isotropic, so that biaxial stress conditions apply, the stress in the film is given by

$$\sigma_0 = E_f\varepsilon_0(1 - v_f), \tag{2.}$$

where E_f and v_f are the Young's modulus and Posisson's ratio of the film, respectively. Here, there is a discontinuity in strain across the interface, as well as a discontinuity in stress. Another source of extrinsic stresses resulting from temperature changes are phase transformations that can have similar effects to those of thermal-expansion mismatches.

Of the stresses that occur during processing, the simplest conceptually are epitaxial strains arising during heteroepitaxial growth. Under some conditions, a film deposited on a substrate will grow so that the atoms have a 1 : 1 correspondence in the plane of the interface with those of the substrate. If there is a mismatch between the lattice parameters of the two components, the atoms in the film will be displaced from their equilibrium position, and a misfit strain will be induced. If the substrate is sufficiently stiff that it controls the overall dimensions of the system, this strain is simply

$$\varepsilon_0 = \Delta a/a_f = (a_s - a_f)/a_f, \quad 3.$$

where a_f and a_s are the relaxed lattice parameters of the film and substrate.

Other stresses arising during processing are sometimes referred to as intrinsic stresses. These can arise from density changes induced by changes in the microstructure after the overall dimensions of the film have been established by bonding to the substrate. Although the distinction between intrinsic and extrinsic stresses is somewhat arbitrary, intrinsic stresses can be considered as irreversible, in contrast to thermal-expansion mismatch stresses that can be reversed by simply changing the temperature. Among the mechanisms that can introduce intrinsic stresses are grain growth (if there is excess volume associated with grain boundaries that can be lost as the grain boundary area is reduced), the introduction of defects and dopants, and the sintering and drying of powder compacts (12–15). Another example of the evolution of intrinsic stresses is provided by the growth of oxide films (7, 8). Oxidation can take place either at the metal/oxide interface or at the surface, depending on whether oxygen ions or metal ions diffuse faster through the oxide. If the reaction takes place at the metal/oxide interface, and the oxide has a different volume from the metal, then residual stresses will be induced. Similar effects occur in other thin-film systems, such as silicides, grown by the reaction of the substrate with an external agent (2), and the formation of intermetallics (16, 17).

MECHANICS OF FILM-SUBSTRATE SYSTEMS

This section reviews some of the basic mechanics pertinent for planar film-substrate systems in order to give insight into the stress distributions and deformations of such systems. For analytical ease, the model system examined here is assumed to be isotopic, to have a rectangular cross-section, and to be under conditions of plane stress. The film, of thickness h, is bonded to a substrate of thickness d, along a planar interface, and there is a mismatch strain of ε_0 that is uniform throughout the thickness of the film.

The stress distribution and deformations of such a composite can easily be obtained by performing the thought experiment illustrated in Figure 1 (18). First, the bonds between the film and substrate are cut, and both components are allowed to relax to their unstressed state (Figure 1*a*). A uniform stress, $\sigma_0 = E_f\varepsilon_0$ is then applied to the ends of the film so that its length matches that of the substrate (Figure 1*b*). At this stage, the stress in the film is uniform, and the substrate remains unstressed. The next step is to bond the film to the substrate with the external stress still applied to the ends of the film (Figure 1*c*). Since the interface is assumed to be planar,

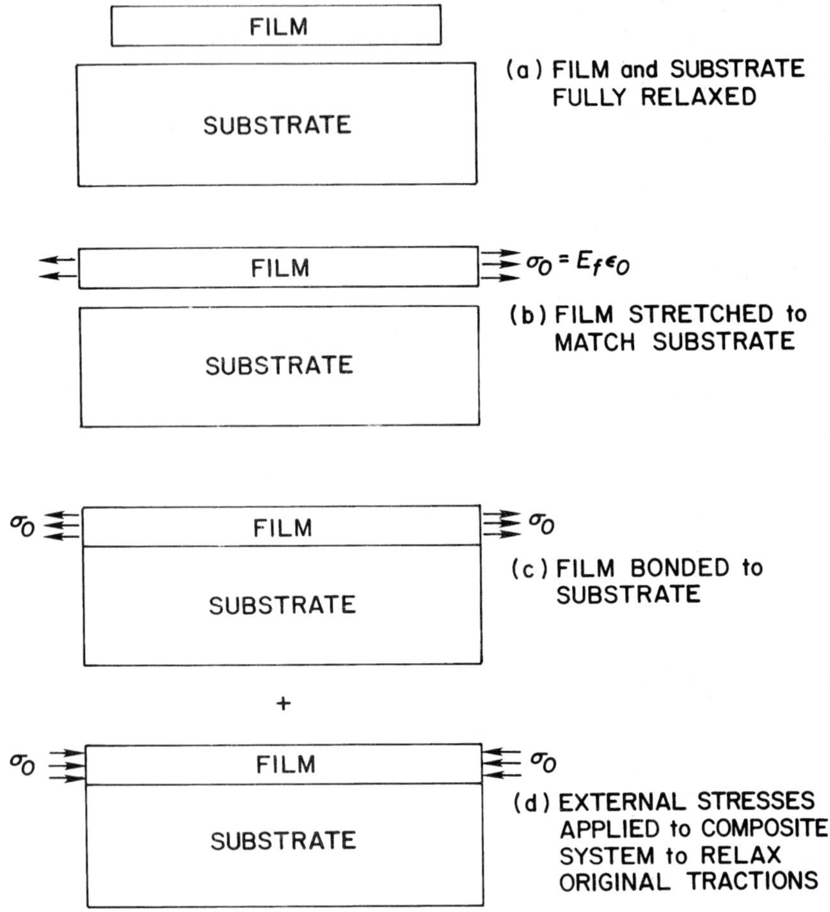

Figure 1 A thought experiment demonstrating the sequence for calculating the stress distribution and deformation of planar film-substrate composite systems. (*a*) The film and substrate are separated so that both are fully relaxed. (*b*) The film is stretched so that its lateral dimensions match those of the substrate. (*c*) The film is bonded to the substrate with the tractions still applied. (*d*) Stresses are applied to the bonded composite so as to relax the surface tractions.

the two surfaces match perfectly at this stage, and the bonding can be done without introducing any further stresses or deformations at the interface. After this bonding step, therefore, there is still a uniform stress σ_0 in the film, no stress in the substrate, and the system remains unbent. The final step is to produce stress-free surfaces by removing the external forces. This can be done by applying equal and opposite forces to the ends of the film in the now bonded composite (Figure 1*d*). The principle of superposition allows the final results of interest to be obtained by adding the stresses and deformations resulting from each of the individual procedures. In particular, attention can be focused on the effect of just the last step, as it is the only one that either produces any deformation of the substrate or introduces stresses at the interface.

Near the free edge, the applied compressive stress introduces normal and shear stresses at the interface. These stresses decay away from the edge in the form shown schematically in Figure 2 (there must be a sign change in the normal-stress distribution so as to maintain equilibrium). An analytical result has been derived for the peak stresses at the free edge in a homogeneous system for the limit where the substrate is infinitely thick (11). The maximum shear stress is

$$\tau_{max} = \sigma_0/\pi = 0.32\sigma_0, \qquad 4a.$$

whereas the maximum normal stress is

$$\sigma_{max} = (12-\pi^2)\sigma_0/(\pi^2-4) = 0.18\sigma_0. \qquad 4b.$$

Note that, while not true in this particular case of a homogeneous system, the interfacial stresses are generally singular at the free edge of a film bonded to a substrate. The strength of the singularities for different

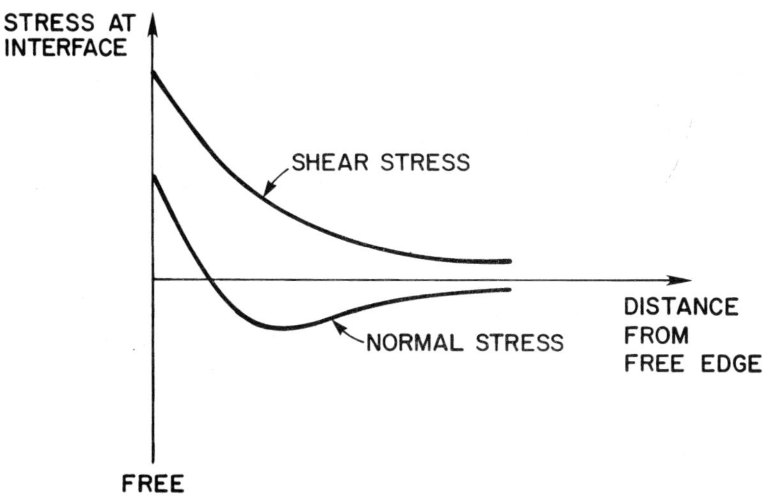

Figure 2 A schematic illustration of the distribution of normal and shear stresses along the interface between a film with a residual mismatch strain and a thick substrate. Equilibrium considerations dictate that the normal stress distribution must exhibit a sign change.

geometries and combinations of materials can be found in a paper by Bogey (19).

Well away from the free edge, the applied compressive stresses can be treated as a point load of $P = -\sigma_0 h$ per unit width acting at the mid-plane of the film. This has the effect on the composite beam of a uniform load P applied to the neutral axis of the beam and a bending moment M per unit width applied to the beam (Figure 3). The bending moment is given by the product of P and the distance between the neutral axis and the mid-plane of the film. Thus, if the modulus of the film and substrate are identical, M is given by

$$M = -Pd/2 = \sigma_0 dh/2. \qquad 5.$$

Simple-beam theory then gives the relationship between the axial stress

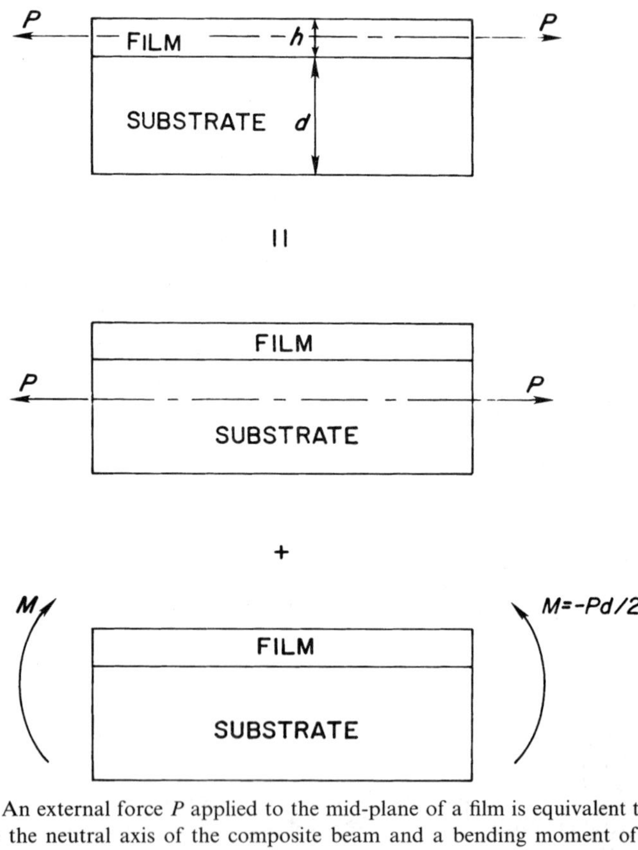

Figure 3 An external force P applied to the mid-plane of a film is equivalent to a force P applied to the neutral axis of the composite beam and a bending moment of magnitude $M = -Pd/2$.

induced by this bending moment and the distance from the neutral axis y measured away from the film:

$$\sigma(y) = My/I, \tag{6.}$$

where I is the moment of inertia which, for the present case of interest, is $(h+d)^3/12$ per unit width. This bending moment also induces bending of the beam with the radius of curvature being given by

$$R = EI/M, \tag{7.}$$

where E is the modulus of the system.

The net stress, well away from any free edges, is given by adding the bending stress (Equation 6) and the uniform axial stress, $P/(h+d)$, arising from the externally applied stress (Figure 1*d*) to the residual stress from the thermal-expansion mismatch (σ_0 in the film and zero in the substrate) (Figure 1*b*). It should be noted that these results for the stress distribution and the bending are valid even if the thermal-expansion mismatch causes the film to exceed its elastic limit. This is because the crucial step in the analysis is the final one, which is always elastic as it involves unloading. If the strain in the film exceeds the elastic limit, the calculation described above is still valid, with σ_0 representing the appropriate yield stress of the film.

Similar approaches can also be used when the film and substrate have different moduli, or the mismatch strain in the film is not uniform. A difference in modulus has two effects in that it changes the moment of inertia and the position of the neutral axis of the composite beam. The technique for determining these two quantities can be obtained from basic textbooks on simple-beam theory that include a discussion on composite beams. The moment of inertia can be shown to be given by

$$I = \{\Sigma[3(\Delta-\xi)^2-3(\Delta-\xi)+1]+3\Delta\xi(\Delta-\xi)+\xi^3\}/3h^3, \tag{8.}$$

where ξ is the ratio of the substrate to film thickness ($\xi = h/d$), Σ is the ratio of the modulus of the film to the modulus of the substrate, and the distance from the bottom of the substrate to the neutral axis is given by

$$\Delta \equiv \delta/h = \frac{\xi^2+2\Sigma\xi+\Sigma}{2(\xi+\Sigma)}. \tag{9.}$$

These equations can then be used directly in the calculations given above. The effect on the calculations of a nonuniform strain that has the form $\varepsilon(x)$ through the film, where x is measured from the interface, is provided in the first step (Figure 1*a*) when the film and substrate are separated. The film will curl and, before reattaching it to the substrate, a bending moment

$$M_f = E_f \int_0^h \left(\frac{h}{2} - x\right) \varepsilon(x)\, dx \qquad 10a.$$

must be applied to straighten it. The load P then required to bring the film into dimensional compatibility with the substrate is

$$P = \frac{1}{h} \int_0^h E_f \varepsilon(x)\, dx. \qquad 10b.$$

In the final step, both P and M_f must be relaxed by the imposition of the appropriate combination of force and moment ($-P$, and $M - M_f$).

Although the analysis given above was limited to plane-stress conditions, other forms of constraint can readily be analyzed. For example, plane-strain results are obtained directly from the plane-stress solutions by replacing all the modulus terms with $E/(1-v^2)$, where v is Poisson's ratio. Of particular importance are calculations for isotropic biaxial stress states where the relationship between the stress in the film and the curvature of a coated substrate can be shown to be

$$\sigma_0 = E_s d^2 / 6hR(1-v), \qquad 11.$$

where E_s is the modulus of the substrate, when the film is much thinner than the substrate. This is the well-known adaptation of Stoney's equation, which is often used to infer the average stress in a film from the curvature of the substrate (20). Some additional results for conditions when the film is not isotropic are given in the review paper by Nix (1).

STRESS RELAXATION IN EPITAXIAL SYSTEMS

An epitaxial film is one for which the lattice of the substrate provides a template for growing the film. The term homoepitaxial is used for a system in which the film and substrate are of the same material; heteroepitaxial is used to describe systems in which the film and substrate are different. In the latter case, the lattice parameters of the film and substrate may not be identical, so if the atomic positions are forced into a 1 : 1 correspondence with the atoms in the substrate, a state of stress will be induced. It has been shown that the total energy of such a system can be reduced by the introduction of misfit dislocations (21–25). Because dislocations can have a serious impact on electronic properties, understanding how to control their formation is of considerable importance. Although a full discussion is beyond the scope of this paper (for some recent reviews on this subject, see References 26–28), the basic concepts of stress relief by the introduction of dislocations are briefly outlined here because they are relevant to topics covered in following sections.

A particularly elegant approach due to Freund (25) for analyzing the introduction of a misfit dislocation is shown schematically in Figure 4. A threading dislocation in the substrate is continued as the film grows. Residual stresses cause the dislocation to move along its slip lane in the film, but it remains stationary in the substrate where there are no stresses. This causes the dislocation to bend and lay down a misfit dislocation in its wake as it progresses through the film (Figure 4). If the length of the dislocation along the interface is much greater than the film thickness, a steady-state analysis comparing the energy of a slice of material far ahead of the dislocation (AA′ in Figure 4) to the energy of a slice of material in the wake of the dislocation (BB′ in Figure 4) provides the criterion for whether the dislocation will grow or retreat. Assuming isotropic elasticity, the equilibrium condition for the stress σ_y required to move the dislocation in a film of thickness h is given by (25)

$$\sigma_y bh \sin\alpha = \frac{E_f b^2}{8\pi(1-v_f^2)}\left[\ln\left(\frac{2h}{r_0}\right) - \frac{1}{2}\cos 2\alpha - \frac{(1-2v_f)}{4(1-v_f)}\right], \qquad 12.$$

where b is the Burgers vector of the edge dislocation, α is the angle the

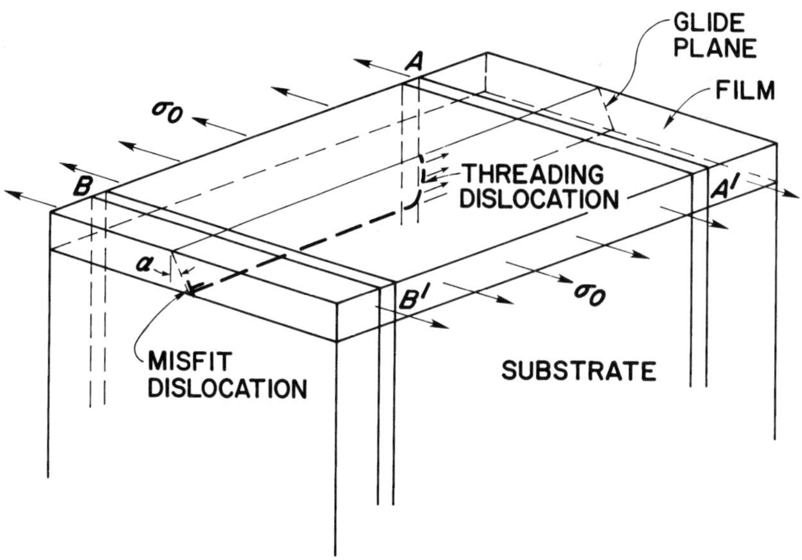

Figure 4 Misfit stresses in an epitaxial film cause a threading dislocation to glide through the film leaving a misfit dislocation in its wake. The glide plane is inclined at an angle α with the normal to the interface (25).

glide plane makes with the normal to the interface, and r_0 is the cut-off radius for the dislocation core, which can be taken to be approximately equal to the Burgers vector. Alternatively, this equilibrium condition can be reexpressed to give the critical thickness h_c required for the formation of a misfit dislocation in a film with a given misfit strain ε_0. If $\alpha = 90°$, and an equi-biaxial stress state exists in the film, the critical thickness is given by the solution to the equation (25)

$$\frac{b}{8\pi(1+\upsilon_f)h_c}\left[\ln\left(\frac{h_c}{b}\right)+\frac{1}{4(1-\upsilon_f)}\right]=\varepsilon_0. \qquad 13.$$

Both of these equations describe the requirements to move a single dislocation under equilibrium conditions. If the film is much thicker than h_c, an array of misfit dislocations can be formed that relaxes a substantial portion of the misfit strain. An approximate analysis to determine the density of the dislocations, which ignores the interaction between them, can be done by comparing the strain energy of a dislocated film to the line energy of the dislocations. If a regular square array of dislocations with a spacing D sits at the interface between a film and substrate, the average strain in the film is approximately $(\varepsilon_0 - b/D)$. The total energy per unit area of interface is then (1)

$$U_{total} = E_f h(\varepsilon_0 - b/D)^2/(1-\upsilon_f) + 2U_{dis}/D, \qquad 14.$$

where U_{dis} is the energy per unit length of a dislocation, given approximately by the right-hand term of Equation 12. If additional dislocations are introduced, the strain energy decreases, but at the cost of creating more dislocations. The equilibrium spacing S of the dislocations can be found by differentiating Equation 14 with respect to D and equating to zero (1, 27):

$$\begin{aligned} b/S &= [\varepsilon_0 - U_{dis}(1-\upsilon_f)/E_f hb] \\ &\simeq \varepsilon_0 - \frac{b}{8\pi h}\ln\frac{h}{b}. \end{aligned} \qquad 15.$$

Taking the limit $b/S \to 0$ gives an approximation for the critical thickness similar to that of Equation 13. Above this critical thickness, the equilibrium calculation predicts that the spacing decreases as the film gets thicker, but the strain can never be completely relaxed except when the film becomes infinitely thick.

Experimental results indicate that the trends predicted by these equilibrium models are followed qualitatively (29). However, both the observed critical thicknesses and the dislocation spacings are generally larger than the predicted values (30–33). Although there is a problem in determining the critical thickness, since strictly speaking it relies on finding a single dislocation in an infinite interface, the discrepancies between theory and experiment are generally attributed to nucleation barriers and kinetic effects (1, 27).

STRESS RELAXATION IN POLYCRYSTALLINE FILMS

Mechanisms

The simplest assumption for modeling the stress relaxation of polycrystalline films is that their behavior is controlled by the same mechanisms that operate in the bulk. Polycrystalline materials flow in response to a shear stress by a variety of thermally activated processes involving the diffusion of atoms and the motion of dislocations (34). At low temperatures, yielding occurs by dislocation glide. The motion of the dislocations is impeded by obstacles such as other dislocations, solute atoms, precipitates, grain boundaries, and the intrinsic resistance of the crystal lattice itself. The yield stress is raised as the spacing of these obstacles is reduced. For example, the Hall-Petch relationship describes an experimentally observed inverse-square root dependence of the yield stress on grain size: As the grain size decreases, the number of grain boundaries and, thus, the obstacles to dislocation motion, increase.

Power-law creep, which has a nonlinear dependence on stress, occurs when dislocations have enough thermal energy to climb and, therefore, escape from pinning sites on the glide planes. Climb is controlled by diffusion through the lattice at high temperatures and by diffusion along the dislocation cores at low temperatures. Often, it is found that a simple power-law relationship between the strain rate and stress does not satisfactorily describe experimental observations at high stresses. Under these circumstances, alternative constitutive equations have to be used that better describe the behavior in this transition regime, where dislocation motion changes from being glide-controlled to climb-controlled (34).

The flow of matter by diffusion in polycrystalline materials under shear stress occurs in response to a difference in chemical potential between neighboring grain boundaries. This difference in potential arises from the variation in normal stresses on boundaries at different orientations to the stress axes. The atoms flow between the boundaries either along the boundaries themselves (Coble creep) (35) or through the lattice (Nabarro-Herring creep) (36, 37). In contrast to dislocation mechanisms, flow by both of these diffusion processes becomes more rapid as the grain size is reduced because the path length over which diffusion has to occur is reduced.

Although all these mechanisms may operate in thin films to some degree, there clearly are significant differences between the thin-film and bulk forms of materials. Films may contain only one grain through their thickness, and the ratio of surface area to volume is relatively large. There will also be, for example, an interface between the film and substrate on one side, and a free surface or an interface between the film and a second layer on the other. These two interfaces are particularly important when there is an equi-biaxial stress in the plane of the film, as would generally be the case for isotropic films under a thermal-expansion mismatch stress. Under these conditions, there is no shear stress within the plane of the film, thus any slip processes must have a component normal to the interfaces. As a particular example, if the film contains only one grain through its thickness, then all the grain boundaries are under an identical normal stress, and Coble and Nabarro-Herring creep, as they are formulated for bulk materials, cannot occur. Instead, diffusion of atoms can only take place between the bulk of the film to the free surface (Figure 5) (38, 39). If no diffusion can occur at the substrate-film interface, the relaxation rate for grain-boundary diffusion controlled creep is (39)

$$\dot{\varepsilon} \simeq 3D_b\delta_b\Omega \frac{\sigma}{kTlh^2}, \tag{16.}$$

where $D_b\delta_b$ is the boundary diffusivity, which is exponentially dependent on temperature, Ω is the atomic volume of the diffusing species, k is Boltzmann's constant, T is the absolute temperature, l is the grain size, and h is the film thickness. As the surface diffusivity decreases, the relaxation rate is decreased owing to the finite rate at which atoms can be transported along the surface, and grain boundary grooves (for a tensile stress), or hillocks (for a compressive stress) will develop (39, 40). In an extreme limit, a coherent passivation or oxide layer could act as a diffusion barrier and completely suppress all diffusional relaxation.[2]

The movement of dislocations within a film is also affected by the interfaces between the film and its neighbors (41–44). If there is no passivation layer on the surface of the film, dislocations can emerge without hindrance from the free surface and produce slip traces and possible rumpling of the film (Figure 6*a*). However, if the surface is covered by a

[2] It should be noted, however, that many practical geometries may not exhibit an equi-biaxial stress state. Diffusional relaxation by flow between neighboring grain boundaries could then occur and would not be impeded by the surface constraints.

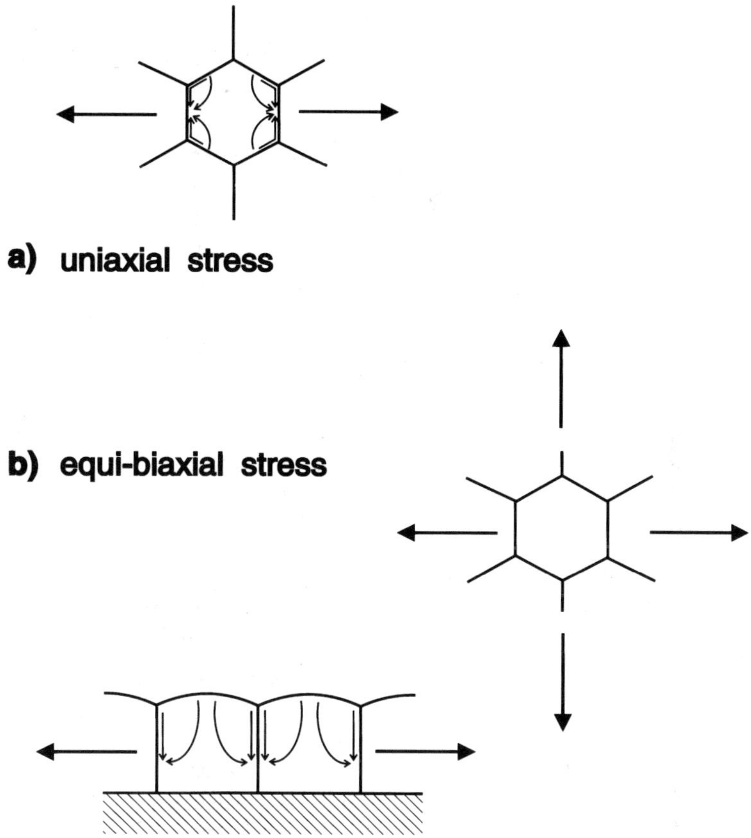

Figure 5 Schematic illustrations of the possible paths along which atoms can diffuse to relax stresses in (*a*) a uniaxial stressed film, (*b*) an equibiaxially stressed film. Diffusion to the surface will, of course, also occur in uniaxially stressed films. (From Reference 54.)

hard oxide, any dislocation motion would have to involve deformation or cracking of the film. In the absence of such deformation, the dislocations would be pinned at the interface, hindering further relaxation (Figure 6*b*) (1). To account for the constraint provided by the substrate and any overlayer such as an oxide or passivation layer, Nix (1) has adapted the model for dislocation motion in epitaxial films by Freund (25). The constraint provides an additional resistance to dislocation motion above that which could arise from any other obstacles within the film such as grain boundaries and precipitates (41–46), and results in a yield stress σ_y of the form (1)

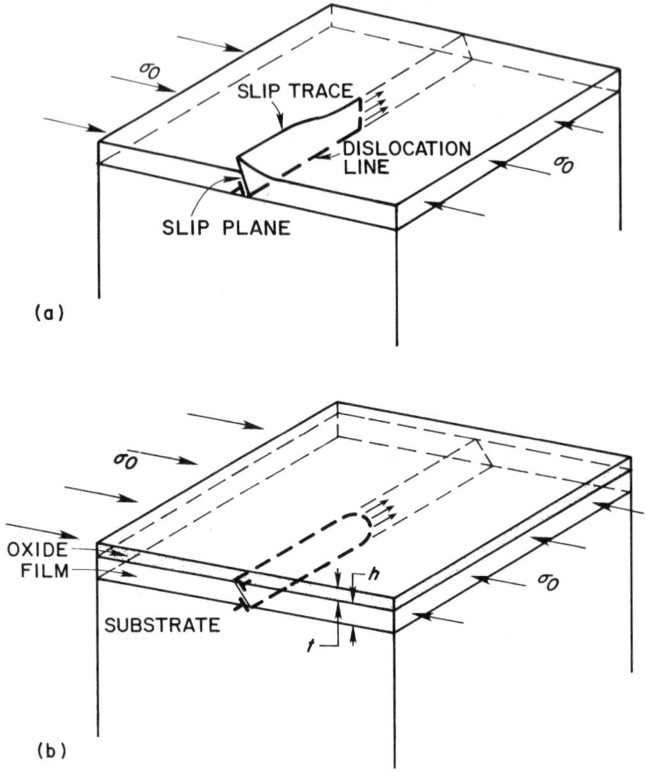

Figure 6 (*a*) Surface step and dislocations formed by slip in a thin film bonded to a substrate. (*b*) Deformation of any overlayer provides an additional constraint to relaxation. (Redrawn from Reference 1.)

$$\sigma_y \propto \frac{b}{h}\left[\frac{\mu_f \mu_s}{\mu_f + \mu_s} \ln\left(2.6\frac{h}{b}\right) + \frac{\mu_f \mu_{ov}}{\mu_f + \mu_{ov}} \ln\left(17.5\frac{t}{b}\right)\right], \quad 17.$$

where μ_f, μ_s, and μ_{ov} are the shear moduli of the film, substrate, and overlayer respectively, t is the thickness of the overlayer, and b is the Burgers' vector for the dislocation. If there is no surface layer, the second term of the expression is dropped. This expression predicts an inverse dependence of the yield stress on film thickness, which is analogous to the grain-size dependence for bulk materials, and is in marked contrast to the dependence expected when diffusion relaxation mechanisms dominate (see Equation 16). Such an inverse dependence has been observed or inferred in some specific systems such as aluminum films at elevated temperatures (43, 44), and lead and tin films at cryogenic temperatures (47, 48). In these systems, it might be expected that diffusional mechanisms are effectively suppressed by the presence of a coherent oxide (for the aluminium) or very low temperatures (for the lead and tin).

Modeling

If constitutive equations are known for the relevant deformation mechanisms, finite-element analyses can be used to predict the development of stresses for arbitrary geometries and temperature cycles. These stresses could then, in principle, be compared with stresses measured experimentally using X-ray techniques (49–51). A geometry for which both the analysis and experimental stress measurements are fairly straightforward is one in which a blanket film is deposited on a wafer (52–54). The biaxial stress in the film can be obtained approximately from Equation 11 by measuring how the curvature of the wafer changes as the system is cycled through a range of temperatures. A particular example is given in Figure 7 for a 1 μm thick copper film deposited on a silicon wafer.

The evolution of stresses in a film during thermal cycling is the result of two competing processes: development of elastic stresses from the thermal expansion mismatch between the film and substrate, and stress relaxation by plastic flow. This competition can be modeled by representing the film as a Maxwell element consisting of a spring in series with linear and

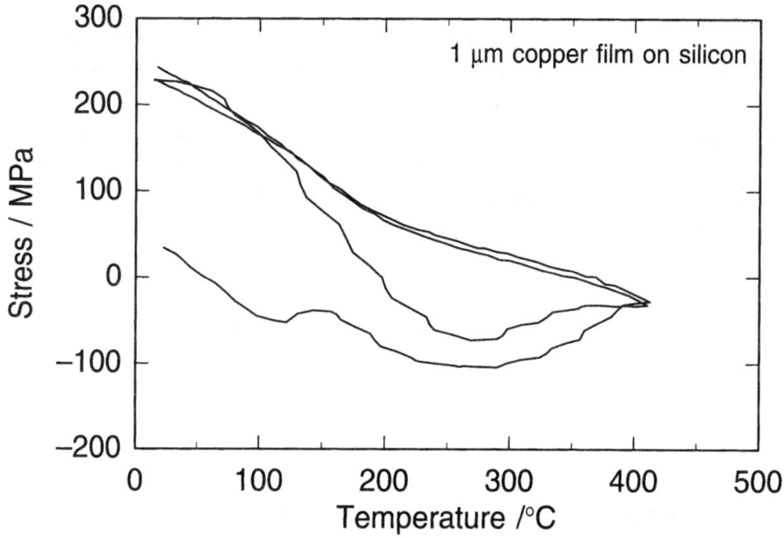

Figure 7 Plot of stress determined as a function of temperature for a 1 μm copper film on silicon. The first heating curve is associated with grain growth; once the microstructure has stabilized, further cycling follows a steady-state loop. (Courtesy of KP Rodbell.)

nonlinear dashpots (Figure 8) (54). The elastic response is determined by the spring, which has a spring constant equivalent to the biaxial modulus of the film, whereas the different dashpots represent the various relaxation mechanisms of the film. If the substrate is much stiffer than the film, an incremental change in temperature ΔT results in an incremental strain of

$$\Delta\varepsilon = \Delta\alpha\Delta T, \tag{18.}$$

where $\Delta\alpha$ is the difference in the coefficients of thermal expansion between the film and substrate. The instantaneous response of the Maxwell element to this strain is completely elastic, so that the elastic strain, $\varepsilon_{\mathrm{elastic}}$, increases by $\Delta\varepsilon$. With time, the elastic strain is converted to a plastic strain, but this results in no net change in the total strain of the film, which is established by the dimensions of the substrate. The stress σ acting on the film is related to the elastic strain by

$$\sigma = E_{\mathrm{f}}\varepsilon_{\mathrm{elastic}}/(1-v_{\mathrm{f}}). \tag{19.}$$

This stress controls the rate at which the dashpot elements extend and, hence, the rate at which elastic strain in the spring is converted to irreversible plastic strain in the dashpots. The general form of this deformation is

$$\dot{\varepsilon}_{\mathrm{plastic}} = \sum_{\mathrm{i}} C_{\mathrm{i}}\sigma^{n_{\mathrm{i}}}, \tag{20.}$$

where C_{i} and n_{i} are material parameters that depend on the appropriate relaxation mechanisms. It is important to note that the plastic-strain rate is not dictated by the heating or cooling rate, rather, it depends on the instantaneous stress in the film and on the constitutive behavior of the relaxation mechanisms. These equations can be solved analytically to show how the stress would vary as a function of temperature for a linear viscoelastic film (55). However, the general non-Newtonian nature of the defor-

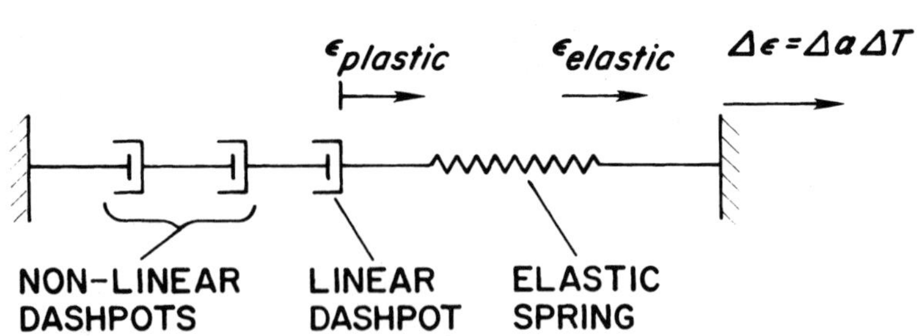

Figure 8 Relaxation of stresses in a film modeled as a Maxwell element consisting of an elastic spring in series with linear and non-linear dashpots.

mation of metals, and its temperature dependence, means that, in general, the response must be computed numerically (54).

If the deformation parameters have no time dependence, which is appropriate in the absence of microstructural changes, several general and important features emerge from the modeling as illustrated in Figures 9 and 10 (MD Thouless, KP Rodbell & C Cabral, paper in preparation). Figure 9 shows how a steady-state hysteresis loop is established after the first cycle, which is followed on all subsequent cycles for the same heating and cooling rates. This is consistent with many experimental observations (52–54). The hysteresis loop is uniquely established by the competition between the strain rate from the thermal-expansion mismatch and the rate of the stress relaxation. It should be noted that the minimum compressive stress is not a yield stress in the conventional sense of being associated with some critical event—it is merely the point at which the relaxation rate equals the applied strain rate. Relaxation processes occur during the entire thermal cycle, even at low temperatures during heating when the curve follows an elastic trajectory (Figure 9). In this region, however, the relaxation rates are so low that they have a negligible influence on the behavior. As the temperature and compressive stresses are increased, the

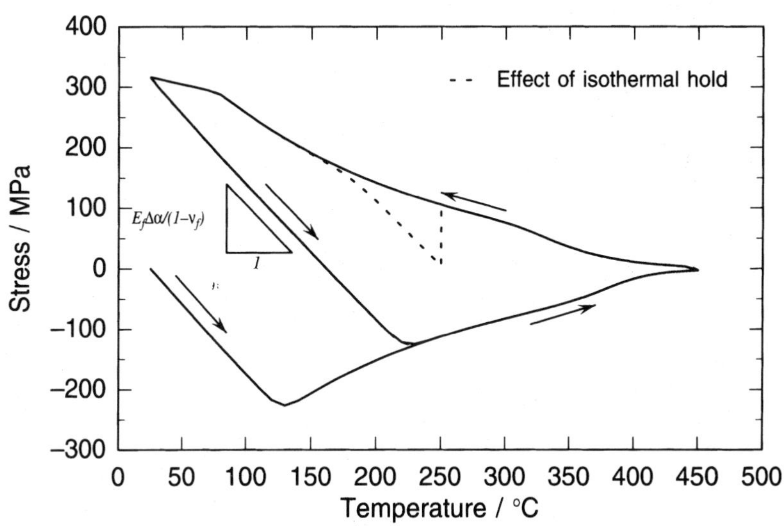

Figure 9 Example of the form of a stress-temperature hysteresis loop that is expected when no microstructural changes occur during cycling. A steady-state curve is generated after the first cycle. During heating, the initial portion of this loop follows an elastic line with a slope $E_f\Delta\alpha/(1-\upsilon_f)$. The predicted effect of an isothermal hold at elevated temperatures is also shown on this figure. If there are no microstructural changes, the curve rejoins the steady-state envelope upon resuming cooling at the original rate.

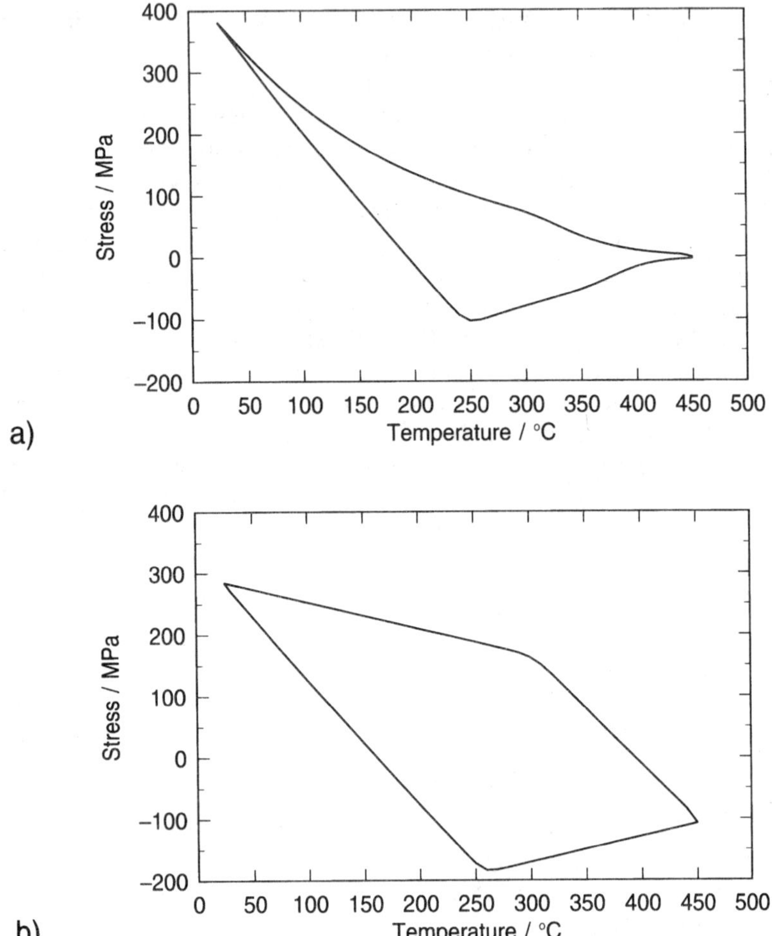

Figure 10 Examples of the steady-state hysteresis loops that would be established when the constitutive equation of the dominant low-temperature relaxation mechanism (*a*) does not and (*b*) does contain an exponential dependence on the stress. (From MD Thouless, KP Rodbell & C Cabral, paper in preparation.)

effects of relaxation start to become more important and to influence the shape of the curve.

The steady-state envelope is followed even after a prolonged hold at elevated temperatures (Figure 9). It can be seen that while relaxation occurs during the hold, further cooling results in the stresses building up elastically until the steady-state curve is reached. This curve is then followed during any subsequent cycling. There is no dependence on prior thermal history (or stress state) because there is nothing in the constitutive equations that carries such information. If the stress is decreased below the level of the steady-state cycle during an isothermal hold, then upon subsequent cycling, the initial relaxation rate will be low (because of the low stress), and the stress will build up elastically until the rates are compatible and the steady-state envelope for the appropriate heating or cooling rate is reached. Such behavior has been observed experimentally (MD Thouless, KP Rodbell & C Cabral, paper in preparation), and it is also indicated by the form of the curves reported for copper films by Flinn (53). These observations imply that the relaxation process often does consist of a simple competition between the relaxation rate and the loading rate without any significant changes in microstructure or dislocation density.

A final general feature of the stress-temperature curves is their behavior on the cooling cycle at low temperatures. This gives some indication of the type of mechanism responsible for relaxation at these lower temperatures. The relaxation rates of thermally activated processes decay exponentially at low temperatures unless the constitutive equations contain an exponential stress dependence. If there is no exponential dependence on stress, the stress-temperature curve will tend to an elastic line as illustrated in Figure 10*a*. Conversely, if there is an exponential stress term in the appropriate constitutive equation (e.g. as for dislocation glide), the cooling curve will bend over in the opposite sense, as indicated in the schematic curve of Figure 10*b*. In both of these examples, the heating portion of the curve follows the elastic line as indicated in Figure 9.

A comparison of the experimental thermal cycling curves with predictions based on constitutive equations developed for bulk metals is illuminating. Using the constitutive equations summarized in Frost & Ashby (34), it can be shown that the behavior of a copper film can be modeled very well (54). However, aluminum behaves in a very anomalous fashion. The experimental stress-temperature curve for aluminium can be reproduced by modeling only if the creep parameters are decreased by orders of magnitude from the ones appropriate for the bulk form (56; MD Thouless, KP Rodbell & C Cabral, paper in preparation). In particular, the diffusional and power-law creep components can be neglected, leaving dislocation glide as the dominant mechanism, as proposed by Flinn et al (52). This is consistent with the shape of the experimental curves, which, in contrast to the curves for copper (53, 54), do not exhibit any elastic response on cooling to room temperature (43, 44, 52). The addition of a passivation layer to copper and silver appears to alter the deformation characteristics (57; MD Thouless, KP Rodbell & C Cabral, paper in preparation). The shapes of the experimental curves suggest that, as with aluminum, glide mechanisms dominate under such conditions.

STRESS RELIEF BY CRACKING

This section considers an additional mechanism of stress relaxation, namely cracking. Since it can compromise the structural integrity of a thin-film system, cracking is almost inevitably identified with failure, and other forms of stress relief are often pursued in an active effort to prevent it. A rich variety of different cracking phenomena can occur (58–60), including failure of the film itself (61–65), delamination at the interface under tensile (66–69) or compressive (70) loads, buckling-driven delamination (71–75), and cracking of the substrate (76–79). For a detailed discussion of these mechanisms, the interested reader is referred to a recent and comprehensive review by Hutchinson & Suo (59) on the fracture of laminated materials. Attention here will be limited to the formation of cracks within the film because it provides some analogies to the dislocation mechanism discussed above.

Film Cracking

A tensile stress in a brittle film, which is bonded to a relatively tough substrate, can be relieved by the formation of cracks that tunnel along the film. Common examples are crazing on glazes and the cracks that can be observed in drying mud. These cracks can be initiated as surface flaws that grow down to the interface and then spread through the film in a steady-state fashion (Figure 11). When this figure is compared to Figure 4, the similarity to the geometry of a propagating dislocation becomes obvious. In both mechanisms, strain energy in the film is relaxed at the cost of the energy associated with introducing a defect. Furthermore, both involve the laying down of an elastic singularity near the interface. In the present case of cracking, the singularity is a crack tip; in the former case of glide, it was a dislocation line. A direct analogy with the calculations leading to Equation 12 gives the result that the equilibrium condition for a single crack growing across a film is (63, 77)

$$\frac{1.98(1-\nu_f^2)\sigma_0^2 h^2}{E_f} = \Gamma_f h, \qquad 21.$$

when the film and substrate have identical elastic properties. The term on the left-hand side of this equation corresponds to the relaxation of the strain energy caused by the introduction of the crack. The term on the

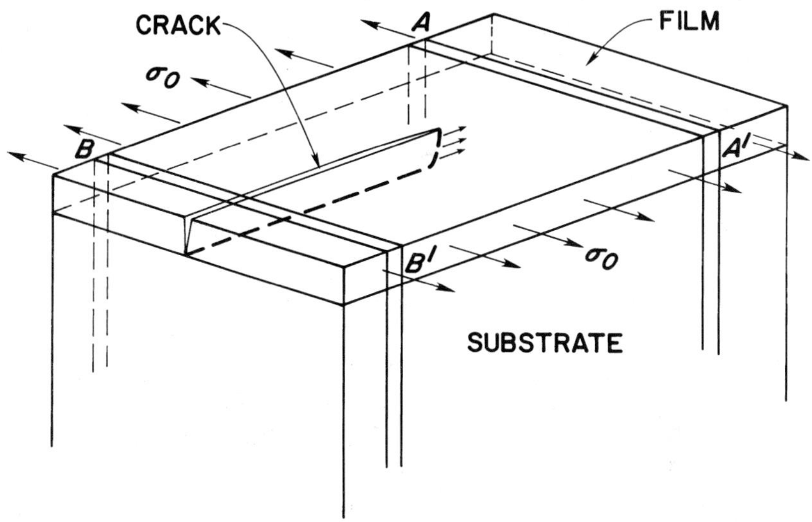

Figure 11 Schematic illustration of the geometry for a steady-state crack propagating through a thin film. The critical condition for propagation is evaluated by comparing the energy of a slice AA′ ahead of the crack with a slice BB′ in the wake.

right-hand side is the energy associated with the formation of the crack.[3] This criterion shows that, as for dislocation motion, there is a critical film thickness below which cracking cannot occur for a given residual strain in the film. This critical thickness is given by (63, 77)

$$h_c = \frac{0.5\Gamma_f(1-v_f)}{(1+v_f)E_f\varepsilon_0^2}. \qquad 22.$$

Similar calculations have been done for the general case when the film and substrate have different elastic constants (65). The resulting plot for the critical thickness as a function of elastic modulus ratio is shown in Figure 12 for the special case when Poisson's ratio of both components is equal to 1/3.

[3] Γ_f is defined as the toughness of the film, which is formally the energy required to create a unit area of new crack surface. At the simplest level of approximation, Γ_f is taken as twice the surface energy of the film 2γ. (The factor two arises because two new surfaces are formed by a crack.) However, in practice, other dissipative processes generally occur during crack advance, so $\Gamma_f \gg 2\gamma$.

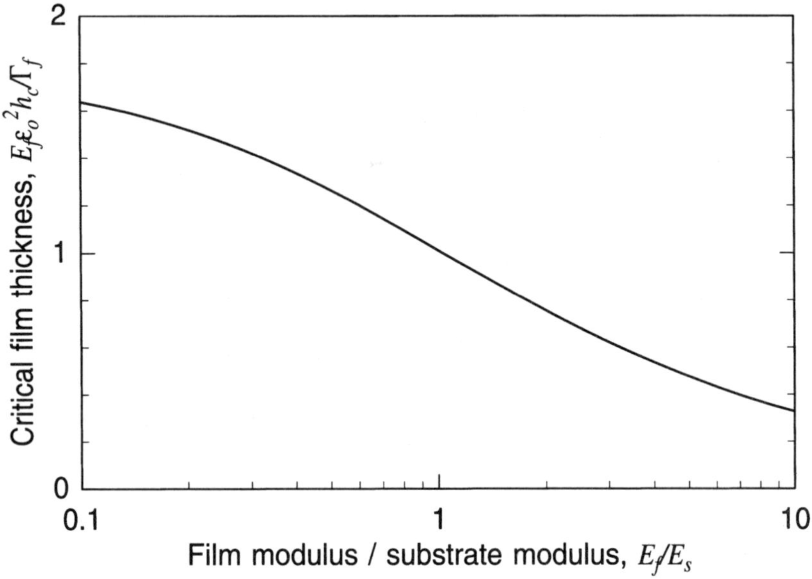

Figure 12 Effect of the ratio of the film modulus to the substrate modulus on the critical film thickness for cracking, assuming Poisson's ratio for both components is 1/3 and an equi-biaxial residual strain of ε_0. (Plotted from results given in Reference 65.)

Crack Spacing

The stress relaxation associated with a crack prevents subsequent cracks from developing unless they are sufficiently distant for the local strain energy to build up to a critical level. This is illustrated in Figure 13, which shows cracks formed in a soda-lime glass cover slip diffusion-bonded to a fused-quartz substrate and cooled to room temperature so as to establish a residual tension in the cover slip (80). No cracks formed until the indent on the right side of the micrograph was introduced by a Vickers indent. The resulting crack then grew to the edges of the film. A second indent was placed in close proximity to the first one. A crack began to grow under the influence of the stress field of the indent but was too close to the initial crack to continue growing. A third indent was then introduced farther away, and the resulting crack grew to the edges of the film.

The relaxation of stresses in the vicinity of a crack and the consequent imposition of a minimum crack spacing is, again, analogous to the above discussion on dislocations. A similar approach can be used to obtain an approximate result for the crack spacing as a function of film thickness and residual strain. If the film and substrate have identical elastic properties, cracks have a negligible interaction if they are at a spacing more than

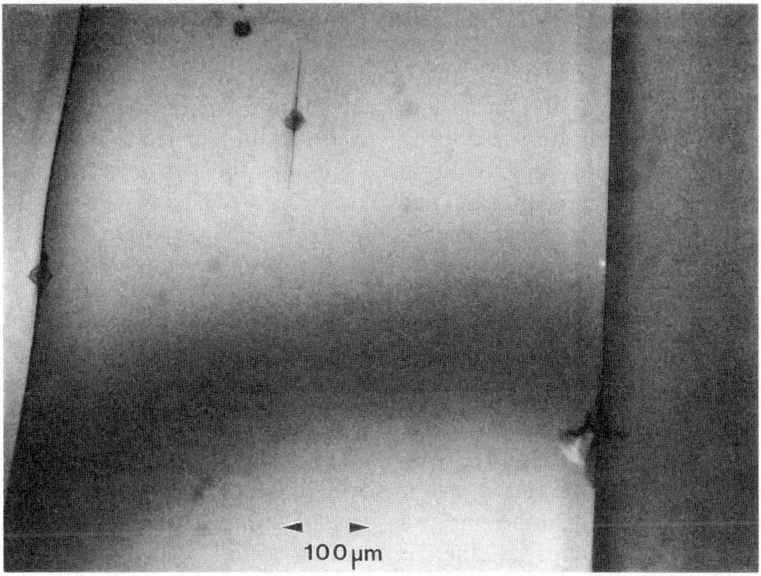

Figure 13 A micrograph showing the growth of cracks from indentation flaws in a soda-lime glass cover slip ($\simeq 140$ μm thick) bonded to a thick-fused-silica substrate. (From Reference 80.)

about eight times the thickness of the film. Beyond this spacing, they are essentially independent. If they are closer than this distance, the total energy per unit area of interface in a film with cracks at an equal spacing D in a one-dimensional array is approximately (63)

$$U_{\text{total}} = [0.5 - 1.98(h/D)](1 + v_f)E_f\varepsilon_0^2 h/(1 - v_f) + \Gamma_f h/D. \qquad 23.$$

The first term is the strain energy of a cracked film, and the second term is the energy associated with the cracks. Differentiating this expression with respect to D and equating to zero gives an equilibrium spacing of (64)

$$S/h \simeq 5.6\sqrt{\frac{\Gamma_f(1 - v_f)}{(1 + v_f)E_f\varepsilon_0^2 h}}. \qquad 24.$$

Thus the crack spacing is predicted to be proportional to the square root of the film thickness. This relationship agrees with the results of a series of model experiments investigating the fracture of $PrBa_2Cu_3O_{7-x}$ films grown by laser ablation on a [110]-$SrTiO_3$ substrate (64).

Although the analysis is similar to that done for determining the dislocation spacing, there are several significant differences between cracks and dislocations. The most important of these is that cracks, once formed, are not mobile. As a result, the crack spacing is often dictated by the stochastic nature of the initial flaw distribution (81), rather than by the energetic considerations described here. For example, in the system illustrated by the micrograph of Figure 13, cracks formed only where they were initiated by an indentation flaw. However, provided there is a suitable density of initial flaws, the general trends are not seriously affected by assumptions about the exact order in which cracks develop, as can be seen by comparing the model given above with one described by Hutchinson & Suo (59).

CONCLUDING REMARKS

Some of the mechanisms by which the generation and propagation of defects can reduce the elastic-strain energy of a film have been reviewed in this paper. Typically, if the energy associated with these defects is less than the strain-energy released, then relaxation occurs. Because the strain-energy scales with the thickness of the film, this balance naturally introduces the concept of a critical film thickness below which relaxation is not energetically favored. Although this paper began with a review of the origin of stresses, it should be noted that the distinction between origins and relaxation is rather arbitrary. In practice, stress development and stress relaxation are interdependent processes. Stresses can evolve because the energy balance favors the elimination of defects at the expense of creating additional strain energy. Similarly, they can relax when the balance is tilted the other way. For example, if it were possible to introduce a misfit dislocation in an epitaxial film below the critical thickness, energetic considerations would then indicate a tendency for the dislocation to be annihilated and the residual strain in the film to be increased. An even clearer example of this dynamic is seen in the phase transformations that may occur in thin-film systems. This paper cites phase transformations as a possible source of stresses in order to provide an analogy to thermal-expansion mismatch strains. However, they can equally well be considered as a mechanism for stress relaxation. In ferroelectric thin films (82, 83), for example, domain formation has been proposed as a relaxation mechanism. The appropriate view depends on the context and on other contributions to the strain.

Finally, a clear understanding of the mechanisms of stress evolution and relaxation is critical to the design of effective thin-film systems. The key issue is to control the stresses that will inevitably occur so that they act in the most benign fashion. Ideally, this will involve designing thin-film systems to enhance certain relaxation mechanisms while impeding others so as to maximize the desired properties of the system at hand.

ACKNOWLEDGMENTS

The author acknowledges helpful discussions with TM Shaw, AD Westwood, and KP Rodbell. Additionally, the help of Y Wu in reviewing the manuscript is greatly appreciated.

Literature Cited

1. Nix WD. 1989. *Metall. Trans. A* 20: 2217–45
2. Tu KN, Mayer JW, Feldman LC. 1992. *Electronic Thin Film Science for Electrical Engineers and Materials Scientists*, New York: Macmillan
3. Guthrie WL, Patrick WJ, Levine E, Jones HC, Mehter EA, et al. 1992. *IBM J. Res. Dev.* 36: 845–57
4. Tummala RR, Knickerbocker JU, Knickerbocker SH, Herron LW, Nufer RW, et al. 1992. *IBM J. Res. Dev.* 36: 889–904
5. Bhushan B. 1990. *Tribology and Mechanics of Magnetic Storage Devices*. New York: Springer-Verlag
6. Takeuchi YR, Kokini K. 1994. *J. Eng. Gas Turbines Power*116: 266–71
7. Whittle DP, Stringer J. 1980. *Philos. Trans. R. Soc. London Ser. A* 295: 309–29
8. Evans AG, Cannon RM. 1989. *Mater. Sci. Forum* 43: 243–68
9. Evans AG, Crumley GB, Demaray RE. 1983. *Oxid. Met.* 20: 193–216
10. Zhang Y, Shores DA. 1993. *Oxid. Met.* 40: 529–53
11. Thouless MD, Evans AG, Ashby MF, Hutchinson JW. 1987. *Acta Metall.* 35: 1333–41
12. Bordia RK, Jagota A. 1993. *J. Am. Ceram. Soc.* 76: 2475–85
13. Heintz J-M, Sudre O, Lange FF. 1994. *J. Am. Ceram. Soc.* 77: 787–91
14. Brinker CJ, Scherer GW. 1990. *The Physics and Chemistry of Sol-Gel Processing*. New York: Academic
15. Chiu RC, Garino TJ, Cima MJ. 1993. *J. Am. Ceram. Soc.* 76: 2257–64
16. Gardner DS, Longworth HP, Flinn PA. 1992. *J. Vac. Sci. Technol. A* 10: 1426–41
17. Rodbell KP, Knorr DB, Mis JD. 1993. *J. Electron. Mater.* 22: 597–606
18. Eshelby JD. 1957. *Proc. R. Soc. London Ser. A* 241: 376–96
19. Bogy DB. 1971. *J. Appl. Mech.* 93: 377–86
20. Stoney GG. 1909. *Proc. R. Soc. London Ser. A* 82: 172–75
21. Frank FC, van der Merwe JH. 1949. *Proc. R. Soc. London Ser. A* 198A: 205–16
22. Frank FC, van der Merwe JH 1949. *Proc. R. Soc. London Ser. A* 198A: 216–25
23. van der Merwe JH. 1963. *J. Appl. Phys.* 34: 123–27
24. Matthews JW, Blakslee AE. 1974. *J. Cryst. Growth* 27: 118–25
25. Freund LB. 1987. *J. Appl. Mech.* 54: 553–57
26. Jain SC, Willis JR, Bullough R. 1990. *Adv. Phys.* 39: 127–90
27. Freund LB. 1992. *MRS Bull.* 17: 52–60
28. Hull R, Bean JC, Logan RA. 1993. *Solid State Phenomena* 32–33: 417–32
29. Petruzzello J, Greenberg BL, Cammack DA, Dalby R. 1988. *J. Appl. Phys.* 63: 2299–303
30. Bean JC, Feldman LC, Fiory AT, Nakahara S, Robinson IK. 1984. *J. Vac. Sci. Technol.* A2: 436–40
31. Tsao JY, Dodson BW, Picraux ST, Cornelison DM. 1987. *Phys. Rev. Lett.* 59: 2455–58
32. Dodson BW, Tsao JY. 1987. *Appl. Phys. Lett.* 51: 1325–27
33. Scott MP, Laderman SS, Kamins TI, Rosner SJ, Nauka K, et al. 1989. *Mater. Res. Soc. Symp. Proc.* 130: 179–84
34. Frost HJ, Ashby MF. 1982. *Deformation-Mechanisms Maps*. Oxford: Pergamon

35. Coble RL. 1963. *J. Appl. Phys.* 34: 1679–82
36. Nabarro FRN. 1948. In *Report of a Conference on Strength of Solids*, pp. 75–90. London: Phys. Soc.
37. Herring C. 1950. *J. Appl. Phys.* 21: 437–45
38. Gibbs GB. 1966. *Philos. Mag.* 13: 589–93
39. Thouless MD. 1993. *Acta Metall. Mater.* 41: 1057–64
40. Génin FY, Mullins WW, Wynblatt P. 1993. *Acta Metall. Mater.* 41: 3541–47
41. Menter JW, Pashley DW. 1959. In *Structure and Properties of Thin Films*, ed. CA Neugebauer, JB Newkirk, DA Vermilyea, pp. 111–50. New York: Wiley
42. Murakami M. 1979. *Thin Solid Films* 59: 105–16
43. Doerner MF, Gardner DS, Nix WD. 1986. *J. Mater. Res.* 1: 845–51
44. Venkatraman R, Bravman JC. 1992. *J. Mater. Res.* 8: 2040–48
45. Sanchez, JE Jr, Arzt E. 1992. *Scripta Metall. Mater.* 27: 285–90
46. Thompson CV. 1993. *J. Mater. Res.* 8: 237–38
47. Kuan TS, Murakami M. 1982. *Metall. Trans.* 13A: 383–91
48. Blumberg RH, Seraphim DP. 1962. *J. Appl. Phys.* 33: 163–68
49. Besser PR, Mack AS, Fraser D, Bravman JC. 1993. *J. Electron. Soc.* 140: 1769–72
50. Besser PR, Brenann S, Bravman JC. 1994. *J. Mater. Res.* 9: 13–24
51. Chidambarrao D, Rodbell KP, Thouless MD, DeHaven PW. 1994. *Mater. Res. Soc. Symp. Proc.* 338: 261–68
52. Flinn PA, Gardner DS, Nix WD. 1987. *IEEE Trans. Electron Devices* 34: 689–99
53. Flinn PA. 1991. *J. Mater. Sci.* 6: 1498–501
54. Thouless MD, Harper JME, Gupta JA. 1993. *J. Mater. Res.* 8: 1845–52
55. Liu HC, Muraka SP. 1992. *J. Appl. Phys.* 72: 3458–63
56. Koleshko VM, Belitsky VF, Kiryushin IV. 1986. *Thin Solid Films* 142: 199–212
57. Vinci RP, Bravman JC. 1993. *Mater. Res. Soc. Symp. Proc.* 308: 337–41
58. Gille G. 1985. In *Current Topics in Materials Science*, ed. E Kaldis, 12: 420–72. Amsterdam/New York: North Holland
59. Hutchinson JW, Suo Z. 1992. In *Advances in Applied Mechanics*, ed. JW Hutchinson, TY Wu, 29: 63–191. San Diego: Academic
60. Thouless MD. 1991. *J. Vac. Sci. Technol.* A9: 2510–15
61. Agrawal DC, Raj R. 1989. *Acta Metall.* 37: 1265–70
62. Hu MS, Evans AG. 1989. *Acta Metall.* 37: 917–25
63. Thouless MD. 1990. *J. Am. Ceram. Soc.* 73: 2144–46
64. Thouless MD, Olsson E, Gupta A. 1992. *Acta Metall. Mater.* 40: 1287–92
65. Beuth JL Jr. 1992. *Int. J. Solids Struct.* 29: 1657–75
66. Kendall K. 1971. *J. Phys. D: Appl. Phys.* 4: 1186–95
67. Thouless MD. 1988. *Acta Metall.* 36: 3131–35
68. Suo Z, Hutchinson JW. 1990. *Int. J. Fract.* 43: 1–18
69. Jensen HM, Hutchinson JW, Kim K-S. 1990. *Int. J. Solids Struct.* 26: 1099–114
70. Thouless MD, Jensen HM, Liniger EG. 1994. *Proc. R. Soc. London Ser. A.* 447: 271–79
71. Nir D. 1984. *Thin Solid Films* 112: 41–49
72. Amaratunga GAJ, Welland ME. 1990. *J. Appl. Phys.* 68: 5140–45
73. Chai H. 1990. *Int. J. Fract.* 46: 237–56
74. Hutchinson JW, Thouless MD, Liniger EG. 1992. *Acta Metall. Mater.* 40: 295–308
75. Thouless MD, Hutchinson JW, Liniger EG. 1992. *Acta Metall. Mater.* 40: 2639–49
76. Cannon RM, Fisher RM, Evans AG. 1986. *Mater. Res. Soc. Symp. Proc.* 54: 799–804
77. Hu MS, Thouless MD, Evans AG. 1988. *Acta Metall.* 36: 1301–7
78. Drory MD, Thouless MD, Evans AG. 1988. *Acta Metall.* 36: 2019–28
79. Suo Z, Hutchinson JW. 1989. *Int. J. Solids Struct.* 25: 1337–53
80. Thouless MD. 1994. *Proc. Int. Conf. Residual Stresses, 4th, Baltimore*, pp. 1088–96. Bethel, MD: Soc. Exp. Mech.
81. Delannay F, Warren P. 1991. *Acta Metall. Mater* 39: 1061–72
82. Kwak BS, Erbil A, Wilkins BJ, Budai JD, Chisholm MF, Boatner LA. 1992. *Phys. Rev. Lett.* 68: 3733–36
83. Pompe W, Gong X, Suo Z, Speck JS. 1993. *Mater. Res. Soc. Symp. Proc.* 310: 179–86

Annu. Rev. Mater. Sci. 1995. 25: 97–127

LASER-BEAM INTERACTION WITH DEFECTS ON SEMICONDUCTOR SURFACES: An Approach to Generation of Defect-Free Surfaces

Noriaki Itoh, Jyun'ichi Kanasaki, Akiko Okano, and Yasuo Nakai

Department of Physics, Nagoya University, Furocho, Chikusa, Nagoya 464–01, Japan

KEY WORDS: defects on surfaces, laser irradiation, electronic excitation, laser-induced surface processes, control of defects on surfaces, laser-induced bond breaking

ABSTRACT

Current studies of laser-induced emissions of semiconductor atoms from semiconductor surfaces, originating from electronic bond breaking, are reviewed. Electronic excitation in solids can lead to atomic emissions by breaking the bonds of the surface atoms. For semiconductor surfaces, this process requires multiple excitations and is defect related. Experimental results of atomic emissions induced by nanosecond laser pulses on Si, GaAs, and GaP surfaces show that the emission originates from surface defects, and that the yield is a superlinear function of the laser fluence. Excitation spectra of the yield confirm that emission originates from bond breaking. Related phenomena, including laser-induced dry etching and laser ablation, are also discussed in the context of bond breaking. We conclude that laser-induced electronic bond breaking may be used to produce defect-free surfaces.

INTRODUCTION

Recent observations by scanning tunneling microscopy (STM) and scanning tunneling spectroscopy (STS) have revealed that the surfaces grown

0084–6600/95/0801–0097$05.00

by epitaxy and those cleaned in a high vacuum contain a substantial number of defects (1–5). The physics and chemistry of surface defects are in the initial stages of development; the electronic and atomic structure of these defects and their effect on the properties of surface-grown devices are problems to be explored (6). For some time, physics and chemistry of defects in the bulk have dealt with their atomic and electronic structure, their thermodynamic properties, and their photophysics and photochemistry (7). The ultimate goal has been the fully controlled introduction of defects into solids whereby a given concentration of specific defects can be introduced and unwanted defects can be eliminated from materials. This same goal holds for the physics and chemistry of surface defects, which are considered to be more easily controlled because they are exposed on the surface and can actually be seen and treated, as demonstrated by the STM technique (8).

Photophysics and photochemistry of defects in the bulk have been well developed in insulating crystals (9) and in semiconductors (10), in which photons below the band-gap energy can penetrate into solids to interact with defects. These studies have made a profound contribution, not only in understanding the properties of defects, but also in developing a method of altering or even passivating the defect properties (11). Photophysics and photochemistry, discussed in this review, are considered to be particularly useful for controlling surface defects because a large number of defects can be treated simultaneously in a short time.

Electronic excitation of a defect in the bulk often alters its structure owing to the coupling between the excited state and the lattice, as depicted using configuration coordinate curves in Figure 1. The abscissa denotes

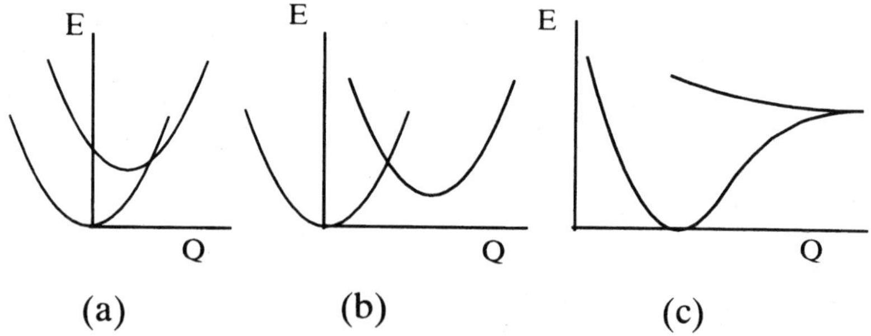

Figure 1 The configuration coordinate curves for the ground state (*lower curve*) and electronically excited state (*upper curve*). The relation between the total electronic and lattice energy E and the lattice configuration Q, describe the lattice relaxation after electronic excitation leading to (*a*) a luminescent state, (*b*) a non-luminescent metastable state, and (*c*) an atomic emission.

lattice deformation and the ordinate is the total electronic and lattice energy. The lower curves of Figure 1 denote the adiabatic potential energy surface (APES) for the ground state, whereas the upper curves show the APES for the excited states. Figure 1 indicates three cases where the equilibrium configuration of the excited state is (*a*) within the parabola for the ground state, (*b*) outside the parabola, and (*c*) at infinity. According to the Franck-Condon principle (12), the electronic excitation initially populates the excited state at the same configuration as the ground state, i.e. vertically in the diagram. Then the lattice relaxes following the APES to (*a*) a luminescent state, (*b*) a metastable state, and (*c*) a configuration with an atom missing from the surface. The last APES is analogous to that of the Menzel-Gomer-Redhead (MGR) process (13, 14) and denotes that for the desorption induced by electronic transition (DIET).

It is the purpose of this review to explore present knowledge of laser interaction with defects on semiconductor surfaces, in particular laser-induced modification of defects on surfaces. First we describe the basic processes for the structural change induced by electronic excitation. Then we present experimental observation of laser-beam interaction with defects on surfaces carried out by the group at Nagoya University. Finally related phenomena and future perspective of these processes is presented. We restrict ourselves to discussions on the effects of laser irradiation primarily with nanosecond laser pulses; picosecond and sub-picosecond laser pulses produce different phenomena, which are reviewed briefly.

LOCAL LATTICE MODIFICATION DUE TO BOND BREAKING IN SOLIDS

Classification of Non-Metallic Materials in Terms of Self-Trapping

For convenience of further discussions, we classify the materials into two groups: type A, in which excitons are not self trapped, and type B, in which excitons are self trapped (15). The typical materials belonging to both types are shown in Table 1.

Bond Breaking in Materials in which Excitons Are Self Trapped (Type B)

Interaction of laser beams of photon energies above the band-gap energy with non-metallic solids induces both one photon band-to-band transitions as well as excitation of defect states, whereas the excited defect states dominate if the photon energy is below the band-gap energy. Multi-photon excitation may occur in the latter case, but its cross section is small

Table 1 Classification of non-metallic materials

Solid type	Self-trapping of excitons	Material
Type A	No	Semiconductors MgO, CaO, ZnO Cu, and Tl halides
Type B	Yes	Alkali halides, alkaline earth fluorides SiO_2

(16) and may be disregarded in the present context. Holes and excitons (composite particles consisting of an electron and a hole interacting through the coulomb force) are self trapped in many insulators (17), i.e. localized by the lattice distortion they produce.[1]

In an ionic solid MX, where M is a cation and X is an anion, the valence band often consists of the *p* orbitals of X, hence a hole is regarded as $X^{-(z-1)}$, where z is the valency. In a monovalent solid, there is no Madelung potential acting on X^0, and hence the X^0 can relax to a halogen molecular ion X_2^-, called a self-trapped hole (Figure 2*a*) (18). An exciton in most alkali halides with a NaCl structure, expressed as $(X^0 + e)$, is self trapped, causing further lattice deformation (Figure 2*b*) (19), which is more likely a pair of an *F* center (a halogen atom vacancy) and an *H* center (a halogen atom interstitial). It follows that putting an electron on the configuration shown in Figure 2*a* induces further lattice distortion by virtue of the energy gain due to the electron relaxation (20, 21). Here the energy needed to deform the lattice from the configuration of Figure 2*a* to that of Figure 2*b* is compensated by the difference in the potential energy for the electron. According to recent theoretical calculation (22, 23), an exciton on the surface leads to the emission of a halogen atom leaving an *F* center behind. The APES for an exciton on the surface is the anti-bonding type (Figure 1*c*), due to the low symmetry for atoms on the surface (24). We note that the lattice modification described above is the result of softening of the ionic bonding caused by the change in the charge state.

[1] The self-trapping of excitons is induced when the electron-lattice relaxation energy is larger than their translational energy (Toyozawa T. 1983. *Physica* 117/118: 23). Thus it occurs in limited insulators such as alkali halides, alkaline earth oxides and SiO_2 (Itoh N, Tanimura K. 1990. *J. Phys. Chem. Solids* 45: 323–40).

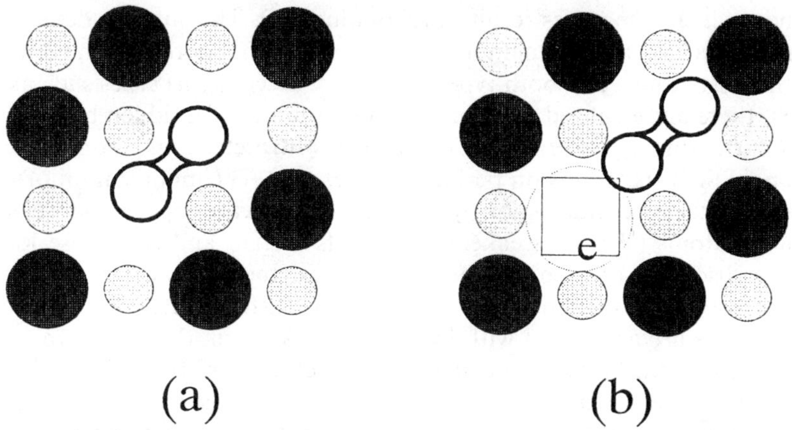

Figure 2 The configuration of (*a*) a self-trapped hole and (*b*) a self-trapped exciton in alkali halides. The self-trapped exciton has several types of configurations; however, the structure is controversial. (See Reference 19, and Suzuki T, Tanimura K, Itoh N. 1993. *Phys. Rev. B* 48: 9298 and references therein for more recent developments.)

Bond Breaking in Materials in Which Excitons Are Not Self Trapped (Type A)

There are two important differences between the type A and type B solids with respect to lattice relaxation at excited states. First, because no self-trapping is induced in type A solids, there is no way for electronic excitation energy to be localized, unless it is localized by defects. Thus any local lattice modification of electronic origin should be related to lattice defects. Second, the local lattice deformation is largely governed by the removal of electrons from the bonding covalent orbital and occupation of the anti-bonding orbital with electrons. Thus both trapping of charge carriers by defects as well as direct electronic excitation of the defects can cause lattice deformation. In this context, the negative-U interaction (25) is of particular importance: When the lattice deformation energy overcomes the coulomb repulsion energy, two holes are localized on a bonding orbital, or two electrons are localized on an anti-bonding orbital, weakening the bond further. In semiconductors, the cause of the lattice modification comes mainly from the change in the charge state; a doubly charged state is particularly effective for lattice modification.

Defects in semiconductors are accompanied by bond disorders, either with a lower coordination number than the host atoms or with the same coordination number, but with different bond strengths. Defects of the former type, for example, vacancies and interstitials, possess dangling bonds, which sometimes result in rebonding (26). The anti-site defects in semiconductors, and often impurities, have the same coordination number as the host atoms. For both types of defects, except some cases such as C in Si, there are localized electronic states within the forbidden band gap, and furthermore, some atoms associated with defects are bonded less strongly than the atoms on perfect surfaces (26). The modification of the charge states by excitation or by trapping charge carriers of these weakly bonded atoms (WBAs) weakens the bonds further and can cause local lattice disorder. On semiconductor surfaces, the presence of dangling bonds produces two-dimensionally extended surface states (27). Defects on surfaces are associated with localized states within the band gap, and the bonds for the WBAs associated with defects on surfaces are considered to be weak and can be broken by modifying the charge state.

Two examples of theoretical calculations of lattice modification due to the change in the charge state or by electronic excitation are (*a*) formation of a metastable state of anti-site defects, and (*b*) bond breaking of an adatom on the GaP(110) surface. Figure 3 shows the ground and metastable state configurations of neutral As anti-site defects (28). In the metastable state, which is higher in energy by about 0.3 eV than the ground state configuration, one of the As-As bonds is broken, and the As anti-site atom is displaced to an interstitial position. The relatively small energy difference in the two configurations compared with the As-As bonding energy (3.96 eV) is due to the energy gain by the rebonding at the metastable configuration. The configuration coordinate diagrams for the metastable formation has been described in Figure 1*b*. Conversion from the

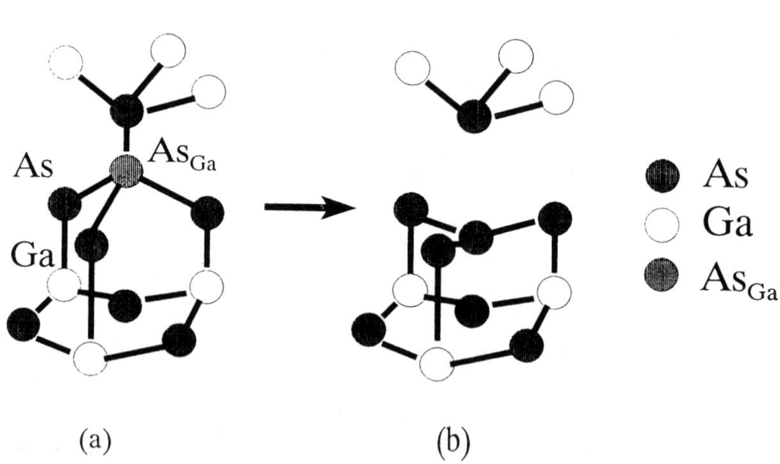

Figure 3 A schematic diagram of the atomic structure of the As anti-site defect in GaAs: (*a*) at the ground state, and (*b*) at a metastable state (28).

stable configuration to the metastable configuration is feasible at an excited state or by trapping charge carriers due to the weakening of one of the As-As bonds.

The effects of electronic excitation of a Ga anti-site defect near the GaP(110) surface have been dealt with by Khoo et al (29, 30). They replaced one of the P atoms in the second layer with a Ga atom and obtained the APESs for the ground state and the excited state, in which an electron contributing to Ga-Ga bonds is excited. The APES at the excited state has a minimum, lower than the ground state minimum by 1.3 eV, at a configuration where the Ga atom on the surface is shifted outward but is still bonded; a metastable state is formed after excitation. Obviously a part of the internal energy of the surface defect is released upon excitation. It has been argued that excitation of another bonding electron induces a negative-U state, in which the surface Ga atom is displaced to infinity.

For a Ga adatom atop a P atom of the GaP(110) surface, it has been shown that the positively charged state of the Ga adatom (in which a hole is localized on the bonding orbital) has a relaxed configuration at a shorter Ga-P distance, and that further excitation from this relaxed configuration leads to the emission of the adatom. Thus it is likely that the bond of WBAs on surface defects can be broken by a sequence of excitation and relaxation.

Electronic excitation produces a metastable state of defects in the bulk and bond breaking of WBAs on the surface. The substantial difference between the metastable state formation in the bulk and bond breaking on surfaces is that the latter requires much more energy than the former because the energy compensation by rebonding is effective for the defects in the bulk, but not for those on surfaces. Thus although the energy difference between the ground state and metastable state in the bulk is often smaller than the band-gap energy E_G, the bonding energy E_B of defect-associated WBAs on semiconductor surfaces is higher than 2 eV (31–33), and hence it exceeds the band-gap energy for many semiconductors. Because the energy quantum that can be transferred to the atomic processes is the band-gap energy[2], single excitation cannot induce atomic emissions if E_B is $> E_G$, but multiple excitation is required (Figure 4). Thus although the yield of the metastable states in the bulk of semiconductors is proportional to the number of excitations, the yield for the atomic emission induced by electronic excitation of semiconductor surfaces should

[2] Of the energy imparted to the generation of electron-hole pairs, that imparted to kinetic energy is dissipated by electron-phonon interaction within 1 ps. Only the internal energy possessed by the electron-hole pairs can be converted to generation of atomic processes.

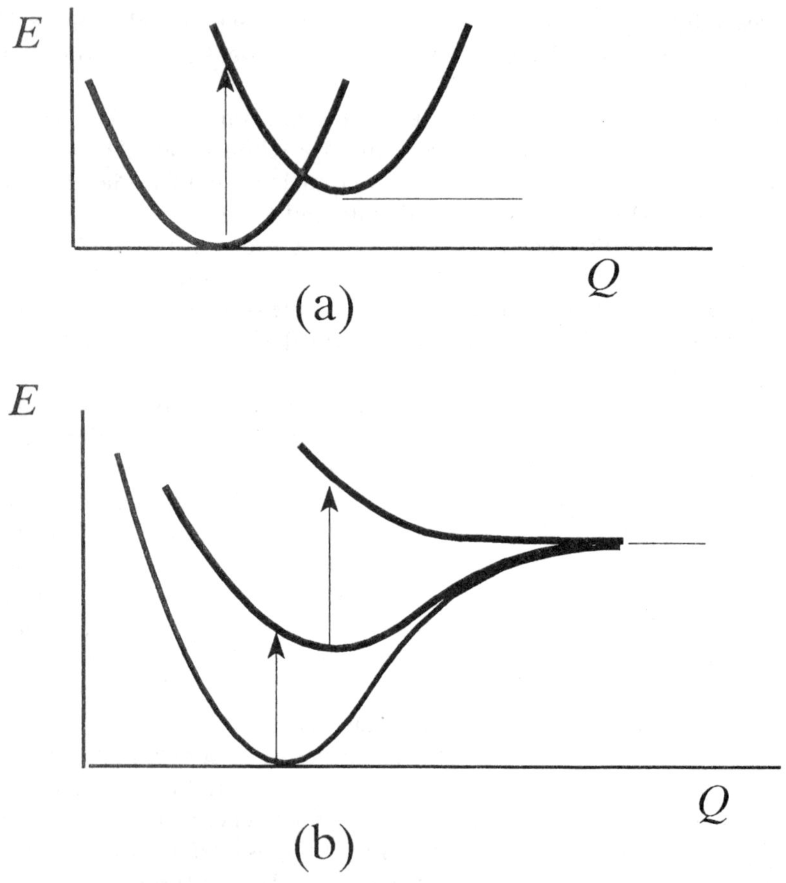

Figure 4 Configuration coordinate diagrams describing (*a*) the formation of a metastable state in the bulk, and (*b*) the emission of an atom from surfaces of type A materials induced by electronic excitation.

be a superlinear function of laser fluence. Summarizing the arguments described above, if DIET of semiconductor surface atoms occurs, it should be defect-related and the yield should be a superlinear function of the number of excitations.

Hattori et al (34) have suggested that a series of excitation-relaxation processes (see Figure 4*b*) can lead to bond breaking. As indicated for the As anti-center, an excitation of a defect induces a metastable state that is subjected to a substantial lattice deformation from the ground state configuration. For a WBA on surfaces, the bond will be elongated by an excitation that weakens the bond. As the bond length becomes longer, the defect energy level will be separated from the valence and conduction bands and can act as a site for localization. It appears that the excitation-relaxation series can take place during a laser pulse.

Itoh & Nakayama have suggested that the negative-U type localization of two holes on the surface of type A solids can induce bond breaking and atomic emission (35, 36). In view of the experimental results described below, the negative-U state on the surface is realized more efficiently on the WBAs than on the perfect sites. Thus generation of the negative-U state on a perfect lattice site by dense electron-hole pairs, as suggested by Itoh & Nakayama (36) and Sumi (37), can be the cause of the bond breaking, but probably not under conditions created by nanosecond laser pulses.

DEFECT-INITIATED ATOMIC EMISSION FROM SEMICONDUCTOR SURFACES BY LASER IRRADIATION

Evidence for Laser-Induced Bond Breaking of WBAs

To observe the emission of atoms induced by electronic excitation of defects, whose concentration is not high, emitted atoms in sub-monolayer quantities have to be detected. In this section we show that laser-induced atomic emissions detected at high sensitivity satisfy the requirements for the DIET of semiconductors, i.e. involvement of surface defects, and necessity of multiple excitation. Because the defect configuration is continuously modified, the laser-surface interaction is expected to change gradually as the laser irradiation is continued. Only the detection of emissions at a sub-monolayer level (while the overall change of surface-defect morphology is extremely small) can give information on the interaction of laser beams with the surface of a given defect structure.

Measurements with sub-monolayer sensitivity of the laser-induced emissions from semiconductor surfaces have been carried out employing the resonance ionization technique (38) for Ga atoms from the GaAs(110) surface (39); GaP($\bar{1}\bar{1}\bar{1}$) (40) and (110) (34) surfaces; and for Si atoms from the Si(100) (41) and (111) surfaces (K Ishikawa et al, unpublished data); all cleaned to the extent that low energy electron diffraction (LEED) patterns are clearly seen. The detection limit of the emitted atoms employed in these experiments (emitted atoms are detected by a micro-channel plate in the current mode) is 10^{-6} monolayer atoms emitted by a laser pulse, but can be reduced to 10^{-10} by employing the counting mode.

In Figure 5, a typical example of atomic emissions at a sub-monolayer level, we show the relation between the yield Y of emitted Ga atoms from the GaP(110) surface and the number n of shots on the same spot of 28 ns laser pulses of photon energy 2.07 eV (below the band-gap energy 2.3 eV

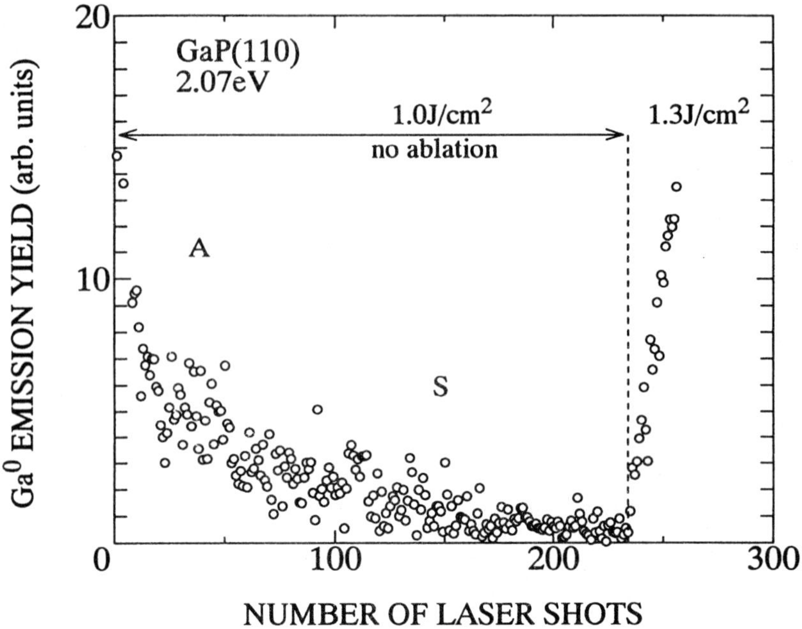

Figure 5 The *Y-n* relation for the GaP (110) surface for 28-ns laser pulses of a photon energy of 2.07 eV and of fluences of 1.0 and 1.3 J/cm^2.

at room temperature). Evidently, for the laser fluence of 1.0 J/cm^2, Y decreases as n increases at the start of irradiation and becomes almost constant. When the laser fluence is higher than a critical value, Y increases with increasing n, as seen from the result for 1.3 J/cm^2. It has been shown that the (1x1) LEED pattern of the surface observed after cleaning is lost by repeated irradiation at 1.3 J/cm^2. We refer to the critical fluence, above which repeated irradiation shows an increasing *Y-n* relation, as F_D. We show below that F_D is closely correlated to the threshold laser fluence that induces surface damage detected by optical microscopy. The fact that Y, detected at a sub-monolayer sensitivity, changes with n indicates that the number of emission sources is altered during irradiation. Based on the arguments above, it is likely that the characteristics shown in Figure 5 are a manifestation of a defect-originated emission; the change in the yield can be ascribed to the change in the defect concentration. It has been suggested that the decreasing part of the *Y-n* relation is due to adatom-type defects, the constant part is due to step-type defects, for example kink sites, and the increasing part results from vacancy-type defects. Elimination of a weakly bonded adatom-type defect leaves a perfect surface site or a more stable site, and elimination of a step-type defect leaves essentially the same defect site. A vacancy-type defect is somewhat different: the atoms neighboring a vacancy are WBAs and elimination of one forms a di-vacancy, resulting in an increase in the number of WBAs. Further successive elimination of WBAs surrounding the di-vacancy induces evolution of a vacancy cluster, which further increases the number of WBAs on the surface. As we explain below, the emission of adatom-type defects occurs at laser fluences lower than those that induce the emission of the step-type defects, and the emission from vacancy-type defects occurs at the highest laser fluence.

The results shown in Figure 5, which have been seen generally in several surfaces, including Si(100), GaAs(110), and GaP(110) and ($\bar{1}\bar{1}\bar{1}$) surfaces, indicate that the surface is progressively modified by laser irradiation: The number of defects is reduced below F_D and is enhanced above F_D. Although the change in Y shown in Figure 5 is a result of modification by successive laser pulses, similar modification should be induced even during a single laser pulse. Of particular importance in terms of laser-induced surface modification is the increasing Y-n relation above F_D, which indicates that the surface is progressively damaged. The final damage seen after the termination of a laser pulse is not the consequence of the interaction of the laser beams with clean surfaces, but is the consequence of the accumulation of the interactions between the laser and damaged surfaces (42, 43). The high sensitivity measurements described in this section can provide information on the interaction of laser beams with the surface of a given morphology.

The Si(111) (7x7) reconstructed surface includes nominally WBAs, i.e. adatoms with a larger bond length. Figure 6*a* shows a typical STM diagram of a Si(111) surface irradiated with a laser beam of 0.5 J/cm^2 (K Ishikawa et al, unpublished data). Evidently, the central two adatoms, which are known to be less strongly bonded (44), are preferentially eliminated by laser irradiation. Further irradiation by laser pulses of slightly higher fluences eliminates the reconstructed pattern, as shown in Figure 6*b*. It appears that the atoms around the vacancies left after the central atoms are removed can then be removed by laser pulses of almost the same fluence. No bond breaking is induced by laser beams of fluences below 0.3 J/cm^2.

Further evidence that defects are the source of laser-induced emissions is given in Figure 7 (T Shibata et al, unpublished data), where the deposition of Ga atoms on the GaP surface is shown to enhance emission. The total number of enhanced emissions is proportional to the number of the deposited atoms, which indicates that isolated deposited atoms can be easily emitted by laser irradiation.

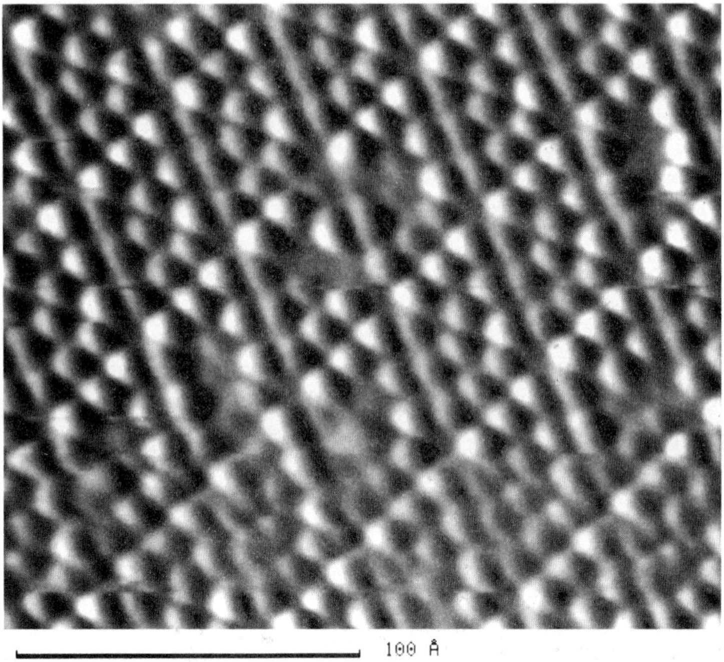

Figure 6(a) A STM diagram of the Si(111) surface after irradiation with laser pulses of fluences of 310 mJ/cm^2.

In summary, defects in semiconductors can act as the source of emissions, and the DIET from these materials is defect related. The experimental observation is consistent with the arguments for type A materials, of which the semiconductors are a typical example. It is expected that similar emission behavior is observed in other type A materials.

Dependence of the Defect-Initiated Emission Yield on Laser Fluence

According to the model of bond breaking by the excitation-relaxation sequence, the yield Y of the emission induced by a laser pulse is a superlinear function of the laser fluence F, unlike the photo-induced emission from type B solids. In this section we describe the Y-F relation obtained with photons below the band-gap energies, for which heating effects are not significant. We shall see that the yield does not change when the photon energy crosses the band-gap energy; hence the same concept is valid for the photons just above the band-gap energy.

B

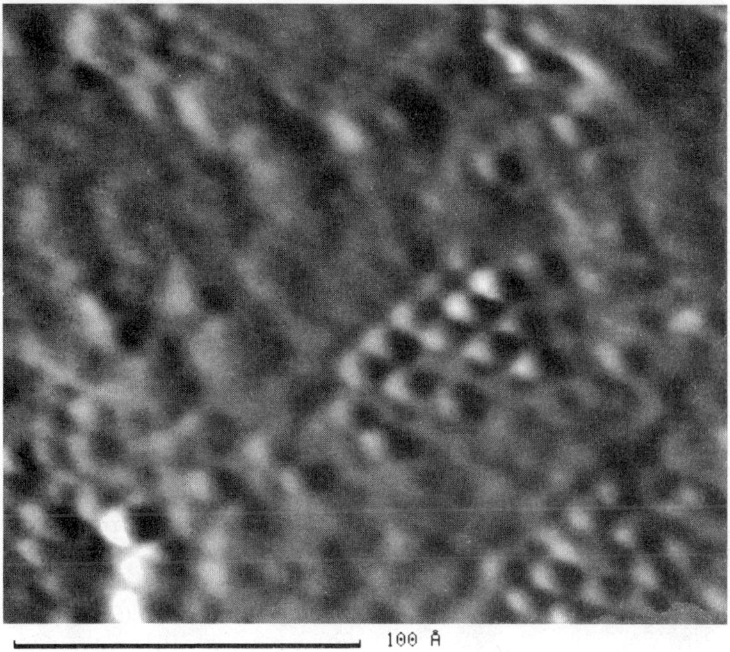

Figure 6(b) A STM diagram of the Si(111) surface after irradiation with laser pulses of fluences of 500 mJ/cm^2.

A typical example of *Y-F* relations is given in Figure 8, which includes the *Y-F* relation for GaAs obtained below F_D, immediately after cleaning and after removal of the adatom-type defects, and that obtained above F_D. The *Y-F* relation can be expressed as a power function F^r, where r is the power index, but the relations often show threshold-type dependence when plotted in linear scales because of the high power and finite sensitivity for detection. Although the threshold-type behavior does not necessarily indicate the presence of any critical process, we define the threshold laser fluence F_A for the adatom-type defects and F_S for step-type defects. F_D represents the threshold laser fluence for the emission of WBAs around the vacancy-type defects and also can be regarded as the ablation threshold detected by LEED. Table 2 summarizes the power indices r and the threshold laser fluences for Si, GaAs, and GaP at several photon energies. The value of r for the vacancy-type defects is half of the exponent for the reason given below.

The emission yield Y of atoms during a laser pulse of fluence F and duration t_L may be given by

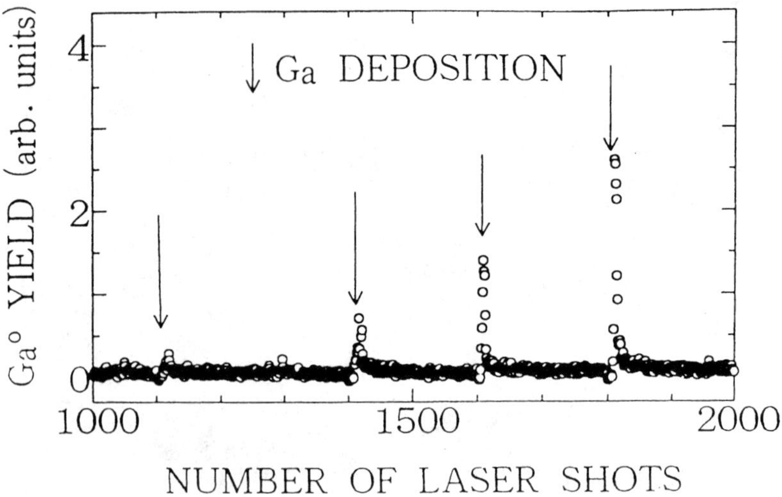

Figure 7 The *Y-n* relation for the GaP(100) surface showing the effects of Ga deposition. Ga atoms are deposited between laser pulses at points indicated by the arrows; the coverage of deposited Ga atoms at each arrow is 1.8, 3.6, 3.6, and 7.2×10^{-3} monolayers, reading from the left.

$$Y = N_D \xi \prod_{i=1}^{r} (\sigma_i \phi t_i) \quad \left(\sum_i t_i = t_L \right),$$

$$= N_D \xi (\sigma F)^r \left(\frac{1}{r} \right)^r \qquad 1.$$

where N_D is the number of WBAs per unit area on the surface; ϕ is the flux; ξ is the probability of the emission after the rth excitation; σ_i is the cross section for the ith excitation; t_i the time interval between (i-1) and ith excitations; t_L is the pulse width; $F = \phi\, t_L$ is fluence; and the product extends to all intermediate metastable states until the bond is broken. σ is the geometrical mean of the cross sections and is given by

$$\sigma^r = \frac{1}{t_D^r} \prod_i (\sigma_i t_i), \quad \sigma^r \left(\frac{t_L}{r} \right)^r = \prod_i (\sigma_i t_i). \qquad 2.$$

The threshold laser fluence F_{th} (F_A, F_S, or F_D) is given by

$$F_{th} = \frac{r}{\sigma} \left(\frac{Y_0}{\xi N_D} \right)^{1/r}, \qquad 3.$$

where Y_0 is the detection limit.

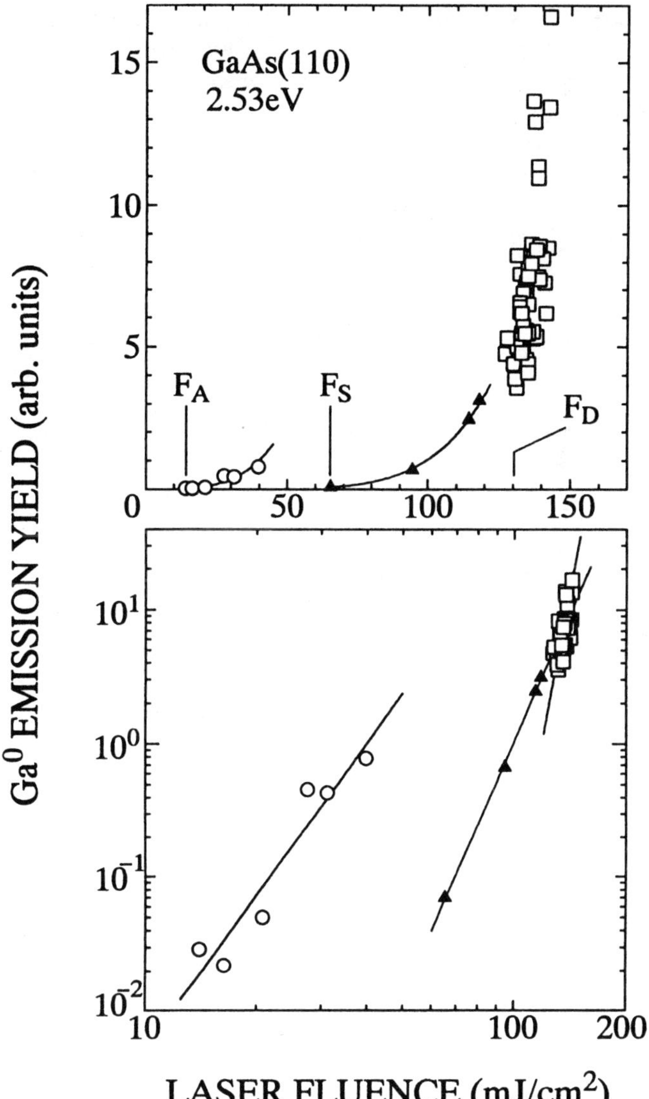

Figure 8 The *Y-F* relation in linear (*upper*) and logarithmic (*lower*) scales for the GaAs(110) surface obtained at a photon energy of 2.53 eV, starting from a clean surface (*circles*), for the surface after the adatom-type defects are removed (*triangles*) and above F_o. F_A and F_S are the threshold laser fluences, respectively, above which the emissions (indicated by circles and triangles) are observed.

In Table 2, the value of r varies from 2 to 7; the lowest index for the adatom-type defects and the highest index for the vacancy-type defects in all materials. The value of r is greater than two for the adatoms, greater than four for steps, and close to seven for vacancies, all of which represent the number of excitation-relaxation sequences that occur before the antibonding state is reached. It has been shown theoretically (29–31) that the bonding strength for WBAs increases progressively from the adatom-, to step-, to vacancy-type defects for all materials, i.e. r is greater as the bonding strength is greater. According to Table 2, the value of F_{th} is the least for the adatom-type defects and increases in the same order as that of r. This is easily understood according to Equation 3, where F_{th} is proportional to r; $[Y_0/(\xi N_D)]^{(1/r)}$ is close to unity; and σ, the geometrical mean of the cross sections for the metastable states, does not depend strongly on the defect types. According to Table 2, F_{th} is the greatest for Si and decreases progressively from GaP to GaAs. This ordering is in accordance with the bond strength of these materials (45): F_{th} is greater as the bond strength increases. It follows that higher laser fluence is needed to break the bonds of more strongly bound atoms. This rule is valid not only for the defect type but also for materials. In view of Equation 2, the

Table 2 The threshold laser fluence and power index for the emission initiated by each type of defect

Defect type	Surface	Threshold: Fluence (F_{th}) (J/cm^2)	Threshold: Ratio	Index (r)
		F_A	F_A/F_o	
Adatoms	Si(100)[a]	0.2	0.1	2–4
	GaP(110)[b]	0.08	0.07	2–4
	GaAs(110)[c]	0.003	0.02	2–4
		F_S	F_S/F_o	
Steps	Si(100)[a]	0.5	0.25	4–7
	GaP(110)[b]	0.5	0.4	4–7
	GaAs(110)[c]	0.05	0.3	4–7
		F_D		
Vacancies	Si(100)[a]	2.0	—	5–7[d]
	GaP(110)[b]	1.2	—	5–7[d]
	GaAs(110)[c]	0.15	—	5–7[d]

[a] Laser photon energy is 2.48 eV.
[b] Laser photon energy is 2.07 eV.
[c] Laser photon energy range is 1.88–2.53 eV.
[d] The value is half of the power index obtained experimentally (see text).

relevance of F_{th} to the bond strength is explained in terms of the correlation between the value of r and the bond strength: A greater number of excitations is needed to break stronger bonds.

Excitation Spectroscopy of Defect-Initiated Emissions

EXCITATION SPECTRA The excitation spectroscopy for Y of laser-induced Ga^0 emissions from the GaAs(110) and GaP(110) surfaces below F_D and for F_D has been carried out (39, 46–48), and the results are depicted in Figure 9. The abscissa is the difference between the photon energy and the band-gap energy. The decrease in F_D corresponds to the increase in Y. In obtaining these results, both the photon energy of the laser beam and the temperature of the specimen were changed. Changing temperature is of particular use for changing the photon energy with respect to the band-gap energy, not only for the bulk, but also for surfaces, because the yield shows extremely weak temperature dependencies unless the wavelength crosses a singular point of either the bulk or surface band structures.

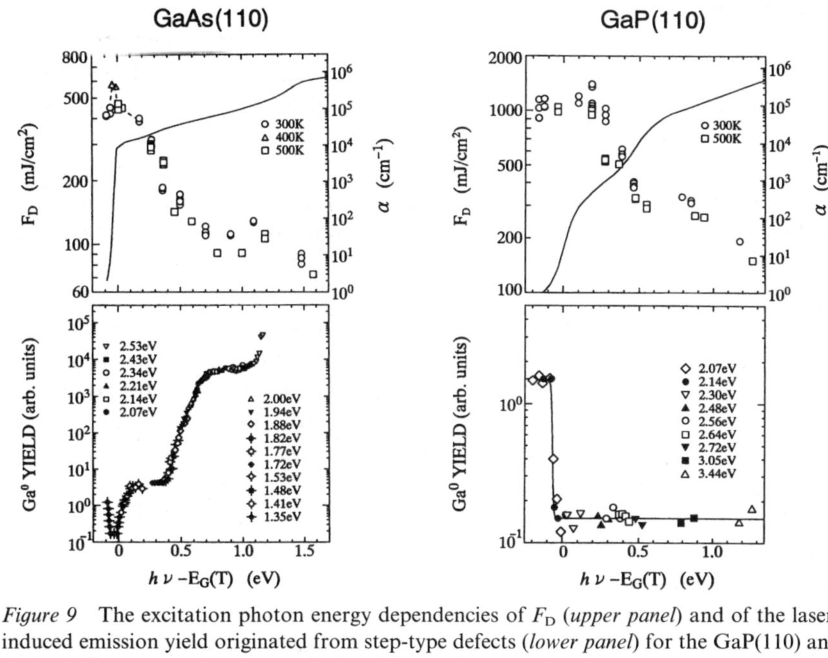

Figure 9 The excitation photon energy dependencies of F_D (*upper panel*) and of the laser-induced emission yield originated from step-type defects (*lower panel*) for the GaP(110) and GaAs(110) surfaces. The abscissa is the difference between the photon energy and the band-gap energy at the temperature where the measurement is made. Each data point has been obtained by laser pulses of the photon energies shown at several temperatures. Experimental data obtained with several photon energies are connected by normalizing for the data points where the abscissa is overlapping.

Changing the temperature rather than the photon wavelength is advantageous because changing the wavelength of dye lasers by using different dyes or by tuning the grating is often accompanied by an undesirable change of the fluence distribution within the laser beam. Although the temperature dependence of the bulk band-gap energies is well documented (46), that of surface band-gap energies is not. However, the assumption that the temperature dependencies of the surface and bulk band-gap energies are the same is appropriate because both depend on the lattice constant. In Figure 9, laser-induced emissions by photons below the band-gap energy are clearly observed for GaAs and GaP. The dependencies above the band-gap energy, however, are remarkably different: Y is substantially reduced almost to the noise level for GaP, but the reduction is of a resonant-type and enhanced at higher photon energies for GaAs.

EXCITATION WITH PHOTONS OF SUB-BAND-GAP ENERGIES The emission induced by photons of sub-band-gap energies can only be ascribed to electronic excitation on defects. The photon energy and temperature dependencies of Y in this energy range are weak (48). This is understood because the optical absorption band from a localized defect state on a surface to the continuum may show a broad band, starting from the depth of the defect level, i.e. the conduction band bottom, (see Figure 10*b*), with little temperature dependence.

In both GaP and GaAs, there is a reduction in the yield at energies several tens of meV below the band-gap energies. This energy difference is slightly larger than the binding energy of excitons in both crystals. Thus it has been suggested (49) that the reduction is the result of the resonance energy transfer from the defect excited state to the bulk exciton perturbed by the defects and then to the bulk exciton, which couples with the perturbed excitons. The difference between GaAs and GaP reductions, resonant-type and stepwise, respectively, has been ascribed to the direct and indirect nature of the semiconductors' band gaps. The optical absorption to form excitons perturbed by defects shows a resonant-type band in GaAs, a direct semiconductor, and a broad band starting from a threshold for GaP, an indirect semiconductor (50) (Figure 10). The resonance energy transfer rate is expressed by the overlap of the optical absorption spectra for the donor and the acceptor of the energy transfer (51). The donor band from the surface defect transition is also broad, thus we expect the defect excitation above the band-gap energy to be greatly suppressed for GaP. A similar resonant-type reduction has been observed for InP, a direct semiconductor (A Okano, unpublished work). The efficient energy transfer described here may be characteristic of defects on surfaces, which are decoupled from the bulk, otherwise there is backward energy transfer. The

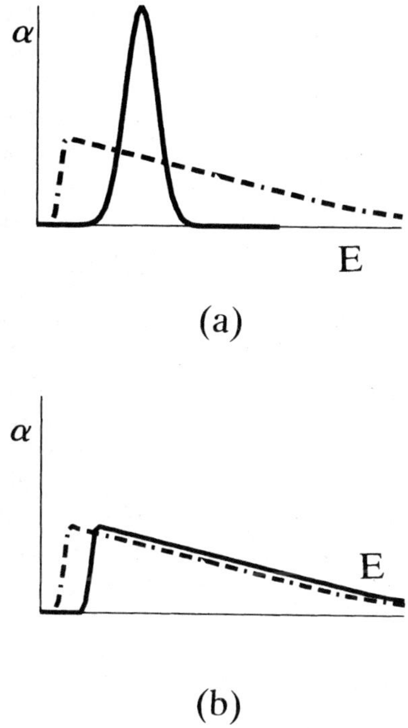

Figure 10 Schematic diagrams describing the transition probability of the resonant energy transfer from the excited states of a defect on the surface to the bulk localized exciton in (*a*) a direct semiconductor, and (*b*) an indirect semiconductor. The solid lines indicate the optical absorption curves resulting from the formation of excitons perturbed by defects and the dashed lines indicate the optical absorption curves resulting from electronic transitions localized on the defects. If the areas of both curves are the same, the transition probability is proportional to the overlap of the two curves.

absence of a substantial change in F_D by crossing the band-gap energy can be ascribed to this backward energy transfer because the vacancies on surfaces may not be completely decoupled from the bulk.

EXCITATION ABOVE THE BAND-GAP ENERGIES The crossing of photon energy across the band-gap energy does not enhance Y below F_D, nor does it suppress F_D for both GaP and GaAs. These results exclude the possibility that heating is the cause of the emission in this fluence range. They also indicate that the electron-hole pairs induced in the bulk make little contribution to the emission.

For GaAs, the photon-energy dependence of Y below F_D shows two stepwise increases above the band-gap energies. These increases have been ascribed (47) to the onset of the transition from the occupied surface states to the bulk conduction band and to the unoccupied surface state (52, 53). In view of Equation 3, the onset of this transition can possibly have three effects: increasing the number of the electron-hole pairs on the surface state, which may be trapped by defects (effectively increasing σ); increasing σ by the configurational interaction between the defect transition and the transitions involving the surface state; and increasing ξ by increasing the probability of the capture of the electrons occupying the surface state by the multiply charged emitted atoms.

Almost the same photon-energy dependencies of F_D are observed for GaAs and GaP. Because the increase in Y corresponds to the decrease in F_D, the energy dependence of F_D in GaAs is analogous to that of the yield below F_D: There are two stepwise changes at nearly the same photon energies. The surface electronic structure of the GaP(110) surface has not been established (27). Recent photoelectron emission experiments have shown that the surface electronic structures of GaAs and GaP are similar (54, 55). Thus the origin of the F_D dependence for GaP is considered to be similar to that for GaAs. Experiments with excitation spectroscopy so far have revealed interesting features of the interaction between decoupled localized states with two-dimensional (surface) and three-dimensional (bulk) continuum: the former enhances emission, and the latter makes no contribution to emission. At present we conclude that the excitations localized near the surface, either on the surface defects or on the surface states, are most effective for inducing emission.

RELATED PHENOMENA

In this section, we deal with the phenomena that are considered relevant to bond breaking, i.e. laser ablation, surface modification on a monolayer scale, and laser-induced dry etching.

Laser Ablation

It is known that irradiation of solid surfaces with laser pulses of fluences above a threshold fluence, denoted by F_{ab}, induces laser ablation, that is, emissions of massive numbers of atoms leaving damaged surfaces that can be seen with an optical microscope (56, 57). The phenomenon of laser ablation has attracted practical interest particularly for thin film growth and patterning. Conventionally, laser ablation is observed in terms of the rapid increase in the atomic emission yield beyond F_{ab}; the removal of many monolayers is required for detection. For many materials, including insulators (58), semiconductors (59–61), and polymers (62), Y increases rapidly above F_{ab}. Because of this rapid increase in Y and high value of F_{ab}, the laser ablation of semiconductor surfaces is considered to be the evaporation resulting from melting. In fact, the thermal model of laser ablation by photons above the band-gap energies satisfactorily explains the laser ablation processes of semiconductors (58, 59).

Ichige et al (63) obtained a simple relation between F_{ab} and the bond strength, as shown in Figure 11. They suggest a model somewhat similar to that suggested by van Vechten (64) and by Combescot & Bok (65): The emission arises from the phase change caused by the electron-lattice coupling in a dense electron-hole plasma. According to their model (63), the localization of electronic excitation energy occurs above a critical concentration of electron-hole pairs by dimerization of nearest neighbor anion atoms in the lattice; the critical concentration depends on the bond strength and the flexibility of the bonding direction.

We have shown that repeated irradiation for laser fluences above F_D on the same spot on a surface enhances Y detected by sub-monolayer sensitivity and leads to the destruction of the surface crystalline order, and have attributed these results to evolution of vacancy clusters on the surface. As shown in Table 3, the photon energy dependence of F_D shows almost the same pattern as that of F_{ab}, the ablation threshold detected by an optical microscope, thus indicating a strong correlation between the two quantities (66). The result strongly suggests that the enhancement of the

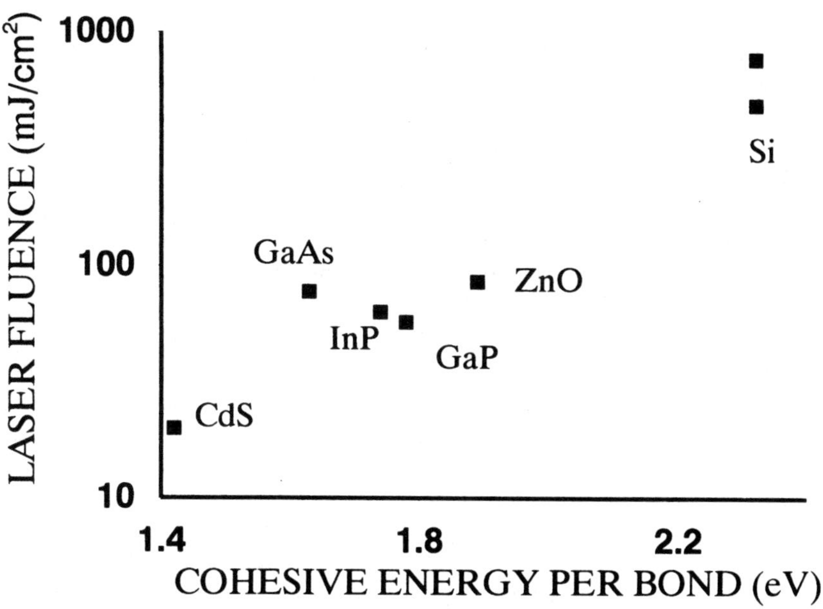

Figure 11 The relation between the macroscopic laser ablation fluence and the bond strength of solids (65).

Table 3 A comparison of the laser ablation threshold F_D determined by high-sensitivity measurements with that of F_{ab} determined by optical microscopy

Material	Photon energy	Ablation laser fluence (mJ/cm^2)	
		F_D	F_{ab}
GaP	2.07	1160	1400
	2.66	400	530
	3.44	190	260
GaAs	1.35	400	460
	1.80	180	250
	2.92	90	130

emission in the sub-monolayer range is a necessary condition for laser ablation.

Based on the results described above, a vacancy model for laser ablation, described schematically in Figure 12, has been suggested (42). According to this model, the laser ablation is initiated if the laser fluence is strong enough to emit WBAs around vacancies. Once this condition is satisfied, the vacancy clusters start to increase in size by emitting the WBAs around them. Furthermore, as the vacancy clusters on the surface top layer evolve, the vacancies that have been in the second layer are exposed on the surface and become the source of vacancy clusters in the second layer. Thus the surface destruction evolves, not only on the surface top layer, but also deep into the bulk. It has been shown that the number N_D of the WBAs produced during a laser pulse due to the evolution of the vacancy clusters is an exponential function of time at the initial stage of growth and then becomes a quadratic function of time. It follows that Y at the later stage of excitation is a power function of fluence with the power index $2r$. The values of r for the vacancy-type defects in Table 2 are the number of excitations evaluated when accounting for the increase in N_D. A Monte Carlo simulation (42) based on the vacancy mechanism leads to the conclusion that Y is a power function of F, with the power index of $2r$ above F_D.

Because of the high-power dependence of Y on F in the sub-ablation stage, the number of emitted atoms per unit area increases dramatically as the laser fluence increases. The randomly generated vacancy clusters over many monolayers will destroy the crystalline order of the surface layers, with accumulating damage to the surface. Furthermore, the decomposition of the surface layers of compound semiconductors appears to occur during the evolution of vacancy clusters. Thus for a laser pulse

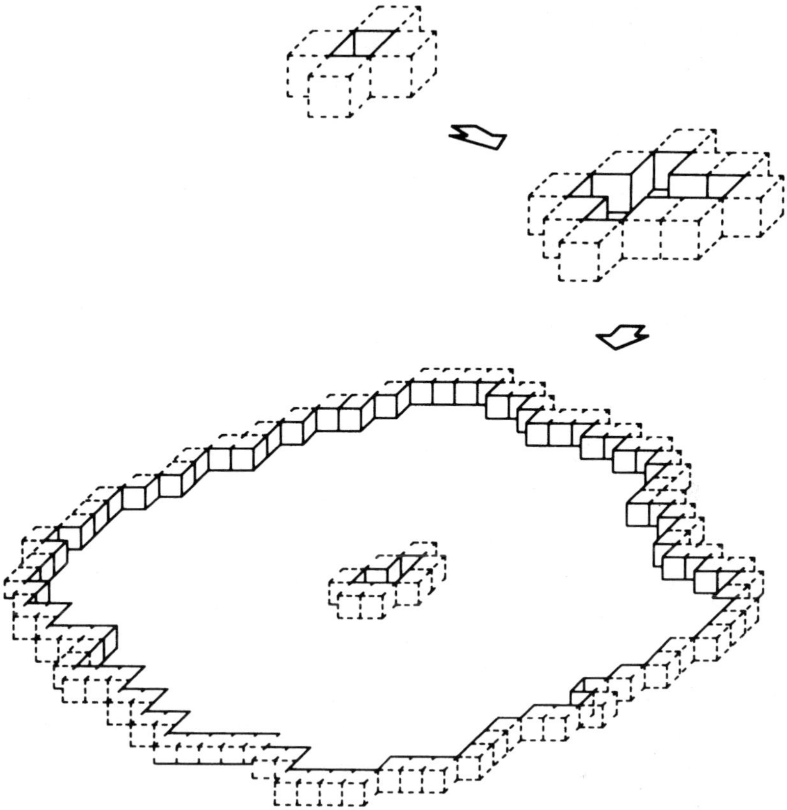

Figure 12 The vacancy model of laser ablation.

of a fluence near F_{ab}, an increase in the optical absorption coefficient and a reduction of the lifetime of the electrons and holes will result. Consequently, the conversion rate of the electronic excitation energy to heat is increased. High electron-hole pair concentration accumulated during a laser pulse may enhance the Auger recombination rate and hence the energy transfer rate to heat at the later stage of a laser pulse. Thus the thermal mechanism will be satisfactory at the later stage for laser pulses of fluence near F_{ab}[3].

[3] One of the problems for the thermal mechanism is the relaxation time required for the transfer of electronic energy to thermal energy. The internal energy possessed by electon-hole pairs is stored during their lifetime and can be converted to the energy required for bond breaking and transferred into the bulk by ambipolar diffusion. Thus we interpret the so-called melt threshold as the calculated value, under the assumption that the relaxation time of the electronic energy transfer to heat is about 1 ps.

In summary, based on the experimental results of the high sensitivity measurement of emitted atoms and Monte Carlo simulation, we suggest that the vacancy mechanism is effective as the primary process of laser ablation of the type A materials. The process should play a major role particularly in laser ablation for photons below the band-gap energies, for which heating by absorption of photons by the bulk of the material is not effective. For photons above the band-gap energies, however, the heating may be the cause of laser ablation, but the similarity of the photon-energy dependence of F_D and F_{ab} suggests that the primary vacancy-related processes make substantial contributions. The correlation between F_{ab} and the bond strength of the materials shown in Figure 11 is in accordance with the correlation between F_D and the bond strength.

Laser-Induced Surface Modification

In this section we review experimental results on monolayer-scale surface modification of semiconductors induced by nanosecond laser pulses. These results involve considerable surface damage and the modification or loss of the surface LEED pattern, therefore we presume that the laser fluence used in these studies is above F_D. The fluence is considered to be below F_{ab}, however, because the damage to the surface is on a monolayer scale.

The reconstructed structures of the Si(111), GaAs(100), InP(100) (67, 68), GaP($\bar{1}\bar{1}\bar{1}$) (69, 70), and CdS(100) surfaces (71, 72) are converted to the (1x1) patterns by irradiation with laser pulses of fluence above a certain threshold value. Further increase in laser fluence converts the (1x1) pattern to the amorphous structure. The conversion of these structures to the (1x1) pattern is considered to be from the loss of the WBAs nominally existing in the reconstructed structures, as is the case for Si(111), or due to the elimination of a substantial fraction of the first layer by the evolution of vacancy clusters. Kumazaki et al (69) have shown that the laser-induced elimination of the structure on the GaP($\bar{1}\bar{1}\bar{1}$) surface is induced by photons below the band-gap energy, which indicates that the modification is not due to heating.

Auger electron spectroscopy of laser-irradiated surfaces has revealed the change in the stoichiometry of the surfaces: above the threshold fluence for the conversion into the (1x1) structure for the GaAs(110) surface (73), and above the threshold for the conversion into the amorphous structure for CdS(100) surface (71). For GaAs, using secondary electron microscopy and photoelectron spectroscopy, it has been shown that Ga islands are formed on surfaces that are otherwise defect free (73, 74). We again presume that this stoichiometric change is induced by laser irradiation above F_D, accompanied by formation of vacancy clusters. The evolution of vacancy clusters requires laser-induced emissions of both component atoms. Thus the emission of either of the two component atoms having higher stability for laser-induced emission will be rate determining. In the intermediate stage of vacancy evolution, the diffusion of atoms may occur from unstoichiometric peripheries of the surface vacancy clusters into the newly exposed surface, thereby inducing overall surface stoichiometry change or even producing metal droplets.

Laser-Induced Dry Etching

In this section we discuss laser-induced (enhanced) dry etching processes of semiconductors, a potential technique in fabrication of electronic devices. Houle (75) has studied the etch rate enhancement by cw-515-nm laser irradiation of Si under XeF_2 exposure, where XeF_2 is known to react spontaneously with Si in the dark, and fluorosilyl layers (SiF, SiF_2, SiF_3), mainly SiF_3, are formed (76). The yield of the photo-enhancement of the etch rate, which is ascribed to photo-induced detachment of SiF_3, is proportional to the 1.6 power of the laser intensity. We suggest that bond breaking by multiple excitation is effective: The reduction of the threshold laser intensity for the bond breaking is due to the weakening of the Si-Si back bond by adsorption of a F atom, which has high electron-negativity.

The influence of temperature, gas pressure, laser repetition rate, and laser fluence on the enhancement of etch rate of silicon by Cl has been studied extensively using pulsed- and cw-lasers. The presence of Cl radicals on the surface appears to enhance the etch rate by 308-nm excimer laser of fluences below the melting threshold (77). Sesselmann et al (78) have found that the etching rate of the surfaces with sub-monolayer chlorine adsorbates by 308-nm laser pulses increases almost linearly above a certain threshold fluence below the melt threshold, although it exhibits a super-linear dependence near the melt threshold. Apparently, the mechanism of the laser-induced enhancement is different, depending on the fluence; at the low fluence range, it is governed by the formation of volatile $SiCl_4$ (78), and at higher fluence range, it is more likely governed by the laser-induced desorption of SiCl (79, 80). Although the process has been ascribed to heating, it is also likely that laser-induced bond breaking is effective (see above). The experimental result (78) that the power index decreases from 6.0 to 1.8 as Cl_2 pressure increases for 308-nm laser irradiation can be explained on the basis of the laser-induced bond-breaking mechanism because higher Cl coverage reduces the strength of the Si-Si back bonds.

The enhancement of dry etching of the compound semiconductor GaAs by 308-nm laser pulses under a Cl_2 atmosphere (81–83) and a HBr atmosphere (84, 85) has also been studied. Brewer et al (84) have suggested that the desorption of the reaction products due to laser-induced excitation of the substrates is a crucial factor for etching rate enhancement. On the other hand, when a large amount of Cl_2 is exposed to a GaAs surface at room temperature, a spontaneous reaction occurs to form a volatile $GaCl_3$ liquid layer on the surface (82, 83), which may be removed thermally by laser irradiation. Thus the laser-induced bond-breaking mechanism may be effective in a limited range of laser fluence and temperature.

So far we have noted several cases where the reaction products of the substrate with halogen gas atoms can be desorbed by laser irradiation in accordance with the concept of laser-induced bond breaking. The question arises whether the reaction occurs in the perfect sites on the surface or at the defect sites. As for the laser ablation, if the WBAs around vacancies or vacancy clusters are eliminated, the surface can be depleted. Similarly, the reduction of the bonding strength of the WBAs around the vacancies by bonding with halogen atoms can cause enhancement of the dry etching.

Kanasaki et al have shown that laser-induced emissions of Si atoms from the Si(100) surfaces (39, 86) and of Ga atoms from GaAs(110) surfaces (87), detected at sub-monolayer sensitivity, are enhanced by deposition of Br atoms. As shown in Figure 13, deposition of sub-monolayer Br atoms on a Si(100) surface from which the adatom-type defects are eliminated induces enhanced emission of Si atoms, which is reduced by

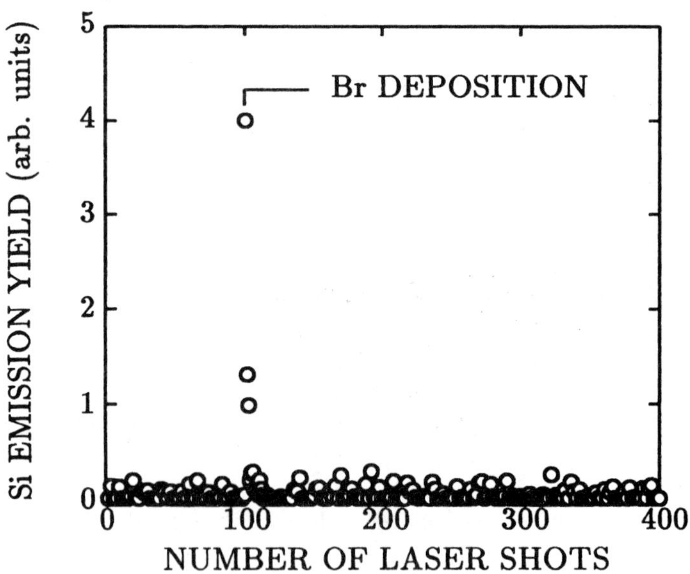

Figure 13 The *Y-n* relation for the Si(100) surface by irradiation with 28-ns 2.48 eV laser pulses of a fluence of 850 mJ/cm^2 before and after deposition of Br_2 at a coverage of 10^{-3} monolayer.

repeated irradiation. The number of atoms emitted after the exposure is proportional to the Br coverage. The absolute number of the Si atoms emitted after coverage is much less than the number of the Br atoms on the surface. The result suggests that only Br atoms deposited at specific sites, namely defect sites, can enhance the emission. Under continuous Br exposure, the enhancement appears to last as long as the exposure is continued. The *Y-F* relation under exposure also shows a superlinear relation, whereas the threshold laser fluence is smaller than that obtained without Br exposure. In an earlier paper (87), it was suggested that the enhancement of the emissions is due to the interaction of Br atoms with steps. Alternatively, the constant value is influenced by Br exposure rate and the enhancement is due to the interaction of Br atoms with vacancies (86).

Further work on how the interaction of halogen gas with surfaces affects the emission on a sub-monolayer scale is necessary to elucidate the mechanism of dry etching. According to the current arguments, the laser-induced dry etching may be the result of the breaking of WBAs on vacancy sites, where the bond is further weakened by bonding with halogen atoms.

PERSPECTIVE

Controlling Defects on Surfaces

In view of the fact that the localization of electronic excitation energy is necessary for inducing atomic processes in semiconductors, wherein no self-trapping of excitons or holes takes place, defects are the only source of laser-induced lattice modification or bond breaking. Thus laser irradiation can be used for controlling the defects on surfaces, i.e. elimination of adatoms and steps on the surface. To eliminate the steps, the atoms at the edge of the steps or at the kink sites should be eliminated, one by one, to remove all terraces on the surface. This process can be enhanced by several orders of magnitude by adsorption of halogen atoms on the surface. The vacancies, which cannot be eliminated by laser irradiation, should be controlled by another method.

Control of defects or production of defect-free surfaces is of fundamental importance for semiconductor device technology. First, the defects existing on the substrate for epitaxial growth cause disorder in an epitaxial layer. Second, the defects on surfaces influence the properties of solids such as surface recombination velocity. Third, defects affect the surface reaction rate. The super-lattice structures grown on a substrate are greatly influenced by the defects on the substrate. Therefore, formation of a step-free surface is essential for obtaining ideal thin layer super-lattices.

Because an increase in temperature induces thermal defects, any method of controlling defects on surfaces should not involve heating processes; the temperature of the surface should be kept below the temperature where the evaporation of any component atoms is negligible, namely a few hundred degrees at the highest. Bond breaking by photons may satisfy this condition. The laser-induced bond breaking technique has several advantages. Yield's superlinear dependence on fluence implies that the laser fluences that are capable of eliminating different types of defects differ considerably. This range allows the choice of an appropriate laser fluence to eliminate specific defects. As we have described, the bonding strength is modified by adsorption of reactive atoms. Therefore, combination of adsorption of reactive atoms and laser irradiation can be utilized to eliminate specific types of defects. Vacancies must be filled by diffusing deposited atoms on the surface prior to laser irradiation.

State-of-the-Art of Growing Defect-Free Epitaxial Layers

Herein we demonstrate the methodology of growing defect-free surfaces and defect-free epitaxial layers. Si is used as an example, but extension to other semiconductors is straightforward. In order to eliminate vacancies from Si, Si atoms or Si compounds such as SiH_4, which are mobile on Si surfaces, are deposited on the surface and allowed to diffuse. Hydrogen elimination by UV laser irradiation is done after SiH_4 is trapped by a vacancy. Termination of the Si dangling bonds on the surface by hydrogen may enhance the diffusion of Si atoms on the surface. If all vacancies are eliminated, the laser-induced emission slightly above F_D does not show any increasing *Y-n* component, which is indicative of the laser ablation detected by sub-monolayer sensitivity. In order to eliminate steps, the surface may be irradiated in a halogen atmosphere with laser pulses at or above F_D, but below the fluence that causes remarkable temperature rise. The irradiation should be continued until all terraces in a given area are eliminated. In this process, the vacancies hidden under the terraces are exposed on the surface. Thus the laser fluence should be either smaller than F_D or laser irradiation and deposition should be repeated alternately. The elimination of step-type defects should lead to the absence of the constant component in the *Y-n* relation. After elimination of the steps, the newly exposed vacancies on the surface should be filled. Any adatoms that are produced in this process can be eliminated by laser-irradiation. The surface, free from defects, will show no laser-induced emission, unless the fluence is high enough to cause melting. We conclude that if the surface includes defects of 10^{-10} in fractional concentration of any type, the emission can be detected using the present technique. Thus the defects on the surface can be reduced below this concentration.

CONCLUSION

We have described the electronic processes for laser-induced atomic emissions. The process involves bond breaking by laser irradiation: Bonds of weakly bound atoms may be broken by multiple excitations during a laser pulse. Inasmuch as fundamental electronic excitation of semiconductors produces electrons and holes, which have extended wave functions, the recombination of these electron-hole pairs involves defects, and the electronic excitation energy can be deposited on defect sites. It is known that the defects in the bulk lead to formation of a metastable state; those on the surface can induce bond breaking that results in atomic emissions. Bond breaking in semiconductor surfaces is induced only if the same defect sites are multiply excited.

The laser-induced bond breaking described herein appears to be the only possible cause of the DIET by nanosecond laser pulses and is the fundamental process for several laser-induced processes, such as laser ablation and photo-induced dry etching. We note, in particular, that the surface is progressively damaged during a laser pulse, and hence studies of laser interaction with defects at various stages of damage evolution are important to elucidate laser-induced processes.

Because laser-induced emission from semiconductor surfaces can occur only via defects, unless heated to the vaporization temperature, laser-induced emission is a potential technique for defect control. The methodology of growing defect-free surfaces and epitaxial layers has been described; however, technological development is needed to achieve practical application.

In view of the present model of bond breaking by the excitation-relaxation sequences, use of sub-picosecond laser pulses is believed to cause excitation from vibrationally excited metastable states. Thus the threshold laser fluence for bond breaking is expected to be lower than for nanosecond laser pulses as, indeed, has been observed for metal-adsorbate systems (88). Recently Wu & Liu (89) observed the emission of an extremely small number of Si ions from cleaned Si (100) surfaces by excimer laser pulses, for which the yield is a superlinear function of laser fluence. The emission is considered to be defect related. The correlation between the emission of neutrals, described in this review, and ions is a topic of future studies.

Literature Cited

1. Biegelsen DK, Bringans RD, Northrup JE, Swartz L-E. 1990. *Phys. Rev. Lett.* 65: 452–55
2. Hamers RJ. 1989. *J. Vac. Sci. Technol.* B6: 1426–67
3. Swartzentruber BS, Mo YW, Webb MB, Lagally MG. 1989. *J. Vac. Sci. Technol.* B7: 2901–5
4. Hamers RJ, Markert K. 1990. *J. Vac. Sci. Technol.* A7: 3524–30
5. Cox G, Szynka D, Poppe U, Graf KH, Urban K, et al. 1990. *Phys. Rev. Lett.* 64: 2402–5
6. Colbourn EA. 1993. *Surf. Sci. Report* 15: 281–319
7. Flynn CP. 1972. *Point Defects and Diffusion.* Oxford: Clarendon
8. Owaman F, Martensson P. 1993. *Surf. Sci.* 303: L367–72
9. Fowler WB. 1968. *The Physics of Color Centers.* New York: Academic
10. Watkins GD. 1989. *Mater. Sci. Forum* 38–41: 39–50
11. Bridge E, Davies G, Robertson J, Stoneham AM. 1990. *J. Phys. Cond. Matter* 2: 2875–2928
12. Hayes W, Stoneham AM. 1984. *Defects and Defect Processes in Nonmetallic Solids.* New York: Wiley
13. Menzel D, Gomer R. 1964. *J. Chem. Phys.* 41: 3311–28
14. Redhead P. 1964. *Can. J. Phys.* 42: 886–905
15. Itoh N, Tanimura K. 1986. *Radiat. Eff.* 98: 269–87
16. Matthias E, Nielsen HB, Reif J, Rosen A, Westin E. 1987. *J. Vac. Sci. Technol.* B5: 1415–22
17. Ueta M, Kanzaki H, Kobayashi K, Toyozawa Y, Hanamura E. 1986. *Excitonic Processes in Solids.* Berlin: Springer-Verlag
18. Kabler MN. 1972. In *Point Defects in Solids.* ed. JH Crawford, LM Slifkin, pp. 291–352. New York: Plenum
19. Song KS, Williams RT. 1993. *Self-Trapped Excitons.* Berlin: Springer-Verlag
20. Itoh N, Tanimura K, Stoneham AM, Harker AH. 1989. *J. Phys. Cond. Matter* 1: 3911–18
21. Shluger AL, Harker AH, Puchin VE, Itoh N, Catlow CRA. 1993. *Modelling Simul. Mater. Sci. Eng.* 1: 673–85
22. Li X, Beck RD, Whetten RL. 1992. *Phys. Rev. Lett.* 68: 3420–23
23. Puchin VE, Shluger AL, Itoh N. 1993. *Phys. Rev. B.* 47: 10,760–68
24. Puchin V, Shluger A, Nakai Y, Itoh N. 1994. *Phys. Rev. B* 49: 11,364–73
25. Anderson PW. 1975. *Phys. Rev. Lett.* 34: 953–55
26. Bourgoin J, Lanoo M. 1981. *Point Defects in Semiconductors I. Theoretical Aspects.* Berlin: Springer-Verlag
27. Hansson GV, Uhrberg RIG. 1988. *Surf. Sci. Rep.* 9: 197-292
28. Chadi DJ, Chang KJ. 1988. *Phys. Rev. Lett.* 60: 2187–90
29. Khoo GS, Ong CK, Itoh N. 1993. *J. Phys. Cond. Matter* 5: 1187–94
30. Khoo GS, Ong CK, Itoh N. 1993. *Phys. Rev. B* 47: 2031–37
31. Ong CK, Khoo GS, Hattori K, Nakai Y, Itoh N. 1991. *Surf. Sci.* 259: L787–90
32. Itoh N, Hattori K, Nakai Y, Okano A, Ong CK, Khoo GS. 1992. *Appl. Phys. Lett.* 60: 3271–73
33. Khoo GS, Ong CK, Itoh N, Kanasaki J. 1994. *J. Appl. Phys.* 75: 255–58
34. Hattori K, Okano A, Nakai Y, Itoh N. 1992. *Phys. Rev. B* 45: 8424–36
35. Itoh N, Nakayama T. 1982. *Phys. Lett. A* 92: 471–75
36. Itoh N, Nakayama K, Tombrello T. 1985. *Phys. Lett. A* 108: 480–84
37. Sumi H. 1991. *Surf. Sci.* 248: 382–410
38. Pellin MJ, Young CE, Calaway WF, Gruen DM. 1986. *Nucl. Instrum. Methods B* 13: 653–57
39. Kanasaki J, Okano A, Ishikawa K, Nakai Y, Itoh N. 1993. *J. Phys. Cond. Matter* 5: 6497–506
40. Hattori K, Nakai Y, Itoh N. 1990. *Surf. Sci.* 227: L115–19
41. Kanasaki J, Yu IK, Nakai Y, Itoh N. 1993. *Jpn. J. Appl. Phys.* 32: 859–62
42. Okano A, Matsuura AY, Hattori K, Itoh N, Singh J. 1993. *J. Appl. Phys.* 73: 3158–62
43. Itoh N, Okano A, Hattori K, Kanasaki J, Nakai Y. 1993. *Nucl. Instrum. Methods B* 82: 310–16
44. Bommer K, Needels M, Larson B, Joannopoules J. 1992. *Phys. Rev. Lett.* 68: 1355–58
45. Phillips JC. 1973. *Bonds and Bands in Semiconductors.* New York: Academic
46. Okano A, Hattori K, Nakai Y, Itoh N. 1991. *Surf. Sci.* 258: L671–75
47. Kanasaki J, Okano A, Ishikawa K, Nakai Y, Itoh N. 1993. *Phys. Rev. Lett.* 70: 2495–98
48. Okano A, Kanasaki J, Nakai Y, Itoh N. 1994. *J. Phys. Cond. Matter* 6: 2697–712
49. Singh J, Okano A, Kanasaki J, Itoh N. 1994. *Phys. Rev. B.* 50: 11730–37
50. Faukner RA. 1968. *Phys. Rev. B* 175: 991–1009
51. Dexter DL. 1953. *J. Chem. Phys.* 21: 836–50
52. Hauijser A, Laar JV, Rooy TLV. 1978. *Phys. Lett. A* 65: 337–39

53. Straub D, Skibowski M, Himpsel FJ. 1985. *Phys. Rev. B* 32: 5237–44
54. Chiaradia P, Fanfoni M, Nataletti P, De Padova P, Brillson LJ, et al. 1989. *Phys. Rev. B* 39: 5128–31
55. Carstensen H, Claessen R, Manzke R, Skibowski M. 1990. *Phys. Rev. B* 41: 9880–85
56. Miller JC, Haglund RF Jr, eds. 1991. *Laser Ablation, Mechanisms and Applications*. Berlin: Springer-Verlag
57. Miller JC. 1994. *Laser-Surface Interaction*. Berlin: Springer-Verlag
58. Dreyfus RW, McDonald FA, von Gutfeld J. 1987. *J. Vac. Sci. Technol. B* 5: 1521–27
59. Stritzker B, Pospieszczyk A, Tagle JA. 1981. *Phys. Rev. Lett.* 47: 356–58
60. Liu JM, Kurz YH, Bloembergen N. 1981. *Appl. Phys. Lett.* 39: 755–57
61. Namiki A, Kawai T, Ichige K. 1986. *Surf. Sci.* 166: 129–40
62. Sutcliffe E, Srinivasan R. 1986. *J. Appl. Phys.* 60: 3315–22
63. Ichige M, Matsumoto Y, Namiki A. 1988. *Nucl. Instrum. Methods B* 33: 820–23
64. van Vechten JA. 1982. *Laser and Electron-Beam Interaction*, ed BA Appleton, GK Celler, pp. 49–57. Amsterdam: North Holland
65. Conbescot M, Bok J. 1982. *Phys. Rev. Lett.* 48: 413–16
66. Okano A. 1994. *Laser-induced atom emissions and laser ablation of surfaces of compound semiconductors*. PhD thesis. Nagoya University. (in Japanese) 88 pp.
67. Moisson JM, Bensussan M. 1983. *Surf. Sci.* 126: 294–300
68. Moison JM, Bensoussan M. 1982. *J. Vac. Sci. Tech.* 21: 315–18
69. Kumazaki Y, Nakai Y, Itoh N. 1987. *Phys. Rev. Lett.* 59: 2883–86
70. Kumazaki Y, Nakai Y, Itoh N. 1987. *Surf. Sci.* 184: L445–51
71. Brewer PD, Zinck JJ, Olson GL. 1991. *Mater. Soc. Symp. Proc.* 201: 543–47
72. Brewer PD, Zinck JJ, Olson GL. 1990. *Appl. Phys. Lett.* 57: 2526–28
73. Goldberg SS, Long JP, Kabler MN. 1991. *Mater. Soc. Symp. Proc.* 201: 519–25
74. Long JP, Goldberg SS, Kabler MN. 1992. *Phys. Rev. Lett.* 68: 1014–17
75. Houle FA. 1988. *Phys. Rev. Lett.* 61: 1871–74
76. McFeely FR, Morar JF, Himpsel FJ. 1986. *Surf. Sci.* 165: 227–87
77. Kullmer R, Bauerle B. 1987. *Appl. Phys. A* 43: 227–32
78. Sesselmann W, Hudeczek E, Bachmann F. 1989. *J. Vac. Sci. Technol. B* 5: 1284–85
79. Boulmer J, Bourguignon B, Budin JP, Debarre D. 1989. *Appl. Surf. Sci.* 43: 424–31
80. Aliouchouche A, Boulmer J, Bourguignon B, Budin JP, Debarre D, Desmur A. 1993. *Appl. Surf. Sci.* 69: 52–57
81. Shih MC, Freiler MB, Haase G, Scarmozzino R, Osgood RM Jr. 1992. *Appl. Phys. Lett.* 61: 828–31
82. Berman MR. 1991. *Appl. Phys. A* 53: 442–48
83. Foulon F, Green M, Goodall FN, De Unamuno S. 1992. *J. Appl. Phys.* 71: 2898–2907
84. Brewer PD, McClure D, Osgood RM Jr. 1985. *Appl. Phys. Lett.* 47: 310–12
85. Brewer PD, McClure D, Osgood RM Jr. 1986. *Appl. Phys. Lett.* 49: 803–5
86. Kanasaki J, Yu IK, Nakai Y, Itoh N. 1994. *Jpn. J. Appl. Phys.* 33: 2255–57
87. Kanasaki J, Matsuura AY, Nakai Y, Itoh N. 1993. *Appl. Phys. Lett.* 62: 3493–95
88. Masewich JA, Heinz TF, Newns DM. 1992. *Phys. Rev. Lett.* 68: 3737–40
89. Liu HT, Wu Z. 1994. *Phys. Rev. Lett.* 72: 3891–94

Annu. Rev. Mater. Sci. 1995. 25: 129–60

MAGNETISM AND GIANT MAGNETO-TRANSPORT PROPERTIES IN GRANULAR SOLIDS

C. L. Chien

Department of Physics and Astronomy, The Johns Hopkins University, Baltimore, Maryland 21218

KEY WORDS: magnetic properties, giant magnetoresistance, Hall resistivity, magneto-thermal conductivity

ABSTRACT

Magnetic granular solids consist of nanometer-sized magnetic particles uniformly dispersed and embedded in an immiscible insulating or metallic medium. They display a rich variety of magnetic properties and giant magneto-transport properties due to their unique nanostructure, which can be usefully described by the metal volume fraction and the metal particle diameter of granular solids. Fabrications and characterizations of granular solids, their enhanced magnetic properties, percolation behavior, single-domain characteristics, superparamagnetic relaxation, and superparamagnetism are reviewed. Giant magneto-transport properties, recently observed in granular solids, including giant magnetoresistance, extraordinary Hall resistivity, magneto-thermal conductivity, and Wiedemann-Franz law are discussed.

INTRODUCTION

Granular metal solids, consisting of ultrafine metallic particles uniformly dispersed and embedded in an immiscible medium, constitute a special class of artificially structured materials (ASM) with low-dimensional entities (1). Multilayers (2) and arrays of nanowires (3) are ASM with quasi two-dimensional and one-dimensional entities respectively; granular solids

0084–6600/95/0801–0129$05.00

have quasi zero-dimensional entities. The novel properties of granular solids arise from the unique nanostructure with which their properties can be altered and tailored. Granular metal solids, pioneered by Abeles et al (4, 5), have recently attracted considerable renewed interest. This is due in part to the current interest in nanophase materials (6–8), and to the unusual physical properties that have been uncovered in granular metal solids. The word granular has also appeared in other contexts, most notably in the studies of fluid flowing through rocks and experiments on sand piles at the threshold of instability. The key difference is the size of the granules, which are of the order of 10 cm in the case of rocks and 100 μm in sand grains. In the present context, granules are referred to as ultrafine solid particles, only a few nms in size, consisting of 10^3–10^6 atoms. Solids of such small sizes are between those of clusters and bulk solids. This is also the size range that is unattainable by usual powder metallurgical and chemical methods. Another important feature of granular solids is that these ultrafine particles are not free standing but are embedded in an immiscible medium. Free-standing ultrafine metallic particles are notoriously susceptible to environmental degradation (e.g. oxidation) and have a strong tendency to conglomerate into larger entities. Both of these difficulties are removed in granular solids, where the ultrafine metal particles are protected by, and uniformly dispersed in, an immiscible medium. The key feature of granular solids is that their properties hinge crucially on their nanostructure, characterized by the metal volume fraction (x_v) and the metal particle size ($2r$). By varying these nanostructure parameters, their physical properties can be greatly altered and tailored.

There are two types of granular metal solids, both of which are two-component systems. Granular metal embedded in an insulating medium, such as SiO_2 and Al_2O_3, have been known since the 1970s (4, 5). More recently, granular metal embedded in an immiscible metallic medium have been realized (9–12). The latter materials have led to the observation of, among other properties, giant magneto-transport phenomena. There is a rich variety of subjects in granular metal solids (4, 5), and we limit our discussions to some essential aspects of their magnetic and magneto-transport properties.

NANOSTRUCTURE

The intricate nanostructure of granular solids can be usefully, although only partially, described by two parameters. The first is x_v whose values can be continuously varied from 0 to 1. The second parameter, is $2r$, which is useful when isolated particles exist for samples with low values of x_v. The particle size can also be experimentally varied to a significant extent through process conditions. In Figure 1, the evolution of the microstructure of granular solids (e.g. $Ag\text{-}Al_2O_3$, $Fe\text{-}SiO_2$) by varying x_v alone is depicted, and only the metal particles are shown. For all granular metal solids with a uniform dispersion of particles, a well-defined percolation volume fraction (x_p) exists, that has been experimentally observed (4, 5, 13) and theoretically understood (14) to be in the vicinity of $x_p = 0.5$–0.6. Isolated metal granules exist for $x_v < x_p$, and a continuous network of connecting granules appear for $x_v > x_p$. All characteristics of the nanostructure of granular metal solids, and indeed the variation of their physical properties, are intimately related to the particle size ($2r$) and the metal volume fraction (x_v).

Once x_v and $2r$ are defined, other structural characteristics of granular solids can be related. In real granular solids, the particles are not spherical but equiaxial, with a narrow size distribution, but not monodispersive. For simplicity, let us consider the ideal situation where all the particles are monodispersive spherical particles of radius r, randomly and uniformly distributed in a medium. Under these simplifications, one can easily calculate that the number of metal particles per volume is

$$n \approx \frac{1}{4}\frac{x_v}{r^3}. \qquad 1.$$

The interface area per volume, which is proportional to the surface to volume ratio of the particles, is

$$S \approx \frac{3x_v}{r}. \qquad 2.$$

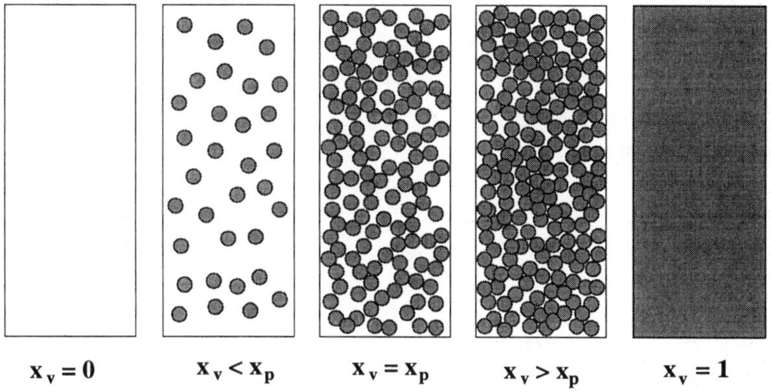

Figure 1 Schematic representations of granular metal-insulator solids as a function of metal volume fraction ($0 \leqslant x_v \leqslant 1$) with a percolation volume fraction (x_p).

The average particle–particle distance (from center to center) is

$$d \approx \left(\frac{16}{x_v}\right)^{1/3} r. \qquad 3.$$

At $x_v = 0.25$, $d \approx 4r$; i.e. the average spacing between the particles is about the size of the particle. These quantities will have relevance to the magnetic properties and the giant magneto-transport properties. It is also straightforward to show that, as long as the particles are randomly distributed, the metal area fraction (x_a) at the surface is the same as the metal volume fraction (x_v).

$$x_a = x_v. \qquad 4.$$

Thus the metal area fraction, as determined from the surface using microscopy, gives the value of the metal volume fraction.

Many physical properties undergo tremendous change as one varies x_v, particularly across x_p. Indeed, all physical properties whose characteristic lengths (e.g. coherent length in superconductors, critical size for single-domain magnetic particles, electron mean free path in transport properties, spin diffusion length in magneto-transport properties) are comparable to, or smaller than, the length scales inherent to a granular solid with certain values of x_v and $2r$ will be unavoidably affected. Most interesting of all, some of the enhanced physical properties can only be captured in granular solids and not in bulk materials. The enhanced magnetic properties and giant magneto-transport properties covered in this article are some of the examples discovered in recent years.

FABRICATION METHODS

Crucial to the formation of a granular solid is the fact that the two components must be immiscible; otherwise, a homogeneous alloy would be formed instead. At present, a number of methods can be used to prepare specimens that contain isolated ultrafine metallic particles of various sizes under 1000 Å. Examples include chemical precipitation (15–17), spark erosion (18), a high-energy ball milling (19), evaporation in an inert gas atmosphere (6, 20), and sol-gel method (21). However, it is a formidable task to prevent the ultrafine metal particles from oxidation and to disperse them uniformly into a suitable and protective medium. Furthermore, none of these methods can fabricate metallic entities in the nm scale with a narrow size distribution. To date, vapor deposition (particularly high-rate sputter deposition) of a metal and an immiscible material has proven to be the most versatile method of fabricating granular metal solids because the formation of the nm-size particles, the protection of the vulnerable metal particles, the uniform dispersion of the particles, and coating onto desired surfaces are accomplished simultaneously (4, 5, 13). Existing results show that the metal particles may be elemental metals (e.g. Fe, Co, Ag) (4, 5, 13), crystalline alloys (e.g. Fe-Ni alloys) (22), metastable crystalline alloys (e.g. Fe-Cu) (22), or amorphous alloys (e.g. Fe-BN).

In the fabrication of granular metal solids using vapor deposition, x_v is determined by the composition. The particle size ($2r$), which depends to some extent on x_v (larger particles for higher values of x_v), can also be altered by processing conditions, of which substrate temperature is one of the most effective (13, 23). There are a number of variations within vapor deposition, and we illustrate these variations by applying to the fabrication of granular Fe-Al_2O_3.

1. Co-deposition from two separate sources (4, 5): The two sources (Fe and Al_2O_3) are directed toward the substrate. The contents of the two components are controlled by the relative deposition rates. With a slight variation in co-deposition, one can also achieve a gradient in composition across a large substrate and effectively obtain many small samples of different compositions [i.e. $Fe_x(Al_2O_3)_{1-x}$] (4). The drawback of co-deposition is that uniform composition exists only over small areas.
2. Tandem deposition from two separate sources (24): In this method, the two sources (Fe and Al_2O_3) are not directed at a common point. Instead, the substrate is moved rapidly over the two separate sources such that the accumulation of each material is small. However, because the deposition is sequential it is in fact the same scheme as that of fabricating a multilayer. Consequently, the composition is intrinsically non-uniform in the growth direction as is the dispersion of the particles. One immediate consequence of the non-uniform dispersion of the particles is that the percolation threshold is often much higher than the value of $x_p \approx 0.55$, which has been generally observed in granular samples with a uniform dispersion of particles (25).
3. Sputtering from a single composition target (4, 5, 13, 22, 23): This method relies on a uniform flux of the same composition over a large substrate area. It can most easily be administered using sputtering by taking advantage of the steady-state sputtering yield (26). Because sputtering sources can be readily scaled up (from a few cms to as large as a meter in size), sputtering is the preferred method for obtaining coatings of homogeneous composition over large surface areas. The drawback of this method is that one composite target has to be prepared for each composition.

4. Phase separation of immiscible metal (9–12): Recently, another method has been devised for the fabrication of metallic particles embedded in a metallic medium by taking advantage of the nearly complete immiscibility between two metals, such as Co-Ag, Co-Cu, and Fe-Ag (27, 28). Using a vapor-quenching method, metastable alloys of immiscible alloys can be readily formed. Annealing a metastable alloy at elevated temperatures induces phase separation, which causes the formation of metallic granules in a metallic medium (9–12). The size of the granules is dictated by the temperature and duration of the anneal. Alternatively, one can also form such granular metal-metal solids by deposition onto a substrate maintained at elevated temperatures.

CHARACTERIZATIONS

Characterization is obviously an integral part of any study of granular solids. Because the intricate structure exists on the nanometer scale, only characterization tools with nm-scale resolution will be useful. The main pieces of information that one wishes to extract from granular solids are purity, structure and phases of the constituent components, $2r$ and its distribution, x_v, and the connectivity of the particles. Some of the methods reported in the literature are X-ray diffraction (XRD), both large-angle and small-angle, transmission electron microscopy (TEM), electron diffraction, X-ray photoelectron spectroscopy (XPS), neutron diffraction (especially for magnetic granular solids), and ^{57}Fe Mössbauer spectroscopy (for granular solids containing Fe). For magnetic granular solids, magnetometery measurements can also shed light on the size of the magnetic particles. Of those, the most useful, and consequently, the most widely used techniques, are XRD and TEM. We briefly describe those results here.

Of the two types of granular solids, the characterization of the nanostructure of granular metal-insulator solids (e.g. Fe-Al_2O_3) are generally more clear-cut than those of granular metal-metal solids. This is because the insulating medium is often amorphous (thus largely featureless in XRD and TEM), and therefore the metal particles stand out (4, 5, 13). Large-angle and small-angle XRD patterns of granular Fe-Al_2O_3 ($x_v = 0.35$) are shown in Figure 2. The large-angle pattern (*left*) clearly shows that the ultrafine Fe particles are crystalline with a body-centered cubic (bcc) structure. The broad diffraction line width is a consequence of the small particle size. In the small-angle XRD pattern shown on the right, there is a distinct minimum at the scattering vector $S = 0.15$ Å due to the particles of size $2\pi/S = 42$ Å. The minimum can only be observed from granular

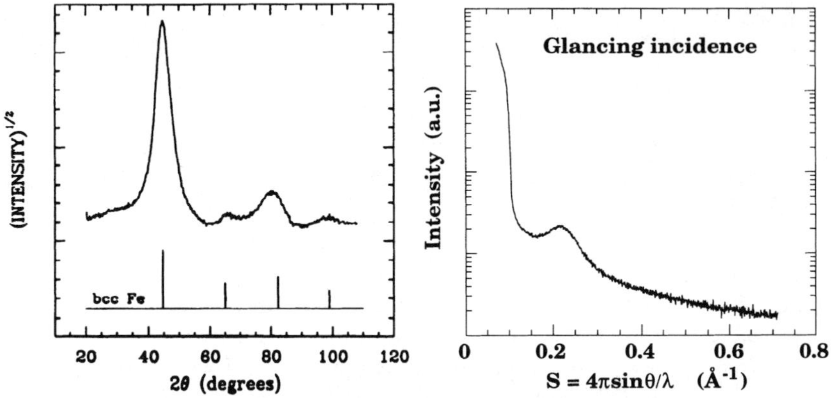

Figure 2 Large-angle and small-angle X-ray diffraction patterns of Fe-Al_2O_3 ($x_v = 0.35$). The large-angle pattern shows the bcc structure of Fe, whereas the small-angle pattern shows a minimum at S ≈ 0.15 Å due to particles with a narrow distribution of size $2\pi/S \approx 42$ Å.

solids with a narrow size distribution (J Jiang & C Chien, unpublished data).

Examples of TEM micrographs of granular metal-insulator solids, specifically ($Fe_{50}Ni_{50}$)-(SiO_2) ($x_v = 0.4$) deposited at $T_S = 150$°C (22) (*upper panel*), and Co-SiO_2 with $x_v = 0.12$, 0.22 and 0.51 (A Tsoukatos & G Hadjipanayis, unpublished data) (*lower panel*) are shown in Figure 3. The

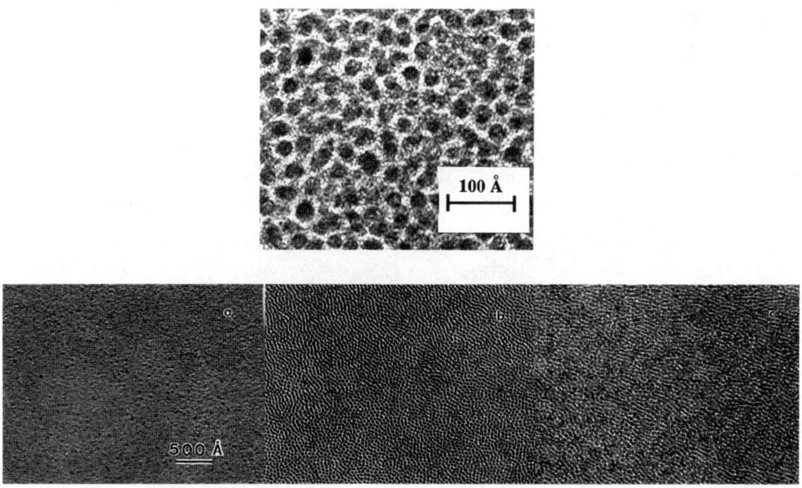

Figure 3 Transmission electron micrographs of $Fe_{50}Ni_{50}$-(SiO_2) ($x_v = 0.4$) deposited at $T_A = 150$°C (*upper panel*), and Co-SiO_2 with (*a*) $x_v = 0.12$, (*b*) $x_v = 0.22$ and (*c*) $x_v = 0.51$ (*lower panel*).

amorphous insulator is the light featureless background, and the dark entities are the metals. Particles of about 40 Å in size (*upper panel*), with a narrow distribution of sizes, are clearly observed. The TEM micrographs on the lower panel also illustrate the fact that the particle size increases with increasing x_v. The TEM micrograph for $x_v = 0.51$ on the lower panel is at the verge of percolation; hence the particles are connected to a large network.

Characterization of granular metal-metal solids (e.g. Co-Cu) are more difficult because both components are crystalline with small grain sizes, and with poorer TEM contrast (10, 29). To complicate matters further, the XRD and electron diffraction patterns of the two components may have overlapping diffraction lines. Bright-field TEM micrographs and electron diffraction patterns of $Co_{20}Ag_{80}$ are shown in Figure 4. The diffraction rings of face-centered cubic (fcc) Co and fcc Ag are also labeled (31). In the as-prepared sample (a), deposited at $T_S = 77$ K, all the diffraction rings match those of bulk Ag, confirming the formation of a metastable fcc alloy. The broad Ag(111) rings result from the small grains, which are about 4 nm in size. After annealing (b), (c), and (d), the grain size becomes larger as expected. No clear-cut diffraction rings corresponding to fcc Co are discernible for the sample annealed at $T_A = 200°C$ [with the caveat that the Co(111) and Ag(200) rings are overlapping]. The results for the samples with $T_A \geqslant 330°C$ show that phase separation has definitely occurred and the particles are larger for higher value of T_A. Converging beam measurements show that each particle is of one element, surrounded by particles of the second element (32).

MAGNETIC PROPERTIES

Strongly ferromagnetic materials may not necessarily have hard magnetic properties that are required for permanent magnet and magnetic recording applications. Metallic Fe has one of the highest saturation magnetizations (M_s) and the second highest Curie temperature (T_C). But the magnetocrystalline anisotropy energy constant (K) is quite small, so that multiple domains with closure-domain structure readily formed in bulk Fe (23). As a result, bulk Fe is magnetically soft with low coercivity (H_c) of the order of 10 Oe. The situation takes a dramatic turn for ultrafine magnetic particles (33, 34). Because the domain walls have a finite thickness (of the order of a few hundred Å), when the magnetic particle is sufficiently small, only one magnetic domain can be formed. In a single-domain magnetic particle, all the magnetic moments are aligned and rotate in unison under an external field as a giant moment, resulting in hard fer-

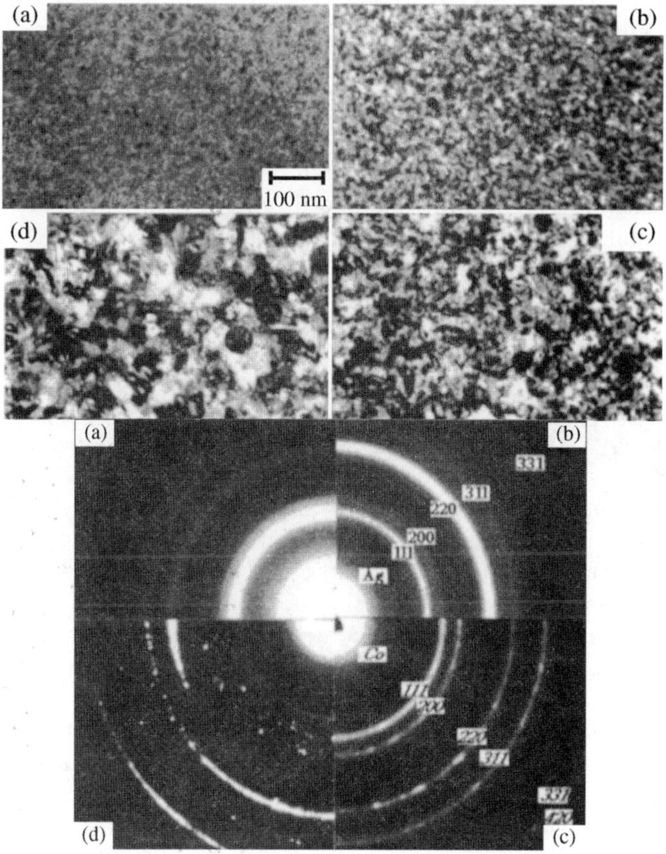

Figure 4 Bright-field transmission electron micrographs (*upper panel*) and electron diffraction patterns (*lower panel*) of $Co_{20}Ag_{80}$ samples deposited at 77 K and after annealing at various temperatures (T_A): (*a*) as-prepared, (*b*) $T_A = 200°C$, (*c*) $T_A = 330°C$, and (*d*) $T_A = 605°C$. In the lower panel, the electron diffraction patterns for fcc Ag and fcc Co are labeled.

romagnetic characteristics. The critical sizes for the formation of single domain for Fe, Co, and Ni are of the order of a few hundred Å, depending on the material and shape of the particle (33, 34). Particles of such small sizes, while challenging for traditional material synthesis, can be readily accomplished in granular magnetic solids. In granular solids, one observes the magnetic behavior of single-domain magnetic particles and the associated phenomenon of superparamagnetism in samples with low volume fractions ($x_v < x_p$). Bulk magnetic properties are recovered in samples with $x_v > x_p$.

Single-Domain Characteristics

In a granular magnetic solid with a low volume fraction, one has a collection of single-domain particles (4, 5, 13, 35, 36), each with a magnetic axis along which all the moments are aligned. In the absence of a magnetic field, parallel and antiparallel orientations along the magnetic axis are energetically equivalent but separated by an energy barrier of CV, where C is the total magnetic anisotropy per volume, and V is the particle volume. Since the size of each single-domain remains fixed, under an external field, only the magnetic axes rotate. Thus, the measured magnetization (M) of a granular magnetic solid with a collection of single-domain particles is the global magnetization

$$M = \frac{\langle \mathbf{M} \cdot \mathbf{H} \rangle}{H} = M_s \langle \cos\theta \rangle, \qquad 5.$$

where θ is the angle between the magnetic axis of a particle, M_s is the saturation magnetization, $\mathbf{H}$ is the external field, and the average $\langle \cos\theta \rangle$ is taken over many ferromagnetic particles. The hysteresis loop of a granular solid is a signature of the rotation of the magnetic axes of the single-domain particles. This should be contrasted with the hysteresis loop of a bulk ferromagnet, in which the sizes and directions of the domains are altered drastically under an external field (33).

An example of a hysteresis loop of a granular magnetic solid at low temperature is shown in Figure 5*a*. In the initial unmagnetized state with $M = 0$ at $H = 0$, the magnetic axes of the particles are randomly oriented, each along its own magnetic axis, which is determined by the total magnetic anisotropy of the particle. The directions of the giant moments are random and static at low temperatures. A saturation magnetization ($M = M_s$) is realized under a large field when all the magnetic axes are aligned. In the remnant state when H is reduced to $H = 0$, one observes the remnant magnetization (M_r), whose value at low temperature is $M_r = M_s/2$. This is because the magnetic axes are oriented only in one hemisphere due to the uniaxial anisotropy of the single-domain particle (13, 36).

It should be noted that the initial $M = 0$ state and the initial magnetization curve in Figure 5*a* does not reappear, whereas the field-cycle part does. The simplest way to recover the initial $M = 0$ state is to heat the sample above the blocking temperature (see below) and cool the sample in zero field back to low temperatures.

Because of the single-domain nature of the magnetic entities, the coercivity (H_c) of the ultrafine particles is expected to be much higher than that in bulk material. For free-standing spherical single-domain Fe particles, theoretical studies indicate a coercivity of $2K/M_s = 600$ Oe (37–39), which

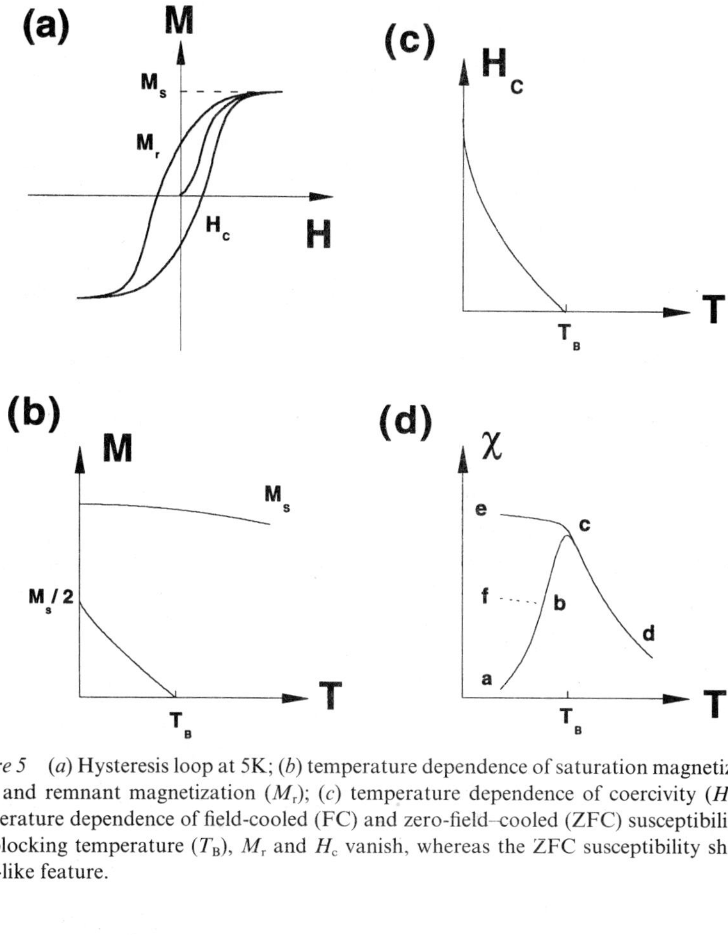

Figure 5 (*a*) Hysteresis loop at 5K; (*b*) temperature dependence of saturation magnetization (M_s) and remnant magnetization (M_r); (*c*) temperature dependence of coercivity (H_c); (*d*) temperature dependence of field-cooled (FC) and zero-field–cooled (ZFC) susceptibility. At the blocking temperature (T_B), M_r and H_c vanish, whereas the ZFC susceptibility shows a cusp-like feature.

is much larger than the 10 Oe value observed for bulk Fe. Many granular Fe-insulator solids actually exhibit even larger coercivities, to as much as 3000 Oe at 5 K and 1100 Oe at 300 K (13, 36, 40, 41). This is due in part, as described below, to the very large total magnetic anisotropy constant (C). The value of C includes magnetocrystalline anisotropy (K) and other contributions that dominate in granular Fe embedded in a medium, so that $C \gg K$. Some researchers have suspected the role of possible metal-oxide shells surrounding the Fe particles to account for the exceptionally large values of coercivity. Evidences of metal oxides have been suggested by XPS and some Mössbauer measurements (25, 42, 43). However, many granular metal–metal systems (e.g. Co-Ag, Fe-Cu, Fe-Ag) also have very large coercivity up to 3000 Oe, but without the presence of oxides (10, 11,

44). The large values of coercivity of granular solids, together with smaller recording noise intrinsic to small particles systems, suggest applications as magnetic recording media (40, 41). However, the anomalously large coercivity, the large total anisotropy constant (C), and the particle size dependence of H_c (H_c increases with particle size) (23) remain as some of the outstanding issues.

Superparamagnetic Relaxation

At sufficiently high temperatures, the energy barrier (CV) is overcome by the thermal energy (4, 5, 13, 33–36, 45–47). Consequently, the magnetic moments within a particle rotate rapidly in unison, exhibiting the well-known superparamagnetic relaxation phenomenon. The simplest form of the relaxation time can be described by the Arrhenius relation (33)

$$\tau = \tau_0 \exp(CV/k_B T), \qquad 6.$$

where τ is the relaxation time, τ_0 is the characteristic time, and CV is the total anisotropy energy. For an instrument that can measure certain magnetic characteristics (e.g. magnetometery, AC susceptibility, Mössbauer spectroscopy) with a measuring time of τ_i, one can define a blocking temperature

$$T_B = \frac{CV}{k_B \ln(\tau_i/\tau_0)}. \qquad 7.$$

At $T < T_B$, τ_i is less than τ and the instrument detects the ferromagnetic nature (e.g. a hysteresis loop) of the system. However, at $T > T_B$, because τ_i is longer than τ, the time averaged values of the ferromagnetic characteristics (e.g. magnetization and coercivity) vanish within the measuring time of τ_i. The system is in an apparent paramagnetic or superparamagnetic state, even though within each particle the magnetic moments remain ferromagnetically aligned. Because of superparamagentic relaxation, the values of remnant magnetization (M_r) and coercivity (H_c) decrease with increasing temperature and vanish at the blocking temperature (T_B) (13, 36, 45). Above T_B, all apparent ferromagnetic characteristics disappear. These aspects are shown in Figure 5*b* and *c*.

Another useful technique for the measurement of superparamagnetic relaxation is low-field susceptibility measurements in both zero-field-cooled (ZFC) and field-cooled (FC) modes (36). As shown in Figure 5*d*, above T_B, the FC and ZFC data are the same and hence reversible. Below T_B, however, the FC and ZFC data are totally different and irreversible. In particular, the ZFC data shows a readily detectable cusp-like feature at

T_B. The requirement of a low field is necessitated by the fact that Equation 6 is defined in the limit of $H = 0$ (36). Low-field susceptibility has been widely used by many people to provide evidence for the existence of ultrafine magnetic particles and/or superparamagnetism. However, it must be emphasized that the irreversibility of the ZFC and the FC branches, and the cusp-like feature of the ZFC branch are also tell-tale signs of spin-glass transitions and often appear even in ferromagnets (e.g. Gd). Another important, but often overlooked technical point is that T_B can be located from the ZFC branch only when the reversible part has been reached (for example, in Figure 5(*d*), $a \to b \to c \to d \to c \to e$). An apparent irreversibility is observed when increasing temperature is reversed to decreasing temperature during a measurement (for example in Figure 5(*d*), $a \to b \to f$), before the actual T_B is reached. This may occur, for example, when the limit of the accessible temperature has been reached. The temperature that corresponds to point *e* is not T_B.

It should be noted that superparamagnetic relaxation is a time effect (4, 5, 13, 33–36, 45–47). Therefore, the best methodology for ascertaining superparamagnetism is to employ two or more instruments with widely different measuring times. The same sample reveals different superparamagnetic relaxation effects when viewed by instruments with different measuring times—one blocking temperature (T_{Bi}) for each instrument with a specific measuring time τ_i, as defined by Equation 7. Thus by using two or more instruments of very different measuring times (36), or by conducting AC measurements over a wide frequency range (46), one can firmly establish superparamagnetism relaxation. For example, using SQUID magnetometery ($\tau_i \approx 10$ s) and Mössbauer spectroscopy ($\tau_i \approx 10^{-8}$ s), for two measured blocking temperatures have the same ratio of 0.35, regardless of the samples (36).

Furthermore, using more than one instrument with very different measuring times, one can also separately determine CV and τ_o. Together with an independent measurement of V (e.g. microscopy), the value of C can be determined (36). It has been found in many ultrafine single-domain magnetic particles that the values of C are much larger than the magnetocrystalline anisotropy energy densities of the constituent magnetic materials in bulk form, some by as much as two orders of magnitude (13, 36, 40, 42, 46, 47). For example, while the magnetocrystalline anisotropy energy constants are $K \approx 10^5$ erg/cm^3 and 10^4 erg/cm^3, respectively, for bulk Fe and $Fe_{50}Ni_{50}$, much larger values of C, in excess of 10^7 erg/cm^3, have been observed in granular Fe-SiO_2 and ($Fe_{50}Ni_{50}$)-(Al_2O_3) (22, 36, 47). Evidently, mechanisms other than magnetocrystalline anisotropy are the dominant factors for the large magnetic anisotropy in ultrafine magnetic particles embedded in a matrix. This enhancement of C in part accounts for the very large coercivities observed in granular magnetic solids as mentioned above.

It may be noted from Equation 7 that the value of T_B is proportional to the volume (V) of the particle, hence proportional to r^3. Even a modest increase in $2r$ (e.g. a factor of 3) can result in hundreds of degrees of increase in the value of T_B (22). Indeed, by controlling the particle size through process conditions, one can vary the value T_B from less than 10 K to several hundred degrees above room temperature.

Finally, we comment on the validity of the Arrhenius relation (Equation 6) for the relaxation time. Because of its simplicity, it has been widely used, especially if one has only one measuring instrument so that the value of τ_0 must be assumed (most often τ_0 is taken to be about 10^{-9} s) in order to extract the value of CV. On the other hand, measurements with two or more instruments, where one can determine both τ_0 and CV by using the Arrhenius law, concluded a wide range of values for τ_0, from 10^{-9} to 10^{-25} s (13, 36, 46). The extremely short values of $\tau_0 < 10^{-15}$ s clearly must be artifacts. This dilemma can be understood as follows. Extensive measurements of the relaxation time using SQUID magnetometery ($\tau_i \approx 10$ s) (36), AC susceptibility ($\tau_i \approx 10^{-4}$–10^{-1} s) (46), Mössbauer spectroscopy ($\tau_i \approx 10^{-8}$ s) (36, 47), and small-angle neutron diffraction ($\tau_i \approx 5 \times 10^{-12}$ s) (48) indicate that the actual relaxation time does not strictly follow an Arrhenius relation. As schematically shown by the plot of $\ln\tau$ vs $1/T$ in Figure 6, the results from magnetometery, AC susceptibility, and Mössbauer measurements fall on a curve rather than a straight line. Small-angle neutron diffraction measurements indicate that the value of τ_0 is likely to be near 10^{-11} s (24). As shown in Figure 6, if SQUID magnetometery and Mössbauer spectroscopy were used, a straight line Arrhenius relation would give $\tau_0 \approx 10^{-13}$ s (36), whereas if AC susceptibility results were used, the straight line would give $\tau_0 \approx 10^{-19}$ s (46).

Superparamagnets with Giant Moments

At $T > T_B$, the magnetic particles exhibit superparamagnetic characteristics of a paramagnetic system of giant moments. The magnetization behavior (i.e. M vs H) follows that of a Langevin function for classical magnetic moments (33),

$$\frac{M}{x_v M_s} = L\left(\frac{\mu H}{k_B T}\right) = \coth\left(\frac{\mu H}{k_B T}\right) - \frac{k_B T}{\mu H}, \qquad 8.$$

where $\mu = M_s V$ is the giant magnetic moment (about 10^4 μ_B) of a single particle with volume V, and H is the external field. In suitable ranges of

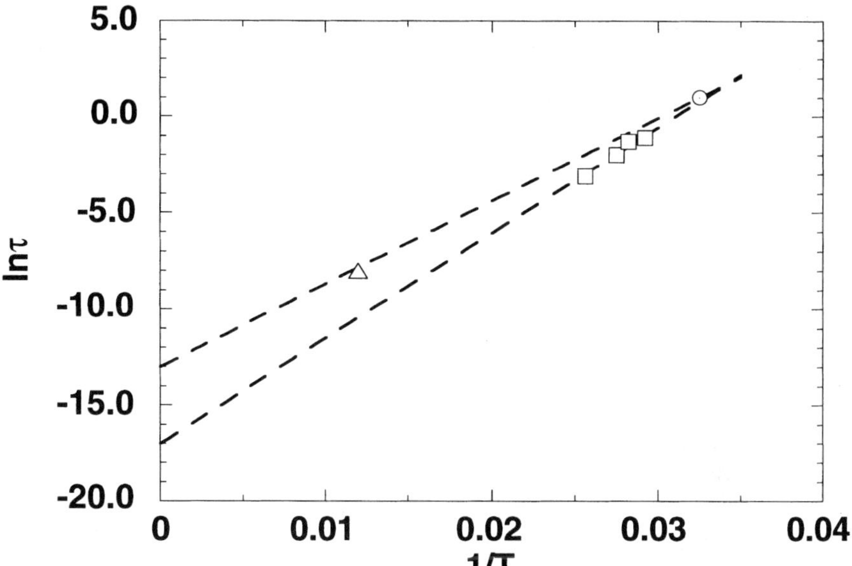

Figure 6 Semi-log plot of relaxation time (τ) as a function of $1/T$. Results from magnetometery (M) and Mössbauer effect (ME) are denoted by circle and triangle respectively. The AC susceptibility (AC) results are denoted by squares. The two dashed straight lines assumed Arrhenius relation linking the results of M and ME, as well as extrapolating linearly the AC results.

temperature and magnetic field, the Langevin function depends sensitively on the value of μ, and hence the particle size, (V), as shown in Figure 7. A detailed analysis of the magnetization behavior permits the determination of the values of the giant moments (49). This information can, in turn, be converted to particle sizes.

At $T > T_B$, the magnetic susceptibility behaves as (49)

$$\chi = \frac{x_v V M_s^2(T)}{3k_B(T - T^*)}. \qquad 9.$$

This form, which follows from Equation 8, is very similar to the well-known Curie-Weiss susceptibility of a paramagnet (33),

$$\chi = \frac{n[g\mu_B J(J+1)]^2}{3k_B(T-\theta)}, \qquad 10.$$

where n is the number of the effective magnetic moment $g\mu_B J(J+1)$, and θ is the Curie-Weiss constant, which is a measure of the strength of the

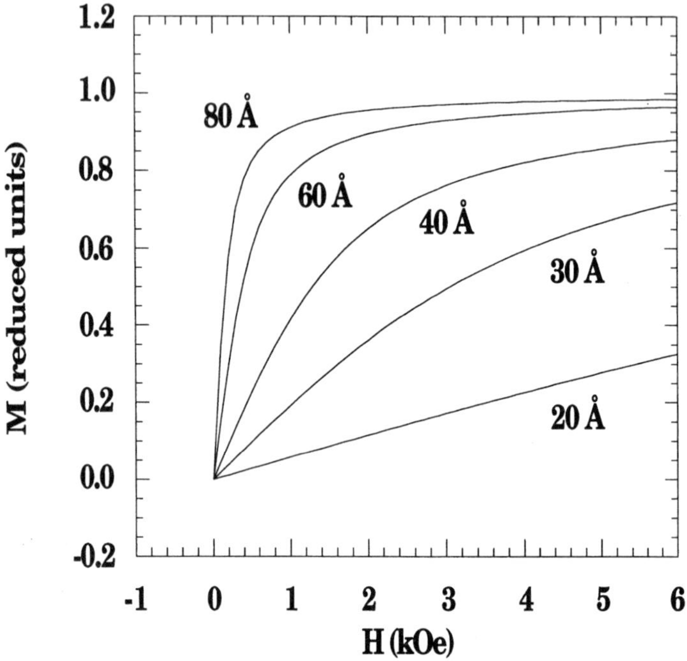

Figure 7 Calculated magnetization behavior at $T = 300$ K of Fe particles of various sizes.

magnetic interactions among the moments (33). The key difference is that the giant moment is about 10^4 times larger than the individual atomic moment. The analyses of the susceptibility data offer still another means of measuring the giant moments and hence the particle sizes. The quantity T^*, the analogue of the Curie-Weiss temperature of a paramagnet, provides a measure of the interaction among the superparamagnetic giant moments (50). In most cases, T^* is of the order of 20 K, suggesting weak inter-particle interactions (49). It should also be noted that the effective magnetic moment in Equation 10 is a constant, thus $1/\chi$ in a paramagnetic system is linear in T. For a granular system, $M_s(T)$ in Equation 9 is temperature dependent, hence the $1/\chi$ plot in a granular system is not linear in T. This fact can be used for the purpose of measuring the temperature dependence of $M_s(T)$ of a superparamagnetic particle, which otherwise is not easily accessible in magnetometery measurements (49).

The giant moments of a superparamagnet can be exploited in magnetocaloric applications. Recently, the utility of granular magnetic solids has been demonstrated in magnetic refrigeration (51), where the giant moments provide higher efficiency and a higher operating temperature range over those of conventional paramagnetic materials.

GIANT MAGNETO-TRANSPORT PROPERTIES

When a material is subjected to an external field, currents of charge or energy may flow. These responses to applied fields are the transport properties, which provide valuable information of the material. The applied fields may be electric, magnetic, thermal gradient, or a combination thereof. Although the specific transport property may involve only one external field, such as electrical resistivity (electric field), more interesting effects may occur when two fields are applied simultaneously, such as magnetoresistivity and Hall resistivity (electric field and magnetic field), and magneto-thermal conductivity (thermal gradient and magnetic field). Each transport property reveals certain information concerning the material in which it occurs. Magnetoresistivity provides a measure of the scattering events contributing to the magnetic resistivity. Thermal conductivity reveals the relative importance of large-angle and small-angle scattering events, as well as that of elastic vs inelastic events.

While the overwhelming majority of studies of giant magneto-transport properties has centered on magnetoresistance, especially in multilayers (52–62), magnetoresistivity is but one of several transport properties, all of which are essential for the understanding of the magneto-transport properties in magnetically heterogeneous systems of multilayers and granular solids. In this section, we describe the main features and the outstanding issues of the giant magneto-transport properties of granular metal systems, including magnetoresistivity, extraordinary Hall resistivity, and magneto-thermal conductivity.

When a specific transport property, for example, electrical resistance (in units of Ω) is measured from a material, its value depends on the resistivity (ρ in units of $\mu\Omega$-cm) as well as the physical dimensions of the specimen. In order to understand the scattering processes that contribute to this transport property, one must determine the absolute value of ρ, by also measuring the physical dimensions of the specimen. If the physical dimensions of the specimen are unknown, one can, at most, measure the changes of the transport property by expressing them in a fraction form (i.e. $\Delta R/R = \Delta\rho/\rho$). Without knowing the value of ρ, one is in no position to seek the physics that causes the observed fractional change. The physics that gives rise to an effect of $\Delta R/R = \Delta\rho/\rho = 20\%$ when ρ is in the $\mu\Omega$-cm range is entirely different from that when ρ is in the mΩ-cm range. In the following descriptions of the transport properties, we discuss the measured transport properties in absolute values of resistivity and conductivity.

MAGNETORESISTIVITY (ρ_{xx})

Magnetoresistance (MR) is the variation of electrical resistance due to an external magnetic field (**H**) (63). Positive or negative MR refers to the increase or decrease of resistance with the magnetic field. For thin-film specimens, where the electric current with a current density (**j**) is usually in the film plane, MR can be measured in three geometrices: $\boldsymbol{\rho}_{\parallel}$ (longitudinal MR, **H**||**j**), $\rho_{\rm T}$ (transverse MR, **H** is in the film plane and perpendicular to **j**), and $\boldsymbol{\rho}_{\perp}$ (perpendicular MR, **H** is perpendicular to the film plane and perpendicular to **j**), as shown in Figure 8. It is generally necessary to measure MR in more than one field orientation to ascertain its characteristics.

For a Drude metal of free electrons, there is no MR, i.e. resistance is independent of *H*, as shown in Figure 9*a*. The characteristics of MR in various real materials at low temperatures can be summarized as follows—roughly in increasing order of the size of MR effect size:

1. Ordinary magnetoresistance (OMR): For ordinary non-magnetic metals, such as Au and Cu, the MR is quite small. Both $\boldsymbol{\rho}_{\parallel}$ and $\boldsymbol{\rho}_{\rm T}$ are

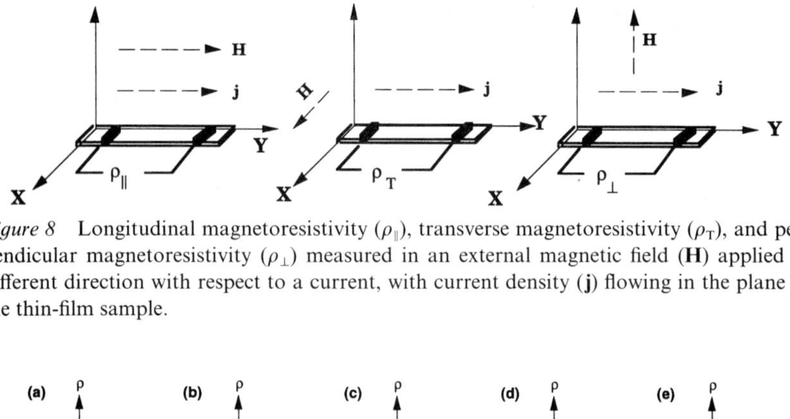

Figure 8 Longitudinal magnetoresistivity ($\rho_{\parallel}$), transverse magnetoresistivity ($\rho_{\rm T}$), and perpendicular magnetoresistivity ($\rho_{\perp}$) measured in an external magnetic field (**H**) applied in different direction with respect to a current, with current density (**j**) flowing in the plane of the thin-film sample.

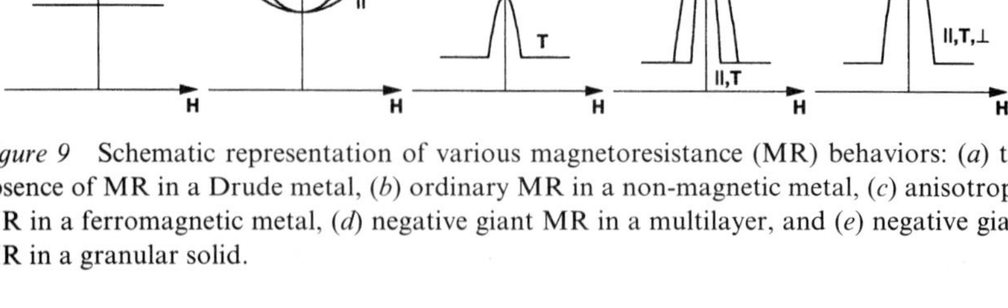

Figure 9 Schematic representation of various magnetoresistance (MR) behaviors: (*a*) the absence of MR in a Drude metal, (*b*) ordinary MR in a non-magnetic metal, (*c*) anisotropic MR in a ferromagnetic metal, (*d*) negative giant MR in a multilayer, and (*e*) negative giant MR in a granular solid.

positive and increase unabated as H^2, with $\rho_T > \rho_\parallel$, as shown in Figure 9*b*.

2. Anisotropic magnetoresistance (AMR): In ferromagnetic metals (e.g. Fe, Co) and alloys (e.g. permalloy), substantial MRs are observed. It is almost always the case that $\rho_\parallel$ is positive and ρ_T is negative, as shown in Figure 9*c*, and $\rho_\parallel > \rho_T$, which is opposite to that of OMR. Both $\rho_\parallel$ and ρ_T can be saturated under a modest field. The AMR depends on the direction of the spontaneous magnetization and is due to the rotation of the ferromagnetic domains under the external field (64). The saturation field (H_s) for the AMR in permalloy is especially small (less than 10 Oe). This, together with the sizable values of $\Delta\rho \approx 0.5$ $\mu\Omega$-cm and $\Delta\rho/\rho_{avg} \approx 2\%$ at room temperature, forms the basis for most low-field magnetoresistive applications (64).
3. Giant magnetoresistance (GMR) in magnetic multilayers: In 1988, Baibich et al (52) discovered giant negative magnetoresistance in Fe/Cr multilayers with certain Cr layers thicknesses through which the interlayer exchange interaction (J) caused antiferromagnetic alignment of adjacent Fe layers, leading to zero magnetization ($M = 0$) (52, 53). Under a sufficiently large magnetic field, a ferromagnetic alignment ($M = M_s$) was achieved. Remarkably, the resistivitiy between the ferromagnetic and the antiferromagnetic alignments differed by as much as a factor of two. Subsequently, such large MR effects and the intricate oscillations in MR, as caused by the oscillatory J, have been observed in a large variety of multilayers (54–62). Because the MR in Fe/Cr was so large, the MR effect has been widely referred to as giant magnetoresistance (GMR). Since the discoveries made in 1988, GMR and interlayer coupling have captured tremendous worldwide interest from many researchers. Prior to 1992, the attention has been focused exclusively on multilayers, upon which most theoretical models have also been based. In 1992, GMR was uncovered in granular magnetic systems, dispelling the notion that the GMR phenomenon is exclusive to multilayers (44, 65).

The GMR effects observed in multilayers and granular solids can indeed be quite large ($\Delta\rho/\rho \geqslant 50\%$) in some cases, but modest ($\Delta\rho/\rho < 5\%$) in others. But what distinguishes the GMR effect from other MR effects is not just the size of the effect, but its distinctively different characteristics. First of all, the GMR is always negative, i.e. resistance decreases with increasing field, irrespective of the direction of the applied magnetic field with respect to the current, as shown in Figure 9*d*. This is qualitatively different from the positive OMR observed in non-magnetic metals, and the positive and negative AMR observed in magnetic metals. Because of the layer structure, GMR in multilayers is not isotropic (52); specifically $\rho_{\perp} > \rho_{\parallel} \approx \rho_{T}$, as shown in Figure 9*d*. Furthermore, other characteristics of GMR in granular solids, as described below, are also different from those of ordinary metals and alloys.

OMR, AMR, and GMR are not isolated effects. AMR is also present in systems that exhibit GMR. This is of particular concern if the GMR is quite small. In such cases, measurements must be extended to several geometries to separate out the AMR contribution (60). Under a large magnetic field, AMR materials often display a weak H^2 dependence due to OMR (64).

The first observation of GMR in granular materials was independently made and simultaneously published by Berkowitz et al (65) and Xiao et al (44), using granular Co-Cu samples after Childress and Chien reported the fabrication of granular Co-Cu (10, 11). Barnard et al (66) observed unexpectedly large values of magnetoresistance in Co-Ag thin films, which probably contained some features of granular solids. Shortly thereafter, GMR was uncovered in a number of granular systems of Co, Fe, Ni, and their alloys embedded in noble metals (Ag, Cu, and Au) (29, 67–75).

In Figure 10*a*, we show the megnetoresitance at 5 K of $Co_{16}Cu_{84}$ deposited at $T_S = 350°C$ (44). The apparent dependence of GMR on H is not particularly revealing because GMR is a direct consequence of the field dependence of the global magnetization (M). The highest and the lowest resistivities are realized respectively at the original unmagnetized state ($M = 0$) and when all the particles are ferromagnetically aligned ($M = M_s$). These two states correspond to the antiferromagnetic ($M = 0$) and ferromagnetic ($M = M_s$) alignments of multilayers. Shown in Figure 10*b* is the hysteresis loop, which closely correlates with the magnetoresistance curves. The increasing-field and the decreasing-field branches of MR correspond exactly to those of the hysteresis loop. The maxima of MR are located precisely at the coercive fields. The correlation between the MR and global magnetization is clearly shown in Figure 10*c*, which is a replot of Figure 10*a*,*b* using the data at the same H. The relationship reveals that the GMR is an even function of the global magnetization (M) (29). It should be noted in Figure 10*a* and *b* that the initial curve ($a \rightarrow b$) is not reproducible under field cycling, and only the field-cycle parts ($b \rightarrow c \rightarrow d \rightarrow e \rightarrow b$) can be used for the correlation shown in Figure 10*c*.

Although GMR data have often been described in percent ($\Delta\rho/\rho$), it is the absolute resistivity values of ρ and $\Delta\rho$ that contain the intricate physics. It has been found in granular magnetic systems exhibiting GMR that the essential physics is contained in the resistivity of the form of (29, 69)

$$\rho(H, T) = \rho_0(T) + \rho_m(T)[1 - F(M/M_s)], \qquad 11.$$

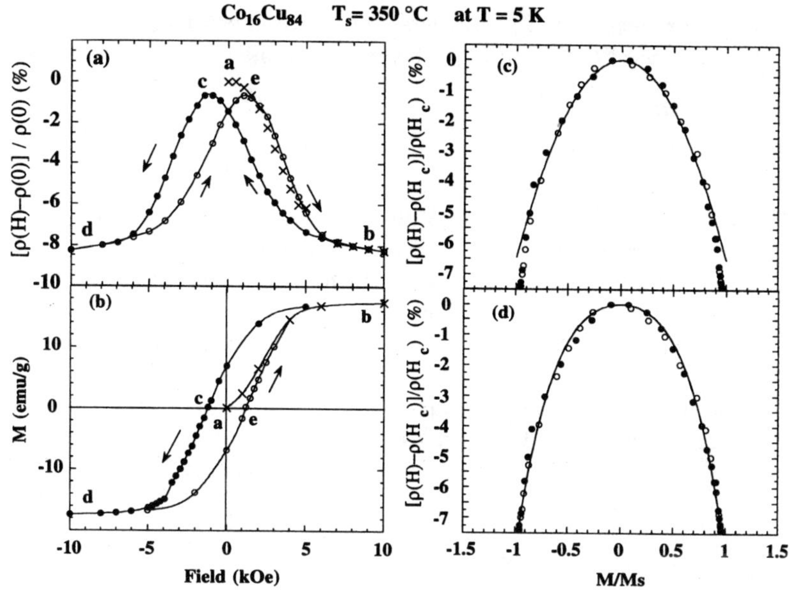

Figure 10 Magnetization (*a*) and hysteresis loop (*b*) of granular $Co_{16}Cu_{84}$ ($T_s = 350°C$) at 5 K. The crosses and closed and open circles denote the initial curve ($a \rightarrow b$), the branch with decreasing field ($b \rightarrow c \rightarrow d$), and the branching with increasing field ($d \rightarrow e \rightarrow b$), respectively. These two plots are combined in (*c*) and (*d*) to show magnetoresistance vs global magnetization. The solid curve in (*c*) is $\Delta\rho/\rho = -0.065\ (M/M_s)^2$, whereas the solid curve in (*d*) is $\Delta\rho/\rho = -0.0824\ [(1-\alpha)\ (M/M_s)^2 + \alpha\ (M/M_s)^4]$ with $\alpha = 0.5$.

where the first term $\rho_0(T)$ is the field-independent part of the resistivity due to the usual contributions from phonon, defect, and impurity scattering. The second term in Equation 11 is responsible for GMR. The value of the magnetic resistivity $[\rho_m(T)]$ in comparison with $\rho_0(T)$ determines the actual effect size of the GMR. The field dependence of $\rho(H, T)$ is through the field dependence of the global magnetization M via the function $F(M/M_s)$, where M_s is the saturation magnetization. The function $F(M/M_s)$ is an even function of M/M_s with the limiting values of $F(M/M_s) \rightarrow 0$ as $M \rightarrow 0$ (when $H = 0$ or H_c, where H_c is the coercive field) and $F(M/M_s) \rightarrow 1$ as $M \rightarrow M_s$ (when $H \geqslant H_s$, where H_s is the saturation field).

The experimentally observed size of the GMR, if expressed in percent and relative to the high resistance state, is just (69)

$$\left(\frac{\Delta\rho}{\rho}\right)_0 = \frac{\rho(0, T) - \rho(H_s, T)}{\rho(0, T)} = \frac{\rho_m(T)}{\rho_0(T) + \rho_m(T)}. \qquad 12.$$

The size so expressed will at most be 100% when $\rho_m(T) \gg \rho_0(T)$. Alternatively, the size of GMR can also be expressed as

$$\left(\frac{\Delta\rho}{\rho}\right)_{H_s} = \frac{\rho(0,T)-\rho(H_s,T)}{\rho(H_s,T)} = \frac{\rho_m(T)}{\rho_0(T)}, \qquad 13.$$

with respect to the low resistance state. The size can now be even greater than 100% when $\rho_m(T) > \rho_0(T)$. Not surprisingly, most results in the literature are expressed in the latter form to maximize the reported size. From Equations 12 and 13, it is clear that the size of the GMR depends on a combination of both $\rho_0(T)$ and $\rho_m(T)$, which must be separately measured.

Experimentally, it has been found that the function $F(M/M_s)$ can be well approximated by $F(M/M_s) \approx (M/M_s)^2$ (44) as shown in Figure 10*c*. In a granular system $(M/M_s)^2 = \langle\cos\theta\rangle^2$, where θ is the angle between the magnetization axis of a particle and the external field, and $\langle\cos\theta\rangle^2$ is averaged over many ferromagnetic particles. The physical significance of $(M/M_s)^2$ is, in the limit of uncorrelated particles $(M/M_s)^2 = \langle\cos\ \theta\rangle^2 = \langle\cos\phi_{ij}\rangle$, where ϕ_{ij} is the angle between the axes of the ferromagnetic particles, which highlights the importance of the relative orientation of the magnetic axes of the ferromagnetic particles (44). This can be readily seen as follows. Consider two ferromagnetic particles of magnetizations $\mathbf{M_i}$ and $\mathbf{M_j}$, then $\mathbf{M_i}\cdot\mathbf{M_j} = M_i\ M_j\cos\phi_{ij}$. Because z is the symmetry axis along which $\mathbf{H}$ is applied, $\langle\cos\phi_{ij}\rangle = \langle\cos\phi_i\cos\phi_j\rangle$. In the limit that the magnetic axes are uncorrelated, $\langle\cos\phi_{ij}\rangle = \langle\cos\phi_j\rangle^2 = (M/M_s)^2$, as experimentally observed. The description of the GMR data can be further improved by using $F(M/M_s) = (1-\alpha)\ (M/M_s)^2 + \alpha\ (M/M_s)^4$, where α is a fitting parameter with a value between 0 and 1, as shown in Figure 10*d*. The dependence of GMR on $(M/M_s)^2$ and other higher power terms accounts for the experimental fact that, under a magnetic field, the global magnetization always saturates faster than the GMR, as shown in Figure 10.

A key feature of GMR in granular systems, distinctively different from that in multilayers and AMR, is that the GMR in granular systems is isotropic (29), whereas both the AMR (64) and the GMR in multilayers (52) are inherently anisotropic. In a granular system that exhibits GMR, $\rho_\perp \approx \rho_\parallel \approx \rho_T$, as schematically shown in Figure 9*e*. Granular magnetic systems are the only magnetic media that show isotropic GMR. When used as a magnetoresistive field-sensing device, the field can be oriented in any direction with respect to the sample, whereas in both AMR alloy films (64) and GMR in multilayers (52), the magnetic field is required to be parallel to the film plane.

The magnetic resistivity (ρ_m), which dictates the size of the GMR as shown in Equations 12 and 13, depends on the particle size and the density of the ferromagnetic particles in a granular solid. The dependence of ρ_m on $2r$ can be revealed in samples with a fixed volume fraction of the magnetic particles, but with different particle sizes. This can be achieved by annealing samples with the same volume fraction to higher annealing temperature (T_A). Experimentally, one finds that $\rho_m \propto 1/r$. Because the interface area/volume in a granular system is $S \approx 3\ x_v/r$, as shown in Equation 4, where x_v is the volume fraction, evidently $\rho_m \propto S$. Thus GMR is due to spin-dependent scattering at the interfaces of the particles (67).

The dependence of ρ_m on x_v is shown in Figure 11, where $\rho(0)$, ρ_m, and

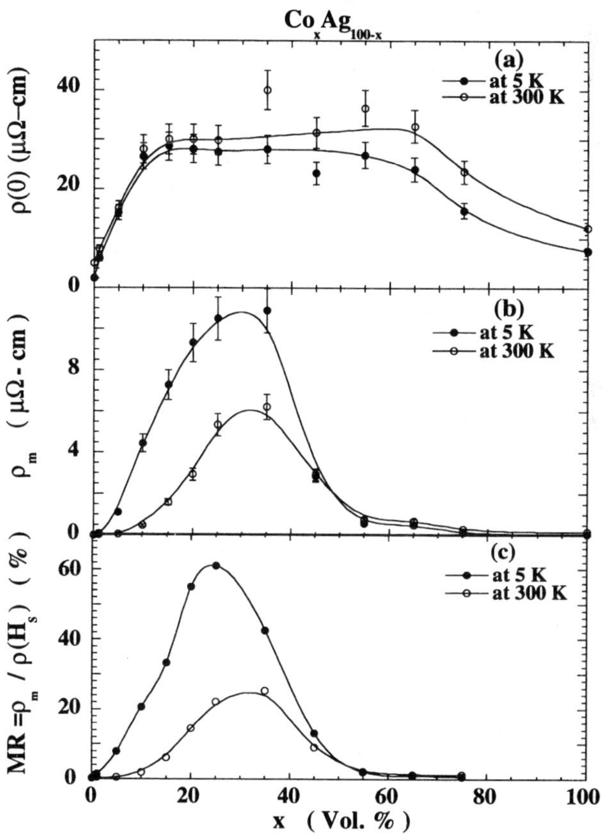

Figure 11 The values of (*a*) resistivity $\rho(0)$, (*b*) magnetic resistivity ρ_m, and (*c*) MR $= \rho_m/\rho(0)$ of Co-Ag as a function of Co content in vol %. The values at 5 and 300 K are denoted by solid and open circles, respectively.

GMR of Co-Ag with $0 \leqslant x_v \leqslant 1$ are shown (32). The value of $\rho(0)$ initially rises when Co is first introduced, reaches a plateau for a broad range of Co contents, and decreases towards the pure Co limit. The value of ρ_m increases from 0, reaching a maximum at about $x_v \approx 25\%$, before decreasing to very small value at $x_v \approx 55\%$, which is just the percolation threshold of a granular solid, beyond which a connecting network of Co is formed and there is no GMR. As shown in Equation 3, at $x_v \approx 25\%$, the spacing between particles is about the size of the particle.

We next compare GMR results of three systems of $Co_{20}Cu_{80}$, $Fe_{30}Ag_{70}$, and $Co_{20}Ag_{80}$ as a function of the annealing temperature (T_A), which, after phase separation, produces particles of larger sizes (76). As shown in Figure 12, in all cases, $\rho(0)$ and ρ_m decrease with increasing T_A. However, because the rates of decrease for $\rho(0)$ and ρ_m are different, the GMR may

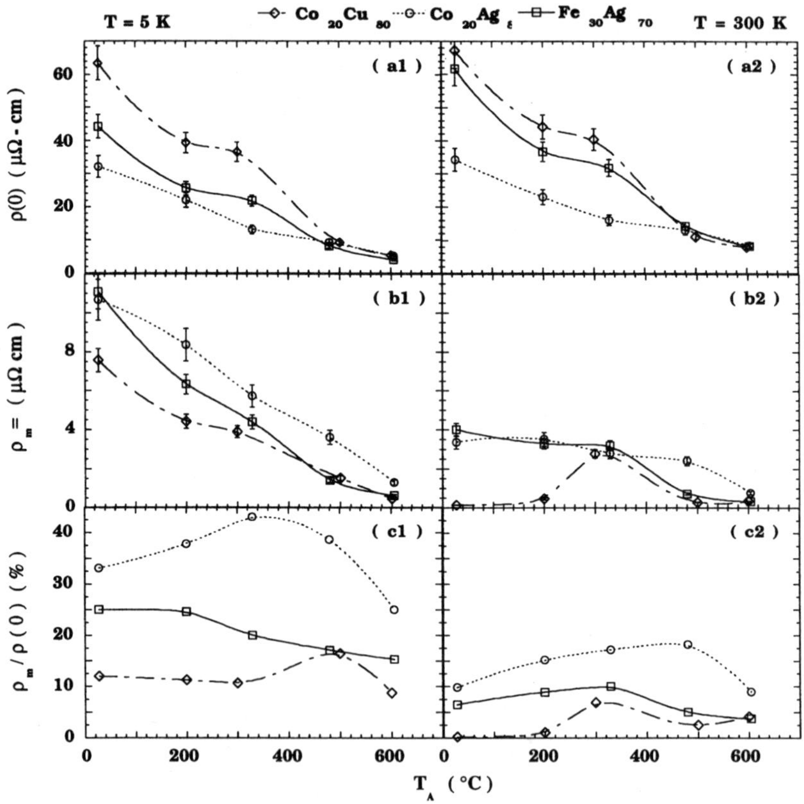

Figure 12 Resistivity [$\rho(0)$], magnetic resistivity (ρ_m) and MR = $\rho_m/\rho(0)$ at 5 and 300 K as a function of annealing temperature (T_A) of $Co_{20}Cu_{80}$, $Fe_{30}Ag_{70}$ and $Co_{20}Ag_{80}$.

be monotonically decreasing (e.g. $Fe_{30}Ag_{70}$ at 5 K), or showing a broad maximum (e.g. $Co_{20}Ag_{80}$ and $Co_{20}Cu_{80}$ at 5 K). It is interesting to note that for large ferromagnetic particles ($T_A = 600°C$), the differences among the three systems become small; in particular, the values $\rho(0)$ and ρ_m are quite similar. As shown in the lowest panel in Figure 12, among the three systems, the GMR effect is the largest in Co-Ag and smallest in Co-Cu. This is because $\rho(0)$ is the smallest and ρ_m the largest in Co-Ag, manifesting the fact that the effect size of GMR is the combination of both $\rho(0)$ and ρ_m.

Although the first observations of GMR in granular solids were made in Co-Cu alloys after phase separation of the metastable alloys by either post-annealing or deposition at elevated temperatures, these granular materials can also be achieved by other techniques. For example, Co-Cu, Fe-Ag, and Co-Ag alloys made by molecular beam epitaxy (MBE) at elevated substrate temperatures (74, 77, 78), and Co-Cu alloys made by melt-spinning technique (75), also have exhibited GMR. In the MBE-made, epitaxially grown granular films, small-angle x-ray diffraction measurements (77, 78) also implied interfacial spin-dependent scattering as the GMR mechanism, in agreement with the conclusion made earlier (67). It should be mentioned that some intended multilayers (e.g. Co/Ag, Co/Cu) may have granular-like entities at the interfaces, resulting in unexpectedly large MR effects. Hylton et al have in fact deliberately disrupted multilayers of permalloy/Ag to capture GMR with a small saturation field (79).

The main technological implications of GMR lie in magnetic field sensing. Of these, the most important are magnetoresistive heads for magnetic recording applications, motion and position sensors, and devices that currently use Hall-effect sensors. Many GMR materials already have sufficiently high sensitivity to displace Hall-effect devices. Magnetoresistive head application requires very small saturation fields, of the order of 10 Oe. The saturation fields for GMR materials have steadily been reduced to values of about 30 Oe (80, 81). The rapid advancement suggests that some or all of the applications mentioned may mature in the near future.

HALL RESISTIVITY (ρ_{xy})

The diagonal element of the magnetoresistivity tensor is the magnetoresistivity (ρ_{xx}), and the off-diagonal element is the Hall resistivity (ρ_{xy}). The Hall resistivitiy of magnetic metals can be expressed as (82)

$$\rho_{xy} = R_0 H + \rho_{xy}^M, \qquad 14.$$

where the first term is the ordinary Hall resistivity, which is linear in H, and the Hall constant $R_0 = 1/nec$ measures the carrier density (n). The second term in Equation 14 is the extraordinary Hall resistivity for magnetic metals

$$\rho_{xy}^{M} = R_s 4\pi M, \qquad 15.$$

which can be readily extracted by removing the linear $R_0 H$ part from ρ_{xy}.

The results of magnetoresistivity (ρ_{xx}), Hall resistivity (ρ_{xy}), extraordinary Hall resistivity (ρ_{xy}^{M}), and global magnetization (M) of granular $Co_{20}Ag_{80}$ ($T_A = 200°C$) are shown in Figure 13. An essential point concerning the extraordinary Hall resistivitiy of a granular solid, for that matter any magnetic material, is its field dependence. As is clearly illustrated in Figure 13, because the reproducible field-cycled results of ρ_{xy}^{M} and M have the same field dependence (i.e. the two hysteresis loops are essen-

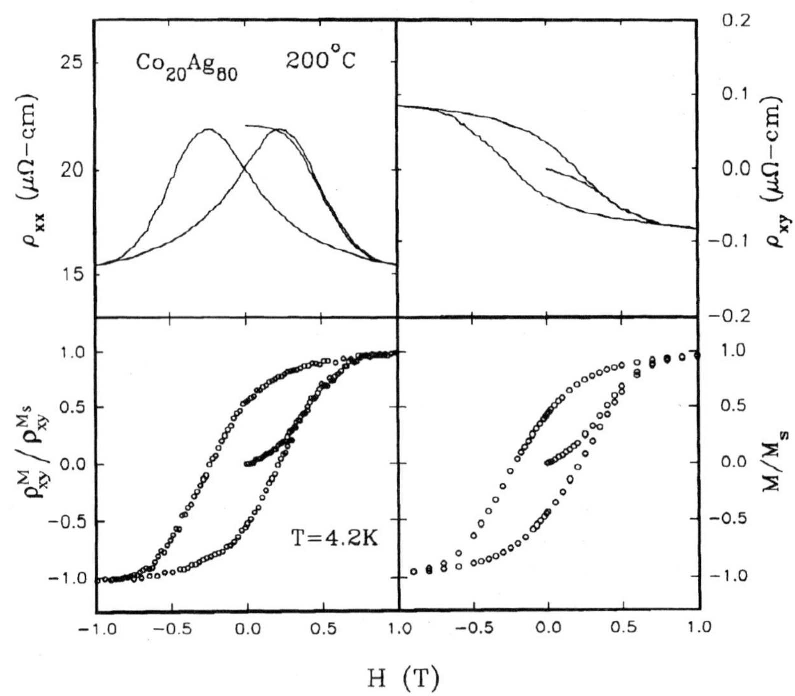

Figure 13 Magnetoresistivity (ρ_{xx}), Hall resistivity (ρ_{xy}), extraordinary Hall resistivity (ρ_{xy}^{M}) and global magnetization (M) of granular $Co_{20}Ag_{80}$ annealed at $T_A = 200°C$. In each case, the initial curve is not reproducible upon field cycling. The results of ρ_{xy}^{M} and M are plotted in reduced units to illustrate that the two hysteresis loops have the same shape. Hence the value of R_s in $\rho_{xy}^{M} = 4\pi R_s M$ is independent of field.

tially identical), ρ_{xy}^{M} and M are proportional to each other, therefore R_s is independent of field (67). It is important to emphasize again that because the initial curves are not reproducible, they cannot be used reliably to determine the correlation among various transport properties, and in particular, to determine the field dependence of R_s, or lack thereof.

In ordinary magnetic metals and alloys, the anisotropic MR is small, both R_s and ρ_{xx} are nearly independent of field. In these cases, the extraordinary Hall constant has been found to be related to the magnetoresistivity,

$$R_s = a\rho_{xx} \quad \text{(skew scattering)}, \qquad 16.$$

and

$$R_s = b(\rho_{xx})^2 \quad \text{(side-jump)}, \qquad 17.$$

according to the different scattering processes (82). There are attempts in the literature to use functions such as $R_s = a\rho_{xx} + b(\rho_{xx})^2$ to describe the extraordinary Hall resistivity of traditional magnetic alloys, multilayers, and granular solids (83), taking advantage of the additional freedom in fitting. However, because a and b are allowed to be of either sign and of any magnitude, the validity of such practices remains dubious. In granular solids, $\rho_{xx}(H)$ has a strong field dependence and R_s is essentially field-independent, thus R_s cannot be any actual function of $\rho_{xx}(H)$. In materials that exhibit GMR, as long as R_s is found to be independent of field, any exercises describing R_s as a function of $\rho_{xx}(H)$ would be futile. However, it has been found in granular Co-Ag that R_s scales with $(\rho_{xx})^{3.7}$ (67), using the values of either $\rho_{xx}(0)$ or $\rho_{xx}(8T)$. Such an unusual scaling is not well understood at present. Recently, very large finite-size effects in ρ_{xy}^{M} have also been observed in granular solids (84). These unusual characteristics of the extraordinary Hall resistivity are some of the outstanding issues in the magneto-transport properties of granular solids.

MAGNETO-THERMAL CONDUCTIVITY (κ)

Virtually all transport studies on multilayer and granular systems have been focused exclusively on electronic transport, one of several transport phenomena, all of which are essential for a complete understanding of the intriguing transport properties in these magnetically inhomogeneous systems. To date, very few studies have been made on thermal conductivity and thermoelectric power. The lack of thermal transport studies of thin films stems in part from experimental difficulties. Although resistivity measurements on thin-film samples can utilize any insulating substrate, such samples are generally unsuitable for thermal transport measurements. Some of the major difficulties lie in the uncertainty in account for heat transfer through substrates much thicker than the films and the requirement of large specimens for thermometry.

Piraux et al reported the first thermal transport measurements on granular solids using large (cms), thick (about 100 μm), and substrate-free samples (85). The temperature dependence of the thermal conductivity (κ) of granular $Co_{20}Ag_{80}$ ($T_A = 605°C$) from 2 K to 300 K has been measured. The well known Wiedemann-Franz (WZ) law states that

$$\frac{\kappa\rho}{T} = L_0 = 2.45 \times 10^{-8}\ V^2/K^2, \qquad 18.$$

where κ is the thermal conductivity, ρ is the electrical resistivity, T the absolute temperature, and L_0 the Lorenz number. For ordinary metals, thermal conduction is purely electronic, and the WZ law is valid at low temperatures where the scattering is static. The WZ is also obeyed at high temperatures (above the Debye temperature), where the phonon energy becomes small compared with k_BT, and the dominating large-angle electron-phonon scattering events are quasi-static. In the intermediate temperature range (e.g. 40–250 K), the WZ law generally breaks down because of the importance of the small-angle electron-phonon scattering and inelastic scattering, and consequently, the values of $\kappa\rho/T$ are much smaller than L_0. These features are shown in Figure 14 by the solid curve. Most surprisingly, in granular Co-Ag, the WZ law is found to hold for the entire temperature range of 2–300 K. These results demonstrate the dominance of electronic thermal conductivity and large-angle scattering events throughout the temperature below the Debye temperature (85, 86).

Very large magneto-thermal conductivity has also been observed (85). For example, at 72 K, upon the application of a magnetic field, the thermal conductivity increases by 28%. The field dependence of the magneto-thermal conductivity is the same as that of the magnetoresistance. While the giant magnetoresistance is negative, the giant magneto-thermal conductivity is positive. Thus the magneto-thermal transport properties also satisfy the WZ law. These unusual thermal transport properties have not been observed in traditional metals and alloys.

CONCLUDING REMARKS

The unique features of granular solids are a collection of ultrafine metal particles with a narrow size distribution uniformly dispersed in an immiscible medium, which may be insulating or conducting, amorphous or

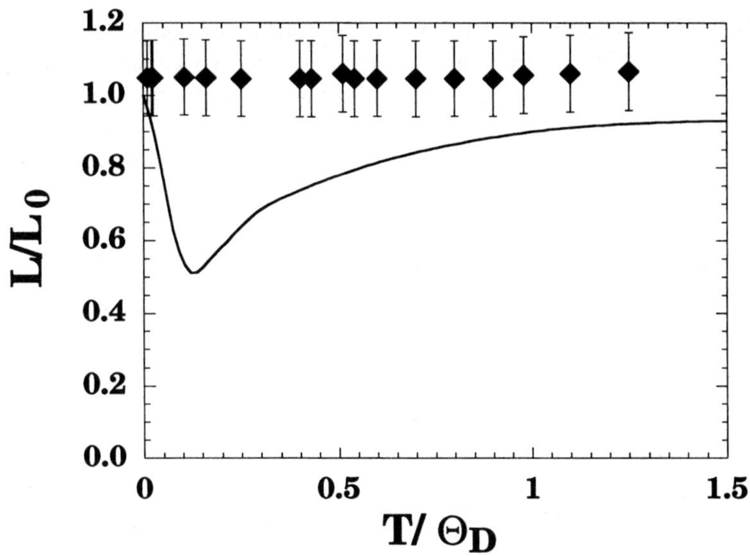

Figure 14 The values of L/L_0 of $Co_{20}Ag_{80}$, with $L = (\kappa\rho/T)$ and L_0 is the Lorenz number, as a function of temperature expressed in terms of the Debye temperature (Θ_D).

crystalline. By varying x_v and $2r$, one can greatly manipulate their physical properties, in general, and their magnetic and magneto-transport properties in particular. The enhanced magnetic and giant magneto-transport properties captured in granular solids cannot be realized in a bulk homogeneous solid. For granular magnetic solids with $x_v < x_p$, the single-domain magnetic properties and the associated superparamagnetism are amply displayed.

Granular magnetic solids also exhibit giant magneto-transport properties. The common feature of magnetic multilayers and magnetic granular solids that exhibit giant magneto-transport properties is that they are magnetically inhomogeneous with well-defined magnetic entities on the nanometer size. The importance of spin-dependent scattering events in giant magneto-transport properties appears to be well accepted (52, 87, 88). In order to quantitatively account for the rich and unconventional magneto-transport properties unveiled, theoretical endeavors should explicitly include these nanometer-size magnetic entities in the calculations. Considerable success has already been realized in magnetic multilayers (88). In the case of granular solids, the randomly dispersed ultrafine magnetic particles present special theoretical challenges. The most impressive advancement to date has been achieved by Maekawa and co-workers (89), who made calculations with the explicit inclusion of magnetic particles. In the case of GMR, their theoretical results have already captured the MR dependence on $\langle \cos\theta \rangle^2$ and $1/r$, two key experimental results of GMR. Other magneto-thermal transport properties, particularly extraordinary Hall resistivity, magneto-thermal conductivity, and the Wiedemann-Franz law, await further theoretical and experimental exploration. One cannot emphasize enough the importance of measuring transport properties in absolute values. Otherwise, much of the discussion in the physics of the magneto-transport properties could not have proceeded.

Granular solids are indeed richly endowed. Even though extensive research began more than two decades ago, interesting new effects continue to be uncovered. As vividly made clear by the progress accumulated over these years, the exploration and the understanding of their unique properties require multi-disciplinary efforts, with theoretical and experimental expertise from fabrication, processing, characterization, measurements, and interpretations of various properties, as well as technological applications. If past experience is any indication, the fruitful study of granular solids may well extend into the next century.

ACKNOWLEDGEMENTS

This work is supported by Office of Naval Research Grant No. N00014-91-J-1633 and National Science Foundation Grant No. ECS92-02222. The author is indebted to colleagues who have contributed to the results mentioned in this paper: they include JQ Xiao, JS Jiang, JR Childress, A Gavrin, SH Liou, KM Unruh, A Tsoukatos, GC Hadjipanayis, G Xiao, P Xiong, JQ Wang, L Piraux and M Cassart.

Literature Cited

1. Falicov LM, Pierce DT, Bader SD, Gronsky R, Hathaway KB, et al. 1990. *J. Mater. Res.* 5: 1299–40
2. Dhez P, Weisbuch C. 1988. *Physics, Fabrication and Applications of Multilayered Structures*, New York: Plenum
3. Whitney TM, Jiang JS, Searson PC, Chien CL. 1993. *Science* 261: 1316–19
4. Abeles B. 1976. In *Applied Solid State Science: Advances in Materials and Device Research*, ed. R Wolfe, pp. 1–117. New York: Academic
5. Abeles B, Sheng P, Couts MD, Arie Y. 1975. *Adv. Phys.* 24: 407–61
6. Karch J, Birringer R, Gleiter H. 1987. *Nature* 330: 556–58
7. Siegel RW, Ramasamy S, Hahn H, Zongquan L, Ting L, Gronsky R. 1988. *J. Mater. Res.* 3: 1367–72
8. Hadjipanayis GC, Prinz GA. 1991. *Science and Technology of Nanostructured Magnetic Materials*, New York: Plenum
9. Kitada M. 1985. *J. Mater. Sci.* 20: 269–73
10. Childress JR, Chien CL, Nathan M. 1990. *Appl. Phys. Lett.* 56: 95–97
11. Childress JR, Chien CL. 1991. *J. Appl. Phys.* 70: 5885–87

12. Liou SH, Malhotra S, Shan ZS, Sellmyer DJ, Nafis S, et al. 1991. *J. Appl. Phys.* 70: 5882–84
13. Chien CL. 1991. *J. Appl. Phys.* 69: 5267–72
14. Sheng P. 1988. *Phys. Rev. Lett.* 45: 60–63
15. Matijevic E. 1989. *Mater. Res. Soc. Bull.* XIV: 18–20
16. Luborsky FE. 1961. *J. Appl. Phys.* 32: 171S–83S
17. Yiping J, Hadjipanayis GC, Sorenson CM, Klabunde KJ. 1989. *J. Magn. Magn. Mater.* 79: 321–26
18. Berkowitz AB, Walter JL, Wall KF. 1982. *Phys. Rev. Lett.* 49: 1484–87
19. Ambrose T, Gavrin A, Chien CL. 1992. *J. Magn. Magn. Mater.* 116: L311–14
20. Tasaki A, Tomiyana S, Iida S. 1965. *Jpn. J. Appl. Phys.* 4: 707–11
21. Shull RD, Ritter JJ. 1988. In *Physical Phenomena in Granular Materials*, ed. GD Cody, TH Geballe, P Sheng, pp. 435–40. Amsterdam: North Holland
22. Gavrin A, Chien CL. 1990. *J. Appl. Phys.* 67: 938–42
23. Liou SH, Chien CL. 1988. *J. Appl. Phys.* 63: 4240–42
24. Logothetis EM, Kaiser WJ, Plummer HK, Shinozak SS. 1986. *J. Appl. Phys.* 60: 2548–52
25. Tsoukatos A, Wan H, Hadjipanayis GC, Papaefthymiou V, Kostikas A, Simopoulos A. 1993. *J. Appl. Phys.* 73: 6967–69
26. Vossen JL, Kern W. 1991. *Thin Film Processes*, Vols. 1, 2. New York: Academic
27. Childress JR, Chien CL. 1991. *Phys. Rev. B* 43: 8089–93
28. Chien CL, Unruh KM. 1983. *Phys. Rev. B* 28: 1214–18
29. Xiao JQ, Jiang JS, Chien CL. 1992. *Phys. Rev. B* 46: 9266–69
30. Deleted in proof
31. Deleted in proof
32. Xiao JQ, Jiang JS, Chien CL. 1993. *Mater. Res. Soc. Proc. Nanophase and Nanocomposite Materials, No.* 286, ed. S Komarneni, JC Parker, GJ Thomas, pp. 197–207. Pittsburgh: Mater. Res. Soc.
33. Morrish AH. 1965. *Physical Principles of Magnetism*, New York: Wiley
34. Jacobs IS, Bean CP. 1963. In *Magnetism III*, ed. GT Rado, H Suhl, pp. 271–350. New York: Academic
35. Gittleman JL, Goldstein Y, Bozowski S. 1972. *Phys. Rev. B* 5: 3609–20
36. Xiao G, Liou SH, Levy A, Taylor JN, Chien CL. 1986. *Phys. Rev. B* 34: 7573–77
37. Kittel C, Galt JK. 1946. *Phys. Rev. B* 70: 965–71
38. Stoner EC, Wohlfarth EP. 1948. *Philos. Trans. R. Soc. London Ser. A* 240: 599–642
39. Kittel C. 1949. *Rev. Mod. Phys.* 21: 541–83
40. Xiao G, Chien CL. 1987. *Appl. Phys. Lett.* 51: 1280–82
41. Liou SH, Chien CL. 1988. *Appl. Phys. Lett.* 52: 512–14
42. Paparazzo E, Dormann JL, Fiorani D. 1983. *Phys. Rev. B* 28: 1154–57
43. Shah SI, Unruh KM. 1991. *Appl. Phys. Lett.* 59: 3485–87
44. Xiao JQ, Jiang JS, Chien CL. 1992. *Phys. Rev. Lett.* 68: 3749–52
45. Kneller EF, Luborsky FE. 1963. *J. Appl. Phys.* 34: 656–58
46. Dormann JL, Fiorani D, Tholence JL, Sell C. 1983. *J. Magn. Magn. Mater.* 35: 117–20
47. Morop S, Topsoe T, Clausen BC. 1982. *Phys. Scr.* 25: 713–19
48. Childress JR, Chien CL, Rhyne JJ, Erwin RW. 1992. *J. Magn. Magn. Mater.* 104–107: 1585–86
49. Xiao G, Chien CL. 1987. *J. Appl. Phys.* 61: 3308–10
50. Chantrell RW, Wohlfarth EP. 1983. *J. Magn. Magn. Mater.* 40: 1–11
51. Bennett LH, McMichael RD, Swartzendruber LJ, Shull RD, Watson RE. 1992. *J. Magn. Magn. Mater.* 104–105: 1094–97
52. Baibich MN, Broto JM, Fert A, Nguyen van Dau F, Petroff F, et al. 1988. *Phys. Rev. Lett.* 61: 2472–75
53. Binasch G, Grünberg P, Saurenbach P, Zinn W. 1989. *Phys. Rev. B* 39: 4828–30
54. Parkin SSP, Bhadra R, Roche KP. 1991. *Phys. Rev. Lett.* 66: 2152–55
55. Pratt WP Jr, Lee SF, Slaughter JM, Loloee R, Schroeder PA, Bass J. 1991. *Phys. Rev. Lett.* 66: 3060–63
56. Chaiken A, Prinz GA, Krebs JJ. 1990. *J. Appl. Phys.* 87: 4892–94
57. Dieny B, Speriosu VS, Parkin SSP, Gurney BA, Wilhoit DR, Mauri D. 1991. *Phys. Rev. B* 43: 1297–1300
58. Johnson MT, Purcell ST, McGee MWE, Coehoorn R, van de Stegge J, Hoving W. 1992. *Phys. Rev. Lett.* 68: 2688–91
59. Shinjo T, Yamamoto H. 1990. *J. Phys. Soc.* 59: 3061–64
60. Chen Y, Florczak JM, Dahlberg ED. 1992. *J. Magn. Magn. Mater.* 104–105: 1907–8
61. Fullerton EE, Kelly DM, Guimpel J, Schuller IK, Bruynseraede Y. 1992. *Phys. Rev. Lett.* 68: 859–62
62. Tsui F, Chen B, Barlett D, Clarke R, Uher C. 1994. *Phys. Rev. Lett.* 72: 740–43
63. Rossiter PL. 1987. *The Electrical Resis-*

tivity of Metals and Alloys, Cambridge: Cambridge Univ. Press

64. McGuire TR, Potter RI. 1975. *IEEE Trans. Magn.* MAG-11: 1018–38
65. Berkowitz A, Young AP, Mitchell JR, Zhang S, Carey MJ, et al. 1992. *Phys. Rev. Lett.* 68: 3745–48
66. Barnard JA, Watnis A, Tan M, Haftek E, Parker MR, Watson ML. 1992. *J. Magn. Magn. Mater.* 114: L230–34
67. Xiong P, Xiao G, Wang JQ, Xiao JQ, Jiang JS, Chien CL. 1992. *Phys. Rev. Lett.* 69: 3220–23
68. Jiang JS, Xiao JQ, Chien CL. 1992. *Appl. Phys. Lett.* 61: 2362–64
69. Chien CL, Xiao JQ, Jiang JS. 1993. *J. Appl. Phys.* 73: 5309–14
70. Carey MJ, Young AP, Starr A, Rao D, Berkowitz AE. 1992. *Appl. Phys. Lett.* 61: 2935–37
71. Tsoukatos A, Wan H, Hadjipanayis GC. 1992. *Appl. Phys. Lett.* 61: 3059–61
72. Xiao G, Wang JQ, Xiong P. 1993. *Appl. Phys. Lett.* 62: 42–44
73. Wang JQ, Xiong P, Xiao G. 1993. *Phys. Rev.* B47: 8341–44
74. Parkin SSP, Farrow RFC, Rabedeau TA, Marks RF, Harp GR, et al. 1993. *Europhys. Lett.* 22: 455–62
75. Dieny B, Chamberod A, Genin JB, Rodmacq B, Teixeira SR, et al. 1993. *J. Magn. Magn. Mater.* 126: 433–36
76. Xiao JQ, Jiang JS, Chien CL. 1993. *IEEE Trans. Magn.* MAG-29: 2688–93
77. Rabedeau TA, Toney MF, Marks RF, Parkin SSP, Farrow RFC, Harp GP. 1993. *Phys. Rev. B* 48: 16810–13
78. Thangaraj N, Echer C, Krishnan KM, Farrow RFC, Marks RF, Parkin SSP. 1994. *J. Appl. Phys.* 75: 6900–2
79. Hylton TL, Coffey KR, Parker MA, Howard JK. 1993. *Science* 261: 1021–24
80. Parkin SSP, Farrow RFC, Marks RF, Cebollada, Harp GR, Savoy RJ. 1994. *Phys. Rev. Lett.* 72: 3718–22
81. Hossain S, Seale D, Qiu G, Barnard JA, Fujiwara H, Parker MR. 1994. *J. Appl. Phys.* 75: 7067–69
82. Berger L, Bergmann G. 1979. In *The Hall Effect and Its Applications*, ed. CL Chien, CR Westgate, pp. 55–76. New York: Plenum
83. Sato H, Kumano T, Aoki Y, Kaneko T, Yamamoto R. 1993. *J. Phys. Soc. Jpn.* 62: 416
84. Wang JQ, Xiao G. 1995. *Phys. Rev. B.* 51:
85. Piraux L, Cassart M, Jiang JS, Xiao JQ, Chien CL. 1993. *Phys. Rev. B* 48: 638–41
86. Piraux L, Cassart M, Bayot V, Jiang JS, Xiao JQ, Chien CL. 1993. *IEEE Trans. Magn.* MAG-29: 2700–4
87. Camley RE, Barnas J. 1989. *Phys. Rev. Lett.* 63: 664–67
88. Valet T, Fert A. 1993. *J. Magn. Magn. Mater.* 121: 378–82
89. Asano Y, Oguri A, Inoue J, Maekawa S. 1994. *Phys. Rev. B* 49: 12,831–34

Annu. Rev. Mater. Sci. 1995. 25: 161–92

SYNTHESIS OF POROUS SILICATES

M. M. Helmkamp and M. E. Davis

Chemical Engineering, California Institute of Technology, Pasadena, California 91125

KEY WORDS: zeolite, structure-directing agent, thermodynamics, kinetics, ordered mesoporous materials

---

ABSTRACT

The issues of importance and future concern in the synthesis of porous silicates and porous materials that contain a large fraction of silica, e.g. zeolites and other crystalline molecular sieves, are reviewed. The thermodynamics of zeolite synthesis are discussed, including a detailed thermodynamic analysis of the synthesis of pure-silica ZSM-5. The kinetics of porous silicate synthesis are reviewed, with particular emphasis on the control of porous structure formation through the use of organic structure-directing agents. Ordered mesoporous materials are discussed in the context of distinguishing their features from zeolites in order to describe further the unique properties of each class of material. Finally, several unresolved issues in the understanding of the synthesis process are outlined, the resolutions of which would aid in the synthesis of porous silicates by design.

INTRODUCTION

The technological importance of silicon dioxide is enormous. Its widespread use in vastly different applications, e.g. electronic materials, catalysts and adsorbents, ranks it as one of the most critical materials for advanced societies. In this paper we discuss one form of silicon dioxide, namely, porous silicon dioxide. We describe the issues of importance and future concern in the synthesis of porous silicas and porous materials that contain a large fraction of silica, e.g. zeolites and other crystalline

0084–6600/95/0801–0161$05.00

molecular sieves. We concentrate on the synthesis of crystalline materials; however, other porous silicates are also discussed, e.g. ordered mesoporous silica, when it is necessary to distinguish their features from zeolites in order to describe further the unique properties of each class of material.

Zeolites are crystalline aluminosilicates that were first discovered in 1756 by the mineralogist AF Cronstedt. He reported a new class of silicate minerals that when heated in a blowpipe flame behaved in a unique manner, i.e. the solid emitted large amounts of water vapor (1, 2). Cronstedt designated these minerals as zeolites—a combination of two Greek words meaning boiling stones. Zeolites are tectoaluminosilicates in that they are composed of TO_4 tetrahedra (T = tetrahedral atom, i.e. Si, Al); each apical oxygen atom is shared between adjacent tetrahedra. Thus the framework ratio of O/T equals 2 for all zeolites (3).

Due to its valency, a silicon atom prefers to make bonds with four nearby oxygen atoms in a tetrahedral geometry. In solution, a SiO_4 group would have a formal charge of -4 because the silicon cation carries a formal charge of $+4$ and each oxygen anion has a formal charge of -2. In a solid, because each oxygen atom forms a bridge between two T atoms and shares electron density with each, the SiO_4 unit is neutral (see Figure 1). Therefore, a pure SiO_2 framework without any defects will be a neutral solid (3). If an aluminum atom is tetrahedrally coordinated to oxygen atoms in a similar manner, the $+3$ valency of aluminum will result in a net formal charge of -1. When silicon and aluminum are both present in a tetrahedral framework, a negative charge is affiliated with every aluminum atom; a cation (M^+ in Figure 1) balances the negative charge. Alkali metal ions, e.g. K^+, Na^+; alkaline earth ions, e.g. Ba^{2+}, Ca^{2+}; and protons, H^+,

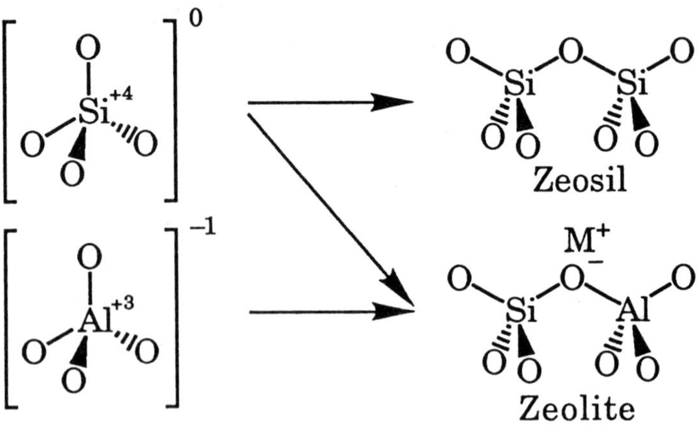

Figure 1 TO_2 units in zeosils and zeolites. (Adapted from 3.)

are typical balancing cations. All tectoaluminosilicates that are synthesized via hydrothermal methods (described below) have ratios of Si/Al $\geqslant 1$. This effect was first explained by Löewenstein (4), who assumed that isolated negative charges ($^{-}$Al-O-Si-O-Al^{-}-O-Si) are more stable than adjacent ones ($^{-}$Al-O-Al^{-}-O-Si-O-Si). This principle is now referred to as Löewenstein's rule.

The modern age of zeolite science and technology really began in the 1950s when the large scale syntheses of several zeolites were accomplished by workers at Union Carbide (5, 6). Since then, the use of zeolites has burgeoned both in size, e.g. over 300,000 metric tons per year for use in powdered laundry detergents as environmentally benign replacements for polyphosphates, and diversity, e.g. catalysts, adsorbents, and MRI contrast agents. As stated by Breck, "Rarely in our technological society does the discovery of a new class of inorganic materials result in such a wide scientific interest and kaleidoscopic development of applications as has happened with the zeolite molecular sieves" (7). Informative, short reviews on zeolites include those of Newsam (8, 9), Smith (10), Davis (3, 11), Gies (12), Davis & Lobo (13), Lobo et al (14), Burkett & Davis (15), and Higgins (16)

Strictly speaking, zeolites are aluminosilicates. In 1978, Flanigen et al (17) announced the synthesis and characterization of a pure-silica polymorph of a zeolite called ZSM-5. [In this paper we discuss numerous zeolites. If the reader is unfamiliar with a particular zeolite, details of the structure can be found elsewhere (18).] Is a pure-silica molecular sieve a zeolite? The answer is really no; however, the term zeolite is used loosely in the literature. Liebau and co-workers have proposed general classifications for materials that are four-connected tetrahedral frameworks (see Table 1) (19, 20). Materials with framework densities (FD), i.e. the number of tetrahedra per nm^3, of less than 21 are arbitrarily denoted as porous (18). Porous frameworks can be classified depending on their pore sizes and

Table 1 Classification of tectosilicates[a]

Porosils, SiO_2		Porolites, $Si_{0.5+x}Al_{0.5-x}O_2$ ($0 \leqslant x \leqslant 0.5$)	
Clathrasils	Zeosils	Clathralites	Zeolites
Silica sodalite	Silica ZSM-5	Sodalite	Faujasite
Silica ZSM-39	Silica ZSM-12	Melanophlogite	Mordenite
DD3R	SSZ-24		ZSM-5
			Zeolite A

[a] Adapted from Reference 20

framework compositions. Materials with frameworks that allow access to the intracrystalline void spaces through rings consisting of 6 T atoms or 6 oxygen atoms (6-membered rings, or 6 MR) or smaller and that tend to trap ions or organic molecules within the intracrystalline space are denoted as clathrasils (SiO_2) or clathralites ($Si_{0.5+x}Al_{0.5-x}O_2$, $0 \leqslant x \leqslant 0.5$). When the access to the crystal interior is by larger rings that permit ion exchange and/or diffusion of organic molecules, these materials are classified as zeosils (SiO_2) or zeolites ($Si_{0.5+x}Al_{0.5-x}O_2$, $0 \leqslant x \leqslant 0.5$). Additionally, frameworks of composition $Si_yM_zO_2$, $y+z = 1.0$ can be formed, e.g. GeO_2, $Si_{0.5}Ge_{0.5}O_2$, $Si_{0.9}Zn_{0.1}O_2$, $Si_{0.95}B_{0.05}O_2$. For these cases, the best way to denote these framework oxides is to call them crystalline molecular sieves.

In this discussion, we are primarily concerned with clathrasils, zeosils, zeolites, and molecular sieves of the form $Si_yM_zO_2$, where $y \gg z$. Pure-silica and high-Si/M materials are somewhat hydrophobic, and this property is important to the discussion presented below. We describe the state-of-the-art in the preparation of porous high-silica materials and outline several unresolved issues that would aid in the synthesis of these classes of materials by design.

PREPARATION OF POROUS SILICATES

Porous silicates are normally prepared by hydrothermal synthesis techniques. A typical crystallization procedure involves the use of water, a silica source, a mineralizing agent, and a structure-directing agent. The sources of silica are numerous and include colloidal silica, e.g. Ludox™ fumed silica, precipitated silica, and even silicon alkoxides such as $Si(OCH_3)_4$ and $Si(OCH_2CH_3)_4$. The typical mineralizing agent is OH^- but F^- can also give good results. (In addition to dissolving the silica, OH^- and F^- can expand the coordination sphere of silicon and produce 5 or 6 coordinate intermediates in solution). The structure-directing agent is a soluble organic species, e.g. a quaternary ammonium ion, that assists in the formation of the silica framework and ultimately resides within the intracrystalline voids (see below). Alkali metal ions such as Na^+ and K^+ can also play a role in the crystallization process. The syntheses can be sensitive to the reagent type, the order of addition, the mixing (or lack of mixing), the crystallization temperature and time, and of course, the composition. This is not surprising since numerous complex reactions, equilibria, and organic-inorganic interactions are occurring during the synthesis process. Additionally, if the synthesis contains a heteroatom (M), then the type and concentration can have an important role in the product formed. To set the stage for further discussions, a typical synthesis of a zeosil, e.g. pure-silica ZSM-5, would be as follows: Tetrapropylammonium (TPA^+) hydroxide in water is combined with fumed silica at a composition of

$$1 \text{ mol } TPA^+OH^- : 10 \text{ mol } SiO_2 : 380 \text{ mol } H_2O$$

and the mixture (pH ~ 12) is heated at 423 K in a closed reaction vessel for 3 days. Upon quenching the reaction, the pure-silica ZSM-5 crystals containing occluded TPA^+ are recovered by filtration from the aqueous solution. The crystals are washed with water and allowed to dry. Below, we discuss the processes that occur during this type of synthesis and the progress that is being made toward rationally designing the crystallization process in order to create new materials or known materials with framework atoms in particular crystallographic positions specified a priori (21).

THERMODYNAMICS

Energetic Stability of Silicates

Because various silicas can be used as reagents for the synthesis of porous, crystalline silicates, what is the driving force for this transformation? Additionally, why do these syntheses not yield quartz, the most stable form of SiO_2? (According to the Ostwald ripening effect, given enough time, metastable phases will be transformed into more stable forms, until eventually the most stable phase will result.) Petrovic et al (22) recently measured the enthalpies of formation of porous, crystalline silicates and compared the values to those of quartz and amorphous silicas; the results begin to partially answer the aforementioned questions. Figure 2 shows the enthalpy of formation of silicates relative to quartz as a function of molar volume (cm^3/mol SiO_2). Note that all of the porous silicates are energetically metastable with respect to quartz and silica glass by at most 14 and 6 kJ/mol, respectively. It is known that amorphous silicas derived from gels are energetically less stable than silica glass by 0–10 kJ/mol (23, 24). Therefore, zeosils and amorphous silicas have very similar energies. At 373 K (close to the lower limit for synthesis), the thermal energy (RT) is 3.1 kJ/mol, so the transformation of amorphous silica into zeosils is not energetically hindered (25).

Navrotsky et al (25) measured the enthalpies of formation for ordered mesoporous and zeolitic silicas as a function of pore diameter (see Figure 3). [It should be noted that while zeolites are microporous crystalline materials, mesoporous silicas are amorphous (see below) (26).] Although the energy changes significantly within the small pore region below 1.0 nm, above a pore size of 1.0 nm, the enthalpy relative to quartz reaches a plateau of approximately 15 kJ/mol. These data illustrate a transition from a domain in which the pores or cages appear to play an integral role

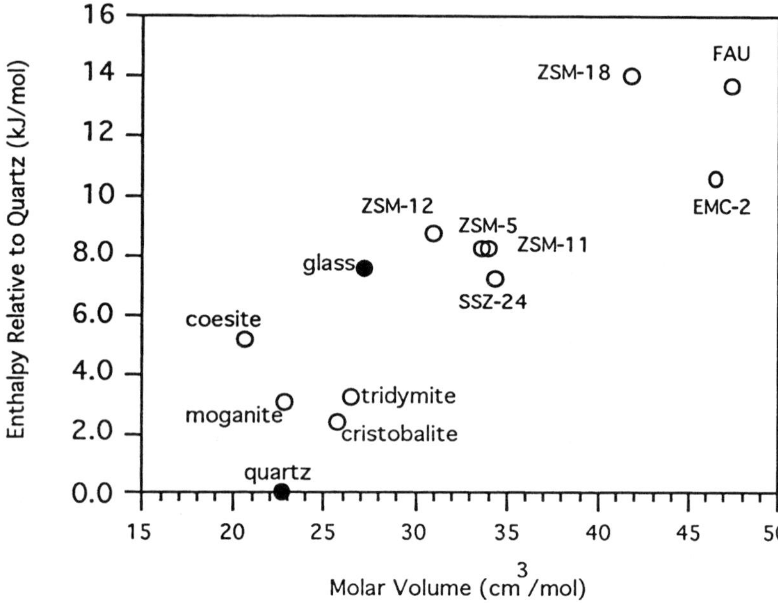

Figure 2 Enthalpy of formation of silicates relative to quartz vs molar volume. (Adapted from 22.)

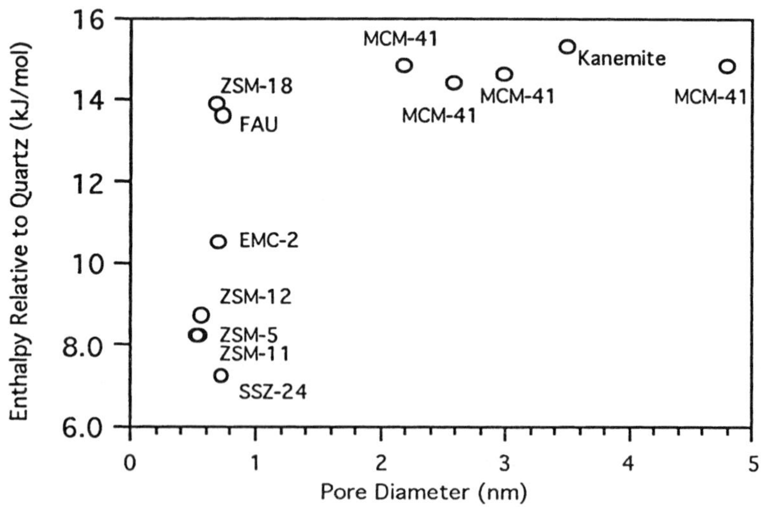

Figure 3 Enthalpy of formation of ordered mesoporous and zeolitic silicas relative to quartz vs pore diameter. (Adapted from 25.)

in the energy of the crystal structure to one where large pores behave energetically as inert volumetric diluents. Because the large pores exist as a second phase, additional increases in pore diameter have little or no effect on the energy of the mesoporous silica. It should be noted that the mesoporous silicas are energetically similar to some high-silica zeolites, in spite of their large pore sizes and lack of crystallinity. It is also significant that the energy plateau is only 15 kJ/mol above quartz.

For crystalline silicas, Petrovic et al (22) found no direct relationship between the enthalpy of formation and the framework density, mean Si-O-Si angle, or molar volume. In silicates, each silicon atom bonds with four oxygen atoms in a tetrahedral fashion; each oxygen atom connects two silicon atoms and the mean Si-O-Si angle is approximately 145°. Quantum mechanical calculations reveal a slight energy barrier to changes in the Si-O-Si angle between 140° and 180°, but angles below 135° experience a marked increase in energy (27). Thus, it is not surprising that all silicates have very similar energies. Petrovic and co-workers indicate that there is a correlation between the enthalpy of formation of silicates and the percentage of Si-O-Si angles below 140°. The correlation is consistent with the potential energy curve of the Si-O-Si angle and implies that the small angles have a more pronounced destabilizing effect than large angles. One note of caution is that numerous bond angles obtained from Rietveld refinements of powder X-ray diffraction data may not accurately represent the angles of the structure. Lobo et al (14) recommend withholding further conclusions until single crystal X-ray data become available, or until other analytical techniques confirm some of the surprising bond angles and lengths reported in the open literature.

Thermodynamic Analysis of Pure-Silica ZSM-5 Synthesis

Zeosils are thermodynamically metastable phases. However, zeosils could conceivably be formed from quartz, e.g.

$$\text{quartz} + 4\text{TPA}^+\text{F}^-(\text{aq}) \rightarrow \text{TPA}^+\text{F}^-\text{-ZSM-5} + \text{H}_2\text{O},$$

where ZSM-5 is the pure-silica version of this zeolite. How could this be? The important issue is that the thermodynamics of the complete synthesis mixture must be considered.

The above reaction can be written as the sum of two reactions:

$$\text{quartz} \rightarrow \text{ZSM-5} \qquad 1.$$

$$\text{ZSM-5} + 4\text{TPA}^+\text{F}^-(\text{aq}) \rightarrow \text{TPA}^+\text{F}^-\text{-ZSM-5} + \text{H}_2\text{O} \qquad 2.$$

[the stoichiometric coefficient for $TPA^+F^-(aq)$ is 4 because 4 TPA^+F^- ion pairs are occluded in each unit cell of pure-silica ZSM-5].

Johnson et al provide thermodynamic properties for Reaction 1 at temperatures of 298.15 and 1000 K (28–31). The values at 298.15 K will be used as a lower bound for the feasibility of the reaction, since the change in Gibbs function is less at 1000 K (typical synthesis conditions involve temperatures of 400–500 K). The thermodynamic parameters for Reaction 1 are

$$\Delta H = 5.5\,\text{kJ/mol}\,SiO_2$$

$$T\Delta S = 1.4\,\text{kJ/mol}\,SiO_2$$

$$\Delta G(\text{reaction 1}) = \Delta H - T\Delta S = 4.1\ \text{kJ/mol}\ SiO_2.$$

Because $\Delta G > 0$, this process cannot spontaneously occur. Although the positive entropy favors Reaction 1, the large positive enthalpy hinders the transformation of quartz to pure-silica ZSM-5.

Partarin et al (32) determined that the change in enthalpy for Reaction 2 is $\Delta H(\text{reaction 2}) = -6.2\,\text{kJ/mol}\,SiO_2$. To estimate the change in entropy for this process, the work of Johnson & Martin (33) can be used. The standard entropy of solution for the salt TPA^+I^- was determined to be $\Delta S(\text{solution}, TPA^+I^-) = 11$ J/K-mol TPA^+I^- for the following process:

$$TPA^+I^-(s) + H_2O \rightarrow TPA^+I^-(aq). \qquad 3.$$

Here, we are concerned with the following reaction:

$$TPA^+F^-(s) + H_2O \rightarrow TPA^+F^-(aq). \qquad 4.$$

To determine the entropy of solution for $TPA^+F^-(s)$, a thermodynamic cycle can be constructed such that $\Delta S(\text{solution}, TPA^+F^-)$ can be estimated from the following equation:

$$\Delta S(\text{solution}, TPA^+F^-) = \Delta S(\text{solution}, TPA^+I^-)$$
$$+ S[F^-(aq)] - S[I^-(aq)] + S[TPA^+F^-(s)] - S[TPA^+I^-(s)].$$

The absolute entropies for aqueous ions were obtained from Marcus (33) and are $S[F^-(aq)] = 8.4$ J/K-mol F^- and $S[I^-(aq)] = 133.5$ J/K-mol I, so that $S[F^-(aq)] - S[I^-(aq)] = -125.1$ J/K-mol halide anion. Johnson & Martin calculated $S[TPA^+I^-(s)]$. We have not found any absolute entropy data for $TPA^+F^-(s)$. As a result, the difference in entropy between $TPA^+F^-(s)$ and $TPA^+I^-(s)$ is estimated using a procedure from Lewis & Randall's text, as revised by Pitzer & Brewer (34). This estimation assumes that the force constants restraining atomic motion are similar for the fluorine and iodine atoms as they exist in the compounds $TPA^+F^-(s)$ and

$TPA^+I^-(s)$ and implies that the compressibility of fluorine is similar to that of iodine in the tetrapropylammonium salts. This approximation is rigorous enough that it does not require similar force constants for elemental (or molecular) fluorine and iodine; only the behavior of these elements in the compounds is important. [Although the validity of this assumption regarding force constants is questionable, it is one of the few, or only, alternatives available in the absence of absolute entropy data for $TPA^+F^-(s)$. The results of such analysis should be evaluated with caution.] The approximation used here applies to compounds differing in the replacement of 1 atom and is

$$S[TPA^+F^-(s)] - S[TPA^+I^-(s)] = 3/2R \ln(M_F/M_I),$$

where R is the universal gas constant, M_F the atomic weight of fluorine, and M_I the atomic weight of iodine. The result from this calculation gives

$$S[TPA^+F^-(s)] - S[TPA^+I^-(s)] = -23.7 \text{ J/K-mol salt.}$$

Substituting the aforementioned values into the thermodynamic cycle used to calculate the entropy of solution for $TPA^+F^-(s)$ gives

$$\Delta S(\text{solution}, TPA^+F^-) = -137.5 \text{ J/K-mol } TPA^+F^-$$

or

$$\Delta S(\text{solution}, TPA^+F^-) = -5.7 \text{ J/K-mol } SiO_2,$$

(the remaining values are expressed in terms of mol SiO_2; there are 96 SiO_2 formula units and 4 TPA^+F^- ion pairs in one unit cell of ZSM-5).

Consider the reverse of Reaction 4:

$$TPA^+F^-(aq) \rightarrow TPA^+F^-(s) + H_2O; \quad \Delta S = 5.7 \text{ J/K-mol } SiO_2. \qquad 5.$$

At 298 K, $T\Delta S$(Reaction 5) = 1.7 kJ/mol SiO_2.

The entropy change for Reaction 5 can be used as an approximation for the change in entropy for Reaction 2 if one assumes that the entropy change for the organic TPA^+F^- going from the aqueous phase into the zeolite is identical to that for transfer from an aqueous phase into a pure crystalline phase of TPA^+F^-. This assumption is subject to criticism, but is preferable to the alternative of statistical mechanical calculations involving other questionable assumptions. The disordering of water and the ordering of TPA^+F^- are both accounted for in this estimate of the change in entropy for Reaction 2. (The concept of water ordering around organic molecules is discussed below.) It is also assumed that the entropy of the ZSM-5 framework is the same whether the TPA^+F^- is occluded or not. Thus any changes in the entropy due to distortions in the crystal structure caused by the occluded TPA^+F^- ion pairs are neglected.

Using the enthalpy and entropy data for Reaction 2, one finds

$$\Delta G(\text{Reaction 2}) = \Delta H - T\Delta S = -7.9 \text{ kJ/mol } SiO_2.$$

The large negative value of ΔH (which favors the process) and the smaller negative value of TΔS (which also favors the process) result in a negative value for the change in Gibbs function. The cause of the large magnitude of the exothermic enthalpy is the contribution of the TPA^+F^- ion pairs to the energetic stabilization of the TPA^+F^--ZSM-5 composite. The tight fit of the TPA^+ ions inside the channels of ZSM-5 and the resulting large number of van der Waals contacts makes the tetrapropylammonium cation a powerful structure-directing agent in the synthesis of ZSM-5.

The increase in entropy accompanying Reaction 2 (and Reaction 5) can be understood in terms of the disordering of water that takes place during this process. Again, consider Reaction 4. As mentioned by Tanford (35), the formation of ionic solutions in water often results in a decrease in entropy, as ion-dipole forces cause an ordering of water molecules around the ions. Frank & Evans (36) explained the entropy decrease that occurs when a hydrocarbon is dissolved in water as a tightening of the structure of water around the hydrocarbon molecules. Lindenbaum & Boyd (37) extended this argument to tetraalkylammonium halides. The nonpolar alkyl groups of the quaternary ammonium cations order the water molecules around the tetraalkylammonium ions (see Figure 4). As the alkyl chains increase in length, the ions become more hydrophobic and further promotes water ordering. Krishnan & Friedman (38) have reported on the

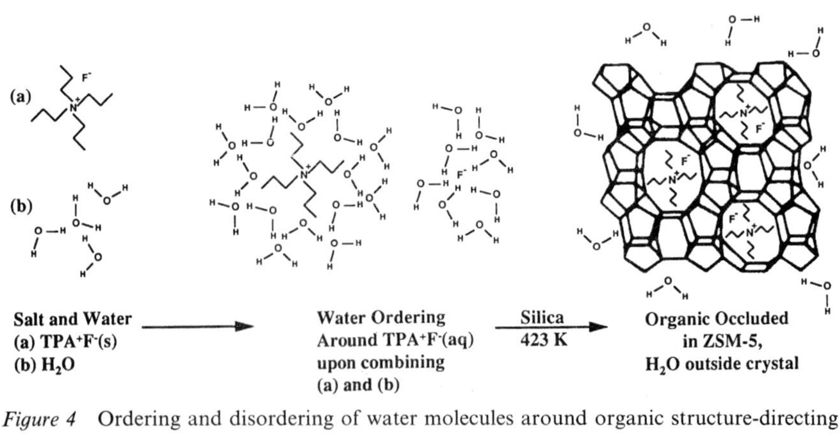

Figure 4 Ordering and disordering of water molecules around organic structure-directing agents.

water structure-making effects of tetrapropylammonium and tetrabutylammonium (TBA^+) cations, and Frank & Wen (39) mention the structure-promoting ability of tetrabutylammonium cations. Based on the above discussion for Reaction 4, the disordering of water in Reaction 5 should result in an increase in the entropy of the system, which would contribute to a more negative change in the Gibbs function. These qualitative arguments agree with the calculations shown above.

Combining Reactions 1 and 2 to achieve the overall zeolite synthesis reaction,

$$\Delta G(\text{overall process}) = -3.8 \text{ kJ/mol } SiO_2.$$

Since $\Delta G < 0$, this process is thermodynamically possible. It is worth noting that Petrovic et al (22) determined the change in enthalpy for Reaction 1 to be $\Delta H = 8.2$ kJ/mol SiO_2 (22), which is 2.7 kJ/mol SiO_2 higher than the value calculated by Johnson et al ($\Delta H = 5.5$ kJ/mol SiO_2). If the value from Petrovic et al is used,

$$\Delta G(\text{overall process}) = -1.1 \text{ kJ/mol } SiO_2.$$

Although this is a low thermodynamic driving force, the change in Gibbs function is still less than zero. Thus the process is thermodynamically feasible.

It should be noted that these calculations involved data from several sources that were working with different systems, as well as some debatable assumptions. The exact numerical values are not as crucial as the trends in ΔG; the important point is that the transformation from quartz to ZSM-5 is thermodynamically possible with the addition of tetrapropylammonium fluoride.

Consider another synthesis of pure-silica ZSM-5, one that is more common than that involving quartz as a starting material:

$$\text{glass} + 4TPA^+F^-(aq) \rightarrow TPA^+F^-\text{-ZSM-5}.$$

This conversion can also be written as the sum of two reactions:

$$\text{glass} \rightarrow \text{ZSM-5} \qquad 6.$$

$$\text{ZSM-5} + 4TPA^+F^-(aq) \rightarrow TPA^+F^-\text{-ZSM-5}. \qquad 7.$$

Referring again to the work of Johnson et al, and using thermodynamic data applicable at 298.15 K, the thermodynamic quantities for Reaction 6 are

$$\Delta H = -3.65 \text{ kJ/mol } SiO_2$$

$$T\Delta S = -0.66 \text{ kJ/mol } SiO_2$$

$$\Delta G(\text{Reaction 6}) = \Delta H - T\Delta S = -3.0 \text{ kJ/mol SiO}_2.$$

From a thermodynamic standpoint, Reaction 6 can spontaneously occur. Pure-silica ZSM-5 is more stable than glass (which is amorphous silica), even without any occluded organic molecules.

Reaction 7 enhances the transformation of glass to ZSM-5. Using the change in enthalpy for Reaction 7 given by Patarin et al ($\Delta H = -6.2$ kJ/mol SiO_2) and the estimate for the change in entropy as described above ($T\Delta S = 1.7$ kJ/mol SiO_2),

$$\Delta G(\text{Reaction 7}) = \Delta H - T\Delta S = -7.9 \text{ kJ/mol SiO}_2.$$

Combining Reactions 6 and 7 to achieve the overall reaction,

$$\Delta G(\text{overall process}) = -10.9 \text{ kJ/mol SiO}_2.$$

Because the change in Gibbs function is well below zero, this process has a strong thermodynamic driving force. As is true for the example involving quartz, these calculations should be used with caution.

One implication for the synthesis of zeolites by design that can be obtained from the results of this analysis is that the "goodness of fit" of the organic structure-directing agent inside the pores of the zeolite is an important component of the thermodynamic driving force for the process. For Reaction 2, $\Delta H = -6.2$ kJ/mol SiO_2. This term relates to the interaction energy between the organic molecule and the inorganic silicate, and its magnitude dominates the entropic contribution. Thus it provides a first-order approximation for the change in Gibbs function of the process. From geometric considerations, it is known that the TPA^+F^- ion pairs fit tightly into the ZSM-5 void space. If the organic molecules do not fit as well inside the void spaces, the number of van der Waals interactions will be reduced, and ΔH should be more positive (less favorable energetics). Therefore, the entropic driving force (release of ordered water) will become more important as the interactions between the organic and silica framework decline.

Burkett & Davis (40) have synthesized pure-silica ZSM-5 in less than a week, using TPA^+OH^- as the structure-directing agent. However, they were unable to prepare ZSM-5 with tetraethanolammonium hydroxide, $(HOC_2H_4)_4N^+OH^-$, even after one month at hydrothermal synthesis conditions. Wen & Saito (41, 42) have shown that the size of the tetraethanolammonium cation is similar to that of the tetrapropylammonium cation. Thus the "goodness of fit" of the organic inside the zeolite pores does not by itself determine whether a particular synthesis can occur (although the sizes are similar, the interaction energies are most likely slightly different). Frank & Wen (39) predicted that solutes with OH groups will not significantly affect the structure of water due to the similar hydrogen-bonding characteristics of water and alcohols. Wen & Saito also claim that quaternary ammonium cations with terminal hydroxyl groups will have less influence on the structure of water than the equivalent tetraalkylammonium cations. Results from Agar (43), Price (44), and H Frank & W Wen (unpublished data) agree with the claim of Wen & Saito. Since $(HOC_2H_4)_4N^+OH^-$ does not order water like TPA^+OH^-, syntheses involving tetraethanolammonium hydroxide will not have the same entropic driving force that is present in syntheses using TPA^+OH^-. Thus on first inspection, the necessity of water ordering may be the critical factor. However, at this time not enough information exists to decouple the enthalpic and entropic driving forces in zeolite syntheses, since the precise interaction energies for tetraethanolammonium ions in ZSM-5 is unknown.

One must exercise caution when attempting to extend the thermodynamic analysis of zeolite synthesis involving fluoride to one using hydroxide anions. With a synthesis of pure-silica ZSM-5 using TPA^+F^- as the structure-directing agent, the fluoride anion is occluded inside the ZSM-5 void space along with the tetrapropylammonium cation. When TPA^+OH^- is used, the TPA^+ cation is occluded but the OH^- anion is not. The positive charge of TPA^+ is balanced by a $Si\text{-}O^-$ defect in the zeolite structure (14). The effect of the ZSM-5 structural defect, i.e. $Si\text{-}O^-$, on the thermodynamics of the synthesis process is unknown at this time and should be investigated.

Energetic Interactions of Organic Structure-Directing Agents with Silicates

Pure-silica zeolites are hydrophobic and have unfavorable interactions with water molecules. Additionally, zeolites in aqueous solution are thermodynamically less stable than dense phases. When organic molecules are occluded in the micropores of the zeolite, water penetration decreases and favorable interactions occur between the silica wall of the zeolite and the occluded organic molecules. It has been speculated that, via these silica-organic interactions, pure-silica zeolites attain enough thermodynamic stability to exist (45). Petrovic et al (22) showed experimentally that this is not the case. Rather, the organic structure-directing agents assist in selecting which zeolite will be formed from a wide range of energetically feasible possibilities (see below). For these favorable interactions to occur, a good fit between the organic molecule and the zeolite void space is required, i.e. the number of van der Waals contacts is maximized. As a result, syntheses involving organic structure-directing agents often yield pure-silica zeolites with pore dimensions and shapes similar to those of the organic molecules.

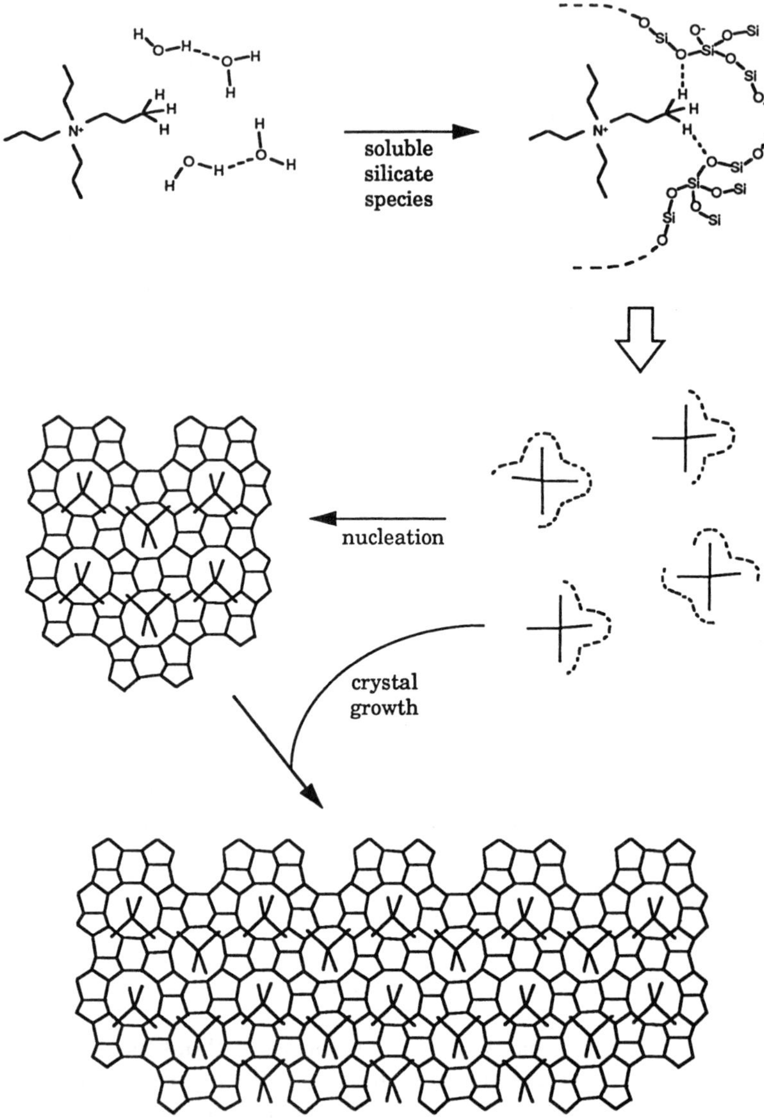

Figure 5 Proposed mechanism of structure-direction and crystal growth involving organic-inorganic composites in the synthesis of pure-silica ZSM-5 using TPA^+ as structure-directing agent. (Adapted from 49.)

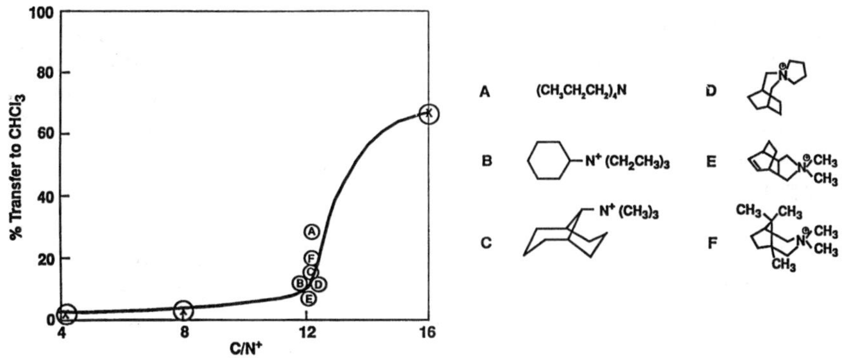

Figure 6 The percent transfer of organic molecules into a chloroform phase from an aqueous phase for varying C/N^+ ratios. (Adapted from 14.)

Effects of Structure-Directing Agent on Crystallization Rate

When an organic molecule is included in a synthesis mixture, otherwise identical synthesis gels with different organic structure-directing agents can yield completely different zeolites. For example, SSZ-24 and ZSM-5 are similar energetically (22), but the former is synthesized using TMADAOH (N,N,N-trimethyl 1-adamantammonium hydroxide) as the organic, whereas the latter is formed using TPA^+OH^-.

The choice of a structure-directing agent can affect the rate of crystallization. Harris & Zones (56) investigated the guest/host energetics for the synthesis of the clathrasil nonasil and the zeolite SSZ-13. They suggest that a rate-determining step for the nucleation of nonasil and SSZ-13 is the interaction of the structure-directing agent with the inorganic gel. Pure-silica clathrasil syntheses where the organic structure-directing molecule is similar in size to the inorganic cage of the final product have faster nucleation rates than those experiments where the size-matching is not as good. This result shows the importance of a tight fit for the structure-directing agent inside the clathrasil cage for fast nucleation rates.

Effects of Alkali Metal Cations on Zeolite Synthesis

The presence of alkali metal cations is required for the vast majority of zeolite syntheses at basic conditions (ZSM-11 and ZSM-5 are exceptions and can be synthesized using TBA^+ and TPA^+ as the structure-directing agents with no alkali metal cations in the synthesis gel). The synthesis of pure-silica ZSM-12 was carried out with and without alkali metal cations at constant hydroxide concentration by Goepper et al (57). Their results indicate that alkali metal ions, e.g. K^+ or Na^+, are required to synthesize pure-silica ZSM-12 in 7 days at 423 K. Although K^+ and/or Na^+ are not needed to synthesize pure-silica ZSM-12, at constant hydroxide ion concentration, the rate of crystallization decreases as the ratio K^+/SiO_2 (or Na^+/SiO_2) declines. In the absence of K^+, and using seed crystals of uncalcined ZSM-12, pure ZSM-12 can be formed in 14 days at 423 K. Usually the use of seed crystals is believed to enhance nucleation alone. If the potassium ions added to the synthesis gel only participate in nucleation, then experiments with $K^+/SiO_2 = 0.3$, and no seeds should yield similar results to experiments with $K^+/SiO_2 = 0$ and 2% ZSM-12 seeds. Since the latter synthesis required twice as much time, it was concluded that the potassium ions participate in nucleation and crystal growth.

Goepper and co-workers eliminated the effects of ammonium cations contained in the Ludox AS-40 silica source by using fumed silica (Cab-O-Sil) in some trials. ZSM-12 was synthesized in 7 days using sodium cations, but the synthesis took $2\frac{1}{2}$ months when no alkali metal ions were included. When ammonium hydroxide was present in the synthesis gel (in a concentration equal to that used in the sodium trials), no zeolite was formed after 14 days. These results indicate that sodium cations enhance the crystallization rate of pure-silica ZSM-12, but ammonium ions do not (at least not as much as sodium ions). It was observed that all of the ZSM-12 samples obtained from gels with different concentrations of potassium ions had similar morphologies; however, these morphologies were different from that of ZSM-12 synthesized in the absence of alkali cations. This result again suggests that alkali metal ions are involved in nucleation and crystal growth (16).

The role of potassium and sodium in zeolite synthesis can be understood further by the results of measuring the rate of dissolution of quartz into aqueous solutions of alkali metal cations. The dissolution of quartz in aqueous solutions of KCl, NaCl, $MgCl_2$, and LiCl was studied by Dove & Crerar (58) and Brady & Walters (59). Small concentrations of electrolytes increased the dissolution rate up to 15 times as much as the rate in deionized water. Higher dissolution rates were obtained for NaCl and KCl compared with $MgCl_2$ and LiCl, which could be explained on the basis of capability for adsorption on the surface of quartz. Because Li^+ and Mg^{2+} have large charge-to-size ratios, they can coordinate well with water molecules and have reduced potentials for adsorption onto quartz. Applying this explanation to zeolite synthesis, one can conclude that potassium and sodium play a role in the polymerization and depolymerization of the silicate present in the synthesis gel. Through this process (as well as other possibilities), they can accelerate nucleation and growth rates of high-silica zeolite crystals.

Zones has studied the effects of sodium concentration on the synthesis of large-pore borosilicates (mainly silica) using endo and exo derivatives of tricyclo(5.2.1.0)decanes as organic structure-directing agents (60). At high alkali-metal ion concentrations, the sodium cations appear to compete with the organics as structure-directing agents. A layered silicate was obtained from high sodium concentrations, whereas zeolites were synthesized with lower amounts of sodium ion. Why do high sodium concentrations result in layered silicates? The answer is currently unknown, although the valence matching principle of Brown may be used to formulate a hypothesis (61). A bond valence approach to chemical structure and bonding would suggest that the most stable structures will be formed when the Lewis base strength of the anion is closest to the Lewis acid strength of the cation. Silicate anions can exist in the form of monomers (Q^0), dimers (Q^1), chain structures (Q^2), or layered structures (Q^3). According to Figure 7 (61), the best valence match for sodium ions occurs with a layered (Q^3) silicate anion. As the ratio Na^+/Si increases, the formation of a layered silicate becomes more likely.

The above examples illustrate the delicate balance that must be maintained in the synthesis gel to achieve the desired product in the shortest possible time. Alkali metal cations accelerate the rate of crystallization of

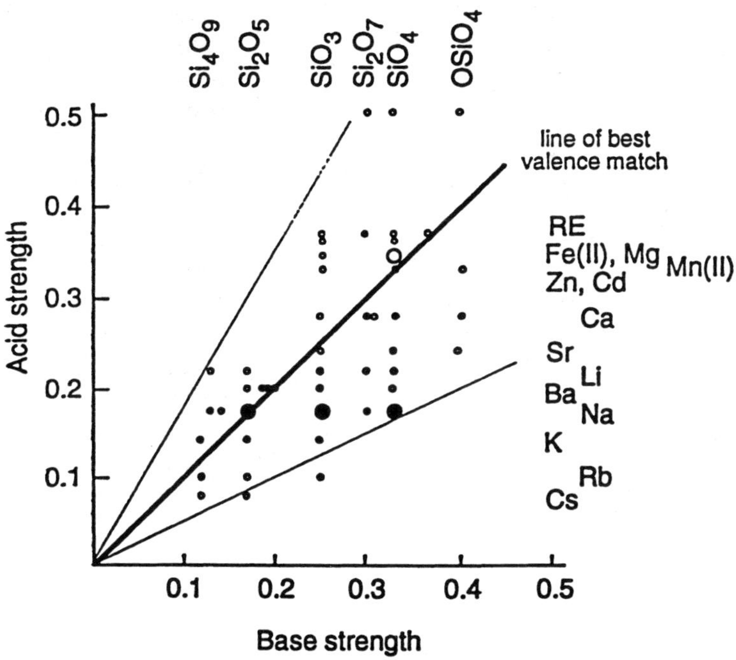

Figure 7 Silicate compounds as a function of the acid strengths and base strengths of the charge balancing cations. (Adapted from 61.)

high-silica zeolites, but too much alkali metal ion may result in competition with the organic structure-directing agent for interactions with silica such that layered products can result.

Effects of O-H (or O-D) Bond Breaking on Nucleation Rate

In addition to the structure-directing agent and alkali metal cations present in the synthesis gel, other factors can influence the nucleation rate. For example, Burkett & Davis (40) have observed isotope effects in the syntheses of pure-silica ZSM-48 and ZSM-5 using fumed silica (Cab-O-Sil). A delay in nucleation is observed when D_2O is used in place of H_2O in the synthesis mixture. When the syntheses are conducted in the presence of Na^+ ions, the rates of nucleation are virtually indistinguishable for D_2O and H_2O and are enhanced relative to the sodium-free trials. Since Na^+ ions have been shown to increase the rate of SiO_2 dissolution, the dissolution of silica is believed to play a role in the delay of nucleation for D_2O compared with H_2O. [Dutta et al (62) also noted a delay in the crystallization rate of zeolite A when D_2O was used instead of H_2O, even though sodium was present. They suggested that an isotope effect with the condensation polymerization reactions required for growth of zeolites may be responsible for this delay.] Because OD^- is a stronger base than OH^-, silica should be dissolved more rapidly in D_2O, and one might expect nucleation to occur more rapidly for D_2O than H_2O. However, the converse was experimentally observed; therefore Burkett & Davis suggest that the breaking of O-H (or O-D) bonds may play a significant role in the nucleation process. This is important for several reasons. First of all, these results provide experimental evidence to suggest that a stronger base does not necessarily cause faster nucleation. Secondly, the O-H (or O-D) bond breaking most likely is a significant factor in the nucleation process and must be considered in addition to the effects of structure-directing agents and alkali metal cations employed.

CONTROL OF POROUS STRUCTURES THROUGH ORGANIC STRUCTURE-DIRECTING AGENTS

The syntheses of high-silica zeolites appear to be the most prone to design via the use of structure-directing agents [this exploits the hydrophobic interactions between silica and the organic molecules and the ability to control the charge and shape of the organic (13)]. Lobo and co-workers (14) indicate that the molecular size (somewhat correlated to the total number of carbon and nitrogen atoms) of a structure-directing agent can strongly influence the number of different zeolites that can be synthesized using a particular organic molecule. (It has been shown experimentally that one organic can be responsible for the synthesis of more than one zeolite). Additionally, Zones noted that this trend is followed as the size of the organic molecule increases until only one zeolite structure is found for large structure-directing agents. Lobo et al recommend that larger organic molecules be investigated to synthesize new zeolite structures; however, the ratio of C/N^+ must be kept somewhat within the previously mentioned limits in order to have the appropriate hydrophobicity/hydrophilicity. The presence of two or more charged nitrogens can circumvent potential problems with hydrophobicity for large organic molecules (14).

Examples of Structure-Direction

The concepts of structure-direction discussed above are now illustrated using five examples. These examples demonstrate the effects of the geometry of the structure-directing agent on the geometry of the zeolite formed.

SSZ-26 is a zeolite with intersecting 10- and 12-ring pores (63). A propellane-based structure-directing agent, **I**, is used in the synthesis of SSZ-26 (64).

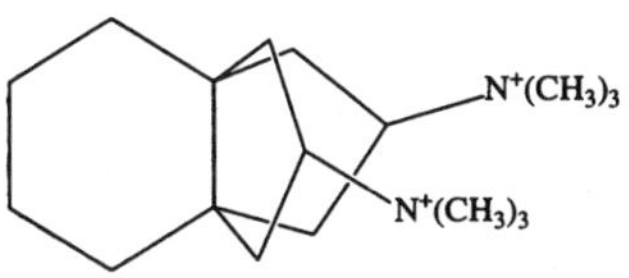

I

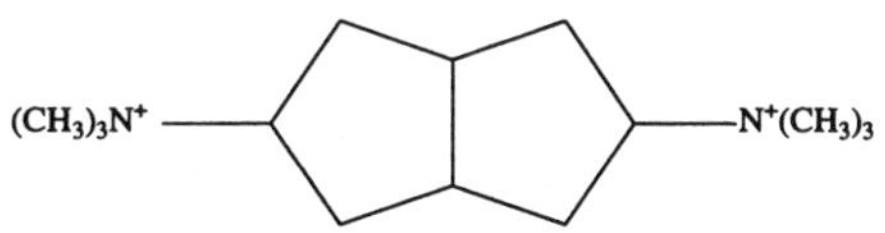

II

Zones & Santilli (64) noted that organic cation **II** directed the synthesis of ZSM-12, a unidimensional zeolite. They speculated that the inclusion of another ring as in **I** would cause an intersecting pore system to form. Using the structure of SSZ-26 (65), chemical and thermogravimetric analyses indicate that one guest molecule is present in each channel intersection (14). Lobo et al (66) performed molecular force-field calculations on cation **I** inside the SSZ-26 pores and noted excellent agreement between the geometry of the organic cation and the geometry of the pore intersection. The conformation of minimum energy for **I** inside the channel intersections bears close resemblance to the conformation suggested by Zones & Santilli (64). SSZ-26 thus appears to be the first zeolite where the pore geometry has been designed a priori (14). This example illustrates how the concepts of structure-direction can be used to synthesize new zeolites.

ZSM-18 has a unique structure in that it is the only aluminosilicate containing 3-membered rings (67). An excellent fit exists between the zeolite cage and the organic structure-directing agent. Hong et al (68) used static and CP MAS ^{13}C NMR to determine that the triquaternary ammonium cation snugly fits inside the cage of ZSM-18. The size, rigidity, and especially the threefold axis of rotation of the organic molecule are factors favoring the formation of the ZSM-18 cage (14). Using this knowledge, Schmitt & Kennedy (69) recently synthesized ZSM-18 using structure-directing agents **III** and **IV**.

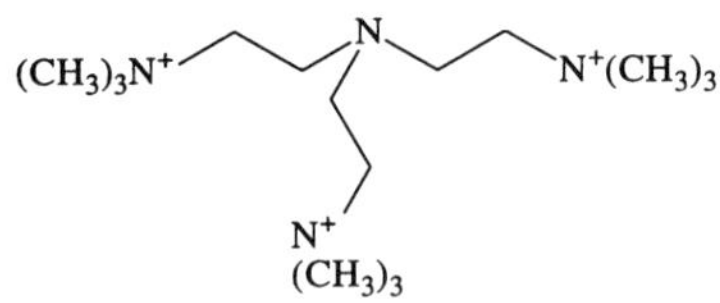

III

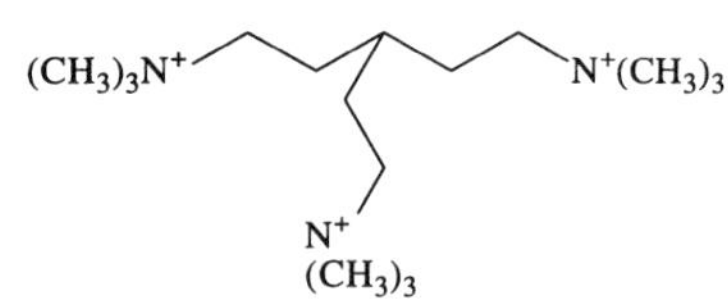

IV

Organic cations **III** and **IV** were chosen by using molecular modeling to help screen for potential replacements of the original ZSM-18 organic structure-directing agent. This is the first published result using molecular modeling to design a structure-directing agent.

Another example of the complementary relationship between the geometry of the organic molecule and the geometry of the zeolite void space is the occlusion of 18-crown-6 in the cages of EMC-2, the hexagonal polymorph of faujasite (14). 18-crown-6 is the only proven structure-directing agent for the formation of EMC-2. It has been shown via X-ray diffraction (70) and solid-state NMR (71) results that 18-crown-6 is located inside the small cages of EMC-2 and roughly conforms to the cage geometry.

Arhancet & Davis (72) provided the first example of exercising the purposeful control of long-range order in zeolite synthesis using structure-directing agents. They were able to control the fractions of cubic and hexagonal polymorphs of faujasite present in intergrowth zeolites by specifying the concentrations of 15-crown-5 and 18-crown-6 in the reaction mixture.

Lobo et al (73) have reported a second example of control of long-range order using structure-directing agents in the synthesis of the related materials CIT-1 and SSZ-33. CIT-1 and SSZ-33 are members of a family of polymorph structures (51). SSZ-33 is an intergrowth of polymorph A (30%) and polymorph B (70%), whereas CIT-1 is the pure polymorph B. By preparing synthesis gels in the presence of the structure-directing agents for SSZ-33 and CIT-1, the fractions of the two polymorphs in the resulting zeolites can be controlled in the range 70% polymorph B (SSZ-33) to 100% polymorph B (CIT-1) (51). This example and the previous example from Arhancet & Davis demonstrate the control of long-range order in zeolites using organic structure-directing agents.

Heteroatoms and Structure-Direction by Organic Molecules

The addition of small quantities of tetrahedral atoms besides silicon (Al, Zn, B, etc) to synthesis gels has dramatic effects on the zeolite structures formed when using identical structure-directing agents (14). Here we outline the changes in zeolite structures observed when aluminum, zinc, or boron is added to an otherwise pure-silica reaction mixture. Heteroatoms other than B, Zn, and Al have been incorporated in high-silica zeolite frameworks (9, 74), but they are not discussed here.

The syntheses of ZSM-12 and zeolite beta using TEA^+ as structure-directing agent are used to illustrate the effect of adding aluminum to the synthesis gels while keeping other factors constant (see Table 2) (14). ZSM-12 is formed when the ratio of SiO_2 to Al_2O_3 is greater than 50. As more aluminum is added and the ratio of SiO_2 to Al_2O_3 becomes slightly less than 50, zeolite beta is formed. If still more Al is included such that $SiO_2/Al_2O_3 \sim 15$, ZSM-20 results. When boron is used in place of aluminum, similar results occur. This trend is also observed for organics other than TEA^+ (see Table 2).

The structure-directing effects of organics that synthesize large-pore, pure-silica zeolites are closely related to the presence of heteroatoms in the synthesis gel (14). Trimethyl-2-adamantylammonium directs the structure of the unidimensional pore system in zeolite SSZ-24. However, the addition of boron to this synthesis results in the formation of SSZ-33, which has a multidimensional pore system (51). As indicated from the results shown in Table 2, some structure-directing agents that make SSZ-31 (unidimensional pore system) with pure-silica reaction mixture also yield SSZ-33 when B is present.

If the organic structure-directing agent listed as the last entry of Table 2 is employed, different zeolites are obtained for pure-silica, aluminum-containing, and boron-containing synthesis gels. When only aluminum is present, SSZ-37 is formed (65). SSZ-37 has a structure similar to that of NU-87 (75). Slightly smaller and larger organics do not direct the structure of SSZ-37. These results show how small changes in composition can lead to significant variations in the structures obtained.

As is true for aluminum and boron, adding zinc to a reaction mixture significantly alters the zeolite formed. Zinc is unique in that VPI-8 is formed in all high-silica syntheses in which Zn is involved, regardless of the choice of structure-directing agent (14). Annen & Davis (76, 77) first

Table 2 Effect of aluminum, boron and zinc on the structure of molecular sieves obtained using organic structure-directing agents[a] (adapted from Reference 14)

	SiO_2	SiO_2/Al_2O_3 <50	SiO_2/B_2O_3 <30	SiO_2/ZnO <100
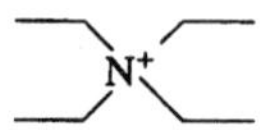	ZSM-12	Zeolite Beta	Zeolite Beta	VPI-8
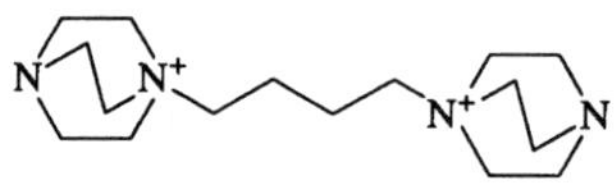	ZSM-12	Zeolite Beta	Zeolite Beta	VPI-8
	ZSM-12	Mordenite	Zeolite Beta	Layered Material
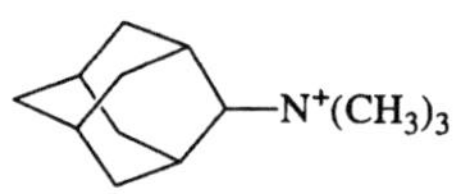	SSZ-24	SSZ-25	SSZ-33	
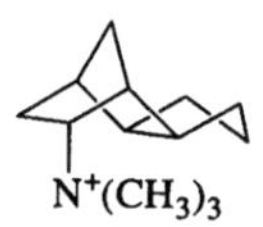	SSZ-31	Mordenite	SSZ-33	VPI-8
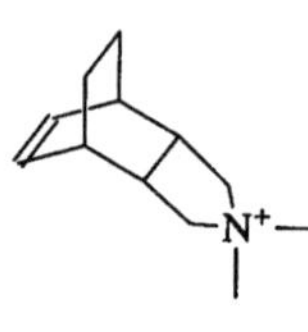	SSZ-31	SSZ-37	SSZ-33	—[b]

[a] From References 77, 79, 92–94; MA Camblor et al, submitted. Some of these results are from experiments in our laboratories and are not yet published.
[b] No product formed after several weeks.

synthesized VPI-8 using TEA^+ as the structure-directing agent and zinc acetate as the source of zinc. Sorption and transmission electron microscopy (TEM) data show that VPI-8, is a one-dimensional large-pore molecular sieve. Because it is synthesized using a wide variety of organic guests, it is probable that a divalent zinc cation determines the structure of the resulting zeolite and that the organic structure-directing agent's primary role is to stabilize the void spaces of the pores (78). Zinc appears to inhibit the formation of other high-silica zeolites besides VPI-8.

Divalent and trivalent T (tetrahedral) atoms in the synthesis gel should exert a strong influence on the formation of high-silica zeolites (14). First of all, substituting Zn, B, or Al for Si in the synthesis gel results in a negatively charged zeolite framework, which will coordinate more strongly with the inorganic cations, e.g. sodium or potassium, present in the reaction mixture. Secondly, the T-O bond lengths and especially the T-O-T angles should affect the formation of particular secondary building units, e.g. 4-rings (14). It is known that Si-O-Al and Si-O-B angles have more limited ranges of variation than Si-O-Si and that the optimum angles are smaller for the heteroatom angles than for Si-O-Si. Thus the addition of framework heteroatoms is likely to cause structural variations.

It is interesting to examine the effects of heteroatoms by comparing the structures of zeolite beta and ZSM-12. The projections along the *a* and *b* axes of zeolite beta are identical to the projections along the 12-ring pores of ZSM-12. The disordered stacking of silicate layers is responsible for the structures of both zeolites (79, 80). There are also more 4-rings present in zeolite beta than in ZSM-12. The additional 4-rings in beta may result from smaller Si-O-T angles, which are promoted by the use of boron or aluminum (14).

Specification of Framework Atoms by Structure-Directing Agents

Recently, Li et al (81) have provided the first example of how to control the placement of framework atoms during a zeolite synthesis. Zeolites ZSM-11 and ZSM-5 were formed using the structure-directing agent *n*-trimethoxysilylpropyltri-*n*-butylammonium (TBA^+ with a silicon alkoxide group covalently attached via a Si-C bond on one of the alkyl groups as a substitute for a methyl group). Thermogravimetric analysis and ^{13}C and ^{29}Si NMR spectroscopy of ZSM-5 proved that the organosilicon structure-directing agent resided intact at the channel intersections in the zeolite void space. It was also shown that the silicon atom in *n*-trimethoxysilylpropyltri-*n*-butylammonium was located in a framework position (81).

The consequences of the work of Li et al are rather broad. First of all, this work shows that there are new opportunities for making microporous materials using organic molecules with at least one silicon atom covalently bonded to a carbon atom in the structure-directing agent. Secondly, the covalent linking of atoms other than silicon to the organic molecules may make it feasible to arrange these heteroatoms in precise locations during the formation of new and previously known zeolites. Eventually it may be possible to distribute these atoms in one particular crystallographic position of a zeolite (81). Controlling atomic arrangements in this way would aid in developing completely uniform properties and in determining intrazeolite catalytic reaction pathways. As an example, the bridging Al-OH-Si groups in ZSM-5 are heterogeneous in terms of acid strength. If all Al atoms in ZSM-5 could be distributed in identical T-sites, a material with homogeneous acid strength could be synthesized (81).

ORDERED MESOPOROUS MATERIALS

From Microporous to Mesoporous Materials

As mentioned previously, Burkett & Davis (49) used the results of intermolecular ^{1}H-^{29}Si CP MAS NMR to propose a mechanism for the synthesis of zeolites with high Si/Al ratios (see Figure 5). This proposed mechanism has several features common to other types of assembly processes, e.g. biological assemblies. These common elements are (*a*) the self-assembly process, which occurs spontaneously with no external mediating factors; (*b*) the organization that involves structure-direction with many noncovalent interactions working together (multipositional molecular recognition events); and (*c*) the process of self-assembly, which occurs such that the structural organization proceeds from molecules to composite structures of increasing size (82). In biological processes, organic materials larger than individual molecules usually act as templates (templating is a more precise form of structure-direction). Therefore one would expect organic molecular aggregates to function as structure-directing agents when synthesizing porous inorganic materials (82). Until the ordered mesoporous materials MCM-41 and MCM-48 were synthesized at Mobil, this had not been accomplished (83). MCM-41 is an (alumino) silicate possessing hexagonally arranged one-dimensional pores with diameters ranging from 15 to 100 Å. MCM-48 is an (alumino)silicate with a three-dimensional pore system and diameters of order 30 Å; its X-ray diffraction pattern has cubic symmetry with space group Ia3d (82–84). It is noteworthy that, unlike zeolites, the inorganic portions of mesoporous materials such as MCM-41 and MCM-48 are amorphous (alumino) silicates (26). This was shown by Raman, IR, and ^{29}Si NMR spectroscopies.

Consider the mixing of water with organic molecules of type $(CH_3)_3N^+(CH_2)_xCH_3$. The molecules should remain isolated in solution if x is small, e.g. $0 \leqslant x \leqslant 2$. When x is larger, e.g. $x = 15$ (hexadecyltrimethylammonium, $C_{16}TMA^+$), spontaneous self-assembly of the molecules can occur so that molecular aggregates are formed. The aggregates (micelles) contain hydrophilic polar groups that align at the water-organic interface as well as hydrophobic hydrocarbon chains that collect in the center (82). For C_{16}TMABr in water at ambient conditions, the critical micelle concentration is 0.03 wt% (85). To investigate if and how molecular aggregation affects synthesis, Davis et al (82) heated a synthesis gel with composition 0.15 mol $C_{16}TMA_2O$: 1.0 mol SiO_2 : 50 mol H_2O at temperatures below and above 423 K. At temperatures below 423 K, MCM-41 was formed (pore diameter ~30 Å), whereas temperatures above 423 K yielded the zeolite ZSM-48 (pore diameter ~5 Å). ^{13}C CP MAS NMR spectroscopy showed that the $C_{16}TMA^+$ remains intact in MCM-41 and ZSM-48. These results are consistent with the idea that organic aggregation causes formation of mesoporous materials, whereas isolated organic molecules direct the structure of microporous zeolites (82).

MCM-41 and MCM-48

MCM-41 and MCM-48 are typically prepared via hydrothermal syntheses with $C_{16}TMA^+$ and silica at temperatures below 423 K (82). MCM-41 is usually formed with ratios of $C_{16}TMA^+/SiO_2 \leqslant 1$ (26, 83, 84). MCM-48 has been more difficult to synthesize than MCM-41. To prepare MCM-48, tetraethylorthosilicate (TEOS) is used as the silicon source and the ratio $C_{16}TMA^+/SiO_2$ is typically 1.1–1.5 (86).

Workers at Mobil suggested that MCM-41 can be obtained via one of two mechanisms (84). In the first, $C_{16}TMA^+$ cations self-assemble into a hexagonal array of rod-like micelles in an aqueous phase (the H_1 phase). After formation of the H_1 phase, silicate species penetrate the region between the rod-like micelles and condense to make the final inorganic-organic material. In the second mechanism, the silicate species affects micelle formation and acts in concert to produce MCM-41. Chen et al (26, 87) studied the synthesis of MCM-41 with thermogravimetric analysis, X-ray powder diffraction, and ^{29}Si NMR spectroscopy of solid samples from throughout the MCM-41 synthesis. More importantly, Chen et al monitored the synthesis with in situ ^{14}N NMR spectroscopy, which can detect the presence of the H_1 phase. The H_1 phase was not observed at any time during the synthesis (87), which indicates that a mechanism where the H_1 phase forms is most likely not correct.

Chen et al (26, 87) suggested a mechanism for the formation of MCM-41 as shown in Figure 8. Rod-like organic micelles initially form with random order. These micelles interact with silicate species through electrostatic interactions between surfactant cations and silicate anions to make micelles coated with silicate (a single layer of silica surrounding a rod-like micelle consists of approximately four silicon atoms per nitrogen atom; each surfactant head group has an approximate area of 50 Å^2) (88).

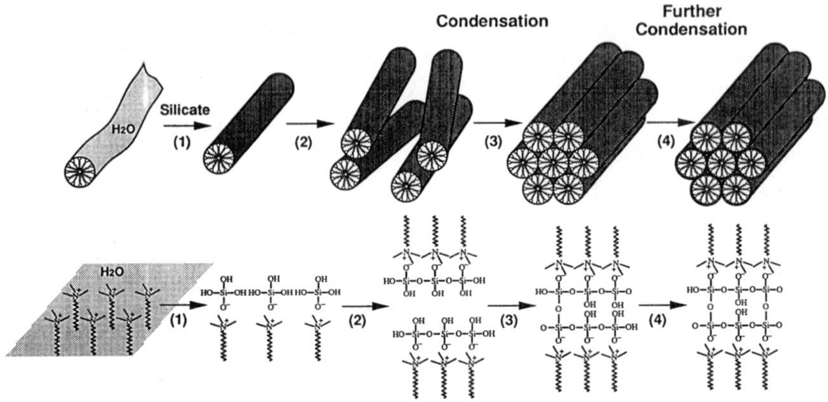

Figure 8 A proposed mechanism for the synthesis of MCM-41. (Adapted from 82.)

Condensation of the silicate species then orders the randomly ordered composite species to form a hexagonal array. Additional silicate condensation occurs while the species is in the hexagonal arrangement. The mechanism proposed by Chen et al suggests that MCM-41 has a ratio of Si/N ~ 4 and a wall thickness of roughly 8 Å. Consistent with this mechanism, MCM-41 has been synthesized with Si/N = 3.5–4.5 (26, 84) and with wall thicknesses of 8–10 Å (84, 89). Beck et al (84) have also demonstrated that the pore diameter of MCM-41 can exceed the value resulting from the exclusive use of $C_{16}TMA^+$ ions by adding hydrophobic organic molecules, e.g. 1,3,5-trimethylbenzene, to the reaction mixture. Using the synthesis mechanism proposed by Chen et al, one can rationalize the mechanism for this enlargement of pore diameter. The small organic molecules are adsorbed into the rod-like micelles and their diameters are then increased. This enlargement of the micelle size leads to the increase in the pore size of the resulting MCM-41 (82). Spegt et al (90) have documented swelling of $C_{16}TMA^+$ micelles, lending further credence to the proposed mechanism. Absorption and swelling of the rod-like micelles may take place at any time before silica encapsulation (82).

Synthesis of Microporous and Mesoporous Materials

A comparison of synthesis routes for zeolites and ordered mesoporous materials reveals that many of the mechanistic steps are similar. Davis et al (82) provided a mechanism for the organic-mediated synthesis of ordered porous materials. This mechanism is discussed below.

Adding an organic molecule to an aqueous liquid-phase can result in the ordering of water in the vicinity of the organic. This "hydrophobic hydration" can take place around isolated molecules or organic aggregates that have self-assembled. Davis and co-workers (13, 49) proposed that molecules which do not order water will not be effective structure-directing agents. Under strongly basic conditions, hydroxyl anions enhance the depolymerization of silica when a non-monomeric silicon source is used, e.g. fumed silica, amorphous silica. The hydrated organic entities have coulombic interactions with the soluble silicate species, and the quaternary ammonium cations promote the condensation of the silicate species. Therefore, the geometry of the organic component influences the geometry of the inorganic species via multipositional molecular-level interactions (van der Waals and coulombic; van der Waals interactions become important after the silicate species condense to make hydrophobic domains) (49). The inorganic-organic composites then merge to create materials with longer-range order. The assembly process has enthalpic and entropic driving forces; the energetically favorable inorganic-organic interactions and the release of ordered water molecules from around the organic species to bulk water in solution favor synthesis of the porous structures (82). The inorganic portion can form crystalline phases in microporous materials, and the properties of the inorganic component play a critical role in determining the final structures formed. If self-assembled organic aggregates behave as structure-directing agents, many silicate anions were needed to balance the charge of the organic cations. As an example, the product ratio Si/N in MCM-41 is approximately 3.5–4.5. This low ratio is likely the reason that materials such as MCM-41 and MCM-48 do not have crystalline walls because the amount of silicon required for charge compensation of the organic cation is too large to permit formation of a crystalline oxide (82). Zeolites typically have Si/N ratios of 7 (beta)–23 (ZSM-5). Finally, the syntheses of ordered mesoporous materials and zeolites can be viewed as examples of a unified strategy of self-assembly whereby structure-direction occurs through a number of noncovalent interactions between inorganic and organic components. These interactions work together to establish order of increasing length scales throughout the synthesis (82). It should also be noted that the inorganic-organic composites are formed by kinetically controlled processes in contrast to mesoscopic organic phases produced in binary water-organic systems (82). These concepts are also important in biological assemblies and in the inorganic-organic composites of biomineralization of pure organic phases (91).

SUMMARY AND FUTURE DIRECTIONS

The mechanisms by which porous silicates are formed still remain unknown. Here we have provided an overview of the types of information that are currently available that enable one to begin to rationalize the assembly processes. Using existing data, partial designs of synthetic strategies are now appearing, e.g. SSZ-26, ZSM-18. Ultimately, the hope is that completely designed syntheses can be accomplished. By comparing and contrasting the assembly processes of porous silicate synthesis to those of other self-assembly processes, e.g. biomineralization, the elucidation of universal rules of construction for inorganic-organic composites may be possible. At this time, we believe that the most critical issues in the area of porous silicate synthesis are investigation of the nucleation process and determination of the precise structures of the nuclei. If these issues can be resolved, we believe that porous silicate synthesis by design will become a reality.

Acknowledgments

The authors acknowledge helpful discussions with Stacey Zones of Chevron Research and Technology Company, John Higgins of Mobil Research and Development Corporation, Kenneth Pitzer of the University of California at Berkeley, and Raul Lobo, Sandra Burkett, Chris Dartt, Cong-Yan Chen, Hubert Koller, George Gavalas, and Zhen-Gang Wang, all from the California Institute of Technology.

Literature Cited

1. Cronstedt AF. 1756. *Svenska Vetenskaps Akademiens Handlingar, Stockholm* 17: 120–23
2. Schlenker JL, Kuhl GH. 1993. *Ninth Int. Zeolite Conference, Montreal*, ed. R von Ballmoos, JB Higgins, MM Treacy, pp. 3–9. Boston: Butterworth-Heinemann
3. Davis ME. 1991. *Ind. Eng. Chem. Res.* 30: 1675–83
4. Löewenstein W. 1954. *Am. Mineral.* 39: 92–96
5. Milton RM. 1959. *U.S. Patent Nos. 2,882,243*; *2,882,244*
6. Milton RM. 1968. *Soc. Chem. Ind.* pp. 199–203
7. Breck DW. 1974. *Zeolite Molecular Sieves*, New York: Wiley & Sons. 771 pp.
8. Newsam JM. 1986. *Science* 231: 1093–99
9. Newsam JM. 1992. In *Solid State Chemistry: Compounds*, ed. AK Cheetham, P Day, pp. 234–80. New York: Oxford Univ. Press
10. Smith JV. 1988. *Chem. Rev.* 88: 149–82
11. Davis ME. 1993. *Acc. Chem. Res.* 26: 111–15
12. Gies H. 1991. In *Inclusion Compounds*, ed. JL Atwood, JED Davies, DD MacNicol, pp. 1–36. Oxford: Oxford Univ. Press
13. Davis ME, Lobo RF. 1992. *Chem. Mater.* 4: 756–68
14. Lobo RF, Zones SI, Davis ME. 1994. In *Inclusion Chemistry with Zeolites. Nanoscale Materials by Design*, ed. N Herron, D Corbin. In press
15. Burkett SL, Davis ME. 1994. In *Comprehensive Supramolecular Chemistry*, ed. JM Lehn. In press
16. Higgins JB. 1994. In *Reviews in Mineralogy: Silica Polymorphs*, ed. PH Ribbe. Vol. 29. Washington DC: Mineral. Soc. Am. In press
17. Flanigen EM, Bennett JM, Grose RW, Cohen JP, Patton RL, et al. 1979. *Nature* 271: 512–16
18. Meier WM, Olson DH. 1992. *Atlas of*

Zeolite Structure Types. London: Butterworth-Heinemann. 200 pp.
19. Liebau F. 1983. *Zeolites* 3: 191–93
20. Liebau F, Gies H, Gunawardane RP, Marler B. 1986. *Zeolites* 6: 373–77
21. Davis ME. 1994. *CHEMTECH* 24: 22–26
22. Petrovic I, Navrotsky A, Davis ME, Zones SI. 1993. *Chem. Mater*. 5: 1805–13
23. Maniar PD, Navrotsky A, Rabinovich EM, Ying JY, Benziger JB. 1990. *J. Non-Cryst. Solids* 124: 101–11
24. Ying JY, Benziger JB, Navrotsky A. 1993. *J. Am. Ceram. Soc*. 76: 2571–82
25. Navrotsky A, Petrovic I, Hu Y, Chen CY, Davis ME. 1994. *Microporous Mater*. In press
26. Chen CY, Li HX, Davis ME. 1993. *Microporous Mater*. 2: 17–26
27. Geisinger KL, Gibbs GV, Navrotsky A. 1985. *Phys. Chem. Minerals* 11: 266–93
28. Johnson GK, Tasker IR, Howell DA. 1987. *J. Chem. Thermodyn*. 19: 617–32
29. Robie RA, Hemingway BS, Fisher JR. 1978. *Thermodynamic Properties of Minerals and Related Substances at 298.15K and 1 Bar (105 Pascals) Pressure and at Higher Temperatures*. Washington DC: US Gov. Printing Office. 456 pp.
30. Richet P, Bottinga Y, Denielou L, Petitet JP, Tequi C. 1982. *Geochim. Cosmochim. Acta* 46: 2639–58
31. Akaogi M, Navrotsky A. 1984. *Phys. Earth Planet. Inter*. 36: 124–34
32. Patarin J, Kessler H, Soulard M, Guth JL. 1989. *Am. Chem. Soc. Symp. Ser*. 398: 221–32
33. Johnson DA, Martin JF. 1973. *J. Chem. Soc. Dalton Trans*. 15: 1585–90
34. Lewis GN, Randall M, eds. 1961. *Thermodynamics*, revis. KS Pitzer, L Brewer, 2nd ed. New York: McGraw-Hill. 723 pp.
35. Tanford C. 1980. *The Hydrophobic Effect: Formation of Micelles and Biological Membranes*. New York: Wiley. 233 pp.
36. Frank HS, Evans MW. 1945. *J. Chem. Phys*. 13: 507–32
37. Lindenbaum S, Boyd GE. 1964. *J. Phys. Chem*. 68: 911–17
38. Krishnan CV, Friedman HL. 1969. *J. Phys. Chem*. 73: 3934–40
39. Frank HS, Wen WY. 1957. *Discuss. Faraday Soc*. 24: 133–40
40. Burkett SL, Davis ME. 1994. *J. Chem. Mater*. In press
41. Wen WY, Saito S. 1964. *J. Phys. Chem*. 68: 2639–44
42. Wen WY, Saito S. 1965. *J. Phys. Chem*. 69: 3569–74
43. Agar JN. 1963. In *Advances in Electrochemistry and Electrochemical Engineering*, ed. P Delahay, pp. 31–121. New York: Wiley
44. Price CD. 1961. PhD thesis. Cambridge Univ.
45. van Santen RA, Ooms G, den Ouden CJJ, van Beest BW, Post MFM. 1989. *Am. Chem. Soc. Symp. Ser*. 398: 617–33
46. Shimizu A, Taniguchi Y. 1990. *Bull. Chem. Soc. Jpn*. 63: 1572–77
47. Shimizu A, Taniguchi Y. 1990. *Bull. Chem. Soc. Jpn*. 63: 3255–59
48. Gies H, Marler B. 1992. *Zeolites* 12: 42–49
49. Burkett SL, Davis ME. 1994. *J. Phys. Chem*. 98: 4647–53
50. Lobo RF, Davis ME. 1993. *U.S. Patent*. Application pending
51. Lobo RF, Pan M, Chan I, Li HX, Medrud RC, et al. 1993. *Science* 262: 1543–46
52. Treacy MMJ, Newsam JM. 1988. *Nature* 332: 249–51
53. Higgins JB, Lapierre RB, Schlenker JL, Rohrman AC, Wood JD, et al. 1988. *Zeolites* 8: 446–52
54. Liebau F, ed. 1985. *Structural Chemistry of Silicates*. Berlin: Springer-Verlag. 347 pp.
55. Griffen DT, ed. 1992. *Silicate Crystal Chemistry*. New York: Oxford Univ. Press. 442 pp.
56. Harris TV, Zones SI. 1994. *Stud. Surf. Sci. Catal*. 94: 29–36
57. Goepper M, Li HX, Davis ME. 1992. *J. Chem. Soc. Chem. Commun*. 22: 1665–66
58. Dove PM, Crerar DA. 1990. *Geochim. Cosmochim. Acta* 54: 955–69
59. Brady P, Walther JV. 1990. *Chem. Geol*. 82: 253–64
60. Zones SI. 1994. *Microporous Mater*. 2: 281–87
61. Brown ID. 1981. In *Structure and Bonding in Crystals*, ed. M O'Keeffe, A Navrotsky, pp. 1–30. New York: Academic
62. Dutta PK, Purl M, Bowers C. 1989. *Am. Chem. Soc. Symp. Ser*. 398: 98–109
63. Zones SI, Olmstead MM, Santilli DS. 1992. *J. Am. Chem. Soc*. 114: 4195–201
64. Zones SI, Santilli DS. 1993. In *Proc. Ninth Int. Zeolite Conf., Montreal* ed. R von Ballmoos, JB Higgins, MMJ Treacy, pp. 171–79. Boston: Butterworth-Heinemann
65. Nakagawa Y. 1993. *U.S. Patent No. 5,254,514*
66. Lobo RF, Pan M, Chan I, Zones SI, Crozier PA, Davis ME. 1994. *J. Phys. Chem*. 98: 12040–52
67. Lawton SL, Rohrbaugh WJ. 1990. *Science* 247: 1319–22

68. Hong SB, Cho HM, Davis ME. 1993. *J. Phys. Chem.* 97: 1622–28
69. Schmitt KD, Kennedy GJ. 1994. *Zeolites* 14: 635–42
70. Baerlocher C, McCusker LB, Chiappetta R. 1994. *Microporous Mater.* 2: 269–80
71. Burkett SL, Davis ME. 1993. *Microporous Mater.* 1: 265–82
72. Arhancet JP, Davis ME. 1991. *Chem. Mater.* 3: 567–69
73. Lobo RF, Zones SI, Davis ME. 1994. *Stud. Surf. Sci. Catal.* 84: 461–68
74. Szostak R. 1992. *Handbook of Molecular Sieves.* New York: Van Nostrand Reinhold. 584 pp.
75. Nakagawa Y. 1994. *Stud. Surf. Sci. Catal.* 84: 323–30
76. Annen MJ. 1992. *Synthesis and characterization of novel molecular sieves containing three-membered rings (zincosilicate gels).* PhD thesis. Virginia Polytech. Inst. 145 pp.
77. Annen MJ, Davis ME. 1993. *Microporous Mater.* 1: 57–65
78. Deleted in proof
79. Fyfe CA, Gies H, Kokotailo GT, Marler B, Cox DE. 1990. *J. Phys. Chem.* 94: 3718–21
80. Newsam JM, Treacy MMJ, Koetsier WT, de Gruyter CB. 1988. *Proc. R. Soc. London Ser. A* 420: 375–405
81. Li HX, Camblor MA, Davis ME. 1994. *Microporous Mater.* 3: 117–21
82. Davis ME, Chen CY, Burkett SL, Lobo RF. 1994. In 1994 *Mater. Res. Soc. Spring Meeting: Better Ceramics Through Chemistry VI, San Francisco.* 346: 831. Pittsburgh, PA: Mater. Res. Soc.
83. Kresge CT, Leonowicz ME, Roth WJ, Vartuli JC, Beck JS. 1992. *Nature* 359: 710–12
84. Beck JS, Vartuli MC, Roth WJ, Leonowicz ME, Kresge CT, et al. 1992. *J. Am. Chem. Soc.* 114: 10834–43
85. Lindblom G, Lindman B, Mandell L. 1973. *J. Colloid Interface Sci.* 42: 400–9
86. Kresge CT, Leonowicz ME, Roth WJ, Schmitt KD, Vartuli JC. 1993. *U.S. Patent No. 5,198,203*
87. Chen CY, Burkett SL, Li HX, Davis ME. 1993. *Microporous Mater.* 2: 27–34
88. Fontell K. 1990. *Colloid Polymer Sci.* 268: 264–85
89. Monnier A, Schuth F, Huo Q, Kumar D, Margolese D, et al. 1993. *Science* 261: 1299–303
90. Spegt PPA, Skoulios AE, Luzzati V. 1961. *Acta Crystallogr.* 14: 866–72
91. Mann S, Archibald DD, Didymus JM, Douglas T, Heywood BR, et al. 1993. *Science* 261: 1286–92
92. Nakagawa Y, Zones SI. 1992. In *Synthesis of Microporous Materials*, ed. ML Occelli, H Robson, pp. 222–39. New York: Van Nostrand Reinhold
93. Zones SI, Nakagawa Y. 1994. *Microporous Mater.* 2: 543–55
94. Kumar R, Reddy KR, Raj A, Ratnasamy P. 1993. In *Proc. Ninth Int. Zeolite Conf. Montreal*, ed. R von Ballmoos, JB Higgins, MM Treacy, pp. 189–96. Boston: Butterworth-Heinemann

Annu. Rev. Mater. Sci. 1995. 25: 193–222

THE NATURE OF GRAIN BOUNDARIES IN THE HIGH-T_c SUPERCONDUCTORS

S. E. Babcock[1,2] *and J. L. Vargas*[1]

[1] Applied Superconductivity Center and [2] Materials Science and Engineering, University of Wisconsin, Madison, Wisconsin 53706

KEY WORDS: weak links, bicrystals, EELS, $YBa_2CU_3O_{7-\delta}$, grain boundary structure

ABSTRACT

Grain-boundary weak links limit the critical current density of polycrystalline high-T_c superconductors. Themes that have evolved from research on these grain boundaries include (*a*) an apparent general and dramatic distinction between the behavior of low-angle and high-angle grain boundaries, (*b*) electrical heterogeneity along as well as among boundaries, (*c*) indications that the macroscopic current path may or may not cross certain high-angle grain boundaries, (*d*) a key role for TEM-EELS in studies of the electronic structure near grain boundaries, and (*e*) evidence that current models for grain boundary structure pertain to the high-T_c superconductors and may eventually help us to understand their electromagnetic character. Results regarding each of these topics are reviewed in this paper.

INTRODUCTION

The mid-to-late 1980s discoveries of superconducting materials with transition temperatures (T_c) above the boiling point of liquid nitrogen caused great excitement among both scientists and the public. The high T_c values of these cuprate superconductors inspired imaginative visions of widespread superconducting technologies. However, it is another critical property, the transport critical–current density (J_c), that really dictates the types

0084–6600/95/0801–0193$05.00

of applications that can be achieved with a particular superconductor. The reality of the low-J_c values of these materials and their unfavorable dependence on magnetic field (H) was soon to quell the euphoria aroused by their discovery. Broad-based efforts to both understand and improve the J_c characteristics of these superconductors have been a focus of high-T_c superconductor research ever since. This paper describes just one aspect of such work, namely the critical role that grain boundaries play in limiting the J_c (H,T) characteristics of polycrystalline high-T_c materials.

THE GENERAL GRAIN BOUNDARY PROBLEM

At the present time, the properties of high-angle grain boundaries are believed to control the macroscopic J_c(H) characteristics of all polycrystalline high–T_c superconductors. This control occurs because most high-angle grain boundaries act like barriers to the current and have electromagnetic properties that often are Josephson junction–like. As a result, the high-angle grain boundaries in the microstructure introduce a network of weak links (reduced J_c regions) into a superconducting path in which the grain interiors may have J_c characteristics that are appropriate for applications.

The characteristic properties of Josephson junction–like weak links include a strongly magnetic field–dependent J_c that can decrease by more than an order of magnitude in weak fields of just a few millitesla. Such J_c(H) characteristics clearly are a serious problem for magnet applications (motors, generators, MRI instrumentation, energy storage systems, etc). As a result, many view high-angle grain boundaries as the immediate obstacle to further development of wire-form materials for such applications. However, polycrystalline Bi-Sr-Ca-Cu-O superconducting tapes with appreciable J_c values (e.g. 10^5 A$\cdot$cm^{-2} at 4.2 K in fields up to 20 T for $Bi_2Sr_2Ca_1Cu_2O_8$) have been fabricated in a number of laboratories (1, 2). Their fabrication suggests that the J_c-limiting effect of weak coupling at grain boundaries can be minimized through either improving the J_c characteristics of individual grain boundaries or texturing of the material so as to minimize the density of weak-link grain boundaries in the current path. Furthermore, direct evidence that supercurrent transport through some high-angle grain boundaries is not limited by weak coupling has been obtained (3–6).

Ironically, the grain boundary behavior that plagues high-field and high-current applications forms the foundation of thin-film and integrated-circuit technologies that are based on the Josephson effect (7, 8). The extreme sensitivity of the J_c of a grain boundary-Josephson junction to weak magnetic fields enables, rather than defeats, applications of high-T_c

materials in electronics. The properties of grain boundary junction-based devices are tied strongly to the grain boundary microstructure because the tunneling phenomenon that results in their magnetic-field sensitivity arises from the presence of a thin interface layer between two superconductors. This layer can have normal metal, insulator, or weak superconducting properties. However, for predictable device performance, it must be of uniform thickness and controlled electrical character. This same barrier layer is responsible for the weak-link character of grain boundaries that is the focus of bulk-scale conductor development.

Thus from both the high-field, flux-pinning viewpoint and the low-field, Josephson junction–based electronics viewpoint, there is strong motivation to develop a detailed picture of the grain boundary structure and microstructure and to describe their effects on the electromagnetic properties of the grain boundaries. Why then, do we not have a more complete understanding of microstructure-property relationships for grain boundaries? Key reasons, described further below, include (*a*) the very high spatial resolution with which all aspects of the boundary must be characterized, (*b*) the sensitivity of at least some of these compounds' properties to oxygen stoichiometry, an element whose distribution is difficult to study with currently available high-spatial-resolution techniques, and (*c*) considerable heterogeneity both among and along high-angle grain boundaries.

The superconducting coherence length ξ defines the grain boundary thickness that can be traversed by the supercurrent. Barriers of thickness up to a few coherence lengths can still show superconductive coupling, albeit of reduced strength. In the layered orthorhombic high-T_c superconductor structures, ξ is on the order of ~ 0.3 nm along the [001] direction and ~ 1.5 nm in (001) planes that carry the current. Thus extremely thin barriers can cause weak-link behavior. High-resolution transmission electron microscope (TEM) imaging, scanning Auger microanalysis of fracture surfaces, and high-spatial-resolution energy-dispersive X-ray spectroscopy in the scanning transmission electron microscope (STEM) have indicated strongly that, at least with respect to the cation distribution, the grain boundaries are defects with thicknesses in the 0.5 to 1 nm range, indeed, approaching ξ (9–12). This result is not surprising because their $\sim$1-nm scale dimensions also approximate the structural widths that are observed for grain boundaries in representative oxide materials (e.g. NiO) that have been studied in considerably more detail (13, 14). The effective electrical width of the boundary appears to be considerably larger, however (see section on High Spatial Resolution EELS).

The known sensitivity of the electrical properties of bulk $YBa_2Cu_3O_{7-\delta}$ (YBCO) to its average oxygen stoichiometry draws attention to the difficult problem of determining the oxygen concentration profile in the grain boundary vicinity with high precision. Undoubtedly, the grain boundary oxygen concentration is important in YBCO because changes in δ as small as ~ 0.1 can result in measurable changes in macroscopic superconducting properties (15–17). Although the effect of oxygen concentration is less well studied in the other high T_c materials, some sensitivity to oxygen stoichiometry appears to be a general feature of the cuprate superconductors (18). It is difficult to measure oxygen concentrations with high precision and high spatial resolution simultaneously. Nevertheless, high-spatial-resolution TEM/electron energy loss spectroscopy (EELS) studies of the oxygen absorption edge have provided some of the most enlightening experimental evidence about the microstructures associated with weak-link character (see section on High Spatial Resolution EELS).

Heterogeneity of the grain boundary structure and properties has arisen as a theme in high-T_c grain boundary studies. The observed general disparity between the behavior of low-angle and high-angle boundaries (see section on Low-Angle/High-Angle Distinction) might have been predicted and is easily rationalized in terms of simple grain boundary structure models. Some variety among boundaries with different crystallographic parameters in the high-angle regime also might be anticipated. However, the collective data suggest that the fundamental mechanism controlling the J_c (i.e. Josephson-coupling or flux-pinning) can vary from boundary to boundary within the high-angle regime and possibly from boundary facet to boundary facet (19–21). Of at least equal significance, weakly coupled, high-angle boundaries show evidence of substantial electrical heterogeneity along the interface (see section on Heterogeneity of Grain-Boundary Weak Links). These grain boundaries appear to be subdivided electrically into sets of parallel junctions with different properties (22–26). The origin of this heterogeneity has yet to be identified, but it may have its roots in either intrinsic or extrinsic features of the grain boundary structure or microstructure (see sections noted above).

The preceding paragraphs have introduced a few of the themes that have evolved from research on grain boundaries in the high-T_c superconductors. Prominent among these are (*a*) an apparent general and dramatic distinction between the behavior of low-angle and high-angle grain boundaries, (*b*) electrical heterogeneity along, as well as among, boundaries, (*c*) indications that the macroscopic current path may or may not cross certain high-angle grain boundaries, (*d*) a key role for TEM-EELS in studies of the electronic structure near grain boundaries, and (*e*) evidence that current models for grain boundary structure pertain to the high-T_c superconductors and may eventually help us to understand their electromagnetic character. Results regarding each of these topics are summarized below.

Because the authors' interests lie in the potential use of superconductors for high-current rather than electronics applications, the following undoubtedly is biased in that direction.

THE LOW-ANGLE/HIGH-ANGLE DISTINCTION

Discussions of the role of grain boundaries in determining critical current characteristics almost universally separate grain boundaries into two classes: those that have small ($< \sim 10°$), and those that have large angles of misorientation θ. The relevance of this partition was first brought to light in the thin-film bicrystal studies of Dimos et al (27, 28), the results of which are often still quoted to represent the essence of the grain boundary weak-link problem. In the Dimos et al study, sets of θ [001] symmetrical tilt, θ [100] symmetrical tilt, and θ [100] twist boundaries were fabricated by epitaxial deposition of YBCO onto bicrystalline $SrTiO_3$ substrates with pre-determined θ values. The J_cs of the boundaries [J_c(gb)] were measured and compared with the J_cs of the neighboring grain interiors J_c(g). In the Dimos et al famous figure, $J_c(gb)/J_c(g)$ decreased rapidly with a θ^{-1}-like dependence for θ values up to about 20°. For $\theta > 20°$, an approximately θ-independent, uniformly low value of $J_c(gb)/J_c(g) \approx 10^{-2}$ was observed. The J_c(gb) and voltage-current (V-I) behavior of boundaries with $\theta > \sim 5$ to 10° mimicked those of a resistively shunted Josephson junction (RSJ), reflecting the presence of weak electromagnetic coupling across the boundary. Later studies of the normal state resistance (R_N) of symmetrical [001] tilt boundaries showed a related dependence of R_N on θ, suggesting (*a*) a correlation between the depression of the boundary J_c value and an increased R_N (29), and (*b*) a general dependence of the electromagnetic properties of the boundary on θ characterized by a θ^{-1}-like dependence for low θs and θ-independence at large θs.

Analogous studies of symmetrical [001] tilt boundaries in other high-T_c superconductors originally suggested that this low-angle/high-angle picture might describe grain boundary behavior in all of the high-T_c superconductors. Epitaxial bicrystal experiments using thin films of $Bi_2Sr_2Ca_1Cu_2O_8$ (BSCCO-2212) (30–32); $TlBa_2Ca_2Cu_3O_x$ (TBCCO-1223) (33); $Tl_2Ba_2CaCu_2O_x$ (TBCCO-2212) (34); $Tl_2Ba_2Ca_2Cu_3O_x$ (TBCCO-2223) (31, 35); and $Nd_{1.85}Ce_{0.15}CuO_4$ (NCCO) (31) all showed a qualitatively similar dependence of J_c(gb) on θ. Such results, as well as the ubiquitous RSJ behavior of weakly coupled boundaries, supported the idea of a common origin to the weak-link character observed in all of these materials (31).

The θ^{-1} dependence/plateau behavior of $J_c(gb)/J_c(g)$ vs θ, led Dimos et al, as well as several others (e.g. 36, 37), to postulate that some aspect of the lattice dislocations [or, in grain boundary nomenclature, primary grain boundary dislocations (PGBDs)] that comprise the grain boundary structure determines its electromagnetic properties. The plateau behavior that initiated at $\theta \sim 20^\circ$ was in good agreement with the misorientation angle at which PGBDs with "conventional" cores of radius $r_o \approx |\mathbf{b}|$ (38) ($|\mathbf{b}| \approx 0.39$ nm in this case) would overlap if they were uniformly distributed along the boundary. However, the transition to Josephson junction–like weak coupling in the $\theta = 5$–10° range for YBCO suggested that the effective size of the dislocation with respect to the local superconducting properties might be somewhat larger if this model were indeed correct. To account for the disagreement, a model was proposed in which the elastic-strain fields of dislocations possessing narrow cores with the correct 1:2:3 cation stoichiometry suppress the local superconducting properties sufficiently to effectively increase the size of the dislocation (37). A key role for the elastic-strain field was contested, however, with arguments that the strain fields near tilt boundaries should have very different characteristics from those near twist boundaries due to the parallel sets of edge dislocations and cross-hatched networks of screw dislocations that comprise their structures, respectively. It was argued that if these strain fields control the properties, distinct electrical characteristics would result, a prediction in disagreement with the observations of Dimos et al (23). Approximately concurrently, PGBDs with radii of ~ 1 nm and a composition different from that of the superconducting grains were observed in low-angle grain boundaries in thin-film samples by high-resolution TEM (36). Their calculated overlap at $\sim 11^\circ$ showed better agreement with the transition to weak-coupling character.

More recent observations have caused reconsideration of the role that PGBD cores and strain fields may play. Evidence that the transition to weak-link behavior can be pushed to $\theta \approx 20^\circ$ in bulk-scale YBCO bicrystals processed by partial melting (6) suggests that it might be possible to reduce the effective size of the dislocation core to the conventional value, if the dislocation model, indeed, pertains. Observations of the strong-to-weak coupling transition at slightly higher θ values in flux-grown, bulk-scale bicrystals relative to the thin-film bicrystals of YBCO is consistent with this idea (3, 39). However, the characteristics of other types of bicrystals have revealed an important variation indicative of a more complex origin for the weak-link behavior than dislocation PGBD overlap. Whereas thin-film BSCCO-2212, TBCCO-2223, and NCCO symmetrical tilt boundaries all showed a continuous evolution of weak-link behavior (31), TBCCO-1223 (33) experiments showed plateau behavior in the low-angle regime and suggested a step function–like decrease in $J_c(\theta)$ to a high-angle value at θ between 15 and 20°. The field dependence of J_c also showed bimodal behavior in these two ranges of θ; the low-angle boundaries had higher, weakly field–dependent J_c values, whereas the high-angle boundaries all showed about the same, stronger field dependence. Evidence for structure in the electrical properties within the high-angle regime also suggests a less clear-cut domination of PGBDs in determining the coupling character of grain boundaries. Asymmetrical [001] tilt boundaries in YBCO (40) showed a linear decline in ln (J_c) throughout the entire range of θ. In fact, the Dimos et al data also show this dependence if J_c(gb) is plotted vs θ (7). Possibly similar behavior was observed for TBCCO-2212 epitaxial bicrystals, although the film quality caused a question as to whether the J_c decrease was continuous or saturated to a plateau (34). The residual current measured for weakly coupled YBCO bicrystals depends on θ for $\theta > 20^\circ$ as well (see below) (26). These results indicate that the structural features of high-angle boundaries may still influence the electromagnetic coupling across the boundary.

Themes similar to those described above appear in the data obtained from grain boundaries in bulk-scale bicrystals. Direct flux imaging, by the magneto-optic technique, at [001] tilt grain boundaries in YBCO crystal clusters suggests a hierarchy of behavior (41). The magnetic flux penetrated all high-angle grain boundaries under the influence of even very low applied magnetic fields. Higher fields were required successively for flux penetration along low-angle grain boundaries, along twin boundaries and, finally, into the grain interiors between the twin boundaries. A plot of the external field required for flux penetration vs θ for nominally [001] tilt boundaries was remarkably similar in shape to the plot of Dimos et al (28). Interestingly, electromagnetic characterization of similar flux-grown bicrystals has revealed evidence for strong coupling at grain boundaries with θ as high as 14°. Strongly coupled boundaries in the high-angle regime (90° [001]) also were first observed in these bicrystals (3).

Studies of bulk-scale θ [001] twist boundaries of BSCCO-2212 tell similar stories. In one study, the critical current density of the weakly coupled boundaries appeared to correlate with the coincidence index, or Σ value, of the boundary (42) which, in turn, depends on θ. Another study described examples of strongly coupled high–angle (23 and 88°) [001] twist boundaries (5). (The 88° boundary could have been considered to be a low-angle boundary if the structural modulation that occurs along the *b*-axis in this compound produced no anisotropy.) An interesting result of the latter study was that strong-link behavior was always correlated with a significantly reduced boundary T_c (up to 30 K), whereas the boundaries with weak-link character had at most very slightly reduced (<3 K) boundary T_c values. This anticorrelation is quite remarkable. It starkly contrasts the behavior of YBCO, for which reduced boundary T_cs always forecast weak coupling (43). All of these points suggest that for many, but possibly not all, cases the partition of boundaries into low-angle and high-angle may be a good approximation, but that the detailed characteristics of grain boundaries are more complicated than a simple dislocation overlap model can describe.

Interestingly, some low-angle grain boundaries appear to pin flux against motion perpendicular to the boundary plane. Evidence for a flux-pinning role of low-angle boundaries has been observed in YBCO (41) and in BSCCO-2212 (44) by magneto-optic imaging; evidence for flux pinning at a boundary with the relatively high value of $\theta = 17°$ has been obtained in the case of BSCCO-2212 (45).

HETEROGENEITY OF GRAIN-BOUNDARY WEAK LINKS

A recurring theme of more recent studies of the electromagnetic character of weakly coupled boundaries in both bulk and thin-film materials is that their superconducting properties vary substantially, and perhaps systematically, with position along the grain boundary. Several types of evidence point to heterogeneity as a general feature of their character, and a number of models and mechanisms have been proposed to describe it. In most models, a subdivided boundary is pictured. Patches of "better" material are separated by weak or nonsuperconducting regions, as shown schematically in Figure 1. A number of possible descriptions of the electrical character and topography of these regions have been proposed, but none has yet emerged as a general mechanism by which local structural or compositional variations affect the order parameter. Local oxygen

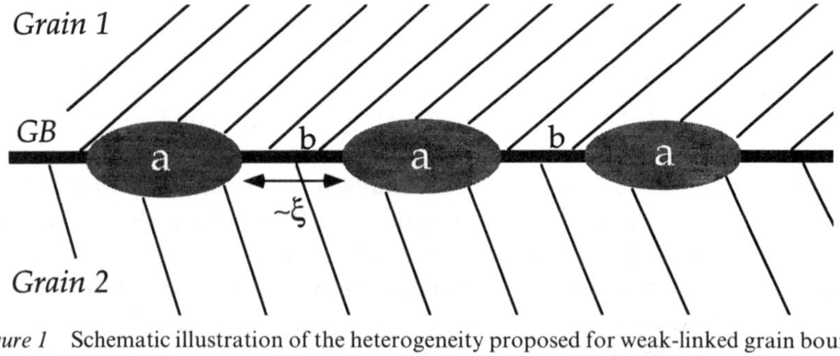

Figure 1 Schematic illustration of the heterogeneity proposed for weak-linked grain boundaries in YBCO. The current flows through microbridges (labeled *b*) that exist between nonsuperconducting regions (labeled *a*). The electrical character of the microbridges is assumed to be either normal, superconducting, or insulating, depending on the model.

depletion and oxygen disorder remain strong candidates for the case of YBCO, and quasi-periodic strain fields and charge trapping centers remain possible sources for all of the high-T_c superconductors.

Very early in the study of sintered YBCO, Larbalestier et al suggested that high-field intergranular transport was limited by the density and character of occasional "microbridges" that spanned otherwise weakly coupled boundaries (43). Studies of the temperature dependence of J_c suggested that these bridges had strong-coupling character but coherence length ξ-scale dimensions (43). The field-independent residual J_c measured for early sintered samples in fields as large as 27 T at 4.2 K (46) could be explained by this model. Evidence for general heterogeneity at all weakly coupled boundaries in the high-T_c materials lies (*a*) in the persistence of small, but finite supercurrents across weakly coupled junctions in the presence of high magnetic fields, (*b*) in the appearance of half-integral multiples of the Shapiro steps in the V-I curves obtained under microwave radiation, and (*c*) in the 1/f noise characteristics of individual grain boundary junctions.

Residual critical currents have been observed consistently in diverse forms of YBCO, including bicrystal thin-film junctions (26, 47) and bicrystalline samples of melt-textured material (48). A microbridge-type model has been proposed by Sarnelli et al to explain observed residual currents across thin-film bicrystal junctions (26). In this model, narrow superconducting channels exist between semiconducting (or insulating) channels in a parallel configuration (26, 49, 50). The superconducting channels have reduced T_c values relative to the grains, resulting in S-S′-S character, where S and S′ indicate superconductors with different properties. Josephson-like behavior is expected if the width of the superconducting channels is of the order of ξ. If the nonsuperconducting channels are associated with the PGBD cores, it is possible to estimate the crossover angle (strong coupling to weak-link behavior) because the PGBD spacing varies approximately with $[\sin(\theta/2)]^{-1}$. The model suggests that the crossover should occur at smaller angles in materials with larger coherence length (e.g. NCCO has $\xi \approx 7$ nm). Preliminary studies appear to confirm this prediction (31).

The presence and density of charge-trapping centers at YBCO grain boundary junctions has been deduced from characterization of the junction noise (23, 51). Measurement of the low-frequency 1/f noise revealed that the density of scattering centers is likely to be larger than 1.3×10^9 cm^{-2}, which corresponds to a spacing of less than 270 nm on average (23). This value is in reasonable agreement with the scale of the heterogeneity that is suggested by the low-field, field dependence of J_c measured for bulk-scale bicrystals (52). Ozone annealing of the thin-film junctions decreased the estimated density of trapping sites, the normal state resistance, and the 1/f noise, but did not eliminate the weak-link behavior (23, 51). These observations suggest that although oxygen deficiency affects the properties of the weak links, some other feature of the grain boundary structure and chemistry must cause it (23).

An alternative role for oxygen has been suggested by Moeckly et al (24). Their measurements of the superconducting properties of grain boundary junctions that have been affected by electromigration currents suggest that oxygen disorder can contribute to, if not cause, weak-link behavior. In their model, the oxygen sublattice is highly disordered and non-uniform near the boundary. Intergranular connections consist of channel-like regions where the oxygen disorder is not as severe as in other regions of the boundary. These channels have lower T_c than the grain interiors but are still superconducting. A large fraction of the filaments from one grain will terminate in strongly disordered regions of the other grain, resulting in a large density of ohmic shunts in parallel with a few strong links. This microstructure is consistent with the transport properties of the tilt boundaries in thin films grown on MgO substrates (22).

A different type of evidence for heterogeneity is found in the V-I characteristic of biepitaxial thin-film boundary junctions (25). The V-I characteristics of these 45° [001] grain boundary junctions display extra steps at half-integral values of the Shapiro steps observed under microwave irradiation. Based on these observations and their analogy to the behavior of two Nb-based junctions in parallel (53), Early et al suggested that biepitaxial boundaries might consist of a one-dimensional array of multiple Josephson junctions in parallel (25).

Understanding the nature of the grain boundary heterogeneity in the high-T_c compounds is a complex problem. Macroscopic electromagnetic measurements that reveal the net effect of microscopically heterogeneous properties must be interpreted through the use of models and analogies. Several other superconductor-normal metal–superconductor (54) and superconductor-insulator–superconductor junction models (55) can describe the data reasonably well. Consider, for example, the resonance measurements in a YBCO thin-film bicrystal (32° [100] tilt boundary) by Winkler et al (56). The grain boundary model used to interpret the measurements is that of a long, uniform, parallel-plate resonator where the grain boundary is the effective dielectric medium and the grains act as the resonator walls. The results are consistent with the presence of an insulating layer of 2 nm thickness. Unfortunately, as the authors remind us, the type of measurements performed cannot distinguish between an insulating tunnel barrier and a barrier containing superconducting filaments, as in the models described above.

A number of possible microstructural sources of the heterogeneity have been observed by electron microscopy studies of weakly and strongly coupled boundaries, including (*a*) wavy boundaries, facet formation, and facet junctions; (*b*) cation composition modulations that occur within the boundary plane; (*c*) strain fields of intrinsic and extrinsic grain boundary dislocations; (*d*) the intersections of the twin planes with the grain boundary plane in the case of YBCO; and (*e*) oscillatory changes in misorientation due to twinning in YBCO. Several of these possibilities are described in other sections of this paper. The wavy nature of the grain boundaries in epitaxial thin-film bicrystals alone illustrates the complexity of identifying the origins of weak-link behavior and its heterogeneity. Thin-film bicrystals often have been perceived as ideal samples with highly controlled boundary crystallography. However, a number of recent investigations, including our own studies of an 18° [001] thin-film junction on a yttria-stabilized zirconia (YSZ) bicrystal substrate, have shown that the grain boundary plane can deviate substantially from the nominal one established in the substrate (57–60; IF Tsu & SE Babcock, unpublished data). Thus epitaxial growth of the thin film does not guarantee that the substrate grain boundary will be reproduced in the superconducting film. These observations highlight the need for coupled electromagnetic measurements and TEM investigations of the same boundaries in order to establish the relevance of some of the models used to interpret electromagnetic measurements.

The role of the small coherence length is a focal point of all of the discussion of weakly coupled high-T_c grain boundaries. In this context, it is interesting to review the behavior of the Chevrel phases (e.g. $PbMo_6S_8$ and $SnMo_6O_8$) that have coherence lengths on the order of 3 nm. Even though their coherence lengths are just twice that of YBCO in the (001) planes, weak-link-free polycrystalline samples have been synthesized (61). Like the high-T_c materials, the unit cells of these compounds are complex. There is little reason to believe that the structural widths of their grain boundaries are considerably different from those of the high-T_c materials. Thus the strong coupling that is observed at Chevrel phase superconductor grain boundaries raises questions about the paradigm that the small coherence length of high-T_c cuprates leads directly to intrinsic weak-link character at grain boundaries.

MACROSCOPIC CURRENT PATHS AND GRAIN BOUNDARIES

The misorientation angle dependence of the grain boundary J_c value in YBCO, together with the rather larger J_c values that have been obtained in silver-sheathed BSCCO tapes (1, 2), has focused attention on the nature of the microscopic and macroscopic current path through polycrystalline bulk materials. Strong alignment of the *c*-axes of the plate-shaped grains that make up the microstructures of these materials is observed in all high-J_c tapes. Texture within the (001) plane is not evident over macroscopic dimensions, however. Light and electron microscopy observations of BSCCO tapes show that the *c*-axis alignment is high near the superconductor/silver sheath interface, but frequently is distributed in the center of the tape (62, 63). This microstructure suggests the presence of three basic types of boundaries. The first type includes the [001] twist boundaries that form between the large faces of grains with perfect (or nearly perfect) *c*-axis alignment. Stacks of [001] twist grain boundaries frequently are observed within the plate-shaped grains that are visible by light microscopy, as well as between well-stacked colonies. The second type of boundary is the [001] tilt boundaries that form where the edges of two plates with well-aligned *c*-axes meet. The third type describes boundaries with the "railway switch" (64) geometry that form between grains with significantly misaligned *c*-axes. These boundaries resemble a railway switch at which the (001) planes form the tracks, an appropriate analogy because these planes appear to carry a large component of the current. The two basic models that have been proposed to describe the J_c characteristics of BSCCO tapes focus on the roles of these types of boundaries (64–66).

It is quite possible that the nature of the current paths differs from material to material among the high-T_c superconductors. Studies of misorientation angles on a grain-by-grain basis in TBCCO-1223 thick films reveal local alignment of the [100] axes of some neighboring grains in tapes that have no apparent in-plane alignment on a macroscopic (X-ray pole-figure) scale (67, 68). More recent work (A Goyal, private communication) indicates that as many as 60 to 80% of the boundaries in the material studied have low θ values. Furthermore, $\sim$40% of the high-angle boundaries that remain most likely correspond to constrained coincidence, site-lattice (CCSL) misorientations (see below). On the basis of these observations, Kroeger et al and Goyal et al have proposed that the current in this TBCCO-1223 material travels along percolative paths composed of only low-angle, presumably strongly coupled, boundaries (67, 69). Coupled studies of the relationship between overall Tl stoichiometry, J_c, and grain boundary stoichiometry led them to conclude that the current travels from grain to grain across low-angle tilt boundaries rather than low-angle twist boundaries. This conclusion is consistent with the general observation that current flows predominantly in the (001) planes of all of the high-T_c materials.

Similarly, the rather weak anisotropy of the normal state resistivity (ρ_n) and J_c measured along vs across a tape section led Cho et al to conclude that the current path in BSCCO-2223 tapes is percolative and involves transport in (001) planes (70). But how does the current travel from grain to grain in these materials? The current path question and the role of grain boundaries in it appear to be rather complicated in the case of BSCCO tapes. In at least BSCCO-2212, the *c*-axis electrical properties vary with oxygen doping in both the normal and superconducting states (71), suggesting that the current path can change as the character of the *c*-axis transport changes from weakly to strongly coupled. In fact, Wang et al have fabricated high-angle [001] twist boundaries of BSCCO-2212 that are strongly coupled in the [001] direction (5). These results (5, 71) suggest that the current path for certain doping states may include a significant *c*-axis component both within and between the grains. Umezawa et al have provided electromagnetic and microstructural evidence that the active current path in polycrystalline BSCCO-2223 tapes indeed includes *c*-axis transport and crosses [001] twist boundaries in low-applied fields (72). However, electron microscopy of these same tapes shows that many of the twist boundaries contain BSCCO-2212 intergrowths that are proximity coupled at low fields. At higher fields, the BSCCO-2212 layers become decoupled, and the twist boundaries that contain them are eliminated from the active current path. As a result, the coupling mechanism across the boundaries that do exist in the current path, as well as the path's geometry, are different in these two field regimes. Comparisons of the distribution of misorientation angles measured on a grain boundary by grain boundary basis with the J_c values measured for BSCCO-2212 and BSCCO-2223 tapes have provided possible evidence that the CCSL-related high-angle boundaries (see below) that form in BSCCO-2212 tapes may be more strongly coupled than other boundaries. However, the observation of Wang et al of weak coupling for near $\Sigma 5$ (36° [001]) and near $\Sigma 17$ (30° [001]) twist boundaries and strong coupling for a near $\Sigma 13$ (23° [001]) twist boundary may provide counter evidence against a direct relationship between the boundary Σ value and the coupling strength (5). ($\Sigma 5$, $\Sigma 17$, and $\Sigma 13$ CSLs form at 37, 28, and 23° [001], respectively, in tetragonal lattices.)

Direct studies of the critical current variation with position in a tape have shown substantial macroscopic heterogeneity in BSCCO tapes. Sectioned BSCCO-2223 tapes showed a factor of three difference in the critical currents measured at the tape edges and at the tape center (Figure 2) (62). Furthermore, the critical current measured near the center of the tape agrees well with the macroscopic critical current density. This conclusion is supported by more recent determinations of the local critical current

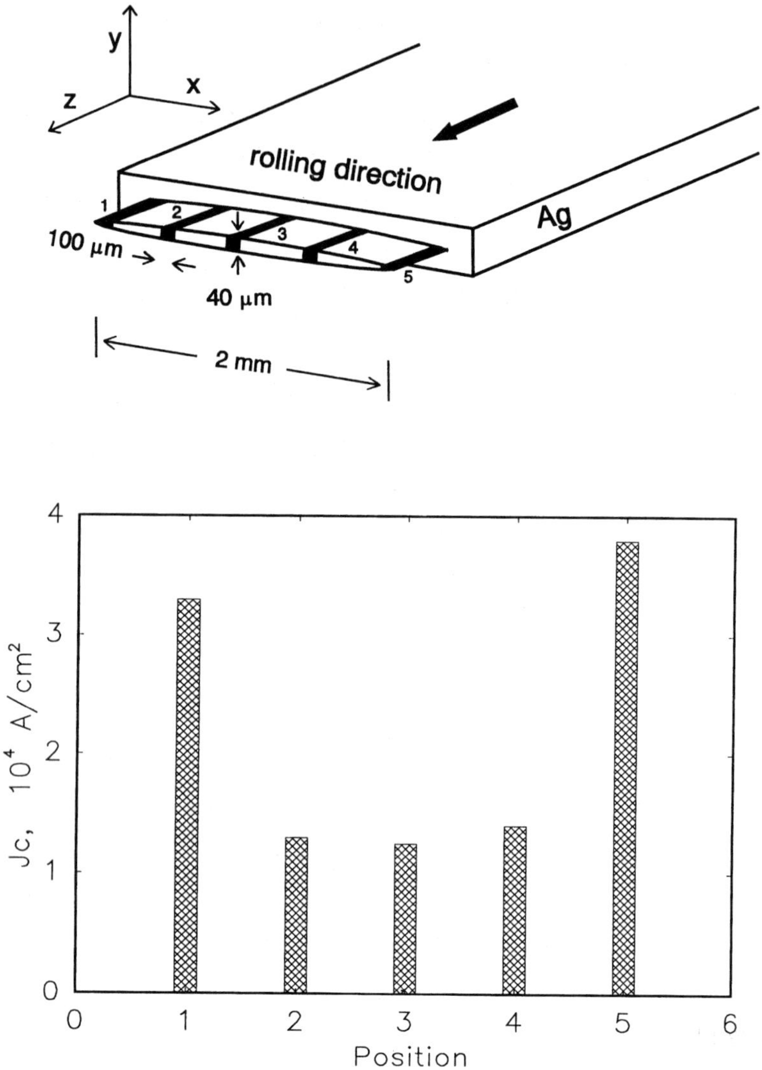

Figure 2 Transport critical current density (zero field, 77 K) values for different sections of a silver-clad BSCCO-2223 tape. Nature of the microslices used is illustrated in top part of figure. (Figure courtesy of DC Larbalestier.)

density via analysis of magneto-optic images of flux penetration into the tape (73). The analysis shows that BSCCO-2223 tapes have regions of very high J_c near the edges of the tape, but these regions may be disconnected from one another. These observations draw attention to the promise of

BSCCO tapes if the microstructure and consequent critical current density can be optimized throughout the tape cross section and length by microstructural control.

The microscopic and macroscopic current path questions are important for the development of high-T_c conductors. The results described above represent some fundamental insights that have been gained but also illustrate the complexity of the problem. It is clear, however, that a thorough understanding of grain boundary properties is one of the keys to the fabrication of high-J_c tapes.

ON THE UTILITY OF MODELS FOR GRAIN BOUNDARY STRUCTURE

Over the past few decades, models for grain boundary structure and related grain boundary phenomena have been developed and tested rather extensively for materials with cubic crystal lattices and rather simple crystal structures. These models have successfully described the types of structural features that are observed in grain boundaries in materials ranging from pure metals like gold and silver to metal alloys, ceramic compounds, and semiconductors. More recently, the basic concepts of these models have been extended to describe grain boundary structural features in other crystal systems (74–78) and even in quasi-crystals (79, 80). This section discusses the application and utility of these models for grain boundaries in the high-T_c superconductors.

Available Models

Two basic approaches are used to describe boundary structures in a systematic way. One, the structural unit approach, focuses on the atomic arrangement at the grain boundary, whereas the other, the grain boundary dislocation (GBD) approach, focuses on the periodic strain field that is observed at many grain boundaries. Both models use a coincidence site lattice (CSL) description of the grain boundary geometry. Although these models are equivalent (81), each one provides a valuable view of the grain boundary structure from a somewhat different perspective. Nevertheless, studies of grain boundaries in the high-T_c superconductors have focused primarily on the GBD description and its relevance to these materials and their properties.

GBD models are based on the idea that the grain boundary free energy is particularly low for a certain set of special misorientation relationships, θ [hkl]. These models assert that boundaries with θ [hkl] values other than special ones relax into a configuration in which patches of a low-energy structure are preserved by a localized rotation of the crystals. The macroscopic θ [hkl] and its deviation from a low-energy misorientation are produced by the introduction of a regular array of dislocations that rotate one crystal relative to the other on a macroscopic scale. On the microscopic scale, these dislocations separate the patches of boundary with low-energy structure. Thus most grain boundaries are envisioned as being subdivided into regions of "good fit," or low-energy atomic structure, separated by a net of "bad fit," or GBDs. As the difference between the actual misorientation relationship and the low-energy one increases, the spacing of the dislocations decreases, the subdivision becomes finer, and the fraction of bad-fitting area increases.

A CSL construction generally is used to identify the low-energy misorientations and describe the structural periodicity of the boundary. The CSL is the superlattice of coincident sites that would arise for a given θ [hkl] if the lattices of each crystal that abut a grain boundary had a fixed common origin and were allowed to interpenetrate. The GBD model assumes that low energy is associated with relatively short-period CSLs or, equivalently, low-coincidence index (Σ) values, where Σ is the fraction of lattice sites of one crystal that are coincident with the other (38). The symmetry and periodicity of a grain boundary are the same as those of the equivalent plane of the CSL that corresponds to its misorientation relationship. Because of the high degree of symmetry of cubic crystal structures, CSLs occur at a high fraction of all possible misorientation relationships.

A principal difficulty with directly extending the GBD model and the CSL construction to other crystal systems is that true CSLs occur for a substantially lower fraction of possible misorientation relationships. CSLs can only form in orthorhombic and tetragonal crystals when the squares of the lattice parameters all form rational ratios (i.e. $a^2 : b^2 : c^2$ is rational) (76). This condition is rarely met in real crystal structures. The approach taken to apply the coincidence concept to these systems is to impose a small, local strain on the lattice in the boundary vicinity. This strain forces the lattice parameters to meet the rational ratios condition, and the lattices can form CSLs. The CSLs that result under strained conditions often are called constrained CSLs (CCSLs). Usually, the additional strain is attributed to additional sets of dislocations that reside in the boundary and become part of the GBD network. Thus the GBD network has two components, one to produce the correct axial ratios, and one to accommodate deviations from low-energy misorientations. An interesting consequence of the dual origin for GBDs in CCSL-related boundaries is that GBD networks are predicted even for boundaries for which the deviation from an exact CCSL misorientation is zero. Furthermore, because the strain-producing and deviation-accommodating GBDs can have opposite signs, the dislocation density may not increase monotonically with the deviation from the exact CCSL misorientation relationship. As a result, the minimum spacing may not occur at the CCSL misorientation, and widely spaced GBD networks may accommodate rather large deviations from coincidence (76), thereby introducing a coarse boundary subdivision into a seemingly nonspecial boundary.

The (C)CSL/GBD model provides a fairly complete picture of several important features of grain boundary structure. The following subsections describe substantial evidence suggesting that this picture pertains to the high-T_c superconductors. However, the degree to which current models describe the details of the grain boundary structural features and the extent to which they can help us understand the superconducting properties of grain boundaries is considerably less clear. At face value, concern with this level of detail may appear excessive; however, there are at least two reasons to explore it. The first is to identify the fundamental origin of the weak-link problem in order to discern whether it is intrinsic to the boundary structure or extrinsic. The dual behavior of the [100] boundaries in step-edge junctions vs bulk bicrystals and planar thin films, and the variation in the transition angle between low-angle and high-angle behavior (see above), suggest that extrinsic control is still a possibility (3, 4). The second reason regards the possible identification of certain families of grain boundaries, other than low-angle boundaries, that have particular properties. The evidence suggesting that θ [001] twist boundaries in BSCCO-2212 are less prone to weak-link behavior (5) draws attention to this latter question.

Evidence that Grain Boundary Models Pertain to Cuprate Superconductors

A variety of experimental evidence suggests that low-energy misorientation relationships exist in the cuprate superconductors and that the CCSL construction explains when they occur. A tendency for boundaries with CCSL misorientation relationships to dominate the microstructure is observed rather generally in samples in which during processing (*a*) the crystals can rotate relative to one another, or (*b*) selective grain growth is possible, as in polycrystalline thin films. Histograms of the distributions of misorientation relationships, usually plotted as frequency vs θ for some fixed rotation axis [hkl], show peaks (implying low grain–boundary energy) at CCSL misorientations in flux-grown YBCO crystals (82), in partial-melt processed BSCCO-2212 tapes (83), in solid-state processed BSCCO-2212 (84), in TBCCO-1223 thick films (A Goyal, private communication), and in YBCO thin films (85, 86). The lack of wetting second-phase layers at the low-Σ boundaries found in non-stoichiometric thin films also is consistent with their presumed low interfacial energy (86). A few examples of improved critical current densities in polycrystalline materials (bulk and thin film) that have a preponderance of low-Σ boundaries indirectly suggest that these boundaries may even have superior superconducting properties (83, 86). The improved mechanical strength and/or critical current densities measured for lower-Σ [001] twist boundaries in synthesized BSCCO-2212 bicrystals are consistent with the notion that such boundaries have stronger bonding and/or better electromagnetic properties (42).

The literature now contains numerous examples of GBD networks in low- and high-angle YBCO grain boundaries. For low-angle boundaries, the observed combinations of Burgers vectors and network geometries appear to be consistent with measured misorientation relationships (36, 88; IF Tsu & SE Babcock, unpublished data). High-resolution TEM investigations of low-angle [001] and [100] tilt boundaries also have shown that the low-energy structure, which is the perfect crystal for the low-angle boundary case, is indeed preserved in the regions between the dislocation cores. However, the dislocation radii themselves can be larger than the magnitude that is often assumed in dislocation models $|\sim\mathbf{b}|$ (38), and their structure appears to be more complicated than just highly distorted YBCO. Extended cores of a different composition and, perhaps crystal structure, with a radius of ~1 nm have been observed for [001] tilt boundaries in thin films (36). The dislocations in [100] tilt boundaries dissociate and can appear to have amorphous cores (89). These results are most interesting when compared with atomic-scale resolution images of high-angle boundaries. Although some high-angle grain boundaries appear to contain regularly spaced amorphous patches (37, 90), the more usual observation is that the boundary structure is highly localized, extending just one or two lattice spacings into the abutting grains (9, 20, 36, 37, 90–93). This observation is consistent with the core widths that are observed in representative ceramic grain boundaries (13). They are not compatible, however, with an oversimplified picture of high-angle grain boundary structure in which the boundary is envisioned as a continuous block of overlapping PGBD cores.

The early observation of GBD networks in high-angle YBCO boundaries provided evidence that the structures of high-angle boundaries also could be described in terms of GBDs (94). Specifically, a concentrated good-fit/bad-fit picture of the boundary as a subdivided defect also pertains to high-angle boundaries. Perhaps more importantly, these dislocations were shown to be sufficiently localized to produce a periodic, imageable, and therefore physically significant, elastic strain field. Of further interest, the dislocations that accommodate the mismatch of the lattice at the Σ3 [100] boundary appear to be partial dislocations that are positioned well away from the boundary in one grain (20, 36, 91). These partial dislocations are connected to the boundary by a stacking fault that most likely introduces excess copper into the boundary region. Nevertheless, this particular boundary appears to act as a strong link, in spite of its extended structure and probable non-stoichiometry (3, 4, 95).

A third implication of the geometrical models for grain boundary structure is facet formation. Because the structures of boundaries that lie parallel to low-index planes of the CCSL have the shortest wavelength periodicity possible for that misorientation relationship, their energies are expected to be minimum, and faceting onto those planes might be expected. Indeed, nanometer-scale faceting is observed in a variety of model grain boundaries (13, 96), but the facet planes often appear to be dictated by the crystal structure rather than the CCSL geometry. Similar effects are observed in the cuprate superconductors. Evidence for macroscopic and microscopic faceting is plentiful (36, 37, 90, 97; IF Tsu & SE Babcock, unpublished data). Facets are observed at boundaries with comparatively simple crystallographies and low Σ values (36) and at boundaries with complex geometries (90). As in the model grain boundaries, the observed facets tend to lie parallel to the lowest index planes of one of the crystals. Nanometer-scale faceting occurs mostly on the (100), (010), and (001) planes. A few examples of symmetrical θ [001] tilt boundaries and [100] tilt boundaries in YBCO are the exceptions to this rule. The most direct and extensive study of macroscopic facets also indicates an insignificant role for the CCSL in determining facet geometries. In this study, the facet planes observed for [001] tilt boundaries in YBCO were independent of θ, and thus the CCSL geometry (97). Observations of faceting add both simplicity and complexity to the grain boundary picture. In principle, modeling of grain boundary structure is simplified in strongly faceting systems because macroscopic boundaries with a variety of boundary planes can be treated as mixtures of just a few boundary facets, each of which has a fixed structure. However, faceting also induces considerable structural heterogeneity into the boundary. The restricted size of microfacets can affect their structure. The facet intersection lines have their own atomic structure, and additional grain boundary dislocations often are associated with them.

Additional Particulars for Orthorhombic Crystal Structures

The fundamental concepts of models for grain boundary structure appear to describe the general character of grain boundary structures in the cuprate superconductors rather well. However, their orthorhombic crystal structures introduce considerable intricacy into the analysis of their grain boundary crystallography. In addition to the length differences in the *a* and *b* lattice parameters, YBCO is heavily twinned, and the BSCCO compounds have an incommensurate modulation along the *b*-axis. These complexities have been only partially considered in studies of the grain boundary structure of these materials. More frequently, approximations like $a = b$ are used to simplify the analysis. However, referral to the tetragonal system ignores the orthorhombic symmetry of the crystal structure and properties. Such approximations may well lead to conclusions that do not represent the boundary crystallography and anisotropy well, even if $a \approx b$ (76, 78). At least two groups have determined CCSL misorientation relationships with Σ less than 50 for the YBCO crystal lattice (78, 98).

Several authors have recognized that the intersection of the twin microstructure with the grain boundary in YBCO results in oscillatory changes in the misorientation relationship along the boundary (12, 75, 85, 94). In at least one case, the geometry of the grain boundary dislocation network proved to be sensitive to the twin microstructure, resulting in a superstructure in the dislocation configuration (dislocation spacings, tangent vectors) with a period equal to two twin thicknesses for that particular boundary (94). This study also provided evidence for the longer-range strain fields with spacings on the order of the twin spacing that might be expected in this situation (52, 75, 94). To our knowledge, the effect on the grain boundary structure of the incommensurate modulation that occurs along the *b*-axis of the BSCCO structures has not been considered much, even though the fundamental superconducting properties are known to differ along the *a* and *b* axes (99).

CCSL descriptions of grain boundary structure often rely on additional grain boundary dislocations to produce the needed strain in the vicinity of the grain boundary. An alternative mechanism to adjust the axial ratios is available in YBCO, to some extent in BSCCO (71), and probably in other compounds with a range of stoichiometry. In YBCO, both the lattice parameters and the superconducting properties vary with oxygen concentration. In particular, the *a* and *c* lattice parameters increase as oxygen is removed from the structure, resulting in an expansion of the unit cell. Zhu et al have explored in considerable detail the possibility that CCSLs are achieved in this compound by local changes in the oxygen stoichiometry (100).

Zhu et al discuss a simplified picture that considers the sets of CCSLs that form for just two constrained axial ratios among a number of possible choices. The lattice parameters needed to achieve one set of CCSLs correspond approximately to fully oxygenated YBCO, whereas those of the other set correspond to strongly oxygen-deficient material. In their model *a* and *b* are crystallographically equivalent; only the $c:a$ ratio differs for the two sets of CCSLs considered. In spite of these simplifications, a strong correlation between the *c* lattice parameter expansion needed (or not needed) to achieve the CCSL and the presence (or absence) of oxygen depletion is observed if rather large deviations from exact coincidence are allowed. Furthermore, simultaneous measurements of the $c:a$ ratio by convergent beam diffraction techniques and of the local electron hole concentration by TEM/EELS suggest a correlation between hole, and presumably oxygen, concentration and lattice expansion to produce a CCSL. These results suggest that oxygen deficiency and its resulting weakening of the superconducting properties may be an intrinsic property of grain boundaries in YBCO. However, in addition to the number of assumptions that are made, the model suggests that a considerable fraction of high-angle boundaries are strongly coupled, a prediction that is in conflict with experimental results. The processing history of the material might explain this discrepancy (100).

CSL and CCSL concepts can be used to rationalize observed grain boundary behavior, but they are not yet predictive. This shortcoming is considerably more pronounced for CCSL descriptions for a number of reasons. First, because the unit-cell dimensions are allowed to vary to achieve coincidence, relatively low-Σ misorientations densely populate θ [hkl] space. As a result, for many observed boundary misorientation relationships there are several nearby CCSLs that could apply to the boundary. Furthermore, the types of relaxations that occur in these boundaries often involve trade-offs among the magnitude of the local strain required to achieve coincidence, the achievement of low Σ values, and the minimization of strain energy due to dislocations. These trade-offs could result in a relatively large fraction of boundaries with more widely spaced dislocations than are observed in cubic crystals (76), a possibility that could produce at least a component of the heterogeneity that is suggested by the electrical properties of these boundaries. The choice of an appropriate CCSL is complicated further by the lack of a suitable estimate for the range of deviations from the exact coincidence misorientation that reasonably can be referenced back to a particular CCSL. This criterion must incorporate the trade-off between strain, low Σ, the mechanism by which the strain is accommodated, and the possible non-monotonic dependence of the dislocation spacing on $\Delta\theta$. It undoubtedly is complicated, and the maximum deviation that was proposed for cubic materials (Brandon criterion) (101) may be inappropriate for quantitative comparisons.

In summary, two themes arise with regard to CCSL models for the grain boundary structure. The first, more negative aspect is that the theory is not yet sufficiently well-developed for predictions nor for quantitative comparisons between model and experiment. CCSL misorientation relationships can be readily computed for any crystal structure; the bigger question is deciding on their physical significance and the appropriateness of any assumptions made. On the other hand, it seems that the general ideas that comprise CCSL/GBD models are obeyed by the cuprate superconductors and that they have made a significant contribution to our understanding of their grain boundaries. The ability and possible propensity for these materials to adjust their lattice parameter via oxygen stoichiometry changes and the existence of physically significant dislocations with associated elastic strain fields in a large fraction of the high-angle grain boundaries could explain important aspects of the weak-link character of the average high-angle boundary. In the long run, such general understanding may be most useful.

HIGH-SPATIAL-RESOLUTION EELS STUDIES OF THE DENSITY OF HOLE STATES NEAR GRAIN BOUNDARIES

There is general agreement that a layer of material with different superconducting properties exists at YBCO grain boundaries. This layer appears to be hole-deficient YBCO (93, 100, 102, 103). However, key questions about the nature of this layer remain unanswered: (*a*) What is the average width of the hole depleted zone, and does it depend on the form of the material (bulk or thin film) or processing method? (*b*) Are the magnitude of the depletion and width of the depleted zone uniform along the boundary? (*c*) What is the microstructural origin of the hole depletion? The experimental approach that has provided the most information about these very fundamental questions is high spatial resolution EELS performed in a (S)TEM. In the case of YBCO, the shape of the oxygen absorption edge (or core-loss near-edge structure) yields important information about the local density of hole states, as described below.

The Electron Energy-Loss Near-Edge Structure (ELNES) of the Oxygen K-Edge and the Density of Hole States

In transmission EELS experiments performed in an electron microscope, a finely focused electron probe is positioned at a region of interest in a TEM sample, and the kinetic energy lost by the electrons that pass through the sample is analyzed. The energetic electrons in the beam excite electrons from the core orbitals of the atoms in the sample into empty states above the Fermi level (E_F). The electrons in the beam lose energy in this process, and absorption edges are recorded at energy losses corresponding to the binding energies of the core electrons of the elements present in the sample. The intensity in the absorption edges is proportional to the concentration of the corresponding element in the volume being analyzed. The fine structure of the absorption edge maps out the density of the empty levels near E_F because the initial states have very specific energies. Therefore, the spectrum reveals the chemical composition and provides electronic band structure information. The spatial resolution of the (S)TEM/EELS technique approaches the diameter of the incident electron beam [typically 1–5 nm, but a diameter as small as 0.2 nm is possible in ultra-high–resolution instruments (104)].

Nücker et al have demonstrated a correlation between the oxygen-absorption-edge fine structure and the electron hole states induced by oxygen doping in YBCO (105). In the undoped state, the valence band, which is mostly of O_{2p} character, is filled and the conduction band, which is mostly of Cu_{3d} character, is empty. When oxygen is added to the chain layer in YBCO, electrons are removed from the electronic states associated with the CuO_2 layers leaving empty levels on the top of the conduction band (electron hole states). The effect of doping is manifested in the appearance of peak-shaped intensity (Figure 3*a*) at an energy loss value slightly smaller than that observed in the undoped state (Figure 3*b*). The intensity in the dopant-induced pre-edge peak relative to that of the main edge is directly proportional to the density of holes in the O_{2p} band because the density of empty states far from the Fermi level is not affected by doping. ELNES and X-ray absorption spectroscopy results are in agreement with this picture of the band structure.

Browning et al have generated a curve of the intensity of the pre-edge peak relative to the intensity of the main edge as a function of overall sample oxygen stoichiometry (106). To do this, the structure observed in the near-edge region was fit by three Gaussian functions centered at each of two possible pre-edge peak energies and at the main edge. A number of researchers have used this approach to quantify their measurements (93, 103, 107, 108). However, it is important to recognize that other features of the band structure can result in pre-edge intensity. For instance, a clear pre-edge peak is observed for nonsuperconducting samples of $(Y_{1-x}Pr_x)Ba_2Cu_3O_7$ (109) and semiconducting samples of $(Tl_{0.5}Pb_{0.5})(Ca_{1-x}Y_x)Sr_2Cu_2O_{7-\delta}$ (110). Similarly, the relative intensity observed in the pre-edge region of spectra from $Nd_{1+x}Ba_{2-x}Cu_3O_{7+\delta}$ samples did not change in samples with large Nd concentrations even though they were not superconducting (107). EELS detects transitions to all empty states with symmetries allowed by the dipole transition rules. Thus localized holes (107, 109) and/or holes associated with the normal state conductivity (110), but

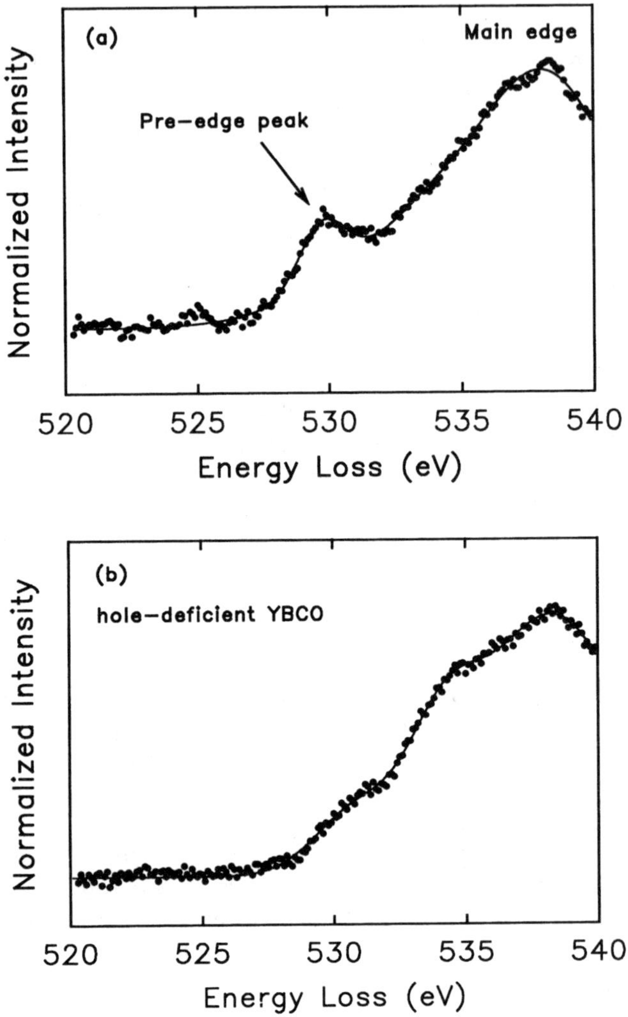

Figure 3 Core-loss (oxygen-K) electron energy-loss spectra obtained in a TEM. (*a*) A pre-edge peak is observed in fully oxygenated YBCO. (*b*) The pre-edge peak is diminished or absent in hole-deficient YBCO.

not the superconducting phase, can be detected. A closer look at the fine structures shows that the pre-edge peaks from nonsuperconducting samples appear at slightly higher energy than does the pre-edge peak associated with superconductivity. This nonsuperconducting pre-edge peak is also slightly sharper (110). However, all reported EELS studies of oxygen-deficient, nonsuperconducting YBCO show either no pre-edge peak or strongly reduced intensity in the pre-edge region. From the standpoint of grain boundary studies, it is true that the simple observation of a pre-edge peak cannot be used as a signature of superconductivity. However, the absence of the pre-edge peak does imply that the material is not superconducting. (S)TEM/EELS measurements of the hole depletion near YBCO grain boundaries are based on the latter observation (93, 100, 102, 111).

(S)TEM/EELS RESULTS FOR GRAIN BOUNDARIES IN YBCO

A number of groups have used the (S)TEM/EELS technique to obtain information about the density of hole states near grain boundaries in several types of samples in the Y-Ba-Cu-O family of superconductors and have used that information to draw conclusions about the character of the boundary. Zhu et al have studied grain boundaries in sintered compacts and have interpreted their results in the context of a CCSL model in an attempt to identify an intrinsic structural origin for hole depletion at grain boundaries (100) (see above). Dravid et al have probed the hole density along a grain boundary in sintered material (103). Browning et al have used a powerful combination of annular dark-field Z-contrast imaging and STEM/EELS at ultra-high ($\sim$0.3 nm) spatial resolution to study the oxygen distribution near [001] tilt grain boundaries in polycrystalline thin films (102). Our TEM/EELS efforts have focused on grain boundaries of known coupling strength (i.e. weakly vs strongly coupled boundaries) by study of grain boundaries in electromagnetically characterized, bulk-scale flux-grown bicrystals of YBCO (93). Wang et al have conducted studies of grain boundaries in sintered $YBa_2Cu_4O_8$ (112). A number of important themes have emerged.

Hole concentration profiles across electromagnetically characterized grain boundaries have revealed a correlation between the magnitude of the hole depletion and the appearance of weak-link characteristics (93). Electron energy-loss spectra and high-resolution TEM images were obtained from bicrystal grain boundaries with and without weak-link character. A broad region of severely hole-deficient material was associated with the boundaries in weakly coupled bicrystals with misorientation relationships of 28 and 31° about [001], respectively. Hole deficiency was either absent or very small in the vicinity of the boundaries in 7 and 12° [001] bicrystals that were not weak linked. Thus severe hole depletion is associated with weakly coupled boundaries.

A striking feature of the hole-deficient zones observed is their breadth.

In fact, a common theme of all the hole-concentration profile studies to date is that the hole depletion zone width exceeds both the superconducting coherence length for YBCO and the apparent structural width of the boundary. Intriguingly, the widths reported vary from one material form to another: Browning et al measured an ~8-nm–wide affected zone at polycrystalline thin-film boundary (102). In contrast, the width of the depletion zone in flux-grown bicrystals is ~60 nm (93). The spectra from Zhu et al's work on sintered materials indicate a width on the order of 10 to 20 nm (100). All of these values indicate that the effect of the boundary on the band structure can be of longer range than the width of the structurally disordered region (~1 nm) or the metal cation non-stoichiometry (~5 nm) (52) and ξ. Unfortunately, neither high-resolution TEM nor the atomic-resolution Z-contrast technique can display the anion lattice, so it is not possible to associate the reduced density of holes with a specific atomic defect such as oxygen non-stoichiometry or oxygen disorder, those being the most likely candidates. It is interesting to note that, unlike the grain boundaries in flux-grown bicrystals, thin-film grain boundaries seem to be free of Cu enrichment (113). This result suggests that there may be important differences in the structure, composition, and electronic properties of grain boundaries in thin films as compared to their bulk counterparts.

Measurements of the oxygen K-edge or oxygen K absorption edge fine structure in a bulk-sintered YBCO sample suggest variations in the hole density along the grain boundary as well (103). This result, obtained for a high-angle grain boundary of undetermined electrical character is, in principle, in agreement with the microbridge models discussed above. Variations in the hole density also were observed in a 12° [001] strongly coupled grain boundary in a flux-grown bicrystal (93). Unfortunately, it is not possible to compare the topography of the observed variations in these two studies. First, not enough information about the length scale of the variations was obtained. Second, somewhat different curve–fitting methods were used to analyze the data, prohibiting a direct comparison of the magnitudes of the depletion. It is important to recognize that the choice of how the Gaussian functions are positioned and the method of establishing their widths can have a large effect on the intensity ratios that are determined by curve fitting.

The variation in the local electronic structure along grain boundaries is consistent with the heterogeneous nature of grain boundary structural features on scales varying from atomic to intermediate to macroscopic dimensions (see above). In particular, variations in the magnitude of cation non-stoichiometry have been observed on a length scale that is at least consistent with the length scales suggested for the heterogeneity in electrical properties (52, 114). It is expected that hole density variations might occur on a number of different length scales in both strongly and weakly coupled boundaries. However, much more complete sets of combined electromagnetic and microstructural data are needed to identify the role of these heterogeneities in determining the macroscopic electromagnetic character of the grain boundaries.

CONCLUDING COMMENTS

Understanding and controlling grain boundary behavior are crucial for both bulk-scale and thin-film applications of high-T_c superconductors. Clearly, much remains to be learned, but important progress has been made. It has become increasingly clear from electromagnetic measurements and microstructural studies that weak-link boundaries are heterogeneous on many length scales. At least some high-angle boundaries are strongly coupled, perhaps leading to the rather good properties of BSCCO tapes. The weak-link character in YBCO is caused by a broad hole-depleted region. This hole depletion appears to be tied to the local oxygen concentration. Considerably less is known about grain boundaries in the other high-T_c materials, some of which may be closer to technological application. Thus it appears that grain boundary behavior will continue to be a focus of high-T_c superconductor research.

Literature Cited

1. Sato K, Hikata T, Mukai H, Ueyama M, Shibuta N, et al. 1991. *IEEE Trans. Magn.* 27(2): 1231–38
2. Heine K, Tenbrink J, Thöner M. 1989. *Appl. Phys. Lett.* 55(23): 2441–43
3. Babcock SE, Cai XY, Kaiser DL, Larbalestier DC. 1990. *Nature* 347: 167–69
4. Eom CB, Marshall AF, Suzuki Y, Boyer B, Pease RFW, Geballe TH. 1991. *Nature* 353: 544–47
5. Wang JL, Cai XY, Kelley RJ, Vaudin MD, Babcock SE, Larbalestier DC. 1994. *Phys. C* 230: 189–98
6. Parikh AS, Meyer B, Salama K. 1994. *Supercond. Sci. Tech.* 7(7): 455–61
7. Gross R. 1993. In *Interfaces in High-T_c Superconducting Systems*, ed. SL Shinde, DA Rudman, pp. 176–209. New York: Springer-Verlag
8. Barone A, Pagano S. 1994. *J. Supercond.* 7(2): 375–80
9. Zandbergen H, Thomas G. 1988. *Acta Cryst. A* 44: 772–75
10. Chiang YM, Ikeda JAS, Roshko A. 1988. In *Ceramic Superconductors*, ed. YF Yan, pp. 607–18. Westerville OH: Am. Ceram. Soc
11. Kroeger DM, Choudhury A, Brynestad J, Williams RK, Padgett RA, Coghlan WA. 1988. *J. Appl. Phys.* 64(1): 331–35

12. Babcock SE, Larbalestier DC. 1989. *Appl. Phys. Lett.* 55(4): 393–95
13. Merkle KL. 1991. *Ultramicroscopy* 37: 130–52
14. Vaudin MD, Lamarre PA, Schmuckle F, Sass SL. 1986. *Philos. Mag. A* 54: 21–35
15. Beyers R, Shaw TM. 1989. *Solid State Phys.* 42: 135–212
16. Vargas JL, Larbalestier DC. 1992. *Appl. Phys. Lett.* 60(14): 1741–43
17. Cava RJ, Hewat AW, Hewat EA, Batlogg B, Marezio M, et al. 1990. *Phys. C* 165: 419–33
18. Kishio K, Shimoyama J, Kotaka Y, Yamafuji K. 1994. In *Proc. 7th Int. Workshop Critical Currents in Superconductors*, ed. HW Weber, pp. 339–42. Singapore: World Scientifc
19. Marshall AM, Eom CB. 1993. *Phys. C* 207: 239–46
20. Gao Y, Bai G, Lam DJ, Merkle KL. 1991. *Phys. C* 173: 487–500
21. Zhu Y, Corcoran YL, Suenaga M. 1993. *Interface Sci.* 1(4): 361–70
22. Russek SE, Lathrop DK, Moeckly BH, Buhrman RA, Shin DH, Silcox J. 1990. *Appl. Phys. Lett.* 57(11): 1155–57
23. Chaudhari P. 1991. *Phys. C* 185–189: 292–96
24. Moeckly BH, Lathrop DK, Buhrman RA. 1993. *Phys. Rev. B* 47(1): 400–17
25. Early EA, Clark AF, Char K. 1993. *Appl. Phys. Lett.* 62(25): 3357–59
26. Sarnelli E, Chaudhari P, Lacey J. 1993. *Appl. Phys. Lett.* 62(7): 777–79
27. Dimos D, Chaudhari P, Mannhart J, LeGoues FK. 1988. *Phys. Rev. Lett.* 61(2): 219–22
28. Dimos D, Chaudhari P, Mannhart J. 1990. *Phys. Rev. B* 41(7): 4038–49
29. Mannhart J. 1990. *J. Supercond.* 3(3): 281–85
30. Amrein T, Seitz M, Uhl D, Schultz L, Urban K. 1993. *Appl. Phys. Lett.* 63(14): 1978–80
31. Kawasaki M, Sarnelli E, Chaudhari P, Gupta A, Kussmaul A, et al. 1993. *Appl. Phys. Lett* 62(4): 417–19
32. Mayer B, Alff L, Träuble T, Gross R, Wagner P, Adrian H. 1993. *Appl. Phys. Lett.* 63(7): 996–98
33. Nabatame T, Koike S, Hyun OB, Hirabayashi I, Suhara H, Nakamura K. 1994. *Appl. Phys. Lett.* 65(6): 776–78
34. Cardona AH, Suzuki H, Yamashita T, Young KH, Bourne LC. 1993. *Appl. Phys. Lett.* 62(4): 411–13
35. Sarnelli E, Chaudhari P, Lee WY, Esposito E. 1994. *Appl. Phys. Lett.* 65(3): 362–64
36. Gao Y, Merkle KL, Bai G, Chang HLM, Lam DJ. 1991. *Phys. C* 174: 1–10
37. Chisholm MF, Pennycook SJ. 1991. *Nature* 351: 47–49
38. Hirth JP, Lothe J. 1982. *Theory of Dislocations*, p. 64. New York: Wiley & Sons. 857 pp.
39. Larbalestier DC, Babcock SE, Cai XY, Field MB, Gao Y, et al. 1991. *Phys. C* 185–189: 315–20
40. Ivanov ZG, Nilsson PÅ, Winkler D, Alarco JA, Claeson T, et al. 1991. *Appl. Phys. Lett.* 59(23): 3030–32
41. Turchinskaya M, Kaiser DL, Gayle FW, Shapiro AJ, Roytburd A, et al. 1994. *Phys. C* 221: 62–70
42. Tomita N, Takahashi Y, Mori M, Ishida Y, 1992. *Jpn. J. Appl. Phys.* 31(7B): L942–45
43. Larbalestier DC, Babcock SE, Cai X, Cooley L, Daeumling M, Hampshire DP, et al. 1989. In *Progress in High Temperature Superconductivity*, ed. S Nakajima, 18: 128–43. New Jersey: World Scientific
44. Nakamura N, Gu GD, Takamuku K, Murakami M, Koshizuka N. 1992. *Appl. Phys. Lett.* 61(25): 3044–46
45. Egi T, Ricketts J, Wen JG. Nakamura N, Unoki H, Koshizuka N. 1994. *Extensive observations of flux pinning at grain boundaries within the bc plane of Bi2211 multi-crystalline samples.* Presented at Int. Workshop Superconductivity, Kyoto, Japan
46. Hampshire DP, Seuntjens, J, Cooley LD, Larbalestier DC. 1988. *Appl. Phys. Lett.* 53(9): 814–15
47. Däumling M, Sarnelli E, Chaudhari P, Gupta A, Lacey J. 1992. *Appl. Phys. Lett.* 61(11): 1355–57
48. Field MB, Cai XY, Babcock SE, Larbalestier DC. 1993. *IEEE Trans. Appl. Supercond.* 3(1): 1479–82
49. Sarnelli E. 1993. *Interface Sci.* 1(4): 287–90
50. Sarnelli E, Testa G, Esposito E. 1994. *J. Supercond.* 7(2): 387–90
51. Kawasaki M, Chaudhari P, Gupta A. 1992. *Phys. Rev. Lett.* 68(7): 1065–68
52. Babcock SB, Larbalestier DC. 1994. *J. Phys. Chem. Solids* 55(10): 1125–36
53. Vannest C, Chi CC, Gallagher WJ, Kleinsasser AW, Raider SI, Sandstrom RL. 1988. *J. Appl. Phys.* 64(1): 242–45
54. Deutscher G, Chaudhari P. 1991. *Phys. Rev. B* 44(9): 4664–65
55. Gross R, Mayer B. 1991. *Phys. C* 180: 235–42
56. Winkler D, Zhang YM, Nilsson PÅ, Stepantsov EA, Claeson T. 1994. *Phys. Rev. Lett* 72(8): 1260–63
57. Rosner SJ, Char K, Zaharchuk G.

1992. *Appl. Phys. Lett.* 60(8): 1010–12
58. Alarco JA, Olsson E, Ivanov ZG, Nilsson PÅ, Winkler D, et al. 1993. *Ultramicroscopy* 51: 239–46
59. Kabius B, Seo JW, Amrein T, Dähne U, Scholen A, et al. 1994. *Phys. C* 231: 123–30
60. Träholt C, Wen JG, Zandbergen HW, Shen Y, Hilgenkamp JWM. 1994. *Phys. C* 230: 425–34
61. Le Lay L, Willis TC, Larbalestier DC. 1992. *Appl. Phys. Lett.* 60: 775–77
62. Larbalestier DC, Cai XY, Feng Y, Edelman H, Umezawa A, et al. 1994. *Phys. C* 221: 299–303
63. Feng Y, Larbalestier DC. 1994. *Interface Sci.* 1(4): 401–10
64. Hensel B, Grivel J-C, Jeremie A, Perin A, Pollini A, Flükiger R. 1993. *Phys. C* 205: 329–37
65. Bulaevskii LN, Clem JR, Glazman LI, Malozemoff AP. 1992. *Phys. Rev. B* 45(5): 2545–48
66. Bulaevskii LN, Daemen LL, Maley MP, Coulter JY. 1993. *Phys. Rev. B* 48(18): 13798–816
67. Kroeger DM, Goyal A, Specht ED, Wang ZL, Tkaczyk JE, et al. 1994. *Appl. Phys. Lett.* 64(1): 106–08
68. Specht ED, Goyal A, Kroeger DM, DeLuca JA, Tkaczyk JE, et al. 1994. *Phys. C* 226: 76–84
69. Goyal A, Specht ED, Wang ZL, Kroeger DM, Sutliff JA, et al. 1994. *J. Electron. Mater.* 23(11): 1191–97
70. Cho JH, Maley MP, Willis JO, Coulter JY, Bulaevskii LN, et al. 1994. *Appl. Phys. Lett.* 64(22): 3030–32
71. Kotaka Y, Kimura T, Ikuta H, Shimoyama J, Kitazawa K, et al. 1995. *Phys. C* 235–36: In press
72. Umezawa A, Feng Y, Edelman HS, High YE, Larbalestier DC, et al. 1992. *Phys. C* 198: 261–72
73. Larbalestier DC, Cai XY, Edelman H, Field MB, Feng Y, Parrell J, et al. 1994. *J. Met.* 46(12): 20–22
74. Grimmer H, Warrington DH. 1987. *Acta Cryst. A* 43: 232–43
75. King AH, Singh A. 1993. *J. Appl. Phys.* 74(7): 4627–30
76. King AH, Singh A, Wang JY. 1993. *Interface Sci.* 4(1): 347–59
77. Chen FR, King AH. 1988. *Philos. Mag. A* 57(3): 431–55
78. Gertsman VY. 1992. *Scripta Met. Mater.* 27: 291–96
79. Koptsik VA. 1993. *Mater. Sci. Forum* 126–128: 13–22
80. Warrington DH. 1993. *Mater. Sci. Forum* 126–128: 57–60
81. Balluffi RW, Bristowe PD. 1984. *Surf. Sci.* 144: 28–43
82. Smith DA, Chisholm MF, Clabes J. 1988. *Appl. Phys. Lett.* 53: 2344–46
83. Zhu Y, Suenaga M, Sabatini RL. 1994. *Appl. Phys. Lett.* 65(14): 1832–34
84. Takahashi Y, Mori M, Ishida Y. 1989. *Appl. Phys. Lett.* 55(5): 486–88
85. Tietz LA, Carter CB. 1991. *Phys. C* 182: 241–51
86. Ravi TS, Hwang DM, Ramesh R, Chan SW, Nazar L, et al. 1990. *Phys. Rev. B.* 42(16): 10141–51
87. Deleted in proof
88. Zhu Y. 1994. *Philos. Mag. A* 69(4): 717–28
89. Chisholm MF, Smith DA. 1989. *Philos. Mag. A* 59(2): 181–97
90. Laval JY, Swiatnicki W. 1994. *Phys. C* 221: 11–19
91. Carim AH, Mitchell TE. 1993. *Ultramicroscopy* 51: 228–38
92. McKernan S, Norton MG, Carter CB. 1992. *J. Mater. Res.* 7(5): 1052–59
93. Babcock SE, Cai XY, Larbalestier DC, Shin DH, Zhang N, et al. 1994. *Phys. C* 227: 183–96
94. Babcock SE, Larbalestier DC. 1990. *J. Mater. Res.* 5(5): 919–28
95. Lew DJ, Suzuki Y, Marshall AF, Geballe TH, Beasley MR. 1994. *Appl. Phys. Lett.* 65(12): 1584–86
96. Carter CB. 1988. *Acta Metall.* 36(10): 2753–60
97. Gayle FW, Kaiser DL. 1991. *J. Mater. Res.* 6(5): 908–15
98. Singh A, Chandrasekhar N, King AH. 1990. *Acta Cryst. B* 46: 117–25
99. Ma J, Quitmann C, Kelley RJ, Berger H, Margaritondo G, Onellion M. 1994. *Phys. C* 235–36: In press
100. Zhu Y, Wang ZL, Suenaga M. 1993. *Philos. Mag. A* 67: 11–28
101. Randle V. 1993. *The Measurement of Grain Boundary Geometry*, pp. 47–48. London: Institute of Physics. 169 pp.
102. Browning ND, Chisholm MF, Pennycook SJ, Norton DP, Lowndes DH. 1993. *Phys. C* 212: 185–90
103. Dravid VP, Zhang H, Wang YY. 1993. *Phys. C* 213: 353–58
104. Chisholm MF, Pennycook SJ, Norton DP, Browning ND. 1993. *Interface Sci.* 1(4): 339–45
105. Nücker N, Romberg H, Xi XX, Fink J, Gegenheimer B, Zhao ZX. 1989. *Phys. Rev. B* 39(10): 6619–29
106. Browning ND, Yuan J, Brown LM. 1992. *Phys. C* 202: 12–18
107. Kramer MJ, Yoo SI, McCallum RW, Yelon WB, Xie H, Allenspach P. 1994. *Phys. C* 219: 145–55

108. Zhu Y, Zuo JM, Moodenbaugh AR, Suenaga M. 1994. *Philos. Mag. A*. 70: 969–84
109. Fink J, Nücker N, Romberg H, Alexander M, Maple MB, et al. 1990. *Phys. Rev. B* 42(7): 4823–26
110. Yuan J, Brown LM, Liang WY, Liu RS, Edwards PP. 1991. *Phys. Rev. B* 43(10): 8030–37
111. Kramer MJ, McCallum RW. 1992. In *Layered Superconductors: Fabrication, Properties and Applications*, ed. DT Shaw, 275: 35–40. Pittsburgh: Mater. Res. Soc.
112. Wang ZL, Brynestad J, Kroger DM, Sun YR, Thompson JR, Williams RK. 1993. *Phys. Rev. B* 48(13): 9726–34
113. Shin DH, Silcox J, Russek SE, Lathrop DK, Moeckly B, Buhrman RA. 1990. *Appl. Phys. Lett.* 57: 508–10
114. Zhu Y, Corcoran YL, Suenaga M. 1993. *Interface Sci.* 1(4): 361–70

Annu. Rev. Mater. Sci. 1995. 25: 223–47

LASER-INDUCED PHASE TRANSITIONS IN SEMICONDUCTORS

Y. Siegal, E. N. Glezer, L. Huang, and E. Mazur

Division of Applied Sciences and Department of Physics, Harvard University, Cambridge, Massachusetts 02138

KEY WORDS: femtosecond laser, bandgap collapse, lattice disordering

ABSTRACT

Optical studies of semiconductors under intense femtosecond laser pulse excitation suggest that an ultrafast phase transition takes place before the electronic system has time to thermally equilibrate with the lattice. The excitation of a critical density of valence band electrons destabilizes the covalent bonding in the crystal, resulting in a structural phase transition. The deformation of the lattice leads to a decrease in the average bonding-antibonding splitting and a collapse of the bandgap. We review the relationship between structural, electronic, and optical properties, as well as the timescales for electron recombination, diffusion, and energy relaxation. Direct optical measurements of the dielectric constant and second-order nonlinear susceptibility are used to determine the time evolution of the phase transition.

INTRODUCTION

For almost two decades the field of laser-induced phase transitions in semiconductors has generated considerable interest. This field arose originally in the context of semiconductor annealing, a technologically important process aimed at repairing the damage to semiconductor crystals caused by dopant atom implantation. The conventional method for this process is thermal annealing—the slow baking of a semiconductor in an oven. When the sample is heated, the increased atomic mobility allows

0084–6600/95/0801–0223$05.00

defects to diffuse to the surface. In the 1970s, a new method for repairing damage to semiconductor crystals, now known as laser annealing, was discovered (1). In laser annealing, a short laser pulse incident on a damaged region of a semiconductor induces structural changes in the material that result in the formation of a defect-free crystal. Although laser annealing of semiconductors has not replaced thermal annealing in industrial semiconductor processing, the discovery of laser annealing did open up a new and exciting chapter in the study of light-matter interactions: the use of light to induce phase transitions in semiconductors. Our focus is not on the thermal transition from solid to liquid (melting) induced by nanosecond and picosecond laser pulses, which has been studied in detail. The details of this transition, including the heating of the solid, thermal diffusion, melting, motion of the interphase boundary, and recrystallization are discussed in Reference 2. Instead, we concentrate in this paper on the highly nonequilibrium electronic and structural dynamics in a semiconductor excited by intense femtosecond pulses.

CHEMICAL BONDS AND ENERGY BANDS

The arrangement of atoms in a crystal is intricately connected with the state of the electrons in the material. On one hand, given a set of atoms and their crystal structure, the electronic properties of the solid can be predicted. On the other hand, the valence electrons of the atoms in the solid determine the most energetically favorable arrangement of the ionic cores of the atoms. The cohesion of a semiconductor crystal is primarily due to covalent bonds between the atoms. Because of the periodicity in the crystal structure, there is a corresponding periodicity in reciprocal or crystal momentum space, which allows a useful alternative description of the state of the electrons. In the momentum representation, the spatial description of a bond is replaced by a delocalized representation of the entire electronic system, where the eigenvalues of the Hamiltonian form energy bands. The bond picture of the spatial distribution of electrons and the energy band picture of electronic states in momentum space are two complementary ways of looking at a solid. Although one description may be more appropriate than the other for a particular discussion, an understanding of both pictures provides additional insight into crystal cohesion, electronic properties, and interactions with light.

Figure 1*a* illustrates schematically a covalent bond between two identical atoms, each having only one electron in a particular energy level (note that each level has two spin states). If the two atoms are brought close enough to each other that their wavefunctions overlap, the energies of their collective state will be different from the energies of the individual

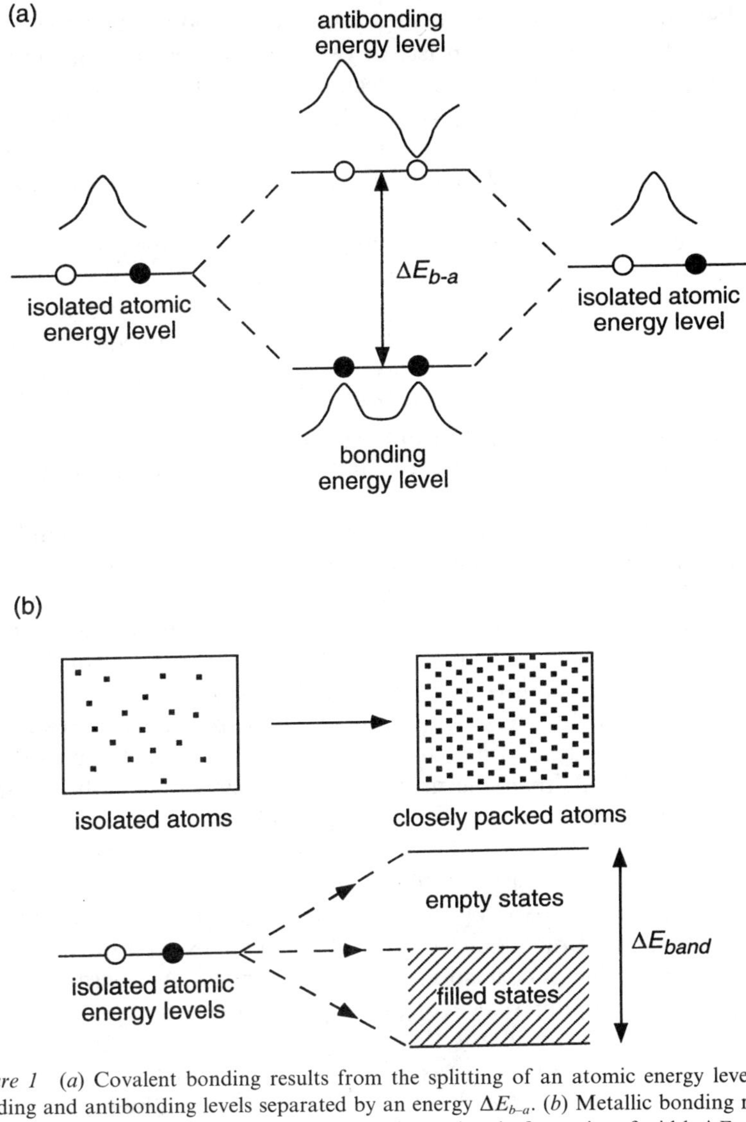

Figure 1 (*a*) Covalent bonding results from the splitting of an atomic energy level into bonding and antibonding levels separated by an energy ΔE_{b-a}. (*b*) Metallic bonding results from the broadening of an atomic energy level into a band of energies of width ΔE_{band}. ●: filled electronic state, ○: empty electronic state.

atoms. The resulting diatomic molecule has a symmetric ground state wavefunction with an energy lower than the initial atomic level, and an antisymmetric excited state wavefunction with an energy higher than the initial atomic level. These are called the bonding and antibonding states. The bound configuration consists of two electrons occupying the lower level, where the energy of the molecule is lower than that of the two individual atoms. The sharing of the electrons results in a covalent bond between the two atoms. If one or both of the electrons are excited to the antibonding level, the molecule no longer has a lower total energy than the two individual atoms.

Metallic bonding arises from a more delocalized overlap of atomic orbitals in a solid. This overlap of the orbitals of a particular atomic energy level leads to a broadening of this energy level into a band of energies. If the atomic energy levels are not fully occupied, then, as the atoms are brought closer together to form a solid, the total energy will be lower than for the isolated atoms, as illustrated schematically in Figure 1*b*. This type of bonding results in a partially filled energy band, which is required for electronic conduction and is characteristic of metals (3).

Semiconductors exhibit characteristics of both covalent bonding and metallic bonding. On the one hand, nearest neighbors form covalent bonds by sharing two electrons, which results in a splitting of atomic levels into bonding and antibonding states. On the other hand, these bonding and antibonding states are broadened into bonding and antibonding energy bands by more delocalized interactions with the other atoms in the solid. When the average bonding-antibonding splitting in the material is larger than the width of the energy bands, an energy gap will exist between the bonding and the antibonding bands (see Figure 2*a*). If the atomic orbitals start out half-filled, then the ground state consists of a fully occupied bonding band and an empty antibonding band. In this case, the material will not conduct electricity. However, when the average bonding-antibonding splitting is smaller than the width of the energy bands, the bonding and antibonding bands overlap (see Figure 2*b*). The resulting partially filled bands lead to metallic electronic characteristics even in the ground state. A semiconductor corresponds to the former case: An energy bandgap separates the bonding (valence) and antibonding (conduction) bands. Semiconductors are insulators in the ground state; however, they start conducting when electrons are excited from the valence to the conduction band. Conductivity in a semiconductor is mediated both by the negatively charged electrons excited to the conduction band and by the positive holes left behind in the valence band by the excited electrons. The conducting electrons and holes are often referred to collectively as free charge carriers or simply free carriers.

Most of the work on laser-induced phase transitions has focused on group IV and III–V semiconductors in the diamond and zincblende crystal structures. These materials are appropriate for studying cohesion because

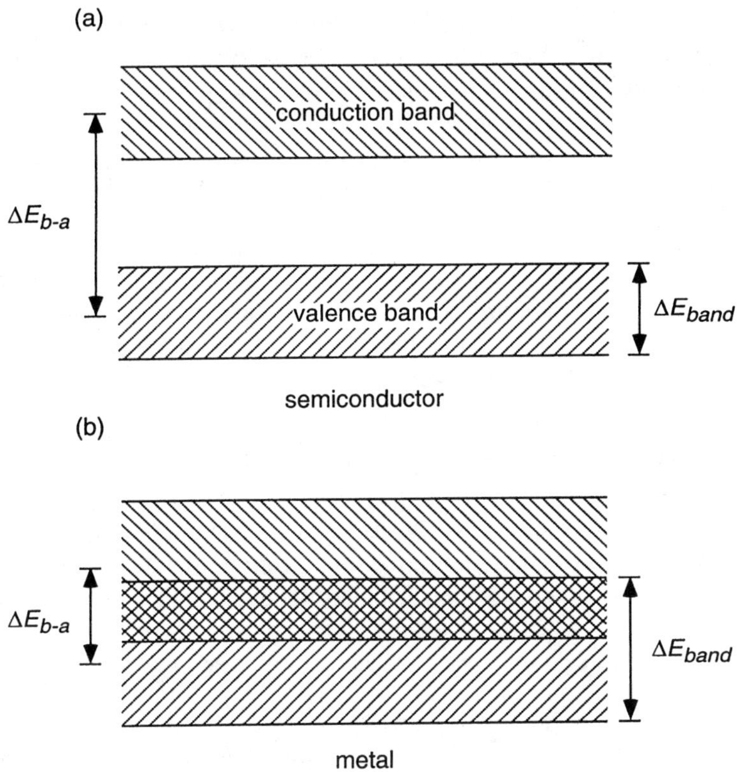

Figure 2 (*a*) Schematic representation of a semiconductor: $\Delta E_{b-a} > \Delta E_{\text{band}}$. (*b*) Schematic representation of a metal: $\Delta E_{b-a} < \Delta E_{\text{band}}$.

of the central role played by covalent bonds in these crystals. Both the diamond and the zincblende structures are built from tetrahedrally coordinated atoms, with each pair of nearest neighbors held together by sharing two electrons. If a significant fraction of these bonding electrons is excited to antibonding states, the cohesion of the crystal will be strongly affected (4).

LIGHT-SEMICONDUCTOR INTERACTIONS

The electronic structure of a material determines its optical properties. Classically, the higher charge-to-mass ratio of the electrons compared with that of the protons in the nucleus accounts for the more important role of the former in the optical response of a material. Quantum mechanically, this dominant role for the electrons arises because optical photon energies and the energy spacings between electronic levels fall in the same range. The energy spacings between nuclear levels are much larger than optical photon energies, resulting in negligible nuclear transition probabilities. Furthermore, the transition energies for the core electrons are also much greater than optical photon energies, and thus only the valence or outer shell electrons significantly contribute to the optical properties. Although electromagnetic radiation can produce collective vibrational excitations of the lattice by exciting phonons, phonon frequencies are generally in the far infrared, well below the frequencies of interest in this discussion. Thus when visible or ultraviolet light is absorbed by a material, most of the energy is initially transferred to the outer shell electrons without directly perturbing the nuclei or the core electrons.

Absorption of Light in a Semiconductor

Absorption of light in a semiconductor can occur either through interband or intraband electronic excitations. However, an electronic transition can only occur between a filled initial state and an empty final state. In the ground state of a semiconductor (i.e. at absolute zero temperature), all bands are either completely filled or completely empty, so intraband transitions cannot occur. Only when a significant number of electrons has been transferred from the valence to the conduction bands does intraband absorption, also known as free-carrier absorption, become an important mechanism for absorption of light. Usually, interband absorption is the principal mechanism for absorption of light in a semiconductor.

An optically induced interband electronic transition must conserve energy and crystal momentum (4). To satisfy the requirement of energy conservation, the difference between the energies of the final and initial electronic states in the transition $E_{\text{final}} - E_{\text{initial}}$ must equal the energy of the absorbed photon $\hbar\omega$. The magnitude of the momentum of the absorbed photon is much smaller than the magnitude of the electron crystal momentum k. In order to satisfy conservation of crystal momentum, an optically induced electronic transition must leave the electron crystal momentum essentially unchanged. Thus a photon of energy $\hbar\omega$ can directly induce an interband electronic transition only between states that satisfy the condition $E_{\text{final}}(\mathbf{k}) - E_{\text{initial}}(\mathbf{k}) = \hbar\omega$. This type of process, known as a direct transition, is a two-body process involving an electron and a photon. Figure 3a illustrates schematically a direct interband transition.

Interband absorption can also occur through indirect electronic transitions, as shown in Figure 3b. In this case, the initial and final values of the electron crystal momentum are not the same. A phonon is required in the process to carry away the excess momentum. A phonon can have a

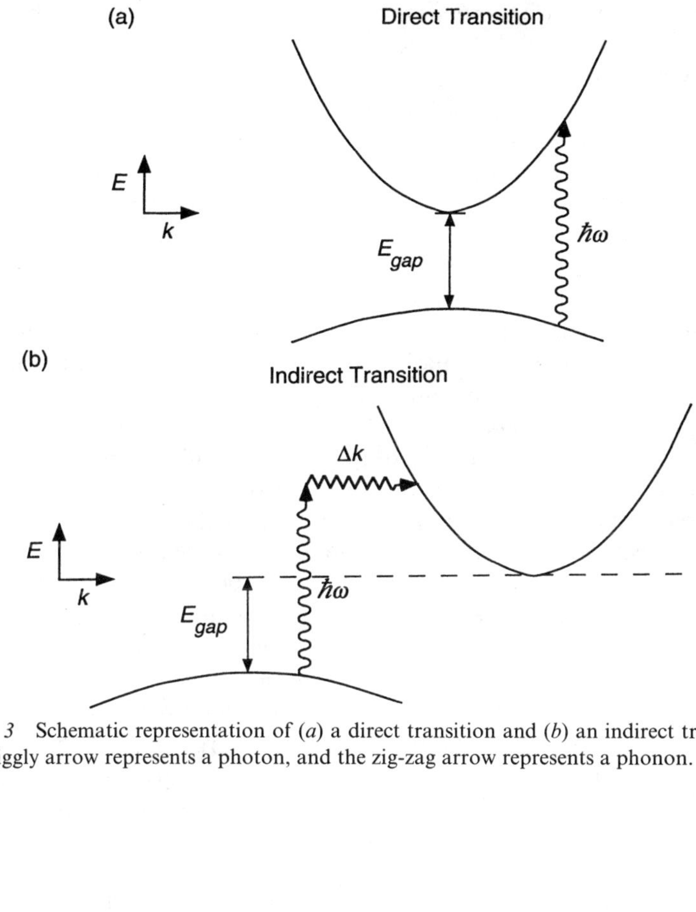

Figure 3 Schematic representation of (*a*) a direct transition and (*b*) an indirect transition. The wiggly arrow represents a photon, and the zig-zag arrow represents a phonon.

large momentum, but its energy is small compared to that of the photon. Thus an indirect transition is characterized approximately by the following two conditions: (*a*) energy conservation, $E_{\text{final}} - E_{\text{initial}} = \hbar\omega_{\text{photon}}$, and (*b*) crystal momentum conservation, $\mathbf{k}_{\text{final}} - \mathbf{k}_{\text{initial}} = \mathbf{k}_{\text{phonon}}$. Because direct transitions are lower-order processes than indirect transitions, direct transitions dominate the absorption spectrum over the range of photon energies in which they are allowed.

There are several common features in the absorption spectra of semiconductors that result from common features in their band structures. One of the most noteworthy features is the fundamental absorption edge at a photon energy corresponding to the minimum energy gap E_{gap} separating the valence and conduction bands. Assuming negligible free-carrier absorption, a semiconductor is transparent to light of photon energy less than E_{gap} because there are no resonant interband transitions in this frequency range, as shown in Figure 3.[1]

At photon energies above E_{gap}, the absorption spectrum will depend on whether a semiconductor has a direct or an indirect minimum bandgap. In a direct-gap semiconductor such as GaAs, the minimum in the conduction band occurs at the same point in k-space (same value of crystal momentum) as the maximum in the valence band. In contrast, the conduction band minimum and the valence band maximum occur at different points in k-space for an indirect semiconductor such as Si. Thus direct transitions dominate the absorption spectrum of direct-gap semiconductors at all photon energies above E_{gap}. Indirect transitions make up the absorption spectrum of indirect-gap semiconductors at photon energies between E_{gap} and the minimum value of the direct energy gap $E_{conduction}(\mathbf{k}) - E_{valence}(\mathbf{k})$ (see Figure 3b).

In general, the conduction band of a semiconductor has more than one relative energy minimum in k-space. The absolute conduction band minimum in GaAs occurs at $\mathbf{k} = 0$, known as the Γ-point. The region around this minimum is called the Γ-valley. GaAs also has relative minima at points symmetrically equivalent to $\mathbf{k} = (111)\pi/a$ and $\mathbf{k} = (100)\pi/a$, where a is the lattice spacing. The eight symmetrical equivalents of $\mathbf{k} = (111)\pi/a$ are known as L-points, whereas the six symmetrical equivalents of $\mathbf{k} = (100)\pi/a$ are called the X-points. The regions around these symmetry points are called the L-valleys and the X-valleys, respectively. In GaAs at room temperature, the direct energy gap at the Γ-point, which is also the minimum energy gap E_{gap}, is equal to 1.42 eV (6). Pseudopotential calculations show the direct gaps at the L- and X-points to be about 3 and 5 eV, respectively (7). Also, the conduction-band minimum at the L-point is 0.29 eV above that at the Γ-point, whereas the X-point conduction-band minimum exceeds that at the Γ-point by 0.48 eV (6). Figure 4 shows the band structure of GaAs between the L- and Γ-points and between the Γ- and X-points.

The optical properties of a semiconductor are quantitatively described by the dielectric function $\varepsilon(\omega)$. The dielectric function is complex and describes the magnitude and phase of the material response to an electromagnetic field. Figure 5 shows both the real and imaginary parts of the dielectric function of GaAs at room temperature (8). The three main features, labeled E_{gap}, E_1, and E_2, are common to the Group IV and III–V semiconductors, although their locations and the relative sizes of these

[1] Discrete absorption lines, however, do appear a few meV below E_{gap} that correspond to the excitation of bound electron-hole pairs of excitons (5).

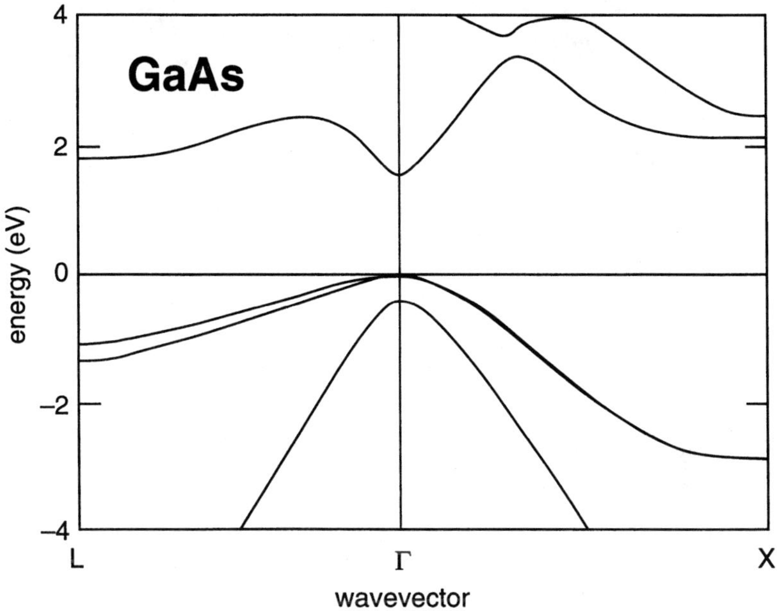

Figure 4 A section of the electronic band structure of GaAs (after Reference 7).

features vary from material to material. The point E_{gap} (at 1.42 eV for GaAs) marks the fundamental band edge below which Im(ε) is zero; E_1 and E_2 (located at 3.0 eV and 4.75 eV, respectively, for GaAs) (9) label the two main absorption peaks in the spectrum. These peaks arise from regions in the band structure (the region around the L-point for the E_1 peak and the region around the X-point for the E_2 peak) in which the valence band is roughly parallel to the conduction band, resulting in a large, joint density of states for direct interband transitions (10). The E_2 peak, which is roughly coincident with the zero-crossing in Re(ε), is the stronger of the two absorption peaks, and its location gives the approximate value of the average bonding-antibonding splitting of GaAs (3).

Much of the work on laser-induced phase transitions discussed in this paper involves the optical excitation of GaAs at photon energies above E_{gap}. Because GaAs is a direct-gap semiconductor, the optical excitation is dominated by direct transitions at all frequencies. Although there are differences in the details of optical excitation in direct-gap vs indirect-gap semiconductors, most of the key issues we address apply to both types of materials. Analysis of the results presented here for GaAs, however, is simplified by the lack of complications arising from indirect transitions.

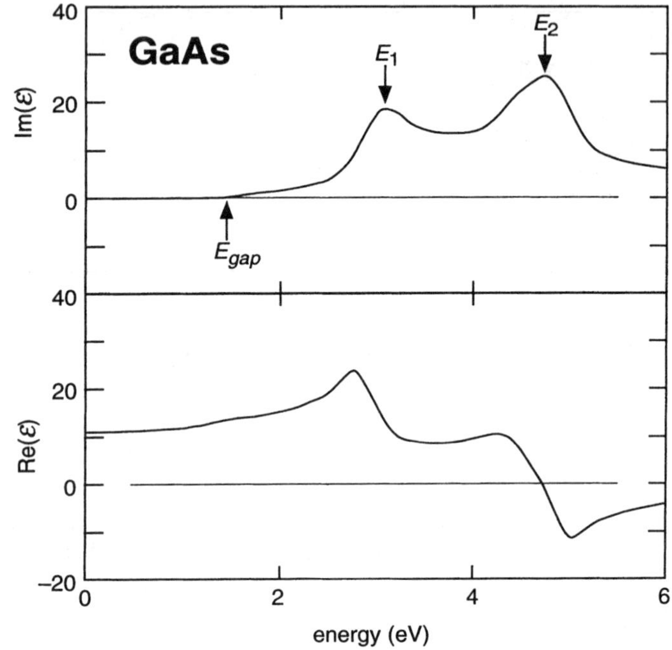

Figure 5 Dielectric function of GaAs (after Reference 8). E_{gap} labels the fundamental absorption edge corresponding to the minimum bandgap. E_1 and E_2 label the two main absorption peaks.

Energy Transfer in a Semiconductor

The optical excitation of electrons from the valence band to the conduction band in a semiconductor leads to a series of processes through which the energy of the excited electrons is distributed to the rest of the electronic system as well as to the lattice. Figure 6 summarizes the main energy transfer processes. Optical excitation of electrons can occur only in regions of k-space satisfying the direct transition condition $E_{\text{final}}(\mathbf{k}) - E_{\text{initial}}(\mathbf{k}) = \hbar\omega$, where ω is within the spectral bandwidth of the incident light. As a result, the incident light creates a nonequilibrium population of free carriers (electrons in the conduction band and holes in the valence band) with a narrow distribution of crystal momentum (localized in k-space).

CARRIER–CARRIER SCATTERING The nonequilibrium free-carrier population quickly spreads out in k-space through carrier–carrier scattering, as indicated by process 1 in Figure 6. The time scale for electron–electron scattering decreases with increasing conduction electron density and is

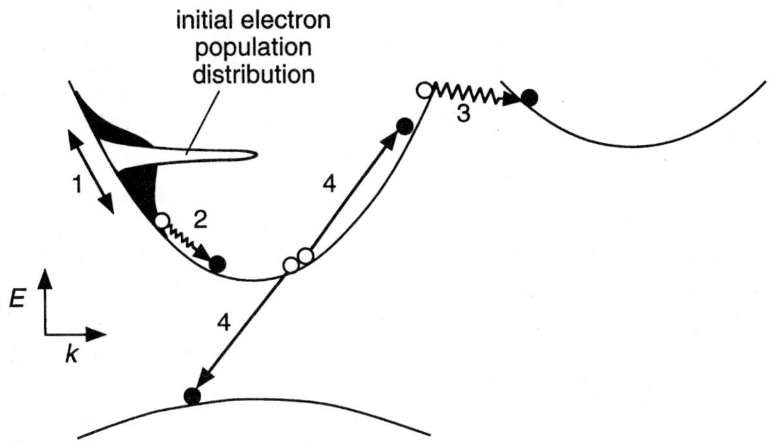

Figure 6 Representation of scattering processes and their corresponding equilibration times following laser excitation of an initial distribution of free carriers. Process 1 indicates the spreading out of the initial carrier distribution in k-space through carrier-carrier scattering, which leads to carrier thermalization. Processes 2 and 3 show intravalley and intervalley carrier-phonon scattering, respectively. The interaction of carriers and phonons leads to electron-lattice equilibration. Carrier-density relaxation occurs through recombination processes such as Auger recombination, shown by process 4. In Auger recombination, an electron and a hole recombine, giving off the excess energy and momentum to a third free carrier.

shorter than 20 fs for electron densities in excess of 10^{18} cm^{-3} (11). Through multiple scattering, the initial nonequilibrium distribution of free carriers thermalizes to a Fermi-Dirac distribution within a few hundred femtoseconds (12). The temperature of this thermal distribution is set by the initial free-carrier excess energy, corresponding to the energy above the conduction band minimum for electrons and the energy below the valence band maximum for holes. Thus carrier–carrier scattering merely redistributes the energy among the excited electrons and holes and does not change the average energy of the excited free carriers.

Carrier–carrier scattering includes three similar but separate processes: electron–electron scattering, hole–hole scattering, and electron–hole scattering. The first two lead to independent thermalization of the excited electron and hole populations, whereas the third leads to equilibration of the two populations with each other. Electron–hole equilibration is driven by the initial difference in average excess energies between excited electrons and holes. In GaAs, for example, a difference between electron and hole excess energies arises during direct interband transitions in the Γ-valley because of the higher curvature of the conduction band compared with that of the valence band (see Figure 3*a*).

CARRIER-PHONON SCATTERING Although the excited free carriers become internally thermalized within a few hundred femtoseconds, they are still far out of equilibrium with the lattice on this time scale. Energy transfer from the excited electrons to the lattice occurs through the slower process of carrier–phonon scattering, shown by process 2 in Figure 6. The time scale for carrier–phonon scattering is longer than that for carrier–carrier scattering because the coupling of free carriers to phonons is weaker than the coupling of free carriers to each other. Typical carrier–phonon scattering times are on the order of a few hundred femtoseconds (13). However, multiple scattering is required to transfer the excess energy from the electrons to the lattice. The phonons primarily responsible for cooling the hot electrons have energies on the order of 30 meV (see below). For initial electron excess energies on the order of 0.5 eV (corresponding to excitation of GaAs with 2-eV photons), the energy transfer process requires over 10 phonon emission events per electron. Thus for excitation significantly above the bandgap, electron-lattice thermalization occurs on a time scale of a few picoseconds (13).

Before describing the interactions between free carriers and phonons, we first examine the phonon spectrum of semiconductors, shown schematically in Figure 7. Because diamond (Si) and zincblende (GaAs) crystal lattices have a two-atom basis, the phonon spectra of these materials split into two main branches: acoustic and optical (4). Each of these branches in turn splits into three polarization vectors for the atomic displacements of a traveling wave: two that are transverse to the direction of propagation and one that is longitudinal. The acoustic branch gets its name from the small wavevector limit in which it has a linear dispersion relation and supports the propagation of sound. The optical branch has higher frequencies than the acoustic branch and a relatively flat dispersion relation. It is called the optical branch because its small wavevector phonons can couple directly to infrared photons. Although free carriers can interact with both optical and acoustic phonons, energy transfer is dominated by optical phonons because of their higher energies.

The nature of the interaction between free carriers and phonons in semiconductors depends on whether there is an ionic component to the crystal bonding. In an elemental semiconductor such as Si, the bonds are purely covalent, and the carrier-phonon interaction is mediated by the deformation potential (5), which is the potential energy associated with the change in electronic band structure arising from small atomic displacements produced by lattice vibrations. This change in band structure introduces an effective force between phonons and free carriers. The deformation potential coupling is independent of phonon wavevector (5), allowing hot free carriers to emit phonons of different wavelengths.

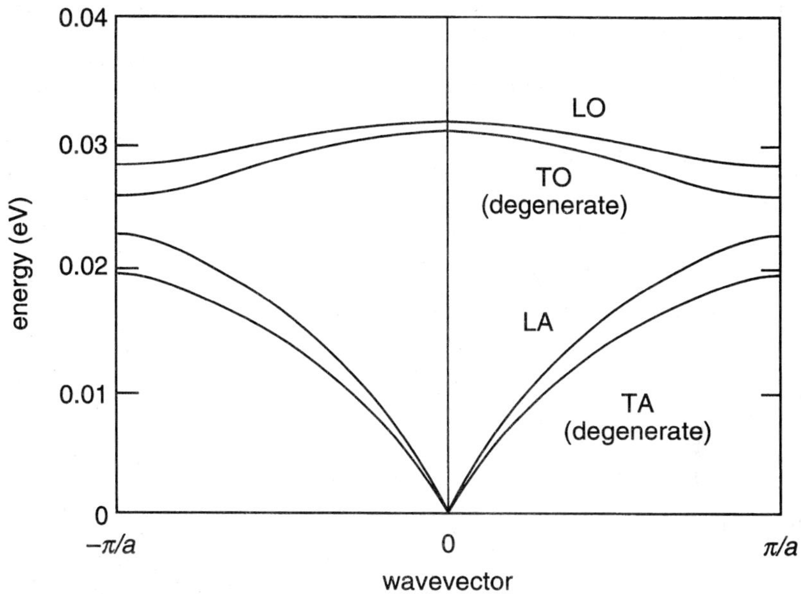

Figure 7 Typical phonon spectrum along a symmetry direction for a crystal with a two-atom basis and cubic symmetry. LO: longitudinal optical modes; TO: transverse optical modes; LA: longitudinal acoustic modes; TA: transverse acoustic modes.

Compound semiconductors such as GaAs, on the other hand, have partially ionic bonds because of the different electron affinities of the two component elements. In this type of material, known as a polar semiconductor, the atomic displacements of a phonon set up an oscillating electric dipole. For longitudinal lattice vibrations, this dipole gives rise to an electric field that couples directly to the charged free carriers (14). The polar coupling is inversely proportional to the square of the phonon wavevector (5), leading to preferential emission of long-wavelength phonons by hot free carriers. Because the longitudinal optical (LO) phonons emitted by electrons through the polar interaction have long wavelengths (small values on q, the phonon wavevector), the change in electron crystal momentum is very small. This small change in crystal momentum means that after emitting an LO phonon, the conduction-band electron remains within the same energy valley as that of its initial state. In GaAs, electronic cooling occurs mainly through intravalley phonon scattering.

Intervalley phonon scattering can also play an important role in energy transfer following the optical generation of free carriers (15). In intervalley scattering, illustrated by process 3 in Figure 6, excited electrons scatter into a different conduction band valley through either phonon emission or absorption. Because the various conduction band valleys are centered at different values of crystal momentum, intervalley scattering requires large q (short wavelength) optical or acoustic phonons to satisfy the requirement of crystal momentum conservation. Thus the deformation potential mediates this type of scattering. Note that intervalley scattering will not occur for an electron if its energy is more than a phonon quantum below the minimum of the destination valley.

By distributing the excited electron population throughout the conduction band, intervalley scattering affects the electron-lattice energy relaxation time. In the experiments on GaAs reviewed in this paper, a laser pulse excites electrons primarily into the Γ-valley. However, intervalley scattering transfers some of these excited electrons into the L- and X-valleys, both of which have higher energy minima than the Γ-valley. Because they do not have as far to decay, the electrons in these side valleys cannot emit as many LO phonons through polar scattering as the electrons in the Γ-valley unless they scatter back into the Γ-valley. In this way, intervalley scattering slows down the transfer of energy from the electronic system to the lattice (15).

CARRIER RECOMBINATION AND DIFFUSION Although the scattering processes discussed so far distribute the initial excitation energy throughout the electronic system and the lattice, none of these processes changes the density of the excited free carriers. Two processes that do change the free-carrier density are carrier recombination and carrier diffusion. Carrier recombination, or electron-hole recombination, refers to processes in which an electron in the conduction band and a hole in the valence band recombine. In other words, an electron previously excited to the conduction band makes a transition back down to the valence band, as shown by process 4 in Figure 6. Carrier diffusion, on the other hand, refers to the average motion of free carriers from high carrier-density regions to low carrier-density regions. Recombination lowers the total number of free carriers in a crystal whereas diffusion redistributes free carriers in real space. However, for the purposes of the work reviewed in this paper, both types of processes reduce the number of free carriers in the laser-excited region.

Time scales for recombination processes depend on the free-carrier density and on the specific type of process. The various types of recombination differ according to how the energy lost by the electron in the downward transition to the valence band is given off. Two of the most important types of recombination are radiative recombination, in which the transition energy is radiated as a photon, and Auger recombination, in which the transition energy is transferred to another free carrier.

A radiative recombination event occurs through the spontaneous emission of a photon when a conduction band electron and a valence band hole annihilate each other. Thus the radiative recombination rate $R_{\text{radiative}}$ is proportional to the product of the conduction electron density times the hole density: $R_{\text{radiative}} = C_{\text{radiative}}np$, where n is the conduction electron density, p is the density of holes in the valence band, and $C_{\text{radiative}}$ is a constant (16). In the case of optical generation of free carriers, $n = p = N$, where N is the density of electron-hole pairs. We can define an instantaneous recombination time $\tau_{\text{radiative}}$ by setting $R_{\text{radiative}} = N/\tau_{\text{radiative}}$. Then for radiative recombination, we have $\tau_{\text{radiative}} = (C_{\text{radiative}}N)^{-1}$.

In contrast, an Auger recombination even involves three free carriers because the electron-hole annihilation transfers energy either to another conduction electron or to a hole in the valence band. Therefore, the Auger recombination rate R_{Auger} in the case of optically generated free carriers with density N is given by $R_{\text{Auger}} = C_{\text{Auger}}N^3$. This expression gives an instantaneous Auger recombination time of $\tau_{\text{Auger}} = C_{\text{Auger}}^{-1}N^{-2}$. In laser-induced structural change experiments using femtosecond-excitation pulses, the femtosecond laser pulse generates initial carrier densities greater than 10^{21} cm^{-3}, a regime where Auger recombination is the dominant mechanism (17).

The above treatment of recombination needs to be modified at high carrier densities where $C_{\text{radiative}}$ and C_{Auger} themselves become functions of N. Once N exceeds roughly 10^{18} cm^{-3}, both of these coefficients will decrease with increasing carrier density due to electronic screening effects. The presence of many mobile charge carriers leads to a partial screening of the Coulomb interaction between free carriers. The weaker interaction reduces the probability of a recombination event, thus lowering the recombination rate compared to the unscreened case. A calculation based on static screening theory for the case of Si suggests that the Auger recombination time saturates at about 6 ps for carrier densities exceeding 10^{21} cm^{-3} (17).

Carrier diffusion is another important mechanism that reduces the free carrier density in the region excited by the laser pulse. The rate at which carriers diffuse out of the laser-excited region is given by $R_{\text{diffusion}} = N/\tau_{\text{diffusion}} = D_a\nabla^2 N$, where the ambipolar diffusion constant D_a takes into account diffusion of both electrons and holes (5). If the laser spot size on the sample surface is much bigger than the absorption depth δ, then the carrier density varies only in the direction normal to the sample surface: $N(z) = N(0)e^{-z/\delta}$. In this case, the diffusion time is given by $\tau_{\text{diffusion}} = \delta^2/D_a$. The diffusion constant D_a is related to the excited electron temperature T_e and to the carrier mobility through the Einstein relation $D_a = k_B T_e \mu_a/e$ (5, 17), where k_B is the Boltzmann constant, μ_a a reduced carrier mobility for conduction electrons and valence holes, and e is the magnitude of the electron charge. If the electron temperature varies in time, such as during and after femtosecond laser-pulse excitation (18), the diffusion constant will also vary in time. The dependence of D_a on electronic temperature is further complicated by the implicit dependence of the carrier mobility μ_a on the electron temperature, which is not easily determined.

Although an increased electron temperature resulting from femtosecond laser-pulse excitation should speed up diffusion, carrier confinement slows it down (18). This effect arises from the reduction in the bandgap in an excited semiconductor due to many-body interactions in the free-carrier system (18, 19). In laser-pulse excitation, the spatial gradient of the light intensity in the material leads to a spatial gradient in the excited carrier density. The carrier-density gradient, in turn, causes a gradient in the bandgap, with a smaller bandgap coinciding with higher carrier density. In the absence of diffusion, free carriers will tend towards the regions with smaller bandgaps, thus this spatial gradient results in carrier confinement that slows down diffusion (18). Simulations and experimental measurements show that carrier-density relaxation through these processes occurs on a picosecond time scale (17, 18, 20, 21).

SUMMARY OF ENERGY TRANSFER The above discussion of energy transfer in a semiconductor is not meant to be a comprehensive treatment of this expansive subject. Many other scattering mechanisms such as scattering with plasmons (collective electronic excitations), impurity scattering, and combinations of all of the above play a role in the distribution of the laser-deposited energy. A full review of these processes can be found in Reference 4. The goal of this section is to provide a sense for the relevant time scales for energy transfer following femtosecond laser-pulse excitation of a semiconductor. In particular, carrier–carrier scattering thermalizes the free–carrier population within a few hundred femtoseconds; carrier-phonon scattering leads to carrier–lattice equilibration on a time scale of a few picoseconds; and carrier recombination and diffusion reduce the free-carrier density in the excited region also within a few picoseconds.

LASER-INDUCED PHASE TRANSITIONS

The strong connection between the electronic system and the lattice structure of a solid suggests that a change in one affects the other. When a laser pulse is incident on a semiconductor, it interacts with the electronic system in the material by exciting electrons from the valence band to the conduction band. If strong enough, this electronic excitation leads to a structural change in the lattice. The central question to address, then, is how the excitation of the electronic system results in atomic motion.

Thermal Model vs Plasma Model

Following the first experiments in laser annealing, two models were proposed to explain the structural change resulting from electronic excitation. One model, known as the thermal model, describes the structural change as a thermal melting process (22–24). The thermal model assumes that the hot electrons rapidly equilibrate with the lattice by exciting lattice vibrations (emitting phonons). With this assumption, the laser energy deposited in the material can be treated as though it is instantly converted to heat. If the incident laser pulse is strong enough, the irradiated part of the sample will heat up to the melting temperature and undergo a transition to the liquid phase as the latent heat of fusion is supplied.

The other model, known as the plasma model, attributes the structural change to destabilization of the covalent bonds resulting directly from the electronic excitation (25, 26). The plasma model assumes a slow rate of phonon emission by the excited electronic system compared with the energy deposition time (i.e. the laser pulse width). This assumption implies that structural change can occur while the electronic system and the lattice are not in thermal equilibrium with each other, although each of these systems may be internally in a quasi-equilibrium. According to this model, the structural change is driven directly by the excited electronic system. If a high enough fraction of the valence electrons is excited from bonding states to antibonding states, the crystal becomes unstable, and a structural phase transition occurs.

Picosecond Pulses vs Femtosecond Pulses

The debate over the explanation for laser-induced structural change prompted experimental tests of these two models. Early results, from experiments involving laser pulses with pulse widths of tens of picoseconds or longer, showed evidence for strong heating of the sample and agreed well with predictions from computer simulations based on the thermal model (22, 27, 28). Later, Raman spectroscopy experiments on semiconductors confirmed that an excited electronic system can equilibrate with the lattice in just a few picoseconds (13), in agreement with the assumption in the thermal model.

The introduction of femtosecond lasers in the mid-1980s led to renewed interest in this topic (29). The development of lasers producing pulses of less than 100 fs duration opened up the possibility of depositing energy in a semiconductor on a time scale that is short compared with the electron-lattice equilibrium time of a few picoseconds. By depositing energy faster than the electrons can transfer it to the lattice or recombine with holes, an intense short pulse can excite a very high density of free carriers before significant heating of the lattice can occur. Due to the high peak intensities in femtosecond pulses, nonlinear absorption further increases the density of excited carriers (30). This suggests that femtosecond laser-pulse excitation of semiconductors can create the conditions necessary for electronically driven structural change. Indeed, recent experimental results show that the response of a semiconductor to a femtosecond laser pulse is fundamentally different from the response to a picosecond or nanosecond laser pulse (29, 31–38). The remainder of this paper focuses on the femtosecond excitation-pulse regime.

Response of the Linear Optical Properties

Determining the behavior of both the electronic and structural properties of a semiconductor is of prime importance to understanding the phase transition induced by a femtosecond optical excitation. However, to probe the structural change directly would require a subpicosecond X-ray or electron probe, which is not currently available. To resolve the response of a material to laser-pulse excitation on such a short time scale, a femtosecond laser pulse is generally used to probe the optical properties of the material over a range of time delays following the excitation pulse. In essence, this pump-probe technique provides a series of snapshots of the material at different times following the excitation, thereby mapping out the time evolution of the optical property being measured. Because the optical properties of a semiconductor are directly related to its electronic structure, as discussed above, optical pump-probe techniques provide a valuable tool for monitoring the changes in the electronic band structure. The band structure in turn carries information about the crystal structure.

Using this pump-probe scheme, a number of researchers have mapped out the time evolution of a semiconductor's reflectivity following intense femtosecond laser-pulse excitation (29, 31–33, 35–38). In both Si (29, 31–33) and GaAs (35–37), the reflectivity shows a sharp rise within roughly 200 fs after the excitation pulse for excitation fluences greater than the threshold for permanent damage. Figure 8 shows a typical example of this reflectivity behavior plotted vs pump-probe time delay. The data points in Figure 8 represent experimental values for the reflectivity of GaAs at a 45° incident angle for *p*-polarized light measured using a 70-fs, 2.2-eV probe pulse following excitation with a 70-fs 1.9-eV excitation pulse. The fluence (energy per unit area) of the excitation pulse in this case is 1.5 kJ/m^2. This is 1.5 times greater than the measured damage threshold of 1.0 kJ/m^2 for GaAs (39). In the past, the increase in reflectivity seen in Figure 8 was

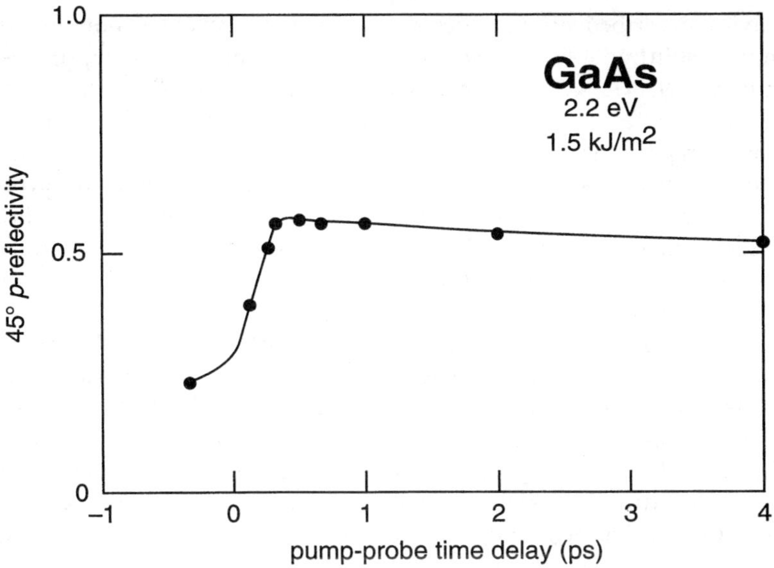

Figure 8 Measured reflectivity of GaAs at 45° incident angle for *p*-polarized, 2.2-eV light following excitation with a 70-fs, 1.5-kJ/m^2, 1.9-eV laser pulse. The reflectivity values are plotted vs the time delay between the arrival of the excitation pulse and the arrival of the probe pulse. The curve through the data is drawn to guide the eye.

attributed to the metallic-like behavior of the excited free carriers. Because the time scale for the reflectivity rise is about an order of magnitude shorter than the time scale for heating of the lattice through phonon emission (see above section), this change cannot be driven by thermal processes.

The reflectivity data alone, however, do not provide a clear link to the changes in the electronic structure induced by the excitation pulse because reflectivity measurements made at a single angle of incidence and polarization do not unambiguously determine the induced changes in the semiconductor's linear optical properties. The ambiguity in interpreting reflectivity data arises because reflectivity is not an intrinsic material property. Rather, it is a function of both the real and imaginary parts of the linear optical susceptibility or, equivalently, the dielectric constant. It is the response of the dielectric constant that provides a clear link to the behavior of the electronic structure. One can unambiguously determine both the real and imaginary parts of the dielectric constant by using two synchronized probe pulses to simultaneously measure the reflectivity at two carefully chosen angles of incidence and polarizations (40).

Recent pump-probe measurements of the dielectric constant made using this type of two-angle reflectivity technique show that the reflectivity behavior described above reflects a drop in the average bonding-antibonding splitting of the semiconductor, leading to a collapse of the bandgap (39, 41, 42). The average bonding-antibonding splitting of a material manifests itself optically as the resonant frequency of the main absorption peak in its spectrum (10). Thus the drop in the average splitting is accompanied by a drop in the resonant frequency of the main absorption peak. As this resonant frequency drops through the frequency of the probe pulse, the imaginary part of the dielectric constant at the probe frequency goes through a peak, while the real part crosses zero. Figure 9, which shows experimental values for the real and imaginary parts of the dielectric constant of GaAs at a photon energy of 2.2 eV plotted vs pump-probe time delay, illustrates this behavior. As in Figure 8, the excitation pulse is a 70-fs 1.9-eV laser pulse with a fluence of 1.5 kJ/m^2. Note that the imaginary part peaks and the real part crosses zero within a few hundred femtoseconds, indicating that within this time, the average bonding-antibonding splitting has dropped from its initial value of about 4.5 eV to below 2.2 eV. This evolution of the dielectric constant produces the rise in

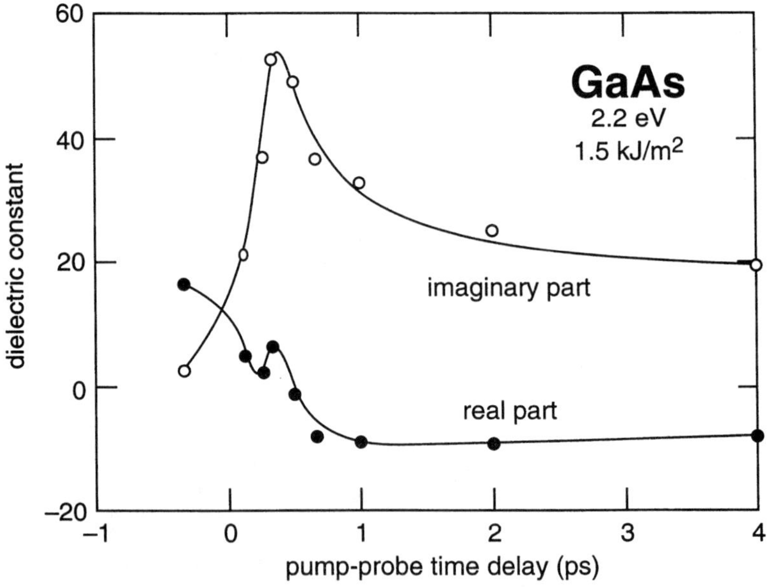

Figure 9 Experimentally determined 2.2-eV dielectric constant of GaAs plotted vs pump-probe time delay. As in Figure 8, a 70-fs, 1.5-kJ/m^2, 1.9-eV laser pulse serves as the pump pulse. The curves through the data are drawn to guide the eye. ●: real part of the dielectric constant, ○: imaginary part of the dielectric constant.

the reflectivity seen in Figure 8. Less intense pulses, down to a fluence of 0.8 kJ/m^2, produce a similar drop in the resonant frequency of the main absorption peak, but on a slower time scale.

The large changes in the electronic band structure evident in the dielectric constant measurements most likely arise from a combination of electronic screening by the excited free carriers and structural deformation of the lattice resulting from an electronically induced lattice instability. The free carriers generated by the laser pulse modify the electronic band structure directly through many-body interactions and by screening the ionic potential. A recent calculation shows that these screening effects cause a 2-eV drop in the direct gap at the X-point in GaAs if 10% of the valence electrons are excited to the conduction band—roughly the free-carrier density attained by femtosecond excitation pulses in these experiments (43). However, the time scale for the drop in the average bonding-antibonding splitting indicates that electronic screening cannot account for the entire bandgap collapse. Screening should be largest immediately following the excitation, i.e. when the free-carrier density is highest. As the free-carrier density relaxes through Auger recombination and diffusion, the changes in the electronic band structure due to screening should also relax. Instead, when excited by pulses of about 1.0 kJ/m^2 fluence, the average bonding-antibonding splitting continues to drop for picoseconds following the excitation. Thus while electronic screening is an important effect immediately following the excitation, the changes in the electronic band structure occurring during the first few picoseconds must be due to changes in the lattice structure itself.

As discussed in the first section of this paper, the covalent bonds that hold a semiconductor together are only stable if the electrons are in the ground state. Excitation of electrons from bonding valence states to antibonding conduction states breaks the covalent bonds. If enough bonds are broken, the lattice structure will become unstable, i.e. certain phonon modes will become soft, and the lattice will begin to deform. Thus a femtosecond laser pulse can induce a lattice instability if it excites a critical density of electrons, which calculations put at roughly 10% of the valence electrons (25, 44, 45). Under these conditions, these ions can move a significant fraction of the bond length within a few hundred femtoseconds (45, 46). Since a 10% change in the average bond length is sufficient for the bandgap to collapse (47), this ionic motion can result in a semiconductor-metal transition on a time scale consistent with the experimental data described above. This interpretation of the behavior of the linear optical properties is further supported by experimental data on the behavior of the nonlinear optical properties, which are discussed below.

Response of the Second-Order Optical Susceptibility

The symmetries of a material play an important role in determining its nonlinear optical properties (48). Thus monitoring the behavior of these properties can provide useful information on changes in the lattice structure. In particular, because of its sensitivity to crystal structure, second-harmonic generation has been used as a probe during laser-induced phase transitions in both the picosecond (49–51) and in the femtosecond regimes (33–37, 52). The central idea in most of this work is that the second-order optical susceptibility, which is responsible for second-harmonic generation in the dipole approximation, is only non-zero in materials that lack inversion symmetry (48). In fact, ion implantation experiments show that a loss of long-range order, which in effect produces inversion symmetry on the scale of the wavelength of light, is enough to make the second-order optical susceptibility vanish (53). Thus in pump-probe experiments on GaAs, which lacks inversion symmetry, a drop to zero in the second-harmonic signal indicates a loss of long-range order (52, 54).[2] It is important to note that extracting the true behavior of the second-order susceptibility from measurements of the generated second-harmonic signal requires a knowledge of the changes in the linear properties (52, 54). For femtosecond laser-pulse excitation of semiconductors at fluences above the damage threshold, second-harmonic generation has been observed to vanish on a time scale of about 100 fs following the excitation (33, 36, 55). This behavior supports the conclusion that a deformation of the lattice causes the collapse of the bandgap within a few hundred femtoseconds. In GaAs, the second-order optical susceptibility has been observed to drop to zero even at fluences below the damage threshold, which indicates that there is a range of fluences at which the laser excitation results in a reversible loss of long-range order (54).

Combining the experimental data on the second-order optical susceptibility with that on the linear optical properties provides a self-consistent picture of the effects of the femtosecond laser-pulse excitation on the semiconductor. The excitation of a critical density of electrons from the valence band to the conduction band destabilizes the covalent bonds in the material. The ionic motion resulting from this laser-induced lattice instability leads to a loss of long-range order in the lattice structure, evident in the drop to zero of the second-harmonic signal. The deformation of the

[2] Although Si has inversion symmetry, second-harmonic generation can still be a useful tool for monitoring long-range order with a proper orientation of the crystallographic axes and probe beam polarization. In this case, second-harmonic generation arises from an order-dependent quadrupole term in the second-order nonlinear polarization (33).

lattice is accompanied by the drop in the average bonding-antibonding splitting seen in the resonance behavior of the dielectric constant.

CONCLUSION

Determining the behavior of the optical properties of semiconductors during laser-induced phase transitions provides a window on the complex interplay between the electronic structure and lattice structure of the material. By exciting a high density of electrons from the valence band to the conduction band, an intense femtosecond laser pulse can destabilize the covalent bonds of the semiconductor. The creation of the required density of free carriers is made possible by the very short time during which the energy is deposited into the electronic system. The deformation of the lattice structure that is driven by the electronic excitation in turn changes the electronic band structure, leading to a drop in the average bonding-antibonding splitting and a transition to metallic behavior.

While much has been learned about laser-induced phase transitions in semiconductors since the discovery of laser annealing, many questions remain to be answered. For example, what are the details of the bandgap collapse? Can we distinguish more clearly the role of electronic screening from that of lattice deformation in modifying the electronic band structure? What does the deformed lattice look like? What are the lattice relaxation dynamics at fluences below the damage threshold? Remaining issues such as these promise to keep this a stimulating field of research.

ACKNOWLEDGMENTS

We appreciate many useful discussions with Professors N. Bloembergen, H. Ehrenreich, E. Kaxiras, and M. Aziz. E.N.G. gratefully acknowledges a Fannie and John Hertz Fellowship. We would also like to acknowledge financial support from the Joint Services Electronics Program under contract number ONR N00014–89-J-1023 and the National Science Foundation under contract number NSF DMR 89–20490.

Literature Cited

1. Shtyrkov EI, Khaibullin IB, Zaripov MM, Galyatudinov MF, Bayazitov RM. 1976. *Sov. Phys. Semicond.* 9: 1309
2. Biegelsen DK, Rozgonyi GA, Shank CV, eds. 1985. *Energy Beam-Solid Interactions and Transient Thermal Processing*. Vol. 35. Pittsburgh: Mater. Res. Soc.
3. Harrison WA. 1989. *Electronic Structure and the Properties of Solids: The Physics of the Chemical Bond*. New York: Dover
4. Ashcroft NW, Mermin ND. 1976. *Solid State Physics*. Philadelphia: Saunders
5. Seeger K. 1991. *Semiconductor Physics: An Introduction*. Berlin: Springer-Verlag
6. Pierret RF. 1987. *Advanced Semiconductor Fundamentals*. Reading, MA: Addison Wesley
7. Chelikowsky JR, Cohen ML. 1976. *Phys. Rev. B* 14: 556
8. Palik ED. 1985. In *Handbook of Optical Constants of Solids*, ed. ED Palik, pp. 429–43. New York: Academic
9. Aspnes DE, Schwartz GP, Gualtieri GJ, Studna AA, Schwartz B. 1981. *J. Electrochem. Soc.* 128: 590
10. Cohen ML, Chelikowsky JR. 1988. *Electronic Structure and Optical Properties of Semiconductors*. Berlin: Springer-Verlag
11. Conwell EM, Vassel MO. 1966. *IEEE Trans. Electron. Devices* 13: 22
12. Goldman JR, Prybyla JA. 1994. *Phys. Rev. Lett.* 72: 1364
13. Kash JA, Tsang JC, Hvam JM. 1985. *Phys. Rev. Lett.* 54: 2151
14. Fröhlich H. 1937. *Proc. R. Soc. London Ser. A* 160: 230
15. Shah J, Deveaud B, Damen TC, Tsang WT, Gossard AC, Lugli P. 1987. *Phys. Rev. Lett.* 59: 2222
16. Landsberg PT. 1991. *Recombination in Semiconductors*. Cambridge: Cambridge Univ. Press
17. Yoffa EJ. 1980. *Phys. Rev. B* 21: 2415
18. van Driel HM. 1987. *Phys. Rev. B* 35: 8166
19. Kalt H, Rinker M. 1992. *Phys. Rev. B* 45: 1139
20. Othonos A, van Driel HM, Young JF, Kelly PJ. 1991. *Phys. Rev. B* 43: 6682
21. Choo HR, Hu XF, Downer MC. 1993. *Appl. Phys. Lett.* 63: 1507
22. Wood RF, White CW, Young RT, eds. 1984. *Semiconductors and Semimetals*. Vol. 23. New York: Academic
23. Malvezzi AM, Kurz H, Bloembergen N. 1985. See Ref. 2, pp. 75–80
24. Malvezzi AM. 1987. In *Excited-State Spectroscopy in Solids*, ed. UM Grassano, N Terzi, pp. 335–54. Amsterdam: North-Holland Physics
25. Van Vechten JA, Tsu R, Saris FW. 1979. *Phys. Lett.* 74A: 422
26. Van Vechten JA. 1984. In *Semiconductors Probed by Ultrafast Laser Spectroscopy*, ed. RR Alfano, p. 95. San Diego: Academic
27. Li KD, Fauchet PM. 1983. *Appl. Phys. Lett.* 51: 21
28. Kurz H, Bloembergen N. 1985. See Ref. 2, pp. 3–13
29. Shank CV, Yen R, Hirlimann C. 1983. *Phys. Rev. Lett.* 51: 900
30. Reitze DH, Zhang TR, Wood WM, Downer MC. 1990. *J. Opt. Soc. Am. B* 7: 84
31. Shank CV, Yen R, Hirlimann C. 1983. *Phys. Rev. Lett.* 50: 454
32. Downer MC, Fork RL, Shank CV. 1985. *J. Opt. Soc. Am. B* 2: 595
33. Tom HWK, Aumiller GD, Brito-Cruz CH. 1988. *Phys. Rev. Lett.* 60: 1438
34. Schröder T, Rudolph W, Govorkov SV, Shumai IL. 1990. *Appl. Phys. A* 51: 1438
35. Govorkov SV, Shumay IL, Rudolph W, Schroeder T. 1991. *Opt. Lett.* 16: 1013
36. Sokolowski-Tinten K, Schulz H, Bialkowski J, von der Linde D. 1991. *Appl. Phys. A* 53: 227
37. Saeta PN, Wang J, Siegal Y, Bloembergen N, Mazur E. 1991. *Phys. Rev. Lett.* 67: 1023
38. Reitze DH, Ahn H, Downer MC. 1992. *Phys. Rev. B* 45: 2677
39. Siegal Y, Glezer EN, Mazur E. 1994. *Phys. Rev. B* 49: 16 403
40. Greenaway DL, Harbeke G. 1968. *Optical Properties and Band Structure of Semiconductors*. Oxford: Pergamon
41. Siegal Y, Glezer EN, Huang L, Mazur E. 1995. In *Ultrafast Phenomena in Semiconductors*, ed. DK Ferry, HM van Driel. Bellingham, WA.: SPIE. In press
42. Siegal Y, Glezer EN, Mazur E. 1995. In *Femtosecond Chemistry*, ed. J Manz, L Wöste. Berlin: Verlag Chemie. In press
43. Kim DH, Ehrenreich H, Runge E. 1994. *Solid State Commun.* 89: 119
44. Biswas R, Ambegoakar V. 1982. *Phys. Rev. B* 26: 1980
45. Stampfli P, Bennemann KH. 1990. *Phys. Rev. B* 42: 7163
46. Stampfli P, Bennemann KH. 1993. *J. Phys. Condens. Matter* 5: A173 (Suppl.)
47. Froyen S, Cohen ML. 1983. *Phys. Rev. B* 28: 3258
48. Shen YR. 1984. *The Principles of Nonlinear Optics*. New York: Wiley & Sons
49. Akhmanov SA, Koroteev NI, Paitan

GA, Shumay IL, Galyautdinov MV, et al. 1983. *Opt. Commun.* 47: 202
50. Akhmanov SA, Emel'yanov VI, Koroteev NI, Seminogov VN. 1985. *Sov. Phys. USP* 28: 1084
51. Liu JM, Malvezzi AM, Bloembergen N. 1985. See Ref. 2, pp. 137–42
52. Siegal Y. 1994. *Time-resolved studies of laser-induced phase transitions in GaAs.* PhD thesis. Harvard Univ., Cambridge. 149 pp.
53. Akhmanov SA, Galyautdinov MV, Koroteev NI, Paityan GA, Khaibullin IB, et al. 1985. *Bull. Acad. Sci. USSR Phys. Ser.* 49: 86
54. Glezer EN, Siegal Y, Huang L, Mazur E. 1995. *Phys. Rev. B.* In press
55. Saeta PN. 1991. *Optical studies of ultrafast carrier dynamics in semiconducting materials.* PhD thesis. Harvard Univ., Cambridge. 167 pp.

Annu. Rev. Mater. Sci. 1995. 25: 249–71

PHASE TRANSITIONS IN III–V COMPOUNDS TO MEGABAR PRESSURES

Arthur L. Ruoff and Ting Li

Department of Materials Science and Engineering, Cornell University, Ithaca, New York 14853

KEY WORDS: X-ray diffraction, transition pressure, solid-state amorphization, bulk modulus, metallization

ABSTRACT

The high pressure behavior of 16 III–V semiconductors, including the nitride, phosphide, arsenide, and antimonide, respectively, of boron, aluminum, gallium and indium, is discussed in a summary of recent research efforts. The emphasis is laid on the structural phase transitions at high pressures utilizing X-ray diffraction technique. The equation of state (EOS) and, hence, the bulk modulus and its pressure derivative at 0 pressure of these compounds is also summarized. Theoretical results are compared with experimental data wherever appropriate to show an ever-increasing reliability of theoretical predictions. Optical absorption, reflectivity, and Raman measurements of these compounds are briefly mentioned. Two closely related IVB elements, Si and Ge, are also discussed.

INTRODUCTION

There has been considerable interest in the high-pressure behavior of III–V compounds, together with Si and Ge, during the last two decades. Technologically, the refinement of diamond-anvil cell (DAC) techniques has made it possible to achieve megabar or submegabar pressures routinely. Although much work on these semiconductor materials has been motivated by possible industrial applications, the theoretical developments, from the quantum dielectric theory by Phillips (1) and Van Vechten

0084–6600/95/0801–0249$05.00

(2–4), to the pseudopotential total-energy approach (5), have served to further interest in this area.

The purpose of this paper is to summarize the experimental results on structural phase transformations of III–V compounds. Comparisons with theoretical calculations are made whenever appropriate. However, no effort has been made to comment on every article of related work; although many advanced techniques are mentioned, the emphasis is on the X-ray diffraction work, especially energy-dispersive X-ray diffraction (EDXD) (6). A summary of the first high-pressure transitions is presented at the end of each section.

Two closely related elements, Si and Ge, are also discussed briefly. We conclude with a look to the future direction in high-pressure studies on III–V compounds.

Basic Characteristics of the First Transition

The first transition in most III–V compounds is from the fourfold structure to quasi-sixfold or sixfold structures, primarily β-Sn and NaCl, with a few exceptions: GaAs (7) becomes orthorhombic and AlAs (8) goes to the sixfold NiAs structure.

An important aspect of this first transition is that it is usually sluggish, and exhibits substantial hysteresis on both uploading and down-loading. This is well illustrated with AlN (9), which transforms to the NaCl structure at 14 GPa but does not revert to the wurtzite structure on down-loading, i.e. it persists as a meta-stable NaCl structure at atmospheric pressure.

Stress Homogeneity and Choice of Pressure Marker

As argued by Besson et al, stress homogeneity is desirable for accurate measurement of pressure in high-pressure experiments (10, 11). However, most pressure-transmitting media transform to the solid state above 10 GPa and thus no longer produce a true hydrostatic or quasi-hydrostatic environment; moreover, trace amounts of water in many commonly used organic pressure media may suffice to contaminate some highly hygroscopic samples (8). On the other hand, shear stress plays an important role in the kinetic process to induce a reconstructive phase transition. Therefore, some degree of stress inhomogeneity is tolerable and, in some cases, favorable.

The secondary ruby pressure scale is currently the most accurate, but in most (EDXD) studies, metal pressure markers, e.g. Au and Mo, are often preferred. Some work has been done to estimate the inaccuracy in the pressure measurements introduced by the presence of large shear stress. For example, Ruoff et al estimated the effect of the presence of a finite yield strength on the measurement of pressure for tungsten (W) (12). In experiments in which the pressure was found by EDXD to be 364 GPa without correction, it was calculated that the actual pressure is 378–395 GPa with a 5–9% correction for nonhydrostaticity. Fortunately, most III–V compounds have a much lower bulk modulus and thus are much softer, so the nonhydrostatic effect is not as manifest as it is in hard metals. In another recent estimate (13), an interesting criterion for choosing a suitable pressure marker is proposed: to minimize the pressure inaccuracy caused by nonhydrostatic effect, the ratio (bulk modulus)/(shear modulus) values for the sample and the pressure marker should be as close as possible. According to this criterion, Mo may be a much better pressure marker than Au for most soft III–V compounds.

Kinetic Effects and Role of the Experimental Technique

In exemplary work on GaAs, Besson et al argued that various physical parameters that characterize the solid in one phase do not simultaneously change in a discontinuous manner at a first-order phase transition (10, 11). For instance, the transition pressure for coordination is generally lower on uploading when observed by incoherent gauges such as X-ray absorption spectroscopy (XAS) or optical opacity (not absorption); the transition pressure for crystal periodicity is usually higher when measured by coherent phenomena such as X-ray or neutron diffraction. Therefore, not only is down-loading highly recommended in order to investigate the hysteresis associated with a solid–solid phase transition, but also a combination of various advanced techniques is desirable to give a complete picture of the kinetics. The thermodynamic equilibrium transition pressure, which is the one theorists can predict, can be reasonably deduced only then.

Solid State Amorphization

In 1965, McDonald et al reported that solid state amorphization (SSA) occurred on heating (at atmospheric pressure) the meta-stable high pressure phase of GaSb that had previously been obtained by rapid quenching to low temperatures before releasing the pressure (14). There are interesting results on two types of pressure-induced amorphization. In GaAs, when the pressure is reduced to 1 atm from 115 GPa, an amorphous phase is formed (15). In the case of BAs, the amorphization occurs when the pressure is increased to 120 GPa (16). Theoretical calculations on BAs (17) had predicted a transition from the zincblende to the NaCl structure at 110 GPa. The experimental result is quantitative evidence that pressure-induced amorphization is basically a kinetically frustrated phase transition.

BN, AlN, GaN, AND InN

BN

BN is known to exist in four crystalline modifications: two graphite-like forms, hexagonal (hBN) and rhombohedral (rBN), and two dense forms, cubic zincblende (cBN) and wurtzite (wBN).

cBN is an extremely hard material with a bulk modulus of 370–390 GPa (18–20). The theoretical calculations show that the transition pressure is as high as 1100 (17) or 1025 GPa (21). The equation of state (EOS) has been studied up to 115 GPa (19), and the bulk modulus was determined to be 369 ± 14 GPa. Good agreement with shock-wave data was found.

wBN has been studied up to 106 GPa, and no transition was observed (22). The bulk modulus obtained therein, 390 ± 3 GPa, is in close agreement with a calculation (23).

Ueno et al (24) have done X-ray diffraction on rBN up to 61 GPa. A transition to cBN takes place at ~8 GPa and is completed above 20 GPa.

AlN

At ambient condition, AlN takes on the wurtzite structure. The experimental and theoretical data for the first high-pressure phase transition in AlN are intriguingly scattered. Vollstädt et al showed that there was a structural phase transformation between 14 and 16.5 GPa, at about 1500°C, to a semiconducting rock-salt structure (25). The pseudopotential calculations by Van Camp et al give a transition pressure of 12.9 GPa (26). Total energy calculations by Christensen & Gorczyca (27) yield a value of 12.5 GPa for the transition pressure, which is lower than their result from linear muffin–tin orbital (LMTO) method (28), but it is argued that the transition pressure in reality is close to 17 GPa, which is in good agreement with their direct observation of appearance of opacity at ~16.5 GPa (28). Xia et al, from their EDXD work, suggested that the equilibrium transition pressure lies on the interval of 0–14 GPa because the rock-salt structure persists all the way through down-loading to atmospheric pressure (9). Another recent angle-dispersive X-ray diffraction study reports a much higher value of 22.9 GPa for the transition pressure during uploading (29).

Therefore, an educated guess on the equilibrium transition pressure, encouraged by the arguments on GaAs of Besson et al could be anywhere below 17 GPa (10). Christensen & Gorczyca (27) also predicted that another transition to NiAs strcuture would occur between 30 and 40 GPa. However, it has not been confirmed up to 65 GPa according to Xia et al (9). Thus another X-ray diffraction study to a higher pressure seems necessary to provide experimental data.

The bulk modulus of the wurtzite phase is around 200 GPa according to the references mentioned above. The frequency of zone-center phonons and the corresponding mode-Grüneisen parameter have been obtained by Raman scattering up to 13 GPa by Perlin et al (30).

GaN

GaN is characterized by its high ionicity of 0.5 (2) or 0.43 (31), the highest among III–V compounds. At ambient pressure, GaN crystallizes in a hexagonal (wurtzite) structure with space group $P6_3mc$ (C^6_{4v}). It has four atoms in the unit cell, and the atomic volume is only about half of that of GaAs. It has a mixture of covalent bonding and ionic tetrahedral bonding and has a direct electronic band-gap of 3.5 eV at the Γ point.

Recent X-ray absorption spectroscopy (XAS) and Raman scattering measurements (32) on GaN indicated a phase transition around 47 GPa. Some recent theoretical calculations based on ab initio pseudo-potential method (33) or LMTO theory (20) predicted that the most favorable high pressure phase of GaN is the rock-salt structure.

Xia et al studied the structure of GaN and discovered a first-order phase transformation from a wurtzite (fourfold) to a rock-salt (sixfold) structure (34). The structural phase transformation was observed beginning at 37 GPa. Ueno et al (35) obtained similar results, but the transition pressure was 52.2 GPa. The earlier X-ray absorption near-edge structure (XANES) study (32) showed the reversal transition begins at 30 GPa. Hence the equilibrium transformation pressure can be anywhere between 30 and 52 GPa.

The bulk modulus was measured to be 245 (32) or 237 GPa (35), which is in good agreement with the theoretical result of 239 GPa by Gorczyca & Christensen (20).

InN

Ueno et al undertook angle-dispersive X-ray diffraction studies on InN and noted that InN transformed from wurtzite to the rock-salt structure at 12.1 GPa upon uploading (22, 35). This value is close to what Xia et al (36, 37) obtained during uploading (10 GPa), but Ueno et al claimed that the equilibrium transformation pressure lies on the interval of 5 GPa to 10 GPa from their down-loading data. These experimental data are close to the theoretical value of 4.9 GPa for the transition pressure by Muñoz & Kunc (38, 39). However, this result is at least disputed by the experimental and theoretical work by Perlin et al (21) and Gorczyca et al (20), which led to a transition pressure of 25.4 GPa (theoretical) or 23.0 GPa (experimental). The rock-salt structure was studied up to 35 GPa, and the equation of state was also obtained by Xia et al (36, 37).

Two theoretical results are very close for the bulk modulus B_0 (165 and 166 GPa) (21, 39); however, the experimental results are much lower [144 ± 4 GPa (36) and 125.5 ± 4.6 GPa (35)]. [The unreasonably high value of the pressure derivative of the bulk modulus ($B'_0 = 12.7 \pm 1.4$) casts some doubt on the validity of the latter because this quantity usually has a value in the neighborhood of four.]

The first- and second-order pressure coefficients of the main band gaps and the crossover pressure from direct to indirect gap of InN, along with other III–V indium compounds, were theoretically studied by Van Camp et al (40).

BP, AlP, GaP, AND InP

BP

BP is known to have a high hardness and a high melting point (41); it is expected to undergo a structural phase transformation from the zincblende

Table 1 Summary of the transition pressures of the first-phase transition in III–V nitrides

Compound	First phase	Second phase	P↑ (GPA)	P↓ (GPa)	$P_{eq.}$ (GPa)	Method	Reference
wBN	Wurtzite	NaCl			1025	LMTO[a]	20
rBN	Rhombohedral	Zincblende	8			ADXD[b]	24
cBN	Zincblende	NaCl			1110	P.P.[c]	17
AlN	Wurtzite	NaCl	14	0	0–14	EDXD	9
			22.9			ADXD	29
					12.9	P.P.	26
					12.5	LMTO	27
GaN	Wurtzite	NaCl	47	30		XAS, Raman	32
			37			EDXD	34
			52.2			ADXD	35
					55.1	P.P.	33
					65	LMTO	20
InN	Wurtzite	NaCl	10	5	7.5	EDXD	37
			12.1			ADXD	35
					4.9	P.P.	38
			23.0			(Experimental)	21
					25.4	P.P.	20

[a] LMTO, linear muffin–tin orbital.
[b] ADXD, angle-dispersive X-ray diffraction.
[c] P.P., pseudopotential.

Table 2 Summary of the bulk moduli and their pressure derivatives at zero pressure of the first two phases in III–V nitrides

	First phase				Second phase			
Compound	B_0 (GPa)	B_0'	$\frac{\Delta V}{V}$ (%)	$\frac{V_N}{V_0}$	B_0 (GPa)	B_0'	Method	Reference
wBN	390	3.5					ADXD	22
	390	6.3					LCAO[a]	23
rBN	33.4	5.25					EDXD	130
cBN	369	4.0					ADXD	19
	382	4.5					ADXD	131
	394	3.57					P.P.	132
	392	3.3					P.P.	18
	387						LMTO	20
	367				425		P.P.	17
	363	3.5					LCAO	19
AlN	185.0	5.7	18.6	0.81	221.0	4.8	EDXD	9
	207.9	6.3	18				ADXD	29
	159.9	5.2					Ultrasonic	133
	207	3.98					LCAO	134
	195.0	3.74			215.0	4.58	P.P.	26
	205		19		270		LMTO	27
GaN	237	4.3	17.9				ADXD	35
	188	3.2	17.0	0.83	248	5.5	EDXD	34
	245	4[b]					EXAFS[c]	32
	190	2.92			223	3.69	P.P.	33
	179	3.93	13.4		244		P.P.	135
	239		12.7		269		LMTO	136
InN	144	4.07	20	0.81	163	3.3	EDXD	36
	143		19.7	0.781			LMTO	39
	155.5	4.02					P.P.	40
	165						LMTO	20

[a] LCAO, linear combination of atomic orbitals.
[b] The value here is arbitrarily assumed.
[c] EXAFS, extended X-ray absorption fine structure.

to the rock-salt structure at a pressure of 160 GPa, according to total energy pseudo-potential calculations (17). A more recent calculation lowered the transition pressure to 139 GPa and predicted that the rock-salt BP would be a semi-metal (42).

EDXD work on BP has only been carried out up to 110 GPa (43), and BP remained stable in its zincblende phase. Thus the pressure-induced transformation remains to be confirmed. The EOS was obtained from the experiment, and the bulk modulus of BP is slightly lower than the theoretical value (44).

AlP

Froyen & Cohen (45) have suggested that AlP should behave as AlAs; thus, following the EDXD work on AlAs, Greene et al performed an EDXD and optical reflectance spectroscopy study on AlP up to 43 GPa (46). Much to their expectation, AlP undergoes a first-order phase transformation from the zincblende to the NiAs structure with a volume collapse of $17 \pm 1\%$. The high-pressure phase was determined to be metallic from reflectivity data, just as predicted by both Froyen & Cohen (45) and Zhang & Cohen (5). This transition is found reversible but with large hysteresis. The equilibrium transition pressure is 9.5 ± 5 GPa, which agrees well with the value of 9.3 GPa of Froyen & Cohen and Zhang & Cohen for their calculated zincblende to rock-salt transition pressure.

GaP

Baublitz & Ruoff first identified the high pressure phase of GaP to be a tetragonal β-Sn-type structure in 1982 (47). The zincblende-to-β-Sn transition began on loading at 21.5 ± 0.8 GPa and room temperature. The new structure was indexed according to a tetragonal unit cell with $a = 4.720$ Å and $c = 2.4687$ Å at $P = 24.0 \pm 1.0$ GPa. The space group of the new phase was not uniquely assigned because there was not sufficient information on the atom ordering. Two Ga and two P atoms were assumed to be either randomly distributed on the sites $(0,0,0)$; $(0,\frac{1}{2},\frac{1}{4})$; $(\frac{1}{2},0,\frac{3}{4})$; $(\frac{1}{2},\frac{1}{2},\frac{1}{2})$ or locally ordered with Ga atoms at $(0,0,0)$; $(\frac{1}{2},\frac{1}{2},\frac{1}{2})$ and P atoms at $(0,\frac{1}{2},\frac{1}{4})$; $(\frac{1}{2},0,\frac{3}{4})$. The existence of one (and possibly a second) forbidden line probably owes its origin to the twins or other planar defects on the (301) planes. This high pressure structure was confirmed by Hu et al (48) with a transition pressure slightly higher due to a more quasi-hydrostatic environment. These results, which are upper bound for the equilibrium transition pressure, are not inconsistent with the calculated transition pressure of 21.7 GPa from the pseudopotential approach (5, 45). The actual transition pressure argued by Besson et al (10), is below 18 GPa based on premonitory variations of the optical properties (49).

The EOS of the zincblende phase was obtained by Itie et al (50) with extended X-ray absorption fine structure (EXAFS). It is also noteworthy that the GaP sample quickly depressurized from 36 GPa was mainly amorphous (50), although no evidence of amorphization was found on release of pressure from ~ 29 GPa (48). Not surprisingly, a sample after depressurization from 32 GPa was found to contain a mixture of polycrystalline and amorphous GaP (49).

The pressure dependence of the fundamental gas of GaP has been determined from optical-absorption measurements up to 23 GPa (49, 51);

frequencies of the zone-center and zone-boundary phonon modes in GaP at high pressures have been theoretically evaluated by Rodriguez & Kunc (52).

InP

A recent high pressure study (53), using angle-dispersive diffraction techniques with an image-plate detector, confirmed that InP underwent a zincblende to NaCl transition at ~10 GPa. This result is slightly lower than the theoretical value, 12.8 GPa, obtained with a pseudopotential approach (5). In an earlier report (54), a further transition to a β-Sn-like structure was found around 19 GPa, and the equations of state for the first two phases were determined.

Among many experimental and theoretical studies on the electronic and phonon properties of InP under high pressures (40–58), the first- and second-order Raman spectra have been measured as a function of hydrostatic pressure up to 10 GPa (59), and the variation of the lowest direct absorption edge with pressure has also been measured in the same pressure range (60). An interesting result stems from the optical-absorption measurements on two alloys $Ga_x In_{1-x}P$ ($x = 0.36$ and 0.5) (51). A linear Vegard law was found for the nominal first transition pressures as a function of composition.

BAs, AlAs, GaAs, AND InAs

BAs

BAs is the most covalent III–V compound because it has the lowest Phillips ionicity, $f_i = 0.002$. In 1987, Wentzcovitch, Cohen & Lam employed the

Table 3 Summary of the transition pressures of the first-phase transition in III–V phosphides

Compound	First phase	Second phase	P↑ (GPA)	P↓ (GPa)	$P_{eq.}$ (GPa)	Method	Reference
BP	Zincblende	NaCl			160	P.P.	17
					139	P.P.	42
AlP	Zincblende	NiAs	14.2	4.8	9.5	EDXD	46
		(NaCl)			9.3	P.P.	45
GaP	Zincblende	β-Sn	21.5			EDXD	47
			<25			EDXD	48
					21.7	P.P.	5
InP	Zincblende	NaCl	10			ADXD	53
			10.8			ADXD	54
					12.8	P.P.	5

For abbreviations see Tables 1 and 2.

Table 4 Summary of the bulk moduli and their pressure derivatives at zero pressure of the first two phases in III–V phosphides

	First phase			Second phase			
Compound	B_0 (GPa)	B'_0	$\frac{\Delta V}{V}$ (%)	B_0 (GPa)	B'_0	Method	Reference
BP	152	4.3				EDXD	43
	173					Brillouin scattering	137
	165					P.P.	44
	166			155		P.P.	17
	160.1	3.77				P.P.	42
AlP	86.0					Empirical estimate	138, 139
	84.5	4.18	18.9[a]			P.P.	5
	86.5		18.9[a]			P.P.	45
GaP	87.4	4.5				EXAFS	50
	88.2	4.78				Ultrasonic	140
	85.8	4.04				P.P.	93
	81	4.4				P.P.	141
	86.8	4.00	15.5[a]			P.P.	5
	90.9					P.P.	69
InP	76	4	16.5	130	1.6	ADXD	54
	71.1	4.59				Ultrasonic	142
	76.1	4.20				P.P.	40
	81			137		LMTO	58
	70.0	4.93	16.6			P.P.	5

[a] The parameters here are calculated assuming rock-salt as the high-pressure structure.
For abbreviations see Tables 1 and 2.

total energy pseudopotential technique to predict that BAs would favor the rock-salt phase and the transition pressure would be 110 GPa (17).

Greene et al recently performed a high pressure EDXD study of the structure of BAs (16). A transformation from the zincblende to an amorphous crystal structure was observed at 125 GPa, slightly above the calculated transition pressure to the rock-salt phase. This was the first quantitative evidence that the pressure-induced amorphization is a kinetically frustrated process. The amorphous state persisted on uploading to 165 GPa, and throughout down-loading to ambient pressure. The first-order Birch equation of state for the zincblende phase was determined with $B_0 = 148 \pm 6$ GPa and $B'_0 = 3.9 \pm 0.3$, which are in good agreement with calculations by Wenzocovitch et al (17).

AlAs

Like many other semiconductors with tetrahedral structures at 1 atm, AlAs is expected to undergo a first-order phase transition from the zincblende structure to a sixfold coordinated structure at high pressure with ~17% volume change. In 1983, Froyen & Cohen (45), on the basis of ab initio pseudopotential calculations, suggested that the high pressure structure could be either rock-salt or NiAs because these two structures had very nearly identical total energy vs volume curves, with NiAs having the slightly lower energy (45). The equilibrium transition pressure was calculated to be 7.6 GPa (5, 45) and 9.0 GPa (61) for zincblende to rock-salt (no other values were given for the NiAs structure). Weinstein et al observed a phase transformation by microscopic examination (62), and later by Raman study (63), at ~12.4 GPa on uploading, but the structure was unknown. A hysteresis of 6–8 GPa was also reported in the latter.

Greene et al (8) did an EDXD study on AlAs recently to 46 GPa and found AlAs transforms to NiAs structure. This was the first experimental observation of a III–V compound transforming to the NiAs structure. The equilibrium transformation pressure was found to be 7 ± 5 GPa, averaging the large hysteresis. The volume collapse on transformation is $17 \pm 1\%$ at 7 GPa. EOS were obtained for both phases, with $B_0 = 74 \pm 4$ GPa for zincblende AlAs. These results are in good agreement with theoretical calculations.

GaAs

The first structural transition in GaAs to an orthorhombic structure was intensively studied by Besson et al with an array of experimental techniques: optical transmittance measurements, Raman scattering, X-ray absorption, X-ray diffraction, and high-resolution electron microscopy (10, 11). By a thorough analysis of the kinetic effects in the transition, they located the thermodynamic transition region at 12 ± 1.5 GPa at 300 K. A consistent picture of what was occurring around the phase transition was presented and that was the first example of combining the strengths of different techniques to achieve a nearly complete understanding of solid–solid phase transition under high pressure.

The orthorhombic structure [GaAs(II)] was determined by Weir et al by EDXD to be space group Pmm2, with atoms at $(0, 0, 0)$, and $(0, \frac{1}{2}, \alpha)$, where $\alpha = 0.35$ at P = 22.9 GPa (7). This structure is identical to InSb(IV) proposed by Yu et al (64). Upon further loading to 24 ± 1 GPa, GaAs(II) transformed to another orthorhombic structure [GaAs(III)], space group Imm2, consisting of a body-centered orthorhombic lattice with a basis of $(0, 0, 0)$ and $(0, \frac{1}{2}, \Delta)$ as shown in Figure 1, where Δ is 0.425 at 28.1 GPa.

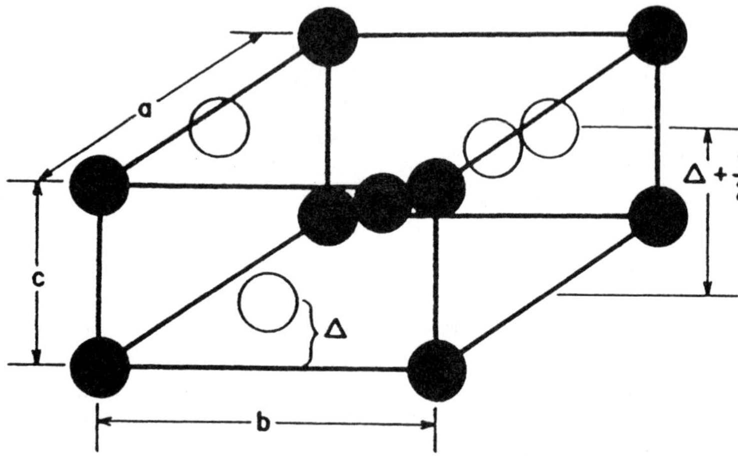

Figure 1 The proposead unit cell of ordered GaAs III. [After Weir *et al* (7)].

This structure can be visualized as an orthorhombic distortion of the simple hexagonal structure with the distortion decreasing as the pressure increases. Although this II-III transition is consistent with the second drop in resistance of a factor of 10–20 observed by Minomura & Drickamer (65), a recent electrical resistance measurement (66) showed a precipitous drop only in the vicinity of 15 GPa and suggested that both GaAs(II) and GaAs(III) are metallic and superconducting with T_c close to 4.5 K.

With increasing pressure, the Δ value for GaAs(III) approached $\frac{1}{2}$ and the GaAs(III) structure gradually assumed the symmetry of the simple hexagonal structure, GaAs(IV). The transition to GaAs(IV) was completed in the vicinity of 60–80 GPa. The structure remained a simple hexagonal up to at least 108 GPa.

In another EDXD and optical reflectivity experiment, Vohra, Xia & Ruoff (15) observed complete amorphization of GaAs at ambient conditions after decompression from 115 GPa. The recompression of the amorphous state produced the orthorhombic GaAs(III) structure at around 27 GPa.

There was strong theoretical interest in the high pressure behavior in GaAs as a prototype solid. For example, Kunc & Martin (67) successfully predicted that the GaAs(II) phase would be orthorhombic and suggested the transition is related to the soft TA(X) phonon; Zhang & Cohen have done first principle, pseudopotential calculations for GaAs(II) structure and found it to be energetically favored over other competing structures (68); García & Cohen used core corrections for exchange and correlation to yield an equilibrium transition pressure of 11.7 GPa for the first transition and, moreover, predicted that a CsCl-type structure should be stable above 125 GPa (69). It is tempting for experimentalists to do another megabar experiment to verify their prediction.

Among various optical studies on GaAs, the refractive index of GaAs was measured between 0.6 and 1.4 eV for hydrostatic pressures up to 8 GPa (70).

InAs

In earlier resistivity measurements (65, 71) and Raman scattering studies (A Jayaraman et al, unpublished) on InAs under hydrostatic pressures, the first transition pressure was found to be around 7 GPa; in a separate X-ray diffraction study, Jamieson reported a phase transition from the zincblende to rock-salt phase at $V/V_0 = 0.926$ with a volume collapse of 18.8% (73).

In 1985, Vohra, Weir & Ruoff extended the high-pressure studies on InAs to 27 GPa with EDXD (74). The zincblende-to-rock-salt transformation was confirmed at 7 GPa. A further transition at 17 ± 0.4 GPa was found with increasing pressure. The β-Sn structure was assigned to the new phase with satisfactory fitting. No volume discontinuity was found for the second transition. The EOS were given up to 27 GPa.

In 1978, Soma did predict the correct transition sequence but overestimated the transition pressures (75). After the EDXD work by Vohra et al, Christensen calculated the EOS of InAs and excellent agreement was found (76).

The first-order Raman scattering by transverse optical (TO) and longitudinal optical (LO) phonons has been measured in InAs under hydrostatic pressures up to its first phase transition (77); a theoretical study on the pressure-dependent phonon properties has also been carried out on InAs, together with some other gallium and indium semiconductors (78).

BSb, AlSb, GaSb, AND InSb

BSb

As the possibility of making BSb is still in doubt (79), there is no high pressure study on this material. It is inferred from the binary phase diagram that as there is some solubility of Sb in B, BSb might form at high pressure in the zincblende structure.

AlSb

As early as 1962, a high-pressure phase transition was reported for zincblende AlSb from electrical resistivity measurements (65). However, controversy surrounds the structure of the new phase because three differ-

Table 5 Summary of the transition pressures of the first-phase transition in III–V arsenides

Compound	First phase	Second phase	P↑ (GPA)	P↓ (GPa)	$P_{eq.}$ (GPa)	Method	Reference
BAs	Zincblende	Amorphous	125	0		EDXD	16
		(NaCl)			110	P.P.	17
AlAs	Zincblende	NiAs	12	2	7	EDXD	8
			12.4	~5.4		Raman	63
		(NaCl)			7.6	P.P.	5
					9.0	P.P.	61
GaAs	Zincblende	Orthorhombic			12	see Reference	10
					16.0	P.P.	5
					11.7	P.P.	69
InAs	Zincblende	NaCl	7			EDXD	74
			8.5			Resistivity	65
			6.9			Resistivity	71
			7.6			Raman	77
					7.8	LMTO	76

For abbreviations, see Tables 1 and 2.

ent types, i.e. β-Sn (80), rock-salt (81), and orthorhombic (82, 83), have been proposed. The reason for this ambiguity may partly lie in the hygroscopicity of the material because the trace amount of water in the pressure medium might suffice to contaminate its chemical structure (82).

Along theoretical lines, Chelikowsky used a total-energy method coupled with an empirical ionicity scale to predict that AlSb would transform to the β-Sn structure at 7.67 GPa (84). Aourag et al (85) arrived at the same structure based on a pseudopotential calculation. Zhang & Cohen showed that the energies of the rock-salt and β-Sn structures were very close, and they predicted a transition pressure of 5.4 GPa for the β-Sn type (5). Singh et al used a three-body-force-potential approach to make a prediction of 6.6 GPa for the transition pressure (86).

The recent EDXD study by Greene et al (87) may serve to clarify all the controversy. Loaded in an argon glove box and with no pressure medium, AlSb was studied up to 50 GPa. The high-pressure phase was indexed to the β-Sn structure. The equilibrium transition pressure, averaging the uploading and down-loading transition pressures, was found to be 5.3 ± 2 GPa, which is in good agreement with Zhang & Cohen's result. EOS were obtained for both phases; the EOS for the low pressure phase agrees with the result by Ves et al (83). Moreover, a further transformation to an unknown phase was reported to start at 43 GPa.

In addition, the dependence of the direct and indirect gap of AlSb on

Table 6 Summary of the bulk moduli and their pressure derivatives at zero pressure of the first two phases in III–V arsenides

Compound	First phase		$\frac{\Delta V}{V}$ (%)	$\frac{V_N}{V_0}$	Second phase		Method	Reference
	B_0 (GPa)	B'_0			B_0 (GPa)	B'_0		
BAs	148	3.9					EDXD	16
	145				143		P.P.	17
AlAs	74	5.0	17	0.831	73	4.6	EDXD	8
	74.1		20.7[a]	0.78[a]			P.P.	45
	71.0	3.26	20.7[a]				P.P.	5
	75						P.P.	143
GaAs	74.7	4.67					Ultrasonic	144
				0.881[b]	57.0[b]	4.80[b]	EDXD	7
	75.6		16.4[a]				P.P.	69
	72.5		17.4[a]	0.81[a]			P.P.	45
	70.9	3.94					P.P.	93
	70.8	3.36	17.4[a]				P.P.	5
InAs	59.2	6.8	18.8	0.86	40.6	7.3	EDXD	74
	58						Ultrasonic	145
	61.9	4.49					P.P.	40
	58.1		17.3				LMTO	76
	58.1	3.60	16.5				P.P.	5

[a] The parameters here are calculated assuming rock-salt as the high-pressure structure.
[b] The parameters here account for GaAs (III, IV).
For abbreviations, see Tables 1 and 2.

pressure was studied by optical absorption measurements up to ~7.9 GPa (88); the first-order Raman scattering by optical phonons was measured for the same pressure range (83).

GaSb

Earlier studies showed that GaSb transformed from the zincblende structure to the β-Sn analogue (65, 81, 89). The transition pressure was found to be 6.2 ± 0.3 GPa (81), which is close to the theoretical result, 6.3 GPa (5). However, the hysteresis accompanying the phase transition is obvious from the fact that holding the sample under 9 GPa at room temperature even for days did not suffice to complete the transition (90).

The relative low ionicity of GaSb suggests that its high-pressure sequence of phase transitions should be similar to those of Ge and Si. In 1987, Zhang & Cohen predicted by ab initio pseudopotential calculations that the next phase in the sequence would be the ordered pseudo-hexagonal structure shown in Figure 2, which they calculated should form at a pressure of 52.8 GPa (5). (It is not strictly a hexagonal structure because it doesn't have sixfold symmetry. The corresponding hexagonal structure had previously been observed in Si and Ge.) This structure, with six unlike and two like neighbors, was found to have the lowest energy among three possible configurations.

Weir et al subsequently studied GaSb with EDXD and found the pseudo-hexagonal structure forms at a pressure of 26.4 GPa (91). This was the first case of finding this pseudo-hexagonal crystal structure in a compound. Due to the fact that certain important peaks were obscured by other peaks, it was not possible to confirm or refute the ordering suggested by Zhang & Cohen.

A further transition occurred on uploading at 61.0 GPa to a structure that was tentatively indexed as a disordered orthorhombic structure with a basis (0, 0, 0) and $(\frac{1}{2}, \frac{1}{2}, 0)$. This orthorhombic structure persisted to 110 GPa. There is a large hysteresis associated with this hexagonal-to-tentative-orthorhombic transition because on down-loading, the hexagonal phase reappeared at 38.0 GPa.

The EOS were given by Weir et al for the three high-pressure phases. The bulk modulus B_0 and its pressure derivative B'_0 for the zincblende phase were given in Reference 92, and a theoretical calculation was done by Van Camp et al for these parameters (93). The first-order Raman

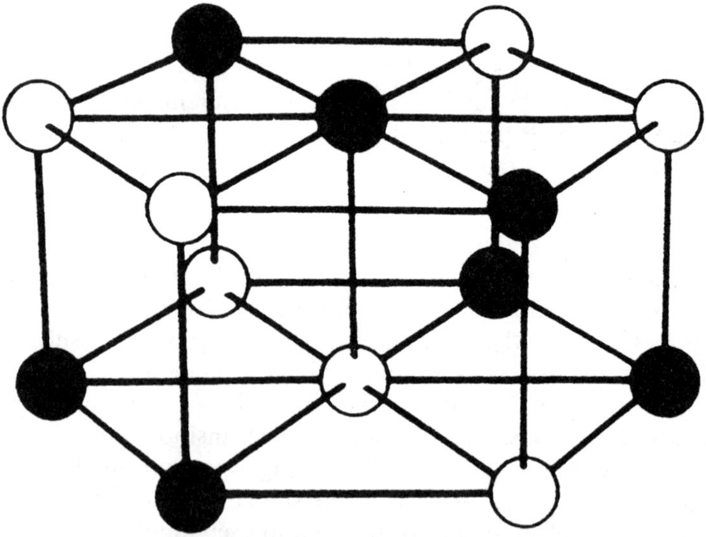

Figure 2 The ordered pseudo-hexagonal structure of GaSb.

scattering by TO and LO phonons has been measured in GaSb under hydrostatic pressures up to the first phase transition (77); a theoretical study also was carried out for pressure-dependent phonon properties (78).

The first SSA was reported on quenched GaSb (14). It is noteworthy that in recent years Russian scientists have shown keen interest in this phenomenon (90, 94–98). For example, Il'ina detected that super-conducting properties arose in amorphous gallium antimonide (a–GaSb) under the nonmetal–metal transition (Anderson transition) at pressures starting at ~ 3.5 GPa (97); Barkalov et al studied the quenched high pressure GaSb β-Sn phase by means of differential scanning calorimetry and dilatometry and found the amorphization process to follow a multistep course (95).

InSb

In 1992, Nelmes et al utilized angle-dispersive X-ray diffraction techniques with an image-plate detector at the Daresbury synchrotron source to examine the high-pressure phases of InSb up to 5 GPa (99, 100). These techniques achieved much greater resolution and sensitivity than had ever been available. Their study successfully clarified the mysterious high-pressure behavior of InSb, which had been the focus of a 30-year-long controversy (64, 80, 101–104).

In their notation, Nelmes et al found two phase transition sequences in IsSb up to 5 GPa, involving three high-pressure phases, **P2**(β-Sn), **P3**(InSb II, orthorhombic), and **P4**(InSb IV, orthorhombic) (99, 100). Either **P1**(zincblende) transforms through a mixture of **P2** and **P3** to **P4** in the pressure range just above 2 GPa, or there is a direct transition from **P1** to **P4** at about 3 GPa. Both **P3** and **P4** were found to be site ordered. **P3** was indicated to be pseudo-A-face centering, with In at $(0, 0, 0)$ and Sb at, or very near, $(0, \frac{1}{2}, \frac{1}{2})$. **P4** has the structure proposed by Yu et al [64] (space group *Pmm*2) with atoms at $(0, 0, 0)$ and $(0, \frac{1}{2}, \alpha)$. Their value of $\alpha = 0.387(1)$ is in good agreement with Vanderborgh et al (104), but **P4** was shown to have a unit cell 12 times larger than previously thought. A mixture of **P2**, **P3**, and **P4** was found to account for the pattern attributed by Vanderborgh et al to a new phase InSb V. In a follow-up study using density functional total-energy calculations by Guo et al (105, 106), the **P2** (β-Sn) phase was found to be unstable compared with **P3** (InSb II, orthorhombic).

In the earlier EDXD work by Vanderborgh et al, InSb was found to experience two further phase transitions: At pressures above ~ 17 GPa, it completed its gradual transition to hexagonal structure (InSb III) (104). InSb later assumed a structure with body-centered symmetry (InSb VI) at pressures above ~ 27 GPa. There was no volume discontinuity associated with these two transitions. The InSb VI phase persisted to at least 66 GPa. Again, the site ordering could not be clarified, but the CsCl structure was hypothesized based on the calculation by Shindo et al (107).

It is interesting to note that the hexagonal (InSb III) phase was also observed in the non-equilibrium phase diagram developed by Vasin et al, during the study of pulsed evaporation and condensation of InSb thin films under ambient pressure (108). Also, Shimomura et al reported a meta-stable NaCl-type phase for amorphous InSb at a pressure of above ~1 GPa (109, 110) and this phase was found to be stable up to 8 GPa (111). This indicates the rock-salt and β-Sn phases are close in energy in InSb.

The pressure dependence of electronic properties of InSb, including bulk modulus and its pressure derivatives, has been summarized (40), and its phonon properties studied (77, 78).

Si AND Ge

High-pressure studies of Si have been performed up to 248 GPa (112). The transition sequence is well established, along with a new high-pressure phase identified recently by McMahon & Nelmes (113). At ~11 GPa, the cubic (diamond) Si (Si I) goes to body-centered-tetragonal (β-Sn) structure (Si II) (73, 114, 115); a further transition occurs above about 16 GPa to simple hexagonal Si V (114, 116), the first observation of such a structure in an elemental system. Angle-dispersive powder-diffraction techniques utilizing an image-plate area detector enabled McMahon & Nelmes to

Table 7 Summary of the transition pressures of the first-phase transition in III–V antimonidesides

Compound	First phase	Second phase	P↑ (GPA)	P↓ (GPa)	$P_{eq.}$ (GPa)	Method	Reference
AlSb	Zincblende	β-Sn	7.2	3.4	5.3	EDXD	87
					7.7	P.P.	84
					5.4	P.P.	5
GaSb	Zincblende	β-Sn	6.2			ADXD	81
			6.7–8.4			Resistivity	89
					6.3	P.P.	5
InSb	Zincblende	β-Sn (**P2**)	2			ADXD	99
		Orthorhombic (**P4**)	3			ADXD	99

For abbreviations, see Tables 1 and 2.

Table 8 Summary of the bulk moduli and their pressure derivatives at zero pressure of the first two phases in III–V antimonides

	First phase				Second phase			
Compound	B_0 (GPa)	B'_0	$\frac{\Delta V}{V}$ (%)	$\frac{V_N}{V_0}$	B_0 (GPa)	B'_0	Method	Reference
AlSb	55.1	4.55					EDXD	83
	57	4.2	15	0.85	59	4.0	EDXD	87
	54.3	4.01	20.6				P.P.	5
GaSb	56.1	4.78					Ultrasonic	92
				0.82	58.8	7.7	EDXD	91
	52.2	4.44					P.P.	93
	55.7	3.83	18.6				P.P.	5
InSb	46.5	3.65					Ultrasonic	146
			17.1	0.80	52.0[a]	4.25[a]	EDXD	104
	45.6	4.88					Ultrasonic	147
	47.7	4.61					P.P.	40
	47.0	5.21	20.7[b]				P.P.	5

[a] The parameters here account for InSb (IV, III, VI).
[b] The parameter here is calculated assuming β-Sn as the high-pressure structure.
For abbreviations, see Tables 1 and 2.

observe a new intermediate body-centered orthorhombic structure (space-group Imma) between Si II and Si V (113); this phase has been confirmed by a follow-up theoretical total-energy study (117). At ~37 GPa, Si V transforms into the intermediate phase Si VI (116), which is found to have the same crystal structure as the X phase of the alloy $Bi_{0.8}Pb_{0.2}$ (112); Si VI transforms to hexagonal-close-packing (hcp) Si VII at 42 GPa (116), which subsequently transforms to face-centered-cubic (fcc) Si at 78 GPa (118). Some theoretical calculations (119–121) have examined the simple hexagonal transition, and it was shown that the transition is driven by a soft phonon mode. On unloading from the β-Sn phase, Si forms a meta-stable BC8 (body-centered cubic) phase (122, 123), a fourfold coordinated structure with a density 9% greater than the diamond cubic phase. Heating Si III above 470 K leads to Si IV, another meta-stable phase, with a hexagonal diamond structure (122, 123). It has been found that Si III is a hole semi-metal and that Si IV, like Si I, is an intermediate-gap semiconductor (124). Two new meta-stable phases, Si VIII and Si IX, both tetragonal, have also been attained from rapid decompression from the high-pressure metallic phases (125).

Ge has a similar transition sequence compared with Si. At around 11 GPa, Ge shows one transition from the cubic diamond structure to the β-

Sn structure (116, 126). Two new phase transitions were observed by Vohra et al (127): the β-Sn to the simple hexagonal phase at 75 ± 3 GPa and to the double hexagonal close-packed structure (dhcp) at 102 ± 5 GPa. The same study also presented results from total-energy pseudopotential calculations, which predicted 84 ± 10 GPa for the β-Sn to simple hexagonal transition and 105 ± 2 GPa for simple hexagonal to hcp (not dhcp) transition (127). The role of 3*d* core electrons in increasing the transformation pressures in Ge, as compared with Si, was also emphasized. Inspired by the recent discovery of the *Imma* phase in Si, Lewis & Cohen did ab initio total energy calculations on Ge and predicted the same *Imma* phase to be a stable high-pressure phase Ge between the transition pressures of the β-Sn and the simple hexagonal phases (128). No experimental verification has been reported. When Ge in the β-Sn structure is depressurized, a fourfold coordinate meta-stable tetragonal structure with a density 11% greater than the diamond cubic phase appears (123, 129).

CONCLUDING REMARKS: WHAT ELSE CAN BE DONE IN THE FUTURE?

Most III–V compounds have been studied only below 50 GPa, which means that high-pressure experimentalists can endeavor to extend the pressure up to megabar range. In some cases, even if a complicated unresolvable phase occurs at a certain pressure range, it is usually desirable to go to higher pressures with the belief that a simple phase might appear.

Angle-dispersive X-ray diffraction techniques, complemented with an imaging plate detector, can achieve much better precision than EDXD. For many complex structures unresolvable by EDXD, angle-dispersive X-ray diffraction is surely worth trying.

Various high-temperature methods have been devised for DAC measurements. Laser heating and/or resistor wire heating can be used to overcome many kinetics-related problems (e.g. disorder, meta-stability). Moreover, to construct the complete phase diagram, the information around the triple point is highly desirable for applying thermodynamic relations.

ACKNOWLEDGMENTS

We acknowledge support by the Department of Energy under Grant No. DE-FG02-87ER-45320. We appreciate helpful discussions with Raymond G Greene and Huan Luo.

Literature Cited

1. Phillips JC. 1968. *Phys. Rev. Lett.* 20: 550–53
2. Van Vechten JA. 1969. *Phys. Rev.* 182: 891–905
3. Van Vechten JA. 1969. *Phys. Rev.* 187: 1007–20
4. Van Vechten JA. 1973. *Phys. Rev. B* 7: 1479–507
5. Zhang SB, Cohen ML. 1987. *Phys. Rev. B* 35: 7604–10
6. Baublitz MA Jr, Arnold V, Ruoff AL. 1981. *Rev. Sci. Instrum.* 52: 1616–24
7. Weir ST, Vohra YK, Vanderborgh CA, Ruoff AL. 1989. *Phys. Rev. B* 39: 1280–85
8. Greene RG, Luo H, Li T, Ruoff AL. 1994. *Phys. Rev. Lett.* 72: 2045–48
9. Xia Q, Xia H, Ruoff AL. 1993. *J. Appl. Phys.* 73: 8198–200
10. Besson JM, Itie JP, Polian A, Weill G, Mansot JL, Gonzalez J. 1991. *Phys. Rev. B* 44: 4214–34
11. Besson JM. 1993. *Jpn. J. Appl. Phys.* (Suppl. 32)1: 11–15
12. Ruoff AL, Luo H, Xia H, Vohra YK. 1991. *High Pressure Res.* 6: 183–86
13. Meng Y, Weidner DJ, Fei Y. 1993. *Geophys. Res. Lett.* 20: 1147–50
14. McDonald TRR, Sard R, Gregory E. 1965. *J. Appl. Phys.* 36: 1498–99
15. Vohra YK, Xia H, Ruoff AL. 1990. *Appl. Phys. Lett.* 57: 2666–68
16. Greene RG, Luo H, Ruoff AL, Trail SS, Disalvo FJ Jr. 1994. *Phys. Rev. Lett.* 73: 2476–79
17. Wentzcovitch RM, Cohen ML, Lam PK. 1987. *Phys. Rev. B* 36: 6058–68
18. Van Camp PE, Van Doren VE, Devreese JT. 1989. *Solid State Commun.* 71: 1055–58
19. Knittle E, Wentzcovitch RM, Jeanloz R, Cohen ML. 1989. *Nature* 337: 349–52
20. Gorczyca I, Christensen NE. 1993. *Physica* B185: 410–14
21. Perlin P, Gorczyca I, Porowski S, Suski T, Christensen NE, Polian A. 1993. *Jpn. J. Appl. Phys.* (Suppl. 32)1: 334–39
22. Ueno M, Yoshida M, Onodera A, Shimomura O, Takemura K. 1993. *Jpn. J. Appl. Phys.* (Suppl. 32)1: 42–44
23. Xu YN, Ching WY. 1991. *Phys. Rev. B* 44: 7787–98
24. Ueno M, Hasegawa K, Oshima R, Onodera A, Shimomura O, et al. 1992. *Phys. Rev. B* 45: 10226–30
25. Vollstädt H, Ito E, Akaishi M, Akimoto S, Fukunaga O, 1990. *Proc. Jpn. Acad.* B66: 7–9
26. Van Camp PE, Van Doren VE, Devreese JT. 1991. *Phys. Rev. B* 44: 9056–59
27. Christensen NE, Gorczyca I. 1993. *Phys. Rev. B* 47: 4307–14
28. Gorczyca I, Christensen NE, Perlin P, Grzegory I, Jun J, Bockowski M. 1991. *Solid State Commun.* 79: 1033–34
29. Ueno M, Onodera A, Shimomura O, Takemura K. 1992. *Phys. Rev. B* 45: 10123–26
30. Perlin P, Polian A, Suski T. 1993. *Phys. Rev. B* 47: 2874–77
31. Camphausen DL, Connell GAN, Paul W. 1971. *Phys. Rev. Lett.* 26: 184–88
32. Perlin P, Jauberthie-Carillon C, Itie JP, San Miguel A, Grzegory I, Polian A. 1992. *Phys. Rev. B* 45: 83–89
33. Van Camp PE, Van Doren VE, Devreese JT. 1992. *Solid State Commun.* 81: 23–26
34. Xia H, Xia Q, Ruoff AL. 1993. *Phys. Rev. B* 47: 12925–28
35. Ueno M, Yoshida M, Onodera A, Shimomura O, Takemura K. 1994. *Phys. Rev. B* B49: 14–21
36. Xia Q, Xia H, Ruoff AL. 1994. *Mod. Phys. Lett. B* 8: 345–50
37. Xia Q, Xia H, Ruoff AL. 1994. *High-Pressure Science and Technology—1993, AIP Conf. Proc.* 309. pp. 307–10
38. Muñoz A, Kunc K. 1993. *Physica* B185: 422–25
39. Muñoz A, Kunc K. 1993. *J. Phys. Condens. Matter* 5: 6015–22
40. Van Camp PE, Van Doren VE, Devreese JT. 1990. *Phys. Rev. B* 41: 1598–602
41. Golikova OA. 1979. *Phys. Status Solidi A* 51: 11–40
42. Van Camp PE, Van Doren VE. 1994. *High-Pressure Science and Technology—1993, AIP Conf. Proc.* 309. pp. 171–74

43. Xia H, Xia Q, Ruoff AL. 1993. *J. Appl. Phys.* 74: 1660–62
44. Wentzcovitch RM, Chang KJ, Cohen ML. 1986. *Phys. Rev. B* 34: 1071–79
45. Froyen S, Cohen ML. 1983. *Phys. Rev. B* 28: 3258–65
46. Greene RG, Luo H, Ruoff AL. 1994. *J. Appl. Phys.* 76: 7296–99
47. Baublitz M Jr, Ruoff AL. 1982. *J. Appl. Phys.* 53: 6179–85
48. Hu JZ, Black DR, Spain IL. 1984. *Solid State Commun.* 51: 285–87
49. Jauberthie-Carillon C, Guillemin C. 1989. *J. Phys. Condens. Matter* 1: 6807–16
50. Itie JP, Polian A, Jauberthie-Carillon C, Dartyge E, Fontaine A, et al. 1989. *Phys. Rev. B* 40: 9709–14
51. Goñi AR, Syassen K Strössner K, Cardona M. 1989. *Phys. Rev. B* 39: 3178–84
52. Rodriguez CO, Kunc K. 1988. *J. Phys. C Solid State Phys.* 21: 5933–41
53. McMahon MI, Nelmes RJ, Wright NG, Allan DR. 1994. *High-Pressure Science and Technology*—1993, *AIP Conf. Proc.* 309. pp. 629–31
54. Menoni CS, Spain IL. 1987. *Phys. Rev. B* 35: 7520–25
55. Leroux M. 1989. *Semicond. Sci. Technol.* 4: 231–32
56. Saxena AK, Benyon RP. 1993. *Phys. Status Solidi A* 136: 529–35
57. Tsai MH, Dow JD, Kasowski RV. 1992. *J. Mater. Res.* 7: 2205–10
58. Gorczyca I, Christensen NE, Alouani M. 1989. *Phys. Rev. B* 39: 7705–12
59. Trommer R, Müller H, Cardona M. 1980. *Phys. Rev. B* 21: 4869–78
60. Müller H, Trommer R, Cardona M, Vogl P. 1980. *Phys. Rev. B* 21: 4879–83
61. Martin RM. 1987. In *Proc. Eighteenth Int. Conf. Physics of Semiconductors, Stockholm*, 1986. ed. O Engström, p. 639. Singapore: World Scientific
62. Weinstein BA, Hark SK, Burnham RD, Martin RM. 1987. *Phys. Rev. Lett.* 58: 781–84
63. Venkateswaran UD, Cui LJ, Weinstein BA. 1992. *Phys. Rev. B* 45: 9237–47
64. Yu SC, Spain IL, Skelton EF. 1978. *J. Appl. Phys.* 49: 4741–45
65. Minomura S, Drickamer HG. 1962. *J. Phys. Chem. Solids* 23: 451–56
66. Zhang SB, Erskine D, Cohen ML, Yu PY. 1989. *Solid State Commun.* 71: 369–73
67. Kunc K, Martin RM. 1981. *Phys. Rev. B* 24: 2311–14
68. Zhang SB, Cohen ML. 1989. *Phys. Rev. B* 39: 1450–52
69. García A, Cohen ML. 1993. *Phys. Rev. B* 47: 6751–54
70. Goñi AR, Syassen K, Cardona M. 1990. *Phys. Rev. B*41: 10104–10
71. Pitt GD, Vyas MKR. 1973. *J. Phys. C* 6: 274–84
72. Deleted in proof
73. Jamieson JC. 1963. *Science* 139: 762–64
74. Vohra YK, Weir ST, Ruoff AL. 1985. *Phys. Rev. B* 31: 7344–48
75. Soma T. 1978. *J. Phys. C* 11: 2681–94
76. Christensen NE. 1986. *Phys. Rev. B* 33: 5096–98
77. Aoki K, Anastassakis E, Cardona M. 1984. *Phys. Rev. B* 30: 681–87
78. Talwar DN, Vandevyver M. 1990. *Phys. Rev. B* 41: 12129–39
79. Clougherty EV, Kaufman L. 1967. *US Govt. Res. Dev. Rep.* 67: 85
80. Jamieson JC. 1963. *Science* 139: 845–47
81. Yu SC, Spain IL, Skelton EF. 1978. *Solid State Commun.* 25: 49–52
82. Baublitz M Jr, Ruoff AL. 1983. *J. Appl. Phys.* 54: 2109–10
83. Ves S, Strössner K, Cardona M. 1986. *Solid State Commun.* 57: 483–86
84. Chelikowsky JR. 1987. *Phys. Rev. B* 35: 1174–80
85. Aourag H, Sellal F, Abid H, Badi N, Mahmoudi A, Khelifa B. 1993. *Mater. Chem. Phys.* 33: 254–59
86. Singh RK, Singh S. 1992. *Phys. Rev. B* 45: 1019–22
87. Greene RC, Luo H, Ghandehari K, Ruoff AL. 1995. *J. Phys. Chem. Solids.* In press
88. Strössner K, Ves S, Kim CK, Cardona M. 1986. *Phys. Rev. B* 33: 4044–53
89. Pitt D. 1969. *High Temp. High Press.* 1: 111–18
90. Degtyareva VF, Belash IT, Ponyatovskii EG, Rashchupkin VI. 1990. *Sov. Phys. Solid State* 32(5): 834–35
91. Weir ST, Vohra YK, Ruoff AL. 1987. *Phys. Rev. B* 36: 4543–46
92. McSkimin HJ, Jayaraman A, Andreatch P Jr, Bateman TB. 1968. *J. Appl. Phys.* 39: 4127–28
93. Van Camp PE, Van Doren VE, Devreese JT. 1988. *Phys. Rev. B* 38: 9906–12
94. Popova SV, Skrotskaya GG, Larchev VI, Zentai G, Pogany L, et al. 1991. *J. Non-Cryst. Solids* 135: 255–58
95. Barkalov OI, Ponyatovsky EG, Antonov VE. 1993. *J Non-Cryst. Solids* 156: 544–46
96. Brazhkin VV, Larchev VI, Popova SV, Skrotskaya GG. 1989. *Phys. Scripta* 39: 338–40
97. Il'ina MA. 1987. *Sov. J. Low Temp. Phys.* 13: 111–13
98. Brazhkin VV, Lyapin AG, Popova SV,

Sapelkin AV. 1994. *JETP Lett.* 59: 130–34
99. Nelmes RJ, McMahon MI, Hatton PD, Crain J, Piltz RO. 1993. *Phys. Rev. B* 47: 35–54
100. Nelmes RJ, McMahon MI, Hatton PD, Piltz RO, Crain J. 1993. *Jpn. J. Appl. Phys.* (Suppl. 32)1: 1–5
101. Jayaraman A, Newton RC, Kennedy GC. 1961. *Nature* 191: 1288
102. Kasper JS, Brandhorst H. 1964. *J. Chem. Phys.* 41: 3768–72
103. Banus MD, Lavine MC. 1969. *J. Appl. Phys.* 40: 409–13
104. Vanderborgh CA, Vohra YK, Ruoff AL. 1989. *Phys. Rev. B* 40: 12450–56
105. Guo GY, Crain J, Blaha P, Temmerman WM. 1993. *Phys. Rev. B* 47: 4841–48
106. Guo GY, Crain J, Temmerman WM. 1993. *Jpn. J. Appl. Phys.* (Suppl. 32)1: 39–41
107. Shindo K, Kobayashi T, Nara N. 1981. *J. Phys. Soc. Jpn.* 50: 2274–78
108. Vasin OI, Petrosyan VI. 1990. *Inorg. Mater.* 26: 769–72
109. Shimomura O, Asaumi K, Sakai N, Minomura S. 1976. *Philos. Mag.* 34: 839–49
110. Asaumi K, Shimomura OO, Minomura S. 1976. *J. Phys. Soc. Jpn.* 41: 1630–35
111. Asaumi M, Minomura S. 1979. *J. Phys. Soc. Jpn.* 46: 1039–40
112. Duclos SJ, Vohra YK, Ruoff AL. 1990. *Phys. Rev. B* 41: 12021–28
113. McMahon MI, Nelmes RJ. 1993. *Phys. Rev. B* 47: 8337–40
114. Hu JZ, Spain IL. 1984. *Solid State Commun.* 51: 263–66
115. Hu JZ, Merkle LD, Menoni CS, Spain IL. 1986. *Phys. Rev. B* 34: 4679–84
116. Olijnyk H, Sikka SK, Holzapfel WB. 1984. *Phys. Lett.* 103A: 137–40
117. Lewis SP, Cohen ML. 1993. *Phys. Rev. B* 48: 16144–47
118. Duclos SJ, Vohra YK, Ruoff AL. 1987. *Phys. Rev. Lett.* 58: 775–77
119. Chang KJ, Cohen ML. 1984. *Phys. Rev. B* 30: 5376–78
120. Chang KJ, Cohen ML. 1985. *Phys. Rev. B* 31: 7819–26
121. Needs R, Martin R. 1984. *Phys. Rev. B* 30: 5390–92
122. Wentorf RM Jr, Kasper JS. 1963. *Science* 139: 338–39
123. Kasper JS, Richards SM. 1964. *Acta Crystallogr.* 17: 752–55
124. Besson JM, Mokhtari EH, Gonzalez J, Weill G. 1987. *Phys. Rev. Lett.* 59: 473–76
125. Zhao YX, Buehler F, Sites JR, Spain IL. 1986. *Solid State Commun.* 59: 679–82
126. Baublitz M Jr, Ruoff AL. 1982. *J. Appl. Phys.* 53: 5669–71
127. Vohra YK, Brister KE, Desgreniers S, Ruoff AL, Chang KJ, Cohen ML. 1986. *Phys. Rev. Lett.* 56: 1944–47
128. Lewis SP, Cohen ML. 1994. *Solid State Commun.* 89: 483–86
129. Bundy FP, Kasper JS. 1963. *Science* 139: 340–41
130. Solozhenko VL, Will G, Hüpen H, Elf F. 1994. *Solid State Commun.* 90: 65–67
131. Aleksandrov IV, Goncharov AF, Stishov SM, Yakovenko EV, 1989. *JETP Lett.* 50: 127–31
132. Nielsen OH, Kunc K. 1991. In *Festschrift in Honor of Rogerio Cerqueira Leite*, ed. M Balkanski, CET Gon calves da Silva, JM Worlock, pp. 112–23. Singapore/New Jersey: World Scientific
133. Gerlich D, Dole SL, Slack GA. 1986. *J. Phys. Chem. Solids* 47: 437–41
134. Ching WY, Harmon BN. 1986. *Phys. Rev. B* 34: 5305–8
135. Muñoz A, Kunc K. 1991. *Phys. Rev. B* 44: 10372–73
136. Gorczyca I, Christensen NE. 1991. *Solid State Commun.* 80: 335–38
137. Wettling W, Windschief J. 1984. *Solid State Commun.* 50: 33–34
138. Wiley JD. 1975. In *Semiconductors and Semimetals*, ed. RK Willardson, AC Beer, 10: 134. New York: Academic
139. Keyes RW. 1962. *J. Appl. Phys.* 33: 3371–72
140. Yoğurtcu YK, Miller AJ, Saunders GA. 1981. *J. Phys. Chem. Solids* 42: 49–56
141. Christensen NE. 1984. *Phys. Rev. B* 30: 5753–65
142. Nichols DN, Rimai DS, Sladek RJ. 1980. *Solid State Commun.* 36: 667–69
143. Chetty N, Muñoz A, Martin RM. 1989. *Phys. Rev. B* 40: 11934–36
144. McSkimin HJ, Jayaraman A, Andreatch P Jr. 1967. *J Appl. Phys.* 38: 2362–64
145. Kunc K. 1973. *Ann. Phys.* (Leipzig) 8: 319–401
146. Slutsky LJ, Garland CW. 1959. *Phys. Rev.* 113: 167–69
147. Bashkin I, Pereseda G. 1975. *Sov. Phys. Solid State* 16: 2058

Annu. Rev. Mater. Sci. 1995. 25: 273–93

FERROELECTRIC THIN FILMS FOR PHOTONICS: Properties and Applications[1]

D. Dimos

Sandia National Laboratories, Materials and Process Sciences Center, Albuquerque, New Mexico 87185

KEY WORDS: electrooptic, optical storage, pyroelectric, waveguide

ABSTRACT

Ferroelectric materials exhibit a combination of electrooptic, nonlinear optical, photo-induced, and pyroelectric properties that make them attractive for a variety of photonic applications. Significant progress has also been made in the development of ferroelectric thin-film fabrication techniques. These advances have made it possible to fabricate optical quality thin films and to integrate these films with functional substrates such as Si and GaAs. At this point, photonic applications for ferroelectric thin films are at either the research or development stage. However, further work and future advances should make ferroelectric films important materials for as yet unknown photonic applications.

INTRODUCTION

In the rapidly developing field of photonics, there is an increasing demand for new devices that are capable of processing large quantities of optical information. These devices include displays, optical memories, image com-

[1] The US Government has the right to retain a nonexclusive, royalty-free license in and to any copyright covering this paper.

parators, optical switches, spatial light modulators, infrared imaging arrays, and frequency doublers. Ferroelectric materials exhibit a combination of properties such as large electrooptic and nonlinear optical coefficients, good transparency to visible light, memory effects, and large pyroelectric coefficients that make them attractive for a variety of photonic applications (1, 2).

Ferroelectric materials can be divided into essentially three material categories. The first and best-studied category is inorganic ferroelectrics. The largest group of materials in this class and the most important for optical applications are oxides that have a distorted oxygen octahedra as the primary structural building unit. This group includes the perovskites such as $BaTiO_3$, $Pb(Zr,Ti)O_3$ (PZT: lead zirconate titanate), and $KNbO_3$; oxides with the tungsten-bronze structure such as $(Sr,Ba)Nb_2O_6$ (SBN), and $Ba_2NaNb_5O_{15}$ (BNN); and the trigonal Li compounds $LiNbO_3$ and $LiTaO_3$. The second category is the ferroelectric polymers, typified by polyvinylidene fluoride (PVDF) and P(VDF-TrFE). Their high pyroelectric coefficients and low permittivities have attracted attention for use in integrated pyroelectric devices (3). However, they do not appear useful for most other electrooptic applications. The last category is the ferroelectric (smetic) liquid crystals. Ferroelectric liquid crystal devices can exhibit large, optical-contrast ratios that make them suitable for spatial light–modulator and display applications. However, they have a limited memory capability, and the fabrication technique used to achieve alignment in displays makes them unsuitable for waveguiding devices. Although both the ferroelectric polymer and liquid crystal films are useful materials for certain applications, this review focuses on the ferroelectric oxides because they have the potential to impact the widest range of photonic applications.

Single crystals and ceramics of ferroelectric oxides have been developed for optical devices such as waveguide switches (e.g. $LiNbO_3$) (4), shutters (e.g. PLZT: lead lanthanum zirconate titanate) (5), and hybrid pyroelectric detectors (6). However, most advanced photonic systems will require integration of the optical components with microelectronic circuits, and to achieve the necessary level of integration, it will be necessary to use thin films. Research over the last five to ten years is now reaching a point where very high-quality thin films of ferroelectric oxides can not only be fabricated by a variety of techniques, but can be processed at temperatures that permit integration with microelectronic circuits. At this point, the progress on ferroelectric thin films has come primarily from work on nonphotonic applications such as nonvolatile memories; however, the work is beginning to provide the basis of an important technology for optical information storage and processing applications.

GENERAL PROPERTIES OF FERROELECTRICS

A ferroelectric can be defined as a material with a spontaneous dipole moment that can be reoriented with an electric field. Therefore, the response that most clearly defines a ferroelectric is the hysteresis in the polarization-voltage behavior, as illustrated in Figure 1. This hysteresis comes from electrical switching of ferroelectric domains, which are regions within the ferroelectric with a unique direction of the spontaneous polarization. The important parameters obtained from the hysteresis loop of Figure 1 are the remanent polarizations $\pm P_R$, which are the intercept points at $E = 0$; the saturation polarizations $\pm P_S$, which are the values at the linear extrema of the loop, and the coercive fields $\pm E_C$, which are the intercept points at $P = 0$. The remanent polarization defined this way is actually the maximum stable value of the polarization in the absence of an applied field because a stable value of the macroscopic polarization can be obtained anywhere between $+P_R$ and $-P_R$ depending on the applied field.

The other critical property is the variation of the spontaneous polarization with temperature. Figure 2 illustrates the simplest case for a material that is paraelectric above a critical temperature (the Curie point), T_C, and that exhibits a second-order phase transition to a ferroelectric phase below T_C. A pyroelectric current, i_{py}, can be derived from a ferroelectric by a change in temperature, where i_{py}–$\propto$-dP/dT. Although a material does not have to be ferroelectric to exhibit pyroelectricity (i.e. a change in its spontaneous polarization with temperature), the best pyroelectric materials are ferroelectric. The transition is also associated with an anomalous increase in the dielectric constant.

Ferroelectric oxides tend to have high refractive indices ($n > 2$) and band gap energies >3 eV, which permit them to be transparent to visible

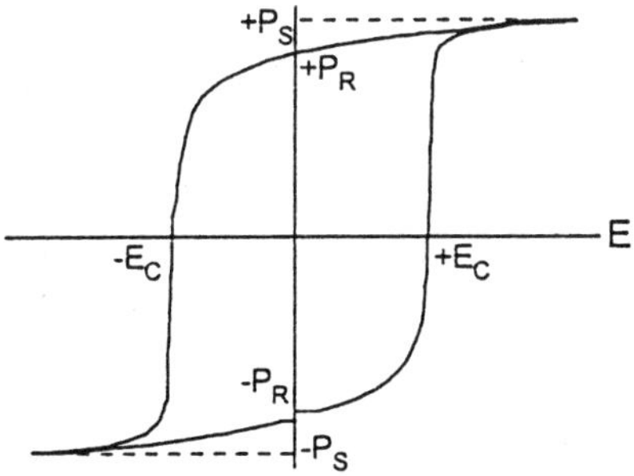

Figure 1 Typical polarization vs electric field behavior for a ferroelectric material.

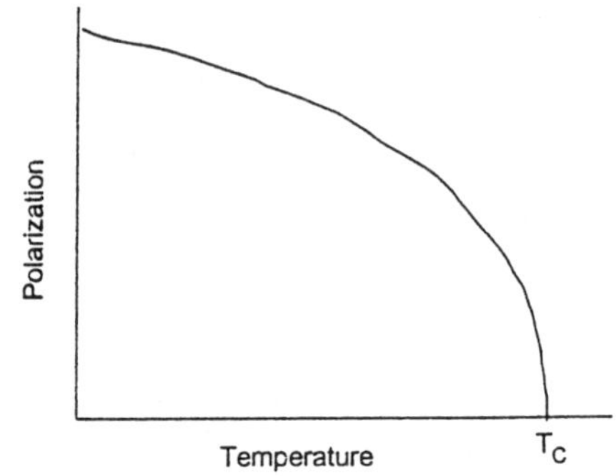

Figure 2 Spontaneous polarization vs temperature for a ferroelectric that exhibits a second-order transition.

light. These materials also tend to have large values of birefringence, which can lead to significant optical scattering, even in materials that are otherwise defect free. In addition, ferroelectric oxides exhibit large second-order susceptibilities that are responsible for their useful electrooptic and nonlinear optical properties, as discussed below.

FABRICATION OF FERROELECTRIC FILMS

The potential for integrating ferroelectric materials with active microelectronics to construct monolithic electrooptic devices has been made possible primarily through the advances in the art of fabricating complex, multicomponent oxides. High-quality, thin films of ferroelectric oxides have been fabricated using a variety of deposition techniques, which can generally be divided into solution, physical vapor–deposition, and chemical vapor–deposition techniques. Because different techniques have inherent advantages and disadvantages, the technique of choice will be dictated by property requirements, as well as by issues of cost, uniformity, and manufacturability. One of the biggest specific concerns for electrooptic applications, except for pyroelectric uses, is scattering losses. Minimizing light scattering generally demands nearly single-crystal thin films, which can be obtained by heteroepitaxial film growth on lattice-matched, single-crystal substrates. Another advantage of thin-film growth techniques is that single-crystal (heteroepitaxial) films can be grown with materials that cannot be grown as bulk single crystals, such as PLZT.

The use of metallo-organic precursors is common to all three of the main solution approaches: sol-gel, metallo-organic decomposition (MOD), and hybrid methods. Sol-gel processing, which is the most common of the solution approaches, generally describes a chemical process in which polymeric gels are formed by hydrolysis of metal alkoxide precursors. In contrast, the MOD method involves direct pyrolysis of the dried film, thereby eliminating the hydrolysis step. Although cracking problems can be encountered during decomposition of the typical carboxylate precursors because of their high organic content, Haertling has developed an alternative MOD approach that exhibits less tendency toward cracking by using acetate precursors, which have smaller organic groups (7). Finally, hybrid methods use small carboxylate and alkoxide precursors and chemical modifiers, such as acetic acid, to reduce hydrolysis rates (8).

Solution-derived films are obtained by spin coating or dip coating a substrate. In general, the ferroelectric films are heat treated using a two-step process in which a low-temperature stage ($<400^\circ$C) is used to pyrolyze the organics and a higher temperature fire ($>600^\circ$C) is used to crystallize the film. Solution techniques have been successfully used to fabricate a wide variety of ferroelectric oxide films. The principal advantages of solution techniques are that they permit excellent compositional control, which is critical for stoichiometric, multicomponent systems, and they provide a relatively simple, low-cost approach to fabricating thin films (9). The primary drawback for optical applications is that solution-derived films are typically polycrystalline and can contain residual porosity, both of which lead to light scattering. However, the ability to prepare highly oriented thin films, which should decrease light scattering, has been reported for sol-gel PZT thin films (10, 11).

The main physical vapor–deposition techniques that have been used for fabricating ferroelectric thin films have been sputtering (rf, dc magnetron, and ion beam) and pulsed laser deposition (PLD). In PLD, a short-wavelength, high-power laser, such as an excimer laser, is used to vaporize small regions of a ceramic target. The vaporized material then condenses onto a substrate that typically is heated. Both sputtering and PLD, which permit layer-by-layer deposition, have been used to fabricate heteroepitaxial thin films of PLZT, $KNbO_3$, $BaTiO_3$, and $LiNbO_3$. One disadvantage of using sputtering techniques to fabricate multicomponent materials is that the sputtering yield and sticking coefficients of the various components are not equal; therefore, precise control of the sputtering rates of the individual components is required (12).

Probably the most promising technique for growing optical quality ferroelectric thin films is metallo-organic chemical vapor deposition (MOCVD). In MOCVD, volatile metallo-organic precursors are vaporized and transported to a reaction chamber where they react and deposit onto a heated substrate. In situ deposition of high-quality, epitaxial PZT (13) and $BaTiO_3$ (14) thin films by MOCVD has been demonstrated. Advantages of MOCVD are that the films also tend to be conformal and that the technique is well-suited to a manufacturing environment. One problem is that stable, volatile precursors for some cation species, such as Zr, are not readily available.

Fabrication techniques for PZT-based thin films have been more widely studied than for the other ferroelectric materials. One of the main issues in fabricating high-quality PZT and PLZT films by any of these techniques is lead oxide volatility. In general, the problem of PbO volatility is overcome by fabricating lead-rich films. For example, 5 to 10% excess Pb is typically added to precursor solutions to compensate for the loss of PbO during firing (8, 15). Similar problems are also encountered for other lead-containing ferroelectric films, such as $Pb(Mg,Nb)O_3$ and $Pb(Sc,Ta)O_3$, and for other materials, such as $KNbO_3$, that contain volatile species like potassium.

Chemical compatibility between the ferroelectric films and appropriate substrates at the required firing temperature is also an issue. In the case of Si and GaAs, buffer layers that act as diffusion barriers are generally required, especially for Pb-containing films. Furthermore, to obtain epitaxial ferroelectric films, epitaxial buffer layers are needed. An example of this approach was demonstrated by Nashimoto et al (16) who fabricated epitaxial $BaTiO_3$ films on GaAs by using an epitaxially grown MgO buffer layer. Similarly, Ramesh has demonstrated the growth of highly oriented PZT films on Si by first depositing a c-axis–oriented bismuth titanate template layer onto a thermally oxidized substrate (17). Another issue for certain thin-film systems is achieving complete crystallization of the desired phase. In the case of PZT, a seed layer approach (using $PbTiO_3$) has been used to enhance perovskite formation in films with high Zr/Ti ratios that otherwise exhibit two-phase microstructures (18).

ELECTROOPTIC PROPERTIES AND APPLICATIONS

The two main electrooptic effects in ferroelectric materials are the linear electrooptic (Pockels) effect and the quadratic electrooptic (Kerr) effect. In the Pockels effect, the refractive indices n vary linearly with the applied field according to relations of the form $\Delta n(E) = -1/2rn^3E$, where r is a linear electrooptic coefficient and E is an applied electric field component. A linear response is only observed in ferroelectrics over a limited applied field range that does not lead to domain switching. In the Kerr effect, n varies quadratically with the applied field according to $\Delta n(E) = -1/2Rn^3E^2 = -1/2gn^3P^2$, where R and g are quadratic electrooptic coefficients. The Kerr effect is generally exhibited by ferroelectrics in the paraelectric phase (above T_C) because the birefringence is a result of a field-induced polarization. Finally, ferroelectric materials can exhibit a hysteretic change of the birefringence with applied field based on the value of the remanent polarization; this effect is primarily useful for optical memories. Values of the linear and quadratic electrooptic coefficients for a variety of ferroelectrics are given in Table 1. In general, ferroelectrics exhibit the largest electrooptic coefficients of any class of material. The change in R over the composition range PLZT 8.5/65/35—PLZT 9.5/65/35, where PLZT x/y/1-y corresponds to $(Pb_{1-x},La_x)(Zr_y,Ti_{1-y})_{1-x/4}O_3$ (22) demonstrates how sensitive the electrooptic reponse is near a phase boundary.

Nonlinear optical effects result from an anharmonic response of the valence electrons to intense light. To account for these effects, the polarizability can be expanded as a power series, such that

$$P_i \chi_{ij}^{(1)} E_j + \chi_{ijk}^{(2)} E_j E_k + \chi_{ijkl}^{(3)} E_j E_k E_l + \ldots,$$

is the linear susceptibility and $\chi^{(2)}$ and $\chi^{(3)}$ are the second- and third-order nonlinear susceptibilities, respectively. The magnitudes of the nonlinear susceptibilities diminish rapidly as the order increases and, thus, useful nonlinear optical effects are generally second or third order. Second-order nonlinear processes occur only in materials such as ferroelectrics that lack a center of symmetry. In fact, the best nonlinear optical materials also tend to be ferroelectrics.

Thin-Film Interference Effects

A critical aspect of working with thin films, either in reflection or transmission, is the important contribution of thin-film interference effects.

Table 1 Values of linear and quadratic electrooptic coefficients for some ferroelectrics

Material	$R \times 10^{12}$ (m/V)	$R \times 10^{18}$ $(m/V)^2$	Reference
$BaTiO_3$ (r33)[a]	820	—	2
$KNbO_3$ (r42)[a]	380	—	2
$KNbO_3$ (r33)[a]	64	—	2
$(Sr,Ba)NbO_3$ (75/25)	1340	—	2
$(Sr,Ba)NbO_3$ (50/50)	210	—	23
$LiNbO_3$	17	—	2, 23
$LiTaO_3$	22	—	23
PLZT (12/40/60)	140	—	2, 23
PLZT (8/40/60)	100	—	23
PLZT (8.5/65/35)	—	3860	23
PLZT (9/63/35)	—	380	23
PLZT (9.5/65/35)	—	150	23

[a] r33 and r42 correspond to the standard contracted notation for a third rank tensor.

Thin-film interference effects, which are caused by multiple reflections due to the index mismatch at the two surfaces, can exert a strong influence on the observed optical and electrooptic properties by changing the effective optical path length of light from its simple geometrical value. It has been shown that these effects can severely complicate the interpretation of simple electrooptic and light-scattering measurements because the effective path length is a nonlinear function of film thickness, indices of the film and substrate, incident angle, and wavelength (19, 20). In fact, these interference effects are typically more pronounced in ferroelectric films due to their high index values.

Electrooptic Properties of Ferroelectric Films

Currently, electrooptic properties have only been evaluated for a small number of ferroelectric thin films. PZT and PLZT compositions remain the most studied ferroelectric films for electrooptic applications. PZT and PLZT thin films have been emphasized partly because they have good electrooptic properties and partly because their processing is reasonably well understood. High-quality thin films of other ferroelectric thin films, such as $KNbO_3$, $BaTiO_3$, $LiNbO_3$, $LiTaO_3$, and SBN have also been fabricated. However, only limited information on the electrooptic properties of these materials has been obtained.

The electrooptic properties of various P(L)ZT compositions have been characterized by field–modulated transmission ellipsometry in which a transverse electric field is applied using interdigitated surface electrodes. These studies have shown that the electrooptic coefficients of thin films are somewhat smaller than for bulk ceramics. For example, Adachi et al (21) showed that PLZT 9/65/35 films exhibited a quadratic electrooptic response with a $R \approx 6 \times 10^{-17}$ $(m/V)^2$, as compared with between 3.8 and 9.1×10^{-16} $(m/V)^2$ for bulk 9/65/35 ceramics (22, 23). Similarly, linear electrooptic coefficients of $r = 4 \times 10^{-11}$ m/V for PLZT 10/20/80 films (24), $r = 2.4 \times 10^{-11}$ m/V for PZT 60/40 films, and $r = 0.8 \times 10^{-11}$ for PLZT 21/0/100 (25) have been reported, compared with values of roughly 1×10^{-10} m/V for PLZT 8/40/60, 12/40/60, and 14/30/70 ceramics. (A direct comparison of identical compositions is not possible because transparent ceramics are not available for some of the film compositions that have been characterized.) It should also be pointed out that these studies have not included the contribution of thin-film interference effects in their calculations.

The electrooptic response of thin films can be smaller than the response for bulk ceramics or single crystals (where appropriate) because of clamping stresses. Both compressive and tensile stresses can be responsible for reducing the magnitude of the field-induced birefringence. Stresses in ferroelectric thin films are primarily due to differences in the thermal expansion coefficient between the film and the substrate. However, internal clamping stresses can also arise during domain reorientation due to the anisotropy of the unit cell. The electrooptic response will also be reduced in films that are not single phase, which is commonly observed for films such as PLZT 9/65/35 that have high Zr/Ti ratios (26). Another consideration specifically important for memory electrooptic effects is whether switching occurs entirely by 180° domain reorientation, as has been suggested for PZT thin films (10). Figure 3 illustrates that the effective macroscopic birefringence (zero bias) of a uniaxial material, such as PZT, does not change if only 180° domain switching occurs (Figure 3*b*). Therefore, 90° switching (or 71 or 109° switching for rhombohedral PLZT compositions) is required to obtain a change in the birefringence at different remanent polarization values, which is necessary for optical memory applications.

Electrooptic responses have also been characterized using ellipsometry in a reflection geometry with a longitudinal applied field (i.e. perpendicular to the plane of the film) (19, 24, 27). However, the thin-film interference (etalon) effects are particularly strong in this geometry, which makes it difficult to calculate index changes from these measurements. On the other hand, the etalon effects do permit optimization of the electrooptic modulation (see below). Probably the most detailed measurements of

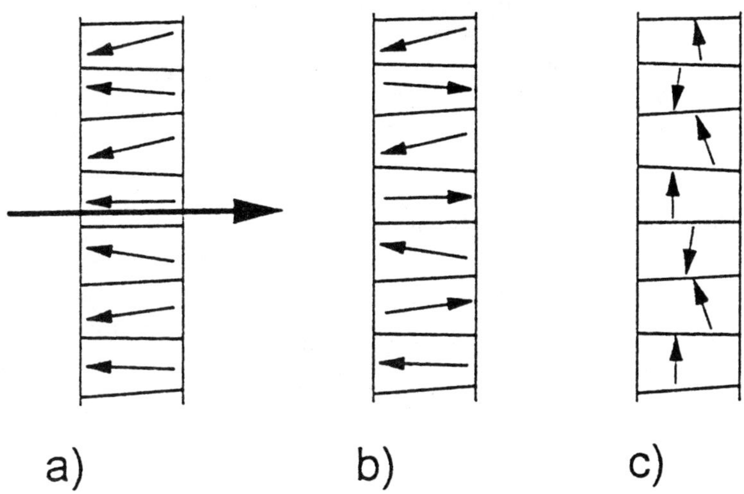

Figure 3 Schematic diagram of columnar thin film. The arrows correspond to the spontaneous dipoles in the grains. (*a*) Poled film with large arrow indicates viewing direction; (*b*) film switched to $P_R \approx 0$ by 180° switching; the index is the same as (*a*); (*c*) film switched to $P_R \approx 0$ by 90° switching. This state shows a maximum birefringence difference compared with (*a*).

longitudinal electrooptic effects in PZT films have been obtained using the prism-coupled refractometry technique. Refractometry is a powerful technique for this type of thin-film study because it permits a quantitative determination of the individual contributions to the total field-induced birefringence, Δn_e and Δn_o, by using transverse-electric (TE) and transverse-magnetic (TM) polarizations (28). As anticipated with this technique, an unpoled PZT 53/47 film was shown to be nearly optically isotropic on a macroscopic scale. Furthermore the ratio of $\Delta n_e/\Delta n_o$ was roughly $-4/1$, which indicates that domain reorientation was not the primary component of the electrooptic response. This ferroelectric composition also exhibited hysteresis in the indices consistent with the electrical hysteresis.

The electrooptic response of some ion beam sputter-deposited $KNbO_3$ films has also been characterized using transmission ellipsometry. It was found that (111)-oriented $KNbO_3$ films grown on (0001) sapphire exhibit a nearly quadratic response with $R = 5.5 \times 10^{-17}$ $(m/V)^2$, whereas the electrooptic response of (100)-oriented $KNbO_3$ grown on (100) MgO was much smaller, with $R = 0.09$–0.3×10^{-17} $(m/V)^2$ (29). This difference was attributed to the fact that the (111) orientation makes use of the shear component (r_{42}) of the electrooptic tensor of $KNbO_3$, which is the largest component for $KNbO_3$. This result shows how tailoring the orientation of the film by choosing the appropriate substrate can be used to optimize the electrooptic response. The good R value for the (111) $KNbO_3$ compares favorably with the measurements on PLZT films. A direct comparison of the thin-film coefficients with single-crystal values has not been made because only linear coefficients have been reported for $KNbO_3$ crystals. The nearly quadratic response is presumably the result of domain reversal in these multidomain films.

Spatial Light Modulators

A spatial light modulator (SLM) is a device that can variably modulate incident light (either in reflection or transmission) over an area. The important performance factors for SLMs are speed, contrast ratio, and light throughput. Most displays, which are a simple type of SLM, require high contrast ratios but not fast response times. Liquid crystal materials are ideally suited to these applications. However, standard liquid crystals are not suitable for high speed SLMs that are required for optical computing applications. For these applications, a variety of materials, including PLZT films (30), ferroelectric liquid crystals (31, 32), and magnetooptic films (33), have been evaluated. Ferroelectric films, such as PLZT 9/65/35, are being considered for use in two-dimensional SLMs because of their large quadratic electrooptic coefficients, fast response, and lack of hysteresis. These films can also be integrated with Si microelectronics to produce monolithic SLMs.

The main drawback to the use of ferroelectric thin films in a two-dimensional array is limited modulation (i.e. poor contrast) due to the short optical path length. One approach to improving the modulation capability of thin films is to optimize constructive interference to increase the effective path length. This type of etalon technique has been used to obtain large phase retardation angles (up to 60°) in relatively thin PZT films ($<1\ \mu$m) (19, 27). As previously discussed, the resonance condition for a thin film depends on the film thickness, wavelength, incidence angle, and electrode thickness and indices. The use of thin-film resonant structures may lead to low-voltage, high-contrast ferroelectric SLMs. The potential drawbacks of this approach are that the increased path length can lead to a loss of lateral spatial resolution and that fairly critical tolerances on the thicknesses and quality of the ferroelectric and electrodes will be required to achieve good modulation in devices.

Waveguide Applications

The other approach to achieving a sufficient optical path length for good optical modulation is through the use of waveguide geometries. Typically, electrooptic single crystals, such as $LiNbO_3$, have been used to fabricate waveguide structures and devices. However, thin-film waveguide devices offer a number of potential advantages over diffused or ion-exchanged single-crystal devices. These potential advantages include compatibility with a variety of substrates, including Si and GaAs, low-cost fabrication over large areas, single-mode operation, and the ability to fabricate materials such as $KNbO_3$ and PLZT that are difficult to grow as single crystals. On the other hand, path lengths on the order of millimeters to centimeters place severe constraints on the optical quality of thin-film materials. Although propagation losses in a waveguide are due to a combination of absorption and scattering, the major component of the propagation loss in most thin-film waveguides is scattering (34). Because scattering is caused by inhomogeneities in the film, minimizing scattering losses requires films that are single phase, fully dense, smooth, and epitaxial. Epitaxial films are required because grain-to-grain misalignment leads to scattering in these birefringent materials.

A standard figure-of-merit for waveguide materials is given by $F = n^3 r/\varepsilon$. Based on this figure-of-merit, the most desirable electrooptic materials for waveguide applications are $KNbO_3$, $BaTiO_3$, $LiNbO_3$, $LiTaO_3$, and SBN (35). Recently, very low propagation losses have been reported for thin films of many of these materials. For example, a propagation loss of only 1.1 dB/cm (TE mode) has been achieved in epitaxial (001)-oriented KNbO3 thin films on (100) Mg_2AlO_4 and (100) MgO prepared by rf sputtering (36). Propagation losses <1 dB/cm have also been reported with epitaxial $LiTaO_3$ thin films fabricated on (0001) sapphire by pulsed laser deposition (37), and epitaxial $LiNbO_3$ thin films fabricated on sapphire by rf sputtering (38). Furthermore, low-loss $LiNbO_3$ thin films (<2 db/cm) have also been grown on thermally oxidized silicon (39). This result demonstrates that highly oriented, but not epitaxial, thin films grown on buffer layers may be a suitable approach toward developing monolithic electrooptic circuits.

Although PLZT films with very low propagation losses (<2 dB/cm) have not been achieved (40, 41), a number of simple waveguide devices have been made using PLZT thin films. Wasa et al demonstrated an optical total internal reflection (TIR) switch with sputtered PLT (28% La) thin film on sapphire, with channel waveguides formed by ion beam etching (42). A schematic diagram of the TIR switch is illustrated in Figure 4. This device redirects light from one arm ($E = 0$) into the other ($E = E_S$) by modulating the in-plane refractive index with an applied voltage. PLZT thin films have also been used to make Bragg acoustooptic deflectors (43) and a Mach-Zender interferometer (44). Based on improvements in the optical quality of ferroelectric thin films and the demonstrated ability to make simple waveguide devices, ferroelectric thin films appear very promising for integrated optics applications.

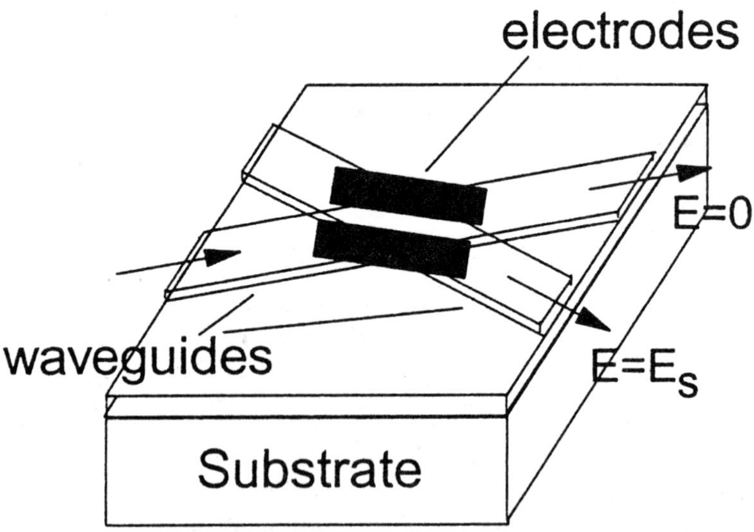

Figure 4 Schematic diagram of thin-film TIR switch.

Second-Harmonic Generation

Next generation optical storage devices will require compact, short-wavelength lasers. One promising approach is through second-harmonic generation using a nonlinear optical material to achieve frequency doubling. Achieving reasonable power for second-harmonic generation requires high conversion efficiency in the nonlinear material. Ferroelectrics are good candidates because they possess some of the largest nonlinear optical coefficients. Ferroelectric thin films are especially attractive for this application because thin-film waveguides can be used to confine the light, thereby increasing the conversion efficiency. An additional advantage of thin-film waveguides is that phase matching can be achieved by designing the waveguide so that the propagation constant of the second-harmonic mode equals twice the propagation constant of the incident light (45). Once again, low propagation losses are critical for such a device. Although second-harmonic generation has not been reported in a ferroelectric thin-film waveguide, good nonlinear optical coefficients have been demonstrated for $BaTiO_3$ thin films (46, 47). Furthermore, corona poling was shown to improve the second-harmonic intensity (46). This effect may not only be the result of domain reorientation, but may also be the result of an induced space-charge field, which could account for the slow decay of the enhanced response.

PHOTO-INDUCED PROPERTIES AND APPLICATIONS

Ferroelectric materials exhibit a variety of photo-induced responses associated with photo-excitation, transport, and trapping of electronic charge carriers. These effects include photorefraction, photovoltaic effects, and photo-induced hysteresis changes. Photo-induced effects have primarily been investigated for use in optical memory and image processing applications. In particular, photo-induced changes in the switching behavior and photovoltaic effects have been used for two-dimensional memory and processing applications, whereas photorefractive effects have been studied for three-dimensional holographic memory and processing applications. In general, the response time and magnitude of these photo-induced effects are best when band-gap illumination is used to create electron-hole pairs. However, good photo-induced responses can also be obtained at longer wavelengths by exciting absorption bands that correspond to specific dopant levels, typically transition metal dopants.

Photo-Induced Hysteresis Changes and Two-Dimensional Optical Storage

PLZT ceramics have been the material of choice for two-dimensional optical memories using ferroelectric materials, because of a combination of their good ferroelectric properties (large P_R, high T_C) and good photosensitivities. One approach to storing optical information in PLZT ceramics is to deposit a photoconductive film on a PLZT ceramic plate to produce a photosensitive medium (48). An alternative approach is to take advantage of the intrinsic photosensitivity of the material by storing information with band-gap light ($E_g \approx 3.4$ eV) (49, 50). For this intrinsic photoferroelectric effect, optical storage is based on the use of light to modify the ferroelectric switching behavior. In this approach, optical storage is achieved by projecting an image with band-gap light onto one surface of the ferroelectric while applying a bias that partially switches the material. Illumination enhances the switching kinetics by introducing free carriers that reduce the electrostatic barrier to domain reorientation by screening internal depolarizing fields; this effect has been termed photo-assisted domain switching. Optical information is stored in the ferroelectric material as a spatial variation in the polarization state, which replicates the spatial variation of the incident (image) intensity. Photo-assisted domain switching also leads to charge trapping that suppresses the switchable polarization in the illuminated regions (51). This photo-induced change in the hysteresis behavior results in a stored image that is not simply erased by repoling the material. Although the appropriate trapping centers have not been identified in PZT, a strong correlation has been made in $BaTiO_3$ crystals between charge trapping at Fe impurities and the amount of switchable polarization (52).

PZT and PLZT thin films exhibit a similar photo-induced suppression of the switchable polarization, which is obtained by illuminating and biasing the film just below the threshold switching voltage, as illustrated in Figure 5*b* (53). It has also been shown that P(L)ZT thin films exhibit a photo-induced translation of the hysteresis loop along the voltage axis (as illustrated in Figure 5*a*) on illumination of a sample that has been fully poled or biased with a saturating voltage (53). These intrinsic effects are distinct but related photo-induced changes in the hysteresis behavior that are reproducible and erasable and, thus, can be used to store optically generated information. Once again, the information would be stored as a spatial variation of the remanent polarization value. Both of these phenomena are due to trapping of photo-generated charge at domain boundaries to minimize internal depolarizing fields. Suppression of the switchable polarization occurs when charge is trapped at internal domain

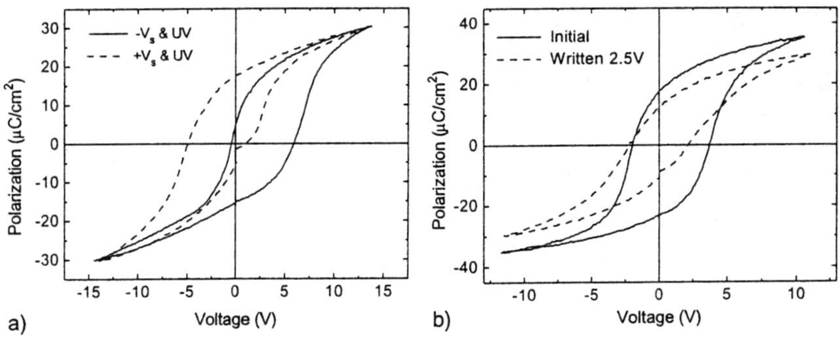

Figure 5 Photo-induced hysteresis changes in PZT and PLZT thin films: (*a*) photo-induced voltage shift for a PLZT 6/20/80 thin film, (*b*) photo-induced suppression of the switchable polarization in a PZT 50/50 thin film.

boundaries, whereas the voltage shift occurs when charge is trapped primarily near the electrode interfaces (53).

Because light can be used to produce changes in the polarization state that can be read optically based on the polarization-dependent birefringence, P(L)ZT films meet the basic requirements for an optical storage medium (54). In comparison to magnetooptic (MO) memories, which are the current state of the art, P(L)ZT memories should have a wider operating temperature range because of their higher Curie temperatures ($T_c \approx$ 350–400°C vs $T_C \approx$ 100–150°C for MO materials). In addition, optical storage in P(L)ZT is an electronic effect as opposed to the thermal effect used in MO memories; hence, ferroelectric memories should be able to use more reliable, low-power lasers. At this point, however, it is not clear whether the photo-induced hysteresis changes in P(L)ZT thin films are sufficiently stable or fast enough to meet the requirements for an optical memory. Furthermore, the use of UV or near-UV light ($\lambda < 400$ nm) is not compatible with current optical memory systems, which are based on diode lasers ($\lambda = 780$–830 nm). Although some improvement in the photosensitivity at sub-band–gap wavelengths has been achieved by thermal diffusion of Fe into PZT films (55), the most promising approach to making PLZT optical memories that are compatible with diode laser wavelengths appears to be by using a photoconductor with the appropriate spectral photosensitivity.

Photovoltaic Effects

In addition to causing photo-induced hysteresis changes, the internal fields that result from the spontaneous polarization also give rise to photovoltaic effects. The sign and magnitude of the photocurrent (or photovoltage) can potentially be used to determine the polarization state. In practice, however, the photocurrent (photovoltage) response in ferroelectric thin films is complicated because photocurrents are also caused by contact potential differences, space-charge fields, and pyroelectric effects. In PZT films, it has been shown that transient photocurrents from band-gap illumination can be used to differentiate the two remanent polarization states (56, 57). This effect was used to operate a simple optical image comparator, in which the photocurrent output from a PZT film was used to compare the position of a projected image with an identical image stored in the PZT film (58). It has also been suggested that photocurrent responses may provide a suitable nondestructive readout technique for ferroelectric memories (59, 60). However, the photocurrent decreases linearly with decreasing sample area; consequently, these techniques are not appropriate for very high-density memories. In addition, the photocurrent response from charge generation (as opposed to the pyroelectric current) exhibits a slowly decaying transient behavior that is not suitable for device applications (56, 60).

Photorefractive Effects

The photorefractive effect refers to a photo-induced change in the refractive index. This effect occurs when photogenerated carriers diffuse from illuminated regions to nonilluminated regions and become trapped. The trapped charge gives rise to internal space-charge fields that cause local changes in the refractive index via the electrooptic effect. After the light is turned off, the space-charge fields and hence the index variations slowly decrease with a relaxation time determined by the dark resistivity of the material. The effect can also be erased by uniform illumination. Photorefractive effects have been observed in $LiNbO_3$, $BaTiO_3$, $LiTaO_3$, $K(Ta,Nb)O_3$, $(Sr,Ba)Nb_2O_6$ (SBN), PLZT ceramics, and $Bi_4Ti_3O_{12}$ (61). Currently, $BaTiO_3$ single crystals doped with Co, Fe, or Rh are the materials of choice for photorefractive applications (62). However, good photorefractive efficiencies have also been obtained with doped semiconductors and quantum well structures.

The primary application area for the photorefractive effect is in image processing, using devices such as phase conjugate mirrors, which are based on short-term storage of three-dimensional holographic images. In addition, volume holography in photorefractive materials has been proposed for high-density optical memories. However, photorefractive holographs are volatile, and the materials that exhibit the best recording sensitivities are the most volatile (61). Although storage of photorefractive gratings in ferroelectric thin films is possible, the diffraction efficiency for thin-films devices would be small. Consequently, little if any work has been done on photorefractive effects in ferroelectric thin films.

PYROELECTRIC PROPERTIES AND APPLICATIONS

Pyroelectric Properties

As discussed above, a pyroelectric current can be derived from a ferroelectric material from the change in spontaneous polarization with temperature (i_{py}–$\propto$-dP/dT). This phenomenon can be used as a sensitive method of detecting incident radiation, particularly at infrared (IR) wavelengths. The other method frequently used to detect IR radiation in ferroelectrics is based on the large change in permittivity that occurs near the transition temperature T_C. This latter method is referred to as the dielectric bolometer mode. Based on the common figure-of-merit for pyroelectric detectors (discussed below), the best dielectric bolometer materials are Pb(Sc,Ta)O_3 (PST), Pb(Mn,Nb)O_3 (PMN), and (Ba,Sr)TiO_3 (BST) because these materials exhibit large dielectric anomalies and can be made with relatively broad transitions near ambient temperatures (63). Similarly, $PbTiO_3$ is an attractive conventional pyroelectric material because of its large pyroelectric coefficient, relatively small dielectric constant ($\varepsilon \approx 300$), and high T_C, which give it a wide operating temperature range.

Good pyroelectric effects have been obtained for PST, $PbTiO_3$, (Pb,La)TiO_3 (PLT), and PZT thin films. Sol-gel derived PST thin films have been prepared with figures-of-merit (11×10^{-5} Pa$^{-1/2}$) comparable to bulk ceramic values (12–15 $\times 10^{-5}$ Pa$^{-1/2}$) (64). PST films with slightly smaller figures-of-merit (7–9 $\times 10^{-5}$ Pa$^{-1/2}$) have also been fabricated by rf sputtering (63). The disadvantage of PST thin films is that they require relatively high processing temperatures ($\approx 900°C$), which makes use with silicon integrated circuits difficult. However, work is being done to decrease the processing temperature by using a high PbO excess (63). Although the figure-of-merit for $PbTiO_3$ thin films (2×10^{-5} Pa$^{-1/2}$) is significantly smaller than for PST thin films (63), $PbTiO_3$, PLT, and PZT thin films have a distinct advantage because they can be fabricated at temperatures that are compatible with integrated circuits. The pyroelectric coefficient can be further optimized by maximizing the component of the spontaneous polarization normal to the plane of the film by fabricating oriented films such as (001)-oriented $PbTiO_3$. Ferroelectric thin films also tend to exhibit broader transitions than bulk ceramics, which should permit a wider operating temperature range. Finally, optical quality thin films are not required for pyroelectric application because light transmission is not an issue. Consequently, high-volume, low-cost fabrication techniques, such as sol-gel methods, are appropriate.

Long-Wavelength Infrared Detectors

The sensing of infrared (IR) radiation in the 8 to 14 μm-range is important for a wide range of applications because this wavelength range corresponds to one of the atmospheric windows of low absorption and to the peak in the black body radiation spectrum for objects near room temperature. Although photoconductive or photovoltaic detectors that utilize III-V or II-VI semiconductors are good for this range, the high cost and cooling requirements for these devices make them unsuitable for many applications, particularly consumer ones. Pyroelectric detectors offer a relatively simple, low-cost alternative to semiconductor detectors. Because pyroelectrics do not require cooling they can be used for untended or portable applications, although pyroelectric detectors have lower resolution than semiconductor detectors. Uncooled infrared detector arrays that utilize bulk ferroelectric ceramics and a hybrid solder-bump technology are now available (63). However, these hybrid detectors are expensive.

Thin-film pyroelectric IR detectors that integrate $PbTiO_3$ thin films with micromachined silicon membranes have now been demonstrated (65). These thin-film devices would not only eliminate the time and expense of the lapping and polishing operations required for ceramic hybrid detectors, but could increase the detector sensitivity by using micromachining techniques to improve the thermal isolation of the detector elements. Figure 6 is a schematic cross section of an integrated thin-film detector element. Thermal isolation, which is one of the most critical aspects to pyroelectric detector design, is achieved by forming an air gap through etch removal of a sacrificial material such as phosphosilicate glass. $PbTiO_3$ thin-film detectors have also been demonstrated using patterned and etched MgO substrates (66).

The standard figure-of-merit for pyroelectric detectors is $F_D = p/c$

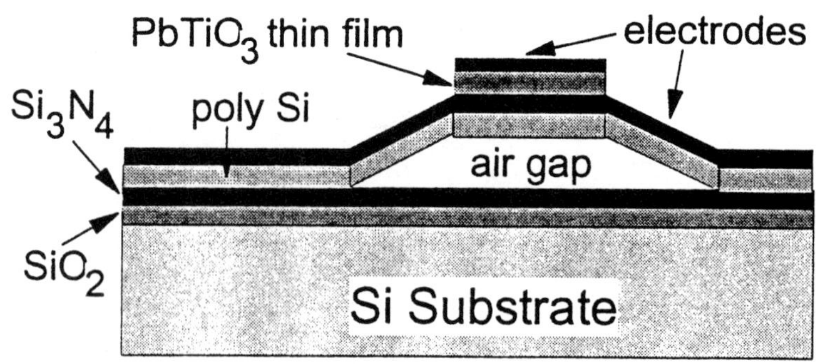

Figure 6 Schematic diagram of a pyroelectric detector element using a $PbTiO_3$ thin film. The air gap is produced by removing a sacrificial layer (65).

$(\varepsilon\varepsilon_o \tan\delta)^{1/2}$, where c is the specific heat, and the pyroelectric coefficient p is given by $p = \mathrm{d}D/\mathrm{d}T$. This figure of merit accounts for the fact that most of the detector noise is due to the dielectric loss of the ferroelectric. One disadvantage of thin films is that they have higher loss tangents than bulk ceramics. It has also been shown that the dielectric bolometer mode of operation gives higher performance figures-of-merit than can be achieved in conventional pyroelectric materials, particularly for large arrays of small detectors. Although thin-film dielectric bolometer detectors have yet to be reported, integrating sol-gel BST thin films with micromachined silicon to make a dielectric bolometer should be straightforward. One disadvantage of the dielectric bolometer mode is that the material must be temperature stabilized in the region just above T_C to gain optimum performance. Achieving temperature stabilization requires that relatively high bias fields ($\sim$50 kV/cm) are applied for long periods of time. However, these fields can lead to degradation of the material properties, particularly for thin films. Nevertheless, ferroelectric thin films seem certain to have a significant impact on pyroelectric detector technology.

CONCLUSIONS

Ferroelectric materials exhibit a combination of electrooptic, nonlinear optical, photo-induced, and pyroelectric properties that make them attractive for a variety of photonic applications. Significant progress has also been made in the development of ferroelectric thin-film fabrication techniques. These advances have made it possible to fabricate optical quality thin films and to integrate these films with functional substrates such as Si and GaAs. At this point, photonic applications for ferroelectric thin films are at either the research or the development stage. However, further work and future advancements should make ferroelectric films important materials for future photonic applications.

Acknowledgments

This work was performed at Sandia National Laboratories, supported by the US Department of Energy under contract # DE-AC04-94AL85000.

Literature Cited

1. Anderson LK. 1974. *Ferroelectrics* 7: 55–63
2. Tuttle BA. 1987. *Mater. Res. Soc. Bull.* 12: 40–45
3. Bauer F, Simonne JJ, Audaire L. 1992. *Proc. 8th Int. Symp. Applications of Ferroelectrics, Greenville, SC.* pp. 27–30.
4. Becker RA. 1988. *Mater. Res. Soc. Bull.* 13: 21–23
5. Cutchen JT. 1980. *Ferroelectrics* 27: 173–78
6. Hanson C, Beratan H, Owen R, Corbin M, NcKenney S. 1992. *Proc. SPIE* 1735: 17–26
7. Haertling GH. 1991. *Ferroelectrics* 116: 51–63
8. Schwartz RW, Bunker BC, Dimos DB, Assink RA, Tuttle BA, et al. 1992. *Integr. Ferroelectr.* 2: 243–54
9. Yi G, Sayer M. 1991. *Ceram. Bull.* 70: 1173–79
10. Tuttle BA, Voigt JA, Goodnow DC, Lamppa DL, Headley TJ, et al. 1993. *J. Am. Ceram. Soc.* 76: 1537–44
11. Barlingay CK, Dey SK. 1992. *Appl. Phys. Lett.* 61: 1278–80
12. Roy RA, Etzold KF. 1992. *J. Mater. Res.* 7: 1433–64
13. De Keijser M, Van Veldhoven PJ, Dormans GJM. 1993. *Mat. Res. Soc. Symp. Proc.* 310: 223–34
14. Bihari B, Kumar J, Stauf GT, Van Buskirk PC, Hwang CS. 1994. *J. Appl. Phys.* 76: 1169–74
15. Klee M, De Veirman A, Taylor DJ, Larsen PK. 1994. *Integr. Ferroelectr.* 4: 197–206
16. Nashimoto K, Fork DK, Geballe TH. 1992. *Appl. Phys. Lett.* 60: 1199–201
17. Ramesh R, Lee J, Sands T, Keramidas VG, Auciello O. 1994. *Appl. Phys. Lett.* 64: 2511–13
18. Swartz SL, Bright SJ, Melling PJ, Shrout TR. 1990. *Ferroelectrics* 108: 71–76
19. Dimos D, Land CE, Schwartz RW. 1992. *Ceram. Trans.* 25: 323–39
20. Sinclair MB, Dimos D, Potter BG Jr, Schwartz RW. 1994. *J. Am. Ceram. Soc.* In press
21. Adachi H, Mitsuyu T, Yamazaki O, Wasa K. 1986. *J. Appl. Phys.* 60: 736–41
22. Land CE, Thacher PD, Haertling GH. 1974. In *Applied Solid State Science*, ed. R Wolfe, 4: 137–233. New York: Academic
23. Haertling GH. 1988. In *Electronic Ceramics*, ed. LM Levinson. pp. 371–492. New York: Dekker
24. Reitze DH, Haton E, Ramesh R, Etemad S, Leaird DE, et al. 1993. *Appl. Phys. Lett.* 63: 596–98
25. Yi G, Wu Z, Sayer M. 1988. *J. Appl Phys.* 64: 2717–24
26. Preston KD, Haertling GH. 1992. *Appl. Phys. Lett.* 60: 2831–33
27. Wang F, Haertling GH. 1994. *Proc. 9th IEEE Int. Symp. Applications of Ferroelectrics, Penn State Univ.* In press
28. Potter BG Jr, Sinclair MB, Dimos D. 1993. *Appl. Phys. Lett.* 63: 2180–82
29. Graettinger TM, Rou SH, Ameen MS, Auciello O, Kingon AI. 1991. *Appl. Phys. Lett.* 58: 1964–66
30. Ersen A, Krishnakumar S, Ozguz V, Wang J, Fan C, et al. 1992. *Appl. Opt.* 31: 3950–65
31. Armitage D, Thackara JI, Clark NA, Handschy MA. 1987. *Mol. Cryst. Liq. Cryst.* 144: 309–16
32. Hudson TD, Worcester RK, Gregory DA. 1991. *Appl. Opt.* 30: 2867–72
33. Hartmann NF, Gaylord TK. 1990. *Appl. Opt.* 29: 4372–83
34. Mir JM, Agostinelli JA. 1994. *J. Vac. Sci. Technol. A* 12: 1439–45
35. Holman RL, Althouse Johnson LM, Skinner DP. 1987. *Opt. Eng.* 26: 134–42
36. Schwyn Thony S, Lehmann HW. 1992. *Appl. Phys. Lett.* 61: 373–75
37. Agostinelli JA, Braunstein GH, Blanton TN. 1993. *Appl. Phys. Lett.* 63: 123–25
38. Huang CH-J, Rabson TA. 1992. *Integr. Ferroelectr.* 2: 13
39. Huang CH-J, Rabson TA. 1993. *Opt. Lett.* 18: 811–13
40. Wood VE, Busch JR, Ramamurthi SD, Swartz SL. 1992. *J. Appl. Phys.* 71: 4557–66
41. Kawaguchi T, Adachi H, Setsune K, Yamazaki O, Wasa K. 1984. *Appl. Opt.* 23: 2187–91
42. Wasa K, Yamazaki O, Adachi H, Kawaguchi T, Setsune K. 1984. *J. Lightwave Technol.* LT-2: 710–13
43. Adachi H, Mitsuyu T, Yamazaki O, Wasa K. 1985. *Jpn. J. Appl. Phys.* 24–2: 287–89
44. Busch JR, Ramamurthi SD, Swartz SL, Wood VE. 1992. *Electr. Lett.* 28: 1591–92
45. Van Buskirk PC, Stauf GT, Gardiner R, Kirlin PS, Bihari B, Kumar J. 1993. *Mater. Res. Soc. Symp. Proc.* 310: 119–24
46. Lu HA, Wills LA, Wessels BW, Lin WP, Zhang TG, et al. 1993. *Appl. Phys. Lett.* 62: 1314–16
47. Bihari B, Kumar J, Stauf GT, Van Buskirk PC, Hwang CS. 1994. *J. Appl. Phys.* 76: 1169–74

48. Fraser DB. 1973. *Proc. IEEE* 61: 1013–18
49. Land CE, Peercy PS. 1977. *Info. Display J.* 13: 20–26
50. Land CE. 1978. *Opt. Eng.* 17: 317–26
51. Chen PJ, Land CE, Madsen MM. 1982. *Acta. Mech.* 43: 61–72
52. Warren WL, Dimos D. 1994. *Appl. Phys. Lett.* 64: 866–68
53. Dimos D, Warren WL, Sinclair MB, Tuttle BA, Schwartz, RW. 1994. *J. Appl. Phys.* 76: 4305–15
54. Dimos D, Potter BG, Sinclair MB, Tuttle BA, Warren WL. 1994. *Integr. Ferroelectr.* 5: 47–58
55. Lin CT, Li L, Webb JS, Leung MS, Lipeles RA. 1994. *Proc. on High Permittivity Materials*, 1993. Pennington, NJ: Electrochem. Soc. In press
56. Dimos D, Schwartz RW. 1992. *Mater. Res. Soc. Symp. Proc.* 243: 73–78
57. Thakoor S. 1992. *Ceram. Trans.* 25: 251–64
58. Martin SJ, Butler MA, Land CE. 1988. *Electr. Lett.* 24: 1486–87
59. Thakoor S. 1992. *Appl. Phys. Lett.* 60: 3319–21
60. Lee J, Esayan S, Prohaska J, Safari A. 1994. *Appl. Phys. Lett.* 64: 294–96
61. Glass AM. 1978. *Opt. Eng.* 17: 470–79
62. Wechsler BA, Klein MB, Nelson CC, Schwartz RN. 1994. *Opt. Lett.* 19: 536–38
63. Watton R. 1994. *Integr. Ferroelectr.* 4: 175–86
64. Patel A, Shorrocks NM, Whatmore RW. 1992. *Mater. Res. Soc. Symp. Proc.* 243: 67–72
65. Polla DL, Ye C-P, Tamagawa T. 1991. *Appl. Phys. Lett.* 59: 3539–41
66. Takayama R, Tomita Y, Iijima K, Ueda I. 1987. *J. Appl. Phys.* 61: 411–15

Annu. Rev. Mater. Sci. 1995. 25: 295–323

THE ULTIMATE STRENGTH AND STIFFNESS OF POLYMERS

Buckley Crist

Departments of Materials Science and Engineering, and Chemical Engineering, Northwestern University, Evanston, Illinois 60208–3108

KEY WORDS: failure, modulus, orientation, chain ends, fracture kinetics

ABSTRACT

Ultraoriented polymer fibers have elastic modulus E as large as 350 GPa and tensile strength σ_b as large as 7 GPa in materials with a density $\rho \approx 1200$ Kg/m^3. Keys to achieving these properties are near perfect orientation of polymer chains along the fiber axis and reduction of the number of chain ends. The two materials that have been most thoroughly studied are polyethylene (PE) and poly(*p*-phenylene terephthalamide) (PPTA). Various schemes for calculating the elastic modulus are reviewed, together with estimates of effects of imperfect chain orientation and the presence of chain ends. Fracture of fibers is treated in terms of covalent bond scission and/or chain slip originating at chain ends. Under laboratory conditions the experimental modulus E can be >90% of the theoretical modulus. It appears that fracture is more sensitive to chain end defects, limiting practical strength to less than 25% of the ultimate strength predicted from bond scission models.

INTRODUCTION

The quest for lightweight materials with great strength and stiffness has led to the development of synthetic polymers having remarkable mechanical properties. With the benefit of hindsight, the continuous crystal model proposed in 1932 by Staudinger (1) has proven to be the basis for high-performance polymer fibers. The structure-property relations conveyed by

0084–6600/95/0801–0295$05.00

the drawing in Figure 1 are obvious; well packed (crystalline) polymer molecules are aligned in a common direction to provide maximum stiffness dictated by covalent bond distortion, whereas chain ends are conspicuous defects for promoting fracture. Hence the greatest stiffness and strength are obtained in the limit of perfectly oriented crystalline materials composed of high molecular weight chains. Secondary bonds between the polymer molecules have energies about 2 to 5% of intramolecular covalent bonds, leading to easier deformation and fracture when the fiber is loaded perpendicular to the chains. It is important to keep in mind that a well-oriented polymer will have good mechanical properties parallel to the axial or fiber direction and be more compliant and weaker in other directions.

Materials approaching the model in Figure 1 first appeared with DuPont's development in the early 1970s of poly(*p*-phenylene terephthalamide) (PPTA, commonly known as Kevlar®) fibers. The following 15 years was a period of immense activity in the area of high-performance polymers (2–4); there are currently a number of commercial products with elastic modulus $E_f = 100–200$ GPa (1 GPa $= 10^9$ N/m^2), breaking strength $\sigma_b = 2–4$ GPa, and density $\rho = 970–1470$ Kg/m^3 (0.97–1.47 g/cm^3). Throughout this review, E_f designates tensile or Young's modulus and σ_b is the tensile strength, both measured in the axial or fiber direction.

Considered here are the tensile properties of uniaxially oriented polymer fibers and films, with emphasis on the fundamental relations between chain architecture, physical microstructure, modulus, and strength. Various structural imperfections, particularly the unavoidable chain ends indicated in Figure 1, are treated as well. We first describe the methods used to process polymers into highly oriented fibers or films and summarize the best mechanical properties (E_f and σ_b) that have been achieved. Fiber modulus is then discussed in depth. It will be shown that experimental moduli can be 80% or more of the ultimate or theoretical values and that shortfalls are reasonably well understood. The macroscopic Young's modulus E_f is thus a quantity that indicates how closely a real fiber approaches the ideal of Figure 1. Tensile strength is the final subject.

Figure 1 Continuous crystal model of a polymer fiber. Molecules are typically ~500 nm long with transverse spacing of 0.5 nm.

The question of theoretical strength is more complex for many reasons, although it appears that currently obtainable σ_bs are at least 10% of the ultimate strength. Not included in the discussion are carbon fibers (4) ($E_f \approx 600$ GPa, $\sigma_b \approx 3.5$ GPa, $\rho \approx 2000$ Kg/m^3), which lack well-defined linear macromolecules. We also omit the technologically important subjects of fatigue and compressive failure of fibers.

FIBER PREPARATION, MORPHOLOGY, AND TENSILE PROPERTIES

It is customary to classify high-strength polymers according to intrinsic chain flexibility in solutions or melts, which in turn dictates the processing methods used to achieve the highly oriented state exemplified in Figure 1. More detailed accounts of the concepts outlined here can be found in Yang (2), Hattery & Hillman (3), and Jiang et al (4). A few representative chemical structures are given in Figure 2. Polyethylene (PE) has the simplest repeat unit of any polymer. Figure 2*a* presents the extended or all trans conformation for the crystalline state; however, chains in solution or melts adopt random conformations by facile rotation about C-C bonds in the backbone. Polyoxymethylene (POM) in Figure 2*b* is another example of a flexible chain polymer. This macromolecule also has random conformations in solutions or melts, but it crystallizes with chains as 9/5 helices (not shown; nine -CH_2O- monomer units execute five turns around the helix axis to establish an equivalent position). Flexible chain polymers have melting temperatures T_m below 400°C and are stable enough to be processed in the melt state. At the other extreme are rigid-rod polymers such as poly(*p*-phenylene benzobisoxazole) (PBO) shown in Figure 2*c*. Although hindered rotation about the C-C single bond in the backbone may be possible, such rotation scarcely alters the chain contour. Lack of conformational entropy in the noncrystalline state raises T_m of such polymers about 500°C. PPTA (Figure 2*d*) has C-N and C-C backbone bonds that are not parallel to the extended chain axis, thus hindered rotation imparts some flexibility to the chain, which is characterized as semi-rigid. Polymers like PBO and PPTA are processed in solution because they degrade before melting. These solutions may have liquid crystalline order; hence the polymers are termed lyotropic liquid crystalline polymers (LLCP). The final structure in Figure 2*e* is a random copolymer of 4-hydroxybenzoic acid (HBA) and 6-hydroxy-2-naphthoic acid (HNA). Acceptably low melting temperatures can be achieved by judicious choice of the mole ratio of HBA to HNA. More significant is the existence of a nematic phase, which greatly facilitates the processing of such thermotropic liquid crystalline polymers (TLCP) into fibers.

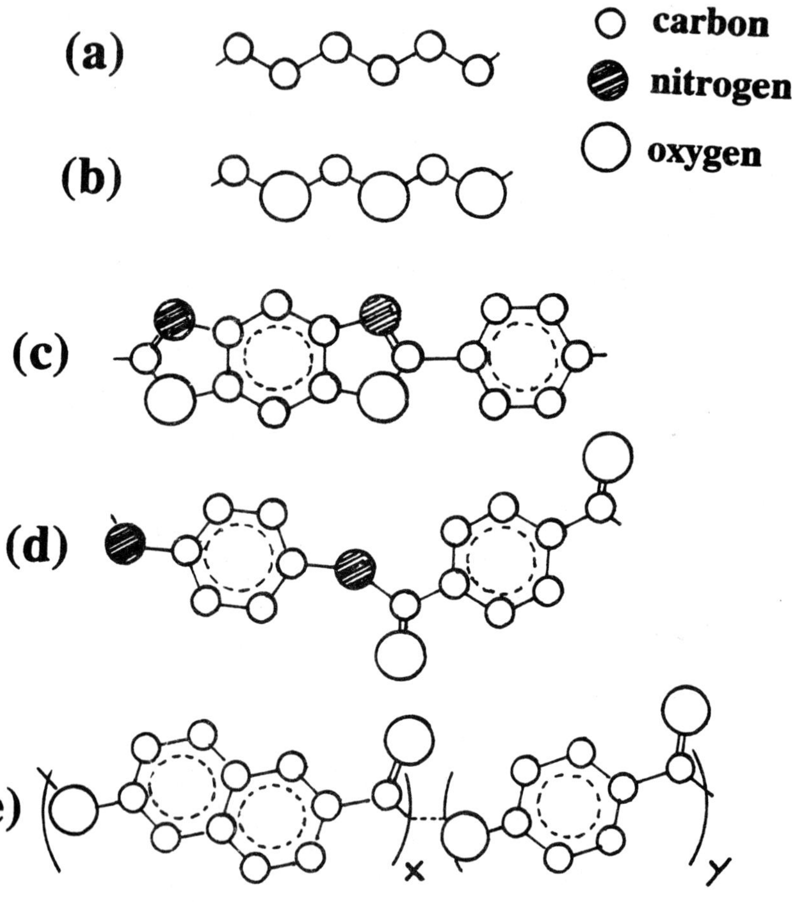

Figure 2 Repeat units for (*a*) PE; (*b*) POM; (*c*) PBO; (*d*) PPTA; (*e*) HBA/HNA random copolymer (y : x ≈ 3 for Vectran). Hydrogen atoms are omitted.

Flexible Chain Polymers

Flexible chain polymers are oriented by plastic deformation (drawing) at temperatures below T_m. Both modulus E_f and strength σ_b increase with the amount of macroscopic stretch, expressed as the draw ratio $\lambda = l/l_0$, and σ_b is seen to be larger for higher molecular weight polymers when compared at the same λ (5–7). Here l_0 is the length of the isotropic sample, and l is the length after drawing. The bottleneck of limited plastic stretch and attendant chain orientation was overcome by Smith & Lemstra (8, 9), who demonstrated that $\lambda > 70$ could be obtained with ultra-high molecular weight PE (UHMWPE, weight average molecular weight $M_w > 10^3$

Kg/mol), which had first been crystallized from dilute solution. The resulting fiber closely resembles the continuous crystal Figure 1; crystallinity is greater than 95% and chain axis orientation is nearly perfect, with misorientation angle $\phi < 4^\circ$. The gel-spinning technique has been developed to the point where UHMWPE fibers such as Spectra® from Allied-Signal and Dyneema® from DSM are commercially available with $E_f \approx 170$ GPa and $\sigma_b \approx 3.0$ GPa (3, 4, 9a). Significantly higher values have been obtained under laboratory conditions, some examples of which are presented in Table 1. Here we list the largest values of room temperature Young's modulus E_f and tensile strength σ_b reported for representative polymers; breaking strain is in the range of $\varepsilon_b = 0.02$–0.04. Multiple entries are given to emphasize the reproducibility of mechanical properties. Each flexible chain polymer except POM was processed by gel spinning or a variation thereof. Ohta (9b) and Porter et al (9c, d) have reviewed the many techniques for orienting flexible chain polymers and summarized tensile properties.

Also in Table 1 are elastic moduli E_x measured by X-ray diffraction

Table 1 Tensile properties of polymer fibers

	E_f (GPa)	σ_b (GPa)	E_x (GPa)	E_c (GPa)	References[a]
Flexible chain polymers					
PE	220	6.5	220	283	10, 11, 12
	264	7.2	235		13, 14
Spectra	170	3.0			3
PVA	74	2.3	230	286	15, 16, 17
PET	34	1.9	110	95	18, 19–20, 21
nylon 6	15	0.9	174	312	22, 23, 24
i-PP	35	2.2	41	40	25, 26, 27
	40	1.5			26
POM	60	—	73	109	28, 28, 29
Lyotropic liquid crystalline polymers					
PPTA Kevlar 49	132	3.3	160	274	30, 31–32, 33
Kevlar 149	179	2.7			30
PBO	370	3.6	477	460	30, 34, 35
	360	5.7			36
PBT	331	4.2	395	405	37, 34, 35
Thermotropic liquid crystalline polymers					
Vectran	65	3.3	126	—	38, 32
Ekonol	138	3.9	130	—	3, 32

[a] References separated by commas in each row are for E_f and σ_b, E_x, and E_c, respectively.

from crystals stressed in the chain axis direction; one expects $E_f = E_x$ for the ideal state in Figure 1. The fifth column in Table 1 presents calculated values of the modulus E_c for perfectly oriented crystalline polymer at room temperature. The relation between fiber modulus E_f, X-ray modulus E_x, and calculated modulus E_c is considered in following sections. Note that for PE, the largest $E_f = 264$ GPa exceeds E_x and is 93% of E_c.

Highly oriented fibers or films of other flexible polymers have been prepared with tensile properties shown in Table 1. Poly(vinyl alcohol) (PVA), polycaprolactam (nylon 6) and poly(ethylene terephthalate) (PET) all crystallize with extended chain conformations and are expected to have high moduli when oriented. Like PE, such chains respond to an axial force by stretching or bending covalent bonds, leading to large stiffness or modulus as discussed below. The low values of E_f for nylon 6 and PET reflect the fact that $\lambda \leqslant 10$ for these polymers, presumably because of intermolecular associations that act in a manner similar to physical entanglements to limit ductility (15).

Isotactic polypropylene (*i*-PP), a saturated hydrocarbon polymer similar to PE, can be drawn to $\lambda > 40$. It is notable that the fiber modulus is equal to the X-ray and calculated moduli, the low values of which reflect the greater compliance and cross-sectional area of the chain, which has a 3/1 helical conformation in the crystalline state. Helical molecules can stretch by rotations about backbone bonds under lower force than that required to deform covalent bonds, reducing axial chain stiffness and modulus. The low modulus $E_f = 60$ GPa of POM is mostly attributed to the 9/5 helical conformation of the crystalline chains.

Of all flexible chain polymers, only PE and *i*-PP have been formed into extremely well-oriented crystalline fibers, with tensile moduli E_f equal to, or nearly equal to, the theoretical E_c. Ultra-drawing of flexible polymers containing atoms other than carbon (C) and hydrogen (H) seems not to be feasible with methods currently available.

Lyotropic Liquid Crystalline Polymers

Stiff-chain polymers like PPTA and rigid-rod polymers like PBO are oriented by extrusion of liquid crystalline solutions (4, 39) and are thus termed lyotropic liquid crystalline polymers (LLCP). PPTA, PBO, and similar materials are processed by dry jet-wet spinning, where the liquid crystalline solution is extruded as filaments and stretched before passing into a nonsolvent bath. The coagulated fibers are washed, dried, and heat treated under tension to improve crystal perfection, chain axis orientation, etc (2–4); this latter step serves to increase appreciably the fiber modulus and to some extent the strength as well (40).

Morphology of LLCP fibers is generally consistent with the continuous crystal model in Figure 1, although there is a definite substructure within a fiber of typical diameter 10–20 μm. Considerable attention has been paid to hierarchical organization within and between microfibrils and larger fibrils within a macroscopic fiber (41, 42). The measured density of PPTA and PBO fibers is about 95% of the crystalline density ρ_c from X-ray structures (4, 39), although the nature of the density defect is not known. Given the nonuniform nature of coagulation, it is likely that dry jet-wet spun fibers contain an appreciable fraction of voids. This may be contrasted to gel spinning, which leads to PE (43) and *i*-PP fibers (26) with macroscopic density $\rho \geqslant 0.99\rho_c$ after drawing.

Tensile properties are presented in Table 1 for PPTA, PBO, and poly(*p*-phenylene benzobisthiazole) (PBT), a heterocyclic polymer similar to PBO, but with sulfur atoms instead of oxygen atoms. Fiber modulus E_f, as well as X-ray and calculated moduli, for PPTA is somewhat less than for PE. The large stiffness of PBO and PBT is consistent with chain architecture; bond angle bending is restricted by ring structures and by bond directions relative to the axial loading direction. Experimental fiber moduli for LLCP polymers are $\sim$80% or more of E_x and E_c. Young's modulus E_f for PBO and PBT are the highest reported for any synthetic polymer, and strength σ_b is in excess of 4 GPa. As with flexible chain polymers, increasing molecular weight leads to higher strength LLCP fibers (4, 36, 44).

Thermotropic Liquid Crystalline Polymers

The molecular structure in Figure 2*d* for the random HBA/HNA copolymer is representative of a large number of aromatic polyesters and similar polymers that display liquid crystallinity between the crystalline and isotropic melt states. A mole fraction of HBA near 0.7 is typical of Vectran®, a product of Hoechst-Celanese (4). Such liquid crystalline polymers can readily be melt spun into fibers having well-oriented chains, relying on the local order and low viscosity of nematic domains to facilitate processing (2, 4, 38). Heat treatment under tension is again used to improve properties. For TLCPs, this processing step may increase tensile strength by as much as threefold, with only a modest (<20%) enhancement of the modulus E_f (2, 45). The origin for this improvement is further polymerization, which decreases the number of chain ends (defects) in the fiber (38, 46).

Morphology of TLCP fibers is not well documented. It is agreed that chain orientation is good, although crystallinity is low, typically below 30%, and "crystals" may not have three-dimensional order (38). Entries in Table 1 are for a HBA/HNA random copolymer (Vectran) and for Ekonol®, a random copolymer from Sumimoto Chemical containing four different monomers, the major one being HBA (4). Although tensile strengths of TLCP fibers are quite respectable, the modulus is less than 150 GPa.

Tashiro & Kobayashi (47) have calculated $E_c = 154$ GPa for HBA/HNA chains like Vectran. This theoretical modulus is similar to E_x but considerably larger than $E_f = 65$ GPa in Table 1.

ELASTIC MODULUS

The experimental fiber modulus E_f in Table 1 is generally less than the X-ray crystal modulus E_x or the theoretical modulus E_c. In this section we consider E_c, E_x and E_f in turn, with an eye to understanding the significance of differences between these measures of stiffness. Calculated modulus E_c clearly represents the theoretical limit for a defect-free material. E_x is an experimental quantity, which is biased toward the most ordered, crystalline regions of the fiber, whereas E_f reflects structural irregularities present in the material.

Calculated Modulus E_c

The axial elastic modulus for a perfectly oriented polymer is usually calculated from the change in internal energy ΔU associated with a length change ΔL in the axial or chain direction. Consider a system of N_0 parallel chains of unspecified but great length (chain ends are ignored), with an area $N_0 A_0$ perpendicular to the chain axis, where $A_0 \approx 0.2$ nm^2 is the effective cross-sectional area of one chain in the equilibrium crystal structure. The equilibrium axial length of a representative chain section is L_0. Elastic strain energy $U(L)$ is by definition at a minimum for the equilibrium length $L = L_0$ and increases when L is changed from L_0. The force required to maintain length L is given by the gradient of the potential $U(L)$:

$$F = \frac{\mathrm{d}U(L)}{\mathrm{d}L}. \tag{1}$$

Young's modulus is given by the ratio of stress $\sigma = F/A$ to strain $\varepsilon = (L - L_0)/L_0$ as $\varepsilon \to 0$, leading to

$$E_c = \left.\frac{\mathrm{d}F/A}{\mathrm{d}L/L_0}\right|_{L=L_0} = \frac{L_0}{A}\left.\frac{\mathrm{d}^2U}{\mathrm{d}L^2}\right|_{L=L_0} = \frac{L_0}{A_0}K. \tag{2}$$

The second derivative of U with respect to L, evaluated at the equilibrium length $L = L_0$ is the force constant K for a single chain; in principle there are both intermolecular and intramolecular contributions to K:

$$K = K_{\text{intra}} + K_{\text{inter}}. \tag{3}$$

One sees that chain area A_0 as well as intrinsic stiffness K contribute to E_c. We have already alluded to the small chain area of PE and the large stiffness of PBO. Length L_0 and area A_0 are established by X-ray crystal structures, whereas K is calculated in some manner.

Methods for evaluating E_c differ according to the type of theory used to evaluate K_{intra} and whether the intermolecular term K_{inter} is included. Certain aspects of these approaches have been reviewed by Grubb (48) and Tashiro (49). One important conclusion is that interchain interactions have effects on axial stretching that are either negligible or small ($K_{inter} \ll K_{intra}$), meaning that E_c can be determined with acceptable accuracy from the response of a single chain (24, 27, 50).

All-important intramolecular energies are handled with either empirical potential functions or quantum mechanical molecular orbital calculations, which are classified as semi-empirical or ab initio. The three methods for calculating single-chain or intramolecular deformation give qualitatively similar results, although E_c from quantum mechanics is generally higher than that from empirical potentials. A comparison can be made with PE, the only polymer that has been treated by ab initio methods (see Table 2). Ab initio results are from Suhai's crystal orbital calculations (51) and the cluster-difference method used in this laboratory (52) and extended to larger basis sets (53). The level of theory increases from STO-3C- to 6–31G** (MP2), which designate basis sets and Møller-Plesset second order perturbation theory to evaluate electron correlations (54). Chain modulus E_c is as much as 18% lower with the crystal orbital method, and further work is required to resolve this discrepancy. For the moment, one must accept that the best ab initio models give $E_c \approx 305 \pm 30$ GPa.

Also shown in Table 2 are single-chain PE moduli from semi-empirical quantum mechanics that are known to be high because a number of approximations have been made in the interest of computational speed (57). The methods are modified neglect of differential overlap (MNDO) and Austin Model 1 (AM1). Recent empirical results are included for comparison; the method uses valence force fields derived from vibrational spectroscopy, and some include additional intramolecular potentials for

Table 2 Single chain elastic modulus of PE (GPa)

Ab initio		Semi-empirical		Empirical
STO–3G	417[a] 404[b]	MNDO[c]	370	315[e]
6–31G	339[a] 360[b]	AM1[d]	400	326[f]
6–31G (MP2)	303[a] 330[b]	—	—	318[g]
6–31G** (MP2)	276[a] 336[b]	—	—	—

[a] Reference 51; [b] Reference 53; [c] References 55, 56; [d] References 57, 58; [e] Reference 24; [f] Reference 59; [g] Reference 50.

nonbonded van der Waals and electrostatic interactions. The empirical calculations consistently give $E_c \approx 320$ GPa, within the ab initio range 305 ± 30 GPa, whereas semi-empirical methods give PE moduli that are $\sim 25\%$ larger.

Single-chain modulus of more complex macromolecules has been calculated with semi-empirical quantum mechanics. Results from the MNDO method are: 68 GPa for *i*-PP, 345 GPa for nylon 6, 670 GPa for PBO, and 500 GPa for PBT (55). More recent AM1 calculations give 690 GPa for PBO and 620 GPa for PBT (60). Comparison with the empirical E_c in Table 1 confirms the differences seen with PE; semi-empirical moduli are from 11% (nylon 6) to 70% (*i*-PP) larger than empirical values. Note also that experimental moduli E_f and X-ray moduli E_x are closer to the empirical than semi-empirical E_c. Ab initio methods, applied only to PE, give reasonably accurate results at high levels of theory.

As stated above, intermolecular bonding makes no sensible contribution to the axial modulus, i.e. $K_{intra} \ll K_{inter}$ in Equation 3. While this is true, there is at least one case where chain packing indirectly affects E_c. Rutledge & Suter (61) found that the calculated E_c of crystalline PPTA was 40% larger than the single-chain value. Subsequent analysis (33) showed that torsion angles of the phenyl rings (see Figure 2*d*) decrease when a single chain is stretched, but this response is prevented in the crystal structure, which stiffens the system. Similar effects may exist in other ring polymers such as PBO.

The preceding discussion is based on internal energy changes associated with chain deformation. Hence these calculations are most appropriate for mechanical response where absolute temperature $T = 0$ K. Somewhat different results are obtained if the free energy, including thermal vibrations and entropy changes, are considered, which is done in the section on Experimental Fiber Modulus.

X-Ray Crystal Modulus E_x

Crystals within the fiber deform elastically when subjected to a tensile stress σ. With X-ray diffraction in the symmetric transmission mode, those crystals with the chain axis or [001] direction parallel to the fiber axis contribute to the observed intensity. Thus the measurement of changes in $00l$ spacing gives a direct measure of crystal strain ε_c in the loading or fiber direction. If uniform stress σ is assumed to exist throughout the fiber, a crystal modulus $E_x = \sigma/\varepsilon_c$ can be evaluated from the ratio of macroscopic stress to crystal strain. This method has met with mixed success as discussed by Grubb (48). The problem stems from the uniform stress (or series loading) assumption, which is thought to be most valid in highly oriented fibers or films. If this condition is not met (effective parallel loading of some crystals), the local stress will exceed the macroscopic stress σ and the apparent $E_x = \sigma/\varepsilon_c$ will be too small.

It is difficult to assess how close E_x is to the true Young's modulus of the crystal. For the most-studied case of PE, there are some reports that E_x is the same in samples of different morphology, suggesting that local stress is constant and that $E_x = E_c$ (15); other work clearly shows that E_x is sample dependent at room temperature (62). Prasad & Grubb were the first to report nonuniform molecular strains in ultra-oriented PE fibers as inferred from shapes of Raman peaks (63) and X-ray peaks (64). Clearly bimodal strain distributions (ε_c differ by up to tenfold!) are observed by Moonen et al (65) in similar PE fibers under stress $\sigma \approx 2$ GPa. Nonuniform crystal strains at large macroscopic stresses certainly raise questions about series loading at lower stresses where E_x is usually evaluated. A similar problem is indicated for PPTA and Ekonol, where X-ray modulus E_x is less than experimental fiber modulus E_f (see Table 1).

Room temperature values of E_x (Table 1) generally track experimental fiber modulus E_f and calculated E_c, being smallest for *i*-PP and greatest for PBO. One desires reliable experimental values of the crystal modulus in order to assess the properties of real fibers (E_f) and to evaluate different theories for calculating E_c. The X-ray modulus E_x must be considered a lower bound to the true elastic stiffness of the crystal. Alternative methods for measuring the axial modulus of polymer crystals rely on Raman or inelastic neutron spectroscopy, which have been reviewed by Grubb (48) and Fanconi & Rabolt (66). Interpretation is not straightforward; the most current analysis of longitudinal acoustic modes gives $E_c = 305$ GPa for PE at room temperature (67), which is consistent with ab initio calculations.

Experimental Fiber Modulus E_f

Even the most carefully prepared fibers represented in Table 1—with the exception of *i*-PP and perhaps PE—have tensile modulus E_f less than ultimate values represented by E_x or E_c. The discrepancy certainly results from defects of various sorts within the fiber. Intercrystalline amorphous regions are always a possibility with flexible chain polymers. Such residual non-crystallinity is assuredly the cause of low E_f in PVA, nylon 6, and POM, which have to date resisted ultra-drawing. There is no evidence, however, for amorphous regions in ultra-oriented PE, *i*-PP, or LLCP fibers. Other defects are known to be present and influence E_f. Perhaps the most significant is chain axis or crystallite orientation, which is never perfect in a real fiber. Intracrystalline features to be considered are chain ends and thermal vibrations.

The best understood structural imperfection is incomplete chain axis orientation. A model for calculating the compliance of an aggregate of imperfectly oriented crystals was developed by Northolt and co-workers (39, 68, 69). A crystal with chain axes inclined to the fiber or stress direction by an angle ϕ responds by chain extension (modulus E_c) and interchain shear; the two strain mechanisms operate in series. With some assumptions appropriate for well-oriented aggregates of anisotropic elastic crystals, the fiber modulus can be written as

$$\frac{1}{E_f} = \frac{1}{E_c} + \frac{\langle \sin^2 \phi \rangle}{2G_c}. \qquad 4.$$

Here $\langle \sin^2 \phi \rangle$ is averaged over all crystals in the fiber and can be determined from X-ray orientation of a (00l) reflection. G_c is the elastic modulus for shear deformation in the [001] direction, which can be obtained from the shear compliance S_{44} calculated for the crystal (48, 49). The analysis embodied in Equation 4 has been applied successfully to PPTA (68) and PE (70), and a related treatment of PBO fiber modulus was done by Jiang et al (4, 71). Figure 3 is an example of the effect of chain misorientation on the compliance $1/E_f$ of PPTA fibers. The intercept gives $E_f = 240$ GPa

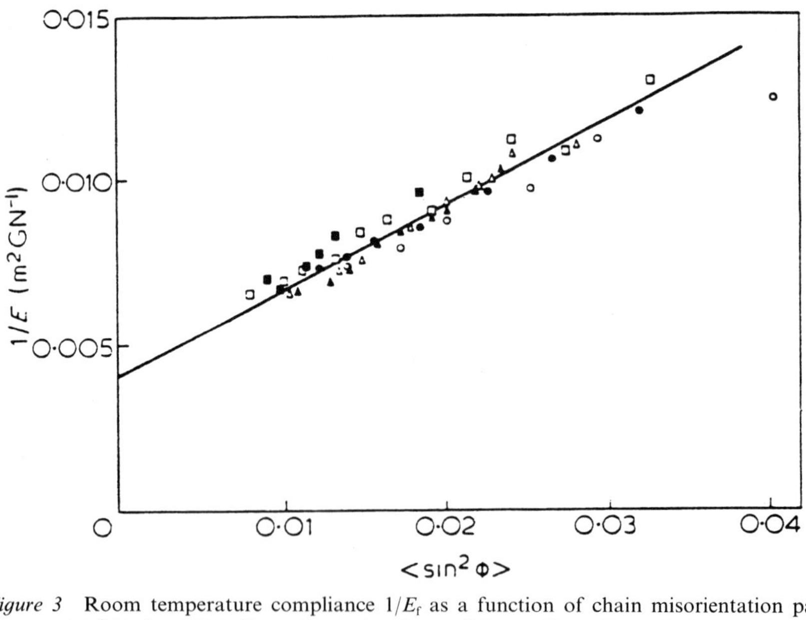

Figure 3 Room temperature compliance $1/E_f$ as a function of chain misorientation parameter $\langle \sin^2 \phi \rangle$ for PPTA fibers. Symbols refer to different fibers. From Reference 68 with permission from Elsevier.

for a perfectly oriented fiber, some 33% larger than the entry for Kevlar 149 shown in Table 1.

Polymers with extended chain conformations have large $E_c \geqslant 200$ GPa, whereas shear moduli, governed by secondary intermolecular bonding, are on the order of $G_c \approx 4$ GPa. This anisotropy results in E_f being very sensitive to chain misorientation, as conveyed by the doubling of the compliance of PPTA when $\langle \sin^2 \phi \rangle = 0.02$, or average misorientation angle $\langle \phi \rangle \approx 8°$ (Figure 3). More nearly isotropic crystals such as *i*-PP are less affected, which probably accounts for $E_f \approx E_c$ for that polymer (Table 1).

A frequently overlooked defect is chain ends. The mean distance between chain end pairs, as seen in Figure 1, is $d_e = (M_n/\rho N_A)^{-1/3}$, where M_n is the number average molecular weight and N_A is Avagodro's number. With representative density $\rho = 1200$ Kg/m^3 and $M_n = 50$ Kg/mol, $d_e = 4$ nm, corresponding to chain end defects occurring every six to eight chains in the transverse direction. These have obvious implications for fracture strength (considered below) but are also points of large, local elastic strain that can reduce the fiber modulus E_f. An analysis similar to shear lag theory was applied to a crystalline fiber with randomly distributed chain ends by Bartenev & Valishin (73), who concluded that E_f is measurably reduced for nylon 6 when $M_n < 10$ Kg/mol.

We are not aware of direct experiments that demonstrate that E_f is a function of chain length, but the deformation-simulation studies by Termonia and co-workers (74–76) are informative in this regard. Three-dimensional coupled oscillators with force constants representing intramolecular (E_c) and intermolecular (G_c) bonds were stretched to evaluate E_f for crystals with different length chains (or M_n). Results are summarized as the relative modulus E_f/E_c vs $1/M_n$ for PE, with $E_c = 300$ GPa and $G_c = 3$ GPa (76), and for PPTA, with $E_c = 200$ GPa and $G_c = 4.1$ GPa (75). Solid lines connecting data points in Figure 4 are for crystals with randomly distributed chain end pairs. PPTA in these simulations is more sensitive to chain ends than PE, particularly at low M_n. Even larger elastic softening is seen if the chain ends are segregated in groups of four, six, eight, etc (76). The dashed line in Figure 4 shows E_f/E_c is reduced further when chain ends are clustered in groups of 18 (nine pairs).

Assuming the simulations are reasonably accurate, the behavior in Figure 4 has obvious importance for the fiber modulus of ultra-oriented polymers. UHMWPE has $M_n \geqslant 150$ Kg/mol, meaning that $E_f \approx E_c$ in the limit of perfect chain orientation. The same can not be said of PPTA, which has number average molecular weight $M_n \leqslant 30$ Kg/mol (44, 76). It is clear that $E_f/E_c < 0.8$ for this range of molecular weights, implying that chain ends reduce the limiting modulus of PPTA by at least 20%.

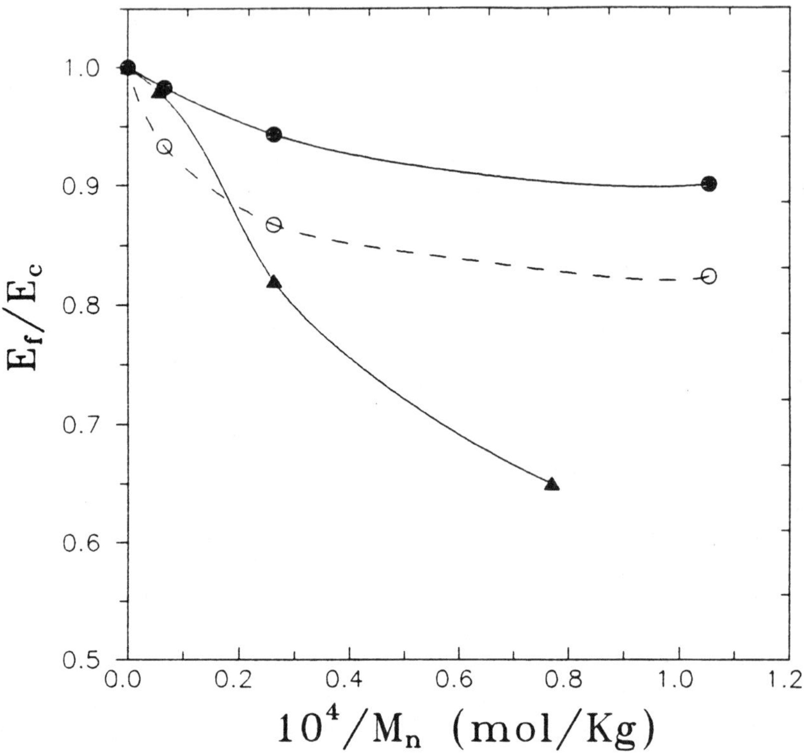

Figure 4 Reduced fiber modulus as a function of reciprocal molecular weight from simulation studies of PE (74) and PPTA (75). Data for randomly distributed chain ends are represented by solid circle (PE) and solid triangle (PPTA). PE crystals with clusters of nine chain-end pairs are represented by open circle.

We conclude this discussion of experimental Young's moduli E_f with the effects of temperature. Most obvious is transverse thermal expansion, which causes A_0 in Equation 2 to increase with temperature. This effect is modest, typically 5% or less between 0 K and 300 K, leading to a commensurate decrease in E_c over the same temperature range. Polymer fibers are viscoelastic, exhibiting both stress relaxation and creep that are more conspicuous at elevated temperatures (4). In ultradrawn PE, the modulus drops by $\sim 15\%$ over 10^4 s (76a) and creep strain at $\sigma = 0.8$ GPa may be as large as 0.04 after 10^6 s (76b) at room temperature. More interesting are effects related to the negative temperature coefficient for length in the chain direction; $\alpha_c = d\ln c/dT \approx -(1\text{–}20) \times 10^{-6}$deg for a number of polymers (16). Nishino et al (77) proposed that thermally generated kink defects contribute to the negative value of α_c and reduce the axial crystal modulus E_c.

Rutledge & Suter (61) and Kluzinger et al (78) independently suggested that thermally activated vibrations, not conformational defects, may influence both α_c and the elastic properties of polymer crystals. Vibrational free energy (enthalpic and entropic) is added to the deformation energy term U of Equation 1. Lacks & Rutledge (12) considered the elastic properties of crystalline (multiple chain) PE. It was shown that vibrational free energy modifies the energy-deformation potential to decrease E_c at room temperature by 29 GPa (about 9%) when compared with a purely static calculation. A similar analysis was done on PPTA crystals by Lacks & Rutledge (33). Here vibrational modification of the potential surface reduces E_c by <5 GPa.

Summary

The tensile modulus of uniaxially oriented polymers is dominated by internal energy changes from deforming intramolecular bonds within the chains. Theoretical modulus E_c based on empirical potentials (entries in Table 1) are reasonable estimates of ultimate stiffness. Semi-empirical quantum mechanics gives values of E_c, which are systematically too large. Given current computing technology, ab initio molecular orbital theory can be applied only to simple systems, e.g. PE. Even the best practical level of ab initio theory gives moduli that are generally 10% too large (79), and methods for simulating long chains may need more study (see Table 2).

Incomplete chain orientation in real fibers is the major reason why $E_f < E_c$, as indicated by Equation 4. UHMWPE can be processed to achieve nearly perfect orientation ($\langle \sin^2 \phi \rangle < 10^{-5}$) (11), hence the room temperature $E_f = 264$ GPa reported by van der Went & Pennings (13) is only 7% below the calculated $E_c = 283$ GPa (12). Chains in PPTA fibers are less well aligned, but Figure 3 gives $E_f = 240$ GPa for the perfectly oriented state. Using the simulation results in Figure 4 to correct for chain end effects, the experimental fiber modulus of PPTA is increased to ~ 290 GPa, in reasonable accord with the most recent empirical $E_c = 274$ GPa (33). Experiments and calculations for other polymers are relatively scarce, but $E_f \approx 350$ GPa for PBO and PBT fibers is consistent with empirical E_cs. The extent to which the fiber modulus of these polymers is reduced by orientation or chain end effects is not known.

TENSILE STRENGTH

Data in Table 1 are clustered about tensile strength $\sigma_b = 3$–6 GPa for the strongest polymer fibers and films at room temperature. The largest strengths are obtained for highly oriented, high modulus materials approximating the ideal continuous crystal as shown in Figure 1. There are two limiting failure modes for such a structure, one involving the scission of every chain in a particular cross section (Figure 5*a*), and the other based on local axial shear or chain slip to separate the crystal into two parts (Figure 5*c*). As indicated in Figure 5, the chain scission mechanism is fairly insensitive to molecular weight, whereas the chain slip process is controlled by chain-end concentration. A combination of scission and slip (Figure 5*b*) would have an intermediate dependence on M_n. The positive correlation between σ_b and molecular weight was mentioned above. Figure 6 gives an example of this behavior for gel spun PE (80). When fibers with the same orientation and Young's modulus, for instance $E_f = 50$ GPa, are compared, strength nearly doubles as M_w is increased from 800 to 4000 Kg/mol.

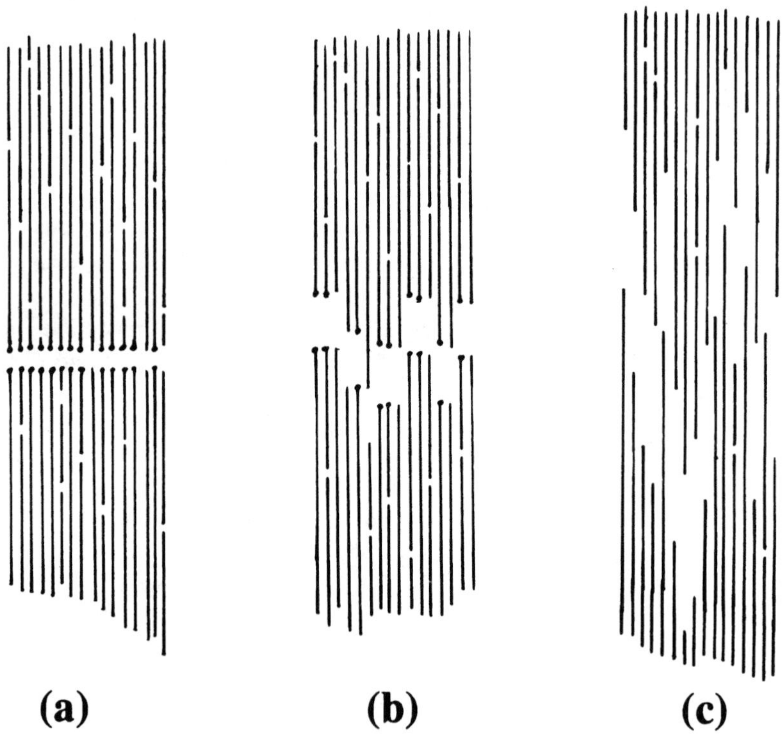

Figure 5 Schematic representations for failure of a continuous crystal by (*a*) chain scission; (*b*) chain scission and chain slip; (*c*) chain slip. Chains terminating with a solid circle indicate point of chain scission.

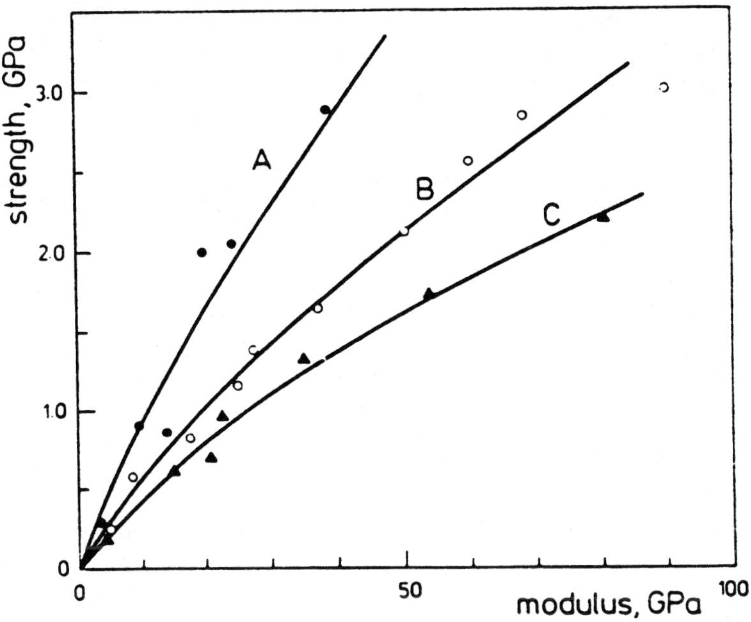

Figure 6 Room temperature tensile strength σ_b as a function of E_f for PE fibers of different molecular weight: (*A*) $M_r = 4000$ Kg/mol; (*B*) $M_r = 1500$ Kg/mol; (*C*) $M_r = 800$ Kg/mol. From Reference 80 with permission from Wiley & Sons.

Strength and failure mechanisms are treated in the following manner. First we review the experimental parameters that affect σ_b, with an eye to information conveyed about elementary fracture processes. Theoretical treatments of chain scission and chain slip, corresponding to failure, as in Figure 5*a* and 5*c*, respectively, are considered next. We conclude with simulation studies that include one or both of the fundamental failure modes.

Experimental Aspects of Polymer Fracture

TEST METHODS AND SIZE EFFECTS A tensile stress σ (static or dynamic) is applied to a fiber or film, and the value at which fracture occurs is defined as σ_b. The most common procedure is the constant strain-rate test. Failure under constant tensile load, also known as static fatigue or stress rupture, is used for establishing the relation between σ_b and the time to failure t_b. Greater polymer strength σ_b is almost always observed for shorter testing times (higher strain rate or lower t_b) or lower temperature. It is important that both temperature and time of the fracture measurement be specified because these parameters have more effect on σ_b than on Young's modulus

E_f. Data in Table 1 are for room temperature and strain rate $\sim 10^{-3}$/s, meaning that fracture occurred in ca 20 s.

Another difference is that fracture is a stochastic process; even with the most careful sample preparation and experimental technique, one observes an appreciable range of σ_b or t_b with replicated measurements. Statistical weak link theories predict that relatively narrow distributions of strength are correlated with size effects, i.e. larger average strength for smaller samples (81, 82). Penning et al (82) report size effects for ultra-oriented PE that can be expressed as $\sigma_b \propto D^{-0.19} l^{-0.01}$ where D is fiber diameter and l is fiber length. Remarkably similar results of $\sigma_b \propto D^{-0.2} l^{-0.06}$ were obtained in Termonia's simulation of fracture in crystalline fibers made of PE-like molecules with clustered or segregated chain ends (83).

High-strength fibers of PPTA, PBO, and PBT have not been studied as extensively but are observed to decrease in strength by $\sim 20\%$ as sample length l is increased by a factor of 10 (84, 85). Strength-size effects in polymer fibers are of practical importance; $\sigma_b \propto D^{-0.2}$ means strength increases by 60% for a tenfold reduction in D. Fundamental interpretation is still lacking, although it is agreed that randomly distributed chain-end defects in an ideal continuous crystal (Figure 1) cause no dependence of σ_b on sample dimensions (81, 83).

TIME AND TEMPERATURE EFFECTS Constant load or creep rupture lifetimes are very sensitive to stress and temperature, as illustrated by the experiments of Wu et al on PPTA fibers (86) shown in Figure 7. Lifetime t_b is reduced exponentially by stress σ; this stress dependence is less at elevated temperatures, and isothermal data extrapolate to a common value $\tau_0 = 0.067$ s at $\sigma = 3.75$ GPa. These data can be summarized in the following expression:

$$t_b(\sigma, T) = \tau_0 \exp\left[\frac{\Delta E_0 - \gamma\sigma}{RT}\right], \qquad 5.$$

$$\sigma(t_b, T) = \frac{\Delta E_0}{\gamma} - \frac{RT}{\gamma}\ln(t_b/\tau_0). \qquad 6.$$

Equation 5 describes creep rupture results; strength σ has been made the dependent variable in Equation 6. Equation 5 is often used to interpret the fracture mechanism through kinetic parameters where τ_0 is the time of a fundamental process, and the apparent activation barrier $\Delta E_a(\sigma) = \Delta E_0 - \gamma\sigma$ decreases linearly with stress. Here ΔE_0 is the stress-free activation energy for the fracture process, and γ is an activation volume that contains a sample-dependent stress concentration factor. The first term $\Delta E_0/\gamma$ of Equation 6 can be thought of as limiting strength at $T = 0$ K, or the stress

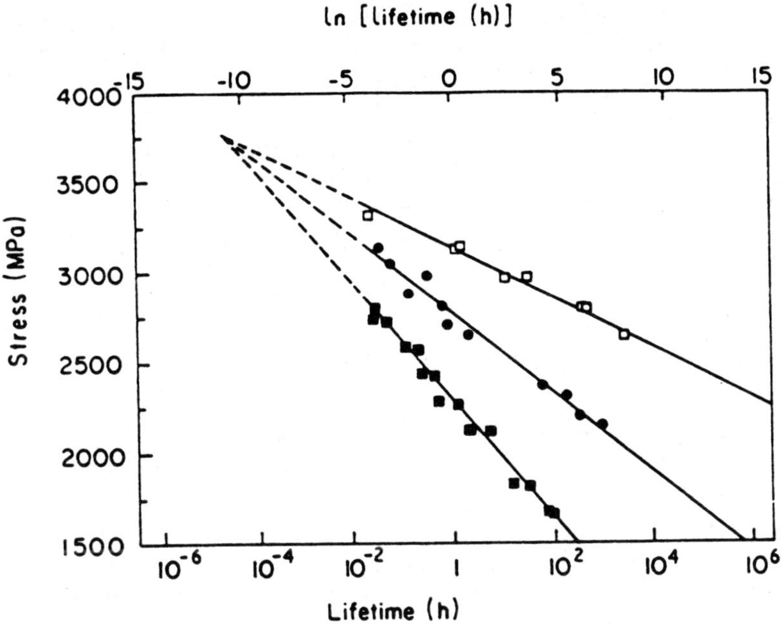

Figure 7 Applied stress vs creep rupture lifetimes for PPTA fibers at 21°C (open square), 80°C (solid circle), 130°C (solid square). From Reference 86 with permission from Chapman & Hall.

required to achieve fracture in the shortest relevant time $t_b = \tau_0$. This limiting strength is inversely proportional to the stress concentration factor, which determines the magnitude of γ. Strength decreases at higher temperatures or longer times as indicated by the second term.

A major advocate of polymer fracture kinetics has been Zhurkov who, with his colleagues, found that creep rupture lifetimes invariably conformed to Equation 5, with $\tau_0 \approx 10^{-12}$ s and ΔE_0 being equal to the activation energy for thermal or mechanical fracture rates of covalent bonds for the polymer (87). Because $\tau_0 \approx \tau_v$, the vibrational period of a covalent bond, and ΔE_0 is identified with chain scission, the macroscopic lifetime t_b was thought to be controlled by the rate of covalent bond rupture in the stressed fiber. A thorough account of the work of Zhurkov and others has been given by Kausch (88). Those earlier experiments were on relatively low-strength samples ($\sigma_b < 1$ GPa with normal tensile testing); quite different results are obtained with ultra-oriented high-strength fibers. Data for PPTA in Figure 7 give $\tau_0 \approx 0.1$ s, clearly incompatible with 10^{-12} s, although $\Delta E_0 = 335$ KJ/mol is near the energy of a C-C or C-N covalent bond. Creep rupture experiments for ultra-oriented

PE give $\tau_0 = 10^{-7}$ s and $\Delta E_0 = 75$ KJ/mol (89) as opposed to 10^{-13} s and 105 KJ/mol for conventional PE (90). To this writer's knowledge, no other high-strength polymer has been studied by creep rupture kinetics at different temperatures.

The limited data available for high-strength UHMWPE and PPTA support the empirical relations in Equations 5 and 6, although with parameters τ_0 and ΔE_0 different from those in low-strength polymers. A preliminary assessment of results for ultra-oriented fibers indicates that high-strength results from a lower γ, which probably reflects small stress concentration factors. The large values of τ_0 are not understood.

Theoretical Approaches to Fracture

STATIC STRENGTH: BOND RUPTURE An absolute upper limit to tensile strength can be estimated by considering bond scission caused by an applied mechanical force in the absence of thermal energy. This ultimate strength may be derived from the first term in Equation 6 as follows. We let ΔE_0 be the bond energy, which is about 400 KJ/mol for covalent C-C or C-N single bonds in the backbone of polymer chains (88). As will be justified by the analysis of kinetic fracture in the following section, we replace γ by the activation volume $v^* = 1.6 \times 10^{-5}$ m^3/mol for covalent bond rupture. The estimated $\sigma_b^* = \Delta E_0/\gamma \approx 26$ GPa establishes the order of magnitude of tensile stress that will break a covalent bond at temperature $T = 0$ K. Similar values from other empirical calculations are summarized by Kelly & MacMillan (91).

Quantum mechanics may be used to obtain less model-dependent estimates of ultimate strength. That procedure is described in the Calculated Crystal Modulus section; however, σ_b^* corresponds to the (nonlinear) elastic stress F/A_0 at the inflection point of the potential surface $U(L)$. Suhai (52) has applied ab initio molecular orbital theory to PE chain fracture. Plotted in Figure 8 is the axial force F required to achieve molecular strain ε by simultaneous stretching of C-C bonds and opening C-C-C bond angles. The force has a maximum of 6.3 nN or $\sigma_b^* = 34$ GPa at a critical strain $\varepsilon^* = 0.3$.

The ultimate strength calculated by Suhai is the most rigorous available. Chain fracture of rigid-rod PBO and PBT was similarly treated with quantum mechanics at the semi-empirical AM1 level by Wierschke et al (60). Molecular fracture occurs at a stress of 59 GPa (PBO) or 55 GPa (PBT) by homolytic scission of the C-C single bond linking the rings (Figure 2*c*). These calculated ultimate strengths are about 70% larger than σ_b^* for PE, although comparing absolute values obtained with different molecules and models may be questioned. Regardless of details, molecular orbital calculations indicate that covalent bond rupture occurs at an

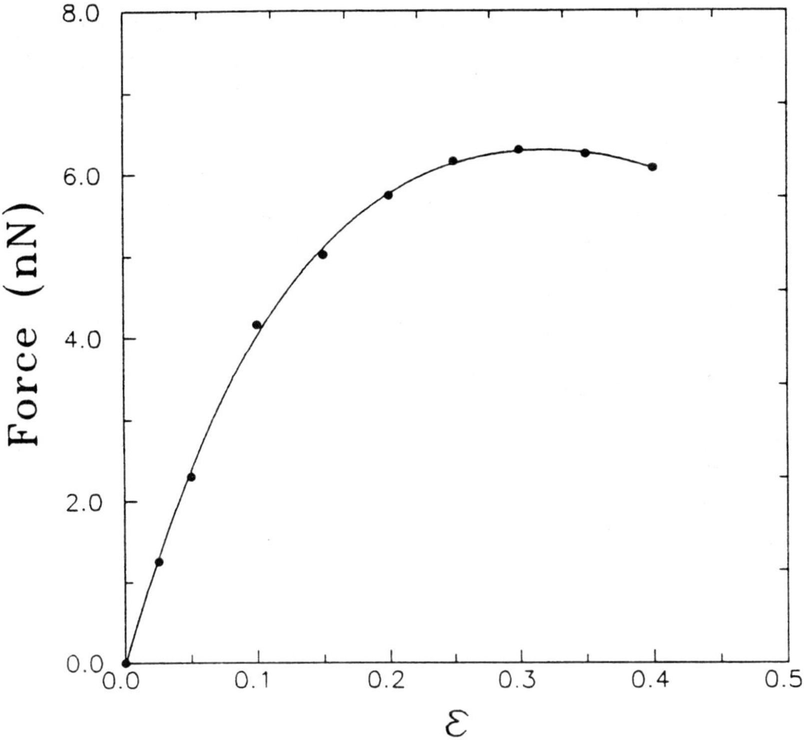

Figure 8 Tensile force F vs molecular strain ε for a single PE chain. Data from Reference 51 were calculated with ab initio methods.

applied stress of 34 to 60 GPa in the absence of thermal energy to assist fracture. This establishes a ceiling for the tensile strength of polymers containing C-C single bonds in the backbone.

STATIC STRENGTH: CHAIN SLIP A continuous crystal of finite length macromolecules may fail without covalent bond rupture if chains slide past one another as in Figure 5. Chain ends modify local stress fields and initiate the chain slip mode of fracture. Yoon (46) employed shear lag theory (elastic load transfer with debonding from the matrix) to calculate the tensile strength σ_{bs} for a perfectly oriented fiber of chains with number average molecular length L_n failing by chain slip:

$$\sigma_{bs} = \frac{2\tau^*}{B \tanh(BL_n/2r_0)}\left(1 - \frac{2r_0}{BL_n \tanh(BL_n/2r_0)}\right). \qquad 7.$$

Here τ^* is the critical shear stress for breaking intermolecular (chain-

chain) bonds, $B \approx (2.3G_c/E_c)^{1/2}$, and r_0 is the chain diameter. This model was applied to HBA/HNA copolymer (Figure 2*e*) fibers as shown in Figure 9 where the dashed line corresponds to Equation 7, and the solid line, conforming well to the data, is from a more complete accounting of the chain length distribution (46). The stress intercept of 4.1 GPa can be thought of as the ultimate strength $\sigma^*_{bs} = 2\tau^*/B$, which is achieved when $L_n \gg 2r_0/B$, and chains are maximally clamped by intermolecular bonding. The rate at which strength decreases with L_n^{-1} here gives $B = 0.016$ (with chain radius $r_0 = 0.5$ nm and monomer length $L_0 = 1$ nm) and hence $\tau^* = 0.03$ GPa.

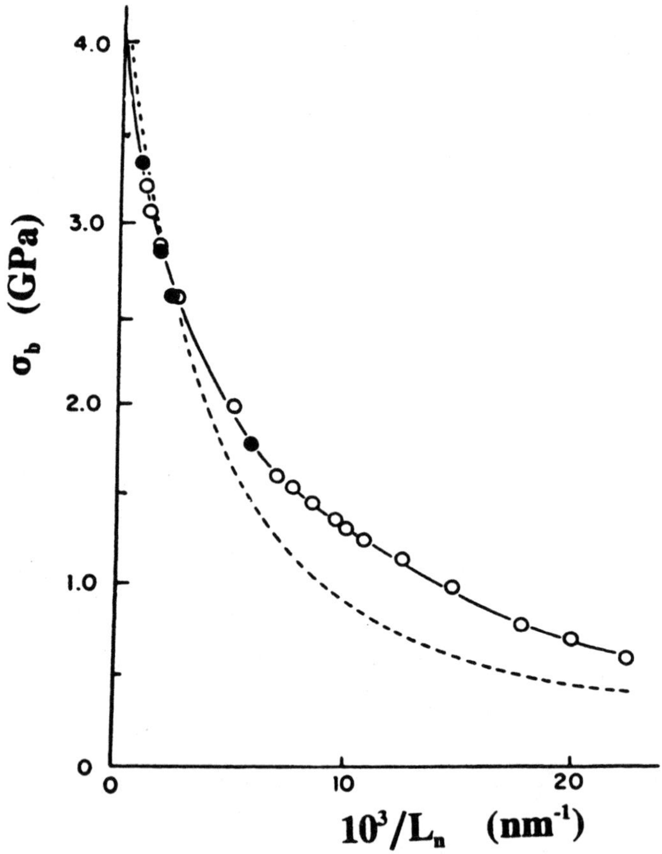

Figure 9 Tensile strength σ_b vs reciprocal chain length for HBA/HNA random copolymers. Open symbols are data for as-spun fibers, solid symbols for heat-treated fibers. Dashed line is Equation 7. The solid line is for the same theory with a more accurate treatment of chain length distribution. After Reference 46 with permission from Steinkopff Darmstadt.

Although molecular weight or chain length effects are well described by Equation 7 (or a more sophisticated variation), the absolute values of B and τ^* seem unrealistic. With typical polymer moduli $E_c = 250$ GPa and $G_c = 3$ GPa, one expects $B \approx 0.2$, and the intermolecular shear strength τ^* can be approximated by $G_c/3 \approx 1$ GPa (91). Each of these estimated parameters is of order 10 larger than those derived from the data in Figure 9. One should, of course, be cautious when applying a continuum theory to molecular events.

Accepting the validity of Equation 7, ultimate strength is $\sigma^*_{bs} = 2\tau^*/B \approx 10$ GPa for typical values for shear strength $\tau^* = 1$ GPa and load transfer coefficient $B = 0.2$. Kausch and co-workers (88, 92, 93) considered a related problem with molecular mechanics, i.e. the maximal axis stress that can be supported by a single unbreakable chain emerging from a semi-infinite crystal. The critical tensile stress for PE is 7.5 GPa and that for nylon 6 is 22.4 GPa; at larger loads chains are pulled through the crystal. Static estimates of fiber strength thus suggest that chain slip will occur at a lower stress (~10–20 GPa) than that required for chain scission (~35–60 GPa). Recall, however, that kinetic aspects have been ignored.

FRACTURE KINETICS Tobolsky & Eyring (94) used absolute rate theory (passage over an energy barrier by thermal fluctuations) to predict the rate of bond breaking and hence the lifetime of a fiber under a constant tensile force or nominal stress σ, as in a creep rupture experiment. Macroscopic lifetime t_b is equal to the time required to break all N_0 bonds initially present in the fracture plane:

$$t_b(\sigma, T) = \theta \exp\left[\frac{\Delta E_0 - \gamma\sigma}{RT}\right]. \qquad 8.$$

The kinetic model in Equation 8 thus accounts for the experimental creep rupture behavior described empirically by Equation 5. Furthermore, the zero-stress activation energy ΔE_0 for macroscopic failure is equal to that for bond rupture, and the parameter γ is identified with gv^*, where v^* is the activation volume for bond rupture and g is an (average) stress concentration factor. The preexponential factor θ can be shown to be $\sim\tau_v/g$, where $\tau_v \approx 10^{-13}$ s is the vibrational period of a covalent bond.

Equation 8 may be rearranged to express the kinetic strength of $\sigma(t_b, T)$ identically to Equation 6. Kinetic strength under typical conditions of $t_b = 10$ s, $T = 300$ K is only ~15% less than the ultimate strength $\sigma^*_b = \Delta E_0/v^* \approx 30$ GPa. The same conclusion, that kinetic factors have second order effects on experimental strength, comes from Figure 7. Breaking stress of PPTA fibers at room temperature, $t_b \approx 50$ sec, is $\sigma_b = 3.3$ GPa, within 13% of the limiting strength of 3.75 GPa.

Model Studies

Molecular dynamics simulations have been used to investigate the fracture of single chains of N-coupled anharmonic oscillators described by Morse or Lennard-Jones potentials. Temperature T and constraints of constant length (95, 96) or force (97) are chosen to achieve fracture in a few hundred vibrational periods τ_v ($t_b < 10^{-9}$ s). Oliveira & Taylor (98) recently completed a study of Lennard-Jones chains at constant length (strain $\varepsilon \approx 0.03$). The average lifetime t_b can be written as

$$t_b(\varepsilon, T) = \frac{10^3 \tau_v}{\mathrm{N}} \exp\left(\frac{\Delta \mathrm{E}_a(\varepsilon)}{RT}\right). \qquad 9.$$

Some of the behavior conveyed by Equation 9 is expected. At constant strain ε the temperature dependence of t_b is of the Arrhenius form, although the apparent activation energy $\Delta E_a(\varepsilon)$ is a nonlinear function of ε (or σ), similar to the finding of Tomashevskii (99) for a Morse chain under constant length. The preexponential factor deserves close attention. Here the lifetime is proportional to N^{-1} because there are N bonds in series, any one of which can fail and cause chain fracture. Tobolsky & Eyring (94) considered only one plane of N_0 bonds loaded in parallel so this size factor is missing in Equation 8. As mentioned above, the prefactor θ in Equation 8 is comparable to τ_v. For the Lennard-Jones chain at constant length, the equivalent prefactor is 10^3 time larger! The reason for this is not clear but presumably reflects cooperative energy pooling associated with the fracture step. Note that the large prefactor ($>10^{-10}$ s) in this molecular dynamics model is in the direction of elementary times seen experimentally for ultra-oriented PE ($\tau_0 \approx 10^{-7}$ s) and PPTA ($\tau_0 \approx 0.01$ s).

Less-detailed models are used for multichain systems. Prevorsek (99a) was among the first to consider both chain slip and chain fracture in a kinetic analysis wherein the stress-dependent activation barrier for failure reflects relative contributions of the two mechanisms. More recently, Termonia and co-workers utilized Monte Carlo methods for calculating rates of chain scission and secondary bond rupture, each governed by Equation 8; chain rupture is irreversible, whereas secondary bonds are allowed to reform, as expected in chain slip. A continuous crystal with finite length chains was stretched at room temperature ($T = 290$ K), at strain rate of 0.017/s as in a conventional tensile test. This simulation with elastic constants and bond parameters appropriate for PE (74) showed that strength

σ_b and fracture strain ε_b increased with molecular weight and gave values consistent with experiment. Fracture occurred exclusively by chain slip (no chain scission) for PE with $M < 100$ Kg/mol. Fibers with $M \geqslant 100$ Kg/mol experience considerable covalent bond rupture before fracture, which reduces molecular weight to the point where chain slip occurs and leads to ultimate failure. A subsequent study (100) showed that strength σ_b was a positive function of strain rate and inverse temperature, as expected for a kinetic fracture model, with relatively more chain scission at high T and/or large strain rate. This logically reflects the fact that chain scission with the larger activation barrier is more sensitive to changes in temperature or time scale.

PPTA fiber-stretching and fracture were simulated in the same manner by Termonia & Smith (75), although the energy $\Delta E_0 = 21$ KJ/mol for interchain (hydrogen) bonds as opposed to 2.7 KJ/mol for van der Waals bonds in PE. Strong interchain bonds in PPTA effectively suppress chain slip and at the same time lead to large stresses near chain ends, initiating chain scission at those locations. One chain rupture microcrack grows and causes fiber fracture. Dominance of chain scission in PPTA failure is consistent with experimental activation energy $\Delta E_0 = 335$ KJ/mol (86), characteristic of covalent bond energy (Figure 7). By the same token, experimental $\Delta E_0 = 75$ KJ/mol for creep rupture of PE fibers (89), a low value that is compatible with chain slip.

Summary

Experimental tensile strengths of ultra-oriented polymer fibers are in the range $\sigma_b = 3$–6 GPa. Creep rupture lifetimes t_b conform to a kinetic model in which the activation energy $\Delta E_a(\sigma)$ decreases linearly with applied stress, although the preexponential time factor τ_0 is many orders of magnitude larger than the classical value of $\sim 10^{-12}$ s. Earlier appraisals of experimental and theoretical strength (87, 88, 90) focused on covalent bond rupture in stressed chains and ignored chain end defects, which are sites for fracture initiation in otherwise perfect fibers. Subsequent advances in polymer synthesis and processing have shown that strength is a positive function of molecular weight (Figures 6 and 9), implying that chain ends limit σ_b.

Only for infinite molecular weight polymers are there no chain ends. Here one may invoke the ultimate strength governed by homolytic scission of C-C bonds, for which $\sigma_b^* \approx 45 \pm 10$ GPA from molecular orbital theory (52, 60). Thermal fluctuations at usual temperatures and time scales will reduce this ideal strength to ca 30–45 GPa. Molecular dynamics studies (98) of single chain fracture confirm the kinetic scheme in Equations 5 and

8, with perhaps larger prefactors that have little bearing on practical strength.

Chain ends associated with finite molecular weight reduce strength by one of two mechanisms. Axial slip originating at chain ends may lead to failure as indicated in Figure 5*c*. The shear lag model written as Equation 7 establishes a (static) strength $\sigma^*_{bs} \approx 10$ GPa, but this continuum mechanics result is only a rough estimate. Even in the absence of chain slip, chain ends cause local stress concentrations that can result in covalent bond scission at applied stress σ much less than σ^*_b. Models involving intermolecular interactions and chain-end defects are necessarily complicated. Monte Carlo simulations of Termonia et al (74, 75) nicely illustrate the interplay between chain scission and chain slip. Weak intermolecular bonds promote chain slip in PE, which fails at $\sigma_b < 15$ GPa, appreciably below the ultimate strength $\sigma^*_b = 45$ GPa for that model. Chain ends have a different role in the simulated fracture of PPTA. Here, stronger intermolecular bonds prevent chain slip, but chain ends cause local stress concentrations that result in fracture by chain scission at $\sigma_b < 12$ GPa, about half of $\sigma^*_b = 24$ GPa in that model. While absolute strengths derived from these simulations may be questioned, they illustrate the profound effect of chain ends on strength.

CONCLUSIONS

Nearly perfect chain orientation can be achieved by gel spinning and drawing PE and *i*-PP, which results in fibers with modulus $E_f \approx E_c$ (~ 270 GPa for PE, 40 GPa for *i*-PP). Semi-rigid or stiff-chain polymers processed by solution spinning are not as well oriented, leading to $E_f \leqslant 0.8\ E_c$, which is described adequately by Equation 4. Empirical calculations of the theoretical modulus E_c base on single chain response are accurate within about 5 to 10% in most cases. On this basis PBO, the stiffest polymer made to date, could have $E_f \approx 450$ GPa with improved processing methods.

The strongest polymer is PE with $\sigma_b = 7.2$ GPa, about 25% of the best estimate of ultimate strength ~ 30 GPa from ab initio calculations of C-C bond scission and adjusted for kinetic effects expressed in Equation 6. Unavoidable chain-end defects are responsible for observed strengths $\sigma_b \approx 5$ GPa in ultra-oriented polymers. For PE with weak intermolecular bonds, chain slip, originating at chain ends, is the likely dominant failure mechanism (74). Stronger intermolecular bonds can indeed suppress slip, but stress concentrations near chain ends cause chain scission and fracture at σ_b that is appreciably less than the ultimate strength σ^*_b (75).

In summary, fiber modulus of ultra-oriented polymers approaches the theoretical modulus E_c as chain orientation is improved, although finite molecular weight may limit E_f if axial load transfer in the region of chain ends is incomplete. Tensile strength is controlled by chain-end defects even when chain orientation is perfect. Better theoretical methods will lead to more accurate predictions of ideal stiffness and strength. Advances in experimental E_f may accrue from improved orientation or synthesis of stiffer macromolecules. Enhanced strength should be realized if molecular weight is increased, but the ideal maximum $\sigma_b^* \approx 45 \pm 1$ GPa should be considered a reference value rather than an obtainable goal.

Literature Cited

1. Staudinger H. 1932. *Die Hochmolecularen Organischen Verbindungen*, p. 111. Berlin: Springer
2. Yang HH. 1989. *Aromatic High-Strength Fibers*. New York: Wiley. 873 pp.

2a. Baer E, Moet A, eds. 1991. *High Performance Polymers*. New York: Hanser. 335 pp.

3. Hattery GR, Hillman MED. 1991. See Ref. 2a, pp. 256–78

3a. Thomas EL, ed. 1993. *Structure and Property of Polymers*. New York: VCH. Vol. 12. 785 pp.

4. Jiang H, Adams WW, Eby RK. 1993. See Ref. 3a, 12: 597–652
5. Sheehan WC, Cole TB. 1964. *J. Appl. Polym. Sci.* 8: 2359–88
6. Samuels RJ. 1974. *Structured Polymer Properties*. New York: Wiley. 251 pp.
7. Barham PJ, Keller A. 1976. *J. Mater. Sci.* 11: 27–35
8. Smith P, Lemstra P. 1980. *J. Mater. Sci.* 15: 505–14
9. Smith P, Lemstra P, Booij HC. 1981. *J. Polym. Sci. Polym. Phys. Ed.* 19: 877–88

9a. Porter RS, Wang LH. 1995. *J. Macromol. Sci. Rev.* C35: 63–115

9b. Ohita T. 1983. *Polym. Eng. Sci.* 23: 697–823

9c. Porter RS, Kanamoto T, Zachariades AE. 1994. *Polymer* 35: 4979–84

9d. Porter RS, Wang LH. 1995. *J. Macromol. Sci. Rev.* In press

10. Savitsky AV, Gorshkova IA, Frolova IL, Shmikk GN, Ioffe AF. 1984. *Polymer Bull.* 12: 195–202
11. Matsuo M, Sawatari C. 1988. *Macromolecules* 21: 1653–58
12. Lacks DJ, Rutledge, GC. 1994. *J. Phys. Chem.* 98: 1221–31
13. van der Wert H, Pennings AJ. 1991. *Colloid Polym. Sci.* 269: 747–63
14. Nakamae K, Nishino T, Ohkubo H. 1991. *J. Macromol. Sci. Phys.* B30: 1–23
15. Schellekens R, Bastiaansen CWM. 1991. *J. Appl. Polym. Sci.* 43: 2311–15
16. Nakamae N, Nishino T. 1991. In *Integration of Fundamental Polymer Science and Technology*, ed. P Lemstra, LA Kleintjens, 5: 121–30. New York: Elsevier
17. Tashiro K, Kobayashi M, Tadokoro H. 1978. *Macromolecules* 11: 914–18
18. Ito M, Takahashi K, Kanamoto T. 1990. *J. Appl. Polym. Sci.* 40: 1257–63
19. Thistlewaite T, Jakeways R, Ward IM. 1988. *Polymer* 29: 61–69
20. Nakamae K, Nishino T, Yokoyama F, Matsumoto T. 1988. *J. Macromol. Sci. Phys.* B27: 407–20
21. Tashiro K, Kobayashi M, Tadokoro H. 1977. *Macromolecules* 10: 413–20
22. Smook J, Vos GJG, Doppert HD. 1990. *J. Appl. Polym. Sci.* 41: 105–16
23. Matsuo M, Sato R, Shimizu Y. 1993. *Colloid Polym. Sci.* 271: 11–21
24. Tashiro K, Tadokoro H. 1981. *Macromolecules* 14: 781–85
25. Kanamoto T, Porter RS. 1989. In *Integration of Fundamental Polymer Science and Technology*, ed. P. Lemstra, LA Kleintjens, 3: 168–77. New York: Elsevier

26. Sawatari C, Matsuo M. 1989. *Macromolecules* 22: 2968–73
27. Tashiro K, Kobayashi M, Tadokoro H. 1992. *Polym. J.* 24: 899–916
28. Nakamae K, Nishino T, Shimizu Y, Hata K. 1990. *Polymer* 31: 1909–18
29. Wu G, Tashiro K, Kobayashi M, Komatsu T, Nakagawa K. 1989. *Macromolecules* 22: 758–65
30. Krause SJ, Haddock TB, Vezie DI, Lenhert PG, Hwang W-F, et al. 1988. *Polymer* 29: 1354–64
31. Ii T, Tashiro K, Kobayashi M, Tadokoro H. 1987. *Macromolecules* 20: 347–51
32. Nakamae K, Nishino T, Shimizu Y, Matsumoto T. 1987. *Polym. J.* 19: 451–59
33. Lacks, DJ, Rutledge GC. 1994. *Macromolecules* 27: 7179–204
33a. Adams WW, Eby RK, McLemore DE, eds. 1989. *Materials Science and Engineering of Rigid-Rod Polymers*. Pittsburgh, PA: Mater. Res. Soc. 704 pp.
34. Lenhert PG, Adams WW. 1989. See Ref 33a, pp. 329–40
35. Tashiro K, Kobayashi M. 1991. *Macromolecules* 24: 3706–8
36. Ledbetter HD, Rosenberg S, Hurtig, CW. 1989. See Ref. 33a, pp. 253–64
37. Wolfe JF. 1988. In *Encyclopedia of Polymer Science and Engineering*, ed. HF Mark, NM Bikales, CG Overberger, G Menges, 11: 601–635. New York: Wiley
38. Donald A, Windle AH. 1992. *Liquid Crystalline Polymers*, Chpt. 8. Cambridge: Cambridge Univ. Press
39. Northolt MG, Sikkema DJ. 1991. *Adv. Polym. Sci.* 98: 115–77
40. Martin DC, Thomas EL. 1991. *Macromolecules* 24: 2450–60
41. Sawyer LC, Jaffe M. 1986. *J. Mater. Sci.* 21: 1897–913
42. Ihm DW, Hiltner A, Baer E. 1991. See Ref. 2a, pp. 280–327
43. Saraf AW, Desai P, Abhiraman AS. 1991. *J. Appl. Polym. Sci. Appl. Polym. Symp.* 47: 67–86
44. Schaefgen JR. 1983. In *The Strength and Stiffness of Polymers*, ed. AE Zachariades, RS Porter, pp. 327–55. New York: Marcel-Dekker
45. Sarlin J, Tormala P. 1993. *J. Appl. Polym. Sci.* 50: 1225–31
46. Yoon HN. 1990. *Colloid Polym. Sci.* 268: 230–39
47. Tashiro K, Kobayashi M. 1991. *Polymer* 32: 454–63
48. Grubb DT. 1993. See Ref. 3a, 12: 301–56
49. Tashiro K. 1993. *Prog. Polym. Sci.* 18: 377–435
50. Karasawa N, Dasgupta S, Goddard WA. 1991. *J. Phys. Chem.* 95: 2260–72
51. Suhai S. 1986. *J. Chem. Phys.* 84: 5071–76
52. Crist B, Ratner MA, Brower AL, Sabin JR. 1979. *J. Appl. Phys.* 50: 6047–51
53. Crist B, Hereña P. 1995. *J. Polym. Sci. B Poly. Phys.* In press
54. Hehre WJ, Radom L, Schleyer PVR, Pople JA. 1986. *Ab Initio Molecular Orbital Theory*. New York: Wiley. 548 pp.
55. Klei HE, Stewart JJP. 1986. *Int. J. Quantum Chem. Quantum Chem. Symp.* 20: 529–40
56. Hong SY, Kertez M. 1990. *Phys. Rev. B* 41: 11368–78
57. Horn T, Adams WW, Pachter R, Haaland P. 1993. *Polymer* 34: 2481–84
58. Meier RJ. 1993. *Macromolecules* 26: 4376–78
59. Sorenson RA, Liau WB, Kesner L, Boyd RH. 1988. *Macromolecules* 21: 200–8
60. Wierschke SG, Shoemaker JR, Haaland PD, Pachter R, Adams WW. 1992. *Polymer* 33: 3357–68
61. Rutledge GC, Suter UW. 1991. *Polymer* 32: 2179–90
62. Clements J, Jakeways R, Ward IM. 1978. *Polymer* 19: 639–44
63. Prasad K, Grubb DT. 1989. *J. Polym. Sci. B Polym. Phys.* 27: 381–403
64. Prasad K, Grubb DT. 1990. *J. Polym. Sci. B Polym. Phys.* 28: 2199–212
65. Moonen JAHM, Roovers WAC, Meier RJ, Kip BJ. 1992. *J. Polym. Sci. B Polym.* 30: 361–72
66. Fanconi B, Rabolt JF. 1985. *J. Polym. Sci. Polym. Phys. Ed.* 23: 1201–15
67. Snyder RG, Strauss HI, Alamo R, Mandelkern L. 1994. *J. Chem. Phys.* 100: 5422–31
68. Northolt MG. 1980. *Polymer* 21: 1199–204
69. Northolt MG, van der Hout R. 1985. *Polymer* 26: 310–16
70. Powell AK, Craggs G, Ward IM. 1990. *J. Mater. Sci.* 25: 3390–4000
71. Jiang H, Eby RK, Adams WW, Lenhert G. 1989. See Ref. 33a, pp. 341–50
72. Deleted in proof
73. Bartenev GM, Valishin AA. 1968. *Polym. Sci. USSR* 10: 26–33
74. Termonia Y, Meakin P, Smith P. 1985. *Macromolecules* 18: 2246–52
75. Termonia Y, Smith P. 1986. *Polymer* 27: 1845–49
76. Smith P, Termonia Y. 1989. *Polymer Commun.* 30: 66–68

76a. Ogita T, Yamamoto R, Suzuki N, Ozaki F, Matsuo M. 1991. *Polymer* 32: 822–34

76b. Ohta Y, Sugiyama H, Yasuda H. 1994. *J. Polym. Sci. B Polym. Phys.* 32: 261–69

77. Nishino T, Ohkubo H, Nakamae K. 1992. *J. Macromol. Sci. Phys.* B31: 191–214

78. Kluzinger PE, Green KA, Eby RK, Farmer BL, Adams WW, Czornyj G. 1991. In *Technical Papers XXXVII*, pp. 1532–35. Brookfield, CT: Soc. Plastics Engineers

79. Pople JA, Scott AP, Wong MW, Radom L. 1993. *Israel J. Chem.* 33: 345–50

80. Smith P, Lemstra PJ. 1981. *J. Polym. Sci. Polym. Phys. Ed.* 19: 1007–9

81. Wagner HD. 1989. *J. Polym. Sci. B Polym. Phys.* 27: 115–49

82. Penning JP, de Vries AA, Van der Ven J, Pennings AJ, Hoogstraten HW. 1994. *Philos. Mag. A* 69: 267–84

83. Termonia Y. 1995. *J. Polym. Sci. B* 33: 147–52

84. Wagner HD, Phoenix SL, Schwartz P. 1984. *J. Composit Mater.* 18: 312–38

85. Im J, Percha PA, Yeakle DS. 1989. See Ref. 33a, pp. 307–12

86. Wu HF, Phoenix SL, Schwartz P. 1988. *J. Mater. Sci.* 23: 1851–60

87. Zhurkov SN, Korsukov VE. 1974. *J. Polym. Sci. Polym. Phys. Ed. A*-2 12: 385–98

88. Kausch HH. 1987. *Polymer Fracture*. New York: Springer. 456 pp. 2nd ed.

89. Smook J, Hamersma W, Pennings AJ. 1984. *J. Mater. Sci.* 19: 1359–73

90. Amelin AV, Pozdnyakov OF, Regel VR, Sanfirova TP. 1971. *Sov. Phys. Solid State* 12: 2034–38

91. Kelly A, MacMillan NH. 1986. *Strong Solids*. New York: Oxford Univ. Press. 423 pp. 3rd ed.

92. Kausch HH, Becht J. 1973. In *The Deformation and Fracture of High Polymers*, ed. HH Kausch, JA Hassell, RJ Jaffee, pp. 317–33. New York: Plenum

93. Kausch HH, Langbein D. 1973. *J. Polym. Sci. Polym. Phys. Ed.* 11: 1201–18

94. Tobolsky A, Eyring H. 1943. *J. Chem. Phys.* 11: 125–34

95. Melker Al, Mikhailin AI. 1981. *Sov. Phys. Solid State* 23: 1016–18

96. Melker AI, Ivanov AV, Govorov SV. 1989. *Mech. Composite Mater.* 6: 746–50

97. Welland RW, Shin M, Allen D, Ketterson JD. 1992. *Phys. Rev. B* 46: 503–6

98. Oliveira FA, Taylor PL. 1994. *J. Chem. Phys.* 101: 10118–25

99. Tomashevskii EE. 1971. *Sov. Phys. Solid State*. 12: 2588–92

99a. Prevorsek DC. 1989. In *Encyclopedia of Polymer Science and Engineering*, ed HF Mark, NM Bikales, CG Overberger, G Menges, Suppl. pp. 803–21. New York: Wiley

100. Termonia Y, Meakin P, Smith P. 1986. *Macromolecules* 19: 154–59

Annu. Rev. Mater. Sci. 1995. 25: 325–56

INTERFACES BETWEEN INCOMPATIBLE POLYMERS

Manfred Stamm and Dirk Wolfram Schubert

Max-Planck-Institut für Polymerforschung, Postfach 3148, 55021 Mainz, Germany

KEY WORDS: polymer interfaces, blends, neutron reflection, interfacial tension, ellipsometry

---

ABSTRACT

The interface between incompatible polymers is not very wide and is typically in the range of 2 to 50 nm, depending on compatibility. This article gives an introduction to the theory of interface formation and interfacial tension based on a free energy functional approach. Effects changing the interface width through finite chain length are also discussed theoretically; different expressions are presented and compared. Experimental methods employed for the investigation of polymer interfaces are reviewed with respect to their accuracy and advantages, with emphasis on neutron- and X-ray reflection, ellipsometry, interfacial tension, and nuclear reaction analysis techniques. Results obtained from these methods reveal information about interface formation, and interaction parameter data are compiled for polymeric systems.

INTRODUCTION

Polymer blends are used in increasing areas of application because they are cost effective and have the potential (1) for design of materials with properties tailored to a specific use. Polymer blends, composites, and lightweight reinforced polymer materials are utilized, for instance, in parts of cars or airplanes with specific demands on material properties, which differ from manufacturer to manufacturer. The materials industry has to meet those demands for an increasing spectrum of applications that is impossible to achieve through the development of chemically new polymers.

0084–6600/95/0801–0325$05.00

Thus existing materials are modified or blended, and understanding microstructures and polymer-polymer interactions gains increasing importance.

In this context, the question of polymer miscibility has great relevance (2, 3). Most polymers are incompatible with each other. They form microphase-separated structures when they are mixed together. To obtain good mechanical properties, good cohesion between phases is required, and different materials have to interpenetrate with each other at the interface (4, 5). This interpenetration depends on the degree of miscibility of components and is generally neither easy to determine experimentally nor easy to predict theoretically.

For most incompatible materials, the interface is not very wide—typically in the range of 2 to 50 nm, depending on compatibility. This situation is schematically shown in Figure 1. If two incompatible polymers are put into contact, some interpenetration will occur when the materials are heated above the glass transition. The interface width increases with time, reaching an equilibrium value, which according to mean-field theory is determined by the Flory-Huggins interaction parameters χ. The quantity χ is a measure of polymer compatibility, although it is not well understood on a molecular basis and is often used as an empirical parameter. However, this parameter allows a quite detailed description of thermodynamics of phase separation for a particular polymer pair within mean-field theory (2, 3, 6). From a more detailed and systematic knowledge of χ for different polymeric materials, one could expect to obtain a better understanding of polymer miscibility. Besides equilibrium thermodynamic effects, other factors originating from sample preparation and experimental conditions also influence interface formation. When two films interface, this includes the influence of initial surface roughness, surface composition, and chain conformation at the surface of the films, which may largely depend on sample history. In specific cases, it is difficult to reach equilibrium, for instance, when segment mobility is slowed by the glass transition T_g. It will be interesting to follow the time dependence of interface formation in order to obtain segment mobility and interdiffusion mechanisms. Thus interface formation can be quite complex in specific cases. Because available information is limited, it is important to develop and utilize techniques that can resolve details of the interface between polymeric components.

In this review we discuss the most common techniques used to measure the interface width of incompatible polymeric materials (see Table 1). The interface width for incompatible polymers is typically smaller than 50 nm; thus the resolution of techniques has to be adapted for a determination of interface width and profile (7–9). Besides resolution requirements, a suitable contrast between components has to be present in order to "see" the interface between the components. This contrast can be generated for

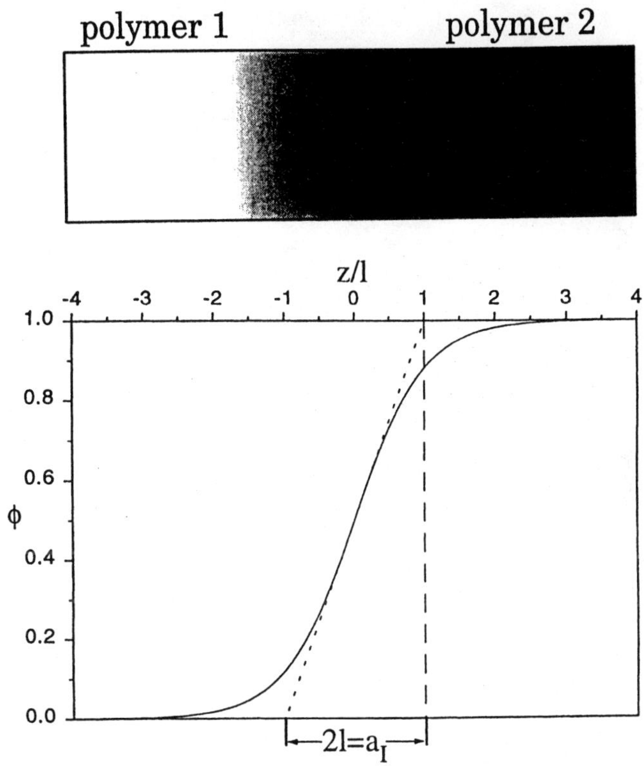

Figure 1 An equilibrium interface is formed during annealing between two incompatible polymers where a particular profile for the volume fraction ϕ exists. z is a coordinate perpendicular to the interface. The origin is located at the inflection point. Shown is the interface width $a_I = 2l$, which may be obtained from a linear approximation in the vicinity of $z = 0$.

neutron reflectometry (NR), for example, by deuteration of one component (10). Thus NR is of general use; and in multicomponent systems, one component and its interfaces with other components can be made visible for neutrons. The application range may be limited by the particular sample geometry needed. Other techniques discussed in more detail below include ellipsometry, small angle scattering, ion techniques, measurement of interface tension or electron microscopy.

We first describe some theoretical aspects of polymer compatibility and interface formation, where in particular a relation between interface width and segment interaction expressed through χ is obtained. Then we compare the most common techniques for the investigation of polymer interfaces between incompatible materials, with emphasis on the measurement of

Table 1 Some common techniques for the investigation of interfaces between polymers with a depth resolution better than 10 nm

Technique	Depth resolution (nm)	Sample	Contrast by	Lateral resolution	Comments
X-ray reflectometry (XR)	0.2	flat, >1 cm	Electron density	—	Requires large contrast
Neutron reflectometry (NR)	0.2	flat, >5 cm	H/D	—	Indirect technique
Ellipsometry (ELLI)	0.1	flat, >100 μm	Refractive index	10 μm	Detailed model assumptions needed
Small-angle X-ray scattering (SAXS)	1	bulk, >1 mm	Electron density	—	Dedicated data analysis necessary
Small angle neutron scattering (SANS)	1	bulk, >5 mm	H/D	—	Dedicated data analysis necessary
Transmission electron microscopy (TEM)	3	thin cut, >1 mm	Staining	3 nm	Dedicated sample preparation necessary
^{15}N-nuclear reaction analysis (NRA)	3–20	flat, >1 cm	H/D	—	Direct technique, depth-dependent resolution
Elastic recoil detection (ERD)	>10	flat, >1 cm	H/D or other elements	—	Direct technique, element specific
Dynamic secondary ion mass spectrometry (DSIMS)	10	flat, >100 μm	Mass difference	10 μm	Matrix effects
Interfacial tension (IT)	—	pendant drop, imbedded fiber, spinning drop	Refractive index	10 μm	Dedicated sample preparation difficult at higher M_r

surface tensions and on NR. Because several recent reviews already describe different techniques (see e.g. 7–10), we only briefly describe those techniques here. There are not many examples of detailed investigations of incompatible polymer systems reported in the literature. However, some recent examples are given and the potential of the techniques illustrated. We generally do not cover the area of diblock copolymers and the effect of compatibilizers at interfaces. When comparing experimentally determined interface widths with calculated ones from known interaction parameters, e.g. from interfacial tension data, one must be cautious, because there are influences from end-group effects, molecular weight distributions, chain orientation, capillary waves, or initial interface roughness that are not very well understood. We also assume that additives like antioxidants or plasticizers are not present, since those materials might migrate to the surface or interface even if present only in trace amounts.

THEORY

The description of phase behavior in polymer blends is of fundamental interest in polymer research. A first approach is based on Flory, Huggins, and Staverman (FHS) theory (11a–e). Using a lattice model, a simple form for the free energy of binary polymer blends including the interaction parameter is derived

$$\frac{F_{FHS}}{kT} = \frac{\phi \ln \phi}{N_1} + \frac{(1-\phi)\ln(1-\phi)}{N_2} + \chi\phi(1-\phi). \qquad 1.$$

N_1 and N_2 are the degrees of polymerization. The condition

$$\phi_1 + \phi_2 = 1 \qquad 2.$$

for the volume fraction of the two polymers is implied by use of $\phi_1 = \phi$.

Many theories have been derived to describe the interaction parameter as a function of temperature to reproduce phase diagrams observed in different experiments. For complicated phase diagrams, it may be necessary to introduce a ϕ-dependence of χ in addition (12a–c). For the following theoretical description of the interface between two incompatible polymers, a volume fraction-independent interaction parameter χ is used. In order to discuss the problem of an interface between two polymers, Equation 1 has to be extended by the so-called square gradient term $\kappa(\nabla\phi)^2$ introduced by Cahn & Hilliard (13) to take fluctuations into account. Several authors (14–18) derive an expression for the square gradient term with a prefactor κ including a ϕ-dependence. Due to different approximations, there are different prefactors. A discussion of the prefactor is also given in Reference 19. For the case, where structures are in the range of the radii of gyration (R_g) of components or smaller ($qR_g > 1$), one obtains a prefactor

$$\kappa = \frac{a^2}{24\phi(1-\phi)}. \qquad \text{3a.}$$

For structures that are large compared to the radii of gyration ($qR_g \ll 1$), a different prefactor is derived

$$\kappa = \frac{a^2}{36\phi(1-\phi)}, \qquad \text{3b.}$$

a is the characteristic segment length assumed to be the same for both polymers, and q is the magnitude of the scattering vector, $q = (4\pi/\lambda)\sin\Theta$ with λ wavelength and 2Θ scattering angle.

Since the interface region between incompatible polymers is mostly of the order of the radii of gyration, we derive the following free energy functional of the mixture augmented by the square gradient term on the basis of Equations 1 and 3a

$$\frac{F}{kT} = \int_V \left\{ \frac{F_{FHS}}{kT} + \frac{a^2}{24\phi(1-\phi)} (\nabla\phi)^2 \right\} dV. \qquad \text{4.}$$

For the case of a planar layer system, the volume fraction varies only in z direction (see Figure 1) with the consequences that

$$\int_V dV \to \int_{-\infty}^{+\infty} dz \quad \text{and} \quad \nabla \to \frac{d}{dz}:$$

$$\frac{F}{kT} = \int_{-\infty}^{+\infty} \left\{ \frac{F_{FHS}}{kT} + \frac{a^2}{24\phi(1-\phi)} \left(\frac{d\phi}{dz} \right)^2 \right\} dz. \qquad \text{5.}$$

Minimization of Equation 5 leads to the equilibrium volume fraction profile. An exact analytical solution for the volume fraction profile Equation 6a, can be calculated for the case of infinite degrees of polymerization ($\chi N \gg 1$) using Euler-Langrange equation (13),

$$\phi(z) = \tfrac{1}{2}\left(1 + \tanh\frac{z}{l}\right) \qquad \text{6a.}$$

with

$$l = \frac{a}{\sqrt{6\chi}}. \qquad \text{6b.}$$

The boundary conditions to be satisfied are

$$\phi(-\infty) = 0 \quad \text{and} \quad \phi(+\infty) = 1. \tag{6c.}$$

One can also find the additional condition of a vanishing derivative for $z = \pm\infty$. With the other prefactor, Equation 3b, only a change in the magnitude of l occurs (20).

$$l = \frac{a}{\sqrt{9\chi}}. \tag{6d.}$$

Following the conventional notation (see also Figure 1), we use the parameter a_I for the total interface width where

$$a_I = 2l. \tag{6e.}$$

The influence of finite chain length on interface formation has been considered by several authors (14, 15, 21–24). Three different methods yield slightly different expressions for the interface width l with finite chain length.

1. A simple way to estimate the influence of finite chain length is possible for the case of two polymers with the same degree of polymerization N. Expanding Equation 1 in the vicinity of $\phi = 1/2$ to terms of second order yields

$$\frac{\phi \ln \phi + (1-\phi)\ln(1-\phi)}{N} + \chi\phi(1-\phi) \approx \frac{\chi}{4} + \frac{\ln\frac{1}{2}}{N} - \left(\chi - \frac{2}{N}\right)(\phi - \tfrac{1}{2})^2. \tag{7.}$$

   Minimization is not influenced by constant terms, and χ can be replaced by $\chi - 2/N$ in Equation 6b to obtain:

$$l = \frac{a}{\sqrt{6\left(\chi - \frac{2}{N}\right)}}. \tag{8.}$$

2. Another approach proceeds by integrating Equation 5 in the vicinity of $\phi = 1/2$, i.e.

$$\int_{-\infty}^{+\infty} dz \rightarrow \int_{1/2-\varepsilon}^{1/2+\varepsilon} 2l \cdot d\phi,$$

   using the linear approximation

$$\phi(z) = \tfrac{1}{2}\left(1 + \frac{z}{l}\right).$$

This is a good approximation of the profile

$$\phi(z) = \tfrac{1}{2}\left(1 + \tanh\frac{z}{1}\right)$$

for small z. One obtains an equation that yields after minimization

$$1 = \frac{a}{\sqrt{6\left(\chi - 2\ln 2\left(\frac{1}{N_1} + \frac{1}{N_2}\right)\right)}}. \qquad 9.$$

This equation is similarly derived by Broseta et al (14).

3. Using Equation 6a as a trial function (21) to minimize Equation 5, one finds

$$1 = \frac{a}{\sqrt{6\left(\chi - \frac{\pi^2}{6}\left(\frac{1}{N_1} + \frac{1}{N_2}\right)\right)}}. \qquad 10.$$

The comparison of these results, Equations 8, 9 and 10, for the particular case of $N_1 = N_2 = N$, reveals that they are different in the prefactor of the $1/N$-term, ranging from 2, $4\ln 2$ (=2.77) to $\pi^2/3$ (=3.29). Equation 8 only proves useful for estimating the influence of finite chain length in a relatively rough manner. Note that Equation 9 is exactly valid only in the vicinity of $\phi = 1/2$. Equation 10, involving the whole z range, seems to be the best way to describe the effect of finite chain length on the interface width l. For most experiments, taking experimental errors into account, the difference between Equations 9 and 10 is not relevant. It is necessary to note that in the derivation of Equations 9 and 10, the same boundary conditions that are used for the case of infinite chain length are utilized. This is still another approximation. To calculate how boundary conditions would have to be modified with chain length, we have to determine the minima of Equation 1 (see Figure 2), which describe the phase equilibrium simplified for the case of $N_1 = N_2 = N$.

With the minima located in the vicinity of $\phi = 0$ and 1, expressions for the modifications of the boundary conditions can be derived (14)

$$\text{for} \quad z \to -\infty \quad \phi = 0 \to \phi = \exp(-\chi N), \qquad 11a.$$

$$\text{for} \quad z \to +\infty \quad \phi = 1 \to \phi = 1 - \exp(-\chi N). \qquad 11b.$$

Equations 11a, 11b, and Figure 2 show clearly that for the case of $\chi N \gg 1$, a modification of the boundary conditions is not essential. Furthermore,

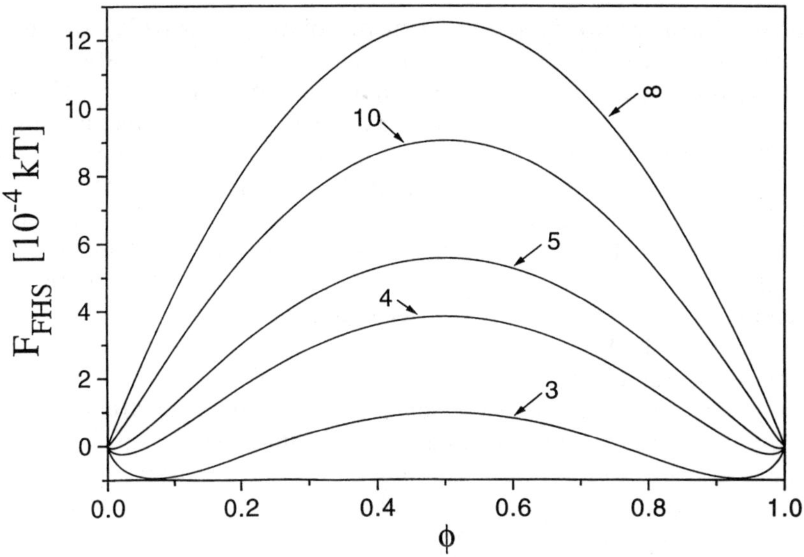

Figure 2 The Flory-Huggins-Staverman free energy F_{FHS} vs the volume fraction ϕ for different χN values, which are indicated as parameters at the individual curves. An interaction parameter of $\chi = 0.005$ is used.

Figure 2 shows that for $\chi N \gg 1$, it is mostly the curvature at $\phi = 1/2$ that changes with variation of χN. This fact can be understood on the basis of the expansion used in Equation 7. Thus, for a not too small chain length, only the influence of finite chain length to the interface width l is significant. The extensions derived for the interface width l for finite chain length in Equations 8, 9, and 10 are equivalent to the introduction of a critical interaction parameter χ_c, which corresponds to the χ-value where l diverges. The condition for the application of the approximations, i.e. $\chi \gg \chi_c$, which is equivalent to $\chi N \gg 1$, still remains.

The fundamental result of the square gradient theory applied to the description of the interface between incompatible polymers is the direct connection of the interfacial width with the interaction parameter χ.

$$l \propto \chi^{-0.5}. \qquad 12.$$

In the vicinity of the critical region, a different exponent of -0.63 is proposed (22). Approaching the spinodal, an interpolation scheme is suggested (15), and deviations from the simple tanh-profile, Equation 6a, caused by limitations of square gradient theory, are discussed (20). Two different length scales are introduced.

Other approaches are based on the lattice theory (23, 24), a random potential treatment (25), the gradient percolation theory (26), or Monte Carlo techniques (27, 28, 30). In particular, computer simulations allow the incorporation of interactions and dynamics in a versatile way (30) and reveal conformation effects as well as the influence of fluctuations or capillary waves, which are usually not included in analytic treatments. An analytical treatment of compositional fluctuations at an interface (29) reveals that these corrections may be significant in some cases, while chain end corrections, disregarding different chemical natures of chain ends, are small for incompatible blend systems (AN Semenov, submitted). Corrections of this sort are, on the other hand, much more severe for block copolymers (31, AN Semenov, submitted). The analysis of experimental data on the basis of Equations 8, 9 and 10 still has to be performed with care because most experimental techniques measure an apparent interface width, which due to lateral averaging, also contains the influence of lateral inhomogeneities, initial interfacial roughness from preparation conditions, and capillary waves or compositional fluctuations (see e.g. 32), as discussed above.

Also not included in most theoretical treatments is the specific effect of chain ends, which may be enriched at the interface and which, in general, have a chemical nature that is different from the rest of the chains, including possibly some groups from the catalyst. Besides having different mobilities, chain ends can change the interfacial tension, which has been demonstrated in some examples (33, 34). The presence of a broad molecular weight distribution will also result in a broadening of the interface, particularly if low-molecular weight components are present. This effect can be estimated on the basis of Equations 8 and 10 and may lead to a fractionation of components at the interface (28).

EXPERIMENTAL TECHNIQUES

There are several experimental techniques that provide sufficient resolution to investigate the relatively narrow interfaces between incompatible polymers. Some of the most common techniques are summarized in Table 1 and are reviewed below.

Interfacial Tension

The appearance of an interfacial tension at the interface between different materials is an expression of different energetics of bulk materials. It thus reflects differences in thermodynamics and can be related to the interaction parameter . A compilation of surface and interfacial tensions of polymers can be found in References 35, 36, and recent discussions are given in References 8, 37. Several authors (14, 15, 19, 38–40) have derived equations that show the connection between interfacial tension γ and the interaction parameter χ, where the effect of finite chain length is also included. The mean-field theory of interfacial tension between two incompatible polymers is similar to the theory of interface width as discussed above. For the case of infinite chain length ($\chi N \gg 1$), the interfacial tension γ is related to the interaction parameter χ and temperature T in the following way:

$$\gamma = \frac{kT}{a^2}\left(\frac{\chi}{6}\right)^{1/2}. \qquad 13.$$

This equation holds for $qR_g > 1$, and similar comments hold as discussed in relation to Equations 3a and 3b. To include the effect of finite chain length, one has to replace χ, as discussed in Equations 8 through 10. This yields the following expressions for γ:

$$\gamma = \frac{kT}{a^2}\left(\frac{\chi - \frac{2}{N}}{6}\right)^{1/2}, \qquad 14a.$$

$$\gamma = \frac{kT}{a^2}\left[\frac{\chi - 2\ln 2\left(\frac{1}{N_1} + \frac{1}{N_2}\right)}{6}\right]^{1/2}, \qquad 14b.$$

$$\gamma = \frac{kT}{a^2}\left[\frac{\chi - \frac{\pi^2}{6}\left(\frac{1}{N_1} + \frac{1}{N_2}\right)}{6}\right]^{1/2}. \qquad 14c.$$

Equations 14b and c have been similarly derived by other authors (14, 15, 40). Restrictions as discussed with respect to Equations 8 through 10 are also applied here. Several other functional forms of interfacial tension have been suggested. Thus lattice theory (23, 24) yields a $\chi^{0.75}$ dependence and density functional theory (15) with a more complex dependence on χ. Experimental results on the other hand, are more consistent with mean-field theory, taking chain end effects into account (40), which yields

$$\gamma = \gamma_\infty\left[1 - \left(\frac{\text{const.}}{\chi N}\right)^{\alpha}\right]. \qquad 15.$$

γ_∞ is the interfacial tension for infinite molecular weight, and $\alpha = 1$ in the asymptotic range. In some experiments, however (19, 41), an exponent of $\alpha = 2/3$ is reported for low-molecular weight samples.

Monte Carlo calculations, on the other hand, show (42) that α increases continuously with chain length and can be approximated at low-molecular weights by $\alpha = 2/3$, while it approaches $\alpha = 1$ at high-molecular weights.

The experimental determination of interfacial tension with polymeric melts is not easy because of problems connected with sample preparation and equilibration. Several techniques have been proposed (for reviews see e.g. 8, 35, 43). An elegant method to determine the interfacial tension is to analyze the shape of a pendant drop (Figure 3*a*) of a polymer in a surrounding melt of another polymer. The equilibrium profile of the pendant drop can be expressed by the set of equations (44),

$$\frac{d\varphi}{dS} = \frac{2}{B} + Z - \frac{\sin\varphi}{X}, \quad \frac{dX}{dS} = \cos\varphi, \quad \frac{dZ}{dS} = \sin\varphi, \qquad 16.$$

which is solved with the boundary conditions applied at the drop apex, $X(0) = Z(0) = \varphi(0) = 0$.

The dimensionless variables are defined as $X = x\sqrt{c}$, $Z = z\sqrt{c}$, $S = s\sqrt{c}$ and $B = b\sqrt{c}$ with $c = (\Delta\rho g/\gamma)$. The shape of the drop is specified in terms of x and z coordinates; the arc length s; the angle ψ between a tangent to the drop profile and the horizontal axis at a point (x, z); the

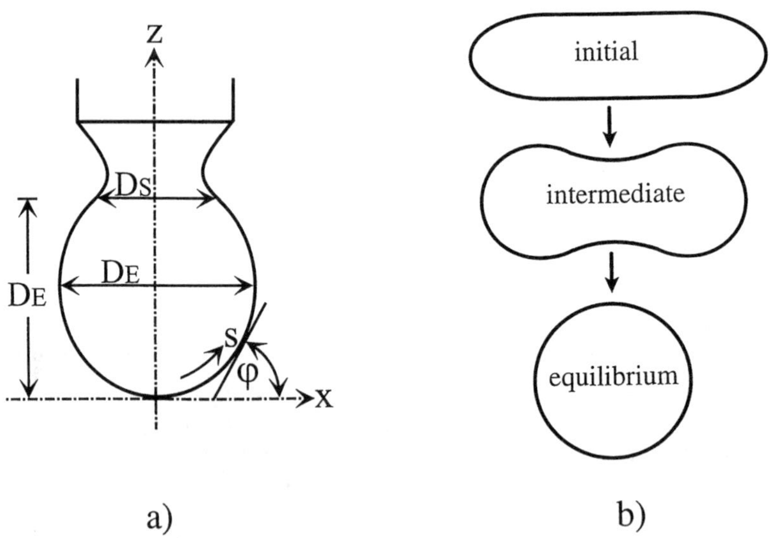

Figure 3 Sketch of (*a*) a pendant drop and (*b*) an imbedded fiber at different relaxation states for interfacial tension measurements. Drop and fiber are embedded in a matrix material. The definition of the indicated parameters is explained in the text.

radius of curvature b at the drop apex; the gravitational constant g; and the mass density difference across the interface $\Delta\rho$.

Different setups have been developed to determine the surface and interfacial tension from the shape of pendant drops. Figure 4 schematically shows a typical experimental setup (45). The video signal of the pendant drop is transmitted to a digital video processor, which performs the frame gripping and digitization of the drop image. A computer is used to acquire the image and perform the image analysis. The accuracy of measurements with the pendant-drop method depends on the algorithm used for the analysis of the drop shape. Using a simple method for the analysis involving only two specific diameters DS and DE (47) (see Figure 3*a*), an accurate measurement of the interfacial tension γ of about 1 mN/m is obtained. The analysis of the full drop profile by fitting the data to Equation 16 gives values with an accuracy of ± 0.1 mN/m.

Classical equilibrium interfacial tension experiments like the pendant-drop technique are frequently too difficult to apply to high polymers because of their high melt viscosities (10^3–10^6 Pa s). The dominant factors limiting equilibrium experiments on high-viscosity materials are (*a*) the inability to effectively ascertain when equilibrium is reached; (*b*) prohibitively long equilibration times; and (*c*) the equilibration time frequently exceeds the thermal-degradation time of the resins. These problems are partially overcome by dynamic techniques such as the spinning top method (48).

The imbedded fiber retraction (IFR) (49) technique is another dynamic method for measuring the interfacial tension between molten polymers. For an IFR experiment, a fiber of a specific polymer is embedded in a matrix of another polymer. Interfacial free energy causes the initial cylin-

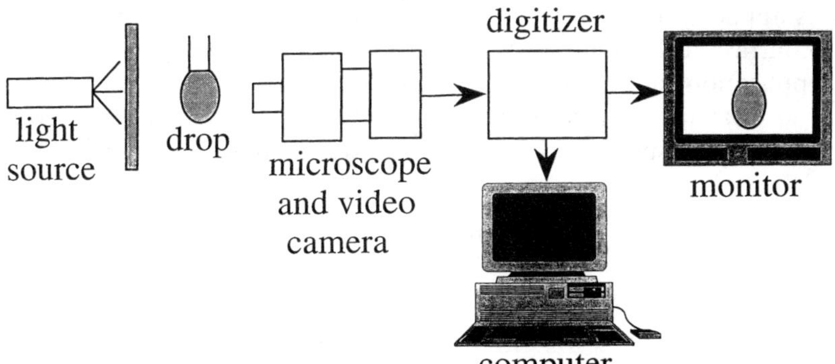

Figure 4 Schematic view of a video-enhanced pendant drop experiment.

drical fiber to retract through a series of shapes (Figure 3*b*). The absolute minimum of the free energy is reached for the spherical shape.

The advantage of IFR lies in its ability to provide measurements of interfacial tension during this minimization process. The experimental set-up is quite similar to the pendant-drop technique. IFR enables determination of interfacial tension much more quickly than equilibrium experiments and allows study of high polymers before they degrade. The fiber material thus may have a relatively high viscosity, whereas matrix material has a reasonably low viscosity. The retracting fiber is characterized at any time t during the retraction by measurement of the fiber length L and the diameter D. These data are then reduced according to the theory developed in Reference 49.

Reflection Experiments with Neutrons and X-rays

NR is probably the most versatile technique used to investigate accurately the interfaces between immiscible polymers. It provides optimal resolution for the interface width a_I, typically of the order of 0.2 nm in the range of $a_I = 2$ to 50 nm. It can also be applied to practically all polymeric systems, provided one component can be deuterated to achieve a contrast at the interface. However, it needs a dedicated sample preparation, where relatively large (typically $5 \times 10\ cm^2$) thin films are deposited on top of each other. Interfacial roughness (typically $\gtrsim 1$ nm) also limits the resolution in some cases. Non-uniqueness of data analysis, which might be a problem in more complicated cases of the application of the technique, is not a problem for the analysis of a double-layer system (Figure 5) of two incompatible polymer films because the starting situation is very well defined. One obtains information on both interface width a_I and form of the profile $\phi(z)$. X-ray reflectometry, on the other hand, is largely limited in its application to polymeric systems because for most polymer pairs, the contrast at the interface is not sufficiently large enough.

Applications of the techniques to polymers have recently been reviewed (10, 50–52) and we describe here only some essential points. Analogous to optics, one can specify a refractive index for neutrons and X-rays:

$$n = 1 - \frac{\lambda^2}{2\pi} b\rho_n - i\frac{\lambda}{4\pi}\mu_n \text{ (neutrons)}, \qquad 17.$$

$$n = 1 - \frac{\lambda^2}{2\pi} a_e\rho_e - i\frac{\lambda}{4\pi}\mu_x \text{ (X-rays)}. \qquad 18.$$

$b\rho_n$ is the neutron scattering length density (b is the scattering length, ρ_n

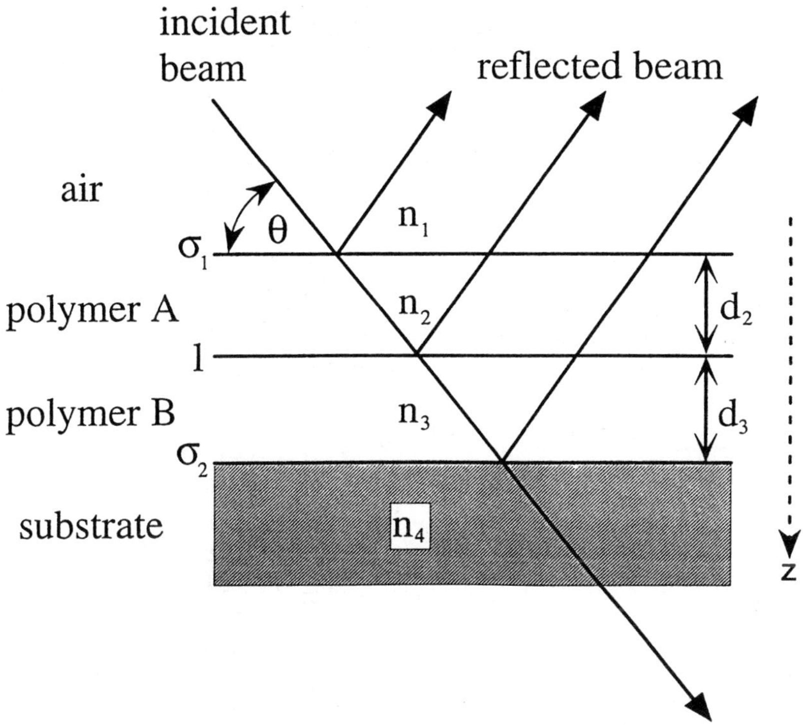

Figure 5 Schematics of a neutron or X-ray reflection experiment. The fit parameters, which define the refractive index profile as explained in the text, are shown.

the particle number density); μ_n is the linear absorption coefficient for neutrons; $a_e\rho_e$ is the electron scattering density (a_e is the classical electron radius, ρ_e the electron density); μ_x is the linear mass absorption coefficient for X-rays; and λ the wavelength. Note that $b\rho_n$, μ_n, ρ_e and μ_x are defined for a specific atomic species, and for the application to a particular material, one has to calculate the mean values involving the macroscopic mass density and the mass of the monomer. The interaction between matter and neutrons (or X-rays) is generally weak, and the refractive index n is very close to $1(\sim 1–10^{-6})$. For neutrons, absorption is negligible. For X-rays it is still small for most polymers ($n \sim 1–10^{-6}-\mathrm{i}10^{-8}$). To better describe the contrast problem, the scattering densities for neutrons and X-rays for some materials are given in Table 2. It is evident from a comparison of the deuterated and protonated materials that a large contrast can be obtained for neutrons by deuteration. In the case of X-rays, contrasts even for significantly different polymers are not large, a situation also depicted

Table 2 Comparison of characteristic data for the reflection of X-rays and neutrons at interfaces[a]

Compounds	Chemical formulas	ρ (g cm^{-3})	$a_e\rho_e$ (10^{10} cm^{-2})	$b\rho_n$(H/D) (10^{10} cm^{-2})	μ_x (cm^{-1})	μ_n(H/D) (cm^{-1})
Air/vacuum		0	0	0	0	0
PS	$(C_8H_8)_n$	1.03	9.5	1.41/6.46	4	3.9/0.6
PMMA	$(C_5H_8O_2)_n$	1.15	10.6	1.06/7.02	7	4.5/0.6
PVC	$(C_2H_3Cl)_n$	1.38	12.1	1.56/5.76	86	3.3/0.5
PBrS	$(C_8H_7Br)_n$	1.57	13.2	1.76/5.58	97	4.0/0.6
Quartz	SiO_2	2.3	19.7	4.20	85	0.2
Silicon	Si	2.33	19.7	2.15	141	0.1
Nickel	Ni(^{58}Ni)	8.91	72.6	9.25 (12.9)	407	2.6
Gold	Au	19.32	131.5	4.48	4170	3.3

[a] Mass density ρ; electron scattering densities $a_e\rho_e$; neutron scattering length densities $b\rho_n$ for protonated (H)/deuterated (D) compounds; linear absorption coefficients for X-rays and neutrons, μ_x ($\lambda = 0.154$ nm) and μ_n ($\lambda = 0.1$ nm). The two values in some of the columns correspond to the H- and D- materials, respectively.

in Figure 6. Thus X-ray experiments are possible for the investigation of some particular incompatible polymer systems, but NR proves to be a much more versatile technique due to the possibility of generating a large contrast between components by deuteration. It should, of course, be kept in mind that deuteration also changes thermodynamics slightly, which for strongly incompatible systems is usually negligible.

A typical sample geometry is shown in Figure 5. A thin film of polymer A has been deposited by the floatation technique (10) on top of a film of polymer B. Typical film thicknesses are in the range from 10 to 300 nm, and to achieve a reasonable smoothness and surface roughness (1–2 nm), spin coating techniques are used. Suitable substrates for film deposition are float glass, silicon wafers, or polished flat surfaces. A typical lateral sample size is from 2 to 10 cm in each direction.

The reflectivity as a function of the angle of incidence θ can be calculated from the refractive index profile $n(z)$, using a matrix formalism (53). Relevant parameters are indicated in Figure 5. Error functions characterized by their variances σ_1^2 and σ_2^2, respectively, are usually employed to describe the roughnesses of the interfaces air/polymer A and polymer B/substrate. A tanh-refractive index profile with the characteristic parameter l (see Equation 6a) is commonly used between polymer A and B. Finally, the thicknesses d_2 and d_3, and the refractive indices n_1 to n_4 are parameters that may be obtained by separate measurements prior to the experiment for the determination of the interface width l.

Because the determination of a reflection curve takes from 2 to 12 h,

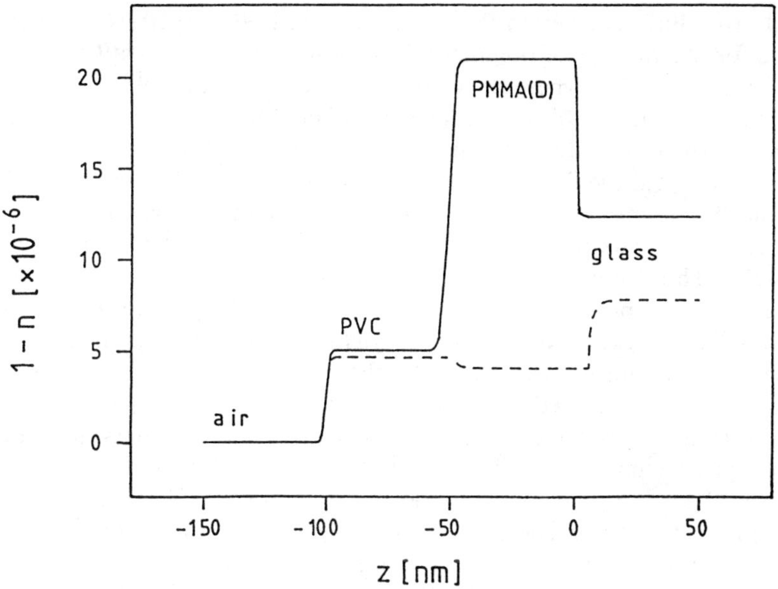

Figure 6 Refractive index profile for the case of a neutron (*solid line*) or X-ray (*dashed line*) reflection experiment according to the sample geometry shown in Figure 5. Typical values for the blend system PVC(H)/PMMA(D) on a float glass substrate are given. The contrast at the interface between the polymers is much larger for neutrons than for X-rays.

samples are mostly annealed outside the reflectometer and then quenched for the neutron or X-ray experiments below the glass transition temperatures of the materials. Because of the slow interdiffusion process of segments at the interface, samples sometimes have to be annealed for several days (depending on segment mobility) before the equilibrium interface width is achieved. It is important to note that (specular) reflectivity techniques do not provide lateral information, and reflection curves are averaged laterally over the sample (lateral coherence length is typically several micrometers). There is in principle the possibility of obtaining lateral information from off-specular scans; however, in most experiments this is not done for intensity reasons. With incompatible materials, one has to be careful with lateral inhomogeneities, for instance, the dewetting phenomenon, which might occur during annealing (see e.g. 54).

Ellipsometry

In ellipsometric experiments, the change of the polarization state of light during the reflection from an interface is analyzed. A review of techniques and procedures is given in References 55, 56. Ellipsometry is successfully used for the analysis of polymer adsorption and surface modification (for reviews see 46, 57), where the change in layer thickness below 0.1 nm can be resolved in favorable cases. A contrast between components is obtained through the difference in the indices of refraction. A principle limitation originates from the fact that only two quantities, the ellipsometric angles Δ and Ψ, are measured mostly at one wavelength and one angle of incidence, which cannot reveal detailed information about the investigated system. Thus monochromatic ellipsometry is relatively insensitive to details of the profile during adsorption from solution (58), and step, parabolic, or exponential profiles yield identical results if the same adsorbed amount is used. Recently, the technique has been employed to investigate the interface width between polymers.

The change in polarization state of polarized light after passing through a sample, which is studied either in reflection or transmission, is characterized by the two ellipsometric angles Δ and Ψ. They describe the complex ratio of the reflectives for *s*- and *p*-polarized light, r_s and r_p, that is for light with its plane of polarization perpendicular (s) and parallel (p) to the plane of incidence, respectively,

$$\frac{r_s}{r_p} = e^{i\Delta} \tan \Psi. \tag{19.}$$

Ellipsometry at fixed angle and wavelength usually yields only a thickness and a refractive index for an assumed uniform layer. At low contrast or very thin samples, only the optical thickness can be obtained. Using the same kind of sample as described for neutron and X-ray reflection experiments (Figure 5), additional information about the original polymer films, i.e. their refractive indices and thicknesses, is necessary for the investigation of the interface width between two polymers. A problem arises if film thicknesses and refractive indices change during annealing at the same time as the interface width changes (59). A maximum sensitivity for the ellipsometric angles is achieved using a wavelength where the refractive indices of the two polymers have a large difference and at an angle of incidence equal to the Brewster angle determined by the first interface between air and polymer. The development of variable-angle spectroscopic ellipsometry (VASE) (60) has provided much more detailed information and yields independent data on the sample and that offers additional means for the examination of multilayer samples, using a fit over the entire wavelength and angular range. Ellipsometry, on the other hand, has the same fundamental restrictions as neutron and X-ray reflectometry, i.e. direct inversion of the measured data is usually not sufficient to obtain the spatial variation of the refractive index profile, and model predictions are required to interpret the data. With a microfocus beam, a lateral resolution of the order of 10 μm can be achieved.

Other Techniques

Several other techniques are also used for the investigation of interfaces between immiscible polymers (see Table 1). Thus from an analysis of the Porod tail in small angle X-ray and neutron scattering (SAXS, SANS) (for reviews, see e.g. 7, 51, 91), information on the width and profile of the interface can be obtained. SAXS and SANS are only applied in specific cases because one generally has to analyze a small signal on a large background. On the other hand, they can be used for interface analysis with standard thick samples, which is similarly true for spin diffusion nuclear magnetic resonance experiments (92).

Transmission electron microscopy (TEM) can be applied to the investigations of interfaces if a thin cut perpendicular to the interface is produced. A contrast between components is then achieved by staining or element-specific discrimination (see e.g. 35, 61, 93). However, resolution for interfaces is also limited due to dedicated preparation and cutting techniques.

For ion beam applications with polymers, the ^{15}N-nuclear reaction analysis technique (7, 62) reaches the best resolution, which is 3 nm at the surface and 12 nm at a depth of 100 nm within the sample. Radiation damage is not a severe problem with most materials, and hydrogen depth profiles are obtained without dedicated data analysis or fitting. Similar to NR application, one component should be deuterated, which then, in a layer system, can be discriminated from the other hydrogenated components. Elastic recoil detection (ERD) (7, 63) is also sensitive to the contrast due to deuteration although it provides a resolution no better than 10 nm. The narrow interface widths between incompatible polymer materials are thus hard to resolve accurately with those ion beam techniques.

The same is true for dynamic secondary ion mass spectrometry (7, 9, 64), where the materials are sputtered at a constant rate and mass fragments are analyzed. Because different materials have characteristic fragmentation spectra, a discrimination between components is achieved without specific tagging. There are still more techniques available, in particular for the analysis of thicker interfaces, that in most cases cannot be applied to the analysis of very thin interfaces between immiscible polymers.

In terms of rank of techniques for the analysis of the very thin interfaces between immiscible polymers, NR is probably most versatile and most accurate. Care has to be taken during sample preparation and data analysis, which is sometimes not straightforward. For low-molecular weight materials, for example, interfacial tension measurements are best suited; however, they do not yield the interface width directly and provide no information on the profile, whereas all the other techniques can be applied only in very specific cases, reveal a limited depth resolution, or are problematic with respect to absolute interface width determination. Specific examples are discussed in the following section.

INTERFACES BETWEEN INCOMPATIBLE POLYMERS

In spite of the widespread use and application of incompatible polymer blends, only a limited number of experimental studies have been reported. This is probably due to the fact that the narrow interface widths are hard to determine experimentally. Values reported from different techniques are then not always compatible with each other and are scattered over a wide range. We discuss some of the more recent investigations in more detail, where the effects of compatibility, glass transition, and temperature on interface width have been studied for several systems. A comparison of data is given in Table 3.

Polystyrene(PS)/Polymethylmethacrylate (PMMA)

One of the most intensively studied incompatible blend systems is PS/PMMA. Values of interface widths $a_I = 2l$ determined with different techniques are compiled in Table 3. The most accurate value has been obtained by Anastasiadis et al (65) with NR. It has been confirmed by these authors from other NR studies with diblock copolymers, where the interface width between lamellae is found to be of the same size. The value of $a_I = 5.0 \pm 0.5$ nm is also consistent with NR data of Fernandez et al (66) after suitable recalculation of their original values (see 65). Also, temperature-dependent ellipsometry investigations (67) and TEM (68) yield similar results within error bars. ERD results (J Kressler et al, submitted), on the other hand, have to be taken with care because first, the profile is convoluted by the resolution function, which actually has a width of typically 12 nm and is thus much larger than the investigated interface width, and second, the oxygen profile is recorded, which might be influenced by contaminations of e.g. water at the interface. One of the problems during a comparison of the data arises from the fact that in some cases the temperature of the experiment is not clearly defined. Samples for NR and TEM, for instance, have been annealed at the indicated temperatures but are cooled down to room temperature for the experiments. Exact procedures are usually not given.

A serious discrepancy, however, arises when the interface width is calculated from interfacial tension data (19, 70) on the basis of the mean-field Equations 6b and 13. Those interface widths are approximately a factor of two smaller than the measured ones. This might indicate that the theoretical description is still not complete, or alternatively, that other effects might significantly contribute to experimental values, which are not properly considered. This could include end-group effects, conformational changes, capillary waves, or surface contaminations. End corrections and concentration-fluctuation effects are considered in a recent calculation of AN Semenov (submitted), who concludes that thermal fluctuations of the concentration at the interface can contribute as much as 2 nm to the interface width. In this approximation, the coherence length of the reflection experiment enters, and results thus depend on experimental conditions. The calculated value of 4.9 nm agrees well with the experimentally determined interface width (Table 3). The agreement is apparently also good with a calculated interface width value obtained from a temperature dependent χ-value from SANS experiments of diblock copolymers (71) on the basis of Equation 10. Here, the much lower molecular weight of the sample has to be taken into account, and the agreement is not as good when it is performed correcting for the higher molecular weights.

PS/Poly(n-butylmethacrylate) (PnBMA)

The temperature dependence of the interface width of PS/PnBMA has recently been investigated by NR (21). Neutron reflection curves are shown in Figure 7 together with best fits based on a simple two-layer model, where the parameters as indicated in Figure 5 are used. The PS-film is deuterated to achieve good sensitivity for the interface between the two polymer films. Because the interface width can be determined with an accuracy of ± 0.2 nm, the temperature dependence of a_I is resolved and can be correlated on the basis of Equation 10 to a temperature-dependent χ-parameter. Data agree well with SANS data from compatible blends of PS with a statistical copolymer of PnBMA and poly(cyclohexylacrylate) of different compositions (21). It is predicted from experiments (72) that PS/PnBMA should be compatible at high temperatures or low-molecular weights.

PS/Poly(styrene-stat-para bromo styrene) (PBr_xS)

The blend system of PS and the statistical copolymer poly(styrene-stat-para bromo styrene) was chosen for systematic investigations of the depen-

Table 3 Experimentally determined interface widths of different incompatible polymer blend systems

	a_I (nm)	T (°C)	$M_r(1)$ (10^3 g mol^{-1})	$M_r(2)$ (10^3 g mol^{-1})	Technique	Comments	Reference
PS/PMMA	5±1	170	220	19	NR	—	66
	5.0±0.5	170	110	107	NR	—	65
	3–9±1	140–170	180	151	ELLI	—	67
	5±3	140	180	151	TEM	—	68
	8	140	180	151	ERD	Error not specified	J Kressler et al, submitted
	(2.6)	190	2–200	24	IT	Calculated from γ_∞	70
	(2.4)	199	2–43	10	IT	Calculated from γ_∞	19
	(4.9)	—	110	107	IT	Calculated from γ_∞ with fluctuation and end corrections	30
	(5.8)	150	13	14	SANS	Calculated from $\chi(T)$ of diblockcopolymer	71
PS/PnBMA	6.4–8.6±0.2	120–156	720	150	NR		21

Table 3—*continued*

PS/PBr$_x$S	4.3–9.6 ± 0.6	120–145	160	160	NR	Degree of bromination $x = 0.16$–1.0	73
	3.9–6.7 ± 0.6	142	100–1450	100–1450	NR, XR	$x = 0.84$–0.97	75
	3.5–4.9 ± 0.4	114–144	117	47	XR	$x = 1.0$	74
PB/PI	260 ± 3	33	58	101	LS		78
PMMA/SAN	5–31 ± 3	130	151	21–259	ELLI	Acrylonitrile content of SAN $w = 0.057$ to 0.55	76
	26–32 ± 3	140–170	151	74–259	ELLI	$w = 0.057$, 0.387	67
	2–62 ± 3	130	14–821	74–259	ELLI	$w = 0.057$ to 0.387	77
	32 ± 3	130	151	74	TEM	$w = 0.387$	68
	49–68	140	151	74	ERD	$w = 0.387$	J Kressler et al, submitted
PC/SAN	4.5	165	19	160	NR	$w = 0.25$	90
Nylon/SMA	12–25 ± 3	170–210	15–20	119–223	ELLI	Maleic anhydride content of SMA $w = 0.08$–0.25	87
Nylon/NBR	48–79 ± 3	250	25	105	ELLI	Hydrogenated nitrile rubber with compatibilizers	88
PMMA/SCPE	< 50	147–181	107	100	RBS	SCPE 66% chlorinated	86
	14–23	120	37	207	NR	SCPE 66% chlorinated	84
PS(H)/PS(D)	80–120	150–170	2800	920	NRA	—	82, 83
PB(H)/PB(D)	50	−30–85	250	227	SANS, LS	—	79, 81

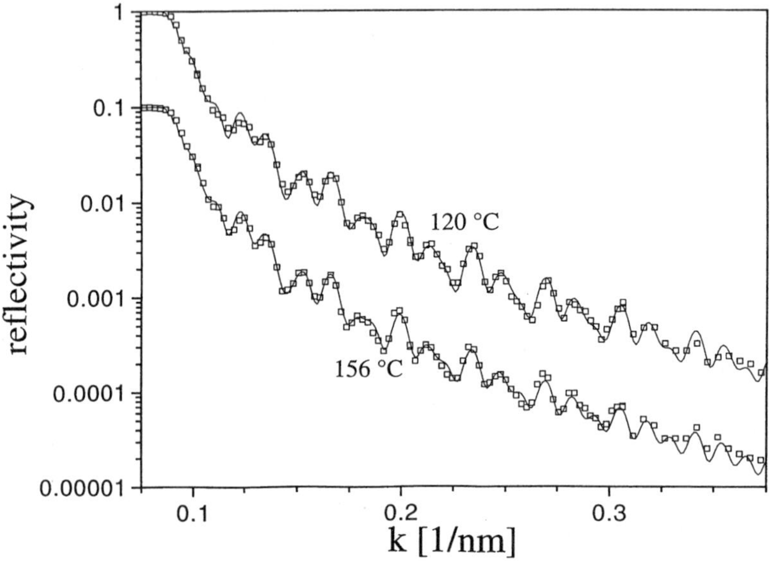

Figure 7 Neutron reflectivity curves of PS(D)/PnBMA(H) on a glass substrate at different temperatures (21). The solid lines correspond to fits assuming an interface width of $a_I = 6.4$ nm (120°C, upper curve) and $a_I = 8.6$ nm after annealing (156°C, lower curve).

dence of interface width on compatibility because the compatibility between components can be easily tuned from completely compatible to highly incompatible by a change in the degree of bromination x. The molecular weight distribution of both partners can be narrow, and at a high degree of bromination, the contrast between components is sufficient also for X-ray reflectivity experiments. At a lower degree of bromination, deuterated PS is utilized, and NR experiments have to be performed.

A systematic study of the change of the interface width with the degree of bromination has been performed by NR (73), and the results are shown in Figure 8. At low x, the interface width diverges because the system becomes compatible. The functional form is fitted by a composition dependent χ-parameter, which contains the concentration-weighted individual segment-segment interaction parameters of the components. Values at large x can be compared to other X-ray and neutron reflectivity data, as shown in Table 3. The a_I-values noted in Table 3 are, for a better comparison, those that have not been corrected for initial surface roughness (see 73, 74). There is good agreement between different data sets if high-molecular weight data of Reference 75 are taken. The strong increase of a_I at lower molecular weights (75) is not well understood and has tentatively been attributed to the influence of capillary waves. Some of the

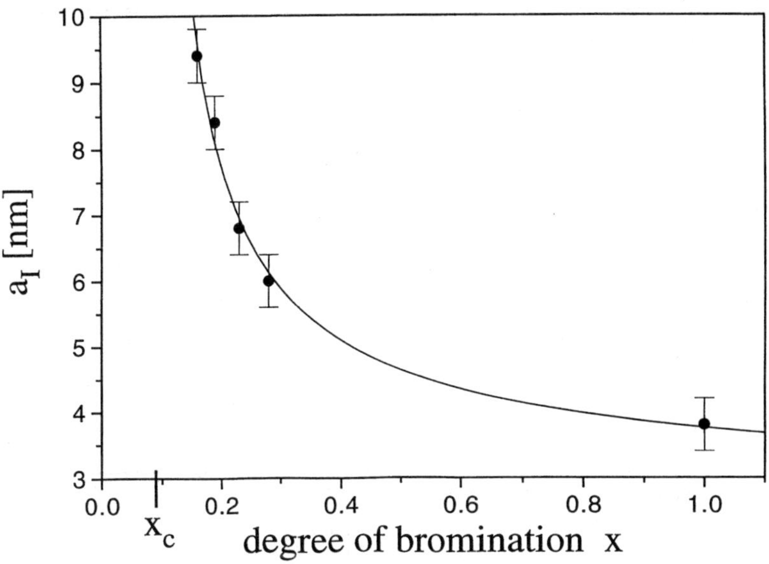

Figure 8 Dependence of interface width a_I on the degree of bromination x for the blend system PS(D)/PBr$_x$S(H) (73). With decreasing x, the compatibility of the blend increases. x_c indicates the critical degree of bromination, where the system becomes compatible and the equilibrium interface width diverges to infinity. The solid line corresponds to a fit assuming a particular composition dependence of χ.

differences between different data sets might be due to different temperatures, molecular weights, weight distributions, and sample preparations.

With some samples, the time dependence of the formation of an equilibrium interface has been followed (73) that is consistent with an initial $t^{0.25}$-behavior, which then levels off to a constant value. Also, the influence of the glass transitions on the interface width has been studied (74). If the sample is annealed between the glass transitions of the components, segment interdiffusion will come to a standstill, whereas at the interface, a concentration is reached at a certain point of the profile, where the local mobility is too low. One thus expects an asymmetric profile, which, however, could not be detected due to the limited resolution of the technique at small interface widths. In this regime, the interface width depends strongly on temperature (Figure 9) until the glass transition temperature also of the second component is reached. At higher temperatures, dewetting phenomena are observed (74) that are a consequence of the strong incompatibility of the components. A comparison of the experimentally determined interface widths with other data is difficult because γ or χ-values for this system are only known for the compatible blend region.

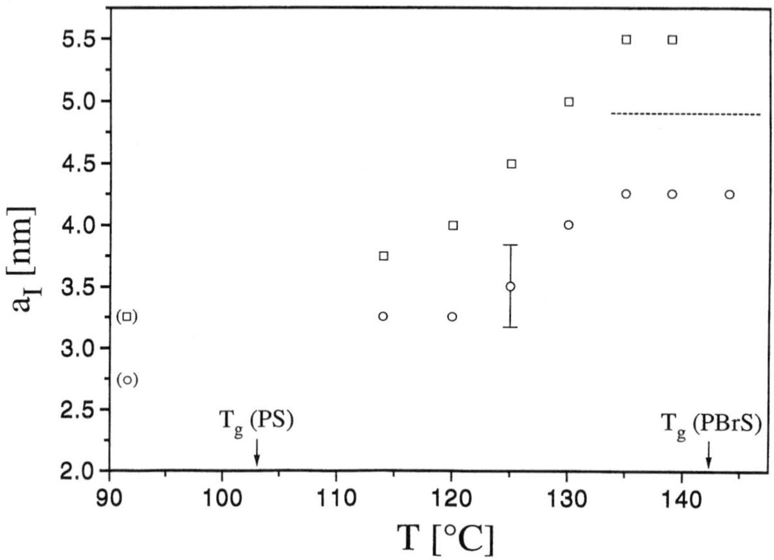

Figure 9 Change of the interface width of PS/PBrS during annealing in the region between the glass transitions of the components determined from X-ray reflectivity experiments (54). The glass transition temperature of PBrS is 142°C. The data points at lower temperatures in brackets indicate the interface width prior to annealing. The symbols (*open box*, *open circle*) indicate two different samples with a different sequence of layers. The dashed line gives the mean equilibrium value of the interface width.

The measured interface widths, on the other hand, may be used to obtain interaction parameters for the incompatible region.

PMMA/Poly(styrene-stat-acrylonitril) (SAN)

With this blend system, compatibility between components can be changed over a wide range by a variation of the acrylonitrile content w in the statistical copolymer poly(styrene-stat-acrylonitrile). At low ($w \lessapprox 0.1$) and high ($w \gtrsim 0.3$, depending on temperature) acrylonitrile content, the blends become incompatible. Molecular weight distributions of SAN are broad ($M_w/M_n \sim 2$), and a temperature range from 120 to 400°C has been used for miscibility studies. A series of experiments mostly with ellipsometry show the influence of temperature, time, and composition on the formation of an interface between the components (67, 76, 77). Ellipsometry can be used because there is a relatively large difference in the indices of refraction between PMMA and SAN ($\Delta n \gtrsim 0.1$). The interface width depends on SAN composition and is for the strongly incompatible materials ($w = 0$ and 0.55) of the order of 5 nm, whereas it increases to $\sim$30 nm for materials approaching the miscibility window (e.g. at $w = 0.057$ and $w = 0.387$) (76). The temperature dependence of a_I in the investigated range of 140 to 170°C is not very strong. Thus, a_I for $w = 0.057$ is nearly constant and changes for $w = 0.387$ in the range 26 to 32 nm with a slight maximum at 150°C (67). Within the accuracy of the technique, the interface width of the incompatible system with $w = 0.057$ does not change with annealing time, whereas it increases with $\sqrt{t}$ for $w = 0.223$ for a miscible system according to a Fickian diffusion behavior (77). From the change of a_I with time, diffusion coefficients are obtained. A correlation with adhesion experiments shows a linear increase of adhesive strength with $\sqrt{a_I}$.

The constancy of a_I with time (76, 77) for the incompatible system is not easy to understand. In one case ($w = 0.55$) (76), the interface width even decreases significantly (from 34 to 8 nm). One would expect an increase from an initial value given by interfacial roughness (of the order of 1–2 nm) to a final value given by the equilibrium interface width after annealing. The upper SAN film is deposited on the PMMA substrate by the floatation technique and is glassy ($T_g \sim 105$°C) at room temperature. This result is in contrast to similar experiments with PS/PBr$_x$S (73) with NR, where initially a $t^{0.25}$ behavior is observed, which levels off to the equilibrium interface width with time. Ellisometry results using the same materials and identical preparation conditions are also not completely consistent; for instance, in one case, $w = 0.057$, an interface width of 19 nm at 130°C is reported (77), whereas in other experiments, this interface width is 32 nm (76). It is interesting to compare ellipsometric experiments with TEM (68) and ERD (69) results. Although the interface widths obtained with TEM are in agreement with ellipsometric data, the ERD values are significantly larger. The reason might be contamination by water at the interface, when the oxygen profile is observed; however, the discrepancy for the nitrogen profile is unresolved. In this case, it is also well outside the resolution limit of the technique, which is of the order of 12 nm. The situation for the interface width of PMMA/SAN is thus not completely resolved, and additional experiments will be helpful.

Other Incompatible and Partially Compatible Blends

Some other incompatible blend systems have been studied. A detailed Porod analysis of time-resolved light scattering experiments on blends of polybutadiene and polyisoprene (PB/PI) (78) gives some information of the time evolution of the interface width at late stages of spinodal decomposition. This system is not far from its critical point ($T - T_S = 7$K) and interfaces are wide. The interface width decreases with time during progressing spinodal decomposition and reaches a constant value of approximately 260 nm (see Table 3) at later times. This interface width thus can be taken as the equilibrium interface width between PB and PI under those conditions.

A similar Porod analysis of SANS experiments is performed with a blend of deuterated and protonated 1,4-polybutadiene (PB(D)/PB(H)) (79). It has been mentioned before that the small differences between deuterated and protonated polymers can give rise to incompatibility at higher molecular weights, first shown for blends of deuterated and protonated high-molecular weight polystyrenes (80). Deuteration gives rise to a slightly different index of refraction, which makes the investigation of spinodal decomposition by light scattering possible (81). Because the interface is relatively wide, a detailed Porod analysis of the SANS data is possible. Thus besides interface width and total surface area in the sample, information on the form of the profile can be obtained as well. The width is determined as $a_I = 50$ nm and the fit is in favor of a tanh-profile, which is also predicted theoretically (see Equation 6a).

An isotopic blend of protonated and deuterated polystyrene (PS(H)/PS(D)) has been investigated by ^{3}He-nuclear reaction analysis (82, 83). At the investigated high-molecular weights, the components are only partially compatible, and during annealing, the concentrations in the lower and upper films change to a value as given by the phase diagram. At the same time, the interface width increases and approaches an equilibrium value in the range 80–120 nm, depending on annealing temperature. Values of the interface width calculated here from the known χ-parameter yield systematically lower values (60–100 nm).

Another statistical copolymer system, where compatibility between components is controlled by the composition of the copolymer, is the blend of polymethylmethacrylate and solution-chlorinated polyethylene (PMMA/SCPE). With a degree of chlorination of $w = 0.63$, the system shows lower critical solution behavior with $T_C \sim 120°C$ (84). It is then annealed near this temperature, where the interface width is determined after different annealing times by NR. The interface width stays approximately constant at $a_I \sim 14$–23 nm, whereas the position of the interface moves toward the PMMA side, indicating a swelling of the SCPE by PMMA. The situation is similar to the case of the miscible-blend system PMMA/PVC, where the mobilities of components are quite different due to different glass transition temperatures, and where a dissolution of the immobile PMMA (being below its glass transition at the utilized temperatures) in the mobile component of PVC is observed by NR (85). In the case of PMMA/SCPE (84), however, both components are mobile and are very close to the critical point. In another investigation of this system by Rutherford backscattering spectroscopy (86), only an upper limit for the interface width can be given due to the limited resolution of this technique. The interface width is smaller than 50 nm at later annealing times. The composition of phases and the interdiffusion coefficients have been determined (86).

Ellipsometry has been used to determine the interface widths for the blend systems amorphous nylon and poly(styrene-co-maleic anhydride) (nylon/SMA) (87); amorphous nylon and poly(styrene-co-acrylonitrile) with a small amount of SMA added (nylon/SAN/SMA) (87); and nylon with hydrogenated nitrile rubber in the presence of different compatibilizers (nylon/NBR) (88). The authors (87, 88) have investigated the time evolution of the interface widths with annealing time and have studied systems with different copolymer compositions, compatibilizers, and at different temperatures. SMA is known to be reactive with the chain end of nylon. The authors attribute the formation of a relatively thick interface to the occurrence of this reaction (87). The addition of compatibilizers causes an increase of the interface width for nylon/NBR (88).

A value of $a_I = 4.5$ nm (from Reference 89) for the interface width between polycarbonate and poly(styrene-co-acrylonitrile) with acrylonitrile content $w = 0.25$ has been reported (cited in Reference 90). The values of the interface widths of incompatible polymer systems are compiled in Table 3 which includes the presently available literature. However, in this review, we generally do not include the work on interfaces in blockcopolymers or the investigations on compatibilizing effects of copolymer addition to incompatible polymer blends.

CONCLUSION

Several incompatible polymer blend systems have been investigated by different techniques and interface widths have been obtained. In particular, NR allows the determination of the interface width with high precision. Only for some systems, where interaction parameters or interfacial tension data are available with sufficient precision is comparison with calculated values meaningful. Those interface widths calculated on the basis of mean-field theoretical expressions and square gradient theory are significantly lower than experimentally determined values. This leads to the conclusion that there must be additional influences not considered in the theory. Several approaches indicate that concentration fluctuations (or capillary waves) at the interface could account for those differences. However, other factors like chain-end effects, distorted chain conformation, molecular weight distribution, trace amount of additives, or initial interface roughness can also contribute. Of course, square gradient theory has its limitations, and detailed computer simulations of incompatible polymer interfaces, for instance, would be helpful. Thus both theoretical and experimental efforts will be necessary to gain a better understanding of incompatible polymer interfaces. One should keep in mind that a large proportion of commercial polymer systems are incompatible and interfacial properties largely determine the application range and materials properties.

ACKNOWLEDGMENTS

We greatly acknowledge the fruitful discussions with Prof. K Binder and Prof. S Anastasiadis. DWS was supported by Bundes Ministerium für Forschung und Technologie (grant FI3HPG) during this work.

Literature Cited

1. Yoda N. 1993. *J. Pol. Sci., Pol. Symp.* 75: 125–36
2. Olabisi O, Robeson LM, Shaw MT. 1979. *Polymer-Polymer Miscibility*. New York: Academic. 370 pp.
3. Utracki L. 1989. *Polymer Alloys and Blends*. München: Hauser. 356 pp.
4. Kausch HH. 1987. *Polymer Fracture*. Heidelberg: Springer-Verlag. 456 pp.
5. Brown HR. 1991. *Annu. Rev. Mater. Sci.* 21: 463
6. Binder K. 1994. *Adv. Pol. Sci.* 112: 181
7. Stamm M. 1992. *Adv. Pol. Sci.* 100: 357
8. Tirrell M, Parsonage EE. 1993. In *Mater. Sci. Tech.*, 12: 653. Weinheim: VCH
9. Allara DL, Zhang P. 1994. In *Mater. Sci. Tech.*, 2BII: 657. Weinheim: VCH
10. Stamm M. 1992. In *Physics of Polymer Surfaces and Interfaces*, ed. IC Sanchez, p. 163. Boston: Butterworth-Heinemann

11a. Flory JP. 1941. *J. Chem. Phys.* 9: 660

11b. Flory JP. 1942. *J. Chem. Phys.* 10: 51

11c. Flory JP. 1944. *J. Chem. Phys.* 12: 425

11d. Staverman AJ. 1941. *Rec. Trav. Chim.* 60: 640

11e. Huggins ML. 1941. *J. Chem. Phys.* 9: 440

12a. Koningsveld R. 1968. *Adv. Colloid Interface Sci.* 2: 151

12b. Koningsveld R, Kleintjens LA, Schoflers HM. 1974. *Pure Appl. Chem.* 39: 1

12c. Dudowicz J, Freed KF. 1991. *Macromolecules* 24: 5076,5112

13. Cahn JW, Hilliard JE. 1958. *J. Chem. Phys.* 28: 258
14. Broseta D, Fredrickson GH, Helfand E, Leibler L. 1990. *Macromolecules* 23: 132
15. Tang H, Freed KF. 1991. *J. Chem. Phys.* 94: 6307
16. Joanny JF, Leibler L. 1978. *J. Phys.* 39: 951
17. de Gennes P-G. 1980. *J. Chem. Phys.* 72: 4756
18. Binder K. 1983. *J. Chem. Phys.* 79: 6387
19. Anastasiadis SH, Gancarz I, Koberstein JT. 1988. *Macromolecules* 21: 2980
20. Binder K, Frisch H. 1984. *Macromolecules* 17: 2928
21. Schubert DW, Abetz V, Hack T, Stamm M, Siol W. 1995. *Macromolecules*. Submitted
22. Fredrickson GH. 1992. *Physics of Polymer Surfaces and Interfaces*, ed. IC Sanchez, pp. 1–28. Boston: Butterworth-Heinemann
23. Roe RJ. 1975. *J. Chem. Phys.* 62: 490
24. Helfand E. 1975. *J. Chem. Phys.* 63: 2192
25. Brosseau C. 1993. *J. Phys. II* 3: 279
26. Wool RP, Long JM. 1993. *Macromolecules* 26: 5227
27. Jilge W, Carmesin I, Kremer K, Binder K. 1990. *Macromolecules* 23: 5001

28. Reiter J, Zifferer G, Olaj OF. 1990. *Macromolecules* 23: 224
29. Semenov AN. 1994. *Macromolecules* 27: 2732
30. Müller M, Binder K, Ded W. 1995. *Sci. Prog.* In press; Müller M, Binder K, Ded W. 1995. *Faraday Discuss.* Submitted
31. Shull KR. 1992. *Macromolecules* 25: 2122
32. Ocko BM, Wu XZ, Sirota BE, Sinka SK, Deutsch M. 1994. *Phys. Rev. Lett.* 72: 242
33. Fleischer CA, Koberstein JT, Krukonis V, Wetmore PA. 1993. *Macromolecules* 26: 4172
34. Fleischer CA, Morales AR, Koberstein JT. 1994. *Macromolecules* 27: 379
35. Wu S. 1982. *Polymer Interface and Adhesion*. New York: Dekker
36. Brandrup J, Immergut EH. 1989. *Polymer Handbook*. New York: Wiley. 3rd ed.
37. Morra M, Occhiello E, Garbani F. 1990. *Adv. Colloid Interface Sci.* 32: 79
38. Helfand E, Tagami Y. 1971. *J. Chem. Phys.* 56: 3592
39. Helfand E, Sapse AM. 1975. *J. Chem. Phys.* 62: 1327
40. Helfand E, Bhattacharjee SM, Fredrickson GH. 1989. *J. Chem. Phys.* 91: 7200
41. LeGrand DG, Gaines GL. 1978. *J. Colloid Interface Sci.* 63: 394
42. Kumar SK, Vocatello M, Yoon DY. 1990. *Macromolecules* 23: 2189
43. Koberstein JT. 1987. *Encyclopedia of Polymer Science and Engineering*, 8: 237. New York: Wiley
44. Harland S, Hartley RW. 1987. *Axisymmetric Fluid-Liquid Interfaces*. Elsevier: Amsterdam
45. Cheng P, Li D, Boruvka L, Rotenberg Y, Neumann AW. 1990. *Colloids Surf.* 43: 151
46. Kawaguchi M, Takahashi A. 1992. *Adv. Colloid Interface Sci.* 37: 219
47. Hansen FK, Rødsrud G. 1991. *J. Colloid Interface Sci.* 141: 1
48. Wagner M, Wolf BA. 1993. *Macromolecules* 26: 6498; Schoolenberg GE, During F. 1994. *Europhys. Conf. Abstr.* 18C: L16
49. Cohen A, Carriere CJ. 1989. *Rheol. Acta* 28: 223; Sammler RL, Dion RP, Carriere CJ, Cohen A. 1992. *Rheol. Acta* 31: 554
50. Russell TP. 1990. *Mater. Sci. Rep.* 5: 171
51. Foster MD. 1993. *Crit. Rev. Anal. Chem.* 24: 179
52. Richards RW, Penfold J. 1994. *Trends Pol. Sci.* 2: 5
53. Lekner J. 1987. *Theory of Reflection*. Dordrecht: Martinus Nijhoff
54. Hüttenbach S, Stamm M, Reiter G, Foster M. 1991. *Langmuir* 7: 2438
55. Azzam RMA, Bashara NM. 1977. *Ellipsometry and Polarized Light*. Amsterdam: North-Holland
56. Riedling K. 1988. *Ellipsometry for Industrial Applications*. Wien/New York: Springer-Verlag
57. Kawaguchi M. 1990. *Adv. Colloid Interface Sci.* 32: 1
58. Motschmann H, Stamm M, Toprakcioglu G. 1991. *Macromolecules* 24: 3681
59. Siqueira D, Schubert DW, Erb V, Amato JP, Stamm M. 1995. *Macromolecules*. Submitted
60. Kim MW, Pfeiffer DG, Chen W, Hsuing H, Rasing T, Shen YR. 1989. *Macromolecules* 22: 2682
61. Machate C. 1989. *Die Polymer-Polymer Interdiffusion an der Grenzfläche des Kunststoff-Kautschuk-Verbundes aus Poly* (2,6-*dimethyl*-1,4-*phenylenether*) *und Poly* (*styrol*-10-*butadien*). PhD thesis. Univ. Münster. 211 pp.
62. Endisch D, Rauch F, Götzelmann A, Reiter G, Stamm M. 1992. *Nucl. Instrum. Methods B* 62: 513
63. Green PF, Doyle BL. 1990. In *New Characterization Techniques for Thin Polymer Films*, ed. HM Tong, LT Nguyen, p. 139. New York: Wiley
64. Sykes DE. 1989. *Methods of Surface Analysis*, ed. JM Walls, p. 216. New York: Cambridge Univ. Press
65. Anastasiadis SH, Russell TP, Satija SK, Majkrzak CF. 1990. *J. Chem. Phys.* 92: 5677
66. Fernandez ML, Higgins JS, Penfold J, Ward RC, Shackleton C, Walsh DJ. 1988. *Polymer* 29: 1923
67. Higashida N, Kressler J, Yukioka S, Inoue T. 1992. *Macromolecules* 25: 5259
68. Kressler J, Higishida N, Inoue T, Heckmann W, Seitz F. 1993. *Macromolecules* 26: 2090
69. Deleted in proof
70. Ellingson PC, Stand, DA Cohen A, Sammler RL, Carriere J. 1993. *Abstr. Pap. Am. Chem. Soc.* 69: 183
71. Russell TP, Hjelm RP, Seeger PA. 1990. *Macromolecules* 23: 890
72. Hammouda B, Bauer BJ, Russell TP. 1994. *Macromolecules* 27: 2357
73. Guckenbiehl B, Stamm M, Springer T. 1994. *Phys. B* 198: 127
74. Hüttenbach S, Stamm M, Reiter G, Foster M. 1991. *Langmuir* 7: 2438
75. Zhao W, Zhao X, Rafailovich MH, Sokolov J, Mansfield T, et al. 1991. *Phys. B* 173: 43
76. Yukioka S, Inoue T. 1993. *Polymer* 34: 1256

77. Yukioka S, Nagato K, Inoue T. 1992. *Polymer* 33: 1171
78. Hashimoto T, Takenaka M, Jinnai H. 1991. *J. Appl. Crystallogr.* 24: 457
79. Bates FS, Dierker SB, Wignall GD. 1986. *Macromolecules* 19: 1938
80. Bates FS, Wignall GD, Koehler WC. 1985. *Phys. Rev. Lett.* 55: 2425
81. Bates FS, Wiltzius P. 1989. *J. Chem. Phys.* 91: 3258
82. Chaturvedi UK, Steiner U, Zak O, Krausch G, Klein J. 1989. *Phys. Rev. Lett.* 63: 616
83. Steiner U, Krausch G, Schatz G, Klein J. 1990. *Phys. Rev. Lett.* 64: 1119
84. Fernandez ML, Higgins JS, Penfold J, Shackleton C, Walch DJ. 1990. *Polymer* 31: 2146
85. Kunz K, Stamm M. 1994. *Macromol. Symp.* 78: 105
86. Rafailovich MH, Sokolov J, Jones RAL, Krausch G, Klein J, Mills R. 1988. *Europhys. Lett.* 5: 657
87. Yukioka S, Inoue T. 1994. *Polymer* 35: 1182
88. Bhowmick AK, Inoue T. 1994. *J. Appl. Pol. Sci.* 53: 77
89. Janarthanan V, Stein RS, Garrett PD. 1994. *Macromolecules* 27: 4855
90. Mansfield T. 1993. *Polymer concentration profiles and methods of surface modification.* PhD thesis. Univ. Mass., Amherst. 116 pp.
91. Stamm M. 1991. *J. Appl. Cryst.* 24: 651
92. Schmidt-Rohr K, Spiess HW. 1994. *Multidimensional Solid-State NMR and Polymers.* London: Academic
93. Shull KR, Winey KI, Thomas EI, Kramer EJ. 1991. *Macromolecules* 24: 2748

Annu. Rev. Mater. Sci. 1995. 25: 357–88

GIANT MAGNETORESISTANCE IN MAGNETIC NANOSTRUCTURES

S. S. P. Parkin

IBM Research Division, Almaden Research Center, 650 Harry Road, San Jose, California 95120

KEY WORDS: multilayers, granular metals, interlayer exchange coupling, magnetic recording, sputtered films

ABSTRACT

This chapter contains a brief review of the giant magnetoresistance (GMR) effect exhibited by magnetic multilayers, granular alloys, and related materials. Subjects covered include a description of the phenomenon, and the related oscillatory interlayer exchange coupling in magnetic multilayers; a simple model of giant magnetoresistance; the inverse GMR effect in spin-engineered magnetic multilayers; structures that display large changes in resistance in small magnetic fields, possibly for use in magnetic field sensors; and the dependence of GMR on various aspects of the magnetic structures.

INTRODUCTION

There has been much interest in recent years in artificially engineered nano-structured materials with novel physical properties. One area of particular interest is that of metal multilayers. These materials have been studied for the past 30 years or more, but it is only relatively recently that detailed expertise has been developed to prepare and sufficiently characterize the structure of such materials (1–3). Metal multilayers have been studied for their novel superconducting properties, for the possibility of creating metals much stronger and tougher than the individual components of the multilayer, for creating super-mirrors for X-ray or neutron scattering monochromators, and for a host of other reasons. Most recently, interest in metal multilayers has centered on multilayers comprised of

0084–6600/95/0801–0357$05.00

thin magnetic layers separated by thin nonferromagnetic layers. These multilayers display unusual magnetic and transport properties that are the subject of this brief review. Note that several excellent comprehensive reviews of the transport and magnetic properties of magnetic multilayers and related materials are available (4–10).

Usually the change in resistance of thin metal films with application of a magnetic field is small (11). However, in 1988 it was found that single crystalline (100)-oriented Fe/Cr/Fe sandwiches (12) and (100)-oriented Fe/Cr multilayers (13) displayed much larger magnetoresistance values than could be accounted for by the magnetoresistance of the individual Fe layers themselves. The resistance of the multilayers was found to decrease by almost a factor of two when a field of $\simeq$20 kOe was applied at low temperatures (4.2 K) (13). The term giant magnetoresistance (GMR) was coined to describe this effect. The thickness of the Cr layers in these structures was chosen to correspond to the thickness that was previously found to give rise to an unusual antiferromagnetic interlayer exchange coupling of neighboring Fe layers (14, 15). These films were prepared by molecular beam epitaxy (MBE) (see, for example, 16–18), which is a sophisticated and expensive ultra-high vacuum deposition technique. Soon it was shown that similar GMR results could be obtained in polycrystalline Fe/Cr sandwiches and multilayers prepared using inexpensive and simpler sputter-deposition methods (19, 20). Sputter-deposition allowed for the study of a wide variety of metal multilayers, and this led to the surprising discovery that GMR was common to many metal multilayers (19, 21). To date, the largest GMR effects at room temperature are found in Co/Cu multilayers comprised of thin Co and Cu layers (21–23).

Interest in transition metal multilayers has also been greatly increased in recent years with the discovery of oscillations in the interlayer magnetic exchange coupling in multilayers comprised of thin ferromagnetic layers of Fe, Co, Ni, and their various alloys separated by thin layers of virtually all the nonferromagnetic transition and noble metals (19, 22, 24, 25). This discovery led to the prospect of artificially engineering multilayered structures with complex magnetic structures, useful both for understanding the physics of these materials as well as for technological applications. Indeed, magnetic thin film structures already have found important applications for storing information in magnetic and magneto-optical disk storage devices (26, 27), for magnetic field sensors in magnetic recording magnetoresistive read heads (28), and for nonvolatile random access memory (29). Applications involving hybrid magnetic/semiconductor structures have been proposed (30), and recently, spin switches, devices based on transport of spin-polarized electrons across magnetic/nonmagnetic interfaces, have been devised and prototypes built (31, 32).

It is likely that devices based on the giant magnetoresistance phenomenon will be used within a few years.

GIANT MAGNETORESISTANCE

Magnetic Multilayers: Polycrystalline Co/Cu and Fe/Cr

Figure 1 shows plots of the change in room temperature resistance ΔR vs in-plane magnetic field for two polycrystalline multilayers of Fe/Cr and

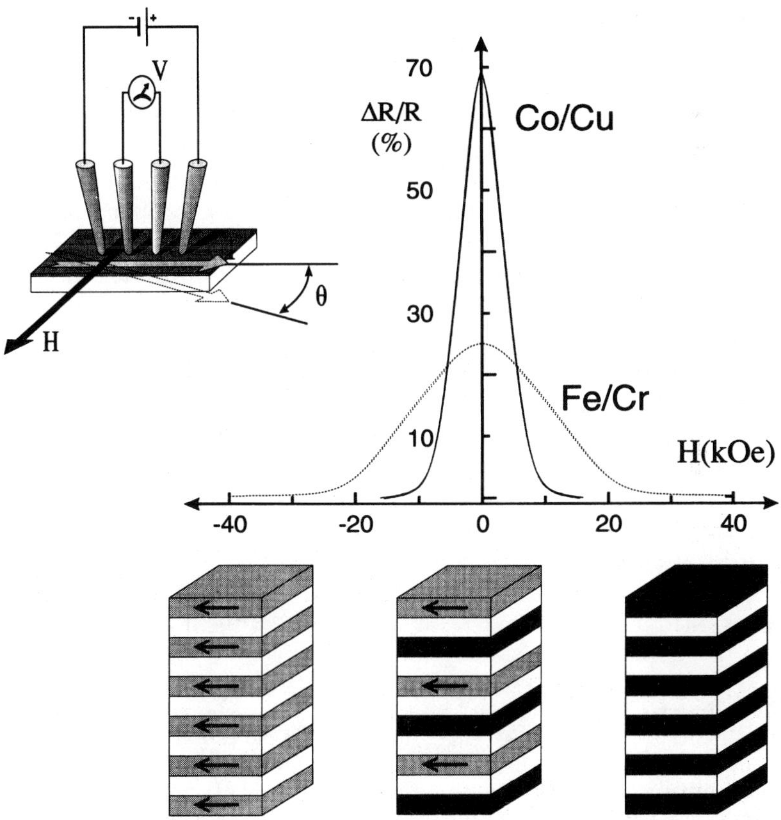

Figure 1 Room temperature resistance vs in-plane magnetic field curves for polycrystalline Fe/Cr and Co/Cu multilayers deposited by magnetron sputtering (for detailed deposition conditions, see 19, 21, but note that the Co/Cu multilayer was deposited at 2.1 mTorr). The measurement geometry is shown schematically in the top left corner. The magnetic state of the antiferromagnetically coupled multilayers is shown schematically in the lower portion of the figure for large negative, zero, and large positive magnetic fields.

Co/Cu, respectively[1]. Note that the resistance is measured with the current in the plane of the layers, and the magnetic field is applied orthogonal to the current also in the plane of the layers. At high magnetic fields, the resistance saturates at some value R, and the sample resistance is normalized to this value. The figure shows that the resistance of the multilayers is significantly higher in small fields as compared with high fields. The variation in resistance is related to a change in the relative orientation of neighboring ferromagnetic Fe or Co layers with applied magnetic field. The resistance is higher when adjacent magnetic layers are aligned antiparallel to one another, as compared with parallel alignment, shown schematically in cartoons of the magnetic multilayer included in Figure 1. In the figure, values of room-temperature saturation magnetoresistance $\Delta R/R$ of more than 25 and 70% are observed for Fe/Cr and Co/Cu, respectively, with corresponding saturation fields of $\simeq$25 and 10 kOe. The corresponding GMR values at 4.2 K are significantly higher at 110 and 130%, respectively.

The GMR of the samples shown in Figure 1 are some of the highest recorded GMR values in magnetic multilayers. Higher values still have been reported in single crystalline (100) Fe/Cr multilayers. GMR values exceeding 150% at 4.2 K have been found in magnetron-sputtered samples (33), and values of more than 220% at 1.5 K have been reported recently in MBE-deposited multilayers (34).

As shown schematically in Figure 1, the magnetic moments of successive magnetic layers in the Co/Cu and Fe/Cr multilayers are arranged antiparallel to one another in small fields. This is a consequence of an antiferromagnetic interlayer exchange coupling J_{AF} propagated through the intervening Cr or Cu layers. As the Cr or Cu layer thickness is varied, the exchange coupling of the magnetic layers is found to vary in sign, oscillating between antiferromagnetic (AF) and ferromagnetic (F) coupling (19, 22). This is manifested, for example, as an oscillation in the magnitude of the GMR effect with increasing separation of the magnetic layers and

[1] The structures of the Fe/Cr and Co/Cu multilayers shown in Figure 1 are as follows: Si/40ÅCr/[8ÅFe/7.5ÅCr]$_{39}$/8ÅFe/15ÅCr, and Si/50ÅFe/[8ÅCo/7.5ÅCu]$_{59}$/8ÅCo/20ÅFe. Both multilayers were deposited on Si(100), which was covered with a thin ($\simeq$10–15 Å thick) native oxide layer. The Fe/Cr multilayer was deposited at 125°C, which was found to give rise to maximal GMR values (20). It was postulated that this was a result of reducing the bulk resistivity of the Fe and Cr layers while limiting intermixing of the Fe and Cr layers. The Co/Cu multilayer was deposited at room temperature because deposition at any higher temperature appears to decrease GMR, presumably because of dissolution or intermixing of the Co and Cu layers (21–23). An Fe buffer layer is used. It was discovered that Fe buffer layers give the largest GMR values in Co/Cu multilayers grown on silicon for thin Cu layers (21, 22).

was first observed in sputtered Fe/Cr multilayers (19). Figure 2 gives results for a series of Co/Cu multilayers in which the magnitude of the saturation magnetoresistance is found to oscillate with increasing Cu spacer layer thickness with an oscillation period of $\simeq 9$ Å. Large GMR values are found for Cu layer thicknesses for which the Co layers are coupled antiferromagnetically (see Figure 2). For strong ferromagnetic coupling of the Co layers, the relative magnetic alignment of the Co layers is unaffected by magnetic field, and consequently there is no GMR effect. Note that for polycrystalline multilayers, as shown in Figure 2, a significant background magnetoresistance (MR) effect is superimposed on the GMR oscillations. This effect is most likely a consequence of the crystallographic structure of the multilayer, which is comprised of small crystalline grains, typically 100 to 200 Å in size (35). Although the grains are preferentially oriented along (111), the texture is poor, and there is a considerable angular dispersion of the ⟨111⟩ axis from grain to grain. In addition, there may also be (100)-oriented grains (36). Because the period of the oscillatory coupling in Co/Cu depends on crystallographic orientation (37, 38), it is possible that GMR oscillations from differently oriented grains in the polycrystalline Co/Cu structures could be superimposed, perhaps accounting for the GMR background shown in Figure 2. For very thick Cu layers, where the interlayer coupling is very weak, GMR is still observed if the magnetic layers break up into magnetic domains, as shown schematically

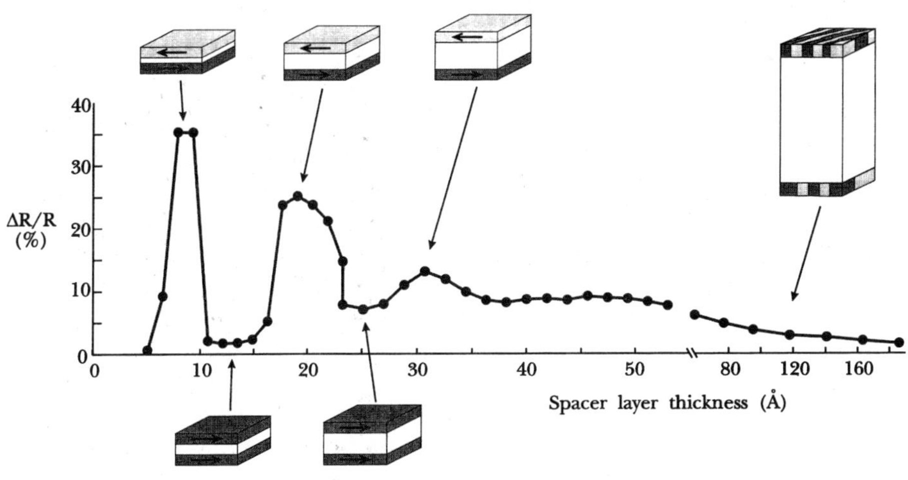

Figure 2 Room temperature saturation magnetoresistance vs Cu spacer layer thickness for a series of polycrystalline sputter-deposited Co/Cu multilayers (from 22). The magnetic state of the multilayer is shown schematically for various Cu layer thicknesses (only two magnetic layers are shown).

in Figure 2. GMR is the result of electrons propagating from one magnetic layer to another magnetic layer for which the magnetic moments are not aligned with one another. Thus there is a certain probability that electrons emanating from a magnetic domain in one magnetic layer will propagate and undergo scattering in a magnetic domain in a neighboring layer (or indeed even in the same layer), which would lead to an increased resistance. The resistance is highest in this case when the net magnetic moment of each layer is close to zero at the coercive field H_c. This leads to twin peaks in resistance at $\pm H_c$.

Granular Alloys—An Example: [111]-Oriented CoCu

The preceding discussion of GMR in uncoupled Co/Cu multilayers indicates that magnetically coupled multilayers are not a prerequisite for the observation of GMR. The layered geometry is also unnecessary, and giant magnetoresistance has been observed in a variety of inhomogeneous magnetic systems. The only requirement is that the system contains a distributed magnetic component in which the relative orientation of the magnetic moments of small magnetic entities will vary with magnetic field or any other variable (such as, for example, temperature or strain). Granular alloys containing magnetic particles in a metallic host bear a strong resemblance to magnetic multilayers, and GMR has been observed in a variety of magnetic granular systems predominantly comprised of Fe, Co, Ni, and their various alloys in Cu, Ag, and Au matrices (39–42) (for a review see C L Chien, this volume). Figure 3 shows typical GMR results for a single crystalline (111)-oriented $Co_{28}Cu_{72}$ granular alloy prepared by MBE (41). Resistance vs field curves are shown for the magnetic field applied both in-plane and orthogonal to the sensing current and perpendicular to the plane of the film. There is a significant magnetic anisotropy, perhaps resulting from the shape of the Co particles, or from the intrinsic magnetic anisotropy of the Co itself. Note also that the resistance of the sample is dependent on the magnetic field history. As a result of the considerable magnetic anisotropy, the coercive fields are very different for field applied in the plane of and perpendicular to the plane of the samples. Just as for the uncoupled Co/Cu multilayers described above, the largest resistance in granular alloys is found when the magnetization is close to zero at $\pm H_c$. This corresponds to the state of maximal anti-alignment of the magnetic moments of neighboring magnetic particles. Thus, because the coercive field depends on field orientation, the sample resistance in zero field can be varied by up to a factor of two, as shown in Figure 3. In this case, the sample was first magnetized perpendicular to the film. After reducing the field to zero, the sample was rotated so that the field was aligned in the plane of the sample. Magnetizing the sample

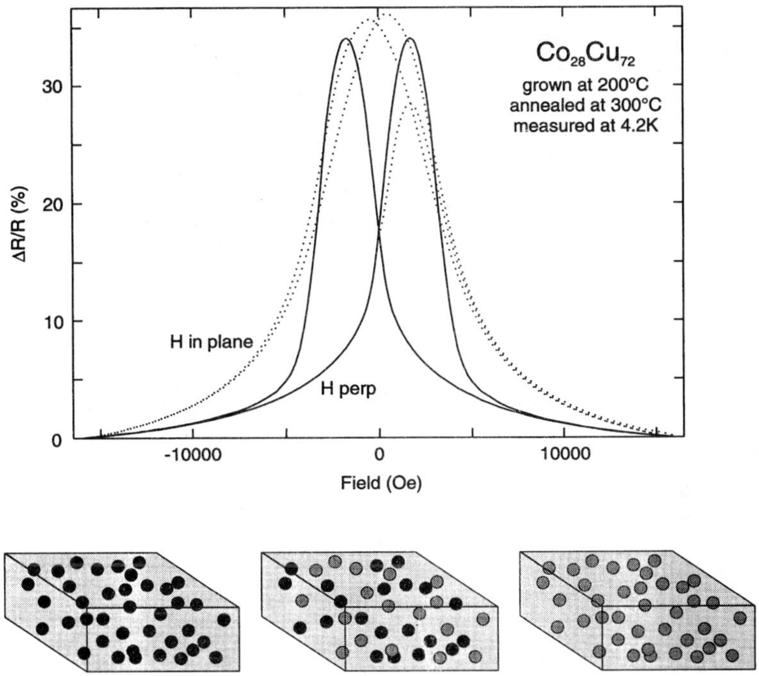

Figure 3 Resistance vs field curves for a MBE-deposited crystalline (111)-oriented $Co_{28}Cu_{72}$ alloy. The film was deposited at 200°C and annealed for 1 h at 300°C (for deposition conditions, see 41). Data are shown at 4.2 K for field applied in-plane (and orthogonal to the sense current) and for field applied along the field normal. The arrangement of the magnetic Co particles is shown schematically in the bottom portion of the figure for large negative, zero, and large positive magnetic fields.

and reducing the field to zero results in an increase of the zero field resistance by 51%!

OSCILLATORY INTERLAYER COUPLING

The Cu or Cr in the layers adjacent to the Co or Fe magnetic layers shown in Figure 1 becomes spin-polarized, and these atoms develop magnetic moments. For Cu, the *d*-spin-induced moment, which is parallel to the Co moment, is very small and is about 100 times smaller than the moment on Co itself, as deduced from X-ray magnetic circular dichroism studies (XMCD) (43). For Cr, the induced *d*-moment, which is antiparallel to the Fe moment (44), is significantly larger and is comparable to that of the Fe moment (45). The induced spin-polarization of Cr has been directly observed by measuring the spin-polarization of secondary electrons excited from a (100)-oriented single crystalline Cr wedge deposited on a single crystal Fe whisker by the electron beam in a scanning electron microscope (SEMPA) (10, 44). The direction of the induced Cr moment at the surface of the Cr layer is observed to oscillate with increasing Cr thickness in a manner consistent with the variation of the interlayer magnetic coupling between Fe layers observed in related structures in which a thin Fe layer is deposited on top of the Cr wedge (46). It is the induced spin density wave in the spacer layer material that mediates the magnetic coupling of the magnetic layers in magnetic multilayers and sandwiches.

The interlayer exchange coupling of thin layers of Fe, Co, Ni, and their alloys via nonmagnetic transition and noble metal spacer layers nearly always exhibits an oscillatory variation with spacer layer thickness (25). The oscillation period varies from metal to metal and lies between $\simeq 8$ Å and 12 Å in most cases, with the exception of Cr, for which it is significantly longer ($\simeq 18$ Å). In some cases, more than one oscillation period has been observed. For Cr, a second short oscillation period just $\simeq 2$ monolayers long has been observed (46, 47). The interlayer coupling has been explored experimentally in many systems using a wide variety of techniques, including SEMPA and XMCD, as well as more conventional techniques including various magnetometries, Brillouin light scattering (see, for example, 48–50) and ferromagnetic resonance (51). A detailed discussion of interlayer coupling can be found elsewhere (7, 10, 38, 52–54).

A SIMPLE MODEL OF GIANT MAGNETORESISTANCE

The detailed origin of GMR has provoked considerable interest. Many theoretical models have been developed, but most of them are based on a model of the electrical conduction in ferromagnetic metals from Mott (55). Mott hypothesized that the electrical current in ferromagnetic metals is carried independently in two conduction channels that correspond predominantly to the spin-up and spin-down *s-p* electrons. These electrons are in broad energy bands with low effective masses. This assumption is believed to be good at temperatures significantly below the magnetic ordering temperature of the magnetic material so that there is little spin-mixing between the two conduction channels. Mott theorized that the conductivity can be significantly different in the two spin channels because the conduction-electron scattering rates in these two channels will be related to the corresponding spin-up or spin-down density of empty states at the Fermi level. These states will be largely of *d* character, and as a result of the exchange split *d* bands, the ratio of spin-up to spin-down density of empty states at the Fermi level can be significantly different in the ferromagnetically ordered states of Fe, Ni, Co, and their alloys. Consequently, this leads to the possibility of substantially different mean-free paths $\lambda^{\pm}$ and conductivities $\sigma^{\pm}$ in the two channels. In Co, for example, the density of states at the Fermi level is ten times higher for down-spin (minority) electrons as compared with up-spin (majority) electrons (56). (For in-depth reviews see, for example, 55–58.)

Detailed theories of the giant magnetoresistance effect in magnetic multilayers have been developed (52, 59–66). However, it is beyond the scope of this short review to discuss these models in any detail, but see the detailed discussion in reference (8). The simplest model for a magnetic multilayer that results in GMR is an equivalent resistor network model (53, 64), shown schematically in Figure 4. In this model, each of the ferromagnetic and nonmagnetic spacer layers consists of two resistors corresponding to the two conductivity channels associated with the up- and down-spin electrons. In the ferromagnetic layers, the resistivity is spin-dependent, $\rho_{\mathrm{M}}^{\pm}$, whereas in the spacer layers, the resistivity in the two channels is identical, ρ_{S}. The resistance of the multilayer is then equivalent to that of a total of eight resistors, with four resistors in each channel. The net resistivities of the two channels can be treated as resistors in parallel. Appropriately summing the resistors within a given channel is more complicated, but there are two simple cases (53). For short mean-free paths compared with the thickness of the layers, the resistors are independent and should themselves be added in parallel. Under these circumstances, it is clear that the resistance in the ferromagnetic and antiferromagnetic configurations is the same, and consequently there is no magnetoresistance defined as $\Delta R/R = (R_{\mathrm{AF}} - R_{\mathrm{F}})/R_{\mathrm{F}}$, where R_{AF} and R_{F} are the resistances corresponding to the AF and F configurations. Another straightforward case is when the mean-free paths are long compared with the layer thicknesses in the multilayer. Then the resistivity is an average of the resistivity of the various layers in the multilayer, in proportion to the thicknesses of the corresponding layers. Note that for the F configuration, only two resistivities must be averaged, but in the AF configuration there are four resistivities to be considered. Taking these averages, and subsequently adding the resistivities of the two spin channels in parallel, leads to $\Delta R/R = [(\alpha - \beta)^2]/[4(\alpha + N/M)(\beta + N/M)]$, where N and M are the thicknesses of the spacer and ferromagnetic layers, and $\alpha = \rho_{\mathrm{F}}^{+}/\rho_{\mathrm{S}}$ and $\beta = \rho_{\mathrm{F}}^{-}/\rho_{\mathrm{S}}$. The magnetoresistance in this model depends on two parameters, α/β and $M\beta/N$. This model shows, not surprisingly, that the magnitude of $\Delta R/R$ is strongly dependent on the scattering asymmetry between the spin conduction channels in the ferromagnetic layers. Of course it is irrelevant in which spin channel the scattering is stronger, and the magnitude of the MR depends on how much the ratio α/β differs from

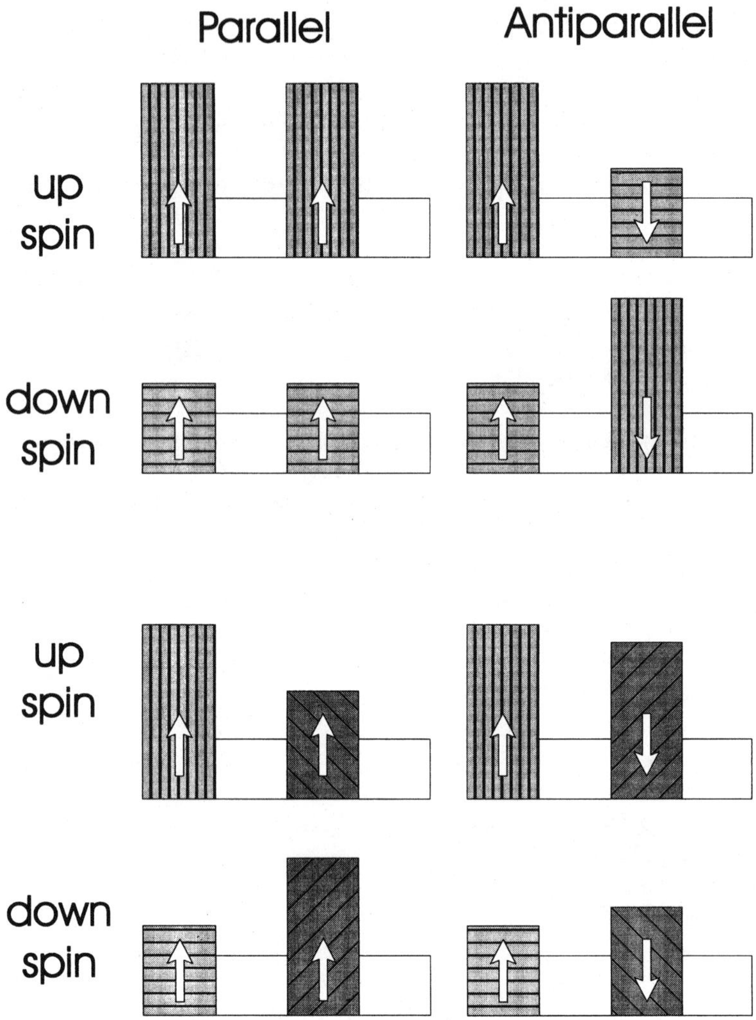

Figure 4 A simple resistor network model of GMR in which the height of the columns is proportional to the resistivity of the magnetic (*arrows*) and spacer layer (*white boxes*) metals for the two independent up- and down-spin channels. The magnetic orientation of the magnetic layers is shown by the direction of the arrow inscribed in the relevant box. Two different multilayers are considered. In the top portion of the figure, a simple multilayer containing one type of magnetic layer is described. In the bottom portion, a resistor network model is shown for a magnetic multilayer containing two different types of magnetic layers separated by the same spacer layer.

1. This highly simplified model also predicts that for a constant ratio α/β, the MR decreases monotonically with increasing spacer layer thickness, falling off as $1/N^2$ for large N. The MR is actually found to decrease exponentially with N for large N (67). The reason for this discrepancy is that the resistor network model is no longer applicable for N large compared to the mean-free path in the spacer layer. Such a simple resistor network model can easily give values of MR exceeding 100% for α/β ratios of $\simeq 8$ to 10 (13, 53), and these ratios are considered reasonable (56, 57). Nevertheless, we note that, as shown by the resistor network model, the magnitude of the MR is expected to be related to the ratio of the scattering rates within the two conduction channels no matter where the spin-dependent scattering takes place. The scattering asymmetries have been indirectly determined from measurements of the resistivity of magnetic ternary alloys (56–58). However, no correlation between the magnitude of the scattering asymmetries from studies of bulk magnetic alloys and the magnitude of the MR in magnetic multilayers has yet been found.

The simple resistor network model discussed above predicts that the resistance of a magnetic multilayer will be higher for antiparallel alignment of the magnetic layers as compared with parallel alignment. This is schematically demonstrated in the upper portion of Figure 4, where it is clear that the lower resistance for parallel alignment of the magnetic layers results from a short circuit of the current through one of the spin conduction channels (shown in the figure as the down-spin channel). A simple extension of this model can be made for the case of a ferromagnetic multilayer containing two different ferromagnetic layers with different ratios (α/β), shown in the cartoon in the lower part of Figure 4. In this case, the resistance of the multilayer will be lower for the antiparallel alignment of neighboring magnetic layers, thereby leading to a positive or an inverse giant magnetoresistance effect. This effect has not yet been unambiguously demonstrated, but the observation of a large inverse GMR effect would be important to clarify theoretical models of GMR.

INVERSE GIANT MAGNETORESISTANCE IN SPIN-ENGINEERED MULTILAYERS

The interlayer exchange coupling strength in antiferromagnetically coupled magnetic multilayers can be readily measured from the field required to rotate the magnetic moments of neighboring magnetic layers parallel to one another (68–71). Indeed, the strength of the AF interlayer coupling J_{AF} is related to the magnetization M_S and thickness t_F of the magnetic layers and the field H_S needed to rotate the magnetic moments of these layers parallel to one another, by the relation, $J_{AF} \simeq H_S M_S t_F/4$

(71, 72). It is more difficult to determine ferromagnetic coupling strengths, although these can be calculated from the stiffness of spin wave modes in ferromagnetic resonance (73) or Brillouin light-scattering studies (50). A simple technique to measure ferromagnetic coupling strengths in multilayers was developed by spin-engineering appropriate structures such that the coupling strength could be inferred directly from the magnetization vs field loop (24). The method is shown in the upper portion of Figure 5. The basic idea is to ensure that the two magnetic layers of interest (the top and middle Co layers in the example shown in Figure 5), which are ferromagnetically coupled, have their magnetic moments arranged antiparallel to one another for some field range. In Figure 5, the two uppermost Co layers are weakly ferromagnetically coupled through a spacer layer of Cu. A third Co layer (the bottom layer in the figure) is designed to be strongly antiferromagnetically coupled to the middle Co layer through a thin layer of Ru. It turns out that Co layers are coupled strongly antiferromagnetically for Ru layers a monolayer or so thick (19). Thus a structure has been spin-engineered in which at low fields the top two ferromagnetically coupled Co layers are aligned antiparallel to the applied field. This occurs because the magnetic moment of the bottom Co layer is designed to be slightly larger than the sum of the magnetic moments of the top two Co layers. Because the ferromagnetic coupling of the top two layers is much weaker in strength than the strong antiferromagnetic coupling between the bottom two Co layers, when a magnetic field is applied, the magnetic moment of the topmost Co layer rotates in the field and becomes aligned with the field for intermediate field strengths. Thus, as shown in Figure 5, the two ferromagnetically coupled Co layers become aligned antiparallel to one another. The field required to rotate the top layer in this way will be proportional to the ferromagnetic interlayer coupling strength (provided the middle Co layer remains blocked). By further increasing the applied magnetic field strength, the middle layer magnetic moment will eventually rotate parallel to the applied field. The field strength required to do this is determined primarily by the strength of the antiferromagnetic coupling of this layer to the bottom Co layer. The magnetization vs field strength of a typical Co/Ru/Co/Cu/Co sample is shown in Figure 5. It behaves exactly as described above. Similar structures have been used to measure oscillations in ferromagnetic interlayer coupling via Ru (24), Cu (74), and Pd (75).

The dependence of the resistance of the spin-engineered structure with field shown in Figure 5 is unusual. The structure does indeed exhibit GMR, but as expected from the unusual variation of the magnetic structure with field, the resistance of the structure increases for small fields. This is simply because the relative orientation of the two Co layers surrounding the Cu

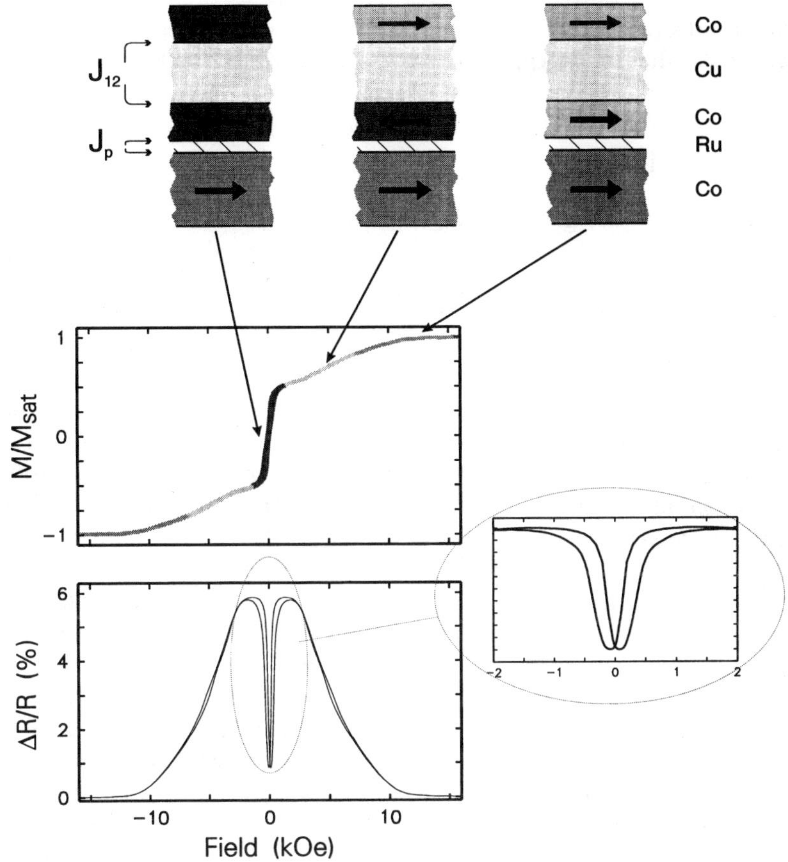

Figure 5 Schematic diagram of a spin-engineered structure in which two Co layers are ferromagnetically exchange coupled through a Cu layer of the appropriate thickness with an interlayer coupling strength J_{12}. The middle Co layer is antiferromagnetically coupled to a third (lower) Co layer through a thin Ru layer, with a coupling strength J_p. The moment of the lower Co layer is set to be equal to or slightly larger than the sum of the moments of the top two Co layers. The magnetic arrangement of the Co layers is shown for small, intermediate, and strong magnetic fields (*left*, *middle* and *right*, respectively). These arrangements correspond to the regions of the magnetization vs field and corresponding resistance vs field curves as indicated by the arrows. The inset at the right of the figure is an enlarged view of the low field region of the resistance vs field curve, which demonstrates an inverse GMR effect (see 24).

layer changes from being parallel to being antiparallel to one another. Thus this structure exhibits an inverse GMR effect, although its origin is simply related to the unusual field dependence of the magnetic configuration of the magnetic layers. At higher fields, the resistance of the structure decreases as the magnetic moments of all three magnetic layers become parallel to the applied field, as shown in the figure.

GIANT MAGNETORESISTANCE AT LOW FIELDS IN MAGNETIC MULTILAYERS

For applications of giant magnetoresistance to magnetoresistive sensors for magnetic recording applications, large changes in resistance at low fields of $\simeq 5$ to 10 Oe are required (28). At present, the magnetoresistive component of such sensors is comprised of a thin film alloy of Ni and Fe, permalloy (Py), $Ni_{81}Fe_{19}$. Permalloy exhibits a reasonably large magnetoresistance at room temperature of nearly 4% for fields of less than 1 Oe for films of thickness $> \simeq 1000$ Å (11). Note, however, that the typical operating temperature in a high-capacity magnetic disk storage drive is likely to be significantly higher than room temperature, perhaps $\simeq 100$°C. An important stimulus to much work on GMR is to develop structures with large GMR values at the lowest possible fields. In addition to large low-field GMR, the structures must possess certain other properties if they are to be useful as MR sensors for magnetic data storage applications. These properties include magnetic stability and reproducibility; stability against the elevated temperatures required for processing of the MR device; low magnetostriction so that the material will be insensitive to strains induced during processing; stability against electromigration due to the very high current densities (10^7 to 10^8 A/cm^2) the material will experience in order to obtain the required signal; and stability against corrosion from the environment within which the material will operate. These conditions are quite stringent and difficult to meet.

Dependence of Giant Magnetoresistance on Spacer Layer Thickness

At first glance, the very large GMR values in Figure 1 are extremely promising for MR read head applications. However, the saturation fields are clearly too high for useful application to MR recording heads. As shown in Figure 2, the magnitude of the GMR decays rather slowly with increasing Cu layer thickness in Co/Cu multilayers. Indeed, the GMR decays approximately as the inverse of Cu layer thickness at low temperatures, where the mean-free path of the conduction electrons within the Cu layers is long compared to the Cu layer thickness. At higher temperatures, where the mean-free path becomes comparable to the range of Cu layer thicknesses shown in Figure 2, the GMR decays more rapidly, approximately as $\Delta R/R \propto 1/t_{Cu} \exp^{-(t_{Cu}/\lambda_{Cu})}$, where t_{Cu} is the Cu layer thickness and λ_{Cu} describes scattering within the Cu layer interior (67). This functional dependence can be understood by considering two effects. First, increasing the Cu layer thickness dilutes the interfacial scattering regions because the measuring current, which is parallel to the layers, is shunted through the bulk of the Cu layers away from these regions. Second, scattering within the interior of the Cu layers will clearly diminish the flow of electrons between Co layers. This scattering can be described by a scattering length λ_{Cu}, which is likely related to the mean-free path in thick Cu layers, where the contribution from scattering at the surfaces of the film is small. Thus, for sufficiently thick Cu layers where most of the current is carried by the Cu layers, we can account for the variation of GMR with Cu layer thickness (67). In contrast, the interlayer exchange coupling decays much more rapidly with increasing Cu layer thickness (22, 74, 76–79). For example, this means that for Cu layers about twice as thick as those contained in the Co/Cu multilayer in Figure 1, the GMR is only reduced by a factor of two, but the saturation field is about 20 times smaller. Thus GMR values of almost 35% for fields of about 100 Oe are possible for Cu layer thicknesses of $\simeq$20 Å (77). Such materials may be useful for magnetic field MR sensors.

Giant Magnetoresistance in Permalloy/Au Multilayers

As mentioned above, oscillatory interlayer coupling has been observed in a wide variety of transition metal multilayers. Whereas the oscillation period is similar for many systems and there is no simple variation of the period with spacer layer material, there is a systematic variation of the interlayer coupling strength with the position of the transition metal spacer element in the periodic table. Indeed, the coupling strength is found to increase exponentially with *d*-band filling across the 3*d*, 4*d*, and 5*d* periods. In addition, the coupling strength increases systematically from the 5*d* to 4*d* to 3*d* elements in any given column of the periodic table (25). Based on these considerations and taking into account that noble metal spacer layers typically give the largest GMR values, good candidates for the large GMR at low saturation fields (low interlayer coupling) are multilayers with Au spacer layers. Indeed, one of the first observations of GMR in magnetic multilayers was in perpendicularly magnetized and magnetically weakly coupled or uncoupled Co/Au/Co sandwiches (80, 81). The two Co layers were sufficiently thin that the magnetic interface anisotropy was able to overcome the demagnetization field from the Co magnetization so that the magnetic moments were perpendicularly magnetized in zero magnetic field. By preparing the two Co layers with slightly different thicknesses, and correspondingly different magnetic perpendicular anisotropies, the moments of the two Co layers switched at different field strengths. Thus there was a certain field range where the Co moments were aligned antiparallel to one another. The resistance of this structure was higher than for fields where the Co moments were aligned parallel to one another. This structure is similar to subsequent observations of GMR in permalloy/Cu/Co/Cu multilayers (82) and exchange-biased sandwiches (83) in which successive magnetic layers are designed to have different in-plane coercivities or switching fields, as discussed further below.

Antiferromagnetic coupling across Au spacer layers and oscillations in the magnetic interlayer coupling with Au thickness have been observed in Fe/Au/Fe sandwiches and Fe/Au multilayers (84–87), although only small GMR values were observed (86). Recently, oscillatory coupling via Au sandwiched between Co layers (88) and permalloy layers (89) has been found. The latter system displays large MR values at very low fields at room temperature, as described below.

The magnitude of the saturation magnetoresistance vs Au layer thickness is plotted in Figure 6*a* for three families of (111)-oriented Py/Au multilayers. The multilayers were prepared by MBE deposition and were grown on (0001) sapphire substrates. The [111] crystalline orientation was established by using a thin Pt seed layer, $\simeq 20$ Å thick, grown at 600°C. Related saturation fields determined from the magnetoresistance curves are shown in Figure 6*b*. They correspond to the field at which the MR is half that of the saturation MR. Four well-defined oscillations in saturation magnetoresistance and saturation field are found as the Au layer thickness is increased (indicated by AFn, $n = 1$–4 in Figure 6*a*). Typical resistance vs in-plane magnetic field curves are shown in Figure 7 for samples corresponding to the four maxima in Figure 6*a* and *b*. Note that a small amount of Au was deliberately inserted in the permalloy layers in two of the families of samples depicted in Figure 6, by coevaporation from the permalloy and Au sources. This was an attempt to take advantage of the likely surface-segregation of the Au from the Py layers during growth so as to reduce the possibility of magnetic pin-holes between the Py layers. Similar samples grown without Au in the Py layers show suppression of the MR peak near $t_{Au} \simeq 11$ Å (AF1). An analogous surfactant effect has been reported for copper surface-segregation through cobalt during epitaxy of exchange-coupled Co/Cu (111) multilayers (90). Note that alloying the permalloy with Au has little effect on the properties of the system for thicker Au layers, except that the resistivity of the multilayers is increased. For example, at the second MR peak, the resistivity is increased by about 30%, consistent with the decreased MR of these samples (see Figure 6, *open squares*) relative to otherwise similar samples.

Typical magnetization vs in-plane magnetic field are shown in Figure 8*a* and *b* for two Py/Au multilayers from the AF1 and AF2 peaks. The magnetization loops are consistent with the resistance curves shown in

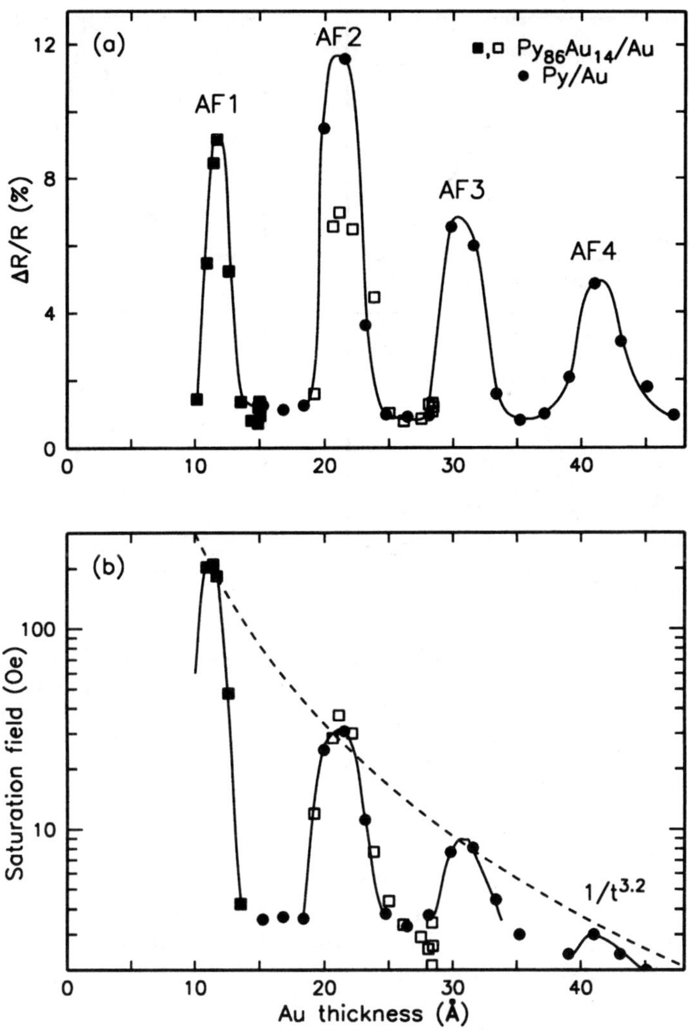

Figure 6 Room temperature saturation magnetoresistance (*a*) and saturation field (*b*) vs Au spacer layer thickness at 295 K for three series of Py/Au multilayers. The open and solid boxes correspond to two samples grown with $Py_{86}Au_{14}$ magnetic layers, and the solid circles correspond to samples prepared with Py layers. Four peaks in MR and saturation field are observed in the Au thickness range spanned, as indicated by AF1, AF2, AF3, and AF4 (from 89).

Figure 7 and indicate antiferromagnetic coupling of the permalloy layers. The degree of antiferromagnetic coupling can be inferred from the remanent magnetization near zero field. Figure 8 demonstrates almost perfect antiferromagnetic alignment at the AF2 peak but incomplete ($\simeq$45%) AF

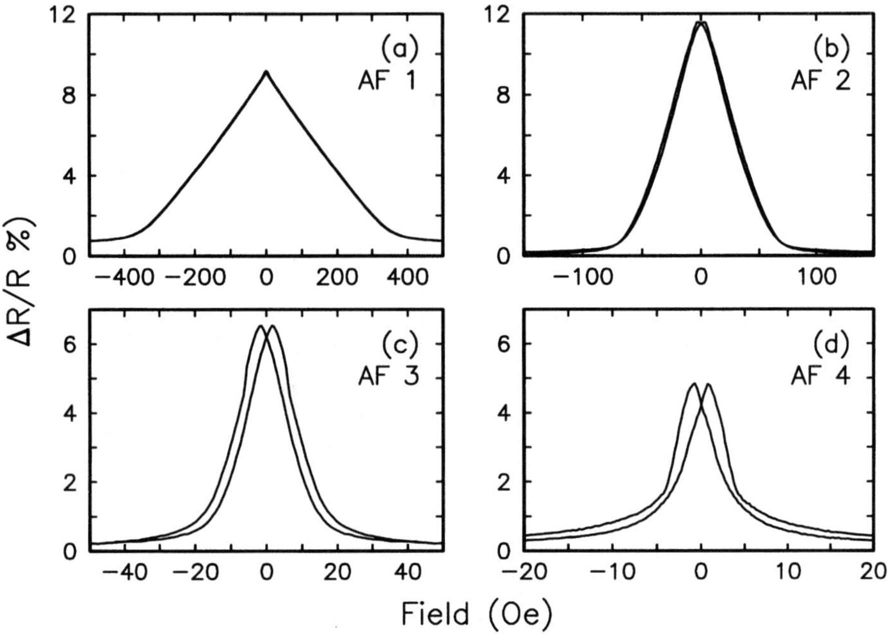

Figure 7 Room temperature resistance vs in-plane magnetic field curves for four samples corresponding to the four maxima in MR shown in Figure 6, with Au layer thicknesses of (*a*) AF1: 11.7 Å, (*b*) AF2: 21.5 Å, (*c*) AF3: 29.8 Å, and (*d*) 41.0 Å. The corresponding resistivities of these samples at room temperature are $\simeq$43, 19, 13, and 10 μOhm-cm, respectively, in fields large enough to saturate the magnetization of the samples.

coupling at the AF1 peak. The detailed relationship of magnetization and MR is explored in Figure 8*c* and *d*. Because permalloy is a soft magnetic material whose magnetization is saturated in small fields, the simplest model of GMR can be used in which the resistance is proportional to the cosine of the angle between the magnetic moments of neighboring layers (91). Then the resistance of the structure is expected to vary as $-M(H)^2$, where $M(H)$ is the magnetization parallel to the applied field H. Figures 8*b* and *d* show a comparison of resistance and $-M^2$ vs field curves. Excellent agreement is obtained confirming the expected relationship. Note that for the AF1 sample, the MR is compared with $-(M-M_0)^2$, where M_0 is the residual magnetization in small fields. M_0 represents the portion of the sample ($\simeq$55%) containing ferromagnetically coupled Py layers due to structural defects. These layers do not contribute to the magnetoresistance. If there were complete AF coupling of the Py layers, these data would suggest that the magnitude of the MR at the AF1 peak would be

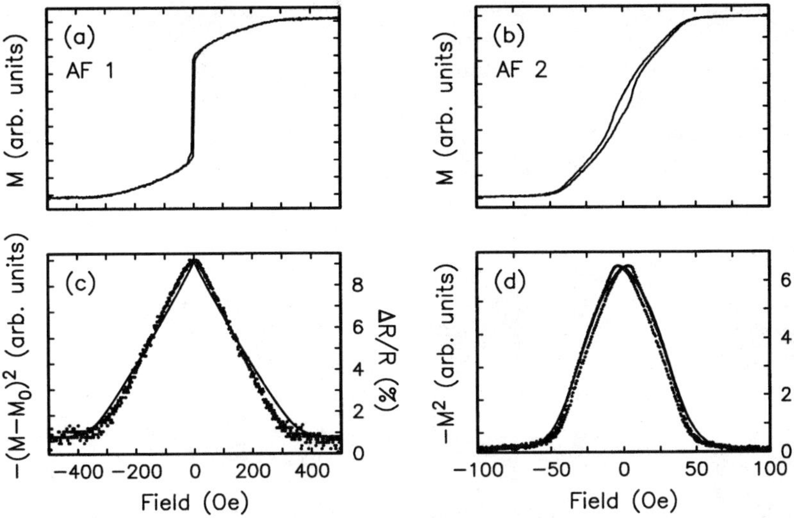

Figure 8 Magnetization vs field curves, (*a*) and (*b*), for two Py-Au/Au multilayers from the AF1 and AF2 peaks shown in Figure 6. In (*c*) and (*d*), corresponding resistance vs field curves (*solid lines*) and plots of $-(M-M_0)^2$ vs field (*solid circles*) are shown. M_0 is the remanent magnetization.

much higher (MR $\simeq$ 20%) for multilayers containing Py-Au magnetic layers and yet higher still for pure Py layers.

The magnetic interlayer coupling in the Py/Au multilayers is relatively small. For example, at the AF1 peak, J_{AF} is $\simeq 0.02$ erg/cm^2. This is approximately five times larger than that found in sputtered (111) textured Py/Au multilayers (92). Although the magnitude is similar to that found in sputtered Py/Cu multilayers (93), if we assume J_{AF} is proportional to M^2, the magnetic coupling through epitaxial (111) Au is about an order of magnitude smaller than the coupling found at AF1 in epitaxial (111) Cu (94). Note that as mentioned above for Co/Cu multilayers, the strength of the coupling falls off rapidly with thicker Au layers. Indeed, as shown in Figure 6*b*, the coupling varies as approximately $1/t^n_{Au}$, where $n = \simeq 3.2$. Such a dependence is more rapid than predicted in simple Ruderman-Kittel-Kasuya-Yosida (RKKY) models of the coupling (95).

Interlayer coupling is usually found to be weaker in polycrystalline multilayers than in similar single crystalline multilayers. Note that the lowest reported saturation fields at room temperature in magnetic multilayers are found in sputtered Py/Au multilayers where changes in resistance of $\simeq 1$ to 2%/Oe can be obtained (92).

Giant Magnetoresistance in Exchange-Biased Sandwiches

An important structure for obtaining GMR at low magnetic fields is shown in Figure 9. This structure, an exchange-biased sandwich (EBS), is in its simplest embodiment, $F_I/S/F_{II}$/FeMn. The EBS contains two ferromagnetic layers, F_I and F_{II}, and a single nonmagnetic spacer layer, *S*, in

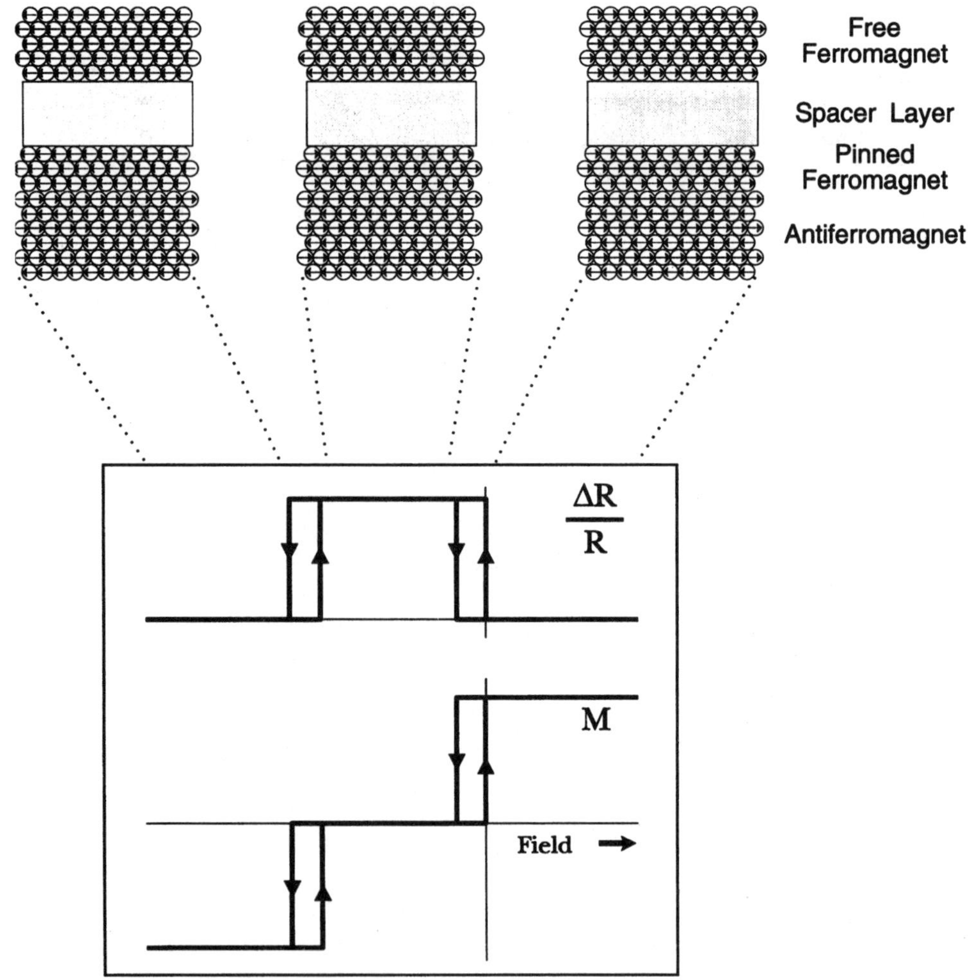

Figure 9 Schematic diagram of an exchange-biased sandwich structure.

which one of the magnetic layers, F_{II}, is exchange coupled to an antiferromagnetic layer. EBS structures with antiferromagnetic exchange bias layers of FeMn (83), NiO (96), and TbFe (97) have been reported. The structure takes advantage of a phenomenon first discovered more than 30 years ago in oxidized Co particles (98) and subsequently extensively studied in a number of thin film systems (99–101). This phenomenon, often referred to as exchange anisotropy, arises from an interfacial magnetic exchange coupling between an antiferromagnetic layer and a ferromagnetic layer. Under appropriate conditions, the exchange anisotropy results in a unidirectional anisotropy in the F layer such that its magnetic hysteresis loop is centered about a non-zero field, a bias field H_B. In contrast, providing that the magnetic coupling of F_I and F_{II} via the spacer layer is sufficiently weak, the magnetic hysteresis loop of F_I is centered close to zero field. The moments of F_I and F_{II} are thus aligned antiparallel for some field range intermediate between zero and H_B. A resistance vs field curve is shown in Figure 10*a* for a typical EBS structure, where F_I and F_{II} are Py and S is Cu. The current and field are aligned along the unidirectional anisotropy direction. The structure displays a giant MR effect exactly analogous to that in multilayers with a higher resistance for fields where F_I and F_{II} are antiparallel. As is found for multilayered structures (22, 93), replacing the Py layers with Co layers of the same thickness increases the magnetoresistance of the structure by approximately a factor of two, as shown in Figure 10*b*.

The quest for structures exhibiting large changes in resistance per unit field has motivated a great deal of work in GMR. The simplest structure with the largest change in resistance per unit field at room temperature is shown in Figure 10*c*. This structure is comprised of elements of the two EBS structures shown in Figure 10*a* and *b*. As can be seen from Figure 10*a*, the Py/Cu/Py EBS has a small switching field and values of $\Delta R/R$ of about 2–4% at room temperature. Similar Co/Cu/Cu EBS have higher MR values, as high as $\simeq 9\%$ at room temperature, but larger switching fields. The increased switching field results from the higher magnetic anisotropy of Co as compared with Py. To obtain the highest values of MR per unit field, EBS sandwiches such as those shown in Figure 10*c* have been developed (102). These structures are essentially comprised of a Py/Cu/Py EBS in which thin Co layers are introduced at each Py/Cu interface. The Co layers need only be approximately 1 atomic layer thick, or just sufficient to completely cover the Py/Cu interface. These structures also dramatically demonstrate the overwhelming importance of spin-dependent scattering at the magnetic/nonmagnetic interfaces (102, 103).

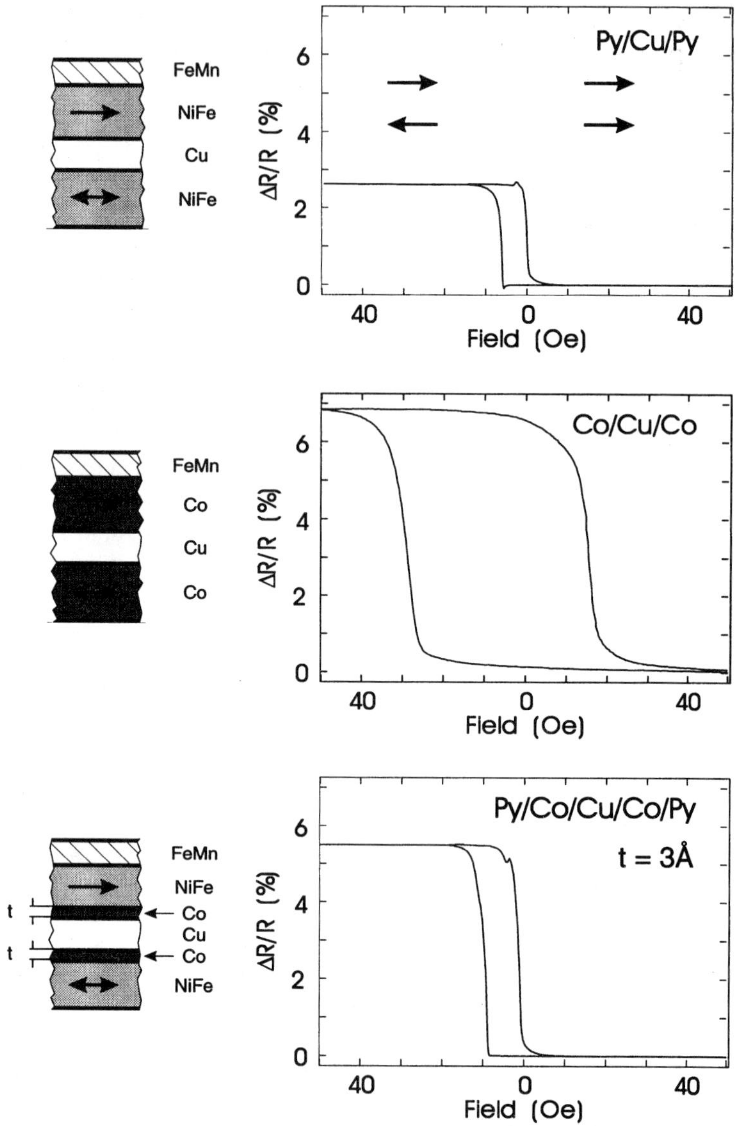

Figure 10 Room temperature resistance vs field curves for (*a*) Si/Py(53 Å)/Cu(32 Å)/Py(22 Å/FeMn(90 Å)/Cu(10 Å); (*b*) the same structure with the Py (permalloy = $Ni_{81}Fe_{19}$) layers replaced by Co; and (*c*) the same structure as in (*a*) but with 3.0 Å-thick Co layers added at each Py/Cu interface. (Note the thicknesses of the Py layers have correspondingly been reduced by 3.0 Å.)

DEPENDENCE OF GIANT MAGNETORESISTANCE ON STRUCTURE

In multilayers, the magnitude of the saturation giant magnetoresistance is strongly dependent on the details of the multilayer structure. In particular, the GMR will depend on the thicknesses of both the magnetic and nonmagnetic layers and the number of bilayers, as well as on the composition and thickness of any seed layers and capping layers. In addition, the GMR is sensitive to the detailed morphology of the structure, which can be changed by varying deposition conditions such as the deposition temperature, the detailed method of preparation, and the substrate. The GMR will also be decreased if the magnetic layers are not completely coupled antiferromagnetically to one another in the high resistance state. This means that it is very easy to obtain different values of GMR for nominally the same structure, and it is very difficult to make meaningful comparisons of results on individual samples from different groups. It is most useful to study sets of samples made under identical conditions where one aspect of the structure is systematically varied.

Dependence of GMR on Number of Bilayers

Perhaps one of the most obvious aspects of the structure of a magnetic multilayer that can be varied is the overall thickness of the multilayer, as determined by the number of bilayers N. The magnitude of the GMR can be, and usually is, strongly dependent on N. Figure 11 shows typical resistance vs field curves for representative samples from two series of Fe/Cr multilayers with different N. The series differ only in the thicknesses of the Cr overlayer t_o and the Cr underlayer t_u. A minimum Cr underlayer thickness of $\simeq 10$ Å was required for the growth of well-layered structures. As can be seen from Figure 11, the measured GMR is increased with increasing N in both series of structures. However, the rate of increase of GMR with N is significantly different for the two series. This is clearly demonstrated in Figure 12, where the detailed dependence of MR on N is displayed for the two families of structures. The magnetoresistance is observed to increase much more rapidly with N for the $[Fe/Cr]_N$ multilayers with very thin Cr under- and overlayers than for the structures with thicker Cr layers. The origin of this effect is similar to that described above for the dependence of GMR on Cu layer thickness in Co/Cu multilayers. The GMR in the Fe/Cr multilayers is reduced in proportion to the measuring current passing through the Cr under- and overlayers. The dilution of the $[Fe/Cr]_N$ portion of the multilayer contributing to the giant MR, by the noncontributing portion of the structure, will be decreased as the relative proportion of the noncontributing portion decreases with increased N.

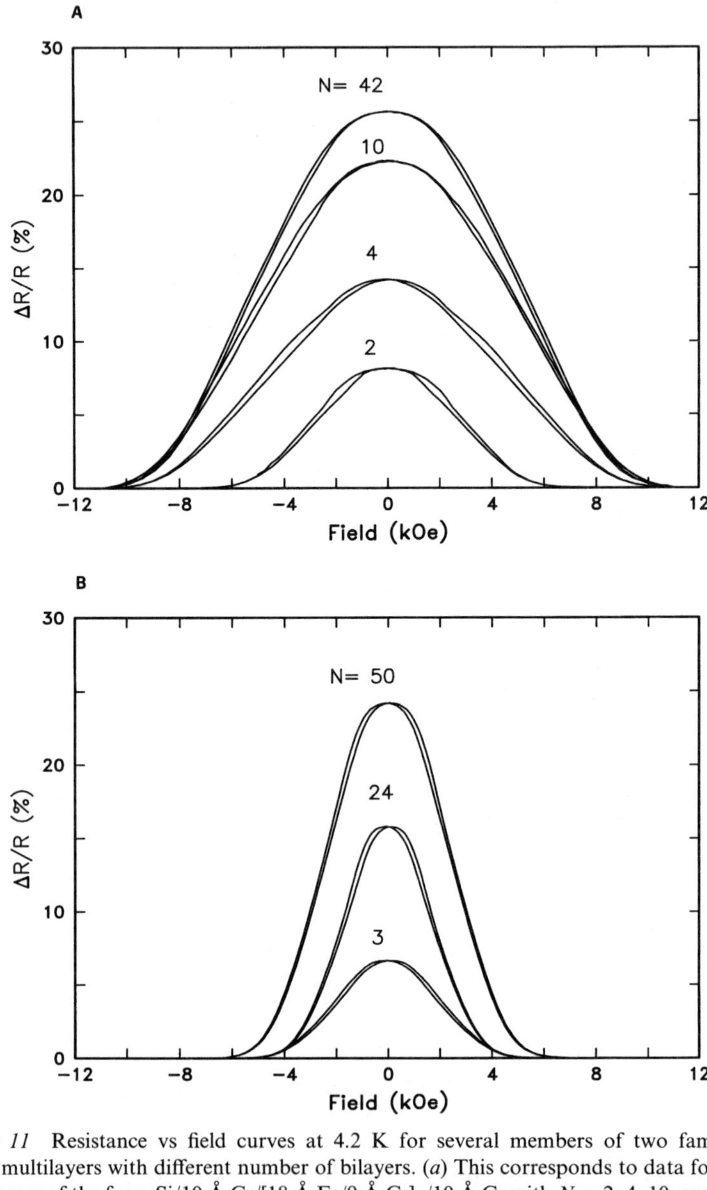

Figure 11 Resistance vs field curves at 4.2 K for several members of two families of Fe/Cr multilayers with different number of bilayers. (*a*) This corresponds to data for Fe/Cr multilayers of the form Si/10 Å Cr/[18 Å Fe/9 Å Cr]$_N$/10 Å Cr with $N = 2$, 4, 10, and 42. (*b*) This corresponds to data for Si/115 Å Cr/[16 Å Fe/11.5 Å Cr]$_N$/115 Å Cr with $N = 3$, 24, and 50.

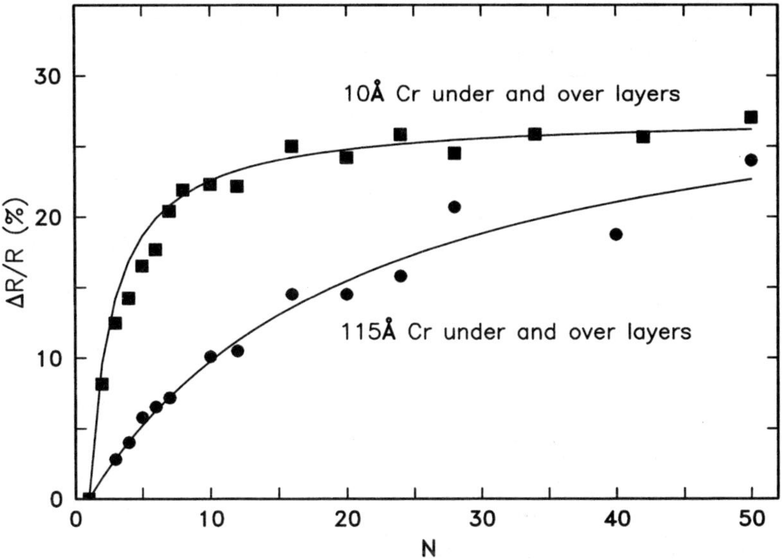

Figure 12 Dependence of saturation magnetoresistance on number of bilayers for two series of Fe/Cr multilayers with structure of the form (*upper line*) Si/10 Å Cr/[18 Å Fe/9 Å Cr]$_N$/10 Å Cr; and (*lower line*) Si/115 Å Cr/[16 Å Fe/11.5 Å Cr]$_N$/115 Å Cr.

The results shown in Figure 12 are rather similar to those found for many magnetic multilayered systems. It is not surprising, therefore, that the largest GMR values in Fe/Cr (and other multilayers) are found in multilayers with large values of N (33, 34). The dependence on N is not completely determined by a simple dilution effect; there are several other factors. First, an important and very simple factor affecting the dependence of GMR on N is that the two magnetic layers at either end of the multilayer contribute to GMR only half as much as the magnetic layers within the interior of the multilayer (assuming diffuse scattering at either extremity of the multilayer). Second, as described in the very first models of GMR (59, 61, 62), the GMR will be increased if the electrons propagate across many magnetic/nonmagnetic interfaces within a conduction carrier mean-free path. When N is small compared with some mean-free path within the multilayer, the GMR is reduced. Thus it follows that the dependence of GMR on N will be temperature dependent, as described by the temperature dependence of the mean-free path. This effect is much more important as the conductivity of the multilayer increases: for example, in multilayer systems containing metals with long mean-free paths such as Cu, Ag, and Au and in single-crystal, defect-free multilayer systems. Third, the resistance of the film will increase as the film thickness is decreased because of scattering at the film surfaces. This leads to a decreased GMR effect.

Finally, it must be noted that the structure of the multilayer itself will vary with total film thickness. For polycrystalline films, the grain size often increases in proportion to the film thickness. The GMR is clearly sensitive to the grain size (35). The variation in number of grain boundaries with film thickness will lead to variations in film resistivity and thus GMR. For single-crystal films with no large-angle grain boundaries, the changes in structure with film thickness may be more subtle. For Fe/Cr, it has been proposed that increases in GMR with N may be attributed to an increase in the roughness of the Fe/Cr interfaces at some length scale (104). Indeed, a particularly important aspect of the origin of GMR, which is not discussed in detail in this article, is the role of the structure of the interfaces, which remains controversial.

Dependence of GMR on Magnetic Layer Thickness

A basic assumption of the resistor network model of GMR described above is that the spin-dependent scattering giving rise to the GMR originates purely within the interior of the magnetic layers, i.e. from bulk spin-dependent scattering. This model can be readily generalized to allow for spin-dependent interfacial scattering by adding additional resistors in the network representing the scattering in the interfacial regions. The relative contribution of spin-dependent scattering from bulk scattering and from spin-dependent scattering at the interfaces between the magnetic and spacer layers is a subject of great current interest.

A simple experiment to examine the role of interfacial vs bulk spin-dependent scattering is to vary the interfacial content of the magnetic multilayer by varying the thickness of the magnetic layers. Figure 13 shows various resistance vs field curves measured at 4.2 K for sets of otherwise identical Fe/Cr multilayers in which the Fe layer thickness is systematically varied from $\simeq 1.5$ to 300 Å. The magnitude of the saturation GMR for these samples is plotted in Figure 14 vs Fe layer thickness. These data show that maximum GMR values are obtained for thin Fe layers $\simeq 8$ Å-thick, which strongly suggests that interfacial scattering is of predominant importance. Note, however, that the GMR is diluted by the presence of necessary underlayers (for growth) and overlayers (for protection against corrosion). This means that the Fe thickness for which maximum GMR is obtained will be affected by the presence of these layers and by the thickness of the Cr spacer layers themselves. Increasing the thickness of these layers will increase the thickness of Fe necessary to obtain maximum

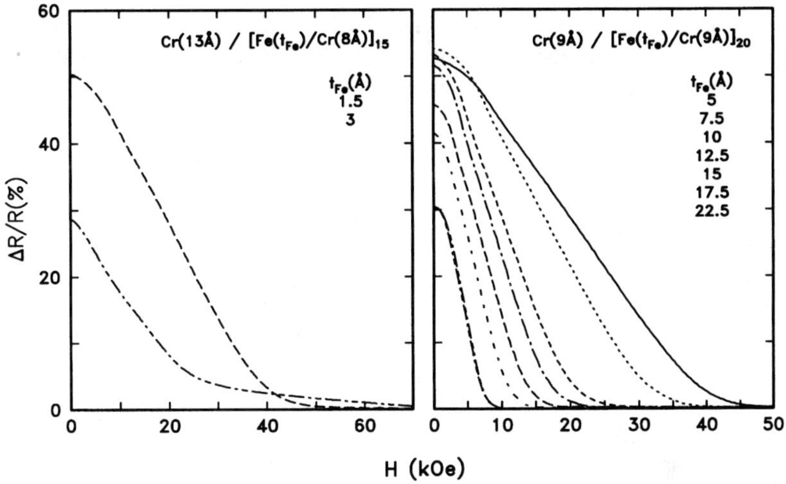

Figure 13 Resistance vs field curves for several Si/Cr(t_u)/[Fe(t_{Fe})/9 Å Cr]$_{20}$ multilayers with varying Fe layer thickness t_{Fe}. The Cr underlayer thickness t_{Cr} is $\simeq 10$ Å. On the left panel, the sample with $t_{Fe} = 1.5$ Å has the smaller GMR. On the right panel, the saturation field systematically increases from the largest to the smallest value of t_{Fe} shown.

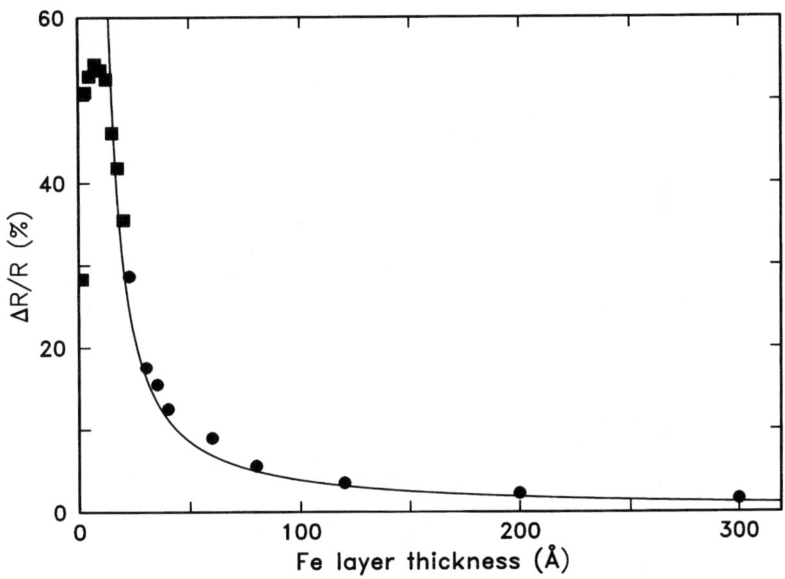

Figure 14 Dependence of saturation magnetoresistance at 4.2 K vs Fe layer thickness for structures Si/9 Å Cr/[Fe(t_{Fe})/9 Å Cr]$_{20}$, for Fe layer thicknesses ranging from 1.5 to 300 Å. For thick Fe layers, the anisotropic magnetoresistance (AMR) becomes a significant fraction of the GMR effect. For these samples, shown as filled circles, the AMR was measured and a correction applied.

GMR. Thus it is not straightforward to interpret such studies without appealing to detailed theoretical models. Note, however, that GMR models predict a maximum in the GMR effect for Fe layers significantly thicker than those found in Figure 14 for any significant spin-dependent bulk scattering (105). Finally, note that the MR decreases approximately as the inverse Fe layer thickness for Fe layers, ranging in thickness from 20 to 300 Å, consistent with dilution of the interfacial scattering regions, and predominant interfacial spin-dependent scattering. Similarly, in granular alloys, the magnitude of GMR decreases approximately as the inverse size of the magnetic particles, as shown in Figure 15.

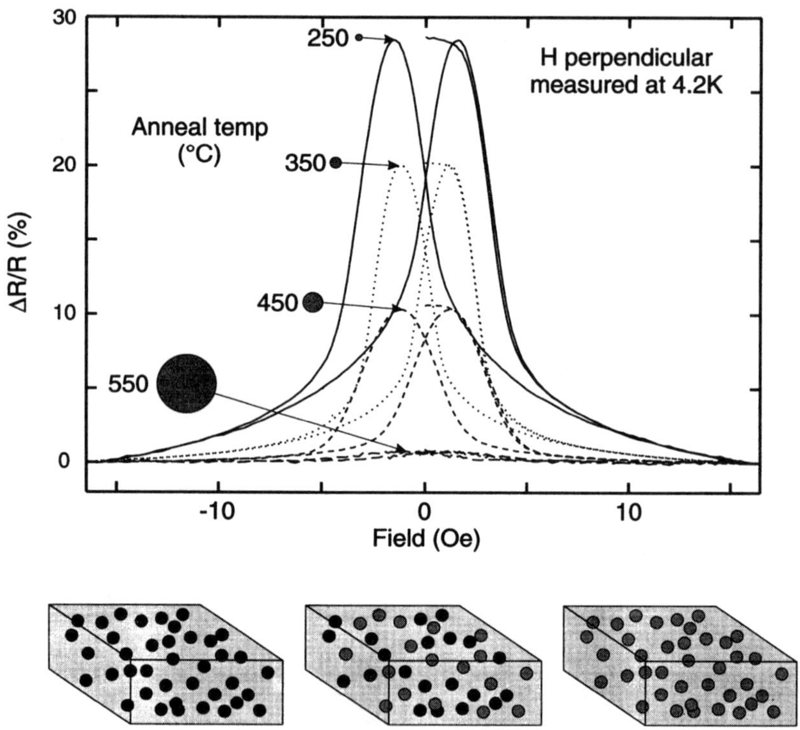

Figure 15 Variation of GMR in a single crystal (111) $Co_{28}Cu_{72}$ granular alloy at 4.2 K with annealing temperature of the alloy. The granular alloy was deposited at 200°C and successively annealed for 1 h at 250, 350, 450, and 500°C. The size of the Co particles as determined from small-angle grazing incidence X-ray scattering is shown by the size of a representative particle. Note that the largest particle size for the sample annealed at 550°C is ≃ 135 Å (see 115 for more details).

CONCLUSIONS

Interest in magnetic nano-structured metals has substantially increased in recent years with the discovery of giant magnetoresistance and oscillatory interlayer exchange coupling in magnetic multilayers and related structures. This brief review has discussed only a small amount of the published work in this field, which continues to grow at an enormous pace. Important omissions include work on detailed models of interlayer coupling and GMR; magnetotransport in magnetic multilayers and sandwiches in which the current flows perpendicular to the layers (31, 32, 106–110); the detailed role of the influence of the structural morphology, especially of the interfaces, on GMR; detailed discussions of magnetic granular alloys and multilayered granular structures (42); spin-dependent quantum confinement of electrons in the magnetic potentials created by the magnetic/nonmagnetic interfaces; and the influence of such potentials on the interlayer coupling and GMR (see, for example, PD Johnson, this volume). Magnetic nano-structures display other anomalous transport properties including giant magneto-thermopower and giant magneto-thermal conductivities (see, for example, 111), which have not been discussed herein.

It seems quite clear that giant magnetoresistance is intimately related to scattering at the interfaces between the magnetic and nonmagnetic components in both magnetic-layered structures and granular alloys. The largest GMR values are usually found for structures in which the interfacial component is maximized, by reducing either the thickness of magnetic layers or the size of the magnetic particles. Similarly, the GMR is extremely sensitive to the detailed chemical nature of the interfaces. Modifications of these interfaces by dusting with just sub-monolayer to monolayer equivalent coverages of impurity elements results in dramatic variations in GMR values (102). Indeed, GMR may thus be used to probe the electronic character of the magnetic/nonmagnetic interfaces, just as oscillatory interlayer coupling can be used to probe details of the Fermi surface topology of the spacer layer material (38, 52) and perhaps the magnetic layer material (112–114). The ability to tailor-make magnetic nano-structured materials with properties especially engineered for particular applications is very attractive and is likely to lead to important applications for such structures in the near future.

ACKNOWLEDGMENTS

I would like to thank many colleagues, especially Drs. R F C Farrow, R Marks, A R Modak and Prof. D J Smith for their contributions to parts of the work described in this paper. I am especially grateful to Kevin Roche for technical support.

Literature Cited

1. Schuller IK, Falco CM. 1982. *Prog. Phys.* 4: 492
2. Schuller IK. 1988. In *Physics, Fabrication, and Applications of Multilayered Structures*, ed. P Dhez, C Weisbuch, p. 139. New York: Plenum
3. Falicov LM, Pierce DT, Bader SD, Gronsky R, Hathaway KB, et al. 1990. *J. Mater. Res.* 5: 1299
4. Mathon J. 1991. *J. Magn. Magn. Matter* 100: 527
5. Fert A, Bruno P. 1994. See Ref. 116, p. 82
6. Hathaway KB. 1994. See Ref. 116, p. 45
7. Heinrich B, Cochran JF. 1993. *Adv. Phys.* 42: 523
8. Levy PM. 1994. In *Solid State Physics*, ed. H Ehrenreich, D Turnbull, 47: 367. New York: Academic
9. Parkin SSP. 1994. See Ref. 116, p. 148
10. Pierce DT, Unguris J, Celotta RJ. 1994. See Ref. 116, p. 117
11. McGuire TR, Potter RI. 1975. *IEEE Trans. Magn.* MAG-11: 1018
12. Binasch G, Grünberg P, Saurenbach F, Zinn W. 1989. *Phys. Rev. B* 39: 4828
13. Baibich MN, Broto JM, Fert A, Nguyen van Dau F, Petroff F, et al. 1988. *Phys. Rev. Lett.* 61: 2472
14. Grünberg P, Schreiber R, Pang Y, Brodsky MB, Sowers H. 1986. *Phys. Rev. Lett.* 57: 2442
15. Carbone C, Alvarado SF. 1987. *Phys. Rev. B* 36: 2443
16. Joyce BA. 1985. *Rep. Prog. Phys.* 48: 1637
17. Ploog K. 1988. *Angew. Chem. Int. Ed. Engl.* 27: 5931
18. Farrow RFC, Marks RF, Harp GR, Weller D, Rabedeau TA, et al. 1993. *Mater. Sci. Eng. Rep.* R11: 155
19. Parkin SSP, More N, Roche KP. 1990. *Phys. Rev. Lett.* 64: 2304
20. Parkin SSP, York BR. 1993. *Appl. Phys. Lett.* 62: 1842
21. Parkin SSP, Li ZG, Smith DJ. 1991. *Appl. Phys. Lett.* 58: 2710
22. Parkin SSP, Bhadra R, Roche KP. 1991. *Phys. Rev. Lett.* 66: 2152
23. Kano H, Kagawa K, Suzuki A, Okabe A, Hayashi K, Aso K. 1993. *Appl. Phys. Lett.* 63: 2839
24. Parkin SSP, Mauri D. 1991. *Phys. Rev. B.* 44: 7131
25. Parkin SSP. 1991. *Phys. Rev. Lett.* 67: 3598
26. Kryder MH. 1993. *Annu. Rev. Mater. Sci.* 23: 411
27. Grochowski E, Thompson DA. 1994. *IEEE Trans. Magn.* 30: 3797
28. Ciureanu P. 1992. In *Thin Film Resistive Sensors*, ed. P Ciureanu, S Middelhoek, p. 253. Bristol: IOP Pub.
29. Daughton JM. 1992. *Thin Solid Films* 216: 162
30. Prinz G. 1990. *Science* 250: 1092
31. Johnson M. 1993. *Science* 260: 320
32. Johnson M. 1993. *Appl. Phys. Lett.* 63: 1435
33. Fullerton EE, Conover MJ, Mattson JE, Sowers CH, Bader SD. 1993. *Appl. Phys. Lett.* 63: 1698
34. Schad R, Potter CD, Belien P, Verbanck G, Moshchalkov VV, Bruynseraede Y. 1994. *Appl. Phys. Lett.* 64: 3500
35. Modak AR, Smith DJ, Parkin SSP. 1994. *Phys. Rev. B*, 50: 4232
36. Parkin SSP, Roche KP, Suzuki T. 1992. *Jpn. J. Appl. Phys.* 31: L1246
37. Johnson MT, Coehoorn R, de Vries JJ, McGee NW, aan de Stegge J, Bloemen PJ. 1992. *Phys. Rev. Lett.* 69: 969
38. Bruno P, Chappert C. 1992. *Phys. Rev. B* 46: 261
39. Berkowitz AE, Mitchell JR, Carey MJ, Young AP, Zhang S, et al. 1992. *Phys. Rev. Lett.* 68: 3745
40. Xiao JQ, Jiang JS, Chien CL. 1992. *Phys. Rev. Lett.* 68: 3749
41. Parkin SSP, Farrow RF, Rabedeau TA, Marks RF Harp GR, et al. 1993. *Eur. Phys. Lett.* 22: 455
42. Hylton TL, Coffey KR, Parker MA, Howard JK. 1993. *Science* 261: 1021
43. Samant MG, Stör J, Parkin SSP, Held GA, Hermsmeier BD, et al. 1994. *Phys. Rev. Lett.* 72: 1112

44. Unguris J, Celotta RJ, Pierce DT. 1992. *Phys. Rev. Lett.* 69: 1125
45. Pappas DP, Prinz GA, Ketchen MB. 1994. *Appl. Phys. Lett.* 65: 3401
46. Unguris J, Celotta RJ, Pierce DT. 1991. *Phys. Rev. Lett.* 67: 140
47. Purcell ST, Folkerts W, Johnson MT, McGee NW, Jager K, et al. 1991. *Phys. Rev. Lett.* 67: 903
48. Demokritov SO, Tsymbal E. 1994. *J. Phys. Conden. Matter* 6: 7145
49. Cochran JF. 1994. See Ref. 116, p. 223
50. Hillebrands B, Güntherodt G. 1994. See Ref. 116, p. 258
51. Krebs JJ, Lubitz P, Chaiken A, Prinz GA. 1989. *Phys. Rev. Lett.* 63: 1645
52. Edwards DM, Mathon J, Muniz RB, Phan MS. 1991. *Phys. Rev. Lett.* 67: 493
53. Mathon J. 1991. *Contemp. Phys.* 32: 143
54. Stiles MD. 1993. *Phys. Rev. B* 48: 7238
55. Mott NF. 1964. *Adv. Phys.* 13: 325
56. Campbell IA, Fert A. 1982. In *Ferromagnetic Materials*, ed. EP Wohlfarth, 3: 747. Amsterdam: North-Holland
57. Dorleijn JW. 1976. *Philips Res. Rep.* 31: 287
58. Rossiter PL. 1987. *The Electrical Resistivity of Metals and Alloys*, Cambridge: Cambridge Univ. Press
59. Camley RE, Barnas J. 1989. *Phys. Rev. Lett.* 63: 664
60. Levy PM, Ounadjela K, Zhang S, Wang Y, Sommers CB, Fert A. 1990. *J. Appl. Phys.* 67: 5914
61. Barnas J, Fuss A, Camley RE, Grünberg P, Zinn W. 1990. *Phys. Rev. B* 42: 8110
62. Barthelemy A, Fert A. 1991. *Phys. Rev. B* 43: 13124
63. Inoue J, Oguri A, Maekawa S. 1991. *J. Phys. Soc. Jpn.* 60: 376
64. Edwards DM, Muniz RB, Mathon J. 1991. *IEEE. Trans. Mag.* 27: 3548
65. Trigui F, Velu E, Dupas C. 1991. *J. Magn. Magn. Matter* 93: 421
66. Zhang S, Levy PM, Fert A. 1992. *Phys. Rev. B.* 45: 8689
67. Parkin SSP, Modak AR, Smith DJ. 1993. *Phys. Rev. B* 47: 9136
68. Barthelemy A, Fert A, Baibich MN, Hadjoudj S, Petroff F, et al. 1990. *J. Appl. Phys.* 67: 5908
69. Dieny B, Gavigan JP, Rebouillat JP. 1990. *J. Phys. Condens. Matter* 2: 159
70. Dieny B, Gavigan JP. 1990. *J. Phys. Condens. Matter* 2: 187
71. Folkerts W. 1991. *J. Magn. Magn. Matter* 94: 302
72. Parkin SSP, Mansour A, Felcher GP. 1991. *Appl. Phys. Lett.* 58: 1473
73. Heinrich B. 1994. See Ref. 116, p. 195
74. Bloemen PJH, van Dalen R, de Jonge WJM, Johnson MT, aan de Stegge J. 1993. *J. Appl. Phys.* 73: 5972
75. Bloemen PJH, de Jonge WJM, Donkersloot HC. 1993. *J. Appl. Phys.* 73: 4522
76. Cebollada A, Miranda R, Schneider CM, Schuster P, Kirschner J. 1991. *J. Magn. Magn. Matter* 102: 25
77. Parkin SSP. 1992. In *Magnetic Surfaces, Thin Films and Multilayers*, ed. SSP Parkin, H Hopster, J-P Renard, T Shinjo, W Zinn, 231: 211. Pittsburgh: Mater. Res. Soc. Symp. Proc.
78. Johnson MT, Purcell ST, McGee NW, Coehoorn R, aan de Stegge J, Hoving W. 1992. *Phys. Rev. Lett.* 68: 2688
79. Qiu ZQ, Pearson J, Bader SD. 1992. *Phys. Rev. B* 46: 8659
80. Velu E, Dupas C, Renard D, Renard JP, Seiden J. 1988. *Phys. Rev. B* 37: 668
81. Dupas C, Beauvillain P, Chappert C, Renard JP, Trigui F, et al. 1990. *J. Appl. Phys.* 67: 5680
82. Shinjo T, Yamamoto H. 1990. *J. Phys. Soc. Jpn.* 59: 3061
83. Dieny B, Speriosu VS, Parkin SSP, Gurney BA, Wilhoit DR, Mauri D. 1991. *Phys. Rev. B.* 43: 1297
84. Fuss A, Demokritov S, Grünberg P, Zinn W. 1992. *J. Magn. Magn. Matter* 103: L221
85. Marty A, Gilles B, Eymery J, Chamberod A, Joud JC. 1993. *J. Magn. Magn. Matter* 121: 57
86. Shintaku K, Hosoito N, Shinjo T. 1993. *J. Magn. Magn. Matter* 121: 413
87. Sato H, Kobayashi Y, Aoki Y, Shintaku K, Hosoito N, Shinjo T. 1993. *J. Phys. Soc. Jpn.* 62: 3380
88. Grolier V, Renard D, Bartenlian B, Beauvillain P, Chappert C, Dupas C, Ferre J, Galtier M, Kolb E, Mulloy M, Renard JP, Veillet P. 1993. *Phys. Rev. Lett.* 71: 3023
89. Parkin SSP, Farrow RFC, Marks RF, Cebollada A, Harp GR, Savoy R. 1994. *Phys. Phys. Lett.* 72: 3718
90. Harp GR, Parkin SSP, Farrow RFC, Marks RF, Toney MF, Lam QH, Rabedeau TA, Savoy RJ. 1993. *Phys. Rev. B* 42: 8721
91. Chaiken A, Prinz GA, Krebs JJ. 1990. *J. Appl. Phys.* 67: 4892
92. Parkin SSP, Rabedeau TA, Farrow RFC, Marks R. 1994. *J. Appl. Phys.* 76: 6617
93. Parkin SSP. 1992. *Appl. Phys. Lett.* 60: 512
94. Parkin SSP, Marks RF, Farrow RFC, Harp GR. Lam QH, Savoy RJ. 1992. *Phys. Rev. B* 46: 9262

95. Roth LM, Zeiger HJ, Kaplan TA. 1966. *Phys. Rev.* 149: 519
96. Anthony TC, Brug JA, Zhang, S. 1994. *IEEE Trans. Mag.* 30: 3819
97. Freitas PP, Leal JL, Melo LV, Oliveira NJ, Rodrigues L, Sousa AT. 1994. *Appl. Phys. Lett.* 65: 403
98. Meiklejohn WH, Bean CP. 1959. *Phys. Rev. B* 102: 1413
99. Yelon A. 1971. In *Physics of Thin Films: Advances in Research and Development*, ed. M Francombe, R. Hoffman, 6: 205. New York: Academic
100. Tsang C, Lee K. 1982. *J. Appl. Phys.* 53: 2605
101. Parkin SSP, Deline V, Hilleke R, Felcher GP. 1990. *Phys. Rev. B* 42: 10583
102. Parkin SSP. 1993. *Phys. Rev. Lett.* 71: 1641
103. Parkin SSP. 1992. *Appl. Phys. Lett.* 61: 1358
104. Fullerton EE, Kelly DM, Guimpel J, Schuller IK, Bruynseraede Y. 1992. *Phys. Rev. Lett.* 68: 859
105. Levy PM, Zhang S, Fert A. 1990. *Phys. Rev. Lett.* 65: 1643
106. Pratt WP, Lee SF, Slaughter JM, Loloee R, Schroeder PA, Bass J. 1991. *Phys. Rev. Lett.* 66: 3060
107. Lee SF, Pratt WP, Loloee R, Schroeder PA, Bass J. 1993. *Phys. Rev. B* 46: 548
108. Yang Q, Holody P, Lee S-F, Henry LL, Loloee R, Schroeder PA, et al. 1994. *Phys. Rev. Lett.* 72: 3274
109. Gijs MA, Giesbers JB, Lenczowski SK, Janssen HH. 1993. *Appl. Phys. Lett.* 63: 111
110. Gijs MA, Lenczowski SK, van de Veerdonk RJ, Giesbers JB, Johnson MT, aan de Stegge JB. 1994. *Phys. Rev. B* 50: 16733
111. Shi J. 1994. *Magneto-transport properties of giant magnetoresistive systems.* PhD thesis. Univ. Illinois, Urbana
112. Bruno P. 1993. *Eur. Phys. Lett.* 23: 615
113. Bloemen PJH, Johnson MT, van de Vorst MT, Coehoorn R, de Vries JJ, et al. 1994. *Phys. Rev. Lett.* 72: 764
114. Okuno SN, Inomata K. 1994. *Phys. Rev. Lett.* 72: 1553
115. Rabedeau TA, Toney M, Marks RF, Parkin SSP, Farrow RFC, Harp G. 1993. *Phys. Rev. B* 48: 16810
116. Heinrich B, Bland JAC, eds. 1994. *Ultrathin Magnetic Structures*, Vol. 2. Berlin: Springer-Verlag

Annu. Rev. Mater. Sci. 1995. 25: 389–415

EPITAXY OF DISSIMILAR MATERIALS

C. J. Palmstrøm

Department of Chemical Engineering and Materials Science,
151 Amundson Hall, 421 Washington Avenue SE,
University of Minnesota, Minneapolis, Minnesota 55455

KEY WORDS: metallic compounds, III-V semiconductors, GaAs, ScErAs, ErAs, NiAl, RE-V/III-V, TM-III/III-V, epitaxial insulators

ABSTRACT

Differences in crystal structure, bonding, and materials properties play important roles in dissimilar materials epitaxy. Emphasis is placed on how these differences affect the growth of dissimilar materials heterostructures. Examples of the growth of elemental metals and semi-metals (Al, Ag, Sb, Bi); metallic and semi-metal compounds (transition-metal group III compounds with CsCl crystal structure, rare-earth group V compounds with NaCl crystal structure, orthorhombic $YbSb_2$, $Fe_3Al_{1-x}Si_x$ with BiF_3 crystal structure); magnetic compounds (tetragonal τ-MnAl, hexagonal α-MnAs); and insulators (rare-earth trifluorides with tysonite crystal structure, metal fluorides with the CaF_2 crystal structure) on compound semiconductors are used to demonstrate the issues involved. The use of template layers is crucial in controlling the orientation and crystal quality of the epitaxial layers.

INTRODUCTION

Devices have been limited, in general, by the available material combinations. Dissimilar materials epitaxy involves combining, through epitaxial growth, materials with distinctly different properties and structure, for example, metals, semi-metals, semiconductors, magnetic materials, insulators, and materials with novel optical properties. This has greatly

0084–6600/95/0801–0389$05.00

expanded the potential for producing novel devices with unique characteristics.

From a fundamental standpoint, the understanding of the dissimilar materials growth mechanisms is still at an early stage. In most treatments of similar materials epitaxy, the major concern has been with the differences in lattice parameters between the two materials. For example, in the growth of similar materials epitaxy such as III-V compound semiconductor heterostructures, the amount of mismatch is considered to be the dominant factor affecting the quality of the grown layers. III-V compound semiconductor/group IV semiconductor structures are also considered as similar materials epitaxy because the structure and bonding of zincblende and diamond structures are very similar. Clearly, the issues involved in similar materials epitaxy are also important in dissimilar materials, but in addition, the issues related to differences in crystal structure, bonding, and properties can play dominant roles.

This chapter emphasizes the relationship of the interface structure to the dissimilar materials epitaxial growth and properties of the epitaxial layers. Although emphasis is made on dissimilar materials growth on III-V compound semiconductors, the issues involved are generic.

MAJOR ISSUES IN DISSIMILAR MATERIALS EPITAXY

Crystal Structure

The easiest epitaxial relationship to consider is one in which the two dissimilar materials have a similar crystal structure, for example, both are cubic. If the lattice parameters are the same, one would expect a cube-on-cube type of epitaxial relationship (Figure 1*a*). For different lattice parameter ratios, other epitaxial relationships can be expected. Examples are shown in Figure 1 for lattice parameter ratios between the grown layer and the substrate of 1, 0.5, $\sqrt{2}$, and $1/\sqrt{2}$. For the latter two examples, a 45° rotation of the cube-on-cube relationship is expected. Clearly, other rotations are also possible for different lattice parameter ratios. It is, however, difficult to achieve arbitrary lattice matching of different materials by a simple rotation in the interface plane (i.e. about the surface normal). By rotations of the overgrown material about other axes, lattice matching may also be accomplished. A small rotation of the epitaxial layer can accommodate some lattice mismatch by forming an asymmetric tilt boundary between the growing layer and the substrates (1, 2). A larger rotation can yield lattice matching with a different crystallographic orientation, such as (110) growth of a dissimilar cubic material on a (100)

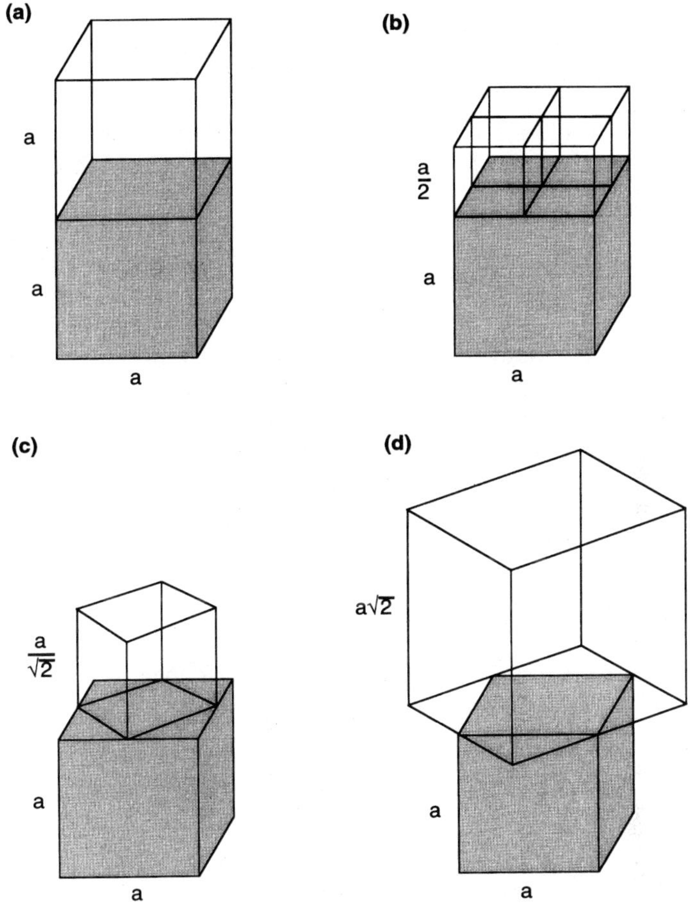

Figure 1 Schematic of simplest potential epitaxial alignments for a cubic material (A) on a cubic substrate (B): (*a*) and (*b*) are referred to as cube-on-cube with (*a*) $a_{0A} = a_{0B}$ and (*b*) $a_{0A} = 0.5a_{0B}$; (*c*) and (*d*) are referred to as 45° rotated alignment with (*c*) $a_{0A} = 1/\sqrt{2}a_{0B}$ and (*d*) $a_{0A} = \sqrt{2}a_{0B}$. (From Reference 52.)

substrate. Obviously, it is desirable to have perfect lattice matching to alleviate problems of stress and relaxation through misfit dislocation formation.

Dissimilar materials with noncubic structure can also be grown epitaxially, but the conditions for lattice matching may require the growth of a different orientation from that of the substrate. The specifics will depend critically on the dissimilar materials system under investigation.

Interface Chemical Bonding and Template Layers

The discussion has been on whether cubes can be aligned on each other in such a way as to produce lattice matching for epitaxial growth. The bonding between these cubes is important in determining whether the proposed alignment can actually occur. For semiconductor heteroepitaxy, elimination of interface charge accumulation is important (3). In the case of Ge growth on GaAs, only the nonpolar (110) substrate can yield an atomically abrupt interface (3). In the case of (100) and (111), where GaAs has polar surfaces, mixed composition interface layers are required for the elimination of interface charge accumulation (3). Electron counting (4) (ensuring that the number of electrons equals the number required for bonding) has been used successfully to design interfaces for semiconductor-on-semiconductor growth without charge accumulation (4–7) and to explain semiconductor surface reconstructions (8, 9). For CuBr/GaAs(100), however, Dassanayake et al (10) found that the interface structure is driven by thermodynamic compound formation rather than by simple electron counting. In general, in dissimilar materials epitaxy, there is electronic bonding mismatch that makes it difficult to apply the electron-counting concept used for covalent bonding to design the surface and interface for epitaxial growth. In practice, techniques for controlling the epitaxial orientation are experimentally determined by trial and error. The use of template layers for controlling the orientation of $NiSi_2$ and $CoSi_2$ epitaxial layers on Si was pioneered by Tung et al (11–13). This concept, as will be shown, is very important in controlling the epitaxial orientation in epitaxy of dissimilar materials.

Surface and Interface Energy Issues

In order for a dissimilar material (A) to grow in a layer-by-layer fashion on another material (B), the growing material must wet the surface of the substrate. For this to occur from a thermodynamic standpoint, the surface energy per unit area of the growing layer γ_A plus the interface energy per unit area γ_{int} must be lower than the substrate surface energy per unit area γ_B (i.e. $\gamma_A + \gamma_{int} < \gamma_B$). This layer-by-layer growth mode is referred to as Frank-van de Merwe growth (14). It follows that if material A wets B, then B will not wet A ($\gamma_B + \gamma_{int} > \gamma_A + 2\gamma_{int}$) and the growth will proceed by a three-dimensional island growth mode called Volmer-Weber (14). This implies that the growth of dissimilar-material multilayer heterostructures is, in principle, impossible. However, considering the case of A wetting B, there are two conceptual ways in which multilayers may be grown: After growing A on B, the first option would be to increase the surface energy per unit area of A (the new substrate) or to reduce the surface energy per unit area of B (the growing layer) by depositing a surfactant layer. This method has been used to suppress island formation for the growth of Ge on Si (15). The second option is to use kinetics to reduce the island formation and form a quasi-two-dimensional growth mode. This can be achieved if the driving force for island formation is small (i.e. $\gamma_B \approx \gamma_A$) and if the growth is done using a relatively high growth rate and/or low temperature.

As in the case of similar materials epitaxy, if there is mismatch, an additional energy term due to the strain must be included. When the strain energy is the dominant term, the growth tends to be intermediate between layer-by-layer and island growth and is described by the Stranski-Krastanov growth mode (14). Initially, layer-by-layer growth occurs until the critical thickness is exceeded and then island growth proceeds on top of the initial wetting layer. However, for dissimilar materials epitaxy the surface and interface energies generally dominate.

Morphological Stability

As pointed out above, kinetics may be used to obtain a quasi-two-dimensional growth mode for dissimilar material A on B even if A thermodynamically should not wet B (i.e. $\gamma_A + \gamma_{int} > \gamma_B$). Clearly, this is a metastable situation. Another metastable situation occurs when A is overgrown by B to yield a B/A/B-layered structure, even in the case where B wets A because the interface energy γ_{int} is generally greater than zero. In both cases, there is a driving force to reduce the surface area of A (i.e. for agglomeration). The rate at which agglomeration can occur depends upon the surface and bulk diffusion coefficients. Whether both these diffusion coefficients are zero-ordered depends upon the melting point of the material. The higher the melting point of material A, the lower the diffusion coefficients, and the higher its morphological stability.

Thermodynamic Stability

Epitaxy does not mean thermodynamic stability. Although *bcc* Fe can be grown epitaxially on (100) and (110) GaAs with cube-on-cube alignment (Figure 1*b*) (16–21), and *bcc* Co(110) can be epitaxially grown on GaAs(110) (17, 22, 23), both Fe (19, 24) and Co (23, 25–29) are not thermodynamically stable on GaAs. During deposition and any subsequent annealing, interface reactions take place. Knowledge of the relevant phase diagram is the only way to predict thermodynamic stability. However, in most cases, the relevant phase diagrams have not been determined. Theoretical calculations and thin film reaction studies may be used to predict the phase diagrams (30–32).

Symmetry Issues

In general, it is easier for a material with a higher symmetry crystal structure to grow on a lower symmetry crystal structure material without the formation of orientation (twin) domain boundaries and translational domain boundaries because of the increased symmetry of the growing material. It follows, therefore, that symmetry-related domain boundaries will be expected when a material of lower symmetry is grown on a material with higher symmetry (33). Materials with the *bcc*, *fcc*, NaCl, and CsCl structures possess the inversion symmetry element. The zincblende structure, however, is noncentrosymmetric (no center-of-inversion symmetry). Consequently, a compound semiconductor (zincblende structure, Figure 2) growing on top of a material with the inversion symmetry element does not "know" which way is up. The resulting two-dimensional defect is known as an inversion domain boundary (commonly referred to as anti-

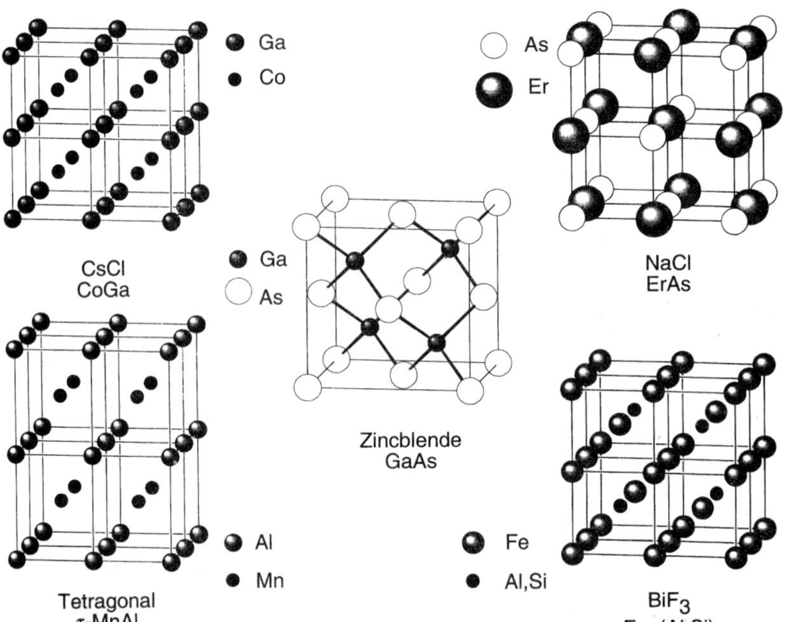

Figure 2 Comparison of different crystal structures with potential epitaxial alignment: CoGa structure (CsCl type); τ-MnAl (tetragonal); ErAs structure (NaCl type); $Fe_3(Al,Si)$ structure (BiF_3 type); and GaAs (zincblende type). Eight unit cells of idealized CoGa (or NiAl) structure, a unit cell of ErAs structure, or $Fe_3(Al,Si)$ structure occupy approximately the same volume as a single unit cell of GaAs. The basal planes of τ-MnAl unit cells are nearly lattice matched to the (100) plane of GaAs. (From Reference 52.)

phase boundary) because it separates two domains of the crystal. Such defects are also common to III-V semiconductor layers grown on Si (the diamond structure of Si possesses a center of inversion symmetry). As in the case of III-V semiconductor growth on Si (34), the symmetry of the substrate surface may be reduced by using vicinal surfaces. Sands et al (33) have discussed these issues for III-V compound semiconductor growth on NiAl(100).

EXAMPLES OF DISSIMILAR EPITAXIAL STRUCTURES

Elemental Metal/III-V Compound Semiconductors

Figure 3*a*,*b* shows the lattice parameters (34a) of metallic elements with *fcc* and *bcc* crystal structure. Only a few elements would be expected to be epitaxial on GaAs. For the *fcc* elements, one may predict that Rh, Ir, Pd, Pt, Al, Au, and Ag should grow with a 45° rotation as shown in Figure 1*c*. Although this orientation relationship has been observed for Al (35–38), it is not observed for Ag (36, 39) growth on GaAs(100). Yb and Ca may be expected to grow with the cube-on-cube relationship (Figure 1*a*). Epitaxially stabilized *fcc* Gd and Y may potentially also form with the cube-on-cube orientation. For the *bcc* elements, Fe, Cr, and V would be expected to grow with the cube-on-cube orientation (Figure 1*b*) and so, potentially, may Rb (Figure 1*a*). This is what is observed for Fe on GaAs(100) (18, 19). Fe (16–18) and metastable *bcc* Co (40) have been grown on (110) GaAs with a cube-on-cube relationship. The initial growth of Fe on (110) GaAs is three dimensional. After island coalescence, the surface becomes atomically flat. For growth on (100), the Fe surface remains atomically rough, even at a substrate temperature of 175°C (19). Fe and Co are not thermodynamically stable on GaAs and thus tend to react with the substrate during deposition (16, 18, 19, 23, 24). Growth on (100) ZnSe results in higher quality films (41, 42), and epitaxial *bcc* Fe_xCo_{1-x} alloy films have also been grown (43). The magnetic properties of epitaxial Co, Fe, and Fe_xCo_{1-x} alloy films have been studied in detail (19, 22, 43–45).

Al/GaAs AND Ag/GaAs Al/GaAs(100) and Ag/GaAs(100) are, in principle, some of the simplest dissimilar heteroepitaxial materials systems to study. From their lattice parameters, Al and Ag would be expected to grow with the 45° rotated alignment, as shown in Figure 1*c*. In practice, the situation is much more complicated. Epitaxy does occur, but the orientation relationships are not straightforward. They depend upon the substrate temperature and GaAs surface structure (reconstruction, composition,

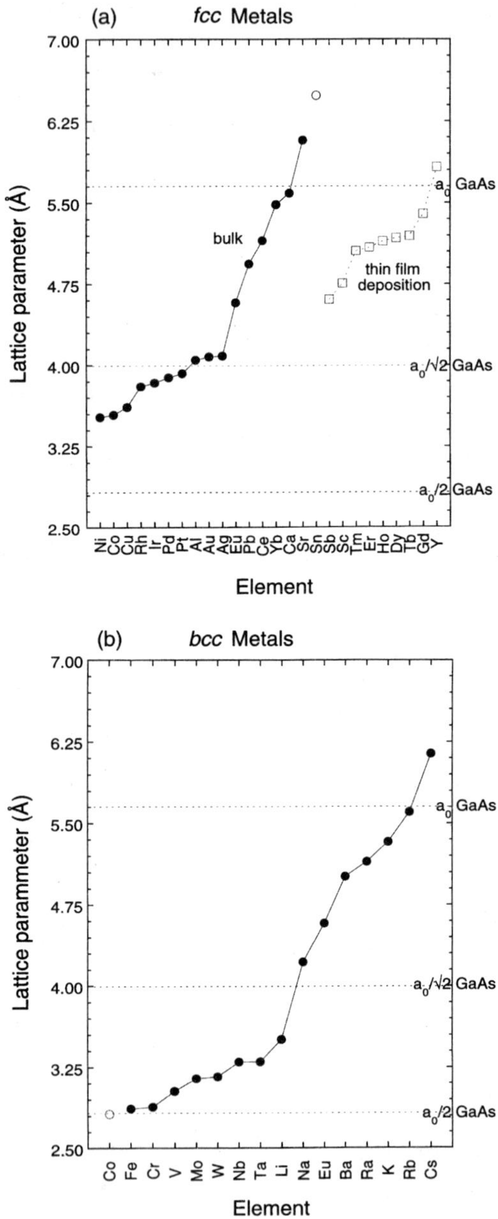
(a) fcc Metals
Lattice parameter (Å)
7.00
6.25
5.50
4.75
4.00
3.25
2.50
bulk
thin film
deposition
a_0 GaAs
$a_0/\sqrt{2}$ GaAs
$a_0/2$ GaAs
Ni Co Cu Rh Ir Pd Pt Al Au Ag Eu Pb Ce Yb Ca Sr Sn Sb Sc Tm Er Ho Dy Tb Gd Y
Element
(b) bcc Metals
Lattice parammeter (Å)
7.00
6.25
5.50
4.75
4.00
3.25
2.50
a_0 GaAs
$a_0/\sqrt{2}$ GaAs
$a_0/2$ GaAs
Co Fe Cr V Mo W Nb Ta Li Na Eu Ba Ra K Rb Cs
Element

step type, and step density) (35–39, 46). In the case of films of Al on (100) GaAs, they nucleate in three different orientations: (100), (110), and (110) R, only one of which is easily rationalized on the basis of lattice matching in the surface plane. Maeda et al (47) report that for Al growth on AlAs(100) at 30°C, Al(100); Al(110)R {Al(110)[110]‖AlAs(100)[011]}; and Al(110) {Al(110)[110]‖AlAs(100)[0$\bar{1}$1]} can all be preferentially grown by using Al- or As-terminated AlAs(100) with and without a 1-monolayer (ML)-thick Ga or In layer deposited prior to the Al growth.

In the case of Ag, the expected 45° rotated orientation relationship (Figure 1*c*) is generally not observed. For growth temperatures less than about 200°C, two orientation variants with (110)Ag‖(100)GaAs are observed (36, 39). At higher growth temperatures, the (100) plane of Ag is parallel to the (100) GaAs surface, but the unit cell of Ag is unrotated, resulting in an apparent lattice mismatch of ~30%! If, however, a 6-ML thick epitaxial film of *bcc* Fe is deposited prior to the Ag (48, 49), it can lead to the expected orientation relationship of Figure 1*c*. This phenomenon is not yet fully understood.

Sb/GaSb(111) AND Bi/CdTe($\bar{1}\bar{1}\bar{1}$) Sb, which has a rhombohedral crystal structure, has been grown epitaxially on GaSb(111) with the following orientation relationship: (111)Sb‖(111)GaSb (50). The in-plane mismatch is only 0.06%. This orientation relationship is equivalent to (0001)Sb‖(111)GaSb, after a rhombohedral to hexagonal coordinate transformation. Bi has been grown in a similar manner on CdTe($\bar{1}\bar{1}\bar{1}$) (51, 51a).

Metallic Compound/Compound Semiconductor Heterostructures

Unlike most elemental metals, the rare-earth monopnictides (RE-V) with the NaCl crystal structure (Figure 2) and the transition-metal group III (TM-III) compounds with the CsCl structure (Figure 2) are generally thermodynamically stable in contact with III-V compound semiconductors (52, 53). By keeping the group III element the same across the

Figure 3 (*a*) Plot of lattice parameter vs element for *fcc* metals (*filled circles* are for the *fcc* metals; the *open circle* is for α-Sn, diamond structure; and the *open squares* are for *fcc*-stabilized deposited thin films). Most *fcc* metals with potential lattice matching to GaAs can be achieved with the metal lattice parameter, $a_{0M} = 1/\sqrt{2}a_{0GaAs}$ (e.g. Rh, Ir, Pd, Pt, Al, Au, Ag), resulting in the expected alignment as shown in Figure 1*c*. Only a few (e.g. Yb and Ca) have the potential of the cube-on-cube alignment (Figure 1*a*). (*b*) Plot of lattice parameter vs element for *bcc* metals [*open circle* is for metastable *bcc* Co grown on GaAs (40)]. The expected epitaxial alignment for *bcc* metals is the cube-on-cube with $a_{0M} = 0.5a_{0GaAs}$ (Figure 1*b* for Co, Fe, Cr and V) and $a_{0M} = a_{0GaAs}$ (Figure 1*a* for Rb). (From Reference 52.)

TM-III/III-V interface and the group V element the same across the RE-V/III-V interface, potential interface exchange reactions can be avoided (52, 53). The TM-Al compounds with the CsCl structure have a higher melting point than the corresponding TM-Ga compounds, which makes them significantly more stable to agglomeration (53). Most RE-As compounds with the NaCl structure have a very high melting point ($\geqslant 2000°C$) and heat of formation, which suggests that they should have good morphological stability (52, 53).

RE-V COMPOUNDS WITH NaCl CRYSTAL STRUCTURE ON III-V SEMICONDUCTORS

The lattice parameters of the RE-V compounds with NaCl crystal structure are shown in Figure 4. The monopnictides of Sc and Y have been included because of their close chemical similarity to the RE elements. The RE-V lattice parameters range from 4.51 Å for ScN to 6.58 Å for LaBi (54), which covers the range of lattice parameters for most group IV, III-V, and II-VI semiconductors. A large number of RE-V compounds with the NaCl structure have been grown on III-V compound semiconductors. These include ErAs (52, 55–63), YbAs (64, 65), LuAs (56), and ScAs (52, 65) on GaAs; ErP (66) on InP; and ErSb (66, 67) on GaSb. The large range in lattice parameters (Figure 4) suggests that ternaries, formed by mixing RE elements ($RE^a_{1-x}RE^b_x - V$) or group V elements ($RE - V^c_{1-y}V^d_y$), or quaternary systems, formed by mixing both ($RE^a_{1-x}RE^b_x - V^c_{1-y}V^d_y$), can be lattice matched to almost all group IV, III-V, and II-VI semiconductors. Palmstrøm et al (52, 53, 68, 69) have used the former approach to grow lattice-matched $Sc_{0.3}Er_{0.7}As$ on GaAs. Guivarc'h and co-workers have used both approaches; they have grown lattice-matched $Sc_{0.2}Yb_{0.8}As$ (70–72), $Sc_{0.3}Er_{0.7}As$ (70, 71), and $ErP_{0.6}As_{0.4}$ (60, 61, 66, 73, 74) on GaAs, and $ErP_{0.47}Sb_{0.53}$ (66) on InP. In the case of RE-As growth on GaAs, direct codeposition of RE and As_4 (or As_2) on the As-terminated (100) GaAs surface yields the expected cube-on-cube orientation (Figure 1*a*) (52, 53, 55, 56). The growth mode is layer-by-layer on the (100) surface, indicating that the RE-V wets the III-V surface (52, 53, 68). Continuous films as thin as three monolayers can be grown easily (52, 53, 75). The effect of strain on the growth mode and film relaxation of ErAs films grown on GaAs(100) (mismatch 1.6%) has been studied by X-ray diffraction (76–79), ion channeling (80), and transmission electron microscopy (81–84). The proposed RE-As/GaAs(100) alignment shown in Figure 5 is consistent with ion-channeling measurements (71, 85) and X-ray photoelectron diffraction measurements (65). The interface structure has also been calculated theoretically (86). The RE-V compounds also grow with the cube-on-cube orientation (Figure 1*a*) on other crystal orientations, including (100), {711}, {511}, {411}, {311}, {211}, {111}, and (110) (52, 57).

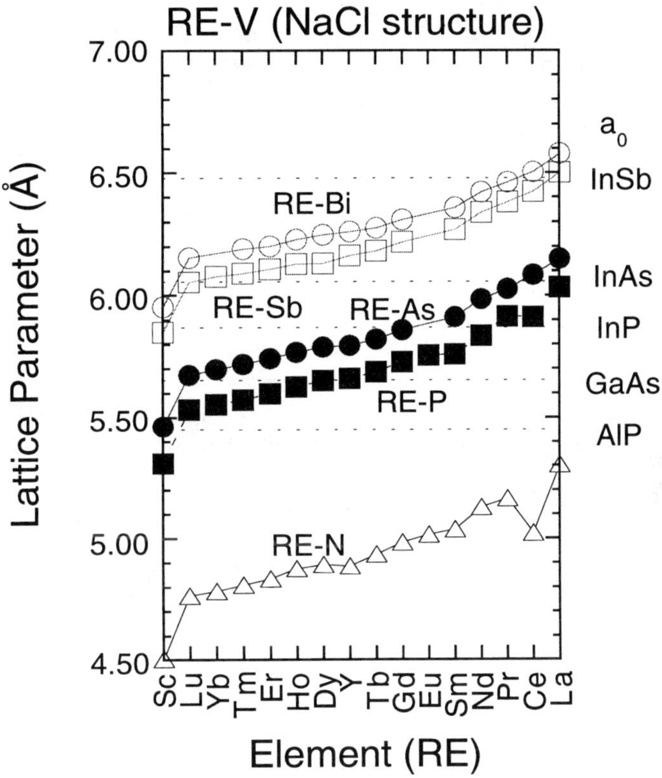

Figure 4 Lattice parameters of RE-V compounds with NaCl structure (data obtained from Reference 54). The RE-V compounds span a large range of lattice parameters ranging from 4.505 Å (ScN) to 6.580 Å (LaBi), which suggests that ternaries, formed by mixing rare-earth elements ($RE^{a}_{1-x}RE^{b}_{x}-V$) or group V elements ($RE-V^{c}_{1-y}V^{d}_{y}$), or quaternary systems, formed by mixing both ($RE^{a}_{1-x}RE^{b}_{x}-V^{c}_{1-y}V^{d}_{y}$), can be lattice matched to almost all group IV, III-V, and II-VI semiconductors. (From Reference 52.)

A number of RE monochalcogenides (RE-VI) also have the NaCl structure. The lattice parameters vary from 5.192 Å (ScS) to 6.595 Å (SmTe) (54), suggesting that these may also be grown lattice matched on most III-V and II-VI compound semiconductors (68).

TM-III COMPOUNDS WITH CsCl CRYSTAL STRUCTURE ON III-V COMPOUND SEMICONDUCTORS The transition-metal group III (TM-III) compounds with the CsCl crystal structure (Figure 2) have lattice parameters of approximately half that of many zincblende III-V compound semiconductors (Figure 6). Hence, the expected epitaxial relationship would be the cube-on-cube relationship shown in Figure 1*b*. A template layer must be used

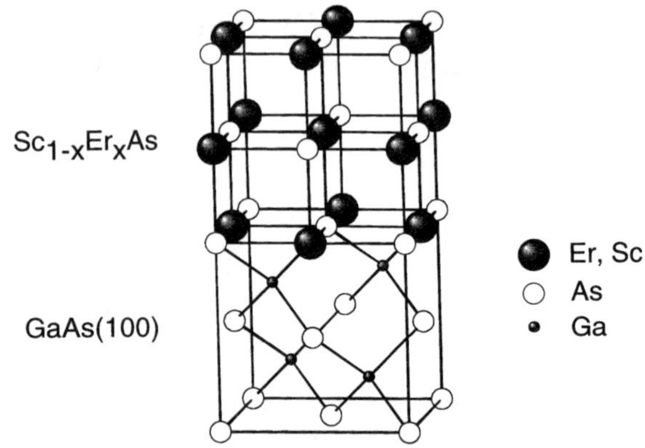

Figure 5 Schematic of a possible $Sc_{1-x}Er_xAs$/GaAs(100) interface alignment. The As-sublattice is continuous across the interfaces. (From Reference 52.)

to achieve this orientation (87–91). In general, a 1–2 ML-thick predeposited layer of the TM is sufficient to ensure TM-III(100) growth on (100) III-V compound semiconductors (87, 88, 90, 91). A number of mechanisms have been proposed (52). One suggests that the TM reacts with the III-V semiconductor to form a thin layer of a TM_x(III-V) compound with pseudo-cubic symmetry, which acts as a template for TM-III cube-on-cube epitaxial growth. For growth of CoGa on $Ga_{1-x}Al_xAs(100)$, a predeposition of $Ga_{1-x}Al_x$ also resulted in (100) CoGa growth (89). If no predeposited template layer is used, mixed (100) and (110) TM-III growth tends to result (33, 53, 87).

A number of TM-III compounds with the CsCl structure have been epitaxially grown on III-V semiconductors in a similar manner utilizing a template layer to ensure the cube-on-cube orientation. These include NiAl (33, 87, 88, 92–95) and CoAl (33, 90, 96) on AlAs(100); NiGa (97–100) on GaAs(100); CoGa (52, 53, 89, 101, 102) on $Ga_{1-x}Al_xAs(100)$; and FeAl (103) on a thin pseudomorphic AlAs(100) layer (used to inhibit interface exchange reactions) grown on InP(100).

OTHER METALLIC COMPOUNDS ON III-V COMPOUND SEMICONDUCTORS $YbSb_2$ has a complex orthorhombic structure ($ZrSi_2$ type with $a_0 = 4.536$ Å, $b_0 = 16.63$ Å, and $c_0 = 4.271$ Å). The lattice parameters a_0 and c_0 are approximately $a_0/\sqrt{2}$ GaSb with lattice mismatch of 5.2% and −0.92%, respectively. Hence, the expected growth alignment for growth on GaSb(100) would be a 45° rotated alignment (Figure 1*c*) with the *b* axis of $YbSb_2$ parallel to the surface normal of (100) GaSb (104). Guivarc'h et al (104) have grown $YbSb_2$ with this alignment on GaSb(100). They have also grown Ni_3Ga_x and Ni_2Ga_3 pseudo-cubic hexagonal phases

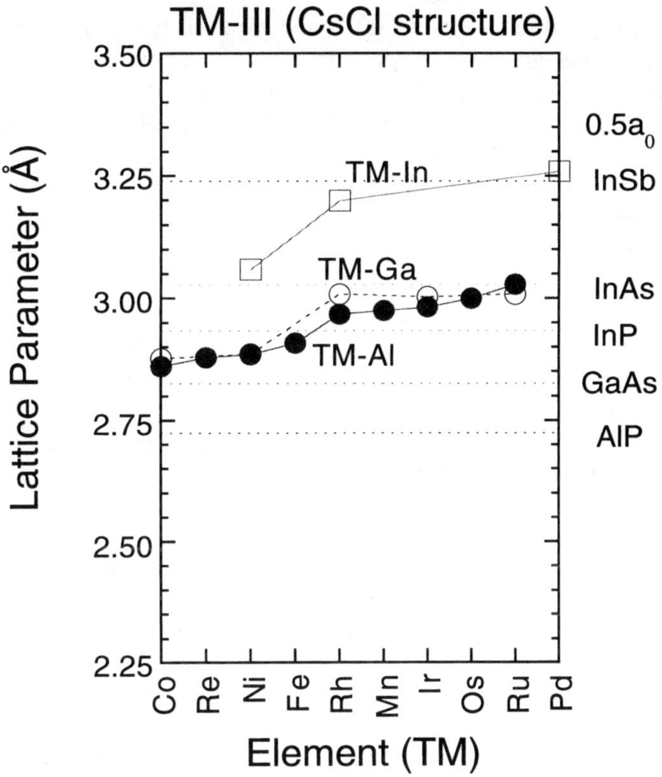

Figure 6 Lattice parameters of TM-III intermetallics with CsCl structure (data obtained from Reference 54). The predicted alignment is the cube-on-cube with $a_{0M} = 0.5a_{0GaAs}$ (Figure 1*b*). Note that all the TM-III compounds have $a_{0M} > 0.5a_{0GaAs}$. (From Reference 52.)

($c_0/a_0 \approx \sqrt{3}/\sqrt{2}$) with effective mismatch of less than 1.5% on GaAs(100) and {111} surfaces (98).

Solid solutions of Fe_3Al ($a_o = 5.74$ Å) and Fe_3Si ($a_o = 5.64$ Å), $Fe_3Al_{1-x}Si_x$ with the BiF_3 structure (Figure 2) have also been grown lattice matched to GaAs with the cube-on-cube relationship without the use of a template layer (Figure 1*a*) (105).

Growth of III-V/Metal/III-V Structures

III-V/ELEMENTAL METAL/III-V STRUCTURES It is clear that the growth of high quality single-crystal Al layers is difficult on GaAs. This fact, together with the low melting point of Al, makes Al a poor choice for the growth of GaAs/metal/GaAs heterostructures. Nevertheless, a few attempts have been made (106–109). The Al grows by the Volmer-Weber mode, and therefore, layer thicknesses >100 Å are required before a smooth film is obtained. For a growth temperature of ~40°C, the Al films were predominantly {110} on GaAs(100) (106, 107). Growth at ~0°C resulted in {111}-textured films (109). Regardless of the Al orientation, the GaAs overgrowth was similar. Growth at 400°C resulted in a completely reacted Al film and no GaAs/Al/GaAs heterostructure. The GaAs overgrowth must be performed at ~200°C to ensure minimal Al-Ga exchange reaction and minimal Al reaction with the impinging As-flux. Conventional MBE was found to yield polycrystalline layers (106). Crystalline layers were obtained by using migration-enhanced epitaxy (MEE) (107–109) but were of poor crystalline quality. These results are a clear indication of the importance of epitaxy, thermodynamic stability, and melting point for the choice of the metallic layer.

DiVenere et al (51, 51a) report the growth of CdTe/Bi/CdTe($\bar{1}\bar{1}\bar{1}$) structures and superlattices. They found that CdTe grown on Bi/CdTe($\bar{1}\bar{1}\bar{1}$) was oriented (111), which indicates that there is a Bi/CdTe($\bar{1}\bar{1}\bar{1}$) interface at both the top and bottom interfaces. Golding et al (50) report the growth on GaSb/Sb/GaSb(111) structures. The properties of Bi/CdTe($\bar{1}\bar{1}\bar{1}$) semi-metal/semiconductor structures indicate a semi-metal to semiconductor transition for Bi layer thickness between 200 and 300 Å (51a).

III-V/METALLIC COMPOUND/III-V STRUCTURES The RE-V compounds are thermodynamically stable in contact with both the group V element and III-V semiconductor. This stability, together with their high melting points, makes RE-V compounds ideally suited for the growth of III-V/RE-V/III-V structures (52, 53). However, because $Sc_{1-x}Er_xAs$ grows in a layer-by-layer manner on GaAs(100) (52, 68, 110), indicating that it wets the GaAs surface, it is not surprising from a surface energy argument that GaAs tends to grow as islands on the $Sc_{1-x}Er_xAs$ surface (52, 53, 111–115). This growth mode results in micro-twins and stacking faults in the overgrown GaAs (52, 53, 111–114, 116). Some improvement in crystalline quality is achieved on vicinal surfaces (52, 72, 114). Palmstrøm et al (52, 110) suggest that a significant bonding mismatch would be expected on the (100) surface (polar growth on a nonpolar surface) and that a significant improvement in the interface bonding configuration would be expected on the {111} surface. The surface energies for GaAs {111} and (110) surfaces are lower than for (100). For NaCl, the (111) surface has about six times higher surface energy than (100) (117). Hence, by growth on {111} surfaces, the $Sc_{1-x}Er_xAs$ surface energy can be increased and the GaAs decreased, making the relative surface energies more equal and the bonding more conducive to GaAs overgrowth. However, although the GaAs wetting of $Sc_{1-x}Er_xAs$ improved dramatically (52, 53, 110, 115), both the $Sc_{1-x}Er_xAs$

and overgrown GaAs were heavily twinned. Palmstrøm et al (52, 110) realized that a compromise between wetting and twinning could be achieved by using surfaces with mixed (111) and (100) character. Growth on $\{h11\}$ surfaces [which consist of steps and ledges with (100) and $\{111\}$ character] enabled substantial improvement in the wetting and crystalline quality of the overgrown GaAs. The overgrown GaAs has a number of micro-twins and stacking faults in the (100) case and only a few stacking faults in the (311) case. Figure 7 shows cross-sectional transmission electron micrographs of GaAs/ErAs/GaAs(100) structures. Ion channeling results showed a dramatic increase in crystalline quality for $\{311\}$ (52, 110). Palmstrøm et al (110) noted that in the growth of GaAs/$Sc_{0.3}Er_{0.7}As$/GaAs($\bar{3}\bar{1}\bar{1}$), the overgrown GaAs orientation was oriented (311). This is similar to the observed growth of CdTe/Bi/CdTe($\bar{1}\bar{1}\bar{1}$) (51).

GaSb wets the surface ErSb(100) better than GaAs wets RE-As (67). Symmetry-related defects were still observed in the overgrown GaSb. By growing GaSb on $YbSb_2$, Guivarc'h et al (104) observed the orientation relationship for GaSb/$YbSb_2$/GaSb$\{100\}$ shown in Figure 8. The GaSb grown on top of the orthorhombic $YbSb_2$ did not have inversion domains and was rotated by 90° compared with the GaSb substrate. This is a result of the symmetry of the $YbSb_2\{010\}$ surface, which is twofold symmetric

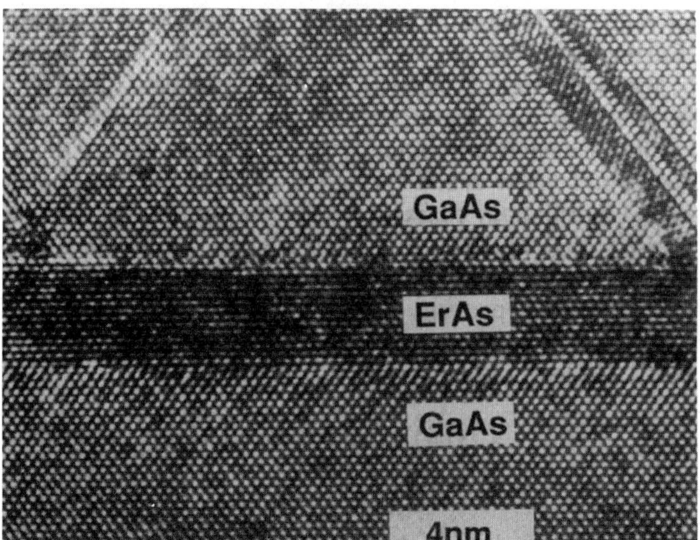

Figure 7 High-resolution cross-sectional TEM micrograph of a GaAs/ErAs/GaAs(100) structure. (From Reference 53.)

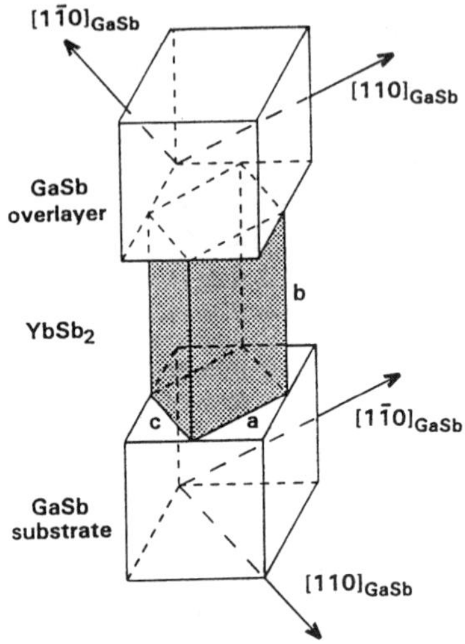

Figure 8 Schematic illustration of the orientation relationships observed between the epitaxial growth of a GaSb/$YbSb_2$/GaSb(001) heterostructure. (From Reference 104.)

with the atomic symmetry being rotated by 90° between the (010) and (0$\bar{1}$0) in the same manner as for zincblende (100) and ($\bar{1}$00) surfaces (the surface dimers are at 90° to each other). However, the crystalline quality of the $YbSb_2$ was not as good as that of ErSb, presumably due to the increased mismatch of the $YbSb_2$/GaSb(100) (mismatch 5.2% and ~0.92% to $a_o/\sqrt{2}$ GaSb for a_o and c_o, respectively) vs ErSb/GaSb(100) (mismatch 0.2%).

Harbison, Sands, and co-workers (33, 53, 87, 88) found that the AlAs interlayer was necessary between GaAs and NiAl to suppress the Ga-Al exchange reaction. For AlAs growth on NiAl, a thin AlAs template layer is initially formed by impinging an As_4 flux onto the NiAl, which reacts with NiAl to form an AlAs template layer for further AlAs and GaAs growth (33, 52, 53, 87, 88, 94). This reaction takes place because NiAl is not thermodynamically stable in contact with arsenic (52, 53, 118, 119). It is controlled by supplying additional Al from an effusion cell. Once a thin AlAs layer has been grown, thermodynamic stability is achieved because NiAl is in contact with AlAs and no longer with elemental arsenic. For growth on thin layers of NiAl, the AlAs and any subsequent GaAs overgrowth must be performed at temperatures below 500°C to prevent agglomeration of the thin NiAl film. An AlAs thickness $\geqslant 20$ Å on either side of the NiAl is sufficient to prevent Al-Ga exchange reactions. Identical growth conditions were found to be appropriate for GaAs/AlAs structures grown on CoAl as well (33, 53). Using this growth procedure, epitaxial GaAs/AlAs/NiAl/AlAs/GaAs heterostructures containing electrically continuous NiAl films as thin as 10 Å have been fabricated (53). The surface morphology of the III-V overlayer can be quite smooth, appearing specular to the eye. The fact that both NiAl growth on GaAs and GaAs growth on NiAl can lead to relatively smooth epitaxial films suggests that the surface energies of NiAl and GaAs (or AlAs) are comparable. Figure 9 shows a high-resolution cross-sectional TEM micrograph of an AlAs/NiAl/AlAs structure. The interfaces are atomically abrupt. Most of the defects in the AlAs result from the symmetry difference between NiAl (CoAl) and AlAs (33, 53).

Properties of Metallic Compound/Compound Semiconductor Heterostructures

IDEAL THERMODYNAMICALLY STABLE CONTACTS The TM-III with the CsCl crystal structure and RE-V compounds with the NaCl structure are ideal for thermodynamically stable contacts to III-V compound semiconductors (52, 53). The RE-V compounds offer the advantage of a larger range in lattice parameter matching (Figures 4 and 6). Figure 10 shows the Schottky barrier height as a function of orientation for lattice-matched $Sc_{0.3}Er_{0.7}As$ layers grown at 500–600°C, and samples grown at 350–400°C and annealed

Figure 9 High-resolution cross-sectional TEM micrograph of an AlAs/NiAl/AlAs(100) structure. (From Reference 53.)

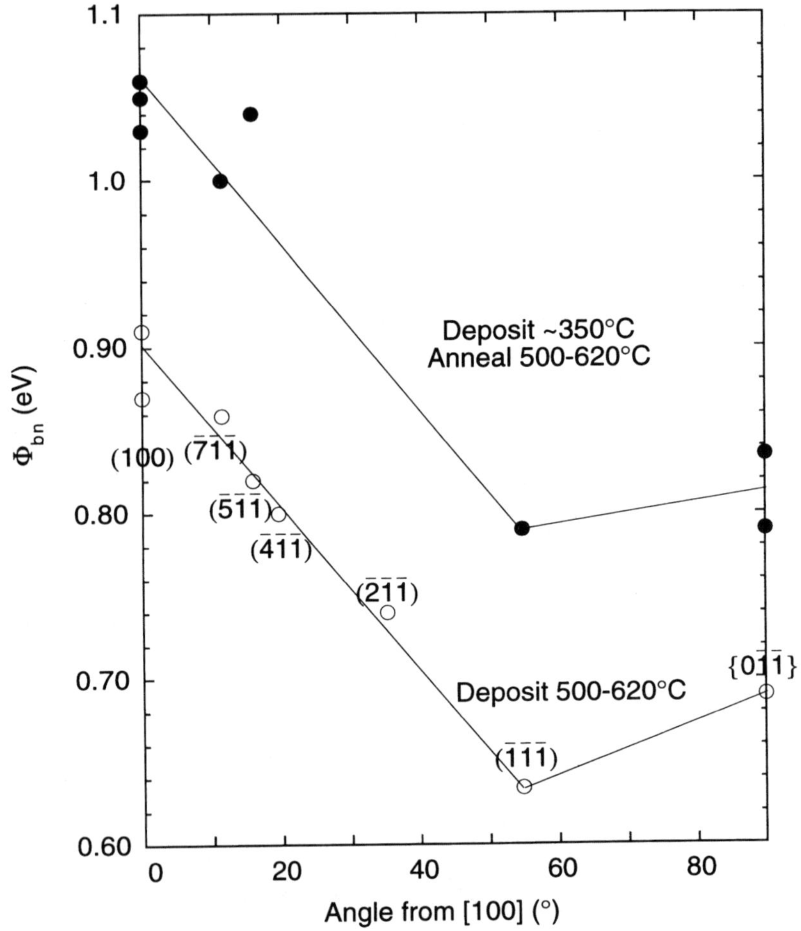

Figure 10 Plot of Schottky barrier height vs substrate orientation for lattice-matched $Sx_{1-x}Er_xAs$/n-GaAs$\{h\bar{1}\bar{1}\}$ diodes. (From Reference 57.)

at temperatures $\geqslant 500°C$ (57). The barrier height is highest for (100), and it decreases approximately linearly with angle towards $(\bar{1}\bar{1}\bar{1})$ for $Sc_{0.3}Er_{0.7}As$ layers grown at 500–600°C (57). As was observed for (100), deposition at $\sim 350°C$ and annealing at ~ 500–600°C results in higher barrier heights than deposition at ~ 500–600°C for all orientations (57). The barrier height variation ranges from ~ 1.03 eV for (100) deposited at 350°C and annealed at $\sim 580°C$ to ~ 0.63 eV for $(\bar{1}\bar{1}\bar{1})$ deposited at $\sim 600°C$, a total variation of ~ 0.4 eV for the same metallization on GaAs. This variation in Schottky barrier height with orientation has been suggested to result from the change in interface bonding configuration, which results in a change in the interface dipole (57). Similar arguments have been used by Tung and co-workers (12, 13, 120) for the changes in the Schottky barrier heights in epitaxial $NiSi_2/Si$ and $CoSi_2/Si$ with Si and silicide orientation. In the specific case of $NiSi_2/Si(111)$, theoretical calculations indicate that the difference in barrier height between type-A (unrotated) and type-B (rotated 180°) results from differences in the interface dipole due to the different bonding configuration (121). This suggests that many of the concepts that have been applied to epitaxial Schottky barriers on Si are also applicable to epitaxial Schottky barriers on III-V compound semiconductors.

ELECTRONIC PROPERTIES OF TM-III AND RE-V COMPOUNDS The electronic properties of bulk TM-III metallic compounds have been well studied (122–125). However, this has not been the case for RE-V compounds. Temperature-dependent lateral magneto-transport measurements of RE-V/III-V semiconductor heterostructures clarified the issue of whether the RE-V compounds with NaCl crystal structures were semiconductors or semi-metals, by demonstrating that the RE-As, RE-P, and RE-Sb are semi-metals with room-temperature resistivities in the range ~30 to 150 $\mu\Omega$.cm (60, 61, 66, 67, 70, 75, 126–131). These data are in agreement with the early band structure calculations by Hasegawa & Yanase (132) for the Gd monopnictides, which suggested that GdN was a semiconductor and that the other monopnictides were all semi-metals, and the recent band structure calculations for ErAs and $Sc_{1-x}Er_xAs$ by Petukhov et al (133). The lattice-matched films were of sufficient quality to show Shubnikov-de Haas (SdH) oscillations in the low-temperature magneto-transport. These oscillations were used to probe the band structure of $Sc_{1-x}Er_xAs$ (127–130). The measurements of $Sc_{1-x}Er_xAs$ are consistent with a semi-metal with an isotropic hole Fermi surface centered at Γ and three equivalent ellipsoidal electron Fermi surfaces at X (127–130, 133). The electron and hole densities were determined to be approximately 3×10^{20} cm^{-3}.

Because the conduction electrons can scatter elastically from the $4f$ Er^{3+} spin and, at low temperature, ErAs orders antiferromagnetically ($T_N \sim 4.5$ K), magneto-transport measurements have been used to measure the anti-ferromagnetic-paramagnetic phase diagram for ErAs (127) and the magnetic properties as a function of film thickness down to three monolayers (~8.5 Å) (52, 75).

The optical properties of ErAs have been studied by Ralston et al (58, 134, 135), and $Sc_{0.2}Yb_{0.8}As$ and $Sc_{0.3}Er_{0.7}As$ have been studied by Neveux and co-workers (131).

QUANTUM CONFINEMENT AND RESONANCE TUNNELING THROUGH METALLIC QUANTUM WELLS Resonant tunneling has been observed in structures utilizing buried NiAl layers (GaAs/AlAs/NiAl/AlAs/GaAs) (136). This has led to the possibility of fabricating three-terminal resonant-tunneling transistor-like structures (53, 139).

From the knowledge of quantum confinement in semiconductors, one may expect that for sufficiently thin films, quantum confinement would result in a semi-metal to semiconductor transition for $Sc_{1-x}Er_xAs$. However, no such transition is observed, and the films remain semi-metallic even for 3 ML–thick films (75). Band structure calculations by Xia et al (140) predicted no semi-metal to semiconductor transition. However, the validity of their predictions remain questionable (133). The fact that this transition is not observed makes these RE-V compounds ideal for ultra-thin, active, conducting materials in novel device structures (141). Zhang et al (142) measured thickness and magnetic field dependence of resonant-tunneling phenomena through ultra-thin ErAs layers. They observed interesting film thickness–dependent tunneling in the range of 5 to 15 ML; the energy of the resonant channel decreased with film thickness. This is opposite to what is observed for resonant tunneling through semiconductor quantum wells. Recent band structure calculations of the quantum confinement suggest that the magnetic field dependence may be consistent with tunneling through a hole state (143). This issue has still to be resolved.

Epitaxial Magnetic Compounds on III-V Compound Semiconductors

τ-MnAl, a ferromagnetic metastable phase with a tetragonally distorted CsCl crystal structure ($a_0 = 2.77$ Å and $c_0 = 3.54$ Å) is nearly lattice matched to half the lattice constant of AlAs (mismatch -2% to $a/2$ of GaAs) when the c-axis is oriented perpendicular to the AlAs(100) surface. The easy magnetization direction is along the c-axis of the tetragonal unit cell. Such a material, if grown on GaAs with the easy magnetization axis parallel to the substrate surface normal, could form the basis for a magnetic or magneto-optic information storage medium integrated with compound semiconductor electronic and photonic devices (144). It has been grown with this orientation on AlAs(100) in a manner similar to the growth of NiAl (144–146). The related phase MnGa, which is a thermodynamically stable ferromagnetic phase with a similar tetragonal crystal structure, has been grown on GaAs with an orientation similar to that observed for τ-MnAl on AlAs(100) (147, 148). $Mn_{1-x}Ni_xGa$ (148) and multilayers of MnGa/NiGa have been grown on GaAs(100) (149) as well as have multilayers of τ-MnAl/NiAl on AlAs(100) (150). The MnGa/NiGa multilayers showed multiple steps in the extraordinary Hall effect, with the number of steps corresponding to the number of superlattice periods in the sample (149). As in the case of TM-III materials with the CsCl structure, the orientation of these tetragonal structures depends critically upon the templates used.

α-MnAs, another ferromagnetic material but with an NiAs-type hexagonal structure with the easy magnetization axis perpendicular to the *c*-axis, has been found to grow epitaxially with either the $(\bar{1}100)$ or the $(\bar{1}101)$ surface parallel to the GaAs(100) surface, depending on the prelayer prior to the growth (Figure 11) (151). Predeposition of As_2 onto the $c(4 \times 4)$ As-terminated surface at 200°C, followed by codeposition of Mn and As_2 resulted in the growth of α-MnAs with the orientation relationship $(\bar{1}100)\langle\bar{1}\bar{1}20\rangle\langle0001\rangle\alpha$-MnAs$\|(100)\langle011\rangle\langle0\bar{1}1\rangle$GaAs (Figure 11*a*). If one ML of Mn was predeposited instead, the resulting orientation was predominantly $(\bar{1}101)\langle\bar{1}\bar{1}20\rangle\langle1\bar{1}02\rangle\alpha$-MnAs$\|(100)\langle0\bar{1}1\rangle\langle011\rangle$GaAs (Figure 11*b*). Note that the α-MnAs $\langle\bar{1}\bar{1}20\rangle$ axis is rotated by 90° on the GaAs(100) surface for the two cases. The lattice mismatch along the $\langle\bar{1}\bar{1}20\rangle$ direction of α-MnAs is -6.8% for both orientations. The mismatches perpendicular to this direction for $(\bar{1}100)$ and $(\bar{1}101)$ grown MnAs are 43 and 71%, respectively. It is interesting to note that the former corresponds to a coincident lattice with two units of α-MnAs to three units of GaAs in the GaAs$\langle0\bar{1}1\rangle$ direction with a mismatch of -4.7%, and the latter with three units of MnAs to five units of GaAs in the GaAs $\langle011\rangle$ direction with a mismatch of 2.4%.

Epitaxial Insulators on III-V Compound Semiconductors

In general, a material with hexagonal symmetry would be more likely to grow epitaxially on a $\{111\}$ than a $\{100\}$ surface of a cubic material. The RE triflourides (LaF_3 and NdF_3) with the hexagonal tysonite crystal structure (152), which is the stable structure at room temperature in the bulk, have been grown on GaAs$(\bar{1}\bar{1}\bar{1})$ with the (0001) RE-$F_3\|(111)$ GaAs and $[1\bar{1}00]$ RE-$F_3\|[\bar{1}10]$ GaAs (153). The mismatch is 3.8 and 1.5% for LaF_3 and NdF_3, respectively (153). Jenkins et al (154) found for growth of GdF_3 on Ge and GaAs$(\bar{1}\bar{1}\bar{1})$ (mismatch of 0.4%) that the tysonite-like structure grows initially. Above a critical thickness, a multidomain orthorhombic structure, which is the stable phase at room temperature in the bulk, forms. Similar behavior was found for HoF_3 grown on Si(111) (154, 155). Strong interface reactions were observed for HoF_3/Si, LaF_3/Si and GdF_3/GaAs but not in the GdF_3/Ge case (154).

Lui et al (156, 157) have grown Er-doped PbF_2 on GaAs(100) (mismatch $\sim 5\%$) and on $(\bar{1}\bar{1}\bar{1})$ with an intermediate SrF_2 (mismatch 2.6%) layer. They observed upconversion luminescence of Er-doped ZnF_2 planar and channel waveguides. The technological drive is for integrating the wave-

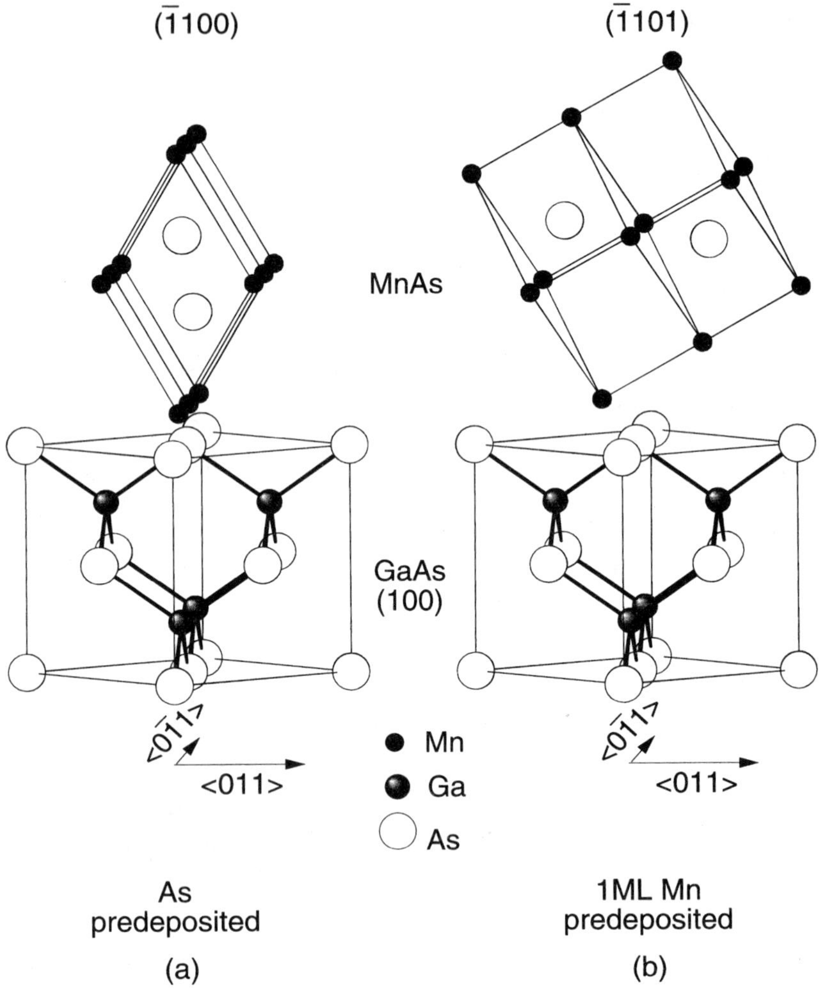

Figure 11 Schematic illustration of the α-MnAs orientation relationships observed on GaAs(100) depending upon the predeposited layer. (*a*) ($\bar{1}$100) MnAs‖GaAs(110). (*b*) ($\bar{1}$101) MnAs‖GaAs(100).

guide laser device with the diode pump laser on the same substrate. Growth on the (100) surface was by island growth. The growth on ($\bar{1}\bar{1}\bar{1}$) resulted in a significant improvement in crystalline quality (157), and it is believed to result from the improvement in wetting of the GaAs (157) due to the lower surface energy of the {111} fluoride surface (158). Details on the growth of group II fluorides (CaF_2 crystal structure) CaF_2, BaF_2, and SrF_2

and lattice-matched $Ca_xSr_{1-x}F_2$ on GaAs are given in a review by Farrow (159). CaF_2 shows promise as a strain relief layer between GaAs and Si (mismatch 4%) in the formation of GaAs/CaF_2/Si(111) structures (160).

CONCLUSIONS

By addressing crystal structure, interface bonding, surface and interface energy, and morphological and thermodynamic stability and symmetry, significant progress has been made in dissimilar materials epitaxy. High-crystalline quality, lattice-matched, thermodynamically stable epitaxial metallic compounds have been grown on III-V semiconductors, which are ideal contacts to III-V semiconductors. The RE-V/III-V dissimilar materials heterostructures have enabled the detailed study and determination of the semi-metallic RE-V compound properties. GaAs/AlAs/metallic compound/AlAs/GaAs structures allow resonant-tunneling metallic and semi-metallic layers to be studied. Further progress is needed to improve the overgrown III-V semiconductor crystal quality to make high-performance device structures with buried metallic layers. Symmetry and interface and surface energies are the major problems. If progress is made on manipulation of surface and interface energies through use of surfactant and template layers, the realization of dissimilar materials epitaxy heterostructures of comparable quality to current semiconductor heterostructures is feasible.

The prospect of combining epitaxial metallic compounds and semiconductors with the properties of epitaxial magnetic compounds and epitaxial optically active materials through dissimilar materials epitaxy opens the possibility of making material structures, such as dissimilar materials superlattices, with novel properties tailored through growth. The generation of a number of new devices with tailored and integrated electronic, optoelectronic, optical, and magnetic properties could be realized.

Literature Cited

1. Dodson BW, Myers DR, Datye AK, Kaushik VS, Kendall DL, Martinez-Tovar B. 1988. *Phys. Rev. Lett.* 61: 2681–84
2. Schowalter LJ, Hall EL, Lewis N, Hashimoto S. 1988. In *Thin Films: Stresses and Mechanical Properties Symposium*, ed. JC Bravman, WD Nix, DM Barnett, DA Smith, pp. 171–77. Pittsburgh: Mater. Res. Soc

3. Harrison WA, Kraut EA, Waldrop JR, Grant RW. 1978. *Phys. Rev. B* 18: 4402–10
4. Harrison WA. 1979. *J. Vac. Sci. Technol.* 16: 1492–96
5. Farrell HH, Tamargo MC, de Miguel JL, Turco FS, Hwang DM, Nahory RE. 1991. *J. Appl. Phys.* 69: 7021–28
6. Farrell HH, Tamargo MC, de Miguel JL. 1991. *Appl. Phys. Lett.* 58: 355–57
7. Maehashi K, Hasegawa S, Nakashima H. 1993. *J. Cryst. Growth* 127: 98–101
8. Frankel DJ, Yu C, Harbison JP, Farrell HH. 1987. *J. Vac. Sci. Technol. B* 5: 1113–18
9. Harbison JP, Farrell HH. 1988. *J. Vac. Sci. Technol. B* 6: 733–35
10. Dassanayake UM, Chen W, Kahn A. 1993. *J. Vac. Sci. Technol. B* 11: 1467–71
11. Tung RT, Gibson JM, Poate JM. 1983. *Appl. Phys. Lett.* 42: 888–90
12. Tung RT, Gibson JM. 1985. *J. Vac. Sci. Technol. A* 3: 987–91
13. Tung RT. 1989. *J. Vac. Sci. Technol. A* 7: 598–605
14. Bauer EG, Dodson BW, Ehrlich DJ, Feldman LC, Flynn CP, et al. 1990. *J. Mater. Res.* 5: 852–94
15. Copel M, Reuter MC, Horn von Hoegen M, Tromp RM. 1990. *Phys. Rev. B* 42: 11682–89
16. Waldrop JR, Grant RW. 1979. *Appl. Phys. Lett.* 34: 630–32
17. Prinz GA, Krebs JJ. 1981. *Appl. Phys. Lett.* 39: 397–99
18. Chambers SA, Xu F, Chen HW, Vitomirov IM, Anderson SB, Weaver JH. 1986. *Phys Rev. B* 34: 6605–11
19. Krebs JJ, Jonker BT, Prinz GA. 1987. *J. Appl. Phys.* 61: 2596–99
20. Koon NC, Jonker BT, Volkening FA, Krebs JJ, Prinz GA. 1987. *Phys. Rev. Lett.* 59: 2463–66
21. Rubenstein M, Rachford FJ, Fuller WW, Prinz GA. 1988. *Phys. Rev. B* 37: 8689–700
22. Prinz GA. 1985. *Phys. Rev. Lett.* 54: 1051–54
23. Xu F, Joyce JJ, Ruckman MW, Chen H-W, Boscherini F, Hill DM, et al. 1987. *Phys. Rev. B* 35: 2375–84
24. Ruckman MW, Joyce JJ, Weaver JH. 1987. *Phys. Rev. B* 33: 7029–35
25. Yu AJ, Galvin GJ, Palmstrøm CJ, Mayer JW. 1985. *Appl. Phys. Lett.* 47: 934–36
26. Palmstrøm CJ, Chang CC, Yu A, Galvin GJ, Mayer JW. 1987. *J. Appl. Phys.* 62: 3755–62
27. Genut M, Eizenberg M. 1987. *Appl. Phys. Lett.* 50: 1358–60
28. Palmstrøm CJ, Chase EW, Hwang DM, Harbison JP, Chang CC, et al. 1988. *J. Vac. Sci. Technol. A* 6: 1456–61
29. Shiau FY, Chang YA, Chen LJ. 1988. *J. Electron. Mater.* 17: 433–41
30. Beyers R, Kim KB, Sinclair R. 1987. *J. Appl. Phys.* 61: 2195–202
31. Schmid-Fetzer R. 1988. *J. Electron. Mater.* 17: 193–200
32. Sands T. 1986. *J. Metal* 38: 31–33
33. Sands T, Harbison JP, Ramesh R, Palmstrøm CJ, Florez LT, Keramidas VG. 1990. *Mater. Sci. Eng. B* 6: 147–57
34. Kroemer H. 1986. *Mater. Res. Soc. Symp. Proc.* 67: 3–14
34a. Villars P, Calvert LD. 1991. *Pearson's Handbook of Crystallographic Data for Intermetallic Phases*, 2nd ed. Metals Park, OH: Am. Metals Soc.
35. Ludeke R, Chang LL, Esaki L. 1973. *Appl. Phys. Lett.* 23: 201–3
36. Ludeke R. 1984. *J. Vac. Sci. Technol. B* 2: 400–6
37. Cho AY, Dernier PD. 1978. *J. Appl. Phys.* 49: 3328–32
38. Luo YS, Yang Y-N, Weaver JH, Florez LT, Palmstrøm CJ. 1994. *Phys. Rev. B.* 49: 1893–99
39. Massies J, Delescluse P, Etienne P, Linh NT. 1982. *Thin Solid Films* 90: 113–18
40. Prinz GA. 1986. *Mater. Res. Soc. Symp. Proc.* 56: 139–44
41. Jonker BT, Prinz GA, Idzerda YU. 1991. *J. Vac. Sci. Technol. B* 9: 2437–44
42. Jonker BT, Prinz GA. 1991. *J. Appl. Phys.* 69: 2938–41
43. Gutierrez CJ, Krebs JJ, Prinz GA. 1992. *Appl. Phys. Lett.* 61: 2476–78
44. Krebs JJ, Jonker BT, Prinz GA. 1989. In *Growth, Characterization and Properties of Ultrathin Magnetic Films and Multilayers*, ed. BT Jonker, JP Heremans, EE Marinero, pp. 23–28. Pittsburgh: Mater. Res. Soc.
45. Prinz GA. 1991. *Science* 250: 1092–97
46. Donner SK, Caffey KP, Winograd N. 1989. *J. Vac. Sci. Technol. B* 7: 742–46
47. Maeda N, Kawashima M, Horikoshi Y. 1993. *J. Appl. Phys.* 74: 4461–71
48. Farrow RFC, Speriosu VS, Parkin SSP, Chien C, Bravman JC, et al. 1989. *Mater. Res. Soc. Conf. Proc.* 130: 281–87
49. Chien CJ, Bravman JC, Farrow RFC. 1990. *J. Appl. Phys.* 68: 4343–45
50. Golding TD, Dura JA, Wang H, Zborowski JT, Vigliante A, et al. 1993. *Semicond. Sci. Technol.* 8: S117–20
51. DiVenere A, Yi XJ, Hou CL, Wang

HC, Ketterson JB, et al. 1993. *Appl. Phys. Lett*. 62: 2640–42
51a. DiVenere A, Yi XJ, Hon CL, Wang HC, Chen J, et al. 1994. *J. Vac. Sci. Technol. B* 12: 1136–39
52. Palmstrøm CJ, Sands T. 1993., In *Contacts to Semiconductors: Fundamentals and Technology*, ed. LJ Brillson, pp. 67–175. Park Ridge, IL: Noyes
53. Sands T, Palmstrøm CJ, Harbison JP, Keramidas VG, Tabatabaie N, et al. 1990. *Mater. Sci. Rep*. 5: 99–170
54. Villars P, Calvert LD. 1985. *Pearson's Handbook of Crystallographic Data for Intermetallic Phases*, Vol. 2, 3. Metals Park, OH: Am. Soc. Metals
55. Palmstrøm CJ, Tabatabaie N, Allen SJ Jr. 1988. *Appl. Phys. Lett*. 53: 2608–10
56. Palmstrøm CJ, Garrison KC, Mounier S, Sands T, Schwartz CL, et al. 1989. *J. Vac. Sci. Technol. B* 7: 747–52
57. Palmstrøm CJ, Cheeks TL, Gilchrist HL, Zhu JG, Carter CB, et al. 1992. *J. Vac. Sci. Technol. A* 10: 1946–53
58. Ralston JD, Fuchs F, Schneider J, Schmälzlin J. 1990. *J. Appl. Phys*. 68: 2176–80
59. Ralston JD, Wagner J, Fuchs F, Hiesinger P. 1991. *J. Cryst. Growth* 111: 989–95
60. Le Corre A, Caulet J, Guivarc'h A. 1989. *Appl. Phys. Lett*. 55: 2298–300
61. Guivarc'h A, Le Corre A, Caulet J, Guenais B, Minier M, et al. 1990. *Mater. Res. Soc. Symp. Proc*. 160: 331–36
62. Yamaguchi H, Horikoshi Y. 1992. *Appl. Phys. Lett*. 60: 2341–43
63. Jourdan N, Yamaguchi H, Horikoshi Y. 1993. *Jpn. J. Appl. Phys*. 32: L1784–87
64. Richter HJ, Smith RS, Herres N, Seelman-Eggebert M, Wennekers P. 1988. *Appl. Phys. Lett*. 53: 99–101
65. Lépine B, Quémerias A, Sébilleau D, Jézéquel G, Agliz D, et al. 1994. *J. Appl. Phys*. 76: 5218–24
66. Guivarc'h A, Caulet J, Le Corre A. 1989. *Electron. Lett*. 25: 1050–51
67. Guivarc'h A, Ballini Y, Toudic Y, Minier M, Auvray P, et al. 1994. *J. Appl. Phys*. 75: 2876–83
68. Palmstrøm CJ, Mounier S, Finstad TG, Miceli PF. 1990. *Appl. Phys. Lett*. 56: 382–84
69. Palmstrøm CJ, Sands T, Harbison JP, Finstad TG, Mounier S, et al. 1990. *SPIE Proc*. 1285: 85–91
70. Guivarc'h A, Guenais B, Ballini Y, Auvray P, Caulet J, et al. 1993. *J. Cryst. Growth* 127: 638–42
71. Guivarc'h A, Ballini Y, Minier M, Guenais B, Dupas G, et al. 1993. *J. Appl. Phys*. 73: 8221–26
72. Guenais B, Poudoulec A, Guivarc'h A, Ballini Y, Durel V, d'Anterroches C. 1992. *Microsc. Microanal. Microstruct*. 3: 299–312
73. Le Corre A, Guenais B, Guivarc'h A, Lecrosnier D, Caulet J, et al. 1990. *J. Cryst. Growth* 105: 234–39
74. Caulet J, Le Corre A, Guenais B, Ropars G, Guivarc'h A. 1990. *J. Cryst. Growth* 102: 309–13
75. Allen SJ Jr, Tabatabaie N, Palmstrøm CJ, Hull GW, Sands T, et al. 1990. *Surf. Sci*. 228: 13–15
76. Miceli PF, Palmstrøm CJ, Moyers KW. 1991. *Appl. Phys. Lett*. 58: 1602–4
77. Miceli PF, Moyers KW, Palmstrøm CJ. 1991. *Mater. Res. Soc. Symp. Proc*. 202: 579–84
78. Miceli PF, Palmstrøm CJ, Moyers KW. 1992. In *Surface X-ray and Neutron Scattering*, ed. H Zabel, IK Robinson 61: 203–8. Heidelberg: Springer Proc. Physics
79. Miceli PF, Palmstrøm CJ, Moyers KW. 1992. *Appl. Phys. Lett*. 61: 2060–62
80. Wilkens BJ, Martinez JA, Mounier S, Palmstrøm CJ. 1991. *J. Appl. Phys*. 70: 4890–93
81. Zhu JG, Palmstrøm CJ, Mounier S, Carter CB. 1990. *Mater. Res. Soc. Symp. Proc*. 160: 325–30
82. Zhu JG, Palmstrøm CJ, Carter CB. 1990. *Inst. Phys. Conf. Ser*. 98: 411–14
83. Zhu JG, Carter CB, Palmstrøm CJ. 1990. *Mater. Res. Soc. Symp. Proc*. 198: 171–76
84. Zhu JG, McKernan S, Palmstrøm CJ, Carter CB. 1990. *Mater. Res. Soc. Symp. Proc*. 159: 89–94
85. Stoffel NG, Palmstrøm CJ, Wilkens BJ. 1991. *Nucl. Inst. Meth. B* 56/57: 792–94
86. Tarnow E, Rappe AM. 1995. *J. Appl. Phys*. In press
87. Sands T, Harbison JP, Chan WK, Schwarz SA, Chang CC, et al. 1988. *Appl. Phys. Lett*. 52: 1216–18
88. Harbison JP, Sands T, Tabatabaie N, Chan WK, Florez LT, Keramidas VG. 1988. *Appl. Phys. Lett*. 53: 1717–19
89. Palmstrøm CJ, Fimland B-O, Sands T, Garrison KC, Bartynski RA. 1989. *J. Appl. Phys*. 65: 4753–58
90. Tanaka M, Sakakibara H, Nishinaga T. 1991. *Appl. Phys. Lett*. 59: 3115–17
91. Sands T, Harbison JP, Palmstrøm CJ, Ramesh R, Keramidas VG 1991. *Mater. Res. Soc. Symp. Proc*. 221: 271–82

92. Sands T, Harbison JP, Tabatabaie N, Chan WK, Gilchrist HL, et al. 1989. *Mater. Res. Soc. Symp. Proc.* 144: 571–82
93. Sands T, Harbison JP, Tabatabaie N, Chan WK, Gilchrist HL, et al. 1990. *Surf. Sci.* 228: 1–8
94. Harbison JP, Sands T, Ramesh R, Tabatabaie N, Gilchrist HL, et al. 1990. *J. Vac. Sci. Technol. B* 8: 242–45
95. Chambers SA. 1989. *J. Vac. Sci. Technol. B* 7: 737–41
96. Tanaka M, Nishinaga T, Ikarashi N, Shimada H. 1994. *J. Appl. Phys.* 75: 885–96
97. Guivarc'h A, Guérin R, Secoué M. 1987. *Electron. Lett.* 23: 1004–5
98. Guivarc'h A, Caulet J, Guenais B, Ballini Y, Guérin R, et al. 1989. *J. Cryst. Growth* 95: 427–30
99. Durel V, Guenais B, Ballini Y, Caulet J, Chomette A, et al. 1990. *Inst. Phys. Conf. Ser.* 112: 129–34
100. Guenais B, Poudoulec A, Caulet J, Guivarc'h A. 1990. *J. Cryst. Growth* 102: 925–32
101. Kuo TC, Kang TW, Wang KL. 1991. *J. Cryst. Growth* 111: 996–1002
102. Baugh DA, Talin AA, Williams RS, Kuo T-C, Wang KL. 1991. *J. Vac. Sci. Technol. B* 9: 2154–57
103. Wowchak AM, Kuznia JN, Cohen PI. 1989. *J. Vac. Sci. Technol. B* 7: 733–36
104. Guivarc'h A, Ballini Y, Auvray P, Caulet J, Minier M, et al. 1993. *J. Appl. Phys.* 74: 6632–35
105. Hong M, Chen HS, Kwo J, Kortan AR, Mannaerts JP, et al. 1991. *J. Cryst. Growth* 111: 984–88
106. Okamoto K, Wood CEC, Rathbun L, Eastman LF. 1982. *J. Appl. Phys.* 53: 1532–35
107. Tadayon B, Tadayon S, Spencer MG, Harris GL, Rathbun L, et al. 1988. *Appl. Phys. Lett.* 53: 2664–65
108. Oh JE, Bhattacharya PK, Singh J, Dos Passos W, Clarke R, et al. 1990. *Surf. Sci.* 228: 16–19
109. Bhattacharya P, Oh JE, Singh J, Biswas D, Clarke R, et al. 1990. *J. Appl. Phys.* 67: 3700–5
110. Palmstrøm CJ, Finstad TG, Hugstad B, Taftø J, Miceli PF, et al. 1993. *Growth of GaAs/$Sc_{1-x}Er_xAs$/GaAs heterostructures on different substrate orientations.* Presented at Spring Mater. Res. Soc. Meet. San Francisco
111. Finstad TG, Palmstrøm CJ, Mounier S, Keramidas VG, Zhu JG, Carter CB. 1991. *Mater. Res. Soc. Symp. Proc.* 202: 413–18
112. Zhu JG, Palmstrøm CJ, Carter CB. 1990. *Mater. Res. Soc. Symp. Proc.* 198: 177–82
113. Zhu JG, Palmstrøm CJ, Carter CB. 1991. *Inst. Phys. Conf. Ser.* 117: 419–22
114. Zhu JG, Palmstrøm CJ, Carter CB. 1995. *J. Appl. Phys.* In press
115. Zhu JG, Palmstrøm CJ, Carter CB. 1995. *J. Appl. Phys.* In press
116. Hugsted B, Taftø J, Finstad TG, Palmstrøm CJ. 1993. *Appl. Phys. Lett.* 63: 2499–501
117. Bokshtein SZ. 1973. *Diffusion and Structure of Metals*, 1: 123. Moscow: Metallurgiya
118. Guérin R, Guivarc'h A. 1989. *J. Appl. Phys.* 66: 2122–28
119. Guérin R, Deputier S, Caulet J, Minier M, Poudoulec A, et al. 1990. *Mater. Res. Soc. Symp. Proc.* 160: 319–24
120. Tung RT. 1984. *Phys. Rev. Lett.* 52: 461–64
121. Das GP, Blöchl P, Andersen OK, Christensen NE, Gunnarsson O. 1989. *Phys. Rev. Lett.* 63: 1168–71
122. Rechtien JJ, Kannewurf CR, Brittain JO. 1967. *J. Appl. Phys.* 38: 3045–50
123. Caskey GR, Franz JM, Sellmyer DJ. 1973. *J. Phys. Chem. Solids* 34: 1179–98
124. Yamaguchi Y, Kiewit DA, Aoki T, Brittain JO. 1968. *J. Appl. Phys.* 39: 231–32
125. Butler SR, Hanlon JE, Wasilewski RJ. 1969. *J. Phys. Chem. Solids* 30: 1929–34
126. Allen SJ Jr, Tabatabaie N, Palmstrøm CJ, Hull GW, Sands T, et al. 1989. *Phys. Rev. Lett.* 62: 2309–12
127. Allen SJ Jr, DeRosa F, Palmstrøm CJ, Zrenner A. 1991. *Phys. Rev. B* 43: 9599–609
128. Bogaerts R, Bockstal LV, Herlach F, Peeters FM, DeRosa F, et al. 1992. *Physica B* 177: 425–29
129. Bogaerts R, Esch AV, Bockstal LV, Herlach F, Peeters FM, et al. 1993. *Physica B* 184: 232–35
130. Bogaerts R, De Keyser A, Herlach F, Peeters FM, DeRosa F, et al. 1994. *Solid State Electron.* 37: 789–92
131. Neveux D, Simon JC, Ballini Y, Guivarc'h A. 1993. *Mater. Lett.* 16: 251–53
132. Hasegawa A, Yanase A. 1977. *J. Phys. Soc. Jpn.* 42: 492–98
133. Petukhov AG, Lambrecht WRL, Segall B. 1994. *Phys. Rev. B* 50: 7800–4
134. Schneider J, Müller HD, Ralston JD, Fuchs F, Dörnen A, Thonke K. 1991. *Appl. Phys. Lett.* 59: 34–36
135. Ralston JD, Ennen H, Wennekers P,

Hiesinger P, Herres N, et al. 1990. *J. Electron. Mater.* 19: 555–60
136. Tabatabaie N, Sands T, Harbison JP, Gilchrist HL, Keramidas VG. 1988. *Appl. Phys. Lett.* 53: 2528–30
137. Deleted in proof
138. Deleted in proof
139. Tabatabaie N, Sands T, Harbison JP, Gilchrist HL, Cheeks TL, et al. 1989. *Tech. Dig. Int. Elect. Dev. Meet., Washington, DC* pp. 555–58
140. Xia J-B, Ren S-F, Chang Y-C. 1991. *Phy. Rev. B* 43: 1692–98
141. Allen SJ Jr, Brehmer D, Palmstrøm CJ. 1993. *Mater. Res. Soc. Symp. Proc.* 301: 307–17
142. Zhang K, Brehmer D, Allen SJ Jr, Palmstrøm CJ. 1995. *Proc. 21st. Int. Symp. Compound Semiconductors*, In press
143. Petukhov AG, Lambrecht WRL, Segall B. 1994. *Quantum confinement effects in ultrathin layers of rare-earth group-V compounds.* Presented at Fall Mater. Res. Soc. Meet. Boston
144. Sands T, Harbison JP, Leadbeater ML, Allen SJ Jr, Hull GW, et al. 1990. *Appl. Phys. Lett.* 57: 2609–11
145. Sands T, Harbison JP, Allen SJ Jr, Leadbeater ML, Cheeks TL, et al. 1992. *Mater. Res. Soc. Symp. Proc.* 231: 341–46
146. Harbison JP, Sands T, Ramesh R, Florez LT, Wilkens BJ, Keramidas VG. 1991. *J. Cryst. Growth* 111: 978–83
147. Tanaka M, Harbison JP, De Boeck J, Sands T, Philips B, et al. 1993. *Appl. Phys. Lett.* 62: 1565–67
148. Tanaka M, Harbison JP, Sands TD, Philips BA, De Boeck J, et al. 1993. *Mater. Res. Soc. Symp. Proc.* 313: 507–12
149. Tanaka M, Harbison JP, Sands T, Philips B, Cheeks TL, et al. 1993. *Appl. Phys. Lett.* 63: 696–98
150. Tanaka M, Harbison JP, Sands T, Cheeks TL, De Boeck J, et al. 1993. *Appl. Phys. Lett.* 63: 839–41
151. Tanaka M, Harbison JP, Park MC, Park YS, Shin T, Rothberg GM 1994. *Appl. Phys. Lett.* 65: 1964–66
152. Gregson D, Catlow CRA, Chadwick AV, Lander GH, Cormack AN, Fender BE. 1983. *Acta Cryst.* B39: 687–91
153. Farrow RFC. 1990. *J. Cryst. Growth* 104: 556–77
154. Jenkins LC, Griffiths C, Hughes A, Richards J, Williams RH 1993. *J. Vac. Sci. Technol. B* 11: 1541–45
155. Griffiths CL, Macdonald JE, Williams RH. 1991. *J. Appl. Phys.* 70: 1858–60
156. Lui M, McFarlane RA, Yap D. 1993. *Mater. Res. Soc. Symp. Proc.* 301: 331–36
157. Lui M, McFarlane RA, Yap D. 1994. *Mater. Res. Soc. Symp. Proc.* 329: 167–72
158. Fathauer RW, Schowalter LJ. 1984. *Appl. Phys. Lett.* 45: 519–21
159. Farrow RFC. 1988. In *The Chemical Physics of Solid Surfaces and Heterogeneous Catalysis*, ed. DA King, DP Woodruff, Vol. 5, chpt. 9. Amsterdam: Elsevier
160. Li W, Anan T, Schowalter LJ. 1994. *Appl. Phys. Lett.* 65: 595–97 and references therein

Annu. Rev. Mater. Sci. 1995. 25: 417–54

GeSi/Si NANOSTRUCTURES

E. A. Fitzgerald

Department of Materials Science, Massachusetts Institute of Technology, 77 Massachusetts Avenue, Cambridge, Massachusetts 02139

KEY WORDS: lattice mismatch, dislocations, quantum wells, electronic devices, two-dimensional gases

---

ABSTRACT

A review of key issues in GeSi epitaxial structures on Si is presented. The review begins with an explanation of why vapor phase techniques are required, and proceeds to describe the mechanical properties of GeSi alloys. Because Ge is a larger atom than Si, GeSi is lattice mismatched to Si substrates. The consequences of this lattice mismatch with regard to defect introduction, degree of relaxation, and band structure are presented. Known methods of dislocation control in GeSi/Si heterostructures are discussed. GeSi/Si epitaxy also can form natural nanostructures, e.g. surface features. Included in the review is a summary of recent progress in artificial nanostructures such as quantum wells and quantum wires. The article finishes with an overview of the most pursued electronic devices to date.

INTRODUCTION

GeSi alloys have been explored for a variety of applications of which one of the first was thermoelectric power generation in portable terrestrial units or satellite systems (1). This interest in GeSi led to many materials studies such as alloy crystal growth, alloy doping, and the determination of the thermal and electrical conductivity of the alloy. In recent history, much of the interest has focused on thin film GeSi, usually deposited by the vapor phase on Si (100) substrates. Interest in GeSi/Si arises predominantly from (*a*) the totally miscible phase diagram, which makes vapor deposition the favorable growth method, (*b*) the dominant Si CMOS industry in which Si (100) is the typical bulk substrate used, (*c*) the favorable electrical

0084–6600/95/0801–0417$05.00

and optical properties of GeSi, and (*d*) the benign nature of Ge in manufacturing and the potential for integrating GeSi with Si. Some of the favorable GeSi electrical and optical properties are higher hole mobility, ability to alter the effective masses of the carriers using strain, ability to employ modulation doping for higher carrier mobility, various advantageous heterojunction band-alignments with Si, and the smaller bandgap adjustable with Ge concentration.

These advantageous properties of GeSi/Si spurred a dramatic increase in research activity from the mid-1980s to the present. Many materials aspects were explored such as accommodating the lattice mismatch between the GeSi alloy and the Si substrate and utilizing lattice-strained GeSi layers on a Si substrate. With this knowledge, prototype devices were fabricated. Much of the early interest in thin film applications of GeSi alloys has been reviewed by People (2). It is not the objective of this review to include detailed repetition of such comprehensive works. In this review, we discuss the thermodynamic information regarding GeSi alloys, followed by the implications of the lattice-mismatch. Methods for reducing the defect densities in relaxed and strained GeSi/Si are discussed as a precursor to descriptions of the nanostructures that have been created in the GeSi/Si system. At all times, this review favors more recent advances in the field and refers the reader to less recent articles that have covered the early work in great detail.

PHYSICAL PROPERTIES OF GeSi ALLOYS

In this section, the phase diagram of GeSi alloys is discussed, and the implications for crystal growth are revealed. Properties such as lattice constant and elastic constants are presented as a function of Ge concentration in the alloy.

Thermodynamic Properties of GeSi and the Phase Diagram

The GeSi phase diagram is shown in Figure 1 (3). Note that across the entire range of alloy compositions, the solid solutions are miscible. The only phases present in such a phase diagram are liquid and solid, with a two-phase region separating these two phases.

For a regular solution, the free-energy of mixing is (4)

$$\Delta G^M = RT(X_A \ln X_A + X_B \ln X_B) + RT\alpha X_A X_B, \tag{1.}$$

where R is the gas constant, T is temperature, X_i is the atomic concentration of component i in the alloy, and α is the regular solution constant. The first term to the left on the right-hand-side of Equation 1 is the entropy

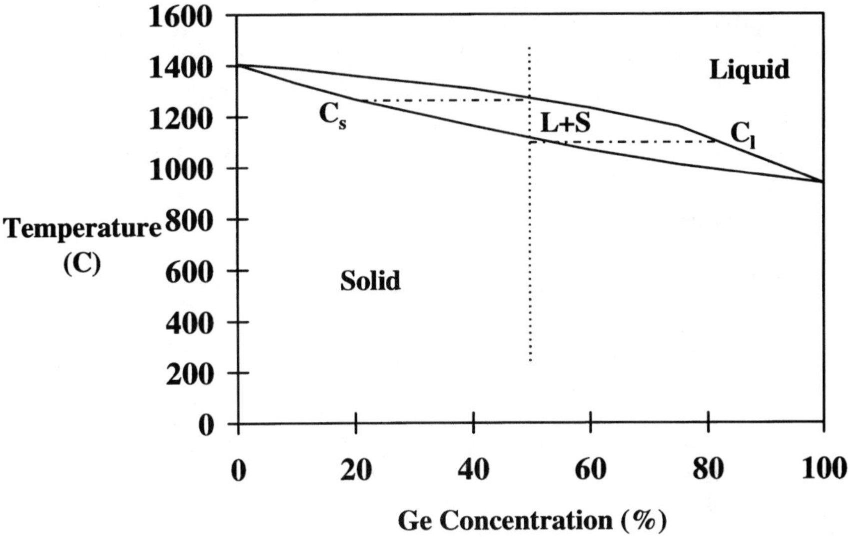

Figure 1 The GeSi phase diagram, showing the first solid composition to form during cooling (C_s) and the final liquid composition just before final solidification (C_l).

contribution to the heat of mixing (ΔS), and the second term on the right-hand-side of the equation is the heat of mixing, or the enthalpy of mixing (ΔH). Because X_i is always less than one for any finite alloy composition, Equation 1 clearly shows that if the heat of mixing is zero (an ideal solution with $\alpha = 0$), the ΔG^M is always less than zero with minimum at $X_i = 0.5$.

A miscible phase diagram like the GeSi diagram in Figure 1 demands that α be small enough such that the ΔS term dominates and ΔG^M has a parabolic dependence on X_i. Microscopically, this implies that Ge-Si bonds in the alloy have energies that are near the average bonding energy of Ge-Ge and Si-Si bonds, i.e. $E_{\text{Ge-Si}} \sim (E_{\text{Ge-Ge}} + E_{\text{Si-Si}})/2$. Small deviations from this relationship will create a small positive or negative heat of formation. Thus despite the much larger size of the Ge atom, there is not a great preference for Ge-Ge, Si-Si, or Ge-Si bonding, and a miscible alloy system is formed.

Implications for Growth Methods

The thermodynamic properties mentioned above have great implications in applications of the alloy, as well as in the production of the alloy. A positive aspect of these properties is that phase separation in the alloys is not a concern; thus, by simply choosing an alloy composition, a robust value of the lattice constant is obtained, and the alloy will retain the crystal structure until the melting point is reached. A negative aspect of this miscible phase diagram is that it is difficult to produce high quality, uniform GeSi crystals using bulk crystal growth techniques. Consider the isocomposition line on the phase diagram in Figure 1 drawn at $X_{Ge} = 0.5$. As the melt of Ge-Si is cooled, the material enters the two-phase liquid + solid region of the phase diagram. Using the phase diagram, we can see that the liquid has a composition of 0.5 when the solid first forms, and the solid has a composition of 0.2. During further solidification, the precipitating solid becomes more rich in Ge, and the liquid becomes less rich in Ge. Before total solidification, the last solid has a composition of 0.5, and the last liquid has a composition of 0.8. Thus in pulling a crystal from a melt using a seed crystal, the solidified crystal will have very different composition from the melt. Because the diffusion of Ge in the melt is relatively slow, a boundary layer forms between the crystal and the melt. To keep the boundary layer thin and uniform, excellent temperature control must be achieved, and impractically slow rates must be used. Under these conditions, fairly homogeneous GeSi alloy crystals with dislocation densities of 10^5 cm^{-2} can be produced (5). Zone-refined cast ingots can be made more quickly and uniformly, but the material is polycrystalline (1).

The combination of the difficulty in bulk growth and the desire to integrate GeSi devices with Si devices has led to vapor-phase epitaxy on Si substrates as the dominant method of GeSi crystal preparation. Molecular beam epitaxy (MBE) and chemical vapor deposition (CVD) are the two most popular methods of GeSi film deposition. MBE growth occurs in ultra-high vacuum, with the substrate rotatively heated in the chamber. The Ge and dopants can be evaporated from a heated boron nitride cell. The temperature needed for significant Si vapor pressure is too high for these cells, so typically an electron beam gun is used to heat a Si charge and produce the Si beam. Ge can also be electron-beam evaporated. The two molecular beams arrive at the heated Si substrate, and epitaxy occurs if the Si surface was properly cleaned.

A variety of CVD techniques, defined by the initial background pressure, the gas pressure during growth and the substrate heating method, have been created for the deposition of the thin films. GeSi has been grown using most of these techniques. The oldest and most utilized CVD techniques are atmospheric and low-pressure CVD, which operate at $\sim$1 atm and 50–100 mtorr, respectively. The reactant gases flow through a quartz tube, which is heated by a conventional furnace. Although well-proven in very large scale integration (VLSI) Si manufacturing, these older CVD techniques have been used to grow only a few GeSi structures (6). Much of the research in the CVD of GeSi has utilized the limited reaction processing CVD (LRPCVD), rapid thermal CVD (RTCVD), and ultra-high vacuum

CVD (UHVCVD). These techniques have been reviewed by others (7–10). The differences in the techniques are that the RTCVD and LRPCVD methods start with a lower background pressure ($\sim 10^{-3}$–10^{-8} torr) than atmospheric CVD and use infrared lamps to heat the wafers. Growth occurs at high flow-rates and high pressures of H_2 gas. Germane (GeH_4) and silane (SiH_4) or dichlorosilane ($SiCl_2H_2$) are mixed in this H_2 at relatively low partial pressures. In UHVCVD, the background pressure is usually in the 10^{-10} torr range, i.e. similar to MBE and lower than RTCVD and LRPCVD. Heating the substrates is accomplished via a standard furnace, and growth is achieved usually by flowing only the reacting gases, e.g. SiH_4 and GeH_4, resulting in a pressure during growth of a few mtorr. Although many researchers have preferred one method over another, this author has experience with MBE, RTCVD, and UHVCVD, and each has strengths and weaknesses. All have been able to grow high-quality GeSi/Si structures.

Lattice Constant and Elastic Constants

Experiments have shown that the GeSi lattice constant has values close to those expected from simple linear interpolation between Si and Ge (1), i.e. $a_{Ge_xSi_{1-x}} = a_{Si} + (a_{Ge} - a_{Si})\,x$, where $a_{Ge} = 5.658$ Å and $a_{Si} = 5.428$ Å. For determining mechanical properties of the alloys, such as in the determination of the critical thickness of GeSi on Si for dislocation introduction, the elastic constants are also usually linearly interpolated:

$$C_{11Si} = 16.57 \times 10^{11}\ \text{dynes/cm}^2 \qquad C_{11Ge} = 12.85 \times 10^{11}$$
$$C_{12Si} = 6.39 \times 10^{11} \qquad C_{12Ge} = 4.83 \times 10^{11}$$
$$C_{44Si} = 7.96 \times 10^{11} \qquad C_{44Ge} = 6.68 \times 10^{11}.$$

After the interpolation between these elastic constants for a particular alloy composition, we can calculate the other relevant elastic constants, such as the Young's modulus equivalent for a biaxial strained layer Y, and the shear modulus G: $Y = C_{11} + C_{12} - 2C_{12}^2/C_{11}$; $G = C_{44} - (2C_{44} + C_{12} - C_{11})/3$.

THE CONSEQUENCES OF LATTICE MISMATCH

As mentioned above, most interest in GeSi deposition involves growth on Si (100) wafers. Since the lattice constant increases with increasing Ge content, GeSi deposition on Si substrates results in a biaxial compressive stress on the GeSi film, assuming the film is much thinner than the substrate. A highly stressed GeSi layer will possess different electrical and optical properties. Also, the degree of stress may influence the quality of the GeSi film; if the stress is too great, one can expect plastic relaxation instead of elastic deformation.

The Critical Thickness

When considering using a highly strained layer in a device, we must know the limits of strain in the film or thickness of the film before plastic relaxation, i.e. dislocation introduction, occurs. The point at which dislocations are introduced into a thin strained film is termed the critical thickness, and this thickness is predominantly determined by the balance between strain energy and dislocation energy. The critical thickness in mismatched epitaxy has become an issue because of the increase in research attempting to utilize strained films (for a review, see 11).

The critical thickness calculation is an equilibrium calculation, i.e. it ignores any kinetic barriers such as dislocation nucleation and dislocation glide. There has been some confusion in the literature over the exact constants in the critical thickness calculation (11). If we use elastic constants, which reflect the anisotropic nature of the Si crystal structure, the equilibrium critical thickness will be

$$h_c = \frac{D(1-\upsilon\cos^2\alpha)(b/b_{\text{eff}})[\ln(h_c/b)+1]}{2Yf}, \qquad 2.$$

where D is an average shear modulus for a dislocation lying at the interface between the film and substrate [$D = G_oG_sb/\pi(G_o+G_s)(1\text{-}v)$], v is the Poisson ratio, α is the angle between the dislocation line and the Burgers vector (b), $f = \text{abs}(a_s - a_o)/a_o$ (a_o is the overlayer lattice constant, a_s is the substrate lattice constant), and b_{eff} is the effective Burgers vector, i.e. the resolved Burgers vector in-plane, which can create a discontinuity at the interface to relieve the mismatch strain. For $\alpha = 60°$, the most common glissile dislocation in the diamond cubic materials b_{eff} equals $b/2$. For a non-glissile edge dislocation ($\alpha = 90°$), which can sometimes form at the interface through other means, $b_{\text{eff}} = b$. For simplication, some researchers use the approximation that the crystal structure is isotropic. This approximation leads to the following expression for the critical thickness for 60° dislocations:

$$h_c = \frac{b(1-(\upsilon/4))[\ln(h_c/b)+1]}{4\pi f(1+\upsilon)}. \qquad 3.$$

Figure 2 is a plot of the critical thickness for both 60° dislocations and edge dislocations in the GeSi/Si system. As the Ge concentration is increased, the mismatch increases, and the strain increases, so the thickness that can be grown without the introduction of dislocations decreases. The

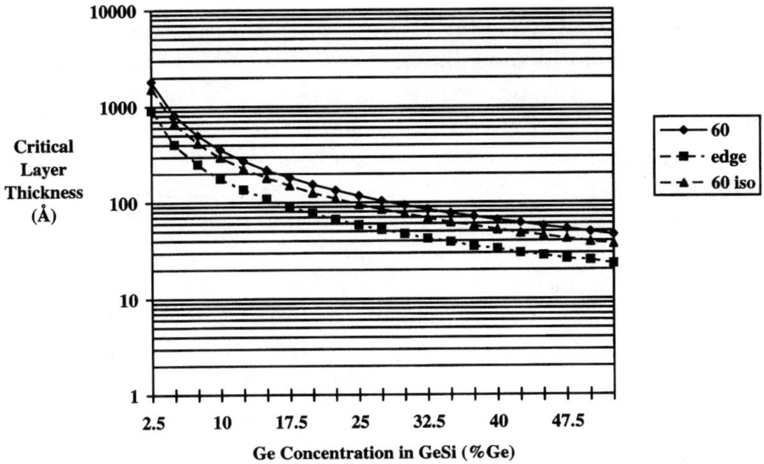

Figure 2 The critical thickness for dislocation introduction vs the Ge concentration in a GeSi film deposited on (001) Si.

60° dislocation curve is more relevant because lower mismatch systems ($f < \sim 1.5\%$) relax by introducing predominantly 60° dislocations. The curve for the isotropic approximation is also shown in Figure 2 for the 60° dislocation only. The difference in the isotropic vs anisotropic curves increases for more anisotropic crystal structures like the III-V and II-VI materials.

Experimental data points can fall above the critical thickness curve (11). This behavior is strong evidence for kinetic barriers and/or problems with experimental sensitivity to small amounts of relaxation. Evidence for metastability due to kinetic barriers has been shown by many authors e.g. Houghton et al. (12). In their experiments, samples grown with MBE at 550° lie above the curve; after sample annealing, the data agree well with equilibrium theory. Some of the discrepancy between experiment and theory may be due to the lack of experimental sensitivity to small amounts of relaxation, as pointed out by Fritz (13).

Effect of Strain on Band Structure

Figure 2 shows that the region underneath the critical thickness curve is accessible for creating strained layers that are in equilibrium, i.e. the strained layers will never have dislocations introduced because it would be energetically unfavorable. Thus we may be able to utilize the advantageous electronic properties of strained GeSi layers on Si.

The two most significant changes that can be expected from straining GeSi are the change in the band-gap energy and the alteration of the band alignment between heterogeneous thin films (i.e. GeSi/Si or $Ge_xSi_{1-x}/Ge_ySi_{1-y}$). Calculations show that due to the nature of the conduction and valence bands, positive or negative biaxial stress decreases the size of the band gap (2, 14, 15). Figure 3 is a summary plot of the band-gap calculations for GeSi alloys grown on either (001) Si (Figure 3*a*) or Ge (Figure 3*b*) substrates. The band gap for the bulk, cubic GeSi alloy is also shown to reveal the additional lowering due to strain. The band gap decreases (Figure 3*a*) with increasing Ge concentration. After an initial increase as the Ge-like L-band increases the gap, the band gap also decreases (Figure 3*b*) as the Ge concentration is decreased. In the case of GeSi on Si, the decrease in band gap is due to the lower band gap in the bulk alloy, and the lowering of the [100] and [010] conduction valleys from increasing compressive stress. For GeSi alloys on Ge, the L-band initially rises rapidly with an increase in Si concentration, but after Δ-L cross over, the strain lowers the [001] conduction valley, decreasing the band gap. Thus both higher Ge content and mismatch strain can be used to lower the GeSi band gap in GeSi/Si devices.

The other key to understanding GeSi heterostructure devices is the alignment of the band edges in the layers. Early work on GeSi/Si modulation-doped hole gases showed that the valence band contained most of the offset due to the band-gap difference (16). Further work showed that tensile-strained Si layers on relaxed GeSi could have both conducting and valence band offsets (type-II alignment), which would confine electrons in the strained Si and holes in the relaxed GeSi (17). Calculations for the extreme cases of Ge strained on (001) Si, and Si strained on Ge (001), verify the duality of the GeSi/Si band alignments (14, 18). In Figure 4, we

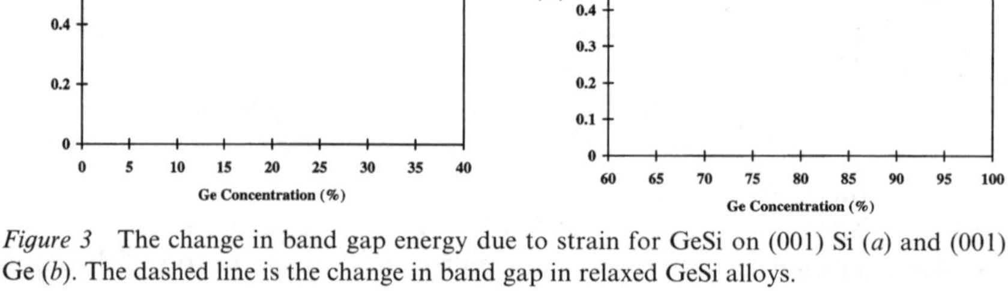

Figure 3 The change in band gap energy due to strain for GeSi on (001) Si (*a*) and (001) Ge (*b*). The dashed line is the change in band gap in relaxed GeSi alloys.

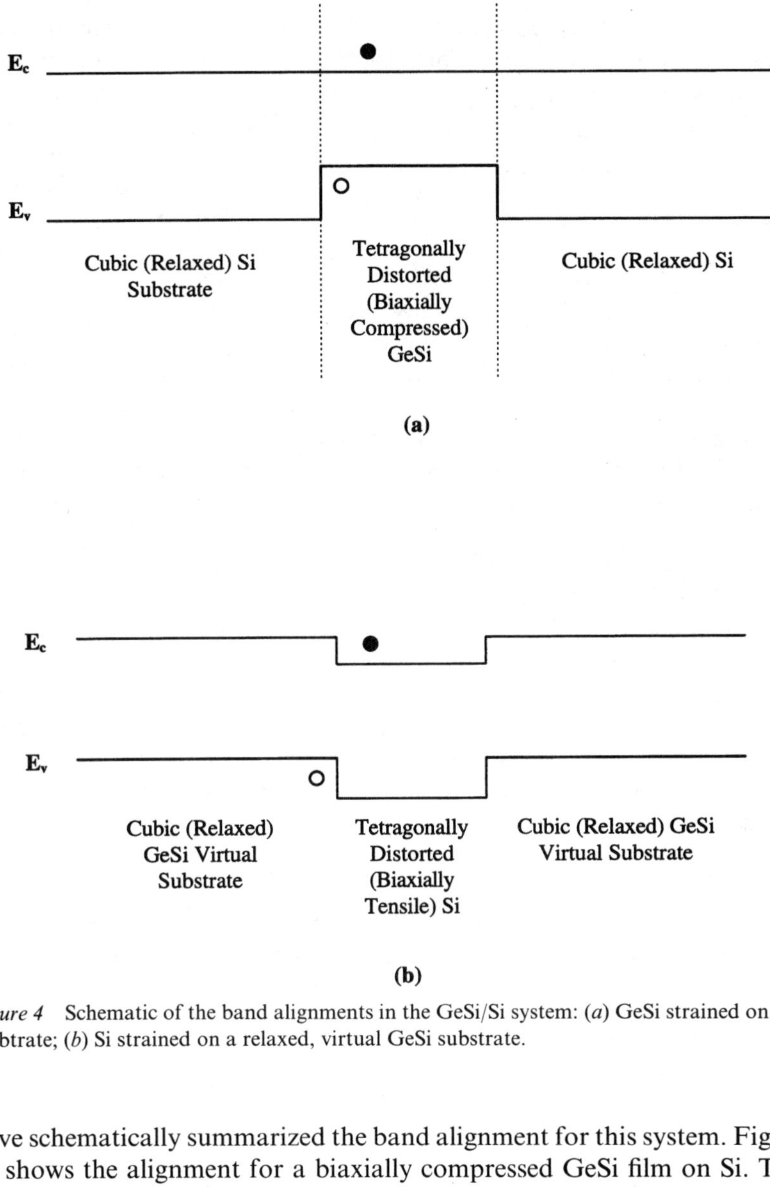

Figure 4 Schematic of the band alignments in the GeSi/Si system: (*a*) GeSi strained on a Si susbtrate; (*b*) Si strained on a relaxed, virtual GeSi substrate.

have schematically summarized the band alignment for this system. Figure 4*a* shows the alignment for a biaxially compressed GeSi film on Si. This type of heterostructure might be used in a quantum well or heterojunction bipolar transistor, as described below. The holes can be confined in such a heterostructure because all of the band-gap difference appears in the valence band. However, electrons are free in this structure, except for their Coulombic attraction to the confined holes.

Now consider the case shown in Figure 4*b*. In this structure, the GeSi alloy is somehow relaxed on a Si substrate (not shown). The thin Si layer is biaxially tensile with electrons being confined in the Si layer and holes confined in the GeSi layer. This structure might be used to confine electrons for a modulation-doped field effect transistor structure. For both cases shown in Figure 4, it should be remembered that the thickness of the strained layers is limited by the critical thickness (Figure 2).

Relaxation of GeSi Alloys on Si

As shown in Figure 4*b*, it can be advantageous to know how to relax GeSi alloys on Si substrates because this can lead to electronic structures that confine electrons instead of just hole-confining structures like those shown in Figure 4*a*. The relaxation of the alloys should follow equilibrium theory, again ignoring kinetic barriers. By rearranging Equation 2 or 3, we can arrive at the predicted strain in the film if it has exceeded the critical thickness:

$$\varepsilon = \frac{D(1-\upsilon\cos^2\alpha)(b/b_{\mathrm{eff}})[\ln(h/b)+1]}{2Yh}. \qquad 4.$$

We can see that the elastic strain in the film, ε, drops quickly with an increase in film thickness h. There is quite a discrepancy between Equation 4 and the elastic strain that is measured experimentally in thin films. For systems with mismatches less than 1.5–2%, the elastic strain left in the layer is higher than the residual strain expected from Equation 4. This difference between equilibrium theory and experiment is believed to be the result of dislocation nucleation, propagation, and interaction.

We have shown that, energetically, dislocations should be introduced above the critical thickness, and then relaxation should occur as the film grows further. As can be seen from Equations 1–4, edge dislocations would be more effective at relaxing the mismatch, and therefore more favorable energetically. However, the (001) diamond or zinc-blend crystals have $\langle 110\rangle\{111\}$ slip systems, i.e. the first dislocations to form and glide to relieve stress will lie on $\{111\}$ plane and have Burgers vectors in the $\langle 110\rangle$ directions. Since edge dislocations, which relieve mismatch, have a glide plane on the $\{001\}$, they are not operating in the primary slip system of the crystal. Therefore, edge dislocations cannot glide from the surface to the interface. The barrier preventing edge dislocations from nucleating and reaching the interface is very large; to activate this process, the secondary slip system would have to become active, which would require such high stresses that the growth mode of the mismatched layer would be three dimensional. Because of this inherent crystallography, only the 60° dislocations can easily glide to form dislocations at the heterointerface.

Figure 5 shows how a 60° dislocation would nucleate and propagate to relieve strain. Because the phenomenon depicted does not require any additional factor to lower the activation energy of this process, this type of nucleation is termed homogeneous nucleation. Calculations have shown that homogeneous nucleation is difficult for systems with a mismatch of less than 6–8%, (11, 19), in agreement with theory, and experimental observations have not proven the existence of homogeneous nucleation. Yet low-mismatched films introduce dislocations.

An obvious solution to this paradox is heterogeneous nucleation, in which some defect in the material aids the nucleation of a dislocation. Using substrates pre-patterned before epitaxial growth to manipulate relaxation, we have shown that dislocations in low-mismatched films are introduced through a variety of heterogeneous sources with varying activation energies (20). The dominant sources observed to date are threading dislocations, particles introduced before or after growth, and edges of the wafer or sample. In the GeSi/Si case, there are few dislocations in the substrates, and therefore only particulates (21), impurities, (22, 23), and edges offer a means of introducting the dislocations into the crystal. Thus a high degree of metastability can result in this mismatched system (24, 25).

After the dislocations have been introduced into the system, the rate of

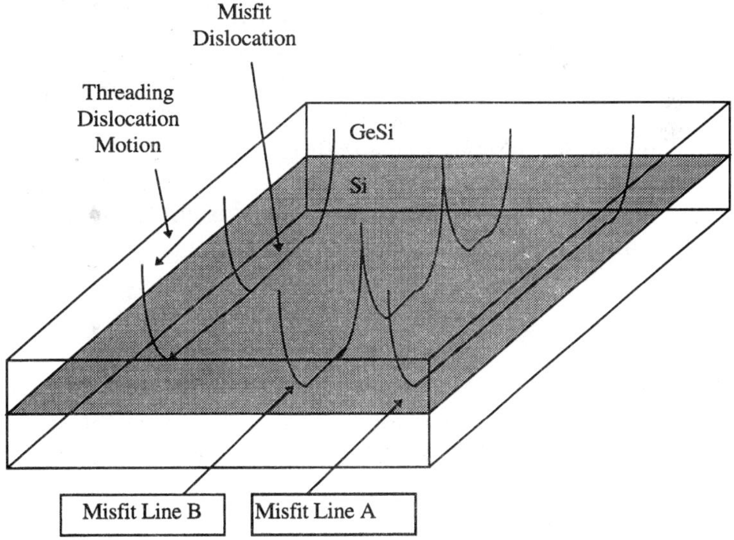

Figure 5 The geometry of misfit and threading dislocation segments in (001) mismatched diamond or zinc-blende structures. Line B occurs when dislocation nucleation is more common; Line A occurs when there is less nucleation than B.

relaxation is also dependent on the propagation of the threading segment shown in Figure 5. Early work (26) in the transmission electron microscope (TEM), and later work using optical microscopy (27), showed that the velocity of dislocations in GeSi/Si can be modeled with the empirical expression for dislocation velocities in bulk crystals:

$$v = B\tau_{\mathrm{res}}^{\mathrm{m}}\,\mathrm{e}^{-U/kT}, \qquad 5.$$

where B is a constant, τ_{res} is the resolved shear stress, m is an exponent approximately equal to 1.1, U is the activation energy for dislocation glide, k is the Boltzman constant, and T is temperature. In principle, with this knowledge we can predict the degree of relaxation above the critical thickness (28). Experiments that use wafer curvature to measure real-time GeSi relaxation show that when there are many artificially made nucleation sources, the relaxation is dominated by dislocation flow, but otherwise, dislocation nucleation controls the degree of relaxation (29). Thus, because heterogeneous nucleation is difficult to model and there are a variety of sources, a universal model to describe relaxation above the critical thickness is difficult to obtain.

Controlling Dislocations in GeSi Nanostructures

Both the relaxed, cubic GeSi and strained Si or GeSi have desirable electronic properties, thus devices may require strained-layers and/or relaxed GeSi layers. In the case of a heterojunction bipolar transistor or a modulation-doped field-effect transistor, the performance of the device increases if more mismatch can be accommodated. Therefore, limiting misfit dislocation introduction is required. For infrared detectors and devices requiring biaxial tensile layers, the GeSi alloy needs to be relaxed to the bulk state. Here, the alloy needs the introduction of dislocations to relieve the mismatch strain. However, the kinetics must be controlled so that few threading dislocations are present in the relaxed layer that would degrade quality and device performance.

LIMITING MISFIT DISLOCATION FORMATION When strained layers are desired, device designers can use the critical thickness curves shown in Figure 2 to keep the mismatch low enough or the thickness small enough so that dislocations will not form. To push device performance even farther, one can take advantage of the kinetic barriers to prevent misfit dislocation formation, thus creating highly strained, metastable layers.

As we discussed above, dislocation nucleation and propagation are key in forming a relaxed layer. These mechanisms can be inhibited by substrate patterning (20, 30, 31). The principle behind substrate patterning is that if the strain is low enough and the growth mode is two dimensional, then dislocation nucleation will be inhibited, and by enclosing the active sources, dislocation propagation will be limited. Consider Figure 6, which is composed of electron beam-induced current (EBIC) images of $Ge_{0.19}Si_{0.81}$ grown on unpatterned (Figure 6*a*) and patterned (Figure 6*b*) (001) Si substrates (32). EBIC images show misfit dislocations running in the two $\langle 110 \rangle$ directions as black lines. The pattern seen in Figure 6*b* was formed in the following way. Before the deposition of the GeSi by MBE at 550°C, the Si substrate was patterned and etched to form an array of square mesas, with edges along the $\langle 110 \rangle$ directions. The mesas were 50–70 μm in size laterally, and were etched 2 μm deep to ensure that the material deposited in the channels did not communicate mechanically with the material deposited on the top of the mesas.

As Figure 6 shows, misfit dislocations still nucleated at certain places on the mesa edges, but overall the misfit dislocation density is much lower on the patterned area. A summary of the degree of metastability introduced is shown in Figure 7. Here we plot the dislocation spacing expected vs Ge concentration. The equilibrium curve is on the graph, as well as data points from the unpatterned area on the samples. As expected, due to the paucity of nucleation sources, even the unpatterned area has a degree of metastability; however the patterning increases this effect dramatically. Thus an active nucleation source will be confined by the mesa; the patterning limits the degree because nucleated dislocations propagate to the edge of the mesa and stop due to the trench. The degree of relaxation is limited because the amount of strain relieved is proportional to the misfit dislocation length.

The patterning technique has been used to fabricate metastable GeSi diodes. Instead of using etched patterns to limit the dislocation glide, selective epitaxy was used inside oxide windows (33). The diodes fabricated from the pre-patterned area showed more ideal diode characteristics, thus implying that the greater number of misfits in the unpatterned led to deleterious characteristics.

We also can use substrate patterning as a diagnostic tool. To determine if homogeneous nucleation will occur even at high temperatures, we performed the same experiments with $Ge_{0.12}Si_{0.88}$ deposited at 900°C. At this elevated temperature, if homogeneous nucleation of dislocations is occurring, then all areas will be heavily dislocated, independently of the patterning. As Figure 8 indicates, patterning is still very effective, with dislocation nucleation dominated by heterogeneous nucleation even at 900°C for low mismatch. Figure 8 is composed of two optical microscope images of modified-Schimmel-etched surfaces (21). Note the small, shallow dishes that are s-pits, believed to be caused by transition metal contamination. Despite these heterogeneities, no dislocations have nucleated from

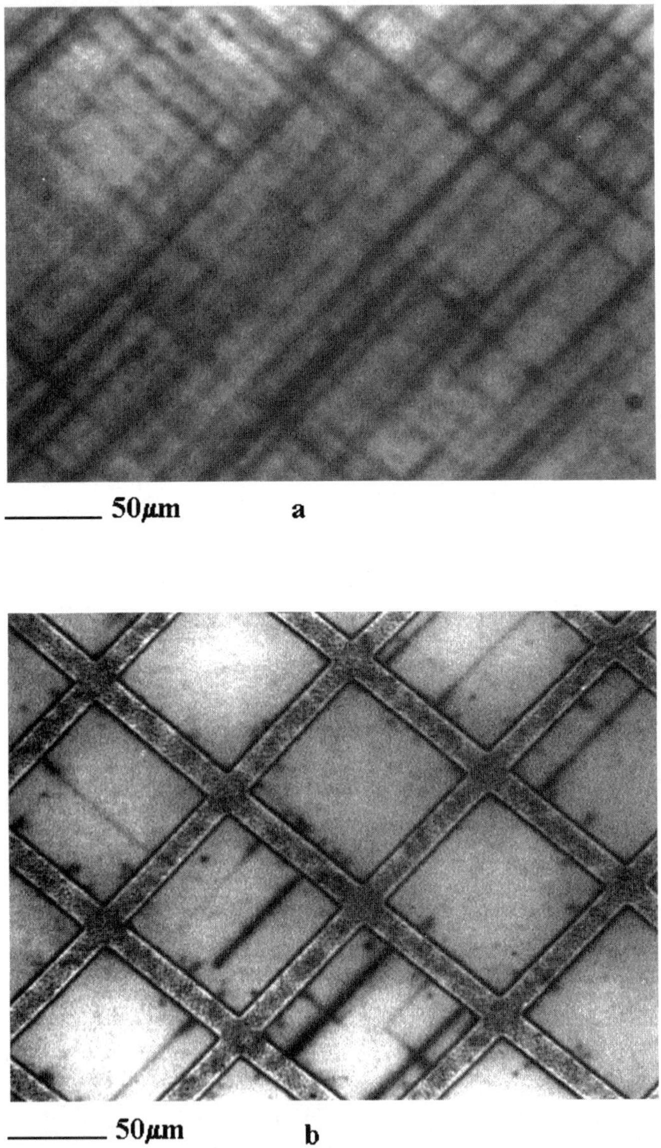

Figure 6 Plan-view EBIC micrographs of (*a*) unpatterned region, (*b*) patterned Si area before the deposition of GeSi.

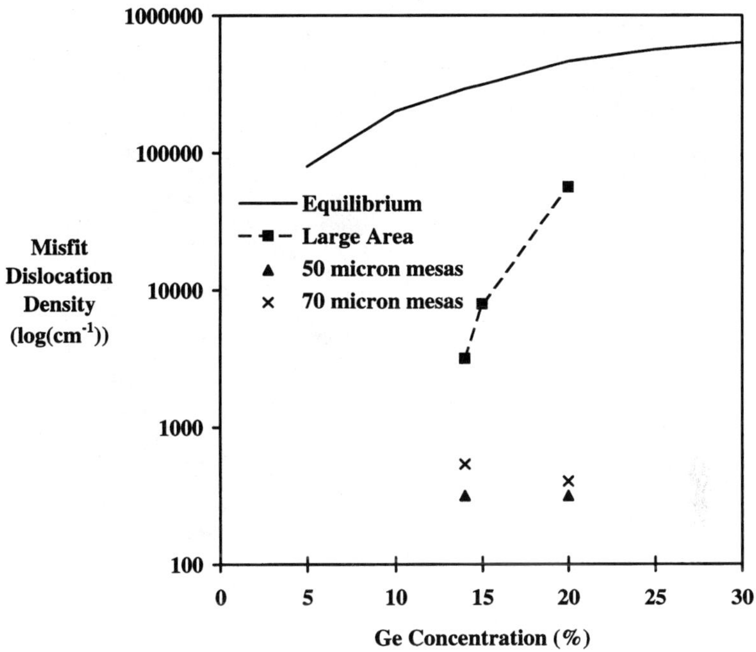

Figure 7 The misfit dislocation density at a mismatched interface for different GeSi concentrations. The curve from equilibrium theory is shown, as well as the curves for large areas and patterned areas.

them. Only a major disruption such as the one shown in the image can nucleate misfit dislocations. Thus the array of misfits that we can see in the unpatterned area is formed by a few dislocation nucleation sources that have produced very long misfit dislocations.

Other methods for reducing misfit dislocation density using patterning have been proposed, such as growing on patterns where the ratio of film thickness to pattern width is much greater (i.e. near 1) than the patterning described above (34–38). All these methods rely on the principle that as the ratio approaches 1, the expansion of the material at the edges allows significant relaxation to occur across the film (39). Thus dislocation introduction does not occur because another avenue is available to reduce some of the mismatch strain. This method is more closely related to cases where complete relaxation of the material is desired (see below). However, we indicate here that a severe disadvantage to this method is that the small structures created have a varying lattice constant across the film; the film is not homogeneously relaxed.

REDUCING THREADING DISLOCATION DENSITIES For some applications, we

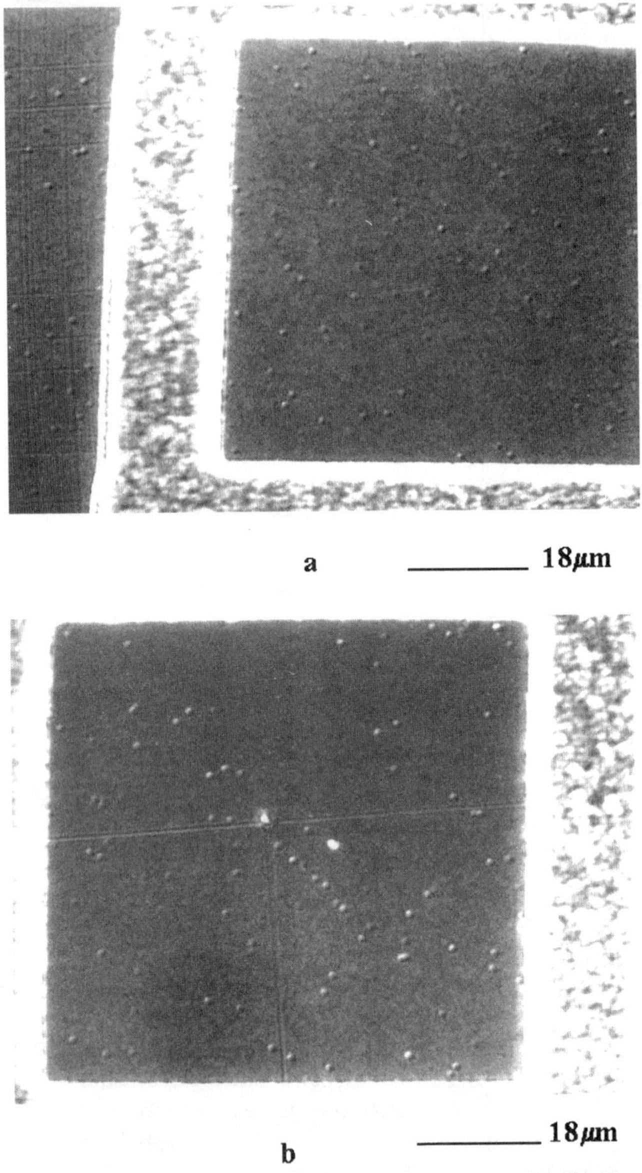

Figure 8 Etched surfaces of $Ge_{0.12}Si_{0.88}$ deposited at 900°C on patterned substrates showing that a large heterogeneity is needed to nucleate dislocations: (*a*) part of the unpatterned area showing misfit dislocations and a mesa without a nucleation source, (*b*) mesa with a rare nucleation source.

need to create a relaxed GeSi alloy on Si. However, until recently, devices made on such material possessed a high threading dislocation density. Typically, a uniform composition GeSi layer is grown directly on Si, and the resulting thickness is so great that substantial relaxation occurs. Figure 9 is a cross-sectional TEM micrograph of such a structure. A 3-μm $Ge_{0.70}Si_{0.30}$ layer was grown directly on Si at 650°C. Note the very high density of dislocations threading through the layer, with a lower density of threading dislocations near the surface. Cross-sectional TEM is actually not sensitive to threading dislocations in low numbers because the TEM sample is so thin. We can see a fair number of dislocations in the sample shown in Figure 9, thus the density must be on the order of 10^8–10^9 cm^{-2}.

Because the amount of strain relieved is proportional to misfit dislocation length and not threading dislocation density, this high density implies a high dislocation nucleation rate. If there were a low dislocation nucleation rate, then a few dislocations would be introduced as shown in segment A in Figure 5, and the few threading dislocations would propagate long distances, relieving the mismatch. For the structure shown in Figure 9, the $Ge_{0.70}Si_{0.30}$ was grown directly on the Si substrate ($\sim$3% mismatch) and a very high stress was applied immediately to the film. This high stress resulted in three-dimensional growth and a high dislocation nucleation rate. Thus many more threading dislocations were created, and each misfit dislocation length was very short, as shown schematically in line B in Figure 5.

For many years, the majority of work has concentrated on reducing the

Figure 9 TEM cross-section of a 3-μm-thick $Ge_{0.70}Si_{0.30}$ layer on Si, revealing many threading dislocations.

high number of threading dislocations after their creation. A common method has been to grow a strained-layer superlattice to try to force threading dislocations to the edge of the wafer. However, calculations show that the threading dislocation density cannot be reduced significantly using this method (11).

One possible way of reducing threading dislocation densities is to use the patterning method described above. Recall that substrate patterning limits the degree of relaxation by limiting the misfit dislocation lengths. A consequence of the limited misfit dislocation length is that the threading dislocation escapes at the side of the mesa. Thus one can imagine a relaxed GeSi layer, in which the threading dislocation density is too high for the desired application. This layer could be patterned, and a strained layer could be applied to push the remaining threading dislocations to the side of the pattern. For typical dislocation densities of 10^5–10^8 cm^{-2}, mesas as large as 10–1000 μm-wide should be large enough to force all the threading dislocations to the edge of the mesa.

When considering device processing, we prefer to have either no patterning or very large patterns if possible. Therefore the initial dislocation density in the relaxed heteroepitaxial layer must be low enough for the application. Based on the information gained from the substrate patterning experiments at 900°C (Figure 8), we know that for low-mismatch (small Ge concentrations), relaxation is still nucleation limited. Equation 5 tells us that the threading dislocation velocity (and therefore the decrease in elastic strain) is very high. Therefore we imagined that if a low-mismatch layer were grown at very high temperature, it might become completely relaxed without having a high threading dislocation density. This relaxed layer could be followed by another layer with the same increase in lattice constant; assuming that no other parameters change, the same threading density already present from the first layer could be used to relax the next layer, and so on. Ultimately, this structure is a linearly graded composition layer because this linear structure is the limit for infinitely small increases in Ge concentrations.

Application of this high-temperature graded layer to produce relaxed GeSi structures with low-threading dislocation densities has been demonstrated with MBE (11, 40). Figure 10 shows cross-sectional TEM micrographs of graded structures grown to a final Ge concentration of 30% (Figure 10*a*) and 70% (Figure 10*b*) using UHVCVD. The samples were grown at 700°C with a pressure of 50 mtorr to accelerate the growth rate. As can be seen, the structures have a slow grading ($< \sim 10\%$ Ge/μm). The top layers in both pictures are uniform 30 and 70% layers. Note that there are no dislocations in the 30% layer, and very few in the 70% layer. Also observe that despite the much greater relaxation present in the 70%

(a) 1μm

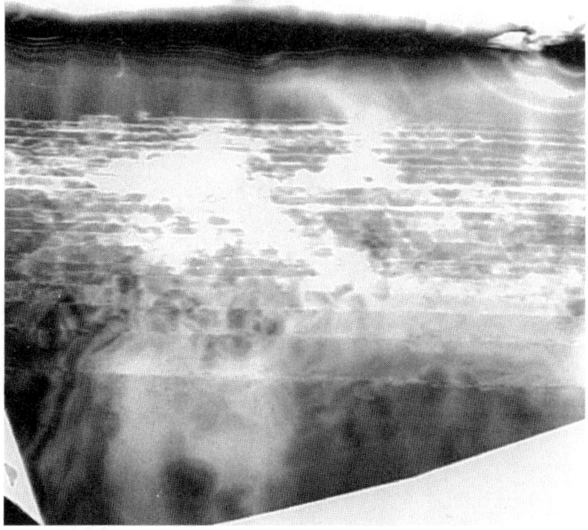

(b) 1μm

Figure 10 TEM cross-sections of compositionally graded structures: (*a*) a 30% Ge final composition, (*b*) a 70% final composition. Note that the images are inverted, i.e. the dislocations are light in color. Also compare the threading dislocation density in the cap layer in (*b*) with Figure 9.

sample, the threading dislocation density has not increased tremendously. This behavior supports the idea that dislocation nucleation is not rampant and that a nearly constant (or slowly increasing) threading dislocation density is present during the formation of the graded region.

The threading dislocation densities are so low that TEM cross-sections are not the best method for determining residual dislocation densities. Figure 11 shows a plan-view EBIC image of a layer graded to 30% Ge, with a uniform 1.5 μm cap layer of 30% Ge. The background doping of the graded region was $\sim 10^{15}$ cm^{-3} using As, and a thin (1000 Å) B-doped layer was grown at the surface of the structure. This type of $p^+/n—/n^+$ structure is ideal for observing threading dislocations in these structures. The black dots seen in Figure 11 are the individual threading dislocations. Because EBIC is a much lower-magnification technique, we can observe larger areas than those seen with TEM micrographs. We arrive at a dislocation density of 7×10^5 cm^{-2} for this sample. Note that this density is two to three orders of magnitude smaller than that expected for uniform layers grown directly on Si substrates.

By growing at high temperatures, we increase dislocation velocities such that despite the low nucleation rate, the layers can be completely relaxed.

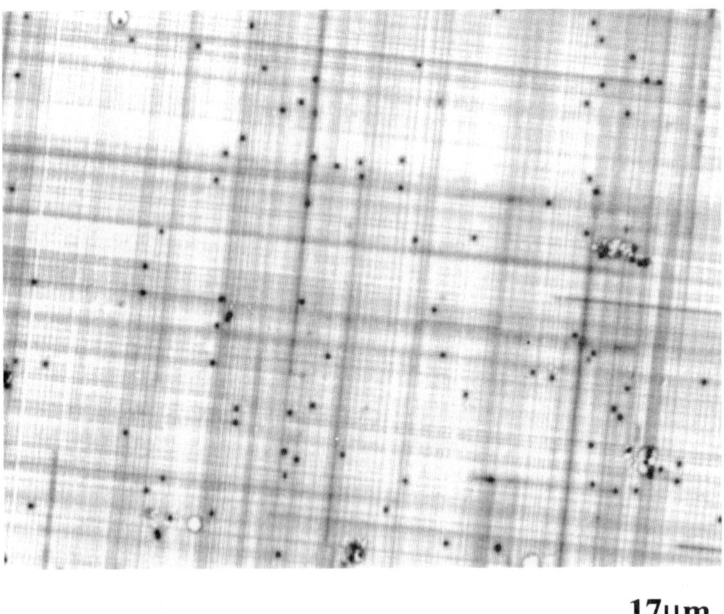

Figure 11 EBIC image of the top uniform layer in a graded, relaxed 30% Ge structure. The black dots are threading dislocations, and the density is 7×10^5 cm^{-2}.

One could also imagine that if some residual strain were present after an individual step in a step-graded structure, one could compensate for that residual stress by applying the next layer with an appropriate lattice constant so that the structure would be in equilibrium (41). A closely related idea was demonstrated in the InGaAs/GaAs mismatched system (42).

UHVCVD with a higher pressure of 50 mtorr was used to grow the samples shown in Figures 10 and 11 because of its high growth rates and ability to produce multiple wafers. Another alternative would be to grade faster, i.e. increase the Ge concentration/thickness at a much higher rate. Higher grading rates for layers grown at lower temperatures have been grown, and very low threading dislocation densities have been claimed (43), but accumulating results suggest that the electronic properties and the overall defect density degrade with increasing grading rate (44–47).

NATURAL NANOSTRUCTURES

Depending on the growth conditions and magnitude of the mismatch between the GeSi alloy and the Si substrate, different nanostructures can naturally occur during GeSi/Si epitaxy. Figure 12 schematically summarizes recent results. Because Si is a desirable substrate, and because GeSi alloys have a larger lattice, most of the structures observed have biaxially compressive stress (shown as $\varepsilon < 0$ in Figure 12). For homoepitaxial growth, the surface remains largely flat under most growth conditions. When GeSi is grown on Si with an intermediate mismatch ($\sim 1\%$), undulations on the order of 1000–2000 Å can form (48). Higher levels of mismatch produce complete islanding without the formation of dislocations (49). Upon further growth, dislocations would eventually be introduced into these structures. But it is noteworthy that small (100–1000 Å) structures can be created without the need for lithography. It was generally believed that the undulations and island growth were solely due to the fact that some strain energy can be released when a three-dimensional structure is formed. However, recent experimental results for the opposite-sign mismatch suggest that another factor may be involved (50).

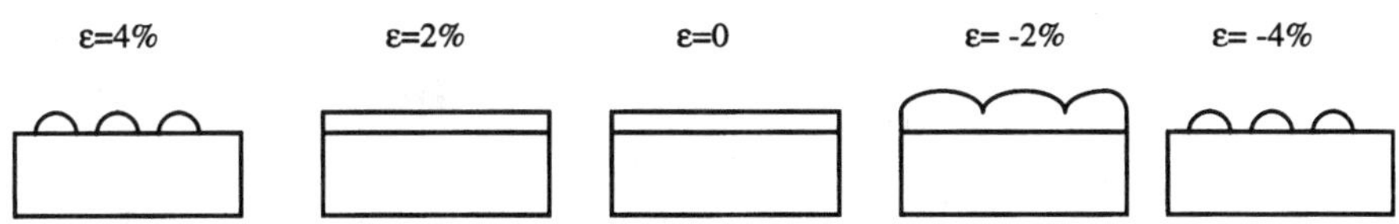

Figure 12 Schematic showing the relationship between the magnitude of strain, the sign of the strain, and the morphology. At intermediate strain, the biaxially compressive layer is more rough than the tensile layer.

Consider Figure 13, which shows two films grown under similar conditions. Figure 13*a* is a TEM micrograph of a biaxially compressed GeSi layer, whereas Figure 13*b* is a TEM micrograph of a biaxial tensile structure. Note that the compressed film forms coherent islands, whereas the tensile film is relatively flat. The numerous lines seen in Figure 13*b*

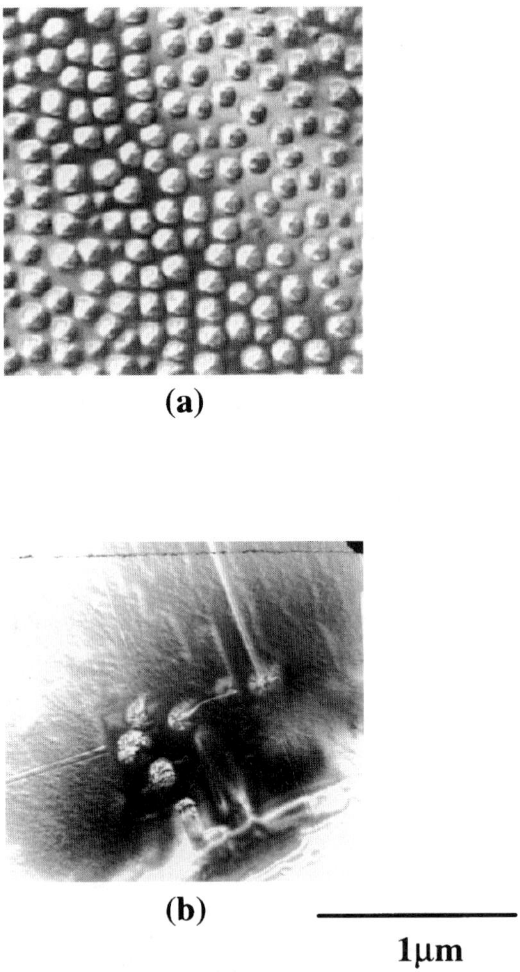

Figure 13 Plan-view TEM images showing (*a*) $Ge_{0.5}Si_{0.5}$ on Si, (*b*) $Ge_{0.5}Si_{0.5}$ on Ge. Both layers were 50 Å-thick and grown at 650°C. The compressive strain produces three-dimensional coherent islands, whereas the tensile strain produces a relatively flat surface except for step bunches. The lines in (*b*) are nucleating misfit dislocations, and the circular features are dislocated three-dimensional features.

are step bunches, as verified by cross-sectional TEM and atomic force microscopy. If the mismatch in the tensile film were much greater ($\sim 3\%$), the film would also form islands (Figure 12). The fact that similar magnitudes of mismatch, but with opposite signs, form islands in one case, but a planar structure in the other, suggests that surface step energies play a role. Calculations show that on a compressed surface, step energies are indeed reduced, allowing the creation of three-dimensional islands at a much lower energy cost than in the tensile case (50).

We emphasize that the tensile and islands discussed are in dislocation-free material. The undulations observed in these cases are at wavelengths that are much shorter than the wavelengths observed in structures where a full array of dislocations is formed. In this case, the wavelengths are on the order of 1 μm and occur because of the influence of the dislocation strain fields on the local growth rate at the surface (20, 51).

ARTIFICIAL NANOSTRUCTURES

We now present brief reviews of the structures created to explore the electronic and optical properties of GeSi/Si structures and begin with the most extensively explored quantum structure, the quantum well. Then laterally patterned wells that form quantum wires are discussed, followed by a brief overview of the GeSi/Si heterojunction bipolar transistor (HBT) and the modulation-doped two-dimensional carrier gas structures.

Quantum Wells

These simple structures were explored first because only a very thin layer was needed, and hole confinement was easily achieved due to the valence band offset. In order to observe quantum effects in GeSi wells, by far the most utilized characterization method is photoluminescence (PL). A laser beam creates electron-hole pairs in the material, which can diffuse and then experience quantum confinement effects in the quantum well. In order to separate the intrinsic effects from the quantum effects, it is necessary to first know the luminescence spectrum from relaxed GeSi alloys. Work by Weber & Alonso revealed the intrinsic spectra of bulk GeSi alloys (52).

When GeSi is deposited on Si, the mismatch affects the quality of the photoluminescence signal. For low Ge concentrations (4%), band-edge luminescence from both the Si substrate and the GeSi alloy are seen at 4.2 K (53). However, as the Ge concentration increases, free excitons cease to exist at 12 K, and deep level luminescence appears (2, 54). The deep luminescence changes character with increasing Ge concentrations: At the lower Ge concentrations, the deep luminescence consists of sharp D-lines, i.e. the luminescence lines associated with dislocation formation in Si; at the higher concentrations, a larger, undefined luminescence occurs. There appears to be a connection of the sharp D-lines with a more ordered misfit array, with dislocations that sometimes propagate into the substrate (55), whereas the broader luminescence is related to the typical high-mismatched structure with many threaded dislocations. Deep luminescence can be observed in strained GeSi structures on Si without the presence of dislocations as well (56, 57). Although still somewhat controversial, it is believed that this luminescence is associated with some sort of point defect or microscopic Ge cluster (58) and has been artificially produced by ion implantation and annealing (59). As expected from the previous section, compositionally graded relaxed GeSi layers grown at high temperatures have few threading dislocations and point defects, and therefore, luminesce as bulk GeSi alloys (40). Figure 14 shows a variety of luminescence spectra due to different material properties that summarize the discussion above.

After this initial work on GeSi alloys, a natural trend was to explore thin strained GeSi layers on Si, capped with another layer of Si. The GeSi layer thickness can then be varied in anticipation of quantum effects, i.e. confining the carrier wavefunctions in very thin layers. Luminescence from such wells was first explored using material grown by RTCVD (60). Soon, other growth techniques led to similar quantum well luminescence from structures grown using atmospheric CVD, gas-source MBE, and solid-source MBE (61–63). Figure 15 shows the typical shifts seen in the PL spectra as a function of layer thickness. Although not always a necessary

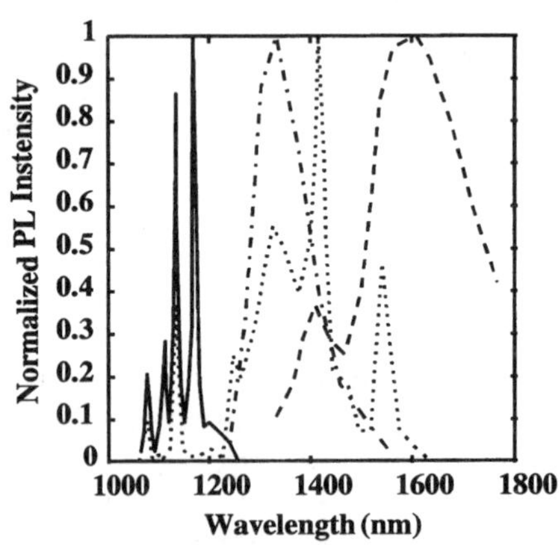

Figure 14 PL spectra from GeSi at ~4 K. Solid line: 4% GeSi, showing Si substrate peak and GeSi peak (54); dotted line: 15% Ge, showing sharp D-lines from 1200–1600 nm (54); dashed line: 38% Ge, revealing that the deep luminescence becomes broad and ill-defined (54); dot-dash line: spectra from implantation and annealing of 20% Ge, simulating luminescent band sometimes appearing in lower temperature MBE material (59).

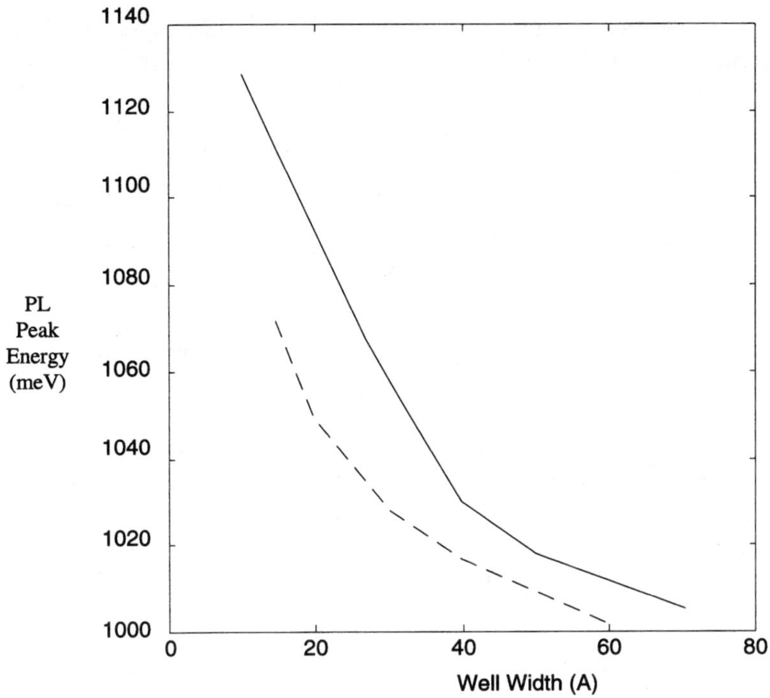

Figure 15 Shift in PL spectra from the confinement in GeSi wells. The solid line represents data from wells grown using conventional MBE, the dashed line shows data from GeSi wells grown by MBE using an Sb layer (70).

condition (63, 64), a band-edge luminescence in solid-source MBE materials without a high degree of deep luminescence usually requires very thin wells or higher growth temperatures (65). A seminal work by Terashima et al compares GeSi quantum wells grown by solid-source MBE, gas-source MBE, and UHVCVD (66). Indeed, they observe that solid-source MBE material exhibits deep luminescence when grown at 400 or 500°C, but after annealing at 800°C, the deep luminescence bands disappear. However, we note that the growth rates were not reported. At 550°C, the growth rate in gas-source MBE and UHVCVD is limited ($\sim < 0.1$ Å/sec) due to the cracking of the source gas. Solid-source MBE is not limited by the cracking of reactant gases, so rates of ~ 3 Å/sec are typically used at most temperatures. Such high growth rates at lower temperatures may result in a higher point-defect background in the material.

High-quality GeSi quantum wells have even demonstrated weak luminescence and electroluminescence at room temperature (67, 68). Although the luminescence efficiency is orders of magnitude lower than III-V materials, room for improvement may exist. For example, cathodoluminescence imaging of GeSi quantum wells has been accomplished, and nonradiative centers can be observed even in the high-efficiency material (69). Also, segregation-assisted growth has been used to improve interface abruptness and improve the quality of the quantum wells (70). The comparison between segregation-assisted quantum wells and conventional quantum wells is shown in Figure 15.

The advent of high-quality relaxed GeSi buffers allows the possibility of growing tensile Si quantum wells. Because of the band alignment, which we have described above, the electrons are confined in the Si layer, and the holes are confined in the GeSi. Thus a type-II band offset occurs and indirect transitions from the conduction band in the Si to the valence band in the GeSi occur. This phenomenon has been recently observed by Nayak et al (71). The energy of the luminescence shifted with Si layer thickness, a strong indication that the type-II recombination was occurring.

Combinations of quantum wells or very thin GeSi have been used to grow strained-layer superlattice structures on Si as well. Early predictions suggested that Ge/Si superlattices may have direct band gaps, and much of the research has concentrated on finding this direct emission. The Ge/Si superlattice research has progressed tremendously (72), although the direct band gap has yet to be observed.

Quantum Wires

The extensive study of GeSi quantum wells in the last few years has led to interest in lower-dimensional structures. The next step in quantum confinement requires growth on patterned substrates or etching of a GeSi quantum well. Electron beam lithography and reactive ion etching in $SiCl_4$ have been used to pattern buried GeSi layers containing 20% Ge into quantum wires. (73) The GeSi layers were quite thick (500 Å), and the vertical confinement was achieved by modulation doping, i.e. the etching was used to further confine a two-dimensional hole gas. A feature in the PL spectra was found to be sensitive to wire width, and therefore was attributed to levels within the quantum wire. However, the spectrum is feature rich. In an effort to avoid dry etching, which may introduce recombination centers, GeSi wells have been grown on v-grooved substrates and through cantilever openings (74, 75). The PL from the quantum wire area in these cases was similar to the unpatterned area except the features were shifted to higher energies, possibly indicating quantum wire confinement. Considering that quantum wires and dots in the III-V material system have been researched for much longer, and that a limitation has been the artifacts resulting from etching, we consider the quantum wire and dot research in GeSi/Si to be in its infancy.

ELECTRONIC DEVICES

In this section we briefly discuss the principles behind the two main devices demonstrated in the GeSi/Si system: the HBT and the modulation-doped field-effect transistor (MODFET).

HBT

The HBT is one of the most simple GeSi devices, and has received much of the recent research effort. Because of this interest, the topic has been extensively reviewed (76, 77). A schematic of the GeSi/Si HBT structure is shown in Figure 16. For this n-p-n structure, the band-gap difference between the GeSi and Si occurs in the valence band, similar to that in the GeSi quantum well structures. In the HBT case, the GeSi layer is thick enough that quantum confinement does not occur.

In the operation of a bipolar transistor, by applying a small current to the base, a large amount of current can flow from the emitter to the collector if the gain is high. In the n-p-n structure, a small current to the base will lower the barrier to electrons in the emitter, allowing the electrons to cross the base and enter the collector. Recalling that a hole flow in the opposite direction to the electron flow will degrade the gain, we must minimize the injection of majority carriers from the base (holes) into the emitter. The HBT has an additional factor in the reduction of hole injection

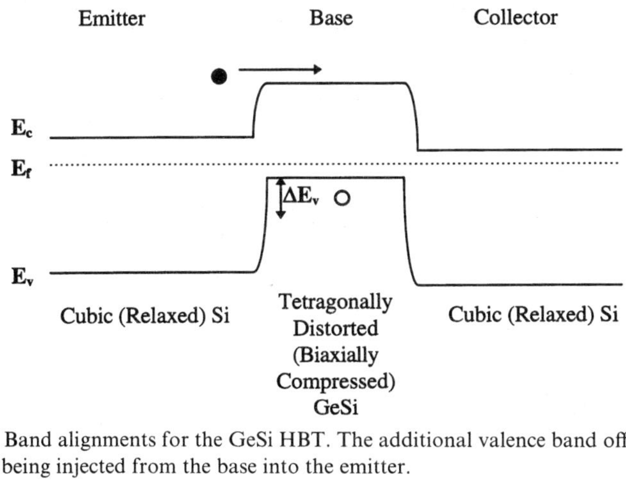

Figure 16 Band alignments for the GeSi HBT. The additional valence band offset prevents holes from being injected from the base into the emitter.

into the emitter: the valence band discontinuity. Contemplating the band structure, we see that the barrier to this hole injection will be exponentially sensitive to the valence band offset ΔE_v.

Mathematically, we can see the advantage of the HBT over the conventional bipolar (BJT) by considering the increase in gain due to the offset, or, equivalently, we look at the ratio of the collector currents for the same base current:

$$\frac{I_{c(\mathrm{HBT})}}{I_{c(\mathrm{BJT})}} = \frac{N_{\mathrm{BJT}} W_{\mathrm{BJT}} D_{\mathrm{HBT}}}{N_{\mathrm{HBT}} W_{\mathrm{HBT}} D_{\mathrm{BJT}}} e^{\Delta E_v/kT}, \qquad 6.$$

where N is the base doping level, W is the base width, and D is the electron diffusivity in the base. If we compare a BJT and HBT with the same base width and electron mobility in the base, we see that low base doping and a large valence band offset in the HBT favors much higher gain in the HBT.

It can now be seen why control of misfit relaxation is desired in this structure. The largest ΔE_v is achieved by incorporating as much Ge in the base as possible; but this requirement is balanced by the critical thickness criterion (Equation 2 and Equation 3). If too much Ge is added to the base, dislocations will exist in the structure, or processing the structure will introduce them, thereby destroying the current gain due to recombination and generation of carriers at the dislocations.

Low doping in the base of the HBT is not desirable if the goal is to have the same drive current but have the transistor operate at a higher frequency. For high frequency operation, the base resistance must be lowered to decrease the RC time constant. Equation 6 shows that an advantage of the HBT is that the base can be doped much higher than a BJT without losing gain due to the exponential factor containing ΔE_v.

Early work using MBE GeSi as the base in a HBT produced transit frequencies as high as 28–40 GHz (78–80). King et al used CVD to produce GeSi HBTs (81), and UHVCVD produced a record transit frequency of 75 GHz for a GeSi HBT (82). Recently, a low-temperature fabrication process was used to achieve a transit frequency of 100 GHz (83). Because of these results, the HBT is considered to be a potentially useful device when integrated with Si for certain applications. The next step is to fabricate HBTs on a large scale in a manufacturing environment.

GeSi MODFET

The GeSi MODFET was actually one of the first devices considered for GeSi/Si heteroepitaxy. The most simple MODFET is the p-channel MODFET, with a structure very similar to the quantum well and HBT structures. A drawing of the p-channel MODFET is shown in Figure 17. At low temperatures, the modulation doping supplies holes to the GeSi layer, and the band structure forces the holes against the GeSi/Si interface, forming a two-dimensional hole gas. This hole gas can have a high mobility because there are no dopants in the conduction channel; the ionized dopant atoms are remotely located across the undoped Si set-back layer. At room temperature, there will be some carriers in the GeSi well, and the modulation doping can be thought of as a threshold adjustment layer because this doping level determines, in part, the voltage needed to turn on the channel. When the channel is turned on, holes can flow from the source to the drain. The speed of this FET is determined by the speed at which holes can cross from the source to the drain and by the parasitic resistances associated with the structure. To obtain a high-speed device, high mobility in the channel is needed, as well as low-resistance source-drain contacts and short channel lengths. When considering real-world applications, we also want the FET to charge other electronic devices quickly; thus a high density of carriers in the channel is also desirable.

Because the performance of the MODFET is closely linked to channel mobility, researchers have attempted to determine the benefit of using a strained GeSi layer as a hole channel. When including deleterious mech-

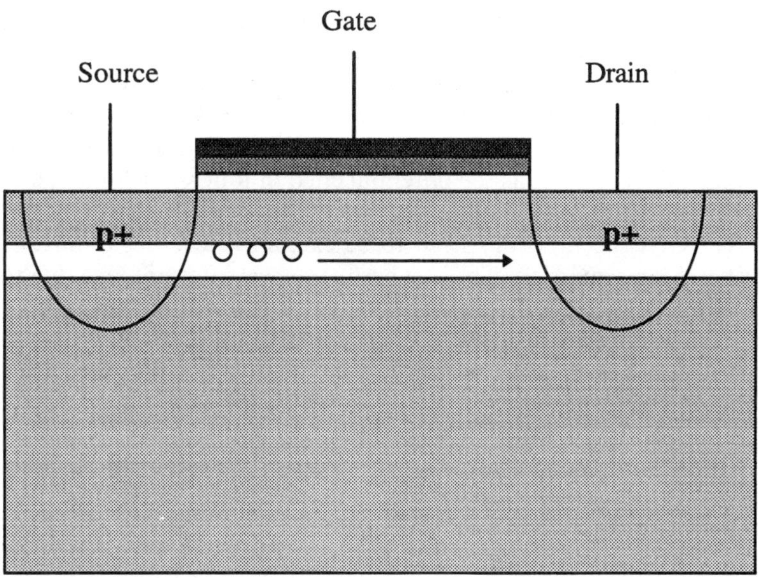

Figure 17 Schematic of the p-channel MODFET. The darker region is Si, the lighter area is the GeSi channel, which holds the confined holes.

anisms such as alloy scattering (84), acoustic-phonon scattering, and Si-Si and Ge-Ge optical phonon scattering, early calculations have shown that the strain and alloying produce a monotonic increase in hole mobility, and greater carrier velocities were found for in-plane transport as well (85). Calculations made explicitly to determine the in-plane effective mass showed that this could be a factor three lighter than the effective mass of holes in Si (86). Bhaumik et al have also shown that the saturation velocity of holes in strained GeSi should be higher, which would translate into higher performance for short channel FETs (87).

Although most calculations support the conclusion that the hole mobility should be higher in GeSi, the experimental evidence shows otherwise. Typical hole mobility in strained $Ge_{0.20}Si_{0.80}$ layers at 5–10 K are 2500–3800 cm^2/V-s (88–90), much lower than expected from the theoretical calculations. Electrical measurements have shown that the hole mobility decreases with increasing Ge concentration (91) and that the drift mobility may be different from the Hall mobility, implying that the ratio of the two mobilities is actually changing with increasing Ge concentration (92). Conversely, cyclotron resonance studies indicate a much lighter hole mass of 0.29 in $Ge_{0.37}Si_{0.63}$, agreeing with at least the effective mass calculations (93).

One possibility to reconcile the effective mass measurement, the mobility measurements, and the theory is that an extrinsic source is limiting the mobility. The fact that the hole mobility increases with higher growth temperatures suggests that transition metal impurities or some other point defect may be limiting the mobility (89). Alternatively, alloy scattering has been proposed as the dominant and limiting scattering mechanism (94, 95). However, hole drift mobility calculations indicate that the alloy scattering potential is not as large as the potential cited in References 93 and 94 (96). Mobility measurements from a number of strained alloy layers with very different Ge concentrations need to be examined in order to answer the alloy scattering question. These experiments were not possible due to the critical thickness limit. With the advance of relaxed, graded layers (40), virtual substrates with different alloy compositions can be created, and strained alloys with 20% more Ge than the virtual substrate Ge concentration can be grown on these relaxed buffers on Si.

To test the potential limit of the hole mobility, relaxed buffers were created on Si with concentrations of 60 and 70% Ge. Pure Ge channels were deposited and hole mobility was measured. Figure 18 shows a cross-sectional TEM image of such a strained Ge layer on a 60% Ge relaxed buffer. Note that the top surface is undulating on a short scale (100s Å). This wavelength of undulation is short enough to cause effective hole scattering and is believed to be responsible for the limiting mobility in these

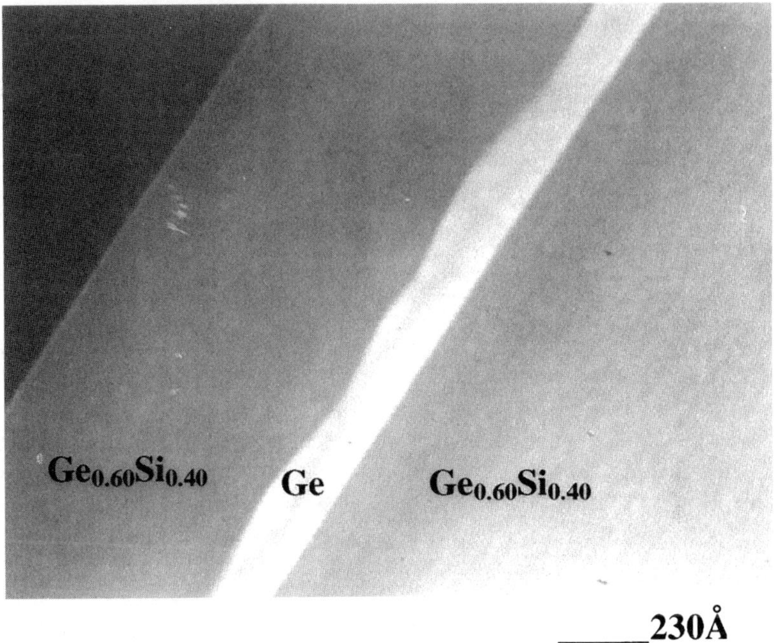

Figure 18 Cross-sectional TEM image of a strained Ge channel grown on a relaxed, graded $Ge_{0.60}Si_{0.40}$ virtual substrate on (001) Si. The compressive strain results in an undulation on the top interface that can degrade hole mobility in the channel.

structures. The observation of this undulation vs the lack of undulation in the tensile-strained Si layer used for the electron gas formed the hypothesis that step energies play a key role in determining interface abruptness (50). Utilizing this information, we placed the hole gas at the bottom interface and increased the thickness of the Ge layer to place the rough interface far from the two-dimensional hole gas. Such a design resulted in the highest hole mobility measured to date in the GeSi/Si system, 55,000 cm^2/V-s at 4 K (97, 98). In agreement with the calculations of the valence band deformations under strain, very light in-plane hole masses were measured (98). The Ge channel results indicate that the hole mobilities in GeSi alloys are likely to be limited by alloy scattering or interface abruptness.

One of the first demonstration devices in GeSi was p-channel GeSi FETs (99). Later work has achieved peak mobilities of 220–240 cm^2/V-s (100, 101) and transconductances of 37 mS/mm at room temperature (102). To try to improve the mobility and performance to their maximum value, pure Ge layers grown on a uniform composition buffer layer (to relax the strain) have been used as the p-channel (103). Room temperature peak mobilities were 400–900 cm^2/V-s and high leakage currents occurred. Both phenomena were probably due to the high defect density in the devices from the use of a uniform GeSi layer to relieve the mismatch strain. With the use of the graded relaxed GeSi layers described above, Ge channels grown on relaxed 70% Ge graded buffers were used for p-channel FET fabrication. Extrinsic transconductance at room temperature of 125 mS/mm was achieved, as well as peak hole mobility of 1300 cm^2/V-s (104). This drastic increase in performance can be attributed to the lower defect density in the device with the implementation of the graded layer buffer.

By raising the mobility and therefore the transconductance of the p-FET, circuit designers hope that the p-channel FET can be fabricated with the same dimensions as the n-channel in a CMOS circuit. Currently, the p-FETs must be larger to correct for the difference in channel mobility, resulting in asymmetric design and larger cells.

One difficulty in any buried-channel device like the p-channel GeSi device is the potential of a second conducting channel at the oxide-semiconductor interface. Essentially, as voltage is applied to the gate, the Fermi level initially crosses the valence band only in the GeSi well. But eventually the Fermi level will cross the valence band at the oxide interface as the voltage is increased (105). This behavior always limits the voltage-range in which these FETs can be used and places severe constraints on device design.

Up to this point, the focus has been entirely on the p-channel layers and devices. Historically, this preference occurred because of the band alignment (Figure 4) and difficulty in obtaining the large amount of lattice relaxation needed to produce a useful conduction band offset. Although type-II alignment had been verified using uniform composition, relaxed GeSi (17), the high threading dislocation density limited the electron mobility.

At the Spring Materials Research Society Meeting in 1991, interest in electron channel devices was renewed with simultaneous announcements of the drastic lowering of threading defects in relaxed GeSi layers and drastic increases in two-dimensional electron gas mobility in strained Si on relaxed GeSi (106–108). Dislocation densities were reduced from greater than 10^8 cm^{-2} to 1×10^6 cm^{-2}, and mobility improved from a ceiling of 17,000–19,000 cm^2/V-s (109, 110) to 96,000 cm^2/V-s at 4.2 K. Soon mobilities were raised to over 120,000 cm^2/V-s (111), and using the graded relaxed buffer, Schaffler et al raised the mobility to over 170,000 cm^2/V-s (44). The AT&T group also achieved 180,000 cm^2/V-s after further optimization (112). Figure 19 is a drawing of the two-dimensional electron gas structure. Note that the structure is more complicated than the GeSi alloy hole-gas

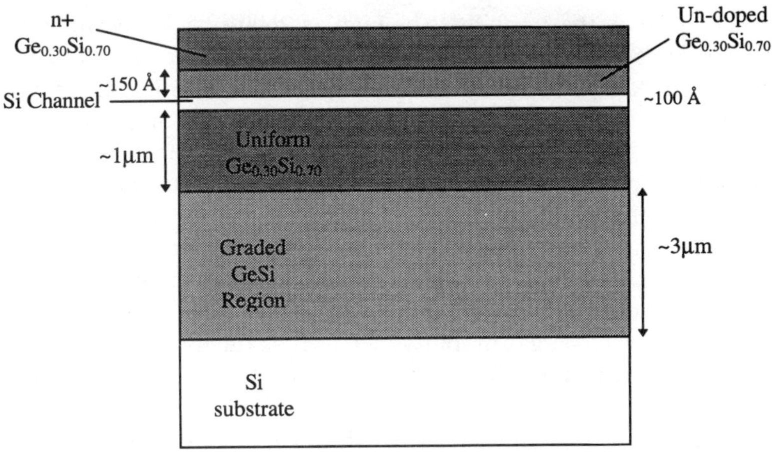

Figure 19 Drawing showing the structure required for testing electron transport in a tensile-strained Si layer.

FET because the GeSi must be relaxed, and then the Si channel can be grown with the tensile strain to achieve electron confinement. Figure 20 shows the electron mobility and sheet carrier density in the two-dimensional electron gas vs temperature.

Nelson et al showed that the room temperature mobility in these tensile Si layers can be quite high as well (113). By correcting for parallel channel

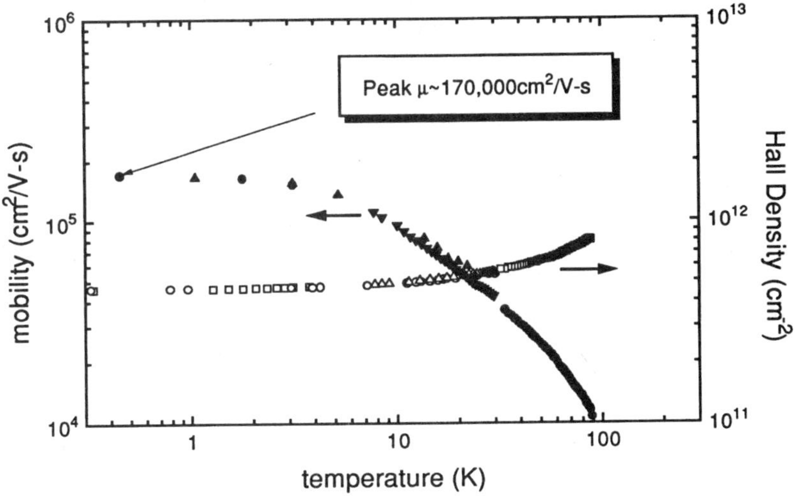

Figure 20 Electron mobility and carrier density vs temperature for an electron gas sample, designed like the structure in Figure 19.

effects, they show that mobilities as high as 2600 cm^2/V-s occur in these structures at room temperature. The same authors have also done a systematic series of experiments to determine the effect of difference materials aspects of the graded region on the mobility in the Si channel (114). Because their UHVCVD is performed at lower temperatures, dislocation nucleation is very difficult during the graded layer growth. Thus the layers are not fully relaxed. On the other hand, faster grading rates can be used because dislocation nucleation is suppressed by the low temperature. To solve the partial relaxation problem, the uniform layer, which is grown after the graded region, can be made very thick to continue to relax the structure. The electron gas mobility increases as the total buffer thickness increases, as expected, but at great thickness, the mobilities decrease; this decrease has been attributed to a roughening of the surface, which appears to be unique to CVD. In the case of slowly graded layers at high temperatures using MBE, the electron mobility and material quality increase with a decreasing grading rate (%Ge/μm) or, equivalent, a thicker total buffer thickness (44–47). However, the cross-hatch pattern on the surface of MBE relaxed GeSi graded layers does not roughen as the top uniform layer thickness is increased.

After such high quality two-dimensional electron gases were produced, the quantum Hall effect and the fractional quantum Hall effect were soon observed (115, 116). Other transport measurements were used to try to determine which factor is limiting the mobility: remote dopants, impurities, interface roughness, alloy scattering, strain fluctuations, cross-hatch morphology, or threading dislocations (117, 118). Although certain mechanisms can be ruled out, remote dopants, impurities, and interface roughness could be the factors limiting mobility. The magnetotransport measurements suggest that the remote dopants are the dominant, limiting scattering mechanism.

The high mobility electron layer can be used as a channel in a FET structure. Unlike the p-channel device layers, where Si is present at the top of the structure, the n-channel device layers terminate at the surface with the relaxed $Ge_{0.30}Si_{0.70}$ alloy. In the p-channel case, Si processes can be used to form a MOSFET in a straightforward fashion. The n-channel requires either a sacrificial Si layer, which can be oxidized at the surface, or a FET must be fabricated with a metal gate. Due to this processing constraint, the first devices to be fabricated with n-channels were metal-gate FETs (MESFETs). The Daimler-Benz group used Pt-Au for the Schottky gate and Au-Sb ohmic contacts for the source and drain (119). Impressive extrinsic transconductances of 340 mS/mm were achieved in this MESFET device. Using a Pt gate, the IBM group produced n-channel MESFET with a 0.25 μm gate (120). The mobility in the channel was 1500

cm^2/V-s and the room temperature transconductance was 330 mS/mm, although gate leakage prevented the device from being turned off at room temperature.

The MESFET results are impressive, but the GeSi FET technology will likely be used in a CMOS environment or integrated with CMOS VLSI. Therefore, another important step in determining the utility of GeSi FETs would be to make MOSFETs from the n-channel material using processes similar to those processes used in CMOS fabrication. Recent results show that both surfaces and buried n-channel MOSFETs can be fabricated (121, 122). The buried channel device had the highest effective mobility (1400–1600 cm^2/V-s) at low electric fields from the source to the drain, but at higher fields, the surface-strained Si device had higher effective mobility (800 cm^2/V-s). The highest transconductance was achieved with the surface-strained Si, reaching 50 mS/mm at 3V.

The high mobilities achieved in Ge p-channel and Si n-channel devices are important when the FETs have relatively low electric fields from the source to the drain. In long-channel FETs ($>=0.5$ μm), the FET performance is expected to dramatically improve because the electrons are being accelerated when traveling from source to drain. However, the continued scaling of Si transistors will make short-channel devices ($<=0.25$ μm) the dominant FETs in the industry. Will the GeSi FET offer advantages at these short-gate lengths? The answer depends on the effective saturation velocity of the electrons, including any velocity overshoot effects at high fields (short channels). Calculations show that electrons in tensile-strained Si will experience a large velocity overshoot (123, 124), and the saturation velocity will be higher than electrons in unstrained Si for all fields except the very highest (5×10^4 V/cm) (125). If these calculations are correct, then the substitution of GeSi FETs for Si FETs will improve device and circuit performance, or allow a larger GeSi FET to operate as fast as a smaller Si FET. The latter may be important as the lithography for fabricating smaller Si FETs becomes prohibitively expensive.

CONCLUSIONS

GeSi/Si materials advancements in the last decade have allowed the fabrication of many GeSi nanostructures. Naturally occurring structures have been discovered, as well as quantum effects. By far, HBTs and FETs have been fabricated more frequently than any other nanostructures because of their more immediate potential applications. Further progress in the processing of GeSi structures should lead to more quantum nanostructures and improve the performance of the HBTs and FETs. Improvement in the control of defect densities will allow more freedom in the design of GeSi heterostructures and allow the integration of GeSi structures of Si electronics.

ACKNOWLEDGMENTS

The author thanks YH Xie, D Monroe, PJ Silverman, GP Watson, TW Sorsch, and JY Cheng for contributions and discussions.

Literature Cited

1. Rosi FD. 1968. *Solid-State Electronics*, 11: 833–68. England: Pergamon
2. People R. 1986. *IEEE J. Quantum Elect.* 22: 1696–710
3. Massalski TB. 1986. *Binary Alloy Phase Diagrams*. Metals Park, Ohio: Am. Soc. Metals
4. Gaskell DR. 1981. *Introduction to Metallurgical Thermodynamics*, pp. 382–88. New York: McGraw-Hill. 611 pp.
5. Kurten M, Schilz J. 1994. *J. Cryst. Growth* 139: 1–5
6. Grutzmacher DA, Sedgwick TO, Northrup GA, Zaslavsky A, Powell AR, et al. 1993. *J. Vac. Sci. Technol. B* 11: 1083–88
7. Gibbons JF, Gronet CM, Williams KE. 1985. *Appl. Phys. Lett.* 47: 721–23
8. Sturm JC. 1991. *J. Metals* 43: 44
9. Meyerson BS. 1992. *Proc. IEEE* 80: 1592–608
10. Greve DW. 1993. *Mater. Sci. Eng. B* 18: 22–51
11. Fitzgerald EA. 1991. *Mater. Sci. Rep.* 7: 87–142
12. Houghton DC, Perovic DD, Baribeau JM, Weatherly GC. 1990. *J. Appl. Phys.* 67: 1850
13. Fritz IJ. 1987. *Appl. Phys. Lett.* 51: 1080
14. Van de Walle CG, Martin RM. 1986. *Phys. Rev. B* 34: 5621–34
15. Hinckley JM, Singh J. 1990. *Phys. Rev. B* 42: 3546–66
16. People R, Bean JC, Lang DV, Sergent AM, Stormer HL, et al. 1984. *Appl. Phys. Lett.* 45: 1231–33
17. Abstreiter G, Brugger H, Wolf T, Jorke H, Herzog HJ. 1985. *Phys. Rev. Lett.* 54: 2441
18. Liu L, Lee GS, Marshak AH. 1993. *Solid State Elect.* 37: 421–25
19. Matthews JW, Blakeslee AE, Mader S. 1976. *Thin Solid Films* 33: 253
20. Fitzgerald EA. 1989. *J. Vac. Sci. Technol. B* 7: 782
21. Tuppen CG. 1990. *Appl. Phys. Lett.* 56: 54
22. Perovic DD, Weatherly GC, Baribeau JM, Houghton DC. 1989. *Thin Solid Films* 183: 141
23. Eaglesham DJ, Maher DM, Kvam EP, Bean JC, Humphreys CJ. 1989. *Phys. Rev. Lett.* 62: 187
24. People R, Bean JC. 1994. *Appl. Phys. Lett.* 47: 322
25. Spitzer J, Thonke K, Sauer R, Kibbel H, Herzog HJ, et al. 1992. *Appl. Phys. Lett.* 60: 1729–31
26. Hull R, Bean JC, Werder DJ, Leibenguth RE. 1988. *Appl. Phys. Lett.* 52: 1605
27. Tuppen CG, Gibbings CJ. 1989. *Thin Solid Films* 183: 133
28. Dodson BW, Tsao JY. 1987. *Appl. Phys. Lett.* 51: 1325
29. Volkert CA, Fitzgerald EA, Hull R, Xie YH, Mii YJ. 1991. *J. Electron. Mater.* 20: 833–7
30. Fitzgerald EA, Watson GP, Proano RE, Ast DG, Kirchner PD, et al. 1989. *J. Appl. Phys.* 65: 2220
31. Fitzgerald EA, Kirchner PD, Proano RE, Pettit GD, Woodall JM, et al. 1988. *Appl. Phys. Lett.* 52: 1496
32. Fitzgerald EA, Xie YH, Brasen D, Green ML, Michel J, et al. 1990. *J. Electron. Mater.* 19: 949
33. Kamins TI, Nauka K, Jacowitz RD, Hoyt JL, Noble DB, et al. 1992. *IEEE Electron. Dev. Lett.* 13: 177–79

34. Luryi S, Suhir E. 1986. *Appl. Phys. Lett.* 49: 140
35. Fitzgerald EA, Chand N. 1991. *J. Electron. Mater.* 20: 839–53
36. van Mieghem P, Jain SC, Nijs J, van Overstraeten R. 1993. *J. Appl. Phys.* 75: 666–68
37. Dietrich B. 1993. *J. Appl. Phys.* 74: 7223–27
38. Fischer A, Richter H. *J. Appl. Phys.* 75: 657–59
39. Suhir E. 1988. *J. Appl. Mech.* 55: 143–48
40. Fitzgerald EA, Xie YH, Green ML, Brasen D, Kortan AR, et al. 1991. *Appl. Phys. Lett.* 59: 811
41. Dunstan DJ. 1991. *Semicond. Sci. Technol.* 6: A76–79
42. Krishnamoorthy V, Lin YW, Park RM. 1992. *J. Appl. Phys.* 72: 1752–57
43. LeGoues FK, Meyerson BS, Morar JF. 1991. *Phys. Rev. Lett.* 66: 2903–6
44. Schaffler F, Tobben D, Herzog HJ, Abstreiter G, Hollander B. 1992. *Semicond. Sci. Technol.* 7: 260–66
45. Watson GP, Fitzgerald EA, Xie YH, Monroe D. 1993. *J. Appl. Phys.* 75: 263–69
46. Dutartre D, Warren P, Provenier F, Chollet F, Perio A. 1994. *J. Vac. Sci. Technol. A* 12: 1009–14
47. Larsen AN, Hansen JL, Jenson RS, Shiryaev Y, Ostergaard PR, et al. 1995. *Phys. Scr.* Submitted
48. Cullis AG, Robbins DJ, Pidduck AJ, Smith PW. 1992. *J. Cryst. Growth* 123: 333–43
49. Eaglesham DJ, Gossman HJ, Cerullo M. 1990. *Phys. Rev. Lett.* 65: 1227
50. Xie YH, Gilmer GH, Roland C, Silverman PJ, Buratto SK, et al. 1994. *Phys. Rev. Lett* 73: 3006–9
51. Fitzgerald EA, Xie YH, Monroe D, Watson GP, Kuo JM, et al. 1994. *Mater. Sci. Forum* 143: 471–82
52. Weber J, Alonso, MI. 1989. *Phys. Rev. B* 40: 5683
53. Terashima K, Tajima M, Tatsumi T. 1990. *Appl. Phys. Lett.* 57: 1925–27
54. Terashima K, Tajima M, Sakai A, Tatsumi T. 1991. *J. Cryst. Growth* 111: 920–24
55. Fitzgerald EA, Ast DG, Kirchner PD, Pettit GD, Woodall JM. 1988. *J. Appl. Phys.* 63: 693
56. Noel JP, Rowell NL, Houghton DC, Perovic DD. 1990. *Appl. Phys. Lett.* 57: 1037
57. Spitzer J, Thonke K, Sauer R, Kibbel H, Herzog HJ, et al. 1992. *Appl. Phys. Lett.* 60: 1729–31
58. Noel JP, Rowell NL, Houghton DC, Wang A, Perovic DD. 1992. *Appl. Phys. Lett.* 61: 690
59. Sturm JC, St Amour A, Lacroix Y, Thewalt LW. 1994. *Appl. Phys. Lett.* 64: 2291–93
60. Sturm JC, Manoharan H, Lenchyshyn LC, Thewalt LW, Rowell NL, et al. 1991. *Phys. Rev. Lett.* 66: 1362
61. Grutzmacher DA, Sedgwick TO, Zaslavsky A, Powell AR, Kiehl RA, et al. 1993. *J. Vac. Sci. Technol. B* 11: 1083–88
62. Fukatsu S, Yoshida H, Usami N, Fujiwara A, Takahashi Y, et al. 1992. *Thin Solid Films* 222: 1–4
63. Wachter M, Thonke K, Sauer R, Schaffler F, Herzog HJ, et al. 1992. *Thin Solid Films* 222: 10–14
64. Schaffler F, Wachter M, Herzog HJ, Thonke K, Sauer R. 1993 *J. Cryst. Growth* 127: 411–15
65. Brunner J, Menczigar U, Gail M, Friess E, Abstreiter G. 1992. *Thin Solid Films* 222: 27–29
66. Terashima K, Tajima M, Ikarashi N, Niino T, Hiroi M, et al. 1992. *J. Electron. Mater.* 21: 1081–85
67. Mi Q, Xiao X, Sturm JC, Lenchyshyn LC, Thewalt MLW. 1992. *Appl. Phys. Lett.* 60: 3177–79
68. Fukatsu S, Usami N, Chinzei T, Shiraki Y, Nishida A, et al. 1992. *Jpn. J. Appl. Phys.* 31: 1015–17
69. Higgs V, Lightowlers EC, Xiao X, Sturm JC. 1994. *Appl. Phys. Lett.* 64: 607–9
70. Usami N, Fukatsu S, Shiraki Y. 1993. *Appl. Phys. Lett.* 63: 388–90
71. Nayak DK, Usami N, Fukatsu S, Shiraki Y. 1993. *Appl. Phys. Lett.* 63: 3509–11
72. Presting H, Kibbel H, Jaros M, Turton RM, Menczigar U, et al. 1992. *Semicond. Sci. Technol.* 7: 1127–48
73. Tang YS, Wilkinson CDW, Torres CMS, Sotomayor CM, Smith DW, et al. 1993. *Solid State Commun.* 85: 199–202
74. Usmai N, Mine T, Fukatsu S, Shiraki Y. 1994. *Appl. Phys. Lett.* 64: 1126–28
75. Brunner J, Rupp TS, Gossner H, Ritter R, Eisle I. 1994. *Appl. Phys. Lett.* 64: 994–96
76. Iyer SS, Patton GL, Stork JMC, Meyerson BS, Harame DL. *IEEE Trans. Electron. Dev.* 36: 2043–64
77. King CA. 1994. *Heterojunction Bipolar Transistors with GeSi Alloys in Heterostructures and Quantum Devices*, pp. 157–87. New York: Academic
78. Temkin H, Bean J, Antreasyan A, Leibenguth R. 1988. *Appl. Phys. Lett.* 52: 1089

79. Tatsumi T, Hirayama H, Aiziki N. 1988. *Appl. Phys. Lett.* 52: 895
80. Patton GL, Iyer SS, Delage SL, Tiwari S, Stork JMC. 1988. *IEEE Electron. Dev. Lett.* 9: 165
81. King CA, Hoyt JL, Gronet CM, Gibbons JF, Scott M, et al. 1989. *IEEE Electron. Dev. Lett.* 10: 52–54
82. Patton GL, Comfort JH, Meyerson BS, Crabbe EF, Scilla GJ, et al. 1990. *IEEE Electron. Dev. Lett.* 11: 171–73
83. Kasper E, Kibbel H, Herzog HJ, Gruhle A. 1994. *Jpn. J. Appl. Phys.* 33: 2415–18
84. Ershov M, Ryzhii V. 1994. *Jpn. J. Appl. Phys.* 33: 1365–71
85. Hinckley JM, Singh J. 1990. *Phys. Rev. B* 41: 2912–26
86. Manku T, Nathan A. 1991. *J. Appl. Phys.* 69: 8414–16
87. Bhaumik K, Ridley BK, Shacham-Diamond Y. 1993. *J. Appl. Phys.* 74: 5546–50
88. Venkataraman V, Schwartz PV, Sturm JC. 1991. *Appl. Phys. Lett.* 59: 2871–73
89. Smith DW, Emeleus CJ, Kubiak RA, Parker EHC, Whall TE. 1991. *Appl. Phys. Lett.* 61: 1453–55
90. Gravesteijn DJ, Mishima T, Fredriksz CW, van de Walle GFA, van den Heuvel RA. 1991. *J. Cryst. Growth* 111: 916–19
91. Manku T, Jain SC, Nathan A. 1992. *J. Appl. Phys.* 71: 4618–19
92. McGregor JM, Manku T, Noel JP, Roulston DJ, Nathan A, et al. 1993. *J. Electron. Mater.* 22: 319–21
93. Cheng JP, Kesan VP, Grutzmacher DA, Sedgewick TO, Ott JA. 1993. *Appl. Phys. Lett.* 62: 1522–24
94. Venkataraman V, Liu CW, Sturm JC. 1993. *Appl. Phys. Lett.* 63: 2795–97
95. Li SH, Hinckley JM, Singh J, Bhattacharya PK. 1993. *Appl. Phys. Lett.* 63: 1393–95
96. Manku T, McGregor JM, Nathan A, Roulston DJ, Noel JP, et al. 1993. *IEEE Trans. Electron. Dev.* 40: 1990–96
97. Xie YH, Fitzgerald EA, Monroe D, Silverman PJ, Watson GP. 1993. *J. Appl. Phys.* 73: 8364–70
98. Xie YH, Monroe D, Fitzgerald EA, Silverman PJ, Thiel FA, et al. 1993. *Appl. Phys. Lett.* 63: 2263–64
99. Pearsall TP, Bean JC. 1986. *IEEE Electron. Dev. Lett.* 7: 308–10
100. Goto K, Murota J, Maeda T, Schutz R, Aizawa K, et al. 1993. *Jpn. J. Appl. Phys.* 32: 438–41
101. Verdonckt-Vandebroek S, Crabbe EF, Meyerson BS, Harame DL, Restle PJ, et al. 1994. *IEEE Trans. Electron. Dev.* 41: 90–101
102. Konig U, Schaffler F. 1993. *Electron. Lett.* 29: 486–87
103. Murakami E, Nakagawa K, Nishida A, Miyao M. 1994. *IEEE Trans. Electron. Dev.* 41: 857–62
104. Konig U, Schaffler F. 1993. *IEEE Electron. Dev. Lett.* 14: 205–7
105. Garone PM, Venkataraman V, Sturm JC. 1991. *IEEE Electron. Dev. Lett.* 12: 230–32
106. Fitzgerald EA, Xie YH, Green ML, Brasen D, Kortan AR, et al. 1991. *MRS Symp. Proc.* 220: 211–15
107. Tuppen CG, Gibbings CJ, Hockly M. 1991. *MRS Symp. Proc.* 220: 187–92
108. Xie YH, Fitzgerald EA, Mii YJ, Monroe D, Thiel FA, et al. 1991. *MRS Symp. Proc.* 220: 413–17
109. Ismail K, Meyerson BS, Wang PJ. 1991. *Appl. Phys. Lett.* 58: 2117–19
110. Schuberth G, Schaffler F, Besson M, Abstreiter G, Gornik E. 1991. *Appl. Phys. Lett.* 59: 3318–20
111. Mii YJ, Xie YH, Fitzgerald EA, Monroe D, Thiel FA, et al. 1991. *Appl. Phys. Lett.* 59: 1611–13
112. Fitzgerald EA, Xie YH, Monroe D, Silverman PJ, Kuo JM, et al. 1992. *J. Vac. Sci. Technol. B* 10: 1807–19
113. Nelson SF, Ismail K, Chu JO, Meyerson BS. 1993. *Appl. Phys. Lett.* 63: 367–69
114. Nelson SF, Ismail K, Jackson TN, Nocera JJ, Chu JO, et al. 1993. *Appl. Phys. Lett.* 63: 794–96
115. Monroe D, Xie YH, Fitzgerald EA, Silverman PJ. 1992. *Phys. Rev. B* 46: 7935–37
116. Nelson SF, Ismail K, Nocera JJ, Fang FF, Mendez EE, et al. 1992. *Appl. Phys. Lett.* 61: 64–66
117. Monroe D, Xie YH, Fitzgerald EA, Silverman PJ, Watson GP. 1993. *J. Vac. Sci. Technol. B* 11: 1731–37
118. Tobben D, Schaffler F, Zrenner A, Abstreiter G. 1992. *Phys. Rev. B* 46: 4344–47
119. Konig U, Boers AJ, Schaffler F, Kasper E. 1992. *Electron. Lett.* 28: 160–62
120. Ismail K, Meyerson BS, Rishton S, Chu J, Nelson S, et al. 1992. *IEEE Electron. Dev. Lett.* 13: 229–31
121. Welser J, Hoyt JL, Gibbons JF. 1994. *Jpn. J. Appl. Phys.* 33: 2419–22
122. Welser J, Hoyt JL, Gibbons JF. 1994. *IEEE Electron. Dev. Lett.* 15: 100–2
123. Miyata H, Toshishige Y, Ferry DK. 1993. *Appl. Phys. Lett.* 62: 2661–63
124. Yamada T, Zhou JR, Miyata H, Ferry DK. 1994. *Semicond. Sci. Technol.* 9: 775–77
125. Vogelsang T, Hofmann KR. 1993. *Appl. Phys. Lett.* 63: 186–88

Annu. Rev. Mater. Sci. 1995. 25: 455–85

SPIN POLARIZED PHOTOEMISSION[1]

P. D. Johnson

Physics Department, Brookhaven National Laboratory, Upton, New York 11973

KEY WORDS: surface magnetism, thin films, multilayers, quantum wells, electronic structure

ABSTRACT

The addition of spin analysis to the photoemission technique allows the investigation of the electronic structure and magnetic properties of clean and adsorbate-covered surfaces, thin films, and multilayers. Deposition of thin films on appropriate substrates allows the investigation of different phases including the *bcc* phase of Ni and *fcc* phases of Fe and Co. Studies of noble metals deposited on ferromagnetic substrates reveal the presence of spin-polarized quantum well states with binding energies reflecting the thickness of the noble metal films. These states sample the Fermi surface with a frequency identical to that characterizing the oscillatory exchange coupling in the associated magnetic multilayers which suggests that they are the electronic states that mediate the coupling.

INTRODUCTION

The requirement for an increased understanding of the magnetic properties of surfaces and thin films is driven by the potential applications in information storage technologies (1) and new device technologies (2). The magnetism in such systems reflects the spin polarization of the electronic structure, which may be directly studied using the photoemission technique. Photoemission is already well established in the study of the electronic structure of atoms and solids (3). In particular, angle-resolved photoemission has been used extensively to map the band structure of clean and adsorbate-covered surfaces, both metal and semiconductor (4). Extending the technique by measuring the spin of the photoemitted electrons allows the possibility of examining the exchange split-band structures that characterize ferromagnetic systems. Because of the typical escape depths of the photoelectrons, spin-polarized photoemission is particularly useful in the study of the magnetic properties of surfaces, thin films, and associated interfaces.

Shown schematically in Figure 1, the photoexcitation of an electron from initial state $|i\rangle$ with binding energy E_i to some final state $|f\rangle$ with energy E_f results in a photocurrent I given by

$$I(hv, E_i) = \sum_i |\langle f|\mathbf{A}\cdot\mathbf{P}|i\rangle|^2 \delta(E_f - hv - E_i), \qquad 1.$$

where **A** is the classical vector potential of the photon, and **P** is the momentum operator, hv represents the energy of the incident light, and the δ-function describes the energy conservation during the excitation.

As written, the intensity in Equation 1 represents excitation from all states at initial binding energy E_i. The summation is over all points in the Brillouin zone. Recognizing that the transitions are momentum conserving, the angle-resolved photoemission experiment evolved as a means of

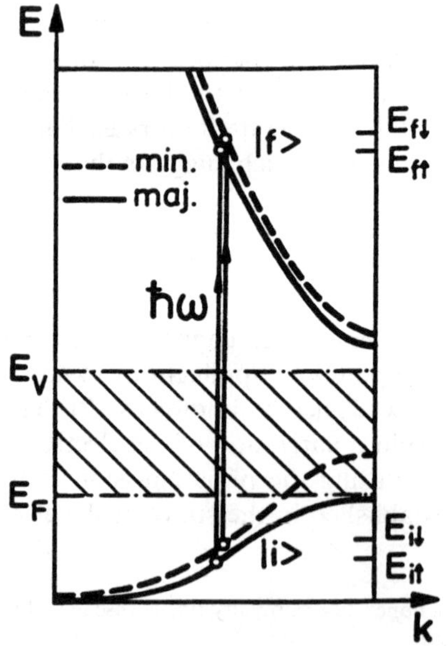

Figure 1 Schematic of the spin-polarized photoemission process. The incident photon excites an electron from an initial state E_i below the Fermi level E_F to some final state E_f with the same spin. Reproduced from Reference 6.

determining the electronic structure at discrete k-points in the Brillouin zone as shown in Figure 1. By measuring the energy and momentum of the photoelectrons emitted at different angles, it is possible to map the entire band structure of the material under investigation. The matrix element of Equation 1 also introduces selection rules that allow the symmetry of the initial state to be determined.

From Equation 1 we note that the photon will couple directly only to the orbital momentum of the electron but not its spin. The spin of the electron will therefore be conserved during the excitation, and by measuring the spin in the final state, we determine the spin in the initial state. This allows the possibility of mapping the spin-dependent band structure of ferromagnetic materials.

A more complete description of the photoemission process allows for the many-body response of the system to the excitation. Here, the δ-function of Equation 1 is replaced by a self-energy correction represented by the function,

$$A(h\nu, E_i) = \frac{1}{\pi}\frac{\operatorname{Im}\Sigma(h\nu, E_i)}{\{[h\nu - E_i - \operatorname{Re}\Sigma(h\nu, E_i)]^2 - [\Sigma(h\nu, E_i)]^2\}}. \qquad 2.$$

This correction results in a broadening and shifting of the peaks that would characterize the one-electron spectrum. In ferromagnetic systems, we anticipate a spin dependence in the response of the system because of the spin polarization of the electronic structure.

Although not discussed in detail in this article, we also mention two further developments in the photoemission field that are adding to our knowledge of ferromagnetic systems. These developments include spin-polarized inverse photoemission and spin-polarized core level photoemission.

Inverse photoemission has developed as a method of examining the unoccupied electronic structure (5, 5a). Electrons incident on the sample make radiative transitions into unoccupied states above the Fermi level E_F. To a first approximation, the technique may be viewed as the time reversal of the photoemission process. The introduction of a spin-polarized electron source allows the study of the spin-dependent unoccupied states in ferromagnetic systems (6).

The second technique, core level photoemission, represents excitation of electrons from the more localized core levels. The experiment has already been well developed as a probe of local site-specific chemical information (7). In recent years, there have been a number of studies examining the possibility of using the spin-polarized counterpart as a probe of the local magnetic information (8–11).

THE EXPERIMENTAL METHOD

Much of the pioneering work in spin-polarized photoemission (12–14) was carried out with spin polarimeters of the high-energy Mott scattering type. Here the electrons to be analyzed are scattered off gold atoms at energies typically of the order of 100 keV. Spin-orbit coupling of the electron in the potential of the gold atom leads to a left-right asymmetry in the scattering, dependent on the spin of the electron. Because of the high energies involved, this type of detector has tended to be large. However, there have recently been a number of successful modifications that have allowed the Mott polarimeter to be scaled down in size.

Figure 2 shows the configuration used in one such device (15, 16). After initially passing through a deceleration lens, the incident beam of electrons is accelerated onto the gold foil at an energy of 20 keV. As in earlier designs (12), those electrons scattered through 120° are detected by two symmetrically opposite channeltrons. Retarding fields in front of these detectors remove any electrons that have undergone inelastic losses. The

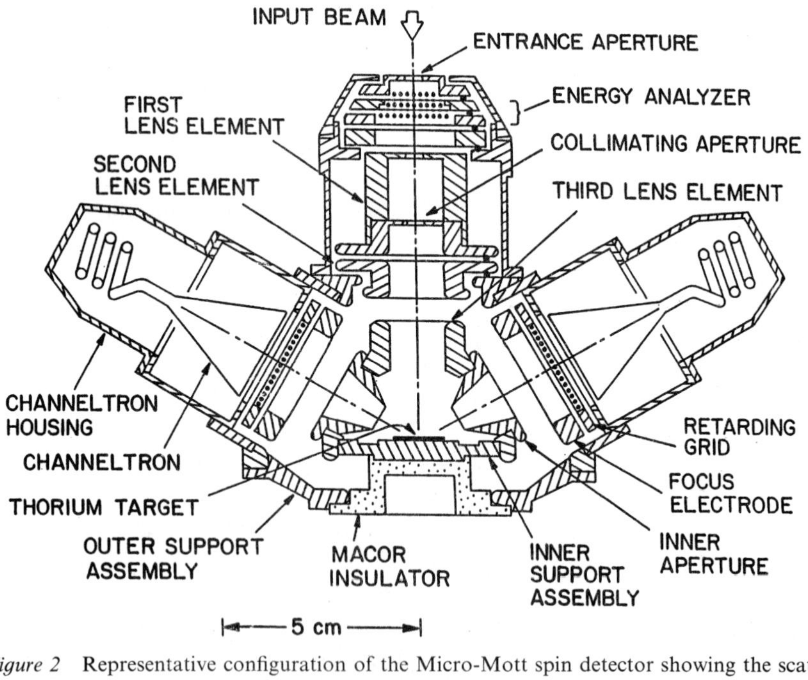

Figure 2 Representative configuration of the Micro-Mott spin detector showing the scattering target and collecting electrodes or channeltrons (17).

polarization P is given by

$$P = \frac{1}{S}\frac{I_A - I_B}{I_A + I_B}, \qquad 3.$$

where S, the Sherman function of the device, is a measure of its ability to distinguish different spins and I_A and I_B represent the intensities measured in opposite channels. Mott polarimeters of this type are now sufficiently small that they can be readily moved inside a vacuum system and thereby allow angle-resolved polarization measurements.

The figure of merit (FOM) used in comparing different polarimeters is defined as (17)

$$\text{FOM} = S^2 \frac{I}{I_0}, \qquad 4.$$

where again S is the Sherman function, I is the sum of the current collection by the two opposite detectors, and I_0 is the incident beam current. The FOM for the traditional Mott polarimeter is typically 10^{-4}; that of the compact low-energy Mott device is approximately 2×10^{-5} (15). However, more recently, the FOM of the compact device has been improved to approximately 1.6×10^{-4} through the use of large-area electron detectors and careful optimization of the electron optics (16).

It was recently demonstrated that the low energy diffuse-scattering type of detector, originally designed at the National Institute of Standards and Technology for use in SEMPA studies (18, 19), also proves an excellent device for energy- and angle-resolved spin polarization studies (20.) Rather than using the high energies characteristic of the Mott devices, the electrons to be spin analyzed are now scattered off a polycrystalline gold surface at the much lower energy of 150 eV, an energy at which the Sherman function of gold exhibits a local maximum. The back-scattered electrons pass through retarding grids to remove low-energy secondary electrons and are then collected by four discrete anodes. This allows a measurement of two components of the polarization to be made in parallel. Shown in Figure 3, this type of detector has an FOM of the order 1.0×10^{-4}, i.e. comparable to the traditional Mott detectors.

If the scattering surface is a single crystal rather than polycrystalline, it is possible to use spin-dependent diffraction to measure the polarization. Detectors of this type include the low energy detector of Kirschner (21), which relies on spin-orbit effects in the diffraction of the electrons from a single-crystal tungsten (001) surface. The intensity of symmetrically opposite diffracted beams is measured with the electrons incident at an energy of 100 eV. The FOM for this detector is reported as 1.6×10^{-4}.

More recently, a second detector employing low-energy diffraction or

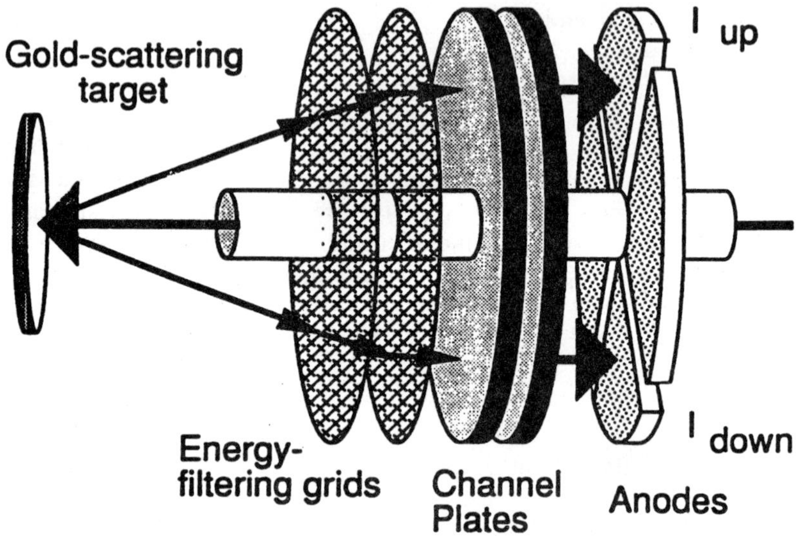

Figure 3 Schematic of the low-energy diffuse scattering spin polarimeter. The incident electrons are back scattered by the target through the energy filtering grids into the discrete anodes.

reflection has been developed (22). However, the device relies on exchange rather than spin-orbit scattering, the scattering surface being a 400Å-thick Fe(001) film grown on a Ag(001) substrate. Reflected intensities are measured when the electrons to be analyzed are reflected from the magnetized iron surface at an energy of approximately 10.0 eV. The asymmetry in the scattering A is given by

$$A = \frac{1}{P}\frac{I^{\uparrow\uparrow} - I^{\uparrow\downarrow}}{I^{\uparrow\uparrow} + I^{\uparrow\downarrow}}, \qquad 5.$$

where P is the spin polarization in the incident beam, and $I^{\uparrow\uparrow}$ and $I^{\uparrow\downarrow}$ are the scattered intensities obtained when the target and primary beam magnetic moments are parallel and anti-parallel, respectively. In this case, the FOM is given by

$$\text{FOM} = A^2 \frac{I_R}{I_P}, \qquad 6.$$

where I_R and I_P are the reflected and primary beam currents. One difficulty with an instrument of this type is that two samples have to be prepared and magnetized. However, with $A = 0.21$ at 10.6 eV, the authors report a FOM for this device of 3.5×10^{-3}.

While the FOM of spin detectors is slowly improving, a large photon flux is still required if a reasonable signal-to-noise ratio is to be achieved. The introduction of undulators into a number of synchrotron storage rings has meant that the quality of the spin-resolved spectra can now be comparable to that obtained in many non-spin-resolved photoemission measurements on normal bending magnet beamlines (20).

In order to define a quantization direction for the spin of the electrons, it is necessary to magnetize the sample under investigation. For elemental metals such as Fe, Co, or Ni, it is often useful to be able to shape the sample in the form of a picture frame. The magnetizing field is then obtained from a solenoid wrapped around one arm of the frame. In the case of thin films, the magnetizing field is obtained by discharging a current pulse through a small coil in close proximity to the film. The photoemission measurements are always made with the sample in a remanent magnet state to minimize problems from stray magnetic fields.

BULK ELECTRONIC STRUCTURE STUDIES

Studies at Room Temperature

Spin-polarized photoemission studies of the elemental materials Fe and Ni have concentrated on the identification of exchange-split bands at low or room temperature and the extension of these studies to the effects of high temperature. It is hoped that the ferromagnetic to paramagnetic phase transition can be observed at high temperature. At low temperatures, the comparison between experiment and theory for Fe has been relatively easy. In the case of Ni, correlation effects have made the interpretation more difficult. To the author's knowledge there have been no detailed studies of hcp Co. However, as described below, fcc Co has been studied in thin-film form.

The bulk electronic structure of Ni was studied extensively with spin-integrated angle-resolved photoemission before any equivalent spin-resolved studies were attempted. However, prior to any of these studies, spin-polarized threshold photoemission experiments were carried out on the Ni(001) surface (23). In agreement with the predictions of Wohlfarth (24, 24a), these studies found a negative spin-polarization at threshold. Moore & Pendry (25) were unable to reproduce these experimental observations in a photoemission calculation taking the local spin-density functional (LSDF) value of 0.6 eV for the exchange parameter ΔE_x. Rather, they found that a value of 0.3 eV for ΔE_x gave a better description of the experiment. This requirement for a reduced value for the exchange splitting was confirmed in subsequent threshold photoemission studies of the Ni(111) surface (26) and angle-resolved photoemission studies of the low-index planes of Ni (27–29).

The study of the Ni(110) surface by Heimann et al (29) revealed a symmetry-dependent, e_g and t_{2g}, exchange splitting. Liebsch (30, 30a) was able to explain the narrowing of the *d* band width in terms of self-energy corrections or correlations among the *d* electrons. His theory also explained the presence of the 6 eV satellite always observed in Ni valence band spectra (31) and further the symmetry dependence of the exchange splitting (29). More recently Victora & Falicov (32), taking full account of both band structure effects and electron-electron interactions, were able to calculate the exact density of emitted states and obtain good agreement with the photoemission experiments.

The first spin, energy, and angle-resolved photoemission study of Ni was carried out by Raue et al (33) on the Ni(110) surface. Using a Ne resonance lamp (16.85 eV) as the excitation source, the authors identified the $X_2^{\downarrow}$ at -0.06 eV and the $X_2^{\uparrow}$ at -0.24 eV, respectively, giving an exchange splitting for the S_4 bands of 0.18 eV. They also located the $X_5^{\uparrow}$ at a binding energy of -0.10 eV. All these results are in excellent agreement with the earlier non-spin-resolved study of Heimann et al (29).

The photoemission spectra from Ni of both the core levels (34) and the valence bands (31) are characterized by a 6 eV satellite. Following a suggestion of Mott (35), this satellite is generally associated with more localized or atomic-like effects (36, 37), particularly as the valence band satellite is known to undergo a resonance at photon energies corresponding to threshold excitation of the 3*p* core level.

The satellite is seen as representing the d^8 configuration with the associated excitation route on resonance being

$$3p^6 3d^9 \rightarrow 3p^5 3d^{10} \rightarrow 3p^6 3d^8 + e^-, \qquad 7.$$

where e^- represents the final state continuum electron. Here the 3*p* core hole is thought to have decayed via a super-Coster-Kronig-type transition. Feldkamp & Davis (38) suggested that the Ni Auger electrons would carry very little spin polarization, whereas the photoelectrons produced in the resonant mechanism would carry spin polarization because of the spin selectivity in the initial excitation at threshold. This prediction was verified in a spin-polarized photoemission study of the satellite on and off resonance by Clauberg et al (39). They found that on resonance, the satellite was spin polarized in excess of the average valence band polarization. However, in a detailed comparison with the results of an Anderson impurity calculation, recent spin-polarized core level photoemission study of the satellites associated with the Ni 3*p* core level assigns the satellites to the d^9 configuration (40).

As in the case of Ni, the exchange split electronic structure of Fe was first examined along the high symmetry directions using angle-resolved photoemission without recourse to the added complication of spin detection. These early angle-resolved photoemission studies with resonance lamps (41, 42) found broad agreement with calculation but were unable to identify specific critical points. Indeed, the authors of these studies concluded that it was not possible to observe band dispersions in Fe. This latter objective was achieved in later studies using synchrotron radiation as the excitation source (43–45). In particular, the study of all three low-index crystal planes by Turner et al (44) found good agreement with first principles calculations by Callaway & Wang (46). However, the experimental study indicated that the presence of surface states complicated the interpretation of spectra recorded from the (001) surface.

The first spin and angle-resolved photoemission study of Fe was undertaken on the (001) surface. Using incident *s*-polarized synchrotron radiation of 60 eV energy, Feder et al (47) were able to identify transitions from the $\Gamma_{25'}$ and Γ_{12} majority spin critical points and from the $\Gamma_{25'}$ minority spin critical point. Because of the polarization of the incident light, these spectra are dominated by the transitions from the $\Gamma_{25'}$ points. As shown in Figure 4, the same group was subsequently able to obtain data from the same crystal with improved resolution (48). The majority spin $\Gamma_{25'}$ and Γ_{12} points are located at binding energies of 2.6 ± 0.2 eV and 1.2 ± 0.2 eV, respectively. The minority spin $\Gamma_{25'}$ point is located at a binding energy of 0.4 ± 0.2 eV. These binding energy values are uniformly higher than the values obtained in the spin-integrated photoemission studies (44), which may simply reflect the choice of position for the Fermi energy. In fact, we note that the value obtained for the exchange splitting in the two studies is nearly identical. The experimental observations may be compared with the theoretical predictions of Hathaway et al (49). The latter study predicted binding energies of 0.69 and 2.51 eV for the two $\Gamma_{25'}$ critical points and 0.99 eV for the Γ_{12} point. In a later spin-polarized photoemission study, Johnson et al (20) were able to enhance the emission from the Γ_{12} critical point by altering the angle of incidence of the light to increase the magnitude of the *p*-polarized component.

Kisker et al (48) were also able to observe the photon energy dependence of the spin-resolved spectra from the Fe(001) surface. With incident *s*-polarized light, they found that at a photon energy below 33 eV, the intensity of the minority spin peak immediately below the Fermi level suddenly dropped as the initial state Δ_5 band crossed the Fermi level at the center of the Brillouin zone. At lower photon energies, the spectra are dominated by majority spin emission.

The only other Fe surface that has been studied by spin-polarized photoemission is an Fe(110) film grown on a GaAs substrate (50). The

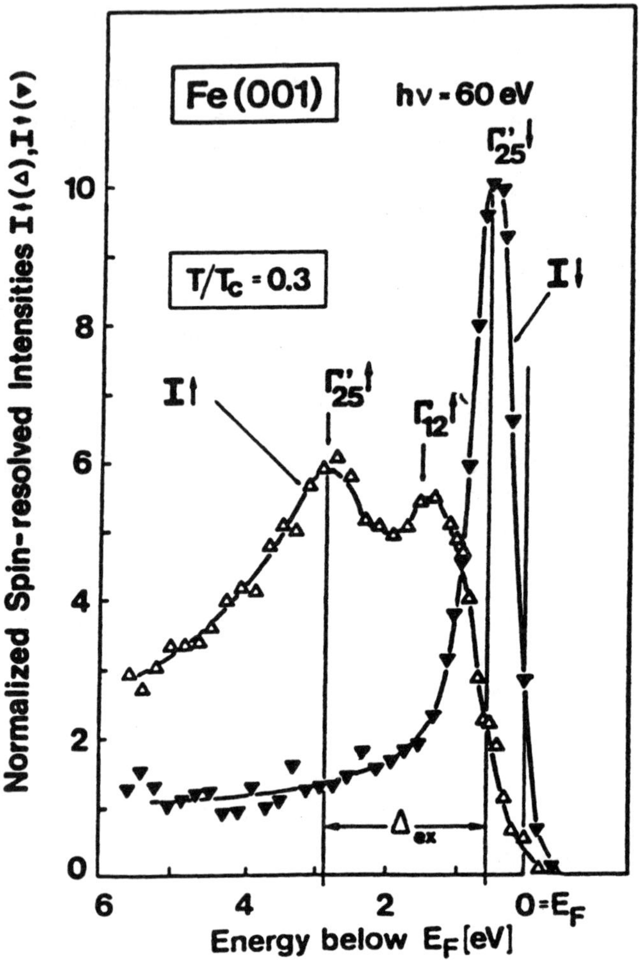

Figure 4 Spin-polarized photoemission spectra recorded from an Fe(001) surface with photons of incident energy 60 eV. The critical points in the spectra are indicated (48).

films that were not grown in-situ were cleaned by Ne ion sputtering. The authors concluded that their study showed good agreement with the results of first principles calculations (49). However, we note that because the experimental study was restricted to the single photon energy of 30 eV, the only new information was the identification of majority spin emission from the band running from Γ_{12} to N_4.

Studies at Elevated Temperatures

An obvious question in the study of magnetism is by what mechanism is the magnetization lost as the temperature is raised above the Curie temperature T_c. In the Stoner model, the spin polarization of the electronic structure scales linearly with the magnetization. Thus it is anticipated that the exchange splitting observed in the photoemission experiment would disappear as the temperature is raised. Alternative models include the local-band model (51) and the disordered local-moment model (52). In the former model, a degree of short-range order of the moments exists above T_c, whereas in the latter model, there is no short-range order. In the local-band model, the exchange splitting above T_c will depend on which k-point is being sampled in the Brillouin zone—any variation reflecting the difference between sampling the short-range and the long-range magnetic order.

Experimental studies have been carried out on Fe and Ni. The first photoemission study of the ferromagnetic to paramagnetic transition was made in a study of Ni(110) by Hopster et al (53). The spectra reproduced in Figure 5 show the exchange splitting disappearing as the temperature approaches T_c. However, the interpretation is complicated by the fact that the line shapes are of the same order as the exchange splitting. In a subsequent analysis of this data by Haines et al (54), it was concluded that the temperature dependence was consistent with a correlation length parameter $L = 5.4$ Å.

A study of the Ni(111) surface found a temperature-dependent exchange splitting at points corresponding to $1/3d$ $(\Gamma - L)$ and $1/2d$ $(\Gamma - L)$, but a temperature-independent splitting at the point corresponding to $2/3d$ $(\Gamma - L)$ (55). However, on the basis of their observations, the authors felt unable to draw any conclusions between the different models.

In a more recent study of the Ni(110) surface, the spin polarization of the 6 eV satellite was found to decrease with increasing temperature (56). This observation was interpreted as evidence of the presence of fluctuating magnetic moments in ferromagnetic Ni.

In a study of the temperature dependence of the exchange splitting in Fe at Γ, the center of the Brillouin zone, no change in the magnitude of the splitting was found on increasing the temperature from 0.3 to 0.85 T_c (57). However, as shown in Figure 6, the intensities of the different peaks in the spectra did show a variation with temperature. As in the case of Ni, modeling of this data led to the conclusion that the spin correlation length must be larger than 4 Å (58).

A similar type of behavior was observed in a temperature-dependent inverse photoemission study of the P point in the Fe Brillouin zone (59).

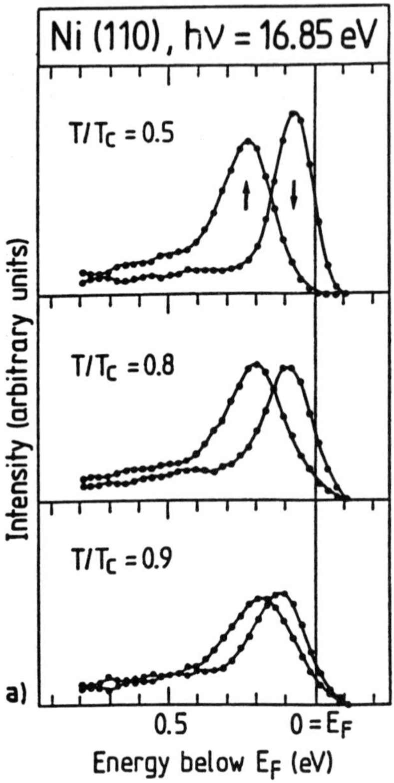

Figure 5 Spin resolved photoemission spectra recorded along the surface normal of Ni(110) at three temperatures ($T = 0.5T_C$, $T = 0.8T_C$, and $T = 0.9T_C$). The incident photon energy is 60 eV and the angle of incidence corresponds to *s*-polarization. The two spin components are indicated (53).

With increasing temperature, no change was observed in the exchange splitting. In contrast, however, in the same inverse photoemission study, it was found that the exchange splitting of a pair of unoccupied states near the $\Gamma - H$ line collapsed as the temperature was raised towards T_c.

A general conclusion drawn from these different studies is that for both Fe and Ni, spin-polarized photoemission indicates that short-range magnetic order on the scale of 4 to 5 Å exists above the T_c.

SURFACE STUDIES

Clean Surfaces

First principles calculations predict that the clean surfaces of many materials will have enhanced magnetic moments (60). As an example, the magnetic moment at the surface of an Fe crystal cut in the (001) plane is predicted to be 2.96 μ_B as opposed to the 2.2 μ_B characteristic of the bulk

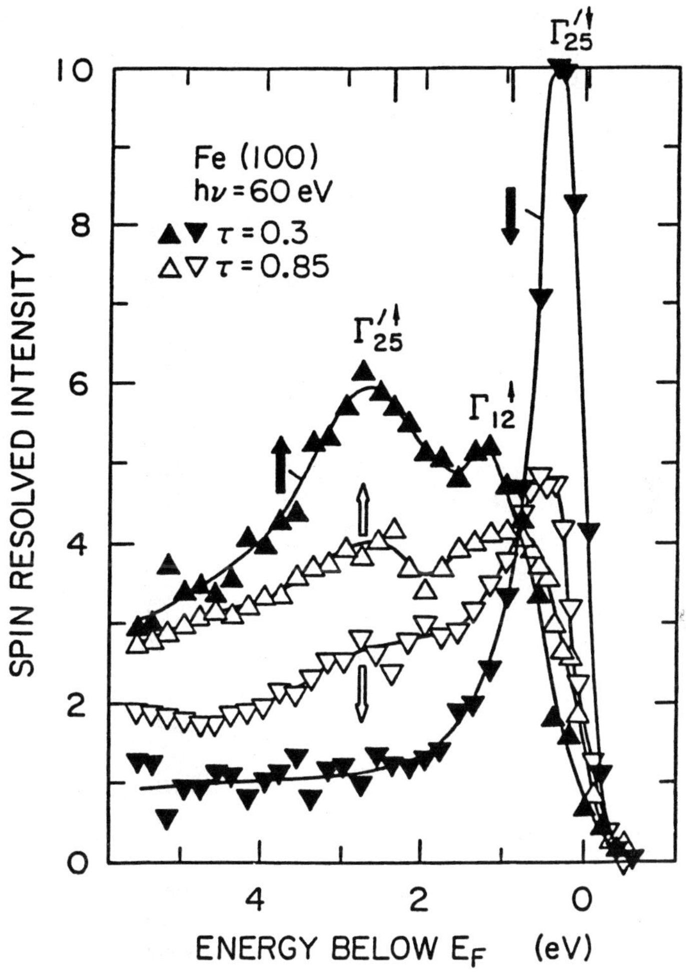

Figure 6 Spin-resolved photoemission spectra recorded along the surface normal of Fe(001) at two temperatures ($T = 0.3T_C$ and $T = 0.85T_C$). The incident photon energy is 60 eV and the angle of incidence corresponds to *s*-polarization. The two spin components are indicated (57).

(61). This enhancement of the moment reflects the coordination number or reduction in the number of nearest neighbors as the dimensionality is reduced. The theoretical predictions have never been directly verified, although several studies of magnetic thin films appear to indicate that the average magnetic moment increases as the film becomes thinner and two-dimensional behavior is approached (62).

Although it is not possible to determine the magnitude of a magnetic moment from angle-resolved photoemission studies, it is possible to verify many other aspects of the calculation. In particular, it is possible to identify and characterize the surface states that are responsible in part for the surface magnetic structure. There have been numerous photoemission studies of surface states on ferromagnetic surfaces. However, to date, the number of spin-polarized photoemission studies has been limited, and herein we concentrate on the latter. We also examine studies of adsorbate systems, with particular reference to the role of the adsorbate in modifying the surface magnetism.

A non-spin-polarized photoemission study of the Ni(001) surface found evidence for a majority spin $\bar{\Gamma}_5$ (Δ_5) surface state near the center of the surface Brillouin zone (63). However, a subsequent spin-polarized photoemission study suggested that the correct spin assignment should be minority rather than majority spin and further suggested that the emission reflected off-normal Δ_2 bulk band contributions rather than a surface state (64).

Two spin-polarized photoemission studies of surface states have been carried out on the Fe(001) surface (65, 66) and one study on the Fe(110) surface (67). The first study, using *p*-polarized incident light, identified a minority spin-surface resonance at the center of the surface Brillouin zone of the Fe(001) surface at a binding energy of 2.4 eV below the Fermi level (65). The symmetry of the state was determined from its sensitivity to the incident light polarization, as shown in Figure 7.

Because of the reduced intensity, there have been only a few examples of band mapping in spin-polarized photoemission studies. However, the study of the Fe(001) minority spin-surface state provides one such example (65). By varying the angle of detection of the photoelectrons, the authors followed the dispersion of the surface state across the zone. Figure 8 shows the comparison of this experimentally determined dispersion as a function of $k_\parallel$ with the results of first principles calculation. The parallel momentum $k_\parallel$ in the figure is given by

$$k_\parallel = \left(\frac{2m}{\hbar^2}\right)^{1/2} (E_f - \phi)^{1/2} \sin\theta, \qquad 8.$$

where ϕ is the work function of the surface, θ is the angle of detection with respect to the surface normal and E_f is the final state energy above the Fermi level. From the figure, we can see that the experimental dispersion follows the calculated dispersion throughout the zone to the boundary $\bar{X}$, where the state falls within a projected bulk band gap and becomes more localized in the surface region.

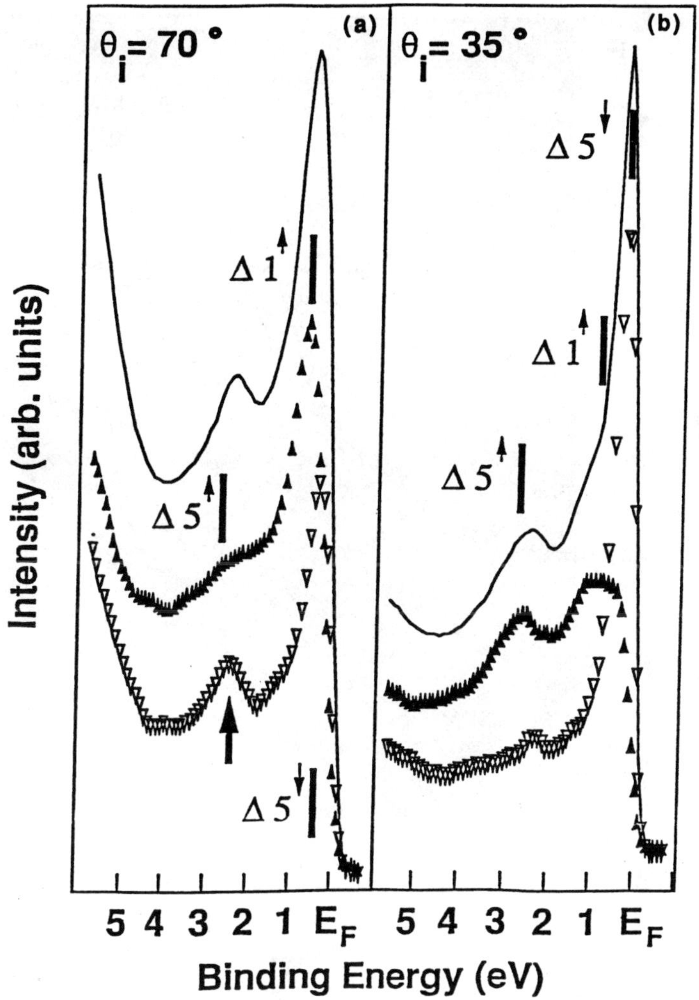

Figure 7 Spin-integrated and spin-resolved photoemission spectra recorded along the surface normal of Fe(001) for photons of incident energy 52 eV. In (*a*) the angle of incidence corresponds to *p*-polarized light (70°) and in (*b*) the angle of incidence corresponds to *s*-polarized light (35°). The filled triangles indicate the majority spin channels, the open triangles indicate the minority spin channel. The surface state at a binding energy of 2.4 eV in the minority spin channel is indicated by the arrow (65).

The second study of the Fe(001) surface used *s*-polarized rather than *p*-polarized incident light (66). For normal emission this changes the sensitivity from Δ_1- to Δ_5-like bands. The study revealed a majority spin

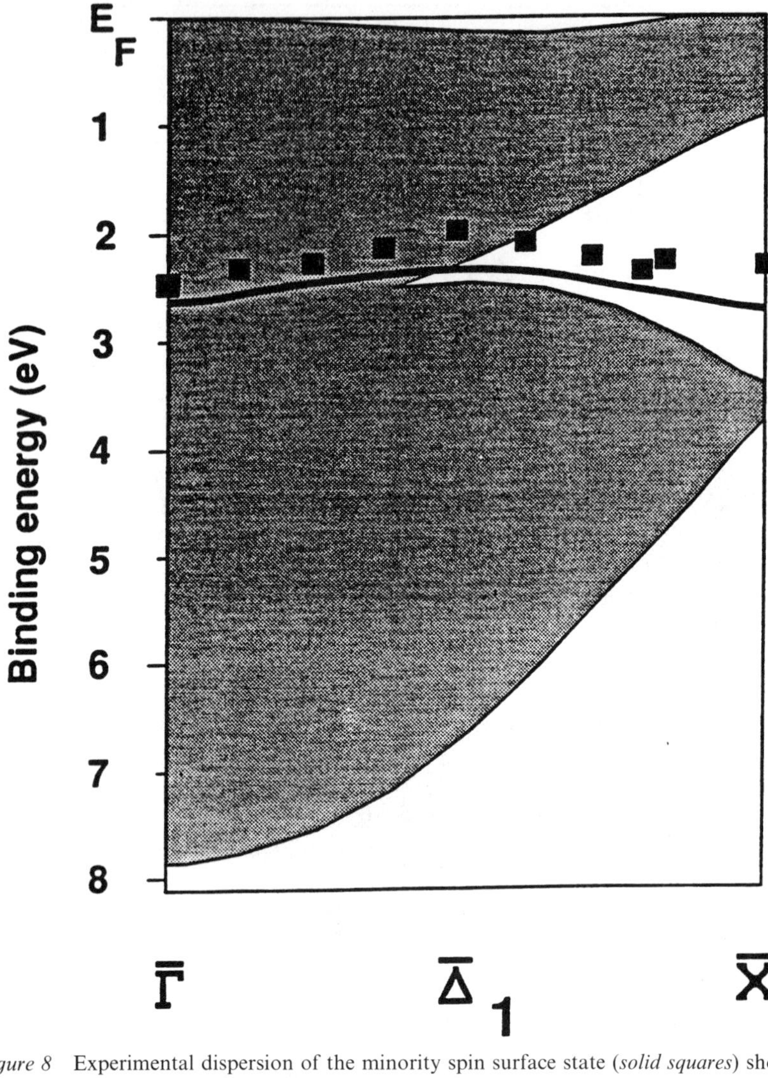

Figure 8 Experimental dispersion of the minority spin surface state (*solid squares*) shown in Figure 7 compared with the theoretical dispersion (*thick solid line*) determined from a 13-layer slab calculation. The calculated minority bulk projected bands are shown by the shaded region (65).

state with the latter symmetry immediately below the Fermi level. The assignment of majority spin again represents a reversal of spin assignment from that suggested by an earlier spin-integrated photoemission study. Finally, a study of the Fe(110) surface (67) using *s*-polarized incident light identified a surface state of minority spin character midway along the $\bar{\Gamma}\bar{H}$ azimuth.

Studies have also been carried out on the surface of the rare earth Gd. A spin-integrated photoemission study identified a surface state 2 eV below the Fermi level (68). A subsequent spin-polarized photoemission study of the same surface state showed that the state carried majority spin, which suggested that the surface magnetic moments were ferromagnetically aligned with the substrate (69). This observation disagreed with the results of a first principles total energy calculation that predicted an antiferromagnetic alignment of the surface and bulk moments (70). Further spin-polarization studies of the secondary electrons and the Gd $4f$ core levels confirmed the ferromagnetic alignment but suggested the possibility of a small canting of the surface moments with a component sticking out of the surface (71).

Adsorbate Covered Surfaces

With the assumption that contamination of a surface leads to the formation of magnetic dead layers, early studies of adsorption concentrated on changes to the substrate electronic structure (72, 73). However, a later study of the ordered (2×1) O structure on Ni(110) found no change in the exchange splitting of the substrate S_4 d-bands (74). In the same paper, a study of the $c(2 \times 2)$ S structure on the same Ni(110) surface found qualitatively similar behavior. Thus the overall conclusion from this study was that adsorption has no strong affect on surface magnetism.

Later studies examined the formation of magnetic moments on the adsorbate atoms themselves. A series of studies identified exchange split adsorbate bands for the systems Fe(001) $p(1 \times 1)$ O and Fe(001) $c(2 \times 2)$ S (75–77). These results, which confirmed earlier theoretical predictions (78–80), provided the first direct evidence of magnetic moments being induced on the adsorbates rather than the adsorbate killing the surface magnetization. In Figure 9 we show the results obtained in the study of oxygen adsorbed on the Fe(001) surface (77). The oxygen-derived bands located 6 eV below the Fermi level show a strong spin polarization indicative of a moment on the oxygen site (75). Again the symmetry of the adsorbate-derived bands are determined by the sensitivity to the incident light polarization. Similar evidence has also been obtained in spin-polarized Auger electron spectroscopy studies of these adsorption systems. Spin-polarized photoemission studies of oxygen adsorbed on Fe(110) (67) and on Co(001) (81) have also identified an exchange splitting in the O $2p_{xy}$ bands. In the case of adsorption on the Fe(110) surface, the exchange splitting was larger than that observed for the other systems, possibly indicative of a stronger bond between the adsorbate and substrate.

The spin-polarized photoemission studies of oxygen adsorption on Fe

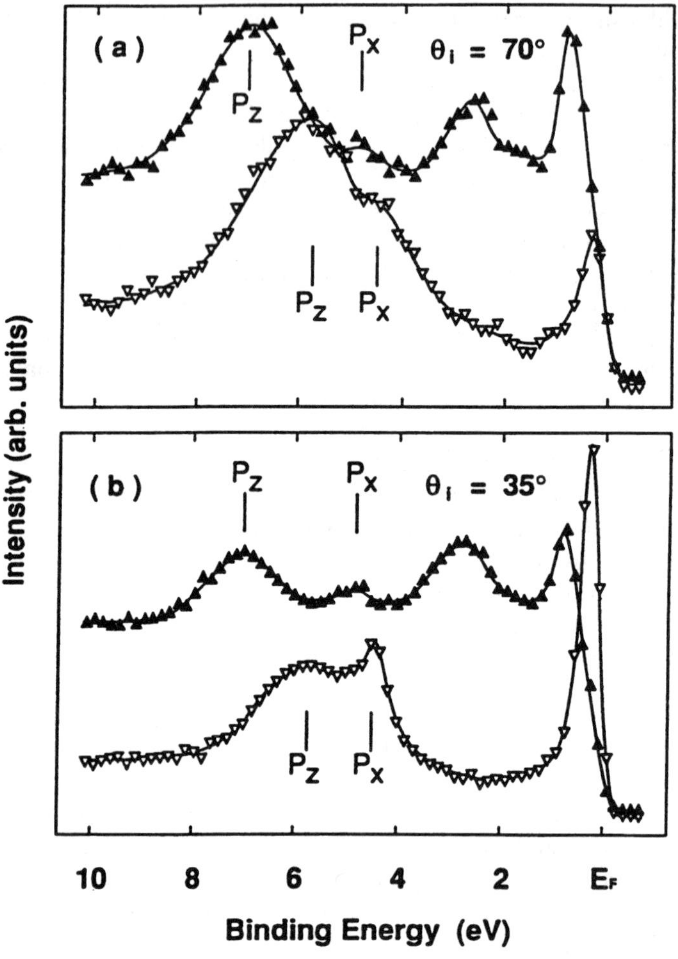

Figure 9 Spin-resolved photoemission spectra for normal emission from Fe(001)-$p(1 \times 1)$ O. Solid upward-pointing triangles correspond to majority spin, and open downward-pointing triangles correspond to minority spin: (*a*) angle of light incidence $\theta_i = 70°$ corresponds to *p*-polarized light; (*b*) $\theta_i = 35°$ giving a larger proportion of *s*-polarization. The symmetry of the different oxygen-derived features is indicated (77).

and Co surfaces have clearly revealed the bonding of the oxygen adsorption to the surface via the interaction of the *p*-orbitals with the substrate *d*-orbitals. Interestingly, a subsequent study of carbon monoxide adsorption on an Fe(001) surface (82) found no spin-polarization evidence for the interaction of the equivalent molecular orbital, the 5σ level, with the substrate *d*-orbitals, even though this is a frequently cited mechanism for a component of the CO bonding (83).

The use of core level photoemission as a means of determining site-specific valency is a well established technique (7). Combining such measurements with spin-polarization analysis allows site-specific magnetic information to be obtained. Such an approach has been used to follow the pathway to oxidation as an Fe(001) surface was exposed to larger amounts of oxygen (9).

Figure 10 shows the spin-resolved Fe 3*p* photoemission spectra recorded from the Fe(001) surface following exposure to 16L of O_2 and after annealing the latter surface to 650°C (9). The main substrate iron emission is represented by the minority spin feature at a binding energy of 52.9 eV. The majority counterpart is at a binding energy 0.5 eV higher. With the initial oxidation (Figure 10*a*), a new majority spin peak appears, shifted to higher binding energy by 4.5 eV with respect to the clean surface peak. The shift in binding energy for this peak suggests that it may be associated with emission from a site of Fe^{3+} valency. This assignment, together with the observation that it is spin polarized, suggests that the initial oxide is of the ferrimagnetic form γ-Fe_2O_3. The authors concluded that the reversal of spin polarization for the initial oxide suggests that the magnetic moment on the outer Fe ions associated with the oxide are antiferromagnetically coupled to the moments of the substrate via a superexchange mechanism involving subsurface atoms (9).

Following annealing (Figure 10*b*), the spectra are dominated by unpolarized emissions shifted approximately 2.0 eV from the main peak at 52.9 eV binding energy. The latter shift in binding energy is indicative of a Fe^{2+} phase, and the oxide following annealing is most probably of the antiferromagnetic Fe_xO form.

THIN FILMS AND INTERFACES

Thin films are particularly interesting because they offer the possibility of studying a range of new phenomena in magnetism. Thus it is possible that normally nonmagnetic materials such as Pd (84–86) or Rh (87) may become magnetic as their dimensionality is reduced. Ferromagnetic films may show marked anisotropies as the film thickness is varied. As an example of the latter phenomenon, the first spin polarized photoemission study of Fe films deposited on a Ag(001) substrate found that the anisotropy resulted in the magnetic moments of the Fe orienting perpendicular to the surface for the thinnest films (88). As the films became thicker the moments aligned in the plane of the film (89). By growing ferromagnetic films on different substrates, it is possible to examine the

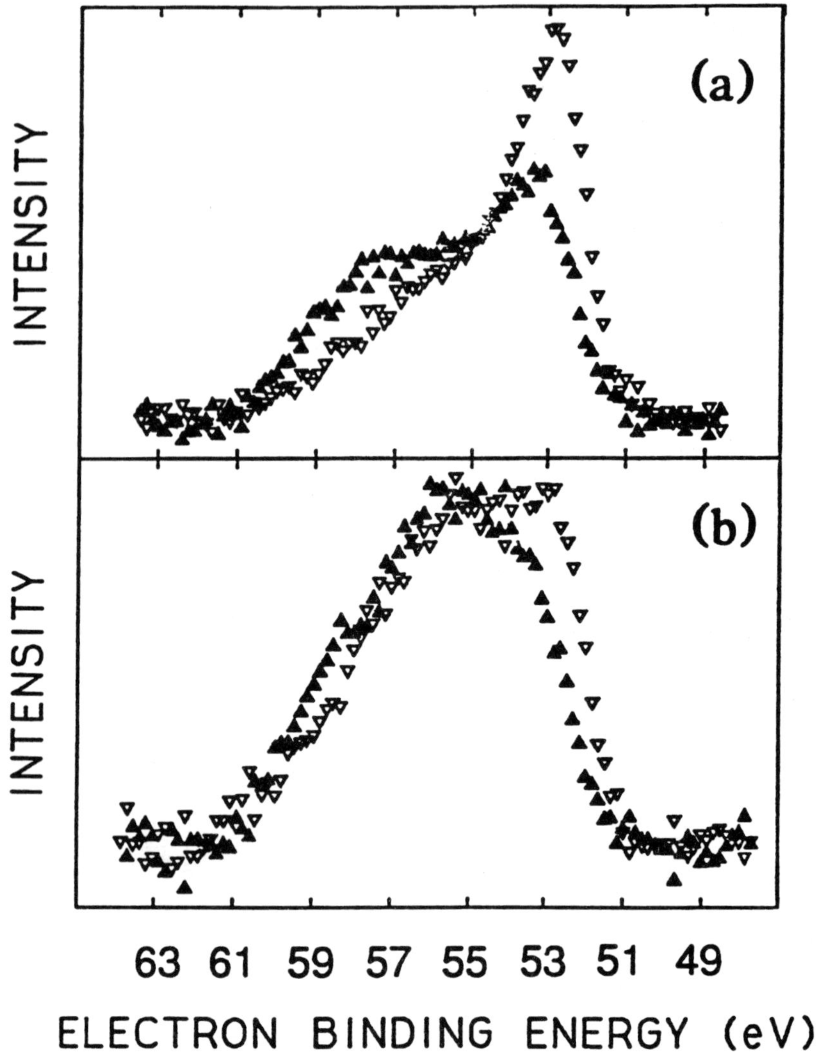

Figure 10 (*a*) Spin-resolved Fe 3*p* spectra showing the initial oxidation of Fe after exposure to 16L of O_2 with the background subtracted. [1*L* (Langmuir) = 10^{-6}]. (*b*) The same as in (*a*) but after annealing to 650°C. The photon energy used in the studies was 90 eV (9).

properties of different phases of a material. Examples here include the fcc phases of Fe (90) and Co (62, 91), both grown by evaporation on a Cu(001) substrate, and the bcc phase of Ni grown by evaporation on an Fe(001) substrate (92).

We note that in all of these systems it is important to examine the magnetic properties of the interface between the thin film and the substrate. Because the symmetry of the system has been broken, we may anticipate the identification of states unique to the metal-metal interface as are found at the solid-vacuum interface.

There have been a number of studies of Pd and Pt interfaces with different materials, and the Pd/Fe interface has been the subject of numerous studies with Fe as the substrate (84–86) and with Pd (93) as the substrate. In both cases, if we consider the (001) direction, calculations predict that the Pd atoms in the interface will have a local moment of the order of 0.3 μ_B (85, 93, 94). A study of Fe films deposited on Pd(001) (93) (Figure 11) revealed that a nonmagnetic surface state associated with the Pd surface evolved into a magnetic interface state localized in the interface between the Fe and Pd. It was shown that the spin polarization of the state, which reflected the hybridization between Pd 4*d* states and Fe 3*d* states, contributed to the development of the magnetic moment on the Pd atoms in the interface.

In the reverse case, Pd films deposited on Fe(001) (85) spin-polarized interface states were identified, as has been the case in all studies of Pd and Pt films deposited on different substrates at the monolayer coverage level (84, 86). It was concluded that the spin polarization reflected hybridization between the overlayer *d*-bands and the exchange-split substrate *d*-bands rather than any lateral interaction in the overlayer. In Figure 12 we show an example of spin-polarized interface states observed in a study of Pd deposited on an Fe(110) surface (84). A similar result was found in a study of a Rh monolayer on Fe(001) (87). The overlayer was ferromagnetic, and as in the case of Pd, it was concluded that the spin polarization reflected the hybridization with the substrate bands. For this system, calculations suggested a moment of 0.82 μ_B on the Rh site.

Chromium films deposited on Fe(001) substrates represent systems that have also been extensively studied in thin-film magnetism. The interest stems from the oscillatory exchange coupling (95–97) and giant magnetoresistance (98) that has been observed in the related Fe/Cr multilayers. Chromium has a lattice structure identical to that of Fe but its half-filled *d*-band results in the antiferromagnetic state rather than the ferromagnetic state representing the ground state. In the bulk crystal, each atom is antiferromagnetically coupled to its nearest neighbor so that every (001) plane is ferromagnetically aligned within the plane but antiferromagnetically aligned to neighboring (001) planes.

Calculations indicate that a Cr monolayer deposited on an Fe(001) substrate will also be ferromagnetically aligned within the plane but antiferromagnetically aligned to the substrate (99–101). The calculations also

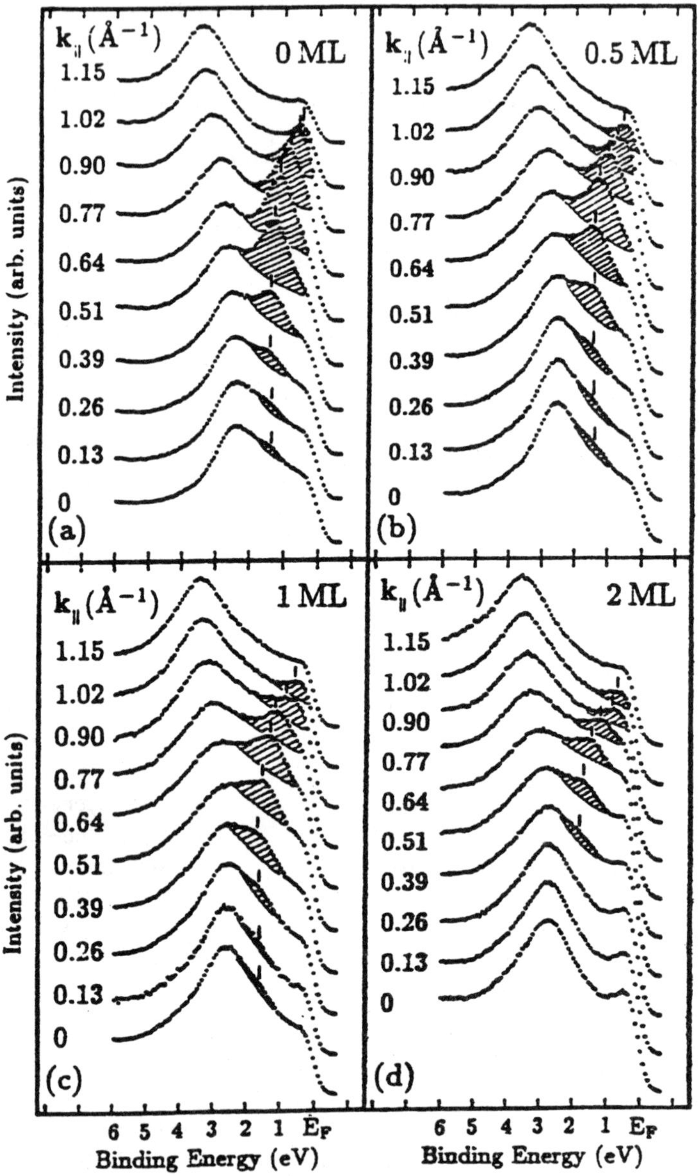

Figure 11 Spin-integrated spectra recorded as a function of $k_{\parallel}$ from Fe films of different thickness deposited on a Pd(001) surface. The hatched area indicates the surface state on the clean surface that evolves into an interface state (93).

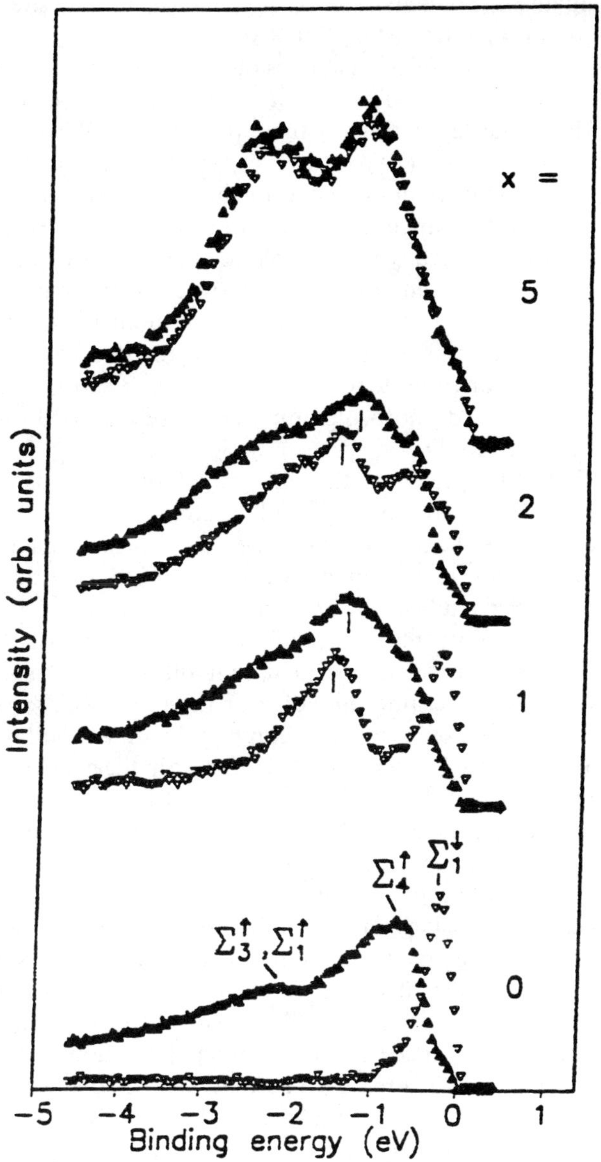

Figure 12 Spin-integrated spectra recorded from different thicknesses (x atomic layers) of Pd deposited on an Fe(110) surface. The incident photon energy is 21.2 eV (84).

predict an enhanced moment of the order of 3.1 μ_B on the Cr site, compared with 0.6 μ_B that is characteristic of bulk Cr.

Spin-polarized photoemission studies have in part confirmed these predictions. A core level study found that the Cr overlayer is indeed ferromagnetic within the layer and, further, that it is antiferromagnetically aligned with the substrate (102). However, an attempt to derive the magnitude of the Cr magnetic moment from the measured spin polarization suggests a value in the range 0.5–1.0 μ_B, i.e. considerably lower than the 3.1 μ_B. A valence band study was unable to conclude anything about the size of the moment but did confirm Cr-derived valence band states in a tight-binding calculation that also predicted a moment of 3.0 μ_B (101). Unfortunately, structural studies indicate that for this system, the overlayer growth mode at room temperature is not necessarily layer by layer (103). Both of the cited photoemission studies were carried out at room temperature.

Stabilization of the fcc phase through growth of thin films on Cu(001) substrates has allowed spin-polarized photoemission studies of Co to be carried out (62, 91). Two studies of thick fcc Co films have identified exchange split bands in agreement with the results of calculation. In particular, a detailed study of the exchange splitting of the Δ_5 bands at the center of the zone (62) confirmed the findings of an earlier study for the thick films (91) but also demonstrated that the exchange splitting was enhanced as the films became thinner. For the monolayer coverage, the splitting was enhanced by 25% when compared with the bulk value. This may be seen as evidence for the predicted increase in the magnetic moments as the dimensionality of the system is reduced (60).

A study of the bcc phase of Ni, stabilized by growth on an Fe(001) surface, provided evidence that the Ni film was ferromagnetic at lattice constants corresponding to that of the Fe substrate (92). By measuring the background spin polarization as a function of the Ni thickness, the authors concluded that the magnetic moment on the Ni sites was of the order of 0.4 μ_B in agreement with theoretical calculations. The study also revealed the presence of a 6 eV satellite identical to that observed in fcc Ni, as described above. Indeed, it was the spin polarization of this localized satellite that confirmed the presence of a magnetic moment on the Ni site.

Finally, we consider a study of a rare-earth transition metal interface, in particular the Sm/Fe(001) interface (104). An earlier spin-polarized Auger electron spectroscopy study of the Gd/Fe(001) interface had shown that at the monolayer coverage, the Gd moments couple antiferromagnetically to the Fe substrate (105). In the Sm/Fe(001) study using spin-polarized photoemission (104), it was found that up to monolayer coverage, the Sm was trivalent, with the in-plane spin moment antiparallel to the Fe 3*d* majority spin direction. With increasing coverage, the Sm developed a divalent surface layer with the underlying layers trivalent. The in-plane ferromagnetic ordering was observed to rapidly decrease with increasing overlayer thickness and to vanish three atomic layers from the interface.

MAGNETIC MULTILAYERS

Because of the potential technological applications in both the recording (1) and device industries (2), magnetic multilayers are currently the subject of an enormous research industry. These multilayers, consisting of ferromagnetic layers interlaced with nonmagnetic layers, have been shown to exhibit an oscillatory exchange coupling between the adjacent ferromagnetic layers dependent on the thickness of the intervening nonmagnetic layer (95, 106). Further, at the thicknesses corresponding to the antiferromagnetic alignment of the ferromagnetic layers, the multilayers have also been shown to exhibit enhanced or giant magnetoresistive properties (98). Examples of systems showing these properties include the Cu/Co and Fe/Ag multilayers, both grown in the (001) direction.

Theories of the oscillatory exchange coupling frequently invoke the bulk Fermi surface of the intervening nonmagnetic layer. In the case of multilayers grown in the (001) direction with copper as the intervening layer, such theories predict two periods of oscillation for the exchange coupling: a long period of oscillation reflecting a spanning vector near the centre of the Brillonin zone, and a short period of oscillation reflecting a spanning vector away from the zone center (107). Both periods of oscillation have been observed in Cu/Co(001) multilayers grown using molecular beam epitaxy for the growth method (108).

The finite thickness of the individual components within the multilayer imposes boundary conditions that produce a quantization of the allowed electronic states within the layers (109). Thus in photoemission studies, rather than seeing electronic transitions or states characteristic of the continuous bulk electronic structure of the material, we anticipate seeing discrete states at binding energies dependent on the thickness of the layer.

This discrete quantization has been clearly observed in photoemission studies of silver films deposited on an Fe(001) substrate (110). Figure 13 shows the photoemission spectra recorded along the surface normal from silver films of different thicknesses as indicated (110). In the region immediately below the Fermi level, a series of peaks or quantum well states were

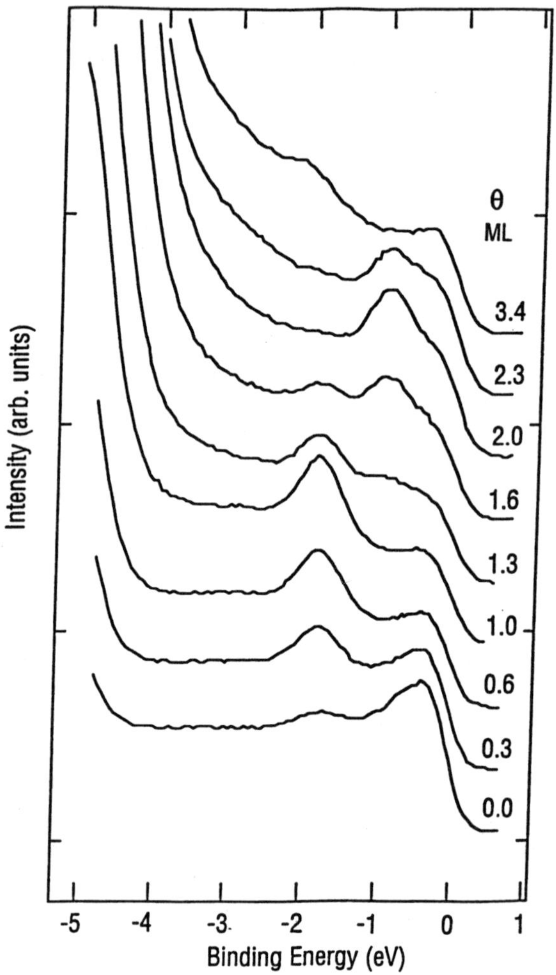

Figure 13 Spin-integrated photoemission spectra recorded from different thickness silver films deposited on a Fe(001) substrate. The different silver coverages are indicated. The spectra correspond to emission along the surface normal, and the incident light is *p*-polarized with an energy of 52 eV.

observed with binding energies dependent on the thickness of the film. In the case of the ultrathin films, these states have been described as interface states because they show a considerable weight in the interface between film and substrate.

A similar quantization of the *sp*-bands has also been observed in studies of copper films grown on a Co(001) substrate (111, 112). Using a tight-binding analysis, Johnson and co-workers (111) were able to show that these states carry a considerable component derived from the Cu *d*-orbitals, a reflection of the strong hybridization between the *sp*-bands and the *d*-bands at the observed binding energies. Such an observation allows for the presence of a magnetic moment of *d* character on the copper sites, as has indeed been observed in recent magnetic circular dichroic studies of Cu/Co multilayers (113). The latter moment was observed to be largest on the interfacial copper sites.

The binding energy separation ΔE between the quantum well states reflects the group velocity $\partial E/\partial k$ of the associated bulk band from which the states are derived and the thickness of the film L (109). Thus

$$\Delta E = \frac{\partial E}{\partial k}\frac{2\pi}{L}. \qquad 9.$$

From such an analysis, we may anticipate that given a typical energy-resolving power in the experiment, quantization of the more slowly dispersing *d*-bands will only be observable in thinner films, as is the case (86, 114).

The spin polarization in the quantum well states (Figure 14) indicates that they are strongly spin polarized predominantly with minority spin, an observation thought to reflect the difference in the reflectivities of the two spin states in the vicinity of the ferromagnetic substrate's bulk band gaps (115, 116). Combined with the observation that they cross the Fermi level with a periodicity identical to the periodicity observed for the oscillatory exchange coupling in the related multilayers, the suggestion has been made that these states represent the electrons that are responsible for the coupling in the multilayers (115). As such, one would anticipate that in the Cu/Co system the quantization would be observed both at the center of the zone and at the point farther out in the zone corresponding to the second period of oscillation (107). To date, the latter quantization has not been observed. However, Johnson and co-workers have shown that in the thinnest Cu films, fewer than seven monolayers, the hybridization in the interface results in an increase in the in-plane effective mass that will in turn lead to a modification of the copper Fermi surface (117). This thin film region may be considered a pre-asymptotic region in which it may not be valid to consider the copper in terms of its bulk electronic structure.

SUMMARY

In this review we have examined the contributions that spin-polarized photoemission has made to our understanding of magnetism. Because of

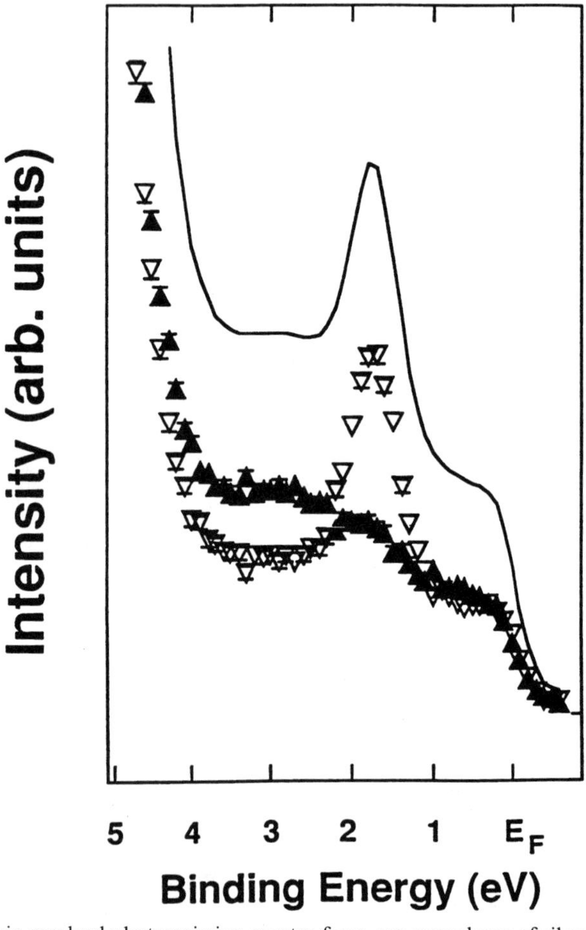

Figure 14 Spin-resolved photoemission spectra from one monolayer of silver deposited on an Fe(001) substrate. The incident light is *p*-polarized with an energy of 52 eV (110, 116).

the sampling depth of the experiment, the biggest impact has clearly been in the field of surface and thin-film magnetism. An understanding of the magnetic properties of these systems is obviously important because we may anticipate the use of magnetic nanostructures in new technologies. Indeed, an improved understanding of the interplay between the microscopic electronic structure and the magnetic properties will allow specific magnetic properties of the nanostructures to be tailored for particular requirements.

We have not considered in any detail the results from spin-polarized core level studies. This site-specific technique, which is still in its infancy, is capable of providing site-specific information that will be an additional resource in the study of magnetic alloys. Neither have we discussed the results that have been obtained using incident circularly polarized light both with and without the added possibility of spin-polarization analysis of the photoemitted electrons (118–120). This valuable technique offers the possibility of examining the electronic structure effects reflecting the spin-orbit interaction as well as the exchange interaction.

ACKNOWLEDGMENTS

The author is pleased to acknowledge many stimulating conversations with NV Smith, B Sinkovic, N Brookes, and M Weinert. This work has been supported by the Department of Energy under contract DE-AC02-76CH00016.

Literature Cited

1. Falicov LM. 1992. *Phys. Today* 45: 46–51
2. Johnson M. 1993. *Science* 260: 320–23
3. Plummer EW, Eberhardt W. 1982. *Adv. Chem. Phys.* 49: 533–656
4. Kevan SD, ed. 1992. *Angle-Resolved Photoemission*. Amsterdam: Elsevier. 615 pp.
5. Smith NV. 1988. *Rep. Prog. Phys.* 51: 1227–94; Johnson PD. 1992. See Ref. 4, Chpt. 13
6. Donath M. 1994. *Surf. Sci. Rep.* 20: 251–316
7. Brundle CR, Chuang TJ, Wandelt K. 1977. *Surf. Sci.* 68: 459–68
8. Carbone C, Kisker E. 1988. *Solid State Commun.* 65: 1107–10
9. Sinkovic B, Johnson PD, Brookes NB, Clarke A, Smith NV. 1990. *Phys. Rev. Lett.* 65: 1647–50
10. Hillebrecht FU, Jungblut R, Kisker E. 1990. *Phys. Rev. Lett.* 65: 2450–53
11. Carbone C, Kachel T, Rochow R, Gudat W. 1988. *Solid State Commun.* 77: 619–22
12. Kisker E, Clauberg R, Gudat W. 1982. *Rev. Sci. Instrum.* 53: 1137–44
13. Clauberg R, Gudat W, Kisker E, Kuhlmann E, Rothberg GM. 1981. *Phys. Rev. Lett.* 47: 1314–17
14. Kisker E, Schroder K, Campagna M, Gudat W. 1984. *Phys. Rev. Lett.* 52: 2285–88
15. Tang FC, Zhang X, Dunning FB, Walters GK. 1988. *Rev. Sci. Instrum.* 59: 504–5
16. Kessler K. 1985. *Polarized Electrons*. Berlin: Springer. 299 pp.
17. Burnett GC, Monroe TJ, Dunning FB. 1994. *Rev. Sci. Instrum.* 65: 1893–96
18. Unguris J, Pierce DT, Celotta RJ. 1986. *Rev. Sci. Instrum.* 57: 1314–23
19. Scheinfein MR, Pierce DT, Unguris J, McClelland JJ, Celotta RJ, Kelley MH. 1989. *Rev. Sci. Instrum.* 60(1): 1–11
20. Johnson PD, Hulbert SL, Klaffky R, Brookes NB, Clarke A, et al. 1992. *Rev. Sci. Instrum.* 63: 1902–8
21. Kirschner J, Feder R. 1979. *Phys. Rev. Lett.* 42: 1008–11
22. Tillman D, Thiel R, Kisker E. 1989. *Z. Phys.* B77: 1–2
23. Eib W, Alvarado SF. 1976. *Phys. Rev. Lett.* 37: 444–46
24. Wohlfarth EP. 1971. *Phys. Lett.* 36A: 131–32

24a. Wohlfarth EP. 1977. *Phys. Rev. Lett.* 38: 524–26

25. Moore ID, Pendry JB. 1978. *J. Phys.* C11: 4615–22
26. Kisker E, Schroder K, Gudat W, Campagna M, Kuhlmann E, et al. 1979. *Phys. Rev. Lett.* 43: 966–69
27. Himpsel FJ, Knapp JA, Eastman DE. 1979. *Phys. Rev.* B19: 2919–27

28. Eberhardt W, Plummer EW. 1980. *Phys. Rev. B* 21: 3245–55
29. Heimann P, Himpsel FJ, Eastman DE. 1981. *Solid State Commun.* 39: 219–23
30. Liebsch A. 1979. *Phys. Rev. Lett.* 43: 1431–34
30a. Liebsch A. 1981. *Phys. Rev. B* 23: 5203–12
31. Guillot C, Ballu Y, Paigne J, Lecante J, Jain RP, et al. 1982. *Phys. Rev. Lett.* 49: 72–76
32. Victora RH, Falicov LM. 1985. *Phys. Rev. Lett.* 55: 1140–43
33. Raue R, Hopster H, Clauberg R. 1983. *Phys. Rev. Lett.* 50: 1623–26
34. Hufner S, Wertheim GK. 1975. *Phys. Lett.* 51A: 299–300
35. Mott NF. 1964. *Adv. Phys.* 13: 325–422
36. Penn DR. 1979. *Phys. Rev. Lett.* 42: 921–25
37. Davis LC, Feldkamp LA. 1979. *J. Appl. Phys.* 50: 1944–49
38. Feldkamp LA, Davis LC. 1979. *Phys. Rev. Lett.* 43: 151–54
39. Clauberg R, Gudat W, Kisker E, Kuhlmann E, Rothberg GM. 1983. *Phys. Rev. Lett.* 51: 829–32
40. Liu Y, Xu Z, Johnson PD, van der Laan G. 1995. *Phys. Rev. Lett.* Submitted
41. Heimann P, Neddermeyer N. 1978. *Phys. Rev. B* 18: 3537–43
42. Schultz A, Courths R, Schultz H, Hufner S. 1979. *J. Phys.* F9: L41–45
43. Eastman DE, Himpsel FJ, Knapp JA. 1980. *Phys. Rev. Lett.* 44: 95–98
44. Turner AM, Donoho AW, Erskine JL. 1984. *Phys. Rev. B* 29: 2986–3000
45. Sakisaka Y, Rhodin T, Mueller D. 1985. *Solid State Commun.* 53: 793–99
46. Callaway J, Wang CS. 1977. *Phys. Rev. B* 16: 2095–105
47. Feder R, Gudat W, Kisker E, Rodriguez A, Schroder K. 1983. *Solid State Commun.* 46: 619–23
48. Kisker E, Schroder K, Gudat W, Campagna M. 1985. *Phys. Rev. B* 31: 329–39
49. Hatthaway KB, Jansen HJF, Freeman AJ. 1985. *Phys. Rev. B* 31: 7603–11
50. Schroder K, Prinz GA, Walker K-H, Kisker E. 1985. *J. Appl. Phys.* 57: 3669–71
51. Korenman V. 1986. In *Metallic Magnetism*, ed. H Capellmann, pp. 109–58. Berlin: Springer
52. Durham PJ, Staunton J, Gyorffy BL. 1984. *J. Magn. Magn. Mater.* 45: 38–42
53. Hopster H, Raue R, Guntherodt G, Kisker E, Clauberg R, Campagna M. 1983. *Phys. Rev. Lett.* 51: 829–32
54. Haines EM, Heine V, Ziegler A. 1985. *J. Phys. F* 16: 1343–64
55. Kamper K-P, Schmitt W, Guntherodt G. 1990. *Phys. Rev. B* 42: 10696–705
56. Kakizaki A, Fujii J, Shimada K, Kamata A, Ono K, et al. 1994. *Phys. Rev. Lett.* 72: 2781–84
57. Kisker E, Schroder K, Campagna M, Gudat W. 1984. *Phys. Rev. Lett.* 52: 2285–88
58. Haines E, Clauberg R, Feder R. 1985. *Phys. Rev. Lett.* 54: 932–34
59. Kirschner J, Globl M, Dose V, Scheidt H. 1984. *Phys. Rev. Lett.* 53: 612–15
60. Freeman AJ, Wu R. 1991. *J. Magn. Magn. Mater.* 100: 497–514
61. Ohnishi S, Freeman AJ, Weinert M. 1983. *Phys. Rev. B* 28: 6741–48
62. Clemens W, Kachel T, Rader O, Vescovo E, Blugel S, et al. 1992. *Solid State Commun.* 81: 739–44
63. Erskine JL. 1980. *Phys. Rev. Lett.* 45: 1446–49
64. Clauberg R, Hopster H, Raue R. 1984. *Phys. Rev. B* 29: 4395–400
65. Brookes NB, Clarke A, Johnson PD, Weinert M. 1990. *Phys. Rev. B* 41: 2643–45
66. Vescovo E, Rader O, Carbone C. 1993. *Phys. Rev. B* 47: 13051–54
67. Vescovo E, Carbone C, Eberhardt W, Rader O, Kachel T, Gudat W. 1993. *Phys. Rev. B* 48: 285–88
68. Li D, Hutchings W, Dowben PA, Hwang C, Wu RT, et al. 1991. *J. Magn. Magn. Mater.* 99: 85–90
69. Mulhollan GA, Garrison K, Erskine JL. 1992. *Phys. Rev. Lett.* 69: 3240–43
70. Wu R, Li C, Freeman AJ, Fu CL. 1991. *Phys. Rev. B* 44: 9400–9
71. Li D, Zhang J, Dowben PA, Garrison K, Johnson PD, et al. 1993. *MRS Symp. Proc.* 313: 451–56
72. Schmitt W, Hopster H, Guntherodt G. 1985. *Phys. Rev. B* 31: 4035–38
73. Seiler A, Feigerle CS, Pena JL, Celotta RJ, Pierce DT. 1985. *Phys. Rev. B* 32: 7776–78
74. Schmitt W, Kamper K-P, Guntherodt G. 1987. *Phys. Rev. B* 36: 3763–68
75. Johnson PD, Clarke A, Brookes NB, Hulbert SL, Sinkovic B, Smith NV. 1988. *Phys. Rev. Lett.* 61: 2257–60
76. Schonhense G, Getzlaff M, Westphal C, Heidemann B, Bansmann J. 1988. *J. Phys. Colloq.* C8: 1643–44
77. Clarke A, Brookes NB, Johnson PD, Weinert M, Hulbert SL, et al. 1990. *Phys. Rev. B* 41: 9659–67
78. Huang H, Hermanson J. 1985. *Phys. Rev. B* 32: 6312–18
79. Chubb SR, Pickett WE. 1987. *Phys. Rev. Lett.* 58: 1248–51

80. Chubb SR, Pickett WE. 1988. *Phys. Rev. B* 38: 10227–43
81. Clemens W, Vescovo E, Kachel T, Carbone C, Eberhardt W. 1992. *Phys. Rev. B* 46: 4198–204
82. Brookes NB, Clarke A, Johnson PD. 1989. *Phys. Rev. Lett* 63: 2746–67
83. Blyholder GJ. 1964. *J. Phys. Chem.* 68: 2772–78
84. Weber W, Wesner DA, Guntherodt G, Linke U. 1991. *Phys. Rev. Lett* 66: 942–45
85. Rader O, Carbone C, Clemens W, Vescovo E, Blugel S, et al. 1992. *Phys. Rev. B* 45: 13823–26
86. Hartmann D, Weber W, Rampe A, Popovic S, Guntherodt G. 1993. *Phys. Rev. B* 48: 16837–40
87. Kachel T, Gudat W, Carbone C, Vescovo E, Blugel S, et al. 1992. *Phys. Rev. B* 46: 12888–91
88. Jonker BT, Walker K-H, Kisker E, Prinz GA, Carbone C. 1986. *Phys. Rev. Lett.* 57: 142–45
89. Qiu ZQ, Pearson J, Bader SD. 1993. *Phys. Rev. Lett.* 70: 1006–9
90. Liu C, Moog ER, Bader SD. 1988. *Phys. Rev. Lett.* 60: 2422–25
91. Schneider CM, Miguel JJ, Bressler P, Schuster P, Miranda R, Kirschner J. 1990. *J. Electron. Spectrs.* 51: 263–74
92. Brookes NB, Clarke A, Johnson PD. 1992. *Phys. Rev. B* 46: 237–41
93. Rader O, Vescovo E, Redinger J, Blugel S, Carbone C, et al. 1994. *Phys. Rev. Lett.* 72: 2247–50
94. Huang H, Hermanson J, Gray JG, Richter R, Smith JR. 1986. *Surf. Sci.* 172: 363–71
95. Grunberg P, Schreiber R, Pang Y, Brodsky MB, Sowers H. 1986. *Phys. Rev. Lett.* 57: 2442–45
96. Parkin SSP, More N, Roche KP. 1990. *Phys. Rev. Lett.* 64: 2304–7
97. Unguris J, Celotta RJ, Pierce DT. 1991. *Phys. Rev. Lett.* 67: 140–43
98. Baibich MN, Broto JM, Fert A, Nguyen Van Dau F, Petroff F, et al. 1988. *Phys. Rev. Lett.* 61: 2472–75
99. Victora RH, Falicov LM. 1985. *Phys. Rev. B* 31: 7335–43
100. Fu Cl, Freeman AJ, Oguchi T. 1985. *Phys. Rev. Lett.* 54: 2700–3
101. Johnson PD, Chang Y, Brookes NB. 1993. *Proc. 10th VUV Conf. Vacuum Ultraviolet Radiation Physics*, ed. FJ Wuilleumier, Y Petrof, I Nenner, pp. 453–64. Singapore: World Scientific
102. Hillebrecht FU, Roth Ch, Jungblut R, Kisker E, Bringer A. 1992. *Europhys. Lett.* 19: 711–16
103. Pierce DT, Stroscio JA, Unguris J, Celotta RJ. 1994. *Phys. Rev. B* 49: 14564–72
104. Vescovo E, Rochow R, Kachel T, Carbone C. 1992. *Phys. Rev. B* 46: 4788–94
105. Taborelli M, Allenspach R, Boffa G, Landolt M. 1986. *Phys. Rev. Lett.* 56: 2869–72
106. Parkin SSP. 1991. *Phys. Rev. Lett.* 67: 3598–601
107. Bruno P, Chappert C. 1991. *Phys. Rev. Lett.* 67: 1602–5
108. Johnson MT, Purcell ST, Mcgee NWE, Coehoorn R, ann de Stegge J, Hoving W. 1992. *Phys. Rev. Lett.* 68: 2688–91
109. Loly PD, Pendry JB. 1983. *J. Phys. C* 16: 423–31
110. Brookes NB, Chang Y, Johnson PD. 1991. *Phys. Rev. Lett.* 67: 354–57
111. Garrison K, Chang Y, Johnson PD. 1993. *Phys. Rev. Lett.* 71: 2801–4
112. Carbone C, Vescovo E, Rader O, Gudat W, Eberhardt W. 1993. *Phys. Rev. Lett.* 71: 2805–8
113. Samant MG, Stohr J, Parkin SSP, Held GA, Hermsmeier BD, et al. 1994. *Phys. Rev. Lett.* 72: 1112–15
114. Li D, Pearson J, Mattson JE, Bader SD, Johnson PD. 1995. *Phys. Rev. B.* Submitted
115. Ortega JE, Himpsel FJ, Mankey GE, Willis RF. 1993. *Phys. Rev. B* 47: 1540–52
116. Smith NV, Brookes NB, Chang Y, Johnson PD. 1994. *Phys. Rev. B* 49: 332–38
117. Bruno P. 1995. *Phys. Rev. Lett.* Submitted
118. Johnson PD, Garrison K, Dong Q, Smith NV, Li D, et al. 1994. *Phys. Rev. B* 50: 8954–56
119. Schneider CM, Hammond MS, Schuster P, Cebollada A, Miranda R, Kirschner J. 1991. *Phys. Rev. B* 44: 12066–69
120. Bansmann J, Getzlaff M, Westphal C, Fegel F, Schonhense G. 1992. *Surf. Sci.* 269/270: 622–26

Annu. Rev. Mater. Sci. 1995. 25: 487–523

FULLERENES AND FULLERENE-DERIVED SOLIDS AS ELECTRONIC MATERIALS

M. S. Dresselhaus

Department of Electrical Engineering and Computer Science, and Department of Physics, Massachusetts Institute of Technology, Cambridge, Massachusetts 02139

G. Dresselhaus

Francis Bitter National Magnet Laboratory, Massachusetts Institute of Technology, Cambridge, Massachusetts 02139

KEY WORDS: graphite, carbon, intercalation, molecular solids, organic magnets

---

ABSTRACT

The status of our present understanding of fullerenes and fullerene-derived solids as electronic materials is reviewed, with particular emphasis given to the strong connection of these electronic materials in the solid state to their molecular fullerene origins. The synthesis, structure, and electronic properties of fullerene solids are considered, and modifications to their structure and properties through doping with various charge transfer agents are reviewed. We conclude with brief comments on potential applications of this unique family of new electronic materials.

INTRODUCTION

Carbon is unique as an electronic material. It can be a good metal in the form of graphite, a wide gap semiconductor in the form of diamond or

0084-6600/95/0801-0487$05.00

fullerenes, a superconductor when intercalated with appropriate guest species, or a polymer when reacted with hydrogen and other species. Furthermore, carbon-based electronic materials provide examples of materials showing the entire range of dimensionalities from fullerenes, which are zero-dimensional quantum dots, to carbon nanotubes, which are one-dimensional quantum wires, to graphite, a two-dimensional layered anisotropic material, and finally to diamond, a three-dimensional wide gap semiconductor. In this family of electronic materials, fullerenes and carbon nanotubes are the newest additions and the subject of this review.

Fullerene solids differ from conventional electronic materials because, in common with most polymeric materials, the fullerene molecule is the fundamental building block of the crystalline phase. This basic difference in structure is reflected in the electronic structure and properties that are summarized here. The modifications to the structure and electronic properties produced by alkali-metal doping are then reviewed. Potential applications for fullerene-based electronic materials are suggested.

In early gas phase work, a molecule with 60 carbon atoms was established experimentally by mass spectrographic analysis (1), and it was conjectured that C_{60} had icosahedral symmetry. The name of fullerene was given to this family of gas phase molecules by Kroto & Smalley (1) because of the resemblance of these molecules to the geodesic dome designed and built by R. Buckminster Fuller (2). The name buckminster fullerene or simply "buckyball" was given specifically to the C_{60} molecule. In the early gas phase work, the fullerene molecules were produced by the laser vaporization of carbon from a graphite target in a pulsed jet of helium (1, 3).

In the fall of 1990, a new type of condensed matter, based on C_{60}, was synthesized for the first time by Krätschmer, Huffman, and co-workers (4), who found a simple carbon arc method for preparing gram quantities of C_{60}, which had previously only been available in trace quantities in the gas phase (1, 5). The availability of large quantities of fullerenes provided a great stimulus to this research field. It was soon found (6) that the intercalation of alkali-metals into C_{60} to a stoichiometry M_3C_{60} (where $M = K$, Rb, Cs) greatly modifies the electronic properties of the host fullerenes and yields relatively high transition temperature ($18 \leqslant T_c \leqslant 40$ K) superconductors (7). The discovery of superconductivity (8, 9) in these compounds further spurred research activity in this field of C_{60}-related materials.

This article reviews synthesis and doping techniques for fullerenes, the structure of fullerene molecules and of their crystalline forms, their electronic structure and optical properties, the results of carrier transport experiments, the superconducting properties of doped fullerenes, and possible applications of fullerenes.

SYNTHESIS OF FULLERENE SOLIDS

Fullerenes are usually synthesized using an ac discharge between graphite electrodes in approximately 200 torr of He gas. The heat generated at the contact point between the electrodes evaporates carbon to form soot and fullerenes, which condense on the water-cooled walls of the reactor. This discharge produces a carbon soot that can contain up to $\sim 15\%$ fullerenes: C_{60} ($\sim 13\%$) and C_{70} ($\sim 2\%$). The fullerenes are next separated from the soot, according to their mass, by use of liquid chromatography (LC) and using a solvent such as toluene.

Property measurements of fullerenes are made either on powder samples, films, or on single crystals. C_{60} powder is obtained by vacuum evaporation of the solvent from the solution. Pristine C_{60} films can be deposited onto a substrate such as a clean silicon (100) surface by sublimation of the C_{60} powder. Single crystals can be grown either from solution using solvents such as CS_2 and toluene or by vacuum sublimation (10), but the sublimation method (in vacuum or in an inert gas such as argon) yields better crystals and is the method of choice.

Doping is used to modify the properties of fullerenes, particularly their electronic properties. Although fullerene solids (called fullerites) can be doped in three ways (endohedrally, substitutionally, and exohedrally), the exohedral doping is of primary interest for fullerenes considered as electronic materials. Endohedral doping denotes the addition of an ion of rare-earth, alkaline-earth, or alkali-metal into the interior of the C_{60} molecule to form an endohedrally doped molecular unit. This configuration has, for example, been denoted by La@C_{60} for one endohedral lanthanum in C_{60}, or Y_2@C_{82} for two Y atoms inside a C_{82} fullerene (11). Thus far, only small quantities of endohedrally doped fullerenes have been prepared, and only limited investigations of endohedrally doped crystalline materials have been reported (12), and almost nothing of relevance to the electronic properties of these materials has been reported.

A second doping method is the substitution of an impurity atom with a different valence state for a carbon atom on the surface of a C_{60} molecule. The only species that can be expected to substitute for a carbon atom on the C_{60} molecule surface is boron, although there has been some discussion of the possibility of nitrogen doping, which might be facilitated by the curvature of the fullerene shell. Nevertheless, substitutional doping has not been widely used for the preparation of electronic materials (13).

The most common method of doping fullerene solids is exohedral donor doping (also called intercalation). In this case, the dopant (e.g. an alkali-metal or an alkaline-earth, M) is introduced into the interstitial positions between adjacent molecules (exohedral locations). Charge transfer can take place between the M atoms and the molecules so that the M atoms become positively charged ions and the molecules become negatively charged with predominately delocalized electrons. With exohedral doping, the conductivity of fullerene solids can be increased by many orders of magnitude (14).

Among the alkali-metals, Li, Na, K, Rb, and Cs and their alloys have been used as exohedral dopants for C_{60} [15, 16]. The doping of C_{60} with alkali-metals can be achieved in a two-temperature oven, similar to the apparatus used to prepare alkali-metal graphite intercalation compounds (14, 17). Some success has also been achieved with the intercalation of alkaline-earth dopants, such as Ca, Sr, and Ba (18, 10, 20). In the case of the alkaline-earth dopants, two electrons per metal atom M are transferred to the C_{60} molecules for low concentrations of metal atoms, whereas for high concentrations of metal atom, the charge transfer is less than two electrons per alkaline-earth ion. In general, the alkaline-earth ions are smaller than the corresponding alkali-metals in the same row of the periodic table. For this reason, the crystal structures formed with alkaline-earth doping are often different from those for the alkali-metal dopants. Doping fullerenes with acceptors has been more difficult than with donors because of the high electron affinity of C_{60} (see Table 1). Although stable compounds with acceptor-type dopants have been synthesized, acceptor doping has not been important for the synthesis of electronic materials.

Fullerene chemistry has become a very active research field, largely because of the uniqueness of the C_{60} molecule and the variety of chemical reactions that appear to be possible (37, 38). Many compounds have already been synthesized and characterized chemically, a few crystal structures have been determined, but almost no study of their electronic properties has been reported until now.

STRUCTURE OF FULLERENES

Because the structure and properties of fullerene solids are strongly dependent on the structure and properties of the constituent fullerene molecules, we first review the structure of the molecules, followed by a review of the structure of these molecular solids.

Structure of C_{60}

To a good approximation, the 60 carbon atoms in C_{60} are located at the vertices of a regular truncated icosahedron. Every carbon site on the C_{60} molecule is equivalent to every other site (see Figure 1*a*), consistent with a single sharp line in the NMR spectrum (39, 40). The average nearest neighbor carbon-carbon (C-C) distance $a_{C\text{-}C}$ in C_{60} (1.44 Å) is almost

Table 1 Physical constants for C_{60} materials

Quantity	Value	Reference
Average C-C distance	1.44 Å	21
fcc lattice constant	14.17 Å	18
C_{60} mean molecule diameter	7.10 Å	22
C_{60}-C_{60} distance	10.02 Å	18
Tetrahedral site radius	1.12 Å	22
Octahedral site radius	2.07 Å	22
Mass density	1.72 g/cm^3	22
Molecular density	$1.44 \times 10^{21}/cm^3$	22
Compressibility ($-d \ln V/dP$)	$6.9 \times 10^{-12} cm^2/dyn$	23
Bulk modulus	14 GPa	24
Structural phase transition	260 K	25
Binding energy per atom	7.40 eV	26
Electron affinity (pristine C_{60})	2.65 eV	27
Ionization potential (1st)	7.58 eV	28
Ionization potential (2nd)	11.5 eV	29
Volume coefficient of thermal expansion	$6.2 \times 10^{-5}/K$	30
Bandgap (HOMO-LUMO)[a]	1.7 eV	31
Spin-orbit splitting of C(2p)	0.00022 eV	32
Velocity of sound v_t	2.1×10^5 cm/s	33
Velocity of sound v_l	3.6×10^5 cm/s	33
Thermal conductivity (300 K)	0.4 W/mK	34
Phonon mean free path	50 Å	34
Static dielectric constant	4.0–4.5	35, 36

[a] Strictly speaking, it is the optical absorption edge that is at ~1.7 eV.

identical to that in graphite (1.42 Å). Each carbon atom in C_{60} (and also in graphite) is trigonally bonded to other carbon atoms, and most of the faces on the regular truncated icosahedron are hexagons. Thus many have considered the C_{60} molecule as a rolled-up graphene sheet (a single layer of crystalline graphite). The regular truncated icosahedron has 20 hexagonal faces and 12 additional pentagonal faces to form a closed shell, in keeping with Euler's theorem, which states that a closed surface consisting of hexagons and pentagons has exactly 12 pentagons and an arbitrary number of hexagons (13).

The symmetry operations of the icosahedron consists of the identity operation: 6 fivefold axes through the centers of the pentagonal faces, 10 threefold axes through the centers of the hexagonal faces, and 15 twofold axes through centers of the edges joining two hexagons. Each of the 60 rotational symmetry operations can be compounded with the inversion operation, resulting in 120 symmetry operations in the icosahedral point group I_h (41) (see Figure 1a). Molecules with I_h symmetry, C_{60} being the

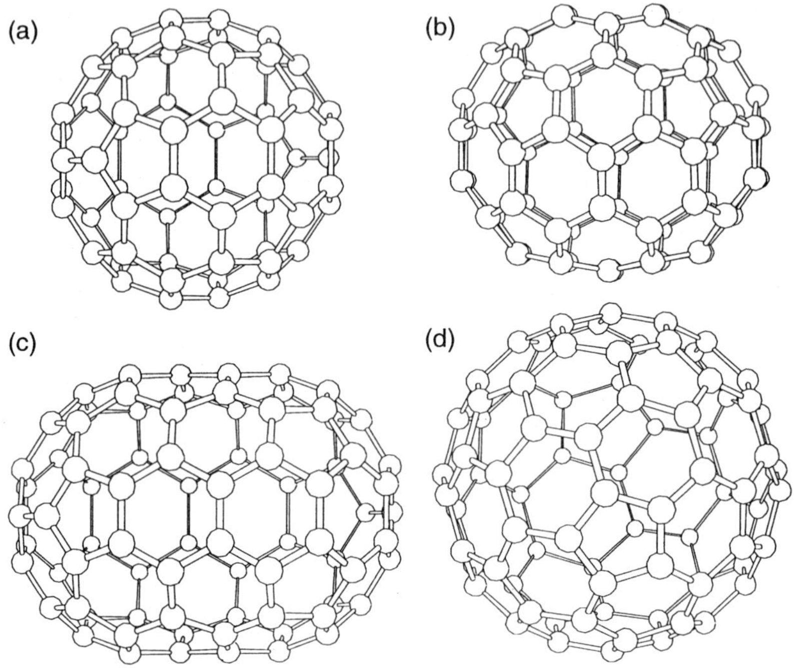

Figure 1 (*a*) The icosahedral C_{60} molecule. (*b*) The C_{70} molecule as a rugby ball-shaped molecule. (*c*) The C_{80} molecule as an extended rugby ball-shaped molecule. (*d*) The C_{80} molecule as an icosahedron.

most prominent example, have the highest degree of symmetry of any known molecule. The diameter of the C_{60} molecule is 7.10 Å (see Table 1), treating the carbon atoms as points (1, 22). When taking account of the size of the π-electron cloud associated with the carbon atoms, the outer diameter of the C_{60} molecule becomes 10.18 Å.

From Euler's theorem it follows that the smallest possible fullerene is C_{20}, which would form a regular dodecahedron with 12 pentagonal faces. However, it is considered energetically unfavorable for two pentagons to be adjacent to each other (referred to as the isolated pentagon rule) as this would lead to higher local curvature and high strain on the fullerene molecule. Because the addition of a single hexagon adds two carbon atoms, all fullerenes must have an even number of carbon atoms, in agreement with the observed mass spectra for fullerenes (3).

Although each carbon atom in C_{60} is equivalent to every other carbon atom, the three bonds emanating from each atom are not completely equivalent. Each of the four valence electrons of the carbon atoms are engaged in covalent bonds, so that two of the three bonds are electron-poor single bonds, and one is an electron-rich double bond. Consistent with X-ray diffraction evidence, the structure of C_{60} is further stabilized by introducing a small distortion of the bond lengths to form the Kekulé structure of alternating single and double bonds around the hexagonal face. We note that the icosahedral I_h symmetry is preserved under these distortions. The single bonds that define the pentagonal faces are increased from the average bond length of 1.44 Å to 1.46 Å, whereas the double bonds are decreased to 1.40 Å (23, 42). Because each carbon atom has its valence requirements fully satisfied, the C_{60} molecules are expected to form a van der Waals bonded solid, which is nonconducting (an insulator or a semiconductor).

Structure of C_{70} and Higher Fullerenes

In the synthesis of C_{60}, larger molecular weight fullerenes C_n ($n > 60$) are also formed, by far the most abundant being C_{70}. However, significant quantities of C_{76}, C_{78}, C_{84}, and higher mass fullerenes have also been isolated and studied in some detail.

C_{70} has been found to exhibit a rugby ball shape (43), and its form can be envisioned by adding either a ring of ten carbon atoms or a belt of five hexagons around the equatorial plane of the C_{60} molecule normal to one of the fivefold axes (see Figure 1*b*). In contrast to the C_{60} molecule with I_h symmetry, the C_{70} molecule has the lower symmetry D_{5h}, which is a subgroup of I (lacking inversion symmetry). Fullerenes often form isomers, because a given number n of carbon atoms C_n can correspond to molecules with different geometric structures (43, 44). For example, C_{80} might be formed in the shape of an elongated rugby ball prepared by adding two rows of five hexagons normal to a fivefold axis of C_{60} at the equator (see Figure 1*c*); an icosahedral form of C_{80} can also be specified, as shown in Figure 1*d*. C_{78}, another example of fullerene isomers, has been shown to have five distinct isomers, one with D_3 symmetry, two with C_{2v} symmetry and two with D_{3h} symmetry (45). In addition, isomers giving rise to right- and left-handed optical activity are expected to occur in molecules belonging to group I (rather than I_h), and not to have inversion symmetry.

CRYSTALLINE STRUCTURE

Much progress has been made in establishing the crystalline structure of C_{60} and C_{70}, and various phases have been identified as a function of temperature. Less is known about the crystal structure of higher fullerenes, and this information is now becoming available. With regard to the doped fullerenes, the most complete information is available for alkali-metal intercalated C_{60}, although structural information on other doped fullerene solids (e.g. alkaline-earth-doped fullerene solids) is also becoming available. Knowledge of the structure is, of course, necessary to gain a microscopic understanding of the electronic properties of fullerene solids.

Crystalline C_{60}

In the solid state, the C_{60} molecules crystallize into a cubic structure with lattice constant of 14.17 Å, a nearest neighbor C_{60}-C_{60} distance of 10.02 Å (22), and a density of 1.72 g/cm^3 (corresponding to 1.44×10^{21} C_{60} molecules/cm^3). At room temperature, the C_{60} molecules rotate rapidly with three degrees of rotational freedom, and the molecules are arranged on a face-centered cubic (fcc) lattice with one C_{60} molecule per primitive fcc unit cell, or four molecules per simple cubic unit cell (see Figure 2*a*). The pertinent space group is O_h^5 or $Fm\bar{3}m$, which has been determined by X-ray and neutron diffraction (30, 47, 48). Relative to the other allotropic

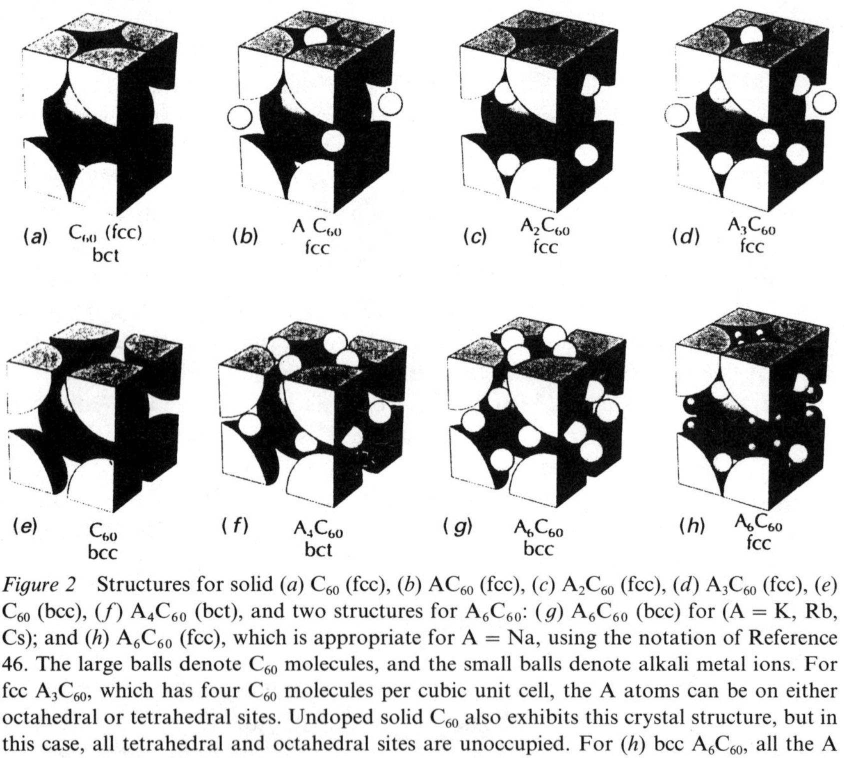

Figure 2 Structures for solid (*a*) C_{60} (fcc), (*b*) AC_{60} (fcc), (*c*) A_2C_{60} (fcc), (*d*) A_3C_{60} (fcc), (*e*) C_{60} (bcc), (*f*) A_4C_{60} (bct), and two structures for A_6C_{60}: (*g*) A_6C_{60} (bcc) for (A = K, Rb, Cs); and (*h*) A_6C_{60} (fcc), which is appropriate for A = Na, using the notation of Reference 46. The large balls denote C_{60} molecules, and the small balls denote alkali metal ions. For fcc A_3C_{60}, which has four C_{60} molecules per cubic unit cell, the A atoms can be on either octahedral or tetrahedral sites. Undoped solid C_{60} also exhibits this crystal structure, but in this case, all tetrahedral and octahedral sites are unoccupied. For (*h*) bcc A_6C_{60}, all the A atoms are on distorted tetrahedral sites. For (*f*) bct A_4C_{60}, the dopant is also found on distorted tetrahedral sites. For (*c*) pertaining to small alkali metal ions such as Na, only the tetrahedral sites are occupied. For (*h*), four Na ions can occupy an octahedral site of this fcc lattice. (Throughout the figure caption A ≡ M.)

forms of carbon, solid C_{60} is relatively compressible, with an isothermal volume compressibility (see Table 1) of 6.9×10^{-12} cm^2/dyn (23).

Below about 260 K, the C_{60} molecules completely lose two of their three degrees of rotational freedom, and the residual rotational motion, described as a ratcheting motion (30, 48–50), for each of the four molecules within the unit cell is about a different $\langle 111 \rangle$ axis. The structure of solid C_{60} below ~260 K thus becomes simple cubic (space group T_h^6 or $Pa\bar{3}$), with a lattice constant $a_0 = 14.17$ Å and four C_{60} molecules per unit cell, as the four molecules within the fcc structure become inequivalent (see Figure 2*a*) (30, 48). Supporting evidence for the 260 K phase transition is provided by many property measurements (13). As the temperature is lowered below 260 K, further ordering of the C_{60} molecules occurs, whereby adjacent C_{60} molecules develop correlated orientations. The degree of order achieved in the crystal structure at low temperature can be understood by referring to Figure 3*a*, where the standard orientation of a C_{60} molecule is shown with respect to a cubic coordinate system (51). Here the [100] axes pass through three, mutually orthogonal twofold molecular axes (the centers of the electron-rich hexagon-hexagon edges), and four $\langle 111 \rangle$ axes pass through the centers of the hexagonal faces. The alternate standard orientation is shown in Figure 3*b*, which is obtained from Figure 3*a* through rotation by $\pi/2$ about the [100] axes. Because of the equivalence of the two standard orientations, the C_{60} molecules have an approximately equal probability to be in either of these orientations,

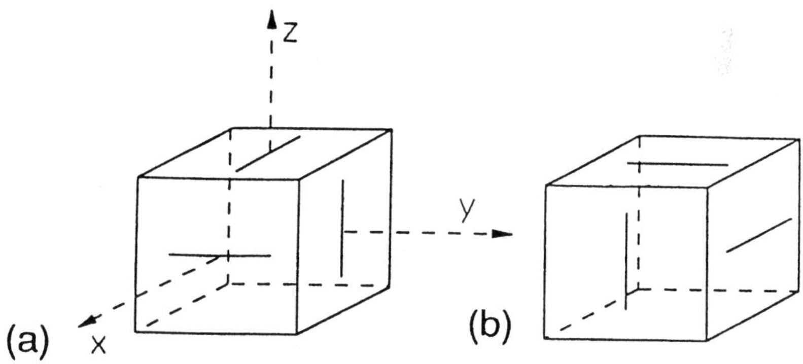

Figure 3 (*a*) One standard orientation for the Cartesian axes in a cubic crystal so that these axes pass through three orthogonal twofold axes. (*b*) The other standard orientation related to that in (*a*) by a 90° rotation about any of the Cartesian axes. The inscribed icosahedron has mirror planes perpendicular to the three $\langle 100 \rangle$ directions. In this orientation, six of the icosahedral edges lie in planes parallel to the mirror planes so that when the icosahedron is circumscribed by a minimal cube, these edges lie in the cube faces, as shown. Each $\langle 111 \rangle$ direction is a threefold axis (51).

leading to what is called merohedral disorder with regard to alignment of the twofold axes. Merohedral disorder persists to the lowest temperatures and contributes to carrier scattering in transport properties. In the low-temperature structure, the relative orientation of adjacent molecules is stabilized by aligning an electron-rich double bond on one molecule opposite the electron-poor pentagonal face of an adjacent C_{60} molecule (see Figure 4*a*) to achieve a minimum in the orientational potential that governs the alignment.

Another structure with only slightly higher energy places the electron-rich double bond on one C_{60} molecule opposite an electron-poor hexagonal face (see Figure 4*b*). This orientation can be achieved from the lower energy orientation described above by rotation of the C_{60} molecule by 60° around a $\langle 111 \rangle$ axis (49). As the temperature T is lowered below 260 K, the probability of occupying the lower energy configuration increases (49). Evidence for residual structural disorder in the low-temperature molecular alignment is found from neutron scattering experiments (49) and many

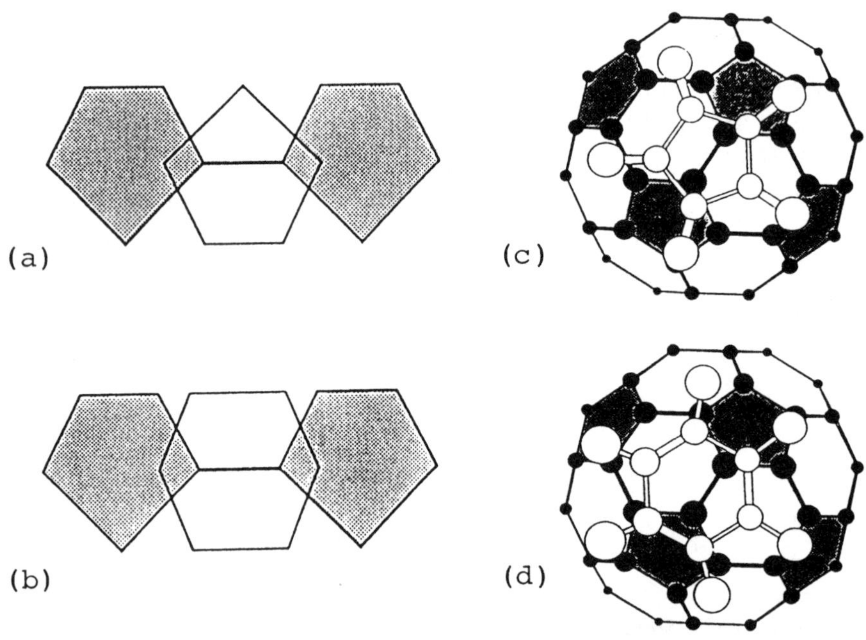

Figure 4 Electron-rich double bond on one C_{60} molecule opposite an electron-poor (*a*) pentagonal face, (*b*) hexagonal face on the adjacent molecule (51): In view (*c*), centers of nearest-neighbor C_{60} molecules follow arrangement (*a*). In view (*d*), rotation of (*c*) by 60° about a threefold axis brings the hexagonal face of one molecule adjacent to the twofold axis of a second molecule (50).

others (13). The mechanism by which partial orientational alignment is achieved is by the ratcheting motion of the C_{60} molecules around the $\langle 111 \rangle$ axes as they execute their hindered rotational motion. The ratcheting motion begins below the 260 K phase transition and continues down to low temperatures.

Crystalline C_{70} and Higher Fullerenes

The crystal structure of C_{70} is more complex than that of C_{60} (52–56), and has been studied in much detail. Less detailed structural information is available for the higher mass fullerenes. At high temperature ($T \gg 340$ K), the fcc phase ($a = 15.01$ Å) of C_{70} with freely rotating molecules is most stable, but because the ideal hexagonal close-packed (hcp) phase with $c/a = 1.63$ is almost equally stable, fcc crystals of C_{70} tend to be severely twinned and show many stacking faults. A transition to another hcp with $a = b = 10.11$ Å and a larger c/a ratio of 1.82 occurs at $\sim$340 K. This larger c/a ratio is associated with the orientation of the C_{70} molecules along their long axis, as the free molecular rotation (full rotational symmetry) that is prevalent in the higher temperature phase goes into a rotation about the fivefold axis of the C_{70} molecule (53, 56). As the temperature is lowered further to $\sim$280 K, the free rotation about the c-axis also becomes frozen, resulting in a monoclinic structure with the unique axis along the c-axis of the hcp structure.

The higher mass fullerenes (C_{76}, C_{84}) have also been shown to crystallize in the fcc structure at room temperature, with an fcc lattice constant that is proportional to n, the number of carbon atoms in the fullerene (57).

Doped C_{60} Crystals

Several stable crystalline phases for exohedrally doped or intercalated C_{60} have been identified. Most widely studied are the crystalline phases formed by intercalation of alkali-metals, although some structural reports have been given for fullerene-derived crystals with alkaline-earth dopants (18). We focus attention only on doped C_{60} materials where charge transfer occurs, and not on clathrate C_{60} compounds where there is no significant charge transfer, and where the electronic properties are of lesser interest.

When C_{60} is doped with the alkali-metals (M = K, Rb, Cs), stable crystalline phases are formed for the compositions M_1C_{60}, M_3C_{60}, M_4C_{60}, and M_6C_{60} (13, 46, 58, 59). These stable phases as illustrated in Figure 5, were experimentally determined for Rb-doped C_{60} (60). This phase diagram shows stability regions for the rock-salt phase of Rb_1C_{60}, the fcc phase of Rb_3C_{60}, the bct phase of Rb_4C_{60} and the bcc phase of Rb_6C_{60}. The stable regions are illustrated in the hatched areas of this binary phase diagram (60).

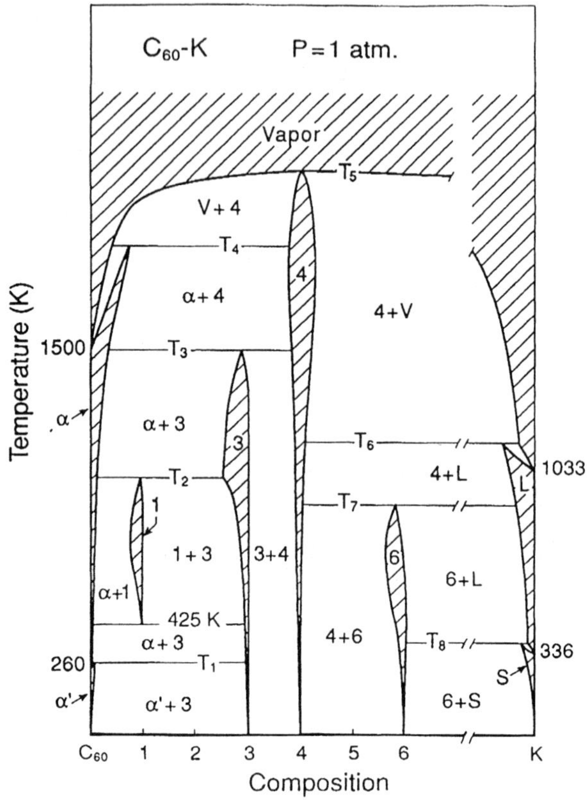

Figure 5 Provisional binary phase diagram for K_xC_{60} at 1 atm pressure. Experimentally determined temperatures are indicated; unknown temperatures of predicted transformations are labeled $T_1, \ldots, T_8$ (60).

Charge transfer of one electron per M atom to the C_{60} molecule occurs, resulting in M^+ ions at the tetrahedral and/or octahedral interstices of the cubic C_{60} host structure. For the composition M_3C_{60}, the resulting metallic crystal has basically the fcc structure (see Figure 2). Within this structure, the alkali-metal ions can sit on tetragonal (1/4,1/4,1/4) sites, which are twice as numerous as the octahedral (1/2,0,0) sites (referenced to a simple cubic coordinate system). The electron-poor alkali-metal ion tends to lie adjacent to a C=C double bond, thereby orienting the C_{60} molecules somewhat and restraining their rotational degrees of freedom (61). From the standpoint of the near neighbors, the K guest species in K_3C_{60} sit between two C_{60} molecules, on any one of four tetrahedral sites or two octahedral sites. For the tetrahedral sites, the radius of the cavity available to a K^+ ion is 1.12 Å, whereas for an octahedral site, the radius is 2.07 Å (22). The structure of the doped C_{60} shows different orientational alignments because of differences in the orientational potentials experienced by the fullerenes due to alkali-metal ions, which are nearest neighbors to the C_{60} ions in the $\langle 111 \rangle$ direction where the nearest neighbor C_{60}-C_{60} approach is along the $\langle 110 \rangle$ direction. The shortest distance between a K^+ ion at a tetrahedral site and a carbon atom on a freely rotating C_{60} molecule is 2.66 Å (22). It is of interest to note that the Cs^+ ion is too large to form a stable Cs_3C_{60} compound with the $Fm\bar{3}m$ structure (47) unless pressure is applied (9).

Because the ionic radii of the alkali-metal ions Na^+ and Li^+ are much smaller than the radii of the K^+, Rb^+, and Cs^+ ions, a number of unusual structural phenomena might be expected. For example, it is possible to fit more than one Na^+ in an octahedral site (15, 62); thus the cubic $Fm\bar{3}m$ structure can be preserved upon adding up to 11 Na atoms, with two Na ions going into tetrahedral sites (see Figure 2*c*) and up to nine going into octahedral sites (see Figure 2*h*) to form $Na_{11}C_{60}$. The maximum stability for four Na atoms on an octahedral site is at the corners of a regular tetrahedron with respect to the center of the octahedral sites (15), and it has been proposed that nine guest Na ions go on the cube edges and the center of an octahedral site. Materials prepared with Na_3C_{60} stoichiometry tend to phase separate (disproportionate) into Na_2C_{60} (Figure 2*c*) and Na_6C_{60} (Figure 2*h*) (15). Also because of the small ionic radius of Na, the Na^+ ions tend to fill the smaller tetrahedral sites when alloys of Na with heavier alkali-metals are made, whereas the larger alkali-metals tend to be on octahedral sites. One unusual effect is that alkali-metal doping with Na_2Cs, Na_2Rb, Na_2K, and Li_2Cs results in a crystal with a closer C_{60}-C_{60} separation than is present in undoped C_{60}, presumably due to the attractive interaction between the metal ions and the fullerene molecules (63).

The structure of the alkaline-earth metal compound Ca_yC_{60} (for $y \leqslant 5$) follows the same space group $Fm\bar{3}m$ as for the heavy alkali-metal M_xC_{60} compounds ($x \leqslant 3$) (18), and the Ca atoms occupy both tetrahedral and octahedral sites. Because of the smaller size of the calcium ion [as noted above for the Na^+ ion (15)], the octahedral sites can accommodate multiple Ca ions, and it is believed that up to three Ca ions can be accommodated in a single octahedral site (18). Also for Ca_5C_{60}, the lattice constant $a_0 = 14.01$ Å is smaller than that for C_{60} prior to doping. The structure of Ba_6C_{60} and Sr_6C_{60} has been investigated by X-ray powder diffraction techniques, showing A15 and other *bcc* phases (64, 19). Unlike alkali-metal-doped C_{60}, where each M atom transfers one electron to C_{60}, the alkaline-earth metal dopants have been reported to transfer two electrons per metal dopant for low metal concentrations but have lower charge transfer values for higher calcium concentrations (18). Band structure calculations of the alkaline-earth compounds (65) have been helpful for determining the charge state of the metal ions in these intercalation compounds as a function of metal concentration.

The high electro-negativity of pristine C_{60} hinders acceptor doping (66). Consequently, relatively few C_{60}-based acceptor compounds have been synthesized, and those that have been synthesized have not yet shown interesting electronic properties.

ELECTRONIC STRUCTURE

Because of weak intermolecular forces, the fullerenes form molecular solids. Thus their electronic structures are expected to be closely related to the electronic levels of the isolated molecules. As seen in Figure 1*a*, each carbon atom in C_{60} has two single bonds along adjacent sides of a pentagon and one double bond between two adjoining hexagons, thereby accounting for the bonding of the four valence electrons for carbon, which indicates that fullerenes should be insulators. If these bonds were coplanar, they would be very similar to the sp^2 trigonal bonding in graphite. The curvature of the C_{60} surface causes the planar-derived trigonal orbitals to hybridize, thereby admixing some sp^3 character to the sp^2 bonding. The shortening of the double bonds and lengthening of the single bonds in the Kekulé arrangement of the C_{60} molecule also strongly influence the electronic structure.

The most extensive calculations of the electronic structure of fullerenes so far have been done for C_{60}. To obtain the molecular orbitals for C_{60}, the multi-electron problem for the free molecule must first be solved (67). Representative results for the free C_{60} molecule are shown in Figure 6*a* (68). Because of the molecular nature of solid C_{60}, the electronic structure for the solid phase would be expected to be closely related to that of the free molecule (69). A molecular approach is especially useful for explaining a large body of experimental observations pertinent to the optical properties of C_{60}.

Nevertheless, until recently more attention had been given to a band structure approach, such as shown in Figure 6*b* for C_{60}, where we see results for fcc solid, based on total energy calculations using norm-conserving pseudopotentials, a local density approximation in density functional theory, and a Gaussian orbital basis set (68). The various band calculations for C_{60} (66, 68) yield a narrow band ($\sim$0.4–0.6 eV bandwidth) solid, with a highest occupied molecular orbital-lowest unoccupied molecular orbital (HOMO-LUMO)-derived direct band gap of $\sim$1.5 eV at the X point of the fcc Brillouin zone. The narrow energy bands and the molecular nature

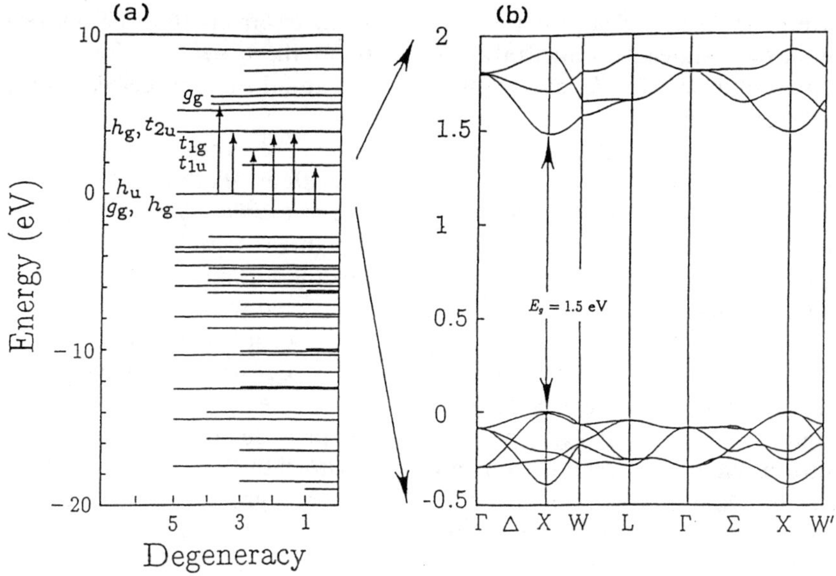

Figure 6 Calculated electronic structure of (*a*) an isolated C_{60} molecule and (*b*) fcc solid C_{60}, where the direct band gap at the *X*-point is 1.5 eV (68).

of the electronic structure of fullerenes are indicative of a highly correlated electron system. Nevertheless, band calculations have been very helpful for describing the electronic properties of fullerene-related solids. Because the HOMO and LUMO levels have the same odd-parity, electric dipole transitions between these levels are symmetry-forbidden in the free C_{60} molecule. In the solid, transitions between the direct bandgap states at the Γ and *X* points in the cubic Brillouin zone are also forbidden, but direct interband transitions are allowed at the lower symmetry points in the Brillouin zone. The allowed electric dipole transitions for the free C_{60} molecule are indicated by arrows in Figure 6*a*.

According to the molecular approach, many-electron-orbitals for the valence and conduction electron states are constructed for the π-electrons on the C_{60} molecule. The lower-lying σ orbitals are filled and are not generally probed by optical experiments. The number of π-electrons on the C_{60} molecule is 60 (i.e. one state per carbon atom), which is exactly the correct number to fully occupy the h_u (HOMO) level and specifies its symmetry in terms of the icosahedral group I_h. In relating the icosahedral levels to those of full rotational symmetry, we note that 50 π-electrons fully occupy the angular momentum states through $l = 4$, and the remaining 10 electrons are available to start filling the $l = 5$ state. In full spherical symmetry, the $l = 5$ state can accommodate 22 electrons, and the splitting of the $l = 5$ state in icosahedral symmetry is into the $h_u + t_{1u} + t_{2u}$ irreducible representations. The h_u level is fully occupied and the two threefold t_{1u} and t_{2u} levels are empty.

Total energy calculations on K_xC_{60} show that the tetrahedral sites fill up before the octahedral sites (68). Doping C_{60} with an alkali-metal transfers electrons to the LUMO levels, which because of their t_{1u} symmetry (Figure 6) can accommodate three spin-up and three spin-down electrons. Assuming one electron to be transferred to the C_{60} molecule per alkali-metal atom dopant, the LUMO levels are expected to be half occupied at the alkali-metal stoichiometry M_3C_{60} and fully occupied at M_6C_{60}, thereby leading to a filled shell configuration. Thus M_6C_{60} would be expected to be semiconducting, with a band gap between the t_{1u} and t_{2g} levels (see Figure 6), whereas M_3C_{60} should be metallic provided that no band gap is introduced at the Fermi level by a Peierls distortion. Because the C_{60}^{3-} ion has fourfold degenerate ground state [double group irreducible representation Γ_8^- (70)] in icosahedral symmetry, a Jahn-Teller distortion would be expected so that a nondegenerate ground state could be achieved. Electron paramagnetic spectra for C_{60}^{3-} molecular ions show an anisotropic g-factor and line width supporting a Jahn-Teller distortion (71). The large increase in optical absorpitivity in M_xC_{60} with alkali-metal doping is observed as $x \rightarrow 3$, and this observation supports the metallic nature of M_3C_{60} (see Optical Properties), in agreement with the transport properties of M_xC_{60} discussed in the section on Electrical Transport. Below ~ 0.5 eV, the free electron contribution to the optical properties is dominant, and above ~ 0.5 eV, interband transitions from the partially occupied t_{1u} level to the higher lying t_{1g} level take place. It should be emphasized that the level-filling arguments for M_xC_{60} pertain to the free molecule as well as to the levels in the solid state. Because of the high electro-negativity of C_{60}, it is not likely that holes would be created in the h_u-derived C_{60} HOMO level, also in agreement with experiment (42).

Fermi surface calculation for K_3C_{60} indicate a hole Fermi surface around the Γ-point that contributes about 12% to the density of states at the Fermi surface (72, 73). These calculations also indicate a larger and more complicated Fermi surface consisting of both electron and hole orbits that contribute the remaining 88% to the density of states. Because of the weak interaction of the molecules with each other and with the alkali-metal dopants, solid M_3C_{60} (although showing transport properties consistent with a one-electron band model) is close to being a molecular solid having energy levels with little dispersion, thus giving rise to a very high density of states near the Fermi level. This property is important in understanding superconductivity in M_3C_{60} (discussed below). The total density of states at E_F is estimated to be $\sim 13/\text{eV}/C_{60}$ molecule per spin state (73, 74). Calculated values for the effective mass m^*, based on a band model, yield $1.3m_e$ (where m_e is the free electron mass) for the conduction band of C_{60} and 1.5 and $3.4m_e$ for the valence bands (66).

Photoemission and inverse photoemission experiments have been especially useful (58) in providing information on the density of states within a few eV of the Fermi level for both undoped and doped fullerenes. Since photoemission experiments are highly sensitive to the first two or three nanometers of material near the surface, the intercalation of the fullerenes is normally done in the measurement apparatus, because cleavage techniques cannot be used for intercalated fullerene samples. This difficulty with sample preparation introduces uncertainties in specifying the exact stoichiometry and homogeneity of the near-surface layers (<20 Å) that are probed in photoemission. Also of concern is verification of the similarity between the near-surface layers and the bulk material with regard to structure and composition, and the formation of spatially separated phases (such as K_3C_{60} and K_6C_{60}), upon addition of K, that are inhomogeneously distributed throughout the sample (see Figure 5).

Typical photoemission spectra ($E < E_F$, where E_F is the Fermi energy) and inverse photoemission (75) spectra ($E > E_F$) are shown in Figure 7 (76) for C_{60} and K_xC_{60} ($0 \leqslant x \leqslant 6$) (77), where the intensity maxima correspond to peaks in the density of states on the basis of a one-electron picture and a constant matrix element approximation. In the photoemission experiments, the density of states for the occupied levels is probed, whereas in the inverse photoemission experiments, the density of unoccupied states is probed. Photoemission data thus provide evidence for charge transfer and for band filling as x in K_xC_{60} increases. Referring to Figure 7, we can see the density of states peak associated with the t_{1u}-derived band just above the Fermi level E_F in the C_{60} trace. As K is added to C_{60}, this peak moves closer to E_F as some K_3C_{60} phase is introduced. Upon further addition of K, the t_{1u}-related peak crosses the Fermi level and eventually, for $5.8 \leqslant x \leqslant 6.0$, the t_{1u}-derived peak falls below E_F, indicating complete filling of the t_{1u}-derived level; the data for this composition further show a small band gap to the next higher lying t_{1g}-derived level (Figure 6).

Whereas photoemission and inverse photoemission (75) studies show strong similarities between the electronic structure of K_xC_{60} and both Rb_xC_{60} and Cs_xC_{60}, differences have also been clearly demonstrated between the density of states for K_xC_{60}, and for Na_xC_{60} and Ca_xC_{60}, wherein multiple metal ions can be accommodated in the octahedral sites (see Doped C_{60} crystals).

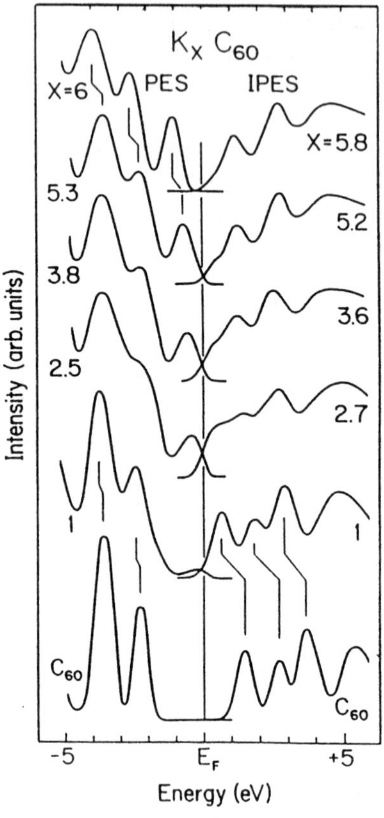

Figure 7 Photoemission (PES) and inverse photoemission (IPES) spectra (76). The spectra show the effects of K incorporation in C_{60} films. The $x = 1$ spectra show emission at E_F from grains of K_3C_{60}. Adding more K results in an increase of emission below E_F and a decrease of intensity above E_F. The top spectra show that E_F shifts into the gap when the LUMO bands are filled.

OPTICAL PROPERTIES

Because the optical transitions near the HOMO-LUMO gap are symmetry-forbidden for electric dipole transitions (see above), and their absorption strengths are consequently very low, study of the absorption edge in C_{60} is difficult from both an experimental and theoretical standpoint. To add to this difficulty, C_{60} is strongly photosensitive, so that unless measurements are made under low-light intensities, photo-induced chemical reactions takes place, in some cases giving rise to irreversible structural changes and polymerization (31). At higher photon energies, dipole-allowed transitions can occur, and the optical absorption increases dramatically. In the higher photon energy regime, typical semiconductor treatments for the optical properties can be used.

Results for the optical absorbance shown as the optical density [which is defined as $\log_{10}$ (1/transmittance)] vs wavelength in the ultraviolet (UV)-visible range (200–700 nm) for C_{60} and C_{70} in hexane (78) are depicted in Figures 8*a* and *b*, respectively. The strong absorption bands in the figure (below ~400 nm in wavelength or above ~2.9 eV in energy) are identified with electric dipole-allowed transitions between occupied (bonding) and empty (antibonding) molecular orbitals (shown in Figure 6 by either the $h_u \rightarrow t_{1g}$ transition or the $h_g \rightarrow t_{1u}$ transition). At longer wavelengths $490 < \lambda < 640$ nm (or $1.9 < E < 2.5$ eV), weak absorption takes place and is associated with electric dipole-forbidden transitions between the one-electron HOMO level with H_u symmetry and the one-electron t_{1u} LUMO level (see above, Electronic Structure). Using many-electron molecular states (67), the fully occupied ground HOMO state is identified with a singlet S_0 state.

In the excited state, we have one electron (t_{1u}) and one hole (h_u^9), each

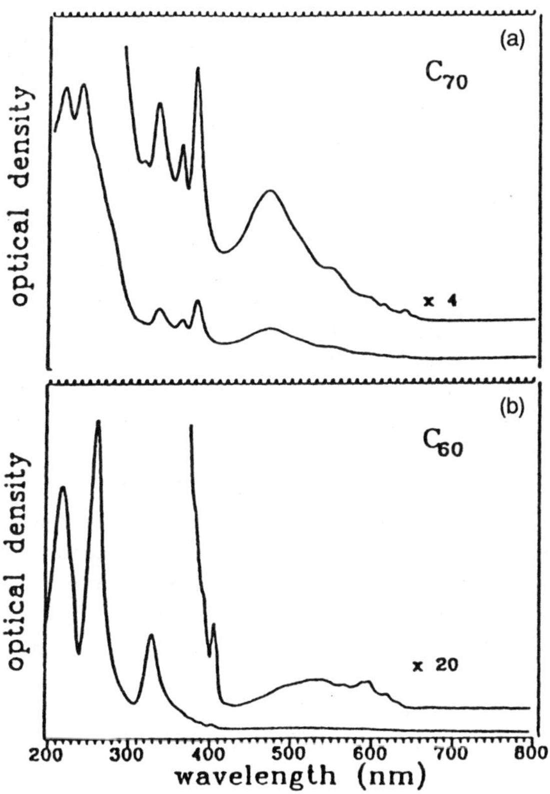

Figure 8 Optical density vs wavelength for (*a*) C_{60} and (*b*) C_{70} in hexane solutions, with the characteristic absorption bands identified (78).

with spin 1/2, so that the interacting electron and hole can either form a singlet S_i state ($S = 0$), or a triplet T_i state ($S = 1$). Because the electric dipole matrix element for optical transitions $\mathscr{H}'_{\text{em}} = -(\vec{p} \cdot \vec{A})/(^e mc)$ does not depend on spin, there is a strong spin selection rule ($\Delta S = 0$) for optical electric dipole transitions. This strong spin selection rule arises from the very weak spin-orbit interaction for carbon, which has a low atomic number ($Z = 6$). To satisfy the electric dipole selection rule, appropriate odd-parity vibrational modes must be admixed with the initial and (or) final electronic states for the optical transition. Thus the weak absorption below 2.9 eV involves optical transitions between appropriate vibronic levels. Optical transitions between the HOMO and LUMO levels can thus occur through the excitation of a vibronic state with the appropriate odd-parity vibrational mode (79–83). Because of the involvement of a vibrational energy in the vibronic state, there is an energy difference between the lowest energy absorption band (78) and the lowest energy luminescence band.

Using these molecular states, the weak absorption observed between 490 and 640 nm for C_{60} in solution (Figure 8*a*) (78) is assigned to transitions between the singlet ground state S_0 and the lowest excited singlet state S_1 (associated with the t_{1u} orbital and activated by vibronic coupling). A schematic diagram showing the various energies important to optical phenomena is presented in Figure 9.

For C_{70}, molecular orbital calculations (68) reveal a large number of closely spaced orbitals both above and below the HOMO-LUMO gap (68). The large number of orbitals makes it difficult to assign particular groups of transitions to structure observed in the solution spectra of C_{70}. UV-visible solution spectra for higher fullerenes (C_n; $n = 76, 78, 82, 84, 90, 96$) have also been reported (43, 44, 85).

Further insight into the electronic structure is provided by pulsed laser studies of C_{60} and C_{70}. Such time-resolved studies of fullerenes in solution have been used to probe the photo-dynamics of the optical excitation/luminescence spectra. The importance of these dynamic studies is to show that photo-excitation in the long-wavelength portion of the UV-visible spectrum leads to the promotion of C_{60} from the singlet S_0 ground state (^1A_g) into a singlet S_1 excited state, which decays quickly with a nearly 100% efficiency (86, 87) via an inter-system crossing to the lowest excited triplet state T_1. The rapid singlet-triplet decay (~ 1.3 ns) (87) is fostered by the overlap in energy between the vibronic manifold of electronic states associated with the lowest singlet S_1 state and the corresponding vibronic manifold for the triplet T_1 state.

Once in the T_1 triplet excitonic manifold, a very rapid transition occurs to the lowest level of the T_1 triplet manifold. It is significant that the lowest

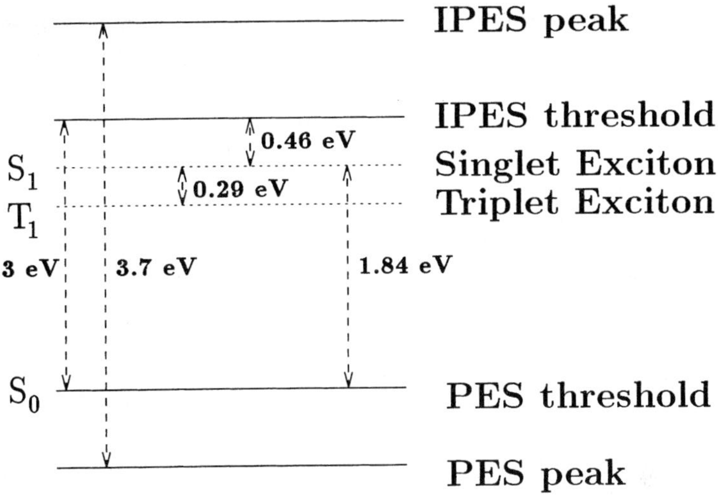

Figure 9 Schematic diagram energy levels close to the Fermi level for C_{60} showing the energy separation between the IPES and PES peaks and thresholds, respectively, associated with the HOMO-LUMO gap between the IPES threshold and the PES threshold. This energy differs from the energy between the IPES and PES peaks by the repulsive energy of two holes on the same C_{60} molecule, denoted by $U = 1.4$ eV. By using a value 1.84 eV for the singlet optical absorption edge, an exciton binding energy of 0.46 eV is obtained. The triplet level at 1.55 eV (84) indicates an exchange energy of 0.29 eV between the singlet and triplet excitons.

energy level of the triplet manifold has been found to lie about 0.3 eV lower in energy than the lowest energy level of the singlet S_1 manifold (87). Also of significance is the long life of the lowest triplet (55 ms) relative to the lowest singlet state (1.2 ns) in the room temperature solution spectra (88). The value of the T_1 triplet lifetime varies widely from one study to another, presumably because of the great sensitivity of this lifetime to the presence of molecular oxygen, which has a ground state with a spin $S = 1$, allowing a large enhancement of the coupling of the fullerene triplet state to the ground state multiplet. Oxygen has been shown to reduce the triplet lifetime from <40 μs to ~ 300 ps for different solutions (87, 89). Despite the difference in the values of the reported triplet, T_1 lifetimes, it is clear that in the absence of oxygen, the triplet T_1 lifetime is much longer than the singlet S_1 lifetime, and therefore a metastable T_1 state can be established. The efficient population of a metastable T_1 level in C_{60} by optical pumping leads to interesting nonlinear optical properties. One practical application that may follow from this nonlinear property is the development of an optical-limiting material, whose absorptivity increases with increasing light intensity (90, 91).

The optical emission spectra for fullerenes have been especially interesting because the emission band generally occurs at lower energies than the absorption band (83). It is interesting to observe that the features in the optical absorption spectra have their counterparts in the photoluminescence spectra but are shifted by twice the frequency of the vibration involved in the vibronic state. Furthermore, a close correspondence is found in the photoluminescence spectra in solution and in solid films (83). For C_{60}, the occupation of the triplet state T_1 has been shown to be quenched with nearly 100% efficiency by dioxygen, which leads to a very short lifetime for C_{60} in the excited T_1 state and a fast, nonradiative relaxation to the ground state in the presence of oxygen (87, 89, 92). However, in the organic glass solvent, the effect of oxygen on the triplet state lifetime is less important because of the lower oxygen diffusion and reduced collision rate. Using photo-acoustic calorimetry (87), the triplet energy $E_T = 36$ kcal/mol or 1.56 eV has been obtained in C_{60}. A similar value of 1.55 eV has been reported for the triplet energy using electron energy loss spectroscopy (EELS), which is not subject to the same selection rules as the optical dipole transitions. The small singlet-triplet exchange splitting ($\sim$0.3 eV; shown in Figure 9) is consistent with the relatively large diameter of a C_{60} molecule. Using pico- and nanosecond flash photolysis, the excited-state optical properties of C_{60} in toluene have been studied to determine the $S_1 \rightarrow S_n$ (88) and $T_1 \rightarrow T_n$ (88, 89) absorption spectra. For studying the $T_1 \rightarrow T_n$ spectrum, the inter-system crossing (discussed above) is used to populate the T_1 level. The $T_1 \rightarrow T_n$ absorption has been reported to be stronger than $S_0 \rightarrow S_n$ absorption, which provides a mechanism (91) to account for the observed nonlinear absorption leading to optical limiting of transmitted, transient, visible light in both C_{60} and C_{70} (7 ns pulses at 532 nm). No phosphorescent emission from the triplet (T_1) vibronic manifold has been observed in either the solution or solid phase emission spectra, presumably because of efficient nonradiative relaxation from the metastable T_1 state to the S_0 ground state.

As shown in Figure 10, a large increase in optical absorption occurs at higher photon energy where electric dipole transitions are allowed. Transmission spectra taken in this range (see Figure 10) confirm the similarity of the optical spectra for solid C_{60} and C_{60} dissolved in decalin (35). The optical properties of solid C_{60} and C_{70} have been studied over a wide frequency range using the variable angle spectroscopic ellipsometry (VASE) technique (35, 93) and electron energy loss spectroscopy (94). From these measurements, the complex refractive index $\tilde{n}(\omega) = n(\omega) + ik(\omega)$ and optical dielectric function $\varepsilon(\omega) = \varepsilon_1(\omega) + i\varepsilon_2(\omega) = \tilde{n}^2(\omega)$ are determined, and the results are shown in Figure 11 for this solid film of C_{60} at $T = 300$ K (35, 93, 97–99). The strong, sharp structure at low

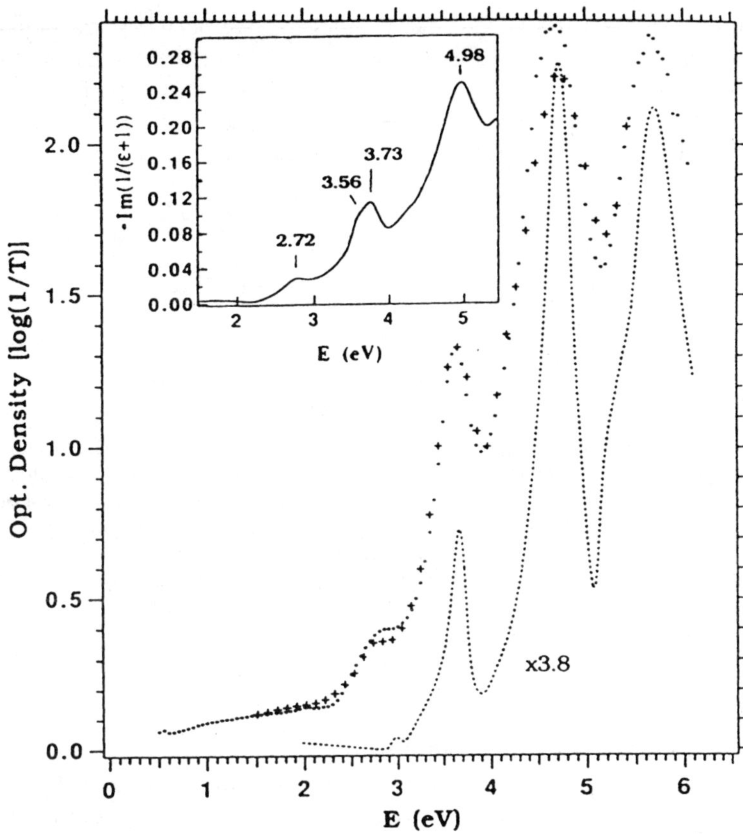

Figure 10 Optical density of solid C_{60} on Suprasil derived from ellipsometry measurements (+) and that measured directly for C_{60} on Suprasil by normal incidence transmission spectroscopy (·). For comparison, the solution spectrum for C_{60} dissolved in decalin (····) is shown. The inset is a plot of $-\mathrm{Im}\,[(1+\varepsilon)^{-1}]$ vs E for comparison with the high-resolution electron energy loss spectra (HREELS) (35).

energy is identified with infrared-active optic phonons. At higher energies the structure is due to electronic transitions. From the VASE studies of C_{60} and C_{70}, similar values for the static dielectric function $\varepsilon_1(0) = [n(0)]^2$ have been obtained: $\varepsilon_1(0) = 3.61$ (for C_{60}) (35) and $\varepsilon_1(0) = 3.94$ (for C_{70}) (93). For the case of C_{60}, this value for $\varepsilon_1(0)$ can be compared with the value $\varepsilon_1(0) = 4.4 \pm 0.2$ obtained from capacitance measurements (96). Fringe structure in the optical transmission data of C_{60} solid films have been used (36) to obtain the value $\varepsilon_1(0) = 4.1$, in better agreement with the capacitance measurements (96).

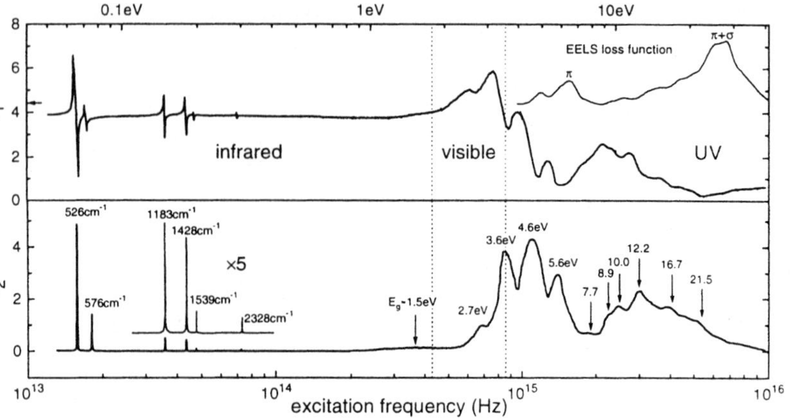

Figure 11 Summary of real $\varepsilon_1(\omega)$ and imaginary $\varepsilon_2(\omega)$ parts of the dielectric function for C_{60} vacuum-sublimed solid films at room temperature over a wide frequency range (97). The data between 0.05 and 0.5 eV (mid- to near-infrared) were collected using the Fourier transform infrared (FTIR) transmission technique (31). The vis-UV range was investigated by VASE (35) and near-normal-incidence reflection and transmission experiments (95). UV data above ~7 eV were obtained using EELS (94) by Kramers-Kronig analysis of the EELS loss function. The arrow at the left axis points to $\varepsilon_1 = 4.4$, the observed low-frequency value of the dielectric constant (96).

Near-normal incidence, transmission/reflection studied on C_{60} and M_6C_{60} (M = K, Rb, Cs) (36) have been carried out in the range 0.5–6 eV to determine the optical dielectric function $\varepsilon(\omega)$ (35). For C_{60}, these studies (36) have assigned absorption structure to transitions between specific, narrow (0.3–0.5 eV) energy bands, using a band model to describe the electronic states. Good agreement was found between the calculated energy band spacing (68, 100) and the experimental peak positions in the absorptive part of $\varepsilon(\omega)$, namely $\varepsilon_2(\omega)$ for energy bands within ~2 eV of the Fermi level.

For alkali-metal-saturated C_{60} solid films (e.g. M_6C_{60}: M = K, Rb, Cs), transmission and reflection spectra largely insensitive to the dopant or intercalate species (M) are obtained (36). These results give strong evidence for only weak hybridization between M and C_{60} states. The optical spectra are thus consistent with a molecular picture in which complete charge transfer of the alkali-metal *s*-electrons takes place to fill a lower lying, sixfold degenerate C_{60} band (t_{1u} symmetry). The narrow spectral features in M_6C_{60} observed in $\varepsilon_2(\omega)$ for $\hbar_\omega < 3$ eV have been assigned to transitions between narrow energy bands or to broadened molecular states whose spacing was found to be in good agreement with theory (73). The energy gap between the t_{1u}- and t_{1g}-derived states is believed to be ~ 1 eV for the M_6C_{60} components.

Using a pulsed Nd:YAG laser, nonlinear optical behavior has been observed in solid C_{60} films at $T = 300$ K (101–103). Using time-resolved four-wave mixing (101, 102), a fast (<35 ps) nonlinear response (including third- and fifth-order contributions) was observed leading to a substantial third-order optical susceptibility $\chi_{xxxx}(3) = 7 \times 10^{-12}$ esu, 60 times greater than that of benzene. The origin of the optical nonlinearity is probably connected to the high efficiency ($\sim 100\%$) in transferring electrons from the excited singlet state manifold S_1 to the T_1 triplet excited states and also to the larger optical matrix elements for triplet-triplet transitions (which are important for high-light power levels) relative to the singlet-singlet transitions that dominate the spectrum at low-light power levels (104–106).

ELECTRICAL TRANSPORT

Because of the high resistivity of undoped C_{60} (10^8 Ω-cm), the doping of C_{60} with alkali-metals creates carriers at the Fermi level and decreases the electrical resistivity ρ of C_{60} by several orders of magnitude. As x in M_xC_{60} increases, the resistivity ρ approaches a minimum at $x = 3.0 \pm 0.05$ (8, 107), corresponding to a half-filled t_{1u}-derived conduction band. Then, upon further increase in x from 3 to 6, the resistivity ρ again increases, as is shown in Figure 12 for various alkali-metal dopants (108). It should be noted that stable M_xC_{60} compounds only occur for $x = 0$, 3, 4, and 6 at room temperature, although the $x = 1$ is a stable phase at elevated temperatures (see Doped C_{60} Crystals). The compounds corresponding to filled molecular levels (C_{60} and M_6C_{60}) exhibit maxima in the resistivity. Furthermore, even at the minimum resistivity in M_xC_{60}, the value of ρ found for K_3C_{60} (2.5×10^{-3} Ω-cm in Figure 12) is high, typical of a high-resistivity metal and here merohedral disorder provides an important scattering mechanism (74, 107).

Studies of the temperature dependence of the resistivity of polycrystalline M_xC_{60} samples in the normal state show that conduction is by a thermally activated hopping process, except for a small range of x near 3, where the conduction is metallic (8, 109). Associated with the hopping process is an activation energy E_a, which increases as x deviates farther and farther from the resistivity minimum at $x = 3$ (107, 108). In the metallic regime, results for $\rho(T)$ for a superconducting single crystal K_3C_{60} sample (see Figure 13) show a linear T dependence of $\rho(T)$ above T_c (109, 110), although the linear T dependence is much weaker than that for the high T_c cuprates.

A number of studies have shown that the temperature dependence of

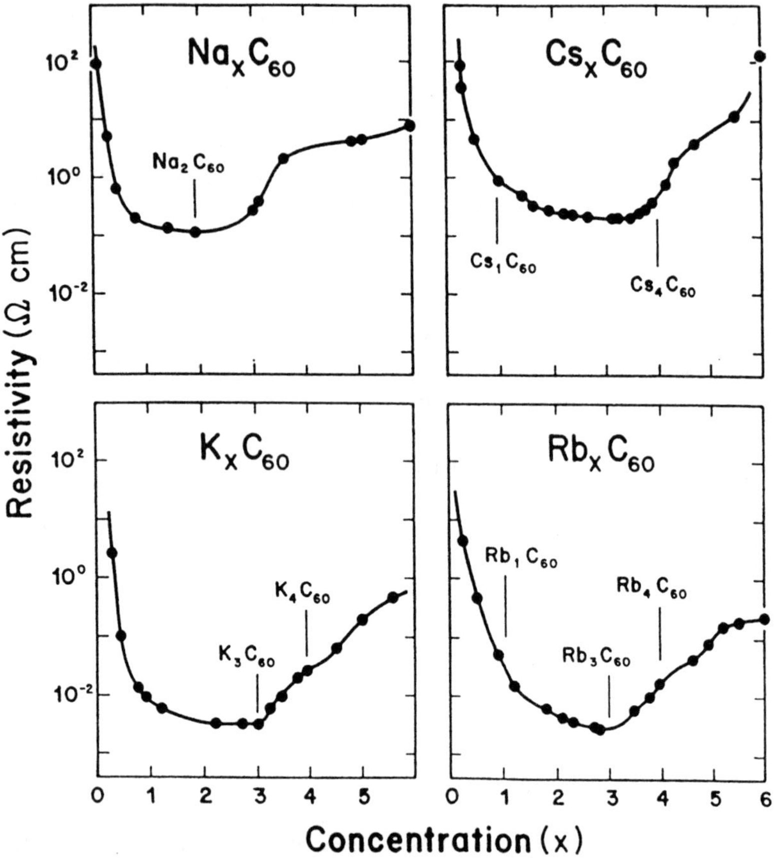

Figure 12 Composition dependence of the resistivity $\rho(x)$ for thick films of C_{60} doped with Na, K, Rb, and Cs. Points indicate where exposure to the alkali-metal source was stopped and X-ray and ultraviolet photoemission spectra were acquired to determine the concentration x. The labels indicate the known fulleride phases at 300 K. The minima in $\rho(x)$ occur for stoichiometries corresponding to Na_2C_{60}, K_3C_{60}, Rb_3C_{60}, and $Cs_{3.5}C_{60}$. Structure in $\rho(x)$ can be associated with the development of different stable phases (108).

the resistivity $\rho(T)$ is strongly dependent on whether the sample is a single crystal or a film (109, 110). Film samples tend to exhibit a negative temperature coefficient of $\rho(T)$ just above the superconducting transition temperature T_c, whereas single crystal samples exhibit a positive $\partial\rho(T)/\partial T$ just above T_c. In general, transport measurements on M_xC_{60} films by different groups are likely to show differences due to small deviations in average stoichiometry and sample homogeneity, leading to differences in the coexisting phases (see Figure 5).

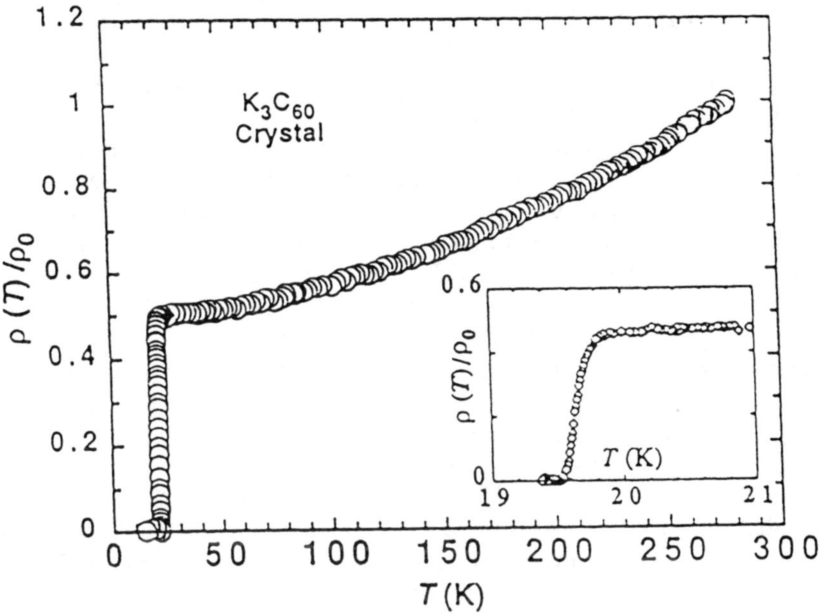

Figure 13 Normalized dc electrical resistivity $\rho(T)$ of single crystal K_3C_{60}. The inset shows the $\rho(T)$ behavior near the superconducting transition temperature $T_c = 19.8$ K (109).

Temperature-dependent Hall effect measurements have also been carried out in the temperature range 30 to 260 K on a K_3C_{60} thin film (111). For three electrons per C_{60}, the expected Hall coefficient R_H based on a one-carrier model would be about one order of magnitude smaller than the experimentally observed value (111). The small value of the observed Hall coefficient suggests multiple carrier types including both electrons and holes. This interpretation is corroborated by the observed sign change in R_H from negative below 220 K to positive above 220 K. Multiple carrier types are consistent with the Fermi surface calculations (73), which also suggest both electron and hole orbits on the Fermi surface.

The high electrical resistivity and the magnitude of the optical bandgap of C_{60} can be reduced by the application of high pressure, with decreases in resistivity of about one order of magnitude observed per 10 GPa pressure (112). However, at a pressure of ~20 GPa, an irreversible phase transition to a more insulating phase has been reported (112). As the intermolecular C-C distance decrease upon application of pressure and becomes comparable to the intramolecular C-C distance, an electronic transition might be expected to occur.

SUPERCONDUCTIVITY

The most striking electronic property of the C_{60}-related materials has been the observation of high-temperature superconductivity ($T_c \sim 40$ K) (9, 59). The first observation of superconductivity in an alkali-metal-doped carbon material occurred in 1965 when superconductivity was seen in the first stage alkali-metal graphite intercalation compound (GIC) C_8K (113). Except for the novelty of observing superconductivity in a compound having no superconducting constituents, this observation did not attract a great deal of attention, because the T_c was very low (~ 140 mK) (14). Later, higher T_cs were observed in GICs using superconducting intercalants (e.g. $KHgC_8$ for which $T_c = 1.9$ K) (114), and in subjecting the alkali-metal GICs to pressure (e.g. NaC_2 for which $T_c \sim 5$ K) (115).

The early observation of superconductivity at 18 K in K_3C_{60} (6) was soon followed by observations of superconductivity at even higher temperatures: in Rb_3C_{60} ($T_c = 30$ K) (8, 116) and $Rb_xCs_yC_{60}$ ($T_c = 33$ K) (16), and finally by applying pressure to stabilize Cs_3C_{60} ($T_c = 40$ K) (9). A large increase in T_c was achieved in the early research by using compounds with larger intercalate atoms, which resulted in unit cells of larger size and with larger lattice constants (117). As the lattice constant increases, the C_{60}-C_{60} coupling decreases, narrowing the bandwidth of the LUMO level, and thereby increasing the corresponding density of states consistent with the Bardeen-Cooper-Schieffer (BCS) expression relating the transition temperature to the density of states $N(E_F)$

$$T_c \sim \omega_{ph} \exp[-1/VN(E_F)], \qquad 1.$$

where V is the electron-phonon coupling energy. Figure 14 shows an empirical, nearly linear relation between T_c and the lattice constant a_0 for superconducting alkali-metal-doped samples (47). This correlation includes compounds derived from alkali-metal dopants, alloys of different alkali metals (62), and samples under pressure (118–120). Because of the close connection between the electronic density of states at the Fermi level $N(E_F)$ and the lattice constant a_0, plots of T_c vs $N(E_F)$ similar to those in Figure 14 have been made (66). The effect of pressure on K_3C_{60} and Rb_3C_{60} (118, 119) reduces T_c in accordance with an approximately linear dependence of T_c on a_0, as shown by the open points in Figure 14.

The reason why the T_c is so much higher in the M_3C_{60} materials relative to other carbon-based materials appears to be closely related to the high density of states that can be achieved at the Fermi level when the t_{1u} LUMO molecular level is half filled with carriers. However, several authors have found an enhancement in the electron-phonon coupling in C_{60}-related materials (121–124). If the electron-phonon interaction provides the pair-

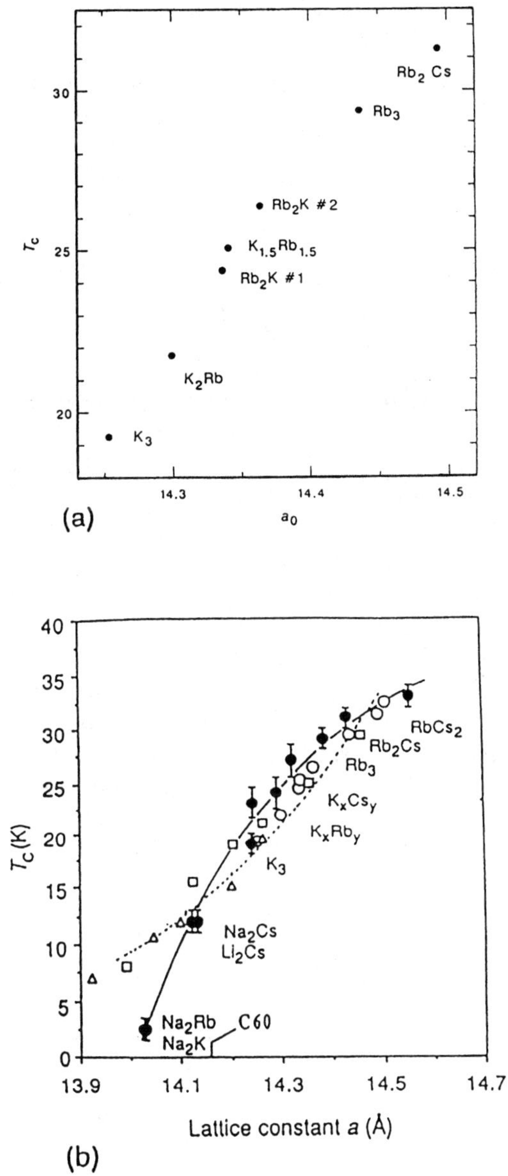

Figure 14 Dependence of T_c for various M_3C_{60} compounds on the lattice constant (47, 62).

ing mechanism for superconductivity in C_{60}-based materials, it is believed (121, 123) that the H_g-derived phonons play the dominant role in the coupling. The observation of broad H_g-derived Raman lines (125, 126) in M_3C_{60} is consistent with a strong electron-phonon coupling. Calculations of the contributions of the various phonon modes to the electron-phonon interaction strength (122) show that the high frequency intramolecular phonons may account for the high T_c values that are observed experimentally.

The magnitude of the superconducting band gap 2Δ has been studied by a variety of experimental techniques, including NMR (117) and optical techniques (127), leading to the conclusion that the superconducting band gap for both K_3C_{60} and Rb_3C_{60} is close to the BCS value of 3.5 kT_c (59, 73, 117, 128). A good fit for the temperature dependence of the bandgap to BCS theory was also obtained using the scanning tunneling microscopy technique (129). Measurements of the isotope effect all suggest that $T_c \propto M^{-\alpha}$. Both small ($\alpha \sim 0.3$–0.4) values (130, 131) and large ($\alpha \sim 1.4$) values (132, 133) of α have been reported. On the basis of the superconducting gap Equation 1, all the reported values of the exponent α suggest that phonons are involved in the pairing mechanism for superconductivity (130, 132). Future work is needed to clarify the experimental picture of the isotope effect in the M_3C_{60} compounds. Also of significance to the pairing mechanism for superconductivity in the doped fullerenes is the muon spin resonance (μSR) experiment, which shows an isotropic superconducting gap (134), as is found in BCS superconductors.

Also of interest is the dependence of the superconducting parameters on pressure and magnetic field. Closely related to the high compressibility of C_{60} (23) and M_3C_{60} (M = K, Rb) (119) is the large linear decrease in T_c with pressure observed in K_3C_{60} (119) and Rb_3C_{60} (119) (see Table 2) for

Table 2 Superconductivity parameters in M_3C_{60} (M = K, Rb)

Parameter	K_3C_{60}	Rb_3C_{60}	Reference
fcc a_0 (Å)	14.253	14.436	18, 135
T_c (K)	19.7	30.0	136
$2\Delta(0)/kT_c$	3.5	3.5	137
$(dT_c/dP)_{p=0}$ (K/GPa)	−7.8	−9.7	119
$H_{c1}(0)$ (T)	13	19	119
$H_{c2}(0)$ (T)	30	55	138
ξ_0 (nm)	3.1	3.0 ± 0.25	138, 139
λ_L (nm)	600[a]	460	127
$\kappa = \lambda_L/\xi_0$	194	153	119

[a] λ_L = 480 nm from μSR measurements (134).

measurements up to a pressure of 1.9 GPa. By attributing the smaller size of the K^+ ion, relative to Rb^+ (0.186 Å), to an effective chemical pressure of 1.06 GPa, the pressure dependence of T_c for both K_3C_{60} and Rb_3C_{60} can be made to coincide, and the result has been fit to the same functional form

$$T_c(P) = T_c(0)\exp(-\gamma P), \tag{2.}$$

with the same value of $\gamma = 0.44 \pm 0.03$ GPa^{-1} (119).

Also listed in Table 2 are values for characteristic superconducting parameters relevant to their magnetic field behavior (118, 119, 140). Based on measurements of the linear temperature dependence of the upper critical field H_{c2} near T_c, a high value for the zero temperature $H_{c2}(0)$ has been estimated (118, 119, 138) using the Werthamer-Helfand-Hohenberg (WHH) formula (141),

$$H_{c2}(0) = 0.69\left[\frac{\partial H_{c2}}{\partial T}\right]_{T_c} T_c, \tag{3.}$$

and high values of $H_{c2}(0)$, exceeding the Pauli-limiting field of 35 T, have been obtained for both K_3C_{60} and Rb_3C_{60} (see Table 2). Measurements of the lower critical field H_{c1}, where the magnetic flux penetration initially occurs (119), show that $H_{c1}(0)$ is very low and that $H_{c1}(T)$ follows the simple empirical formula

$$[H_{c1}(T)/H_{c1}(0)] = 1-(T/T_c)^2. \tag{4.}$$

Values for $H_{c1}(0)$ are also given in Table 2. Based on the relations

$$H_{c2}(0) = \frac{\phi_0}{2\pi\xi_0^2}, \tag{5.}$$

where ϕ_0 is the magnetic flux quantum and

$$H_{c1}(0) = \frac{\phi_0}{4\pi\lambda_L^2}\ln(\lambda_L/\xi_0), \tag{6.}$$

the zero temperature superconducting coherence length ξ_0 and the London penetration depth λ_L have been estimated (see Table 2) (119), showing values of ξ_0 only slightly larger than a lattice constant for the fcc unit cell. The thermodynamic critical field listed in Table 2 is found from the relation

$$H_c^2(0) = H_{c1}(0)H_{c2}(0)/\ln\kappa \tag{7.}$$

where $\kappa = \lambda_L/\xi_0$ (119). The large values of κ listed in Table 2 confirm that the M_3C_{60} compounds are strongly type II superconductors.

MAGNETIC PROPERTIES

Materials with closed shell configurations are diamagnetic. Thus C_{60} itself is expected to be diamagnetic because by Hund's rule, the ground state for the C_{60} molecule is a nondegenerate $J = 0$ state, and paramagnetic contributions from permanent moments are not expected (70). However, the magnitude of the diamagnetic susceptibility is quite low (-0.35×10^{-6} emu/g for C_{60} and -0.59×10^{-6} emu/g for C_{70}) (142, 143) in comparison to that for graphite (-21.1×10^{-6} emu/g) (144) because of the approximate cancellation of ring currents in the hexagons (diamagnetic contribution) and in the pentagons (paramagnetic contribution) (142, 143, 145). Under certain doping conditions, fullerene-derived materials that result in a delocalized metallic state should show Pauli paramagnetism. Other dopants with unfilled *d*- or *f*-shells are expected to be paramagnetic. Thus the use of endohedral dopants could provide one route for the preparation of magnetic fullerenes.

The most extensively studied magnetic fullerene system is that with the organic donor tetrakisdimethyl-aminoethylene (TDAE) complexed with C_{60} (146), which is interesting as the organic ferromagnet (with only *s*- and *p*-orbitals) having the highest Curie temperature of 16.1 K. Some parameters relevant to the magnetization of this organic ferromagnet are listed in Table 3. The compounds TDAE-C_{70}, TDAE-C_{84}, TDAE-C_{90}, and TDAE-C_{94} have been prepared, characterized, and investigated magnetically, but none shows a phase transition to a magnetically ordered state down to 5 K (153).

APPLICATIONS

Although research on solid C_{60} and related materials is still at an early stage, these materials are already beginning to show many exceptional properties, some of which may lead to practical applications.

One promising application for C_{60} is as an optical limiter. Optical limiters are used to protect materials from damage by high-light intensities through a saturation of the transmitted light intensity with increasing incident intensity. Outstanding performance for C_{60} as an optical limiter relative to presently used optical-limiting materials has been observed at 5320 Å for 8 ns pulses using solutions of C_{60} in toluene and in chloroform (CH_3Cl) (91). Although C_{70} in similar solutions also showed optical-limiting action, the performance of C_{60} was found to be superior. The proposed mechanism for the optical limiting is that C_{60} and C_{70} are more absorptive for molecules in the triplet-excited state than for the ground state (see Optical Properties). In this process, the initial absorption of a photon

Table 3 Parameters relevant for the magnetism for TDAE-C_{60}

Property	Value	Reference
T_c (K)	16.1	147
M (5 K) (μ_B/C_{60})	0.1	146
χ_0 Pauli paramagnetization (emu/g)	3.6×10^{-7}	148
Remanent magnetization (5 K) (10^{-3} emu)	0.3226	149
Coercive field (5 K) (G)	1.6	149
Spin concentration (mol-TDAE-C_{60})	4.9×10^{20}	148
dT_c/dH (K/kOe)	0.17	150
dT_c/dp (K/kbar)	−8.8	150
Spin	1/2	150
g-factor (paramagnetic phase)	2.0003–2.0008	151, 150
μ_{eff} (μ_B/C_{60})	1.72	150
θ_{cw} (K)	−22.5, −58	151, 150
Spin concentration (cm^{-3})	8.3×10^{22}	148
Lattice constants[a]	$a = 15.849$ Å $b = 12.987$ Å $c = 9.965$ Å $\beta = 93.31°$	152

[a] C-centered monoclinic structure.

(which takes an electron from the singlet S_0 state to an excited singlet state) is followed by an inter-system crossing from the singlet to a triplet state. Because of the higher excitation cross-section for electrons in the metastable triplet state (relative to the ground state), the population of the metastable triplet state promotes further stronger absorption of photons. (91).

Another interesting application is the photoconducting device that can be fabricated using fullerene-doped polymers. Polyvinylcarbazole (PVK) and other selected polymers, such as poly (paraphenylene-vinylene) (PPV) and phenylmethylpolysilane (PMPS), doped with a mixture of C_{60} and C_{70}, has been reported to exhibit exceptionally good photoconductive properties (154–156) that may lead to the development of future polymeric photoconductive materials. Small concentrations of fullerenes (e.g. ~3% by weight) lead to charge transfer of the excited electrons in the polymer to the fullerenes, thereby separating the holes and the electrons, preventing rapid recombination, and thereby promoting photoconduction by the mobile holes (157).

Fullerene-doped polymers also have significant potential for use in applications as photodiodes, photovoltaic devices, and as photorefractive materials.

By utilizing the strong bonding of fullerenes to clean silicon surfaces, application of a monolayer of C_{60} as a bonding agent between thin silicon wafers has been indicated (156). This strong bonding property, together with the low chemical reactivity of fullerenes, has been utilized in the passivation of reactive surfaces by the adsorption of monolayers of C_{60} on aluminum and silicon surfaces (156).

As research on these materials proceeds, it is expected that other interesting physics and chemistry will be discovered (because of the extreme properties exhibited by carbon-based materials generally) and that promising applications will be found for C_{60}- and C_{60}-related materials.

ACKNOWLEDGMENTS

The authors acknowledge fruitful discussions with Professors M Endo, PC Eklund, R Saito, and RA Jishi. The authors gratefully acknowledge support from the National Science Foundation Grant #DMR-92-10878 and from Air Force Office of Scientific Research grant F49620-93-1-0160.

Literature Cited

1. Kroto HW, Heath JR, O'Brien SC, Curl RF, Smalley RE. 1985. *Nature* 318: 162–63
2. Fuller RB. 1984. In *The Artifacts of R. Buckminster Fuller: A Comprehensive Collection of His Designs and Drawings*, ed. W Marlin. New York: Garland
3. Rohlfing EA, Cox DM, Kaldor A. 1984. *J. Chem. Phys.* 81: 3322
4. Krätschmer W, Lamb LD, Fostiropoulos K, Huffman DR. 1990. *Nature* 347: 354–58
5. Curl RF, Smalley RE. 1988. *Science* 242: 1017–22
6. Haddon RC, Hebard AF, Rosseinsky MJ, Murphy DW, Duclos SJ, et al. 1991. *Nature* 350: 320–22
7. Hebard AF, Rosseinsky MJ, Haddon RC, Murphy DW, Glarum SH, et al. 1991. *Nature* 350: 600–1
8. Hebard AF. 1992. *Phys. Today* 45: 26–32
9. Palstra TTM, Zhou O, lwasa Y, Sulewski P, Fleming RM, Zegarski B. 1995. *Solid State Commun.* 93: 327
10. Meng RL, Ramirez D, Jiang X, Chow PC, Diaz C, et al. 1991. *Appl. Phys. Lett.* 59: 3402–3
11. Smalley RE. 1992. *Acc. Chem. Res.* 25: 98–105
12. Bethune DS, Johnson RD, Salem JR, de Vries MS, Yannoni CS. 1993. *Nature* 366: 123–28
13. Dresselhaus MS, Dresselhaus G, Eklund PC. 1993. *J. Mater. Res.* 8: 2054–97
14. Dresselhaus MS, Dresselhaus G. 1981. *Adv. Phys.* 30: 139–326
15. Rosseinsky MJ, Murphy DW, Fleming RM, Tycko R, Ramirez AP, et al. 1992. *Nature* 356: 416–18
16. Tanigaki K, Ebbesen TW, Saito S, Mizuki J, Tsai JS, et al. 1991. *Nature* 352: 222–23
17. Hérold A. 1979. In *Physics and Chemistry of Materials with Layered Structures*, ed. F Lévy, pp. 323–400. New York: Dordrecht Reidel
18. Kortan AR, Kopylov N, Glarum SH, Gyorgy EM, Ramirez AP, et al. 1992. *Nature* 355: 529–32
19. Kortan AR, Kopylov N, Özdas E, Ramirez AP, Fleming RM, Haddon RC. 1995. *Chem. Phys. Lett.* In press
20. Chen Y, Stepniak F, Weaver JH, Chibante LPF, Smalley RE. 1992. *Phys. Rev. B* 45: 8845–48
21. David WIF, Ibberson RM, Matthewman JC, Prassides K, Dennis TJS, et al. 1991. *Nature* 353: 147–49
22. Stephens PW, Mihaly L, Lee PL, Whetten RL, Huang SM, et al. 1991. *Nature* 351: 632–34
23. Fischer JE, Heiney PA, McGhie AR,

Romanow WJ, Denenstein AM, et al. 1991. *Science* 252: 1288–90
24. Woo SJ, Lee SH, Kim E, Lee KH, Lee YH, et al. 1992. *Phys. Lett. A* 162: 501–5
25. Fischer JE, Heiney PA. 1993. *J. Phys. Chem. Solids* 54: 1725–57
26. Saito S, Oshiyama A. 1991. *Phys. Rev. B* 44: 11532–35
27. Wang LS, Conceicao J, Jin C, Smalley RE. 1991. *Chem. Phys. Lett.* 182: 5–11
28. de Vries J, Steger H, Kamke B, Menzel C, Weisser B, et al. 1992. *Chem. Phys. Lett.* 188: 159–62
29. Steger H, de Vries J, Kamke B, Kamke W, Drewello T. 1992. *Chem. Phys. Lett.* 194: 452–56
30. Heiney PA, Vaughan GBM, Fischer JE, Coustel N, Cox DE, et al. 1992. *Phys. Rev. B* 45: 4544–47
31. Rao AM, Zhou P, Wang KA, Hager GT, Holden JM, et al. 1993. *Science* 259: 955–57
32. Dresselhaus G, Dresselhaus MS, Mavroides JG. 1966. *Carbon* 4: 433–40
33. Shi XD, Kortan AR, Williams JM, Kini AM, Saval BM, Chaikin PM. 1992. *Phys. Rev. Lett.* 68: 827–30
34. Yu RC, Tea N, Salamon MB, Lorens D, Malhotra R. 1992. *Phys. Rev. Lett.* 68: 2050–53
35. Ren SL, Wang Y, Rao AM, McRae E, Hager GT, et al. 1991. *Appl. Phys. Lett.* 59: 2678–80
36. Wang Y, Holden JM, Rao AM, Lee WT, Hager GT, et al. 1992. *Phys. Rev. B* 45: 14396–99
37. Taylor R, Walton DRM. 1993. *Nature* 363: 685–93
38. Olah GA, Bucsi I, Aniszfeld R, Surya Prakash GK. 1992. *Carbon* 30: 1203–11
39. Taylor R, Hare JP, Abdul-Sada AK, Kroto HW. 1990. *J. Chem. Soc. Chem. Commun.* 20: 1423–25
40. Johnson RD, Meijer G, Bethune DS. 1990. *J. Am. Chem. Soc.* 112: 8983–84
41. Dresselhaus G, Dresselhaus MS, Eklund PC. 1992. *Phys. Rev. B* 45: 6923–30
42. Fischer JE, Heiney PA, Smith AB III. 1992. *Acc. Chem. Res.* 25: 112–18
43. Diederich F, Whetten RL. 1992. *Acc. Chem. Res.* 25: 119–26
44. Kikuchl K, Nakahara N, Wakabayashi T, Suzuki TS, Shiramaru T, et al. 1992. *Nature* 357: 142–45
45. Manolopoulos DE, Fowler PW. 1991. *Chem. Phys. Lett.* 187: 1–7
46. Fleming RM, Rosseinsky MJ, Ramirez AP, Murphy DW, Tully JC, et al. 1991. *Nature* 352: 701–3
47. Fleming RM, Ramirez AP, Rosseinsky MJ, Murphy DW, Haddon RC, et al. 1991. *Nature* 352: 787–88
48. Heiney PA, Fischer JE, McGhie AR, Romanow WJ, Denenstein AM, et al. 1991. *Phys. Rev. Lett.* 66: 2911–14
48a. Heiney PA, Fischer JE, McGhie AR, Romanow WJ, Denenstein AM, et al. 1991. *Phys. Rev. Lett.* 67: 1468
49. David WIF, lbberson RM, Dennis TJS, Hare JP, Prassides K, 1992. *Europhys. Lett.* 18: 219–25
50. Axe JD, Moss SC, Neumann DA. 1994. See Ref. 158, pp. 149–224
51. Harris AB, Sachidanandam R. 1992. *Phys. Rev. B* 46: 4944–57
52. Vaughan GB, Heiney PA, Fischer JE, Luzzi DE, Ricketts-Foot DA, et al. 1991. *Science* 254: 1350
53. Verheijen MA, Meekes H, Meijer G, Bennema P, de Boer JL, et al. 1992. *Chem. Phys.* 166: 287
54. Van Tendeloo G, Amelinckx S, Verheijen MA, van Loosdrecht PHM, Meijer G. 1992. *Phys. Rev. Lett.* 69: 1065–68
55. Vaughan GBM, Heiney PA, Cox DE, McGhie AR, Smith AL, Strongin RM, et al. 1995. *Chem. Phys.* In press
56. McGhie AR, Fischer JE, Stephens PW, Cappelletti RL, Neumann DA, et al. 1994. *Phys. Rev. B* 49: 12614–18
57. Armbruster JF, Romberg HA, Schweiss P, Adelmann P, Knupfer M, et al. 1994. *Z. Phys. B* 95: 469–74
58. Weaver JH, Poirier DM. 1994. See Ref. 158, pp. 1–108
59. Lieber CM, Chen CC. 1994. See Ref. 158, pp. 109–48
60. Poirier DM, Owens DW, Weaver JH. 1995. *Phys. Rev. B* 50: 1830–43
61. Fleming RM, Siegrist T, Marsh PM, Hessen B, Kortan AR, et al. 1991. In *Clusters and Cluster-Assembled Materials, MRS Symposia Proceedings*, ed. RS Averback, J Bernholc, DL Nelson, 206: 691–96. Pittsburgh: Mater. Res. Soc.
62. Tanigaki K, Hirosawa I, Ebbesen TW, Mizuki J, Shimakawa Y, et al. 1992. *Nature* 356: 419–21
63. Stephens PW. 1992. *Nature* 356: 383–85
64. Kortan AR, Kopylov N, Thiel FA. 1992. *J. Phys. Chem. Solids* 53: 1683–88
65. Saito S, Oshiyama A. 1993. *Phys. Rev. Lett.* 71: 121–24
66. Oshiyama A, Saito S, Hamada N, Miyamoto Y. 1992. *J. Phys. Chem. Solids* 53: 1457–71
67. Herzberg G. 1945. *Molecular Spectra and Molecular Structure*. Princeton: Van Nostrand
68. Saito S, Oshiyama A. 1991. *Phys. Rev. Lett.* 66: 2637–40

69. Negri F, Orlandi G, Zerbetto F. 1992. *J. Am. Chem. Soc.* 114: 2909–13
70. Saito R, Dresselhaus G, Dresselhaus MS. 1992. *Phys. Rev. B* 46: 9906–9
71. Khaled MM, Carlin RT, Trulove PC, Eaton GR, Eaton SS. 1994. *J. Am. Chem. Soc.* 116(8): 3465–74
72. Erwin SC, Pederson MR. 1991. *Phys. Rev. Lett.* 67: 1610–13
73. Pickett WE. 1994. See Ref. 158, pp. 226–347
74. Gelfand MP. 1994. *Supercond. Rev.* 1: 103–50
75. Jost MB, Troullier N, Poirier DM, Martins JL Weaver JH, et al. 1991. *Phys. Rev. B* 44: 1966–69
76. Weaver JH. 1992. *Acc. Chem. Res.* 25: 143–49
77. Benning PJ, Poirier DM, Ohno TR, Chen Y, Jost MB, et al. 1992. *Phys. Rev. B* 45: 6899–913
78. Ajie H, Alvarez MM, Anz SJ, Beck RD, Diederich F, et al. 1990. *J. Phys. Chem.* 94: 8630–33
79. Pichler T, Matus M, Kürti J, Kuzmany H. 1992. *Solid State Commun.* 81: 859–62
80. Sibley SP, Argentine SM, Francis AH. 1992. *Chem. Phys. Lett.* 188: 187–93
81. Nakayama A, Suzuki K, Enoki T, di Vittorio SL, Dresselhaus MS, et al. 1993. *Synth. Met.* 55–57: 3736–41
82. Matus M, Kuzmany H, Sohmen E. 1992. *Phys. Rev. Lett.* 68: 2822–27
83. Wang Y, Holden JM, Rao AM, Eklund PC, Venkateswaran U, et al. 1995. *Phys. Rev. B* 51: 4547–56
84. Sohmen E, Fink J. 1993. *Phys. Rev. B* 47: 14,532–40
85. Diederich F, Ettl R, Rubin Y, Whetten RL, Beck R, et al. 1991. *Science* 252: 548
86. Birks JB. 1970. In *Photophysics of Aromatic Molecules*. London: Wiley
87. Hung RR, Grabowski JJ. 1991. *J. Phys. Chem.* 95: 6073–74
88. Ebbesen TW, Tanigaki K, Kuroshima S. 1991. *Chem. Phys. Lett.* 181: 501–4
89. Arbogast JW, Darmanyan AP, Foote CS, Rubin Y, Diederich FN, et al. 1991. *J. Phys. Chem.* 95: 11–12
90. Tutt LW, Boggess TF. 1993. *Prog. Quantum Electron.* 17: 299–338
91. Tutt LW, Kost A. 1992. *Nature* 356: 225–26
92. Nissen MK, Wilson SM, Thewalt MLW. 1992. *Phys. Rev. Lett.* 69: 2423–26
93. Ren SL, Wang KA, Zhou P, Wang Y, Rao AM, et al. 1992. *Appl. Phys. Lett.* 61: 124–26
94. Sohmen E, Fink J, Krätchmer W. 1992. *Z. Phys. B* 86: 87–92
95. Wang Y, Holden JM, Rao AM, Lee WT, Bi XX, et al. 1992. *Phys. Rev. B* 45: 14396–99
96. Hebard AF, Haddon RC, Fleming RM, Kortan AR. 1991. *Appl. Phys. Lett.* 59: 2109–11
97. Ecklund PC, Rao AM, Wang Y, Zhou P, Wang KA, et al. 1995. *Solid Thin Films* 257: 211–32
98. Pevzner B, Hebard AF, Haddon RC, Senturia SD, Dresselhaus MS. 1995. In *Science and Technology of Fullerene Materials: MRS Symposia Proceedings*, ed. P Bernier, T Ebbesen, DS Bethune, R Metzger, LY Chiang, JW Mintmire. Pittsburgh: Mater. Res. Soc. 359:
99. Romberg H, Sohmen E, Merkel M. 1993. *Synth. Met.* 56: 3038–43
100. Ching WY, Huang MZ, Xu YN, Harter WG, Chan FT. 1991. *Phys. Rev. Lett.* 67: 2045–48
101. Kafafi ZH, Lindle JR, Pong RGS, Bartoli FJ, Lingg LJ, Milliken J. 1992. *Chem. Phys. Lett.* 188: 492–96
102. Flom SR, Pong RGS, Bartoli FJ, Kafafi ZH. 1992. *Phys. Rev. B* 46: 15598–601
103. Hoshi H, Nakamura N, Maruyama Y, Nakagawa T, Suzuki S, et al. 1991. *Jpn. J. Appl. Phys.* 30: L1397–98
104. Prasad PH, Williams DJ. 1991. *Introduction to Nonlinear Optical Effects in Molecules and Polymers.* New York: Wiley-Interscience. 307 pp.
105. Tang N, Partanen JP, Hellwarth RW, Knize RJ. 1993. *Phys. Rev. B* 48: 8404–8
106. Lindle JR, Pong RGS, Bartoli FJ, Kafifi ZH. 1993. *Phys. Rev. B* 48: 9447–51
107. Kochanski GP, Hebard AF, Haddon RC, Fiory AT. 1992. *Science* 255: 184–86
108. Stepniak F, Benning PJ, Poirier DM, Weaver JH. 1993. *Phys. Rev. B* 48: 1899–906
109. Xiang XD, Hou JG, Briceño G, Vareka WA, Mostovoy R, et al. 1992. *Science* 256: 1190–91
110. Maruyama Y, Inabe T, Ogata H, Achiba Y, Suzuki S, et al. 1991. *Chem. Lett.* (Japan) 10: 1849–52
111. Palstra TTM, Haddon RC, Hebard AF, Zaanen J. 1992. *Phys. Rev. Lett.* 68: 1054–57
112. Nuñez Regueiro M, Monceau P, Rassat A, Bernier P, Zahab A. 1991. *Nature* 354: 289–91
113. Hannay NB, Geballe TH, Matthias BT, Andres K, Schmidt P, MacNair D. 1965. *Phys. Rev. Lett.* 14: 225–26
114. Chaiken A, Dresselhaus MS, Orlando

TP, Dresselhaus G, Tedrow PM, et al. 1990. *Phys. Rev. B* 41: 71–85
115. Belash IT, Bronnikov AD, Zharikov OV, Pal'nichenko AV. 1990. *Synth. Met.* 36: 283
116. Holczer K, Klein O, Grüner G, Huang SM, Kaner RB, et al. 1991. *Science* 252: 1154–57
117. Dresselhaus MS, Dresselhaus G, Saito R. 1994. In *Physical Properties of High Temperature Superconductors IV*, ed. DM Ginsberg, Vol. IV, Chpt. 7. Singapore: World Scientific
118. Sparn G, Thompson JD, Huang SM, Kaner RB, Diederich F, et al. 1991. *Science* 252: 1829–31
119. Sparn G, Thompson JD, Whetten RL, Huang SM, Kaner RB, et al. 1992. *Phys. Rev. Lett.* 68: 1228–31
120. Schirber JE, Overmyer DL, Wang HH, Wang JM, Carlson KD, et al. 1991. *Phys. C* 178: 137
121. Varma CM, Zaanen J, Raghavachari K. 1991. *Science* 254: 989–92
122. Schlüter MA, Lannoo M, Needels M, Baraff GA, Tománek D. 1992. *Phys. Rev. Lett.* 68: 526–29
123. Johnson KH, McHenry ME, Clougherty DP. 1991. *Phys. C* 183: 319
124. Jishi RA, Dresselhaus MS. 1992. *Phys. Rev. B* 45: 2597–600
125. Eklund PC, Zhou P, Wang KA, Dresselhaus G, Dresselhaus MS. 1992. *J. Phys. Chem. Solids* 53: 1391–413
126. Duclos SJ, Haddon RC, Glarum SH, Hebard AF, Lyons KB. 1991. *Science* 254: 1625
127. Degiorgi L, Wachter P, Grüner G, Huang SM, Wiley J, Kaner RB, 1992. *Phys. Rev. Lett.* 69: 2987–90
128. Gelfand MP, Lu JP. 1993. *Appl. Phys. A* 56: 215–17
129. Zhang Z, Chen CC, Lieber CM. 1991. *Science* 254: 1619–21
130. Ramirez AP, Kortan AR, Rosseinsky MJ, Duclos SJ, Mujsce AM, et al. 1992. *Phys. Rev. Lett.* 68: 1058–60
131. Chen CC, Lieber CM. 1992. *J. Am. Chem. Soc.* 114: 3141–42
132. Ebbesen TW, Tsai JS, Tanigaki K, Tabuchi J, Shimakawa Y, et al. 1992. *Nature* 355: 620–22
133. Zakhidov AA, Imaeda K, Petty DM, Yakushi K, Inokuchi H, et al. 1992. *Phys. Lett. A* 164: 355–61
134. Uemura YJ, Keren A, Le LP, Luke GM, Sternlieb BJ, et al. 1991. *Nature* 352: 605–7
135. Zhang Z, Chen CC, Kelty SP, Dal H, Lieber CM. 1991. *Nature* 353: 333–35
136. Cohen ML. 1993. *Mater. Sci. Eng. B* 19: 111–16
137. Kiefl RF, MacFarlane WA, Chow KH, Dunsiger S, Duty TL, et al. 1993. *Phys. Rev. Lett.* 70: 3987–90
138. Foner S, McNiff EJ Jr, Heiman D, Huang SM, Kaner RB. 1992. *Phys. Rev. B* 46: 14936–39
139. Johnson RD, Bethune DS, Yannoni CS. 1992. *Acc. Chem. Res.* 25: 169–75
140. Holczer K, Klein O, Grüner G, Thompson JD, Diederich F, Whetten RL. 1991. *Phys. Rev. Lett.* 67: 271–74
141. Werthamer NR, Helfand E, Hohenberg PC. 1966. *Phys. Rev.* 147: 295–302
142. Ruoff RS, Beach D, Cuomo J, McGuire T, Whetten RL, Diederich F. 1991. *J. Phys. Chem.* 95: 3457–59
143. Haddon RC, Schneemeyer LF, Waszczak JV, Glarum SH, Tycko R, et al. 1991. *Nature* 350: 46–47
144. Ganguli N, Krishnan KS. 1941. *Proc. R. Soc. London Ser. A* 177: 168–82
145. Prato M, Suzuki T, Wudl F, Lucchini V, Maggini M. 1993. *J. Am. Chem. Soc.* 115: 7876–77
146. Allemand PM, Khemani KC, Koch A, Wudl F, Holczer K, et al. 1991. *Science* 253: 301–3
147. Awaga K, Maruyama Y. 1989. *Chem. Phys. Lett.* 158: 556
148. Okahara K. 1994. *Endohedral fullerenes*. PhD thesis. Kyoto Univ., Japan
149. Suzuki A, Suzuki T, Whitehead RJ, Maruyama Y. 1994. *Chem. Phys. Lett.* 223: 517–20
150. Wudl F, Thompson JD. 1992. *J. Phys. Chem. Solids* 53: 1449–56
151. Tanaka K, Zakhidov AA, Yoshizawa K, Okahara K, Yamabe T, et al. 1993. *Phys. Rev. B* 47: 7554–59
152. Stephens PW, Cox DE, Lauher JW, Mihaly L, Wiley JB, et al. 1992. *Nature* 355: 331–32
153. Tanaka K, Okada M, Okahara K, Yamabe T. 1993. *Chem. Phys. Lett.* 202: 394–98
154. Wang Y. 1992. *Nature* 356: 585–87
155. Orczyk ME, Swedek B, Zieba J, Prasad PN. 1994. *Nonlinear Optical Properties of Organic Materials VII*, ed. GR Mohlmann, 2285: 166–77. Bellingham, WA: SPIE Opt. Eng. Press
156. Tong QY, Eom CB, Gösele U, Hebard AF. 1994. *J. Electrochem. Soc.* 141: L137–38
157. Sariciftci NS, Heeger AJ. 1994. *Int. J. Mod. Phys. B* 8: 237–74
158. Ehrenreich H, Spaepen F, eds. 1994. *Solid State Physics: Advances in Research Applications*. New York: Academic

Annu. Rev. Mater. Sci. 1995. 25: 525–46

METAL-ORGANIC CHEMICAL VAPOR DEPOSITION OF FERROELECTRIC OXIDE THIN FILMS FOR ELECTRONIC AND OPTICAL APPLICATIONS

Bruce W. Wessels

Department of Materials Science and Engineering, and Materials Research Center, Northwestern University, Evanston, Illinois 60208

KEY WORDS: ferroelectrics, dielectrics, epitaxy, barium strontium titanate

ABSTRACT

Thin films of ferroelectric oxides, particularly $BaTiO_3$ and its solid state solution with $SrTiO_3$, hold the possiblity of use in microelectronic, electro-optic, and nonlinear optic applications. Many of these applications require thin films that are textured or epitaxial. Synthesis of these ferroelectrics has centered on physical vapor deposition techniques such as magnetron sputtering, molecular beam epitaxy, laser ablation, and activated reactive evaporation. An alternative, powerful approach is metal-organic chemical vapor deposition (MOCVD). This deposition method has produced films of $BaTiO_3$ with ferroelectric, electro-optic, and nonlinear optical properties approaching that of bulk.

INTRODUCTION

The potential of ferroelectric oxides for a variety of microelectronic, electro-optic, and nonlinear optic applications has resulted in a resurgence in interest in their thin film synthesis and properties. Of particular interest are $BaTiO_3$ and its solid solution with $SrTiO_3$. $BaTiO_3$ has excellent electronic properties with a dielectric constant in excess of 3000 and a large ferro-

0084–6600/95/0801–0525$05.00

electric effect with a spontaneous polarization of 26 $\mu C/cm^2$. These properties, along with those of $SrTiO_3$, make it especially attractive for dynamic random access memories (DRAM). The use of $BaTiO_3$, however, will expand beyond electronics and into photonics. Barium titanate's large linear electro-optic coefficient r_{42} of 820 pm/V makes it of interest for electro-optic modulators and couplers for integrated optic devices. Its high nonlinear optical susceptibility also makes it attractive for nonlinear optical applications such as second harmonic generation or frequency doubling. Many of these properties can be enhanced by forming a solid solution between $BaTiO_3$ and $SrTiO_3$. For example, the room temperature dielectric constant for ceramic $Ba_{1-x}Sr_xTiO_3$ (BST) with $x = 0.3$ is 10,000.

For many of these applications, thin films are required that are either textured or epitaxial. Initial studies on the synthesis of these ferroelectrics centered on physical vapor deposition techniques such as magnetron sputtering, molecular beam epitaxy, laser ablation, and activated reactive evaporation. An alternative approach is metal-organic chemical vapor deposition (MOCVD), which has a number of potential advantages. These include the ability to deposit high quality, ultra-thin layers on three-dimensional complex geometries, excellent composition control, and the amenability to large-scale processing. The success of an MOCVD process depends critically on the availability of volatile and stable precursor materials. Although suitable metal-organic precursors of titanium are readily available, this is not the case for the alkaline earth metals Sr and Ba. To date, precursors containing these elements remain somewhat problematic. Nevertheless, excellent progress has been made in MOCVD of both $BaTiO_3$ and $SrTiO_3$ using β-diketonate precursors. In this article, the MOCVD of $BaTiO_3$, $SrTiO_3$, and $Ba{x}Sr_{1-x}TiO_3$ films is reviewed. Discussion of their dielectric, ferroelectric, and nonlinear optical properties is presented.

PRECURSORS

For deposition of Sr- and Ba-containing compounds, volatile, stable precursors are required. Initial studies indicated that alkoxides were unsuitable because they were found to have insufficient vapor pressure and/or decomposed upon heating (1). The metal β-diketonates, on the other hand, are volatile, thermally stable, and do not require special handling procedures. Consequently, the volatile metal β-diketonates have been exclusively used for Ba and Sr deposition. Their relatively high vapor pressure results from the structure of the β-diketonate ligand itself. The ligand shields the metal from interactions with other molecules that would decrease its volatility. These ligands contain either hydrocarbons or fluorocarbons. To maximize the volatility, the metal ion should be encapsulated in a sterically saturated nonpolar environment. This minimizes the tendency to form adducts. A schematic of the β-diketonate precursor is shown in Figure 1.

For the deposition of $BaTiO_3$ and $SrTiO_3$, the most widely used β-diketonates are $Ba(dpm)_2$ and $Sr(dpm)_2$ (dpm = dipivaloylmethanate; $C_{11}H_{19}O_2$). These precursors have sufficient volatility and thermal stability to be used as sources. However, to achieve sufficient transport rates, these precursors must be heated in excess of 200°C.

The necessity for high source temperatures constrains the overall system design. Furthermore, the dpm precursors, upon repeated thermal cycling near the melting point, decrease their volatility with time. The decrease in volatility of the source presumably results from oligomerization/decomposition processes. Indeed, it has also been shown that the volatilization of $Ba(dpm)_2$ is sensitive to the atmosphere (2). The volatility of the compound decreased when stored in air or in a desiccator. The decrease was associated with the interaction of the ligand with carbon dioxide gas and/or moisture.

A strategy to improve the volatility and thermal stability of the β-diketonates is to substitute fluorocarbons for the hydrocarbons in the outer shell of the complex (3). The presence of fluorine reduces the van der Waals interactions and hydrogen bonding between molecules, thereby leading to reduced oligomerization and higher volatility. It has previously been shown that by increasing amounts of fluorine substitution in the

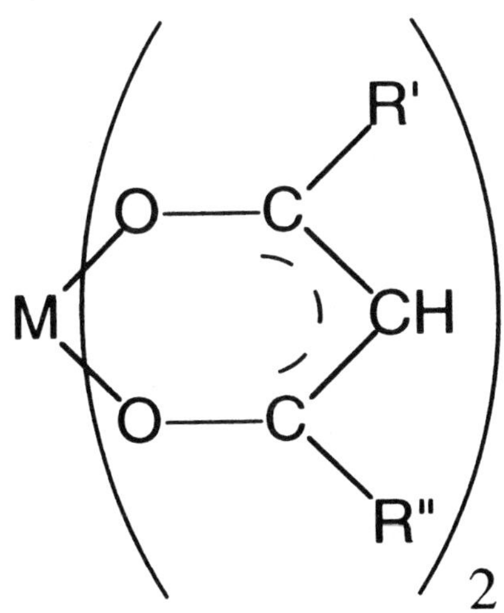

Figure 1 Schematic of a β-diketonate precursor (1).

ligand, the volatility is directly increased. Moreover, these factors lead to reducec air and moisture sensitivity.

To further increase the volatility, the introduction of neutral donor ligands such as polyethers can be considered. One such implementation of the molecular design criteria is the compound $Ba(hfa)_2$ (tetraglyme) (hfa = hexafluoro-aceytlacetonate), which combines fluorinization with the addition of nonpolar molecules (4, 5). The nominal source temperature for this compound is 105°C, which is considerably lower than the 200°C required for $Ba(dpm)_2$ volatilization.

Other improvements in the design of precursor molecules using this strategy have been noted by Hitchman's group (6); they increased the amount of fluorine content of the ring substituents so that all four groups contain C_3F_7 i.e. $Ba(TDFND)_2$. In addition, by forming the tetraglyme complex of $Ba(TDFND)_2$, a compound with a melting point (mp) of 70°C (having the highest volatility under atmospheric conditions yet reported) has been prepared.

Recently, Marks' group synthesized new fluorine-free Ba precursors based upon an encapsulating β-ketoiminate ligands labeled $Ba(diki)_2$ and $Ba(triki)_2$ (7). Both complexes could be sublimed with some decomposition at 10^{-5} torr. However, the compounds are less volatile than the highly volatile fluorinated β-diketonate precursors. Both $Ba(diki)_2$ and $Ba(tiki)_2$ are slightly air sensitive.

DEPOSITION

Initial studies of MOCVD centered on the synthesis of $SrTiO_3$ using the reactants titanium isopropoxide (TPT), $Sr(dpm)_2$, and/oxygen and/or water (1, 8). TPT is a liquid, whereas $Sr(dpm)_2$ is a solid. Because of the low volatility of the Sr precursor, a source temperature in excess of 200°C is required to deposit a film at reasonable growth rates. This requirement of a high source temperature leads to complications in the delivery of the precursor to the reaction chamber. Care must be taken so as not to condense the precursor prior to the reaction zone. In order to prevent this from occurring, initial MOCVD reactors were designed that minimized the condensation of the solid source Sr precursor. Figure 2 shows a schematic of a low pressure MOCVD reactor for the utilization of a solid source. The reactor has two zones consisting of a source zone and a reaction zone. The titanium precursor and oxygen and/or water vapor enter the reactor in separate inlet tubes immediately upstream of the reaction zone. Both the titanium source and the titanium line leading to the reactor are heated to avoid condensation. The $Sr(dpm)_2$ source is placed in a boat in the source zone of the reactor and heated separately.

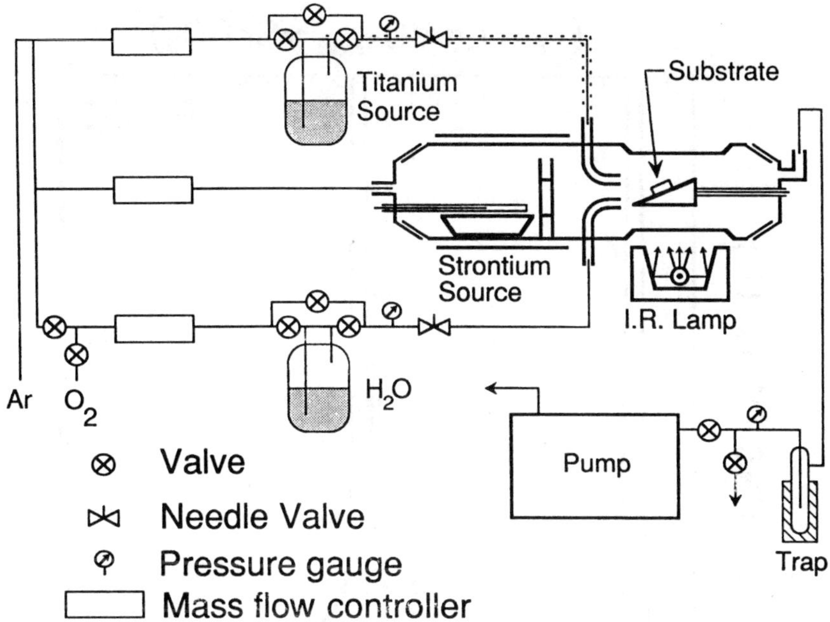

Figure 2 Low-pressure reactor for use with low-volatility precursors (1).

A baffle is placed between the source and the reaction zone. This configuration minimizes condensation of the Sr(dpm)$_2$ as well as pre-reaction with the other reactants. The samples, which were placed on a SiC-coated graphite susceptor, were heated with an infra-red lamp. The system was operated at 5 torr. Deposition was carried out at 800°C.

At 800°C, $SrTiO_3$ epitaxial layers were obtained for deposition on single crystal $LaAlO_3$ substrates (9). Figure 3 and Figure 4 show the X-ray θ-2θ) and ϕ angle scans, respectively, for epitaxial $SrTiO_3$.

An alternative approach to system design is shown in Figure 5 (10). In this case, a separate bubbler is used to contain the solid source Sr(dpm)$_2$. The gas lines between the sources and the reaction chamber are kept at 220°C to prevent condensation. Metal-organics are transported separately and mixed just prior to their introduction into the reaction chamber. In this case, the substrate was resistively heated.

Recently $SrTiO_3$ thin films have been deposited using plasma-assisted MOCVD in a commercial reactor (11). The reactants, again, were TPT and Sr(dpm)$_2$. Deposition was carried out at a system pressure of 10 to 30 torr. A microwave oxygen plasma of 300 W at 2.45 GHz was used. This enabled the lowering of the deposition temperature to 650 to 750°C.

Although electronic grade $SrTiO_3$ films can be deposited using the

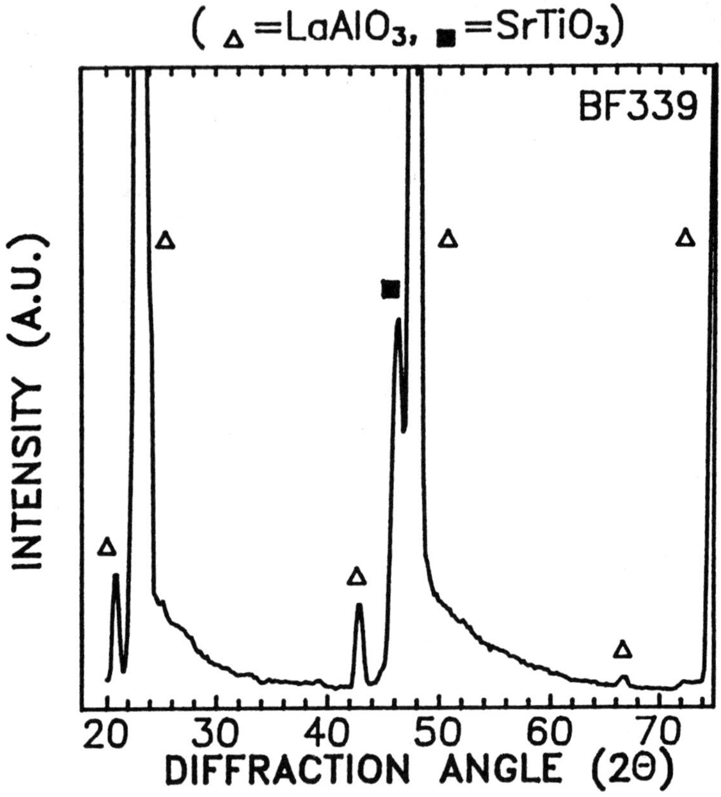

Figure 3 X-ray scan of epitaxial $SrTiO_3$ deposited on $LaAlO_3$.

$Sr(dpm)_2$ precursors, their relatively low volatility and their air sensitivity lead to complexity in system design as well as in long-term reproducibility of the composition and deposition rate.

To improve the reproducibility and minimize system complexity, second generation precursors were developed and utilized (12). $SrTiO_3$ was deposited using the fluorinated β-diketonate $Sr(hfa)_2$ (tetraglyme), along with titanium isopropoxide and oxygen as reactants. The $Sr(hfa)_2$ source temperature was 100–115°C, which is some 100°C lower than that used for deposition using $Sr(dpm)_2$. The reaction occurs as follows:

$$Ti(OC_3H_7)_4 + Sr(hfa)_2(tetraglyme) + O_2 + H_2O \rightarrow SrTiO_3 + HF + \ldots.$$

The deposition temperature ranges from 650 to 850°C. In order to not form SrF_2, gaseous H_2O is added to the reactor by passing O_2 through an H_2O bubbler. Fluorine reacts with water to form HF. For deposition

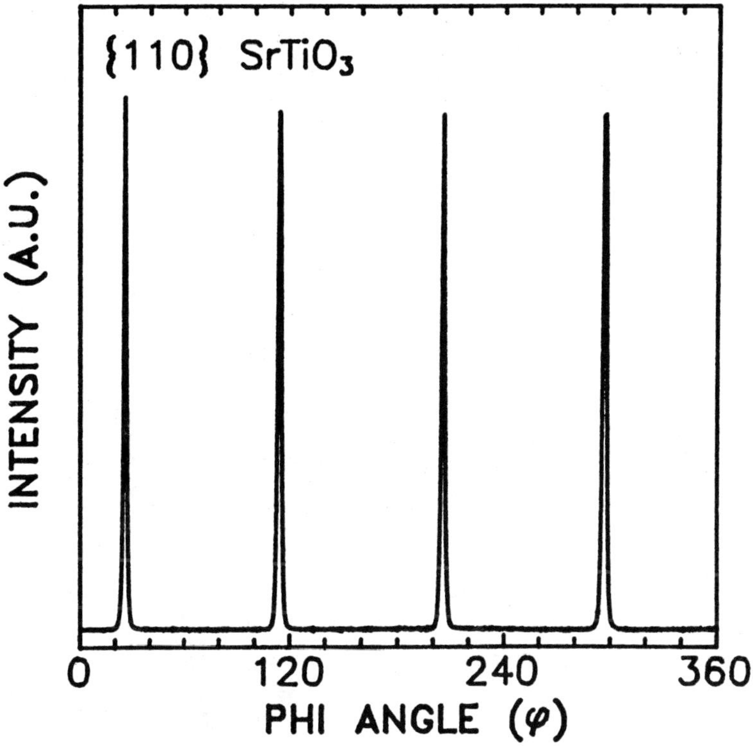

Figure 4 X-ray ϕ-scan of epitaxial $SrTiO_3$ (9).

temperatures of 700°C, SrF_2 is stabilized even in the presence of water. At 800°C, however, phase pure $SrTiO_3$ is deposited. Thus contamination by fluorine impurity phases can be eliminated by appropriate growth conditions.

$BaTiO_3$ Deposition

Following the initial work on $SrTiO_3$ thin-film synthesis using β-diketonates, Kwak et al deposited polycrystalline $BaTiO_3$ thin films using the reactants $Ba(dpm)_2$, $Ti(OC_3H_7)$, and O_2 (13). The source temperatures were 24°C for the $Ti(OC_3H_7)$ and 245°C for $Ba(dpm)_2$. To obtain highly crystalline thin films, oxygen was added to the argon carrier gas. The deposition is performed in an inverted vertical reactor at substrate temperatures of 550 to 800°C and pressure of 160 torr. Films deposited on fused quartz were partially textured with the [110] direction perpendicular to the substrate surface. However, for thicker films, no texture was observed. In contrast, films deposited on Pt sheet exhibited a [111] pre-

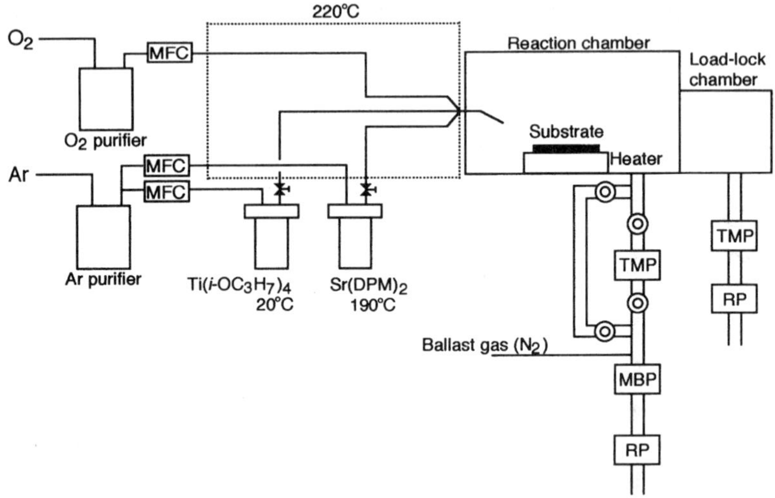

Figure 5 Low-pressure reactor using bubblers for solid sources (10).

ferred orientation. Deposition of $BaTiO_3$ onto $Al_2O_3(1102)$ and Si(100) yielded a texture very similar to that obtained on fused silica.

Subsequent studies by Nakazawa et al of $BaTiO_3$ on (100) MgO indicated that the films were highly textured with a predominant [100] orientation (14). Only a trace of the (101) oriented grains was observed. For these films, the *a*-axis was perpendicular to the substrate. The full width at half maximum (FWHM) of the (200) reflection was 0.18°, indicating good crystallinity. X-ray pole figure analysis indicated that the films had an epitaxial relationship to the MgO (100) substrates for films deposited at 900 and 1000°C.

An alternative approach for the MOCVD of $BaTiO_3$ is to use the fluorinated β-diketonates of barium in order to lower source temperature (15–18). $BaTiO_3$ has been deposited using $Ba(hfa)_2$ (tetraglyme), titanium isopropoxide, oxygen, and H_2O as reagents. The source temperature was 44°C for the titanium isopropoxide and 110°C for the tetraglyme. The reaction occurs as follows:

$$Ti(OC_3H_7)_4 + Ba(hfa)_2(tetraglyme) + O_2 + H_2O \rightarrow BaTiO_3 + HF + \ldots.$$

The deposition temperature ranges from 650 to 850°C. Here again, water vapor is added into the gas stream to eliminate the incorporation of fluorine into the films, by forming gaseous HF as a reaction by-product. At low temperatures (i.e 700°C), the formation of BaF_2 is the predominant crystalline phase, whereas at 750°C, the predominant phase is $BaTiO_3$.

At 800°C, the fluoride impurity phase is eliminated and only $BaTiO_3$ is observed.

Using these precursors, epitaxial $BaTiO_3$ was obtained for deposition onto $LaAlO_3$(100) substrates. X-ray diffraction measurements indicated that the films were [100] oriented with the *a*-axis perpendicular to the substrate. X-ray diffraction rocking curve measurements indicated the FWHM was 0.4° for the (200) reflection, comparable to that obtained using the $Ba(dpm)_2$ precursor. X-ray measurements using Φ scans indicated that the films were epitaxial. This was supported by high resolution transmission electron microscopy (HRTEM). Figure 6 shows a HRTEM image of an epitaxial layer of $BaTiO_3$ on a $LaAlO_3$ substrate. The films are semicoherent in that an array of misfit dislocations are formed at the interface. The interface between the two oxides is nearly atomically abrupt and devoid of extensive reconstruction. Electron energy loss spectroscopy (EELS) performed on the thin films did not indicate the presence of impurity phases such as fluorides in the $BaTiO_3$.

$BaTiO_3$ deposition using either the $Ba(dpm)_2$ or $Ba(hfa)_2$ (tetraglyme) precursors require deposition temperatures of the order of 800°C. To lower the deposition temperature, Chern et al as well as Van Buskirk et al have examined plasma chemical vapor deposition for synthesis of $BaTiO_3$ thin films (20, 21). Chern et al used a commercial-scale system modified

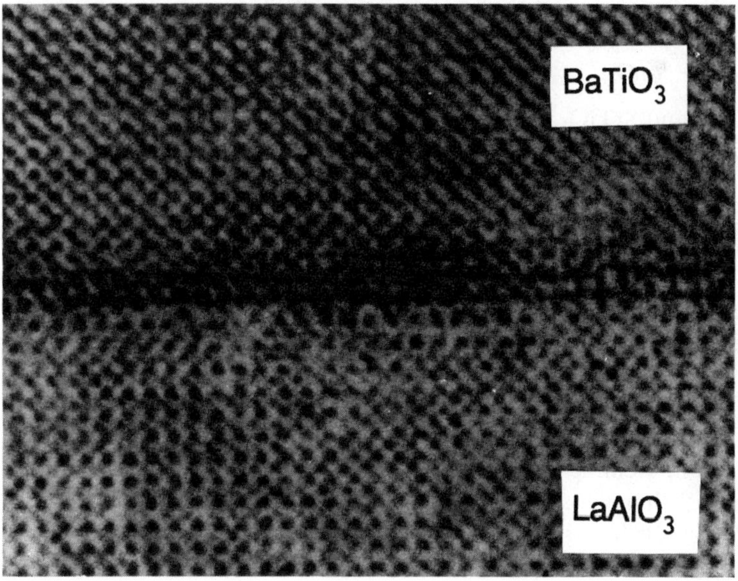

Figure 6 HRTEM image of epitaxial $BaTiO_3$ on $LaAlO_3$ (19).

to handle low volatility precursors as described for $SrTiO_3$ deposition. For the deposition of $BaTiO_3$ thin films, Ba(dpm)$_2$ and titanium isopropoxide were used as source where the Ba(dpm)$_2$ source was maintained at 215°C and the titanium isopropoxide was held at 23.5°C. Reactor chamber pressure was 5 torr. Oxygen was added to the system to enhance deposition. The oxygen was activated by microwaves (2.45 GHz and 300 W) to form an oxygen plasma. For deposition on $LaAlO_3$ substrates, epitaxial $BaTiO_3$ thin films were obtained at a substrate temperature of 680°C at a growth rate of 200 nm/h. X-ray rocking curves indicated a (FWHM) of 0.25 and 0.4° for films deposited on $NdGaO_3$ and $LaAlO_3$, respectively.

To further reduce the growth temperature so as to enable integration of $BaTiO_3$ with semiconductors, Kim et al explored the utilization of strong oxidants (22). For these applications, low temperatures are desired to minimize decomposition of the semiconductor. Using Ba(dpm)$_2$, titanium isopropoxide and N_2O, they were able to deposit $BaTiO_3$ thin films at temperatures as low as 300°C. Films deposited at 450°C on Si(100) had a (110) preferred orientation. Electron diffraction patterns indicated, however, that the films were polycrystalline. HRTEM indicated the presence of a thin interfacial layer between the $BaTiO_3$ and the Si, which presumably resulted in the film's polycrystallinity. Subsequently, Kim et al used these reactants to deposit $BaTiO_3$ on InSb (23). $BaTiO_3$/InSb heterostructures were grown at 300°C. The films were apparently crystalline. Partial epitaxy was observed as indicated by HRTEM for the thin films.

$Ba_{1-x}Sr_xTiO_3$

$BaTiO_3$ and $SrTiO_3$ form a continuous solid solution over the entire composition range. The Curie temperature of $Ba_{1-x}Sr_xTiO_3$ solid solution varies linearly with composition. The solid solution composition corresponding to the phase boundary between ferroelectric and parelectric phases is approximately $x = 0.30$ at 295 K.

MOCVD has been used to deposit thin $Ba_{1-x}Sr_xTiO_3$ films over the entire solid-solution range. Initial synthesis was centered on using the fluorinated β-diketonates (12, 25). The metal-organic precursors used were liquid titanium isopropoxide, Ba(hfa)$_2$ tetraglyme and Sr(hfa)$_2$ tetraglyme. Oxygen saturated with water (17°C) was used as the reactant gas. Films were deposited in a low pressure system at 4 torr, at growth rates from 0.3 to 1.8 μm/h, yielding film thicknesses of 0.3 to 2.0 μm. Films deposited at rates less than 0.7 μm/h were generally single phase and exhibited smooth surface morphologies. Single-phase films deposited on SiO_2/Si and bare Si substrates showed no texture, which is consistent with results obtained previously for $BaTiO_3$ using these precursors. The growth of textured films on silicon is hindered by the formation of an amorphous SiO_2 layer at the interface.

Solid solutions over the entire composition range $x = 0$ to $x = 1$ were deposited at 800°C. A linear dependence of x on Sr source temperature was observed for a constant barium source temperature.

For layers deposited on $LaAlO_3$ single-crystal substrates, epitaxial films were obtained as indicated by in-plane ϕ scans. The films had their a-axes perpendicular to the substrate. The X-ray rocking curves of the (200) reflections range from 0.47 to 1.90°. The chemical purity of the films was determined from Auger analysis. In single-phase films, fluorine was not detected. However, for multiphase films, a fluoride impurity phase was observed.

$Ba_{1-x}Sr_xTiO_3$ epitaxial thin films have also been recently deposited by plasma MOCVD by Chern et al (26). The β-diketonate $Ba(dpm)_2$ and $Sr(dpm)_2$, as well as titanium isopropoxide, were used as sources. A microwave oxygen plasma (2.45 GHz, 300 W) was used to lower deposition temperature to 680°C. Growth rates were of the order of 100 nm/h.

Deposition of barium strontium titanate epitaxial layers was demonstrated using alternative substrates including YBaCuO/$LaAlO_3$ and Pt/MgO substrates. Integration of barium strontium titanate with YBaCuO is important for superconductor device applications. Pt is potentially a bottom electrode suitable for storage capacitors for high density memory devices.

For growth on c-axis oriented YBaCuO films, the BST Φ-scan X-ray patterns indicate no in-plane misorientation. The epitaxy results with YBaCuO a- and b-axes parallel to the BST in plane $\langle 100 \rangle$. A cross-section HRTEM image of BST/YBCO heterostructure on a $LaAlO_3$ substrate indicates excellent heteroepitaxy. Atomically abrupt interfaces were observed. However, in some areas, a high density of stacking faults was observed leading to surface roughness of the epitaxial barium strontium titanate.

DIELECTRIC PROPERTIES

For thin film storage capacitors for one gigabit DRAM applications, high dielectric constant materials with dielectric constants in excess of 300 are required. $BaTiO_3$ has a dielectric constant of 300 in the c-direction and 3000 in the a-direction. $SrTiO_3$, on the other hand, is a paraelectric with a dielectric constant of 320. For BST, dielectric constants as high as 10,000 have been reported in bulk ceramic material. Because the Curie temperature depends on the Ba/Sr ratio, the maximum in the room temperature dielectric constant can be tailored by choosing the appropriate composition.

$SrTiO_3$

Measurements of the dielectric properties of the MOCVD thin films have been made. Table 1 gives the reported dielectric constants for $SrTiO_3$ including dielectric constant, dielectric loss, and resistivity of the thin films prepared by MOCVD.

Initial studies of the dielectric constant of the $SrTiO_3$ thin films were reported by Feil et al, who indicated that the dielectric constant ranged from 25 to 100 for films deposited on (100) Si as indicated in Table 1 (27). The dielectric constant was comparable to that reported for sputtered films. However, the value was considerably lower than the bulk value of 320 and was attributed to the presence of a thin silicon dioxide layer at the interface between the $SrTiO_3$ and silicon, which forms during deposition.

Yamaguchi et al subsequently used $Pt/TaO_x/Si$ as a substrate to minimize the influence of the low dielectric constant interfacial material (10). The $SrTiO_3$ films, which were 100-nm thick, had dielectric constants of 100 to 150 prior to annealing. After annealing in oxygen at 600°C the dieletric constants increased to 180 to 210. For a 40-nm thick film, a capacitance of 2.7 nf and a tan δ of 0.0l were measured at 10 kHz. The dielectric constant of this structure was 140. This has an equivalent SiO_2 thickness of 1.1 nm and a charge capacity of 31 fF/μm^2. The leakage current was of the order of 6×10^{-8} $\mu A/cm^2$ at 1 V.

There is also considerable interest in using epitaxial strontium titanate as the dielectric in superconducting device structures. Liang et al evaluated the dielectric properties of $SrTiO_3$ epitaxially deposited on *c*-axis-oriented YBaCuO (28). For 150-nm thick films, the dielectric constant was in the range of 250 to 300 with a tan δ of 0.1 at 1 kHz. The leakage current was of the order of 1$\mu A/cm^2$.

As can be seen in Table 1, MOCVD $SrTiO_3$ meets the requirements for dielectrics for DRAM applications of leakage currents of less than 10^{-6} A/cm^2 and a capacitance 30 fF/cell.

Table 1 Dielectric properties of $SrTiO_3$ thin films prepared by MOCVD

Substrate	ε	tan δ	ρ	I$\mu A/cm^2$	Reference
(100) Si	25–100	—	4×10^9	—	1
$Pt/TaO_x/Si$	60–210	0.01	—	0.06	10
(001) YBCO	250–320	0.10	—	1	11

$BaTiO_3$

The dielectric properties of the films depend strongly on microstructure and substrate. Initial reports on $BaTiO_3$ thin films prepared by MOCVD were encouraging. For deposition onto NiCr/$BaTiO_3$ substrates, films with relative dielectric constants as high as 250 were realized (13). Nakazawa et al observed dielectric constants as high as 1040 for films deposited on Pt(100)/MgO substrates with a random orientation (14). For comparison, Lu et al found a dielectric constant of the order of 100 for films deposited on similar substrates but having *a*-axis orientation (29).

Yoon et al noted that polycrystalline $BaTiO_3$ films deposited at low temperatures on Si(100) had a dielectric constant of 1147 as determined by C-V measurements (22). The barium titanate films had a (110) texture, which may be the reason for the high values of the dielectric constant observed. For films deposited on p-InSb (111), a dielectric constant of 743 was measured (23). Table 2 summarizes the observed properties of the dielectric films.

$Ba_xSr_xTiO_3$

The dielectric properties of $BaSrTiO_3$ using Au/$Ba_{1-x}Sr_xTiO_3$/p-Si MIS structures were studied (12). The dielectric constant was found to depend upon solid solution composition x, with a maximum value occurring within the region of $x = 0.35$ to 0.45. For measurements at 1 MHz, a film with $x = 0.44$ had a dielectric constant of 220.

Using YBCO/$LaAlO_3$ (100) as the substrate, Chern et al determined the dielectric constant of $Ba_{1-x}Sr_xTiO_3$ for a composition of $x = 0.25$ (26). Dielectric constants were measured over the temperature range of -50 to 100°C. At 100 kHz the room-temperature dielectric constant was 400, increasing to 480 at 1 kHz. The peak in the capacitance was at 34°C, which corresponds to the ferroelectric transition temperature. From capacitors

Table 2 Dielectric properties of $BaTiO_3$ thin films prepared by MOCVD

Substrate	ε	tan δ	ρ Ωcm	Reference
NiCr/$BaTiO_3$	250	—	10^9	13
Pt/MgO	1040	—	—	14
Pt/Ti/ONO/Si	300	0.02	5×10^8	21
Si	—	—	10^4	16
Pt/MgO	100	0.03	—	29
InSb	743	—	—	22
P-Si	1157	—	—	23

Table 3 Dielectric properties of $BaSrTiO_3$ thin films

Substrate	ε	ρ Ωcm	Reference
SiO/Si	220	2×10^9	25
YBCO/$LaAlO_3$	400	—	26

fabricated using Au or Pt sputtered contacts the dc leakage current was about 1×10^{-8} A/cm^2 at 2 V for a film with a thickness of 100 nm, as reported in Table 3.

FERROELECTRIC PROPERTIES

$BaTiO_3$ in bulk form is a ferroelectric at room temperature with a spontaneous polarization of 26 μC/cm^2. In the as-deposited state, the MOCVD films are only weakly ferroelectric (30). In order to have appreciable ferroelectric properties, the films need to be textured and preferably epitaxial. Although the epitaxial deposition of $BaTiO_3$ on oxide substrates has been achieved, the attainment of textured or epitaxial $BaTiO_3$ on metal films has been limited. Lu et al have studied the ferroelectric properties of polycrystalline $BaTiO_3$ thin films prepared by MOCVD on Pt-coated (100) MgO substrates (30).

For the *a*-axis textured films deposited on the highly (100) textured Pt layer, the as-deposited $BaTiO_3$ films had an anomalously low dielectric constant $\varepsilon \sim 10$ and a dissipation factor of 0.03. Furthermore, the as-deposited films exhibited no observable ferroelectricity. Upon application of a dc bias, the dielectric constant remained low until a threshold field E_t was reached. For threshold fields of $\sim$65 kV/cm, the dielectric constant increased an order of magnitude to $\sim$100 for a 100 kHz test frequency as shown in Figure 7. The increased dielectric constant remained after cessation of the dc bias. Concurrently, ferroelectric hysteresis was observed. Figure 8 shows the ferroelectric hysteresis of a textured film observed after application of a dc field greater than the threshold E_t. As a result of the large coersive field required, saturation was not observed. The saturation polarization P_s was $\geq 15\mu$C/cm^2, and the coersive electric field E_c was 220 kV/cm. This can be compared to the reported value for single crystal $BaTiO_3$ of 26μC/cm^2 and $E_c \sim 1.5$ kV/cm.

The measured ferroelectric properties depend upon the microstructure of the $BaTiO_3$ thin films. The as-deposited films are highly *a*-axis textured. The lattice dipoles, which are the origin of the ferroelectricity, prior to poling are aligned in-plane along one of four crystallographic directions

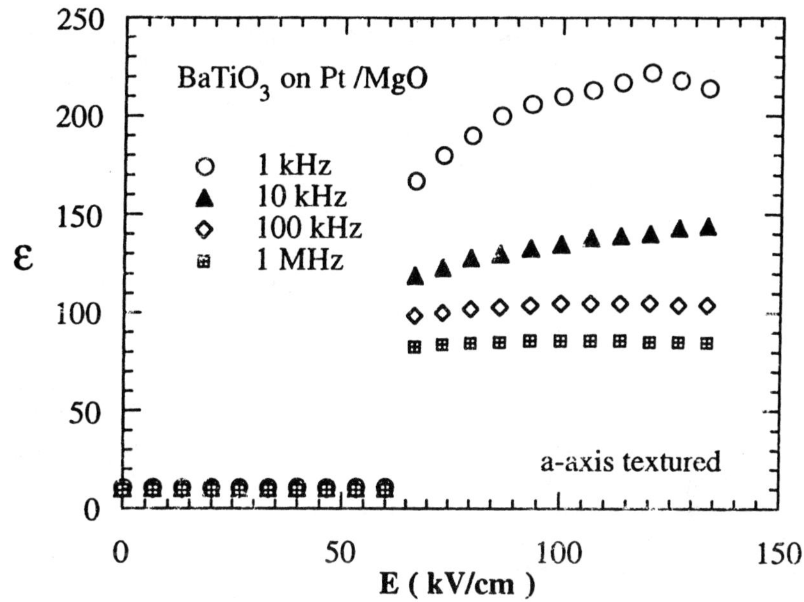

Figure 7 Electric field response of the dielectric constant for a $BaTiO_3$ thin film (29).

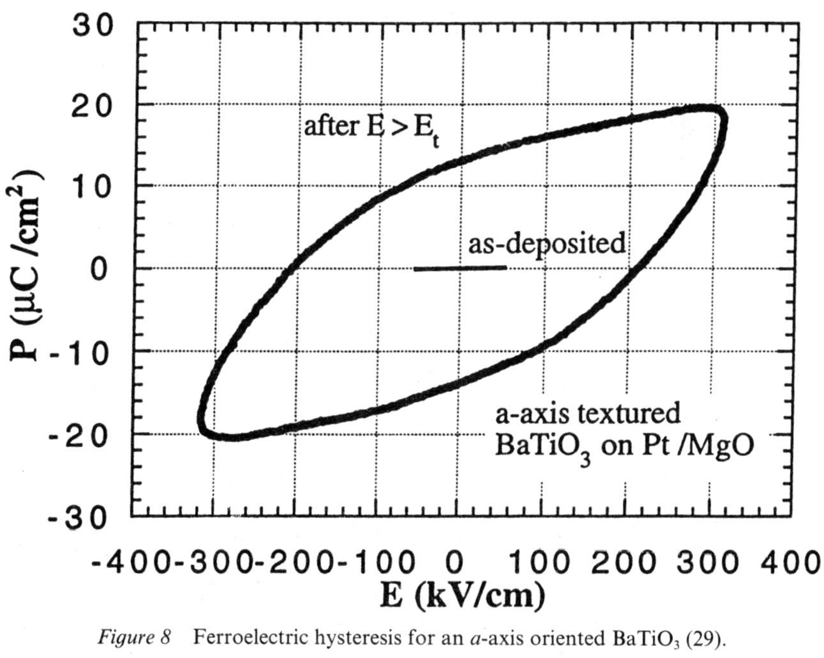

Figure 8 Ferroelectric hysteresis for an *a*-axis oriented $BaTiO_3$ (29).

determined by the cubic lattice frame. This in-plane *c*-axes configuration presumably is responsible for the lack of ferroelectric hysteresis observed in the as-deposited films.

By poling with an external electric field, the lattice dipoles are aligned in the field direction. The sharp change in the dielectric constant observed upon application of an electric field shown in Figure (7) is attributed to the poling of the film. At room temperature, coersive fields of 50 kV/cm are required. This is to be compared to a field of 0.5 to 2 kV/cm required for poling of bulk $BaTiO_3$ at the Curie temperature. In the case of the thin film, upon application of the threshold field, the lattice dipoles are aligned with the *c*-axis now being normal to the substrate. The dielectric constant consequently increases and ferroelectric hysteresis is observed. The polarization of the poled, textured film approaches that of bulk $BaTiO_3$ along the *c*-axis. The extremely high coersive fields required presumably result from residual strain in the film due to the lattice mismatch between film and substrate. This strain provides additional resistance to the reorientation of the lattice dipoles.

Experimental evidence indicates that the ferroelectric properties of the thin films depend on the detailed microstructure (29). For polycrystalline films lacking texture, hysteresis measurements of the poled film gave a $P_s \sim 16\ \mu C/cm^2$ and a $E_c \sim 70$ kV/cm. The shape of the hysteresis curve was also noticeably different in that a sharp transition in ferroelectric behavior was noted at the coersive field. The coersive field was substantially lower than that for textured films. However, the dielectric constant was approximately the same. The differences in the coersive field may be attributed to different states of strain for the two cases. The strain for the textured case would be expected to be larger as a result of coherency strain and clamping effects.

NONLINEAR OPTICAL PROPERTIES

Bulk $BaTiO_3$ has been found to be an excellent nonlinear optical (NLO) material. The non-zero, second-order nonlinear optical coefficients d_{ij} of bulk $BaTiO_3$ are $d_{33} \sim 6.8$, $d_{31} \sim 15.7$, and $d_{15} \sim 17.0$ pm/V, which are nearly comparable to $LiNbO_3$. For a strong second-harmonic generation (SHG) response, the nonlinear optical susceptibilities should be large. For this to be the case, the ferroelectric material must be at least textured or epitaxial and preferably single crystalline.

For as-deposited epitaxial $BaTiO_3$ films prepared by MOCVD, no second-harmonic generation (SHG) frequency doubling signal was observed for Nd : YAG laser excitation at 1.06 μm. However, upon corona poling, whereby voltage of the order of 10 kV is applied to the film at temperatures near the Curie temperature, SHG is observed (31, 32). Figure 9 shows the measured SHG intensity for the $BaTiO_3$ thin film with respect to quartz for different poling conditions. For point A, where the film is at room temperature and no field is applied, the SHG signal is zero. Upon heating the film to just below the Curie temperature and applying a field of 6 kV to the sample, the SHG signal begins to increase B. Upon increasing the field to 8 kV, a strong SHG signal is observed. The maximum SHG response is observed at C yielding a d_{eff} of $\sim$10.3 $d_{11,Q}$ of quartz, which corresponds to a value of $\sim$3.5 pm/V. Upon cooling the film with the corona field on, a decrease is noted due to the temperature dependence of the nonlinear optical susceptibility. Immediately after poling, a d_{eff} of $\sim 7.9 \times d_{11,Q}$ ($\sim$2.7 pm/V) was measured at room temperature.

The low-effective, nonlinear optical coefficient for as-deposited films has been attributed to its microstructure whereby the *a*-axis is perpendicular to the substrate (32). The *c*-axis lies in the film plane with the lattice dipole randomly aligned. This leads to a low SHG response prior to poling. Upon poling, the *c*-axis orients perpendicular to the substrate because of the interaction between the lattice dipole and the electric field. This alignment of the *c*-axis can be observed by X-ray diffraction and is consistent with the observed behavior of the ferroelectric properties upon poling.

The resultant alignment results in an increase in the SHG response

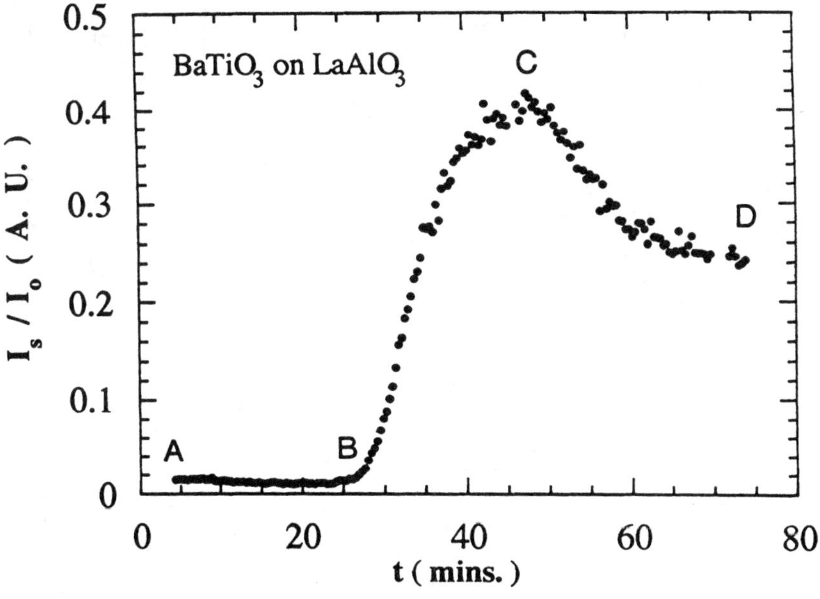

Figure 9 SHG intensity from a $BaTiO_3$ thin film measured in situ during poling (31).

because the ferroelectric domains are aligned perpendiclar to the substate. In-plane, however, poling is not expected to introduce any alignment of the *a*-axis. This may explain the low measured value for d_{eff} in the plane of the film.

From SHG measurements using polarized light, the symmetry of the films was studied (32). In this case, the incident polarization was rotated in the plane perpendicular to the incidence plane. Figure 10 shows the *p*-polarized SH intensity measured for a poled film. The SH signal intensity was plotted as a function of the angle Φ between the incidence plane and the incident polarization with $\Phi = 0$ corresponding to *p*-p and $\Phi = 90^\circ$ corresponding to s-p. Excellent agreement between the experiment and theory is obtained for the measured polarization dependence whereby

$$I_s = (A\cos^2\Phi + B\sin^2\Phi + C\cos\Phi\sin\Phi)^2 I^2(\omega)$$

is obtained when $C = 0$. This indicates that there was no in-plane symmetry for the poled film. A ratio of $(I_p \rightarrow {}_p/I_s \rightarrow {}_p) \sim 14$–17 was measured for the poled films, which corresponds to a ratio of $d_{33}/d_{31} \rightarrow 3.6$ for the nonlinear optical coefficients.

A measured $d_{eff} \sim 2.7$ pm/V for the film immediately after poling corresponds to a d_{33} of ~ 10 pm/V and a $d_{31} \sim 2.9$ pm/V. Upon comparison

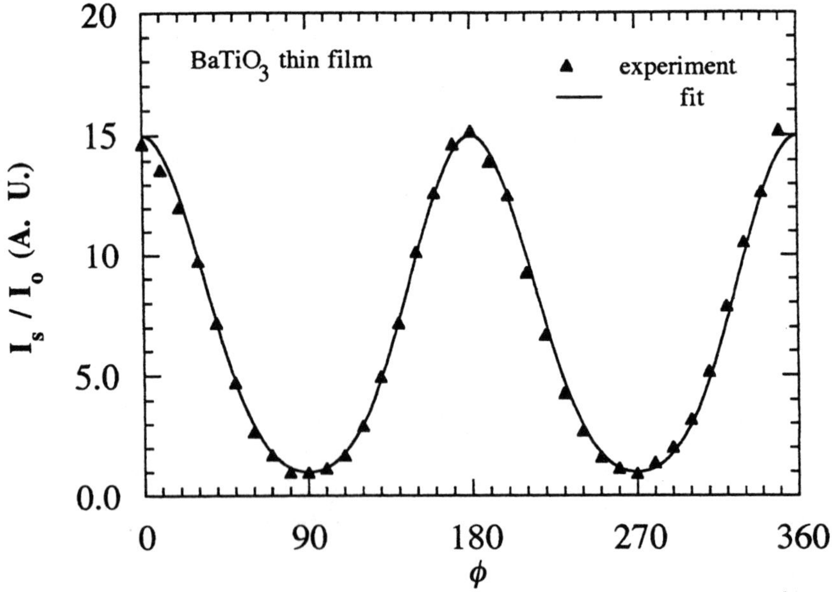

Figure 10 P-polarized SH intensity measured from a poled $BaTiO_3$ film plotted as a function of the incident fundamental polarization angle (32).

of these values with the bulk values, it is concluded that $d_{33}^{film} \sim d_{33}^{bulk}$, whereas $d_{31}^{film} \ll d_{31}^{bulk}$. This is consistent with a film microstructure with grain alignment in the *c*-direction but random alignment of the *a*-axes in-plane.

As to the stability of the NLO properties, the SHG of the poled $BaTiO_3$ films initially decreased by 25%, after which a slow decay was noted with time.

LUMINESCENT PROPERTIES

There has been considerable interest in the utilization of thin-film oxides as an optical gain media. Of particular interest are the rare earth–doped ferroelectrics, which offer the possibility of enabling novel devices that combine their electro-optical and nonlinear optical properties, as well as optical gain associated with rare earth ions (33). Devices such as self-frequency doubled, self-Q-switched, and self-modulated lasers have been proposed. In addition, optical amplifier circuits with no insertion loss are potentially possible. A number of rare earth ions, including Nd, Er and Tm, have been proposed to form the optical gain media. Of particular interest is Er because lasers have been prepared in other ferroelectrics that operate near 1.54 μm.

Synthesis of rare-earth-doped $BaTiO_3$ MOCVD thin films has been only recently reported (33–35). Initial studies were centered on the rare-earth dopant La, which behaves like a donor; however, it is not optically active (SR Gilbert et al, unpublished data). Er-doped $BaTiO_3$ films have been reported that are optically active (33, 35). The films were synthesized by incorporating tris-tetramethylheptanedionate erbium [Er(thd)$_3$] into the gas stream during deposition. Er incorporation as high as 5×10^{21} cm^{-3} was measured using energy dispersive X-ray analysis.

Strong photoluminescence from the Er-doped $BaTiO_3$ was observed under optical excitation as indicated in Figure 11. A peak at 0.800 eV (1.54 μm) is observed that corresponds to the intra-f shell transitions between the first excited state ($^4I_{13/2}$) and the ground state ($^4I_{15/2}$) of the Er^{3+} ion. Strong photoluminescence was also observed at room temperature. The intensity of the room-temperature emission was a factor of two lower than that at 16 K. The FWHM of the peak is 5 meV at 16 K and 30 meV at 300 K, both of which are instrumentally broadened. Time-resolved measurements of the photoluminescence decay indicate the lifetime is 8 ms at 16 K and 7 ms at 300 K (Figure 12). For comparison, the Er^{3+} center PL lifetimes are typically 1 to 10 ms in insulating crystals and 5 to 15 ms in glasses. The long lifetimes and high concentrations of Er^{3+} centers

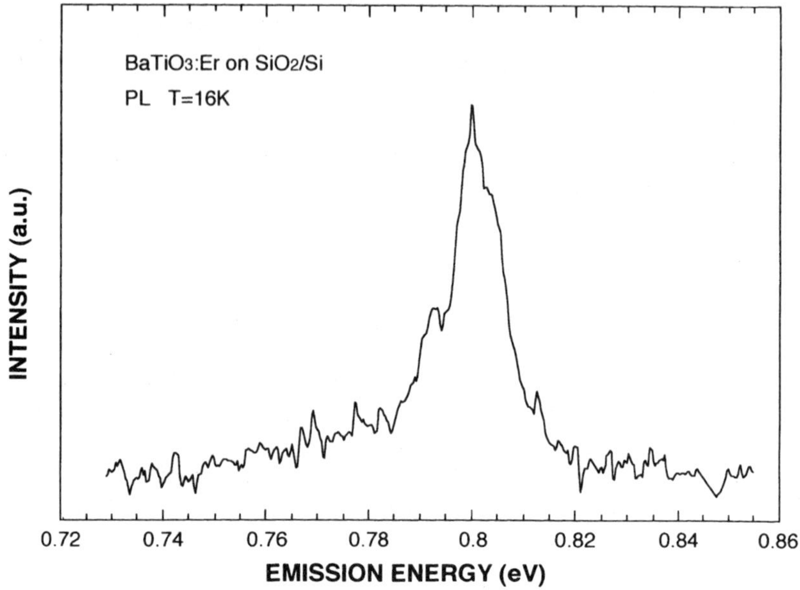

Figure 11 Photoluminescence spectra at 16 K of Er-doped $BaTiO_3$ (33).

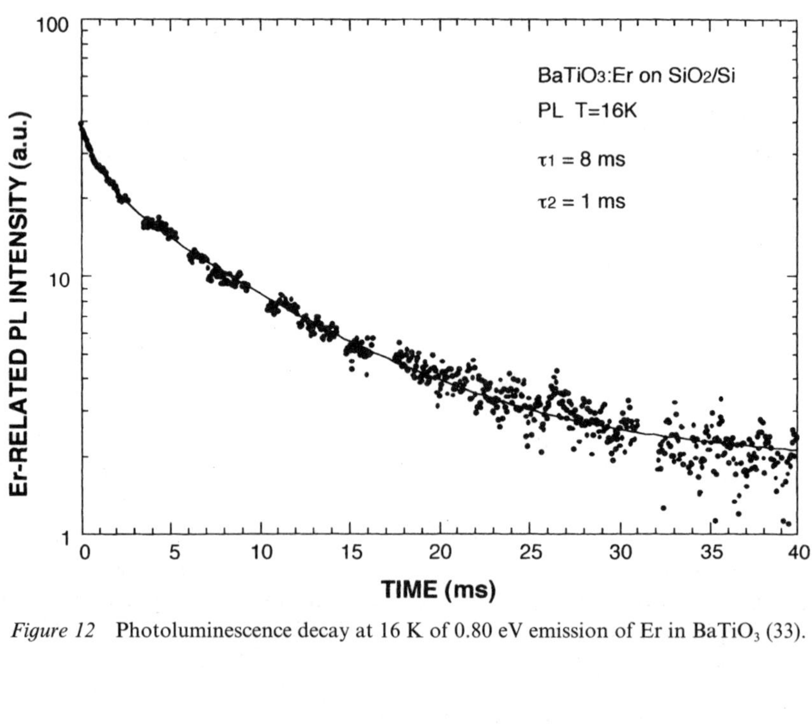

Figure 12 Photoluminescence decay at 16 K of 0.80 eV emission of Er in $BaTiO_3$ (33).

that can be incorporated suggest that $BaTiO_3$: Er has excellent potential as a ferroelectric laser-gain medium.

CONCLUSIONS

Metal-organic chemical vapor deposition is a powerful technique for the synthesis of ferroelectric thin films. Using this deposition method, films of $BaTiO_3$, $SrTiO_3$, and their solid solution can be deposited with properties approaching that of the bulk. The recent attainment of epitaxial thin films of these materials potentially enables their use in a wide variety of applications based on their ferroelectric, electro-optic, and nonlinear optical properties. The prospects of integrating these functions through thin-film technology are closer at hand. Future widespread usage will depend upon continued improvements in compositional and thickness control, conformal coating, epitaxial film quality, and optical quality.

ACKNOWLEDGMENTS

The Office of Basic Energy Sciences of the Department of Energy and the National Science Foundation Materials Research Center at Northwestern University supported the work at Northwestern University. Extensive collaborations with Professors TJ Marks and VP Dravid and members of the Electronic Materials Group at Northwestern are greatly appreciated.

Literature Cited

1. Feil WA, Wessels BW, Tonge LM, Marks TJ. 1990. *J. Appl. Phys.* 67: 3858–61
2. Sato R, Takahashi K, Yoshino M, Kato H, Ohshima S. 1993. *Jpn. J. Appl. Phys.* 32: 1590–94
3. Sievers RE, Sadlowski JE. 1978. *Science* 201: 217–23
4. Van der Sluis P, Spek AL, Timmer K, Meinema HA. 1990. *Cryst. Struct. Commun. C* 46C: 1741
5. Malandrino G, Richeson DS, Marks TJ, DeGroot DC, Schindler JL, Kannewurf CR. 1991. *Appl. Phys. Lett.* 58: 182–84
6. Shamlian SH, Hitchman ML, Cook SL, Richards BC. 1994. *J. Mater. Chem.* 4: 81–85
7. Schulz DL, Hind BJ, Stern CL, Marks TJ. 1993. *Inorg. Chim.* 32: 249–50
8. Wills LA, Feil WA, Wessels BW, Tonge BW, Marks TJ. 1991. *J. Cryst. Growth* 107: 712–15
9. Feil WA. 1991. *Electrical and optical characterization of defects in strontium titanate thin films grown by organometallic chemical vapor deposition.* PhD thesis. Northwestern Univ., Evanston, IL. 338 pp.
10. Yamaguchi H, Lesaicherre P, Sakuma T, Miyasaka Y, Ishitani A, Yoshida M. 1993. *Jpn. J. Appl. Phys.* 9B: 4069–73
11. Liang S, Chern C, Shi ZQ, Lu P, Keer BH, Safari A. 1994. *Mater. Res. Soc. Proc.* 335: 113–16
12. Wessels BW, Wills LA, Lu HA, Gilbert SR, Neumayer DA, et al. 1993. *Chemical Vapor Deposition XII. Electrochemical Society, Honolulu.*

13. Kwak BS, Zhang K, Boyd EP, Erbil A, Wilkens BJ. 1991. *J. Appl. Phys.* 69: 767–72
14. Nakazawa H, Yamane H, Hirai T. 1991. *Jpn. J. Appl. Phys.* 30: 2200–3
15. Feil WA, Wills LA, Wessels BW, Schulz DL, Richeson DS, Marks TJ. 1991. *Growth of strontium, titanate and barium titanate thin films by organometallic chemical vapor deposition.* Presented at TMS Electron. Mater. Conf. Boulder, CO. (Abstr.).
16. Wills LA, Wessels BW, Schulz DL, Marks TJ. 1991. *$BaTiO_3$ thin films prepared by organometallic chemical vapor deposition.* Mater. Res. Soc. Boston. (Abstr.)
17. Wills LA, Wessels BW, Schulz DL, Marks TJ. 1992. *Mater. Res. Soc. Proc.* 243: 47–52
18. Wills LA, Wessels BW, Richeson DS, Marks TJ. 1992. *Appl. Phys. Lett.* 60: 41–43
19. Chen J, Wills LA, Wessels BW, Schulz DL, Marks TJ. 1992. *J. Electron. Mater.* 22: 701–3
20. Chern CS, Zhao J, Luo L, Li YQ, Norris P, et al. 1992. *Appl. Phys. Lett.* 60: 1145–46
21. VanBuskirk PC, Gardiner R, Kirlin PS, Krupanidhi S. 1992. *J. Vac. Sci. Tech. A* 10: 1578–83
22. Yoon YS, Kang WN, Shin HS, Yom SS, Kim TW, et al. 1993. *J. Appl. Phys.* 73: 1547–49
23. Kim TW, Jung M, Kim HJ, Yoon YS, Kang WN, Yom SS. 1993. *Appl. Phys. Lett.* 62: 1788–90
24. Deleted in proof
25. Gilbert SR, Wessels BW, Neumayer DA, Marks TJ, Schindler JL, Kannewurf C. 1994. *Mater. Res. Soc. Proc.* 335: 41–46
26. Chern CS, Liang S, Shi ZQ, Safari A, Lu P, Kear BH. 1994. *Mater. Res. Soc. Proc.* 335: 29–34
27. Feil WA, Wessels BW, Tonge LM, Marks TJ. 1990. *Mater. Res. Soc. Proc.* 168: 375
28. Liang S, Chern CS, Shi ZQ, Lu Y, Lu P, et al. 1994. *Mater. Res. Soc. Proc.* 335: 113–16
29. Lu HA, Wills LA, Wessels BW. 1994. *Appl. Phys. Lett.* 64: 2973–75
30. Lu HA, Wills LA, Wessels BW, Zhan X, Helfrich JA, Ketterson JB. 1994. *Mater. Res. Soc. Proc.* 310: 319–23
31. Lu HA, Wills LA, Wessels BW, Lin WP, Zhang TG, et al. 1993. *Appl. Phys. Lett.* 62: 1314–16
32. Lu HA, Wills LA, Wessels BW, Lin WP, Wong GK. 1993. *Opt. Mater.* 2: 169–73
33. Block BA, Wessels BW. 1994. *Appl. Phys. Lett.* 65: 25–27
34. Deleted in proof
35. Block BA, Wessels BW. 1995. *Integr. Ferroelectr.* 25: 25–32

Annu. Rev. Mater. Sci. 1995. 25: 547–600

LOW-TEMPERATURE GROWN III-V MATERIALS

M. R. Melloch, J. M. Woodall, and E. S. Harmon

School of Electrical Engineering and the MRSEC for Technology-Enabling Heterostructure Materials, Purdue University, West Lafayette, Indiana 47907–1285

N. Otsuka

School of Materials Science, Japan Advanced Institute of Science and Technology, Hokuriku, Nomigun, Ishikawa 923–12, Japan

Fred H. Pollak

Department of Physics, Brooklyn College of the City of New York, Brooklyn, New York 11210

D. D. Nolte

Department of Physics and the MRSEC for Technology-Enabling Heterostructure Materials, Purdue University, West Lafayette, Indiana 47907–1285

R. M. Feenstra and M. A. Lutz

IBM Research Division, T. J. Watson Research Center, Yorktown Heights, New York 10598

KEY WORDS: low-temperature growth, molecular beam epitaxy, GaAs : As, metal/semiconductor composite

ABSTRACT

A new kind of semiconductor composite material is demonstrated with a dispersion of semimetallic particles with properties common to high-

0084–6600/95/0801–0547$05.00

quality single crystals. These materials are arsenides such as GaAs, AlGaAs, and InGaAs containing arsenic clusters. The composites are formed by incorporating excess arsenic in the semiconductor, which percipitates in the anneal. The incorporation of the excess arsenic is accomplished by molecular beam epitaxy at low substrate temperatures. The cluster density can be controlled with the coarsening annealing. The positioning of the clusters can be controlled with heterojunctions and doping. These composites exhibit several interesting properties, including high-resistivity, appreciable optical absorption below the band gap of the semiconductor matrix material, a large electro-optic effect, and very short carrier lifetimes.

INTRODUCTION

Molecular beam epitaxy (MBE) is a film growth technique where atomic or molecular beams are thermally evaporated onto a heated substrate in an ultra-high vacuum (1). The ultra-high vacuum allows monitoring of the growth with techniques such as reflection high-energy electron diffraction (RHEED) (2) and desorption mass spectroscopy (3). By far the largest application of MBE has been for the growth of III-V materials, especially the arsenides. Typical MBE systems have eight ports for effusion furnaces for evaporation of source material. Typical group III sources are gallium, aluminum, and indium. The dominant group V source is arsenic, although antimony and phosphorous sources are not atypical. Arsenic sublimes as a tetramer molecule As_4. Often the arsenic effusion furnace is equipped with a cracking zone that results in dimer molecules of As_2. Silicon is used for the *n*-type dopant and beryllium for the *p*-type dopant almost exclusively for III-V MBE. The typical conditions for MBE of high-quality GaAs are a substrate temperature of 600°C, an As_4-to-Ga beam equivalent pressure (bep) ratio of 15–20, and a growth rate of 1 μm/h. (The bep is measured with an ion gauge in the growth position.) With the Ga effusion furnace at a temperature near 900°C to obtain a 1 μm/h growth rate, every Ga atom that impinges on the substrate at a temperature of 600°C will remain and have enough thermal energy to find a lattice site. The arsenic molecules, originating from an effusion furnace at a temperature of about 250°C, will only result in the incorporation of an arsenic atom if there is a Ga atom to bond to. Therefore, the MBE of arsenides readily lends itself to the growth of highly stoichiometric material. During the last six years there has been considerable interest in the MBE of arsenides at lower substrate temperatures—as low as 200°C. As described below, this has made possible the growth of nonstoichiometric material of high crystal quality, which has many interesting and useful properties.

One of the first applications of a GaAs epilayer grown at low substrate temperatures by MBE was by Murotani et al (4). They were investigating the MBE of GaAs on substrates with a temperature range of 400 to 600°C and found that the material became semi-insulating below a critical substrate temperature—which had obvious applications for buffer layers for MESFETs—so they placed such a buffer layer under a GaAs power metal-semiconductor field-effect transistor (MESFET) (4). This application received little notice until 1988 when Smith et al (5) investigated the use of GaAs grown in the neighborhood of 200°C as a buffer layer under a MESFET and found that it completely eliminated the problem of sidegating. Melloch et al (6) then showed that state-of-the-art high electron mobility transistor (HEMT) material could be grown on top of GaAs buffer layers grown at extremely low substrate temperatures (250°C) and that such buffer layers might have the beneficial effect of slowing diffusion of impurities from the substrate toward the active layers.

There has been an enormous amount of research since 1988 on the MBE of arsenides grown at low substrate temperatures. These materials are usually referred to in the literature as (low-temperature) LT-GaAs. The adoption of the term LT-GaAs is unfortunate because those not familiar with the field obviously assume one is referring to properties of GaAs at low temperatures, not to new materials that exhibit new physics with exciting electronic and optoelectronic device possibilities. The term LT-GaAs also has been used to refer to a broad range of materials whose only thread of commonality is the initial processing step of MBE at low substrate temperatures. As-grown materials are highly nonstoichiometric with the excess arsenic in the form of point defects (7). With sufficient anneal, the nature of the material changes completely, with the excess arsenic precipitating to form semi-metallic arsenic clusters in a GaAs semiconductor matrix (8). Therefore, better terminology would be low-temperature grown (LTG) GaAs to refer only to the as-grown material, and GaAs:As or metal/semiconductor composite to refer to the annealed material.

There have been a number of controversies surrounding LTG-GaAs. The initial controversy was concerned with what happened to the excess arsenic after anneal. First reports of transmission electron microscopy (TEM) of these materials indicated that after anneal the LTG-GaAs epilayers were devoid of any obvious defects (9). However, it is now known that the excess arsenic precipitates during annealing (8, 10–19). Another controversy surrounding LTG-GaAs was a report that the material was superconducting (20). However, it is believed this report was related to the indium used for mounting the substrate during MBE.

Some controversy still remains about the role of residual defects vs arsenic clusters in controlling the material properties (21–26). Part of the debate is whether the arsenic clusters act as embedded Schottky contacts. There is considerable supporting evidence from measurement techniques such as transmission electron microscopy (TEM), scanning tunneling microscopy (STM), and others that the arsenic clusters are indeed embedded Schottky contacts. Transmission electron diffraction clearly indicates the crystal structure of the arsenic clusters as hexagonal (8), the semi-metallic form for arsenic. STM current-voltage spectroscopy reveals a depletion region around the arsenic precipitates in annealed LTG-*n*-GaAs, indicating that the arsenic clusters are acting as buried Schottky barriers (27). Obviously, the properties of as-grown material are dominated by point defects, but as one anneals, the role of defects (which are disappearing) will decrease, whereas the role of the arsenic clusters (which are precipitating) will increase. Another part of the controversy of the role of point defects vs arsenic clusters arises because of this gradual transition and from drawing global conclusions without investigating a spectrum of anneals that take the material from the point-defect-dominated form to the two-phase composite form.

In this article, we describe the fabrication of these materials, review the material properties as one transitions from the as-grown point-defect-dominated material to the annealed composite form, and look at many of the device applications that these new materials have already enabled.

ARSENIC CLUSTER ENGINEERING

Low-Substrate-Temperature Molecular Beam Epitaxy

In 1969, Cho (28) investigated MBE of GaAs at room temperature and found that the deposited material was amorphous, but that if the layer was thin enough, it would turn crystalline with annealing. Murotani et al (4) investigated the MBE of GaAs down to a substrate temperature of 400°C and found that the material remained crystalline and became semi-insulating below a critical temperature. Smith et al (5) investigated the MBE of GaAs in the neighborhood of 200°C, finding that it remained crystalline and that epilayers suitable for MESFETs could be grown on top of these LTG-GaAs epilayers. Eaglesham et al (29) have demonstrated epitaxy of GaAs down to substrate temperatures around 140°C. However, they found there is a critical thickness beyond which the film turns amorphous. This critical thickness is a function of the amount of excess arsenic incorporated into the epilayer, which is a function of the substrate temperature (12, 30); the arsenic species-to-Ga flux ratio (29); and the arsenic species (the dimer As_2 or the tetramer As_4) used (10). GaAs epilayers as thick as 1 μm with 1.5% excess arsenic have been demonstrated (7, 8). Ibbetson et al (31) have determined through their observation of RHEED oscillations that even at these low temperatures, MBE of GaAs is a layer-by-layer process. To observe the RHEED oscillations at these low substrate temperatures requires a smooth surface free of excess arsenic. Below 200°C the RHEED oscillations are very sensitive to the arsenic-to-Ga-flux ratio, the strongest oscillations being obtained under stoichiometric conditions.

The excess arsenic incorporated into GaAs during low-substrate-temperature MBE is in the form of point defects, such as arsenic antisites and possibly arsenic interstitials (7), and the epilayer is strained (7, 32, 33). The amount of excess arsenic can be controlled with the substrate temperature during MBE. The lower the substrate temperature, the greater is the amount of excess arsenic that is incorporated (12, 34). The double crystal rocking curve for a LTG-GaAs epilayer containing 1.5% excess arsenic is shown in Figure 1—a perpendicular mismatch of 0.133% is observed. What is remarkable about the rocking curve in Figure 1 is that although the epilayer was grown at a substrate temperature of 225°C and contains 1.5% excess arsenic, there is a clear presence of Pendellösung fringes. The Pendellösung fringes are caused by interference between the X-ray fields from the epilayer and the substrate. To observe these fringes, there must be a high degree of coherence between the two X-ray fields, indicating a high degree of structural quality for the epilayer (35). Most of the strain in LTG-GaAs is relaxed with modest anneals, e.g. 600°C for 10 to 30 min or 700°C or higher for 30 s or less (7, 32, 33). This strain relaxation is accompanied by precipitation of the excess arsenic (8). Figure 2 shows TEM images of two samples from a LTG-GaAs epilayer, one that was annealed for 30 s at 550°C and one that was annealed for 30 s at 900°C. Clearly seen in Figure 2 is a coarsening of the clusters with annealing; the sample that was annealed at 550°C for 30 s has a much higher density of smaller arsenic clusters than the sample annealed at 900°C for 30 s. This coarsening occurs because the two-phase system of the arsenic clusters in the GaAs matrix is minimizing its free energy with annealing by reducing the interfacial area between the arsenic precipitates and the GaAs matrix—an Ostwald ripening process (36). This process of incorporating excess arsenic followed by an anneal also has been used to obtain arsenic precipitates in AlGaAs (12) and InGaAs (17–19) semiconductor matrices. The wide range of control that one has over the structure of these composites is illustrated in Figure 3, where the average density and diameter of the arsenic clusters is shown as a function of temperature for a 30 s anneal for four different epilayers. Three of these epilayers are GaAs, one with 0.9% excess arsenic, one with 0.7% excess arsenic, and one with 0.3% excess arsenic. The fourth epilayer is $Al_{0.25}Ga_{0.75}As$ containing 0.3% excess

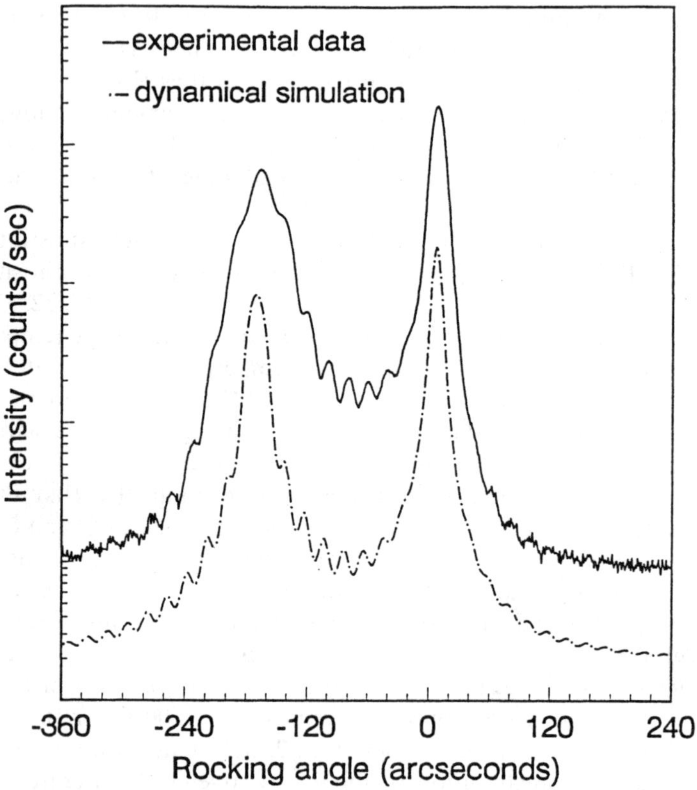

Figure 1 Experimental and calculated (004) double-crystal rocking curves for LTG-GaAs sample containing 1.5% excess arsenic.

arsenic. The amount of the excess arsenic in these epilayers was fixed by the substrate temperature during MBE.

Taking advantage of the strain relaxation that occurs with anneal, Melloch et al (15) have demonstrated a technique to grow arbitrarily thick LTG-GaAs layers. After growth of 2 μm of material, the substrate temperature was cycled to 600°C and then back to the growth temperature of 250°C. With this approach the strain is relaxed before the critical thickness is reached and LTG-GaAs with arbitrary thickness can be grown.

Extrinsic Control of Arsenic Clusters with Doping

When dopants are incorporated into LTG-GaAs, the arsenic cluster coarsening processes can be significantly affected. In MBE, the typical *n*-type dopant is Si and the typical *p*-type dopant is Be. For normal substrate

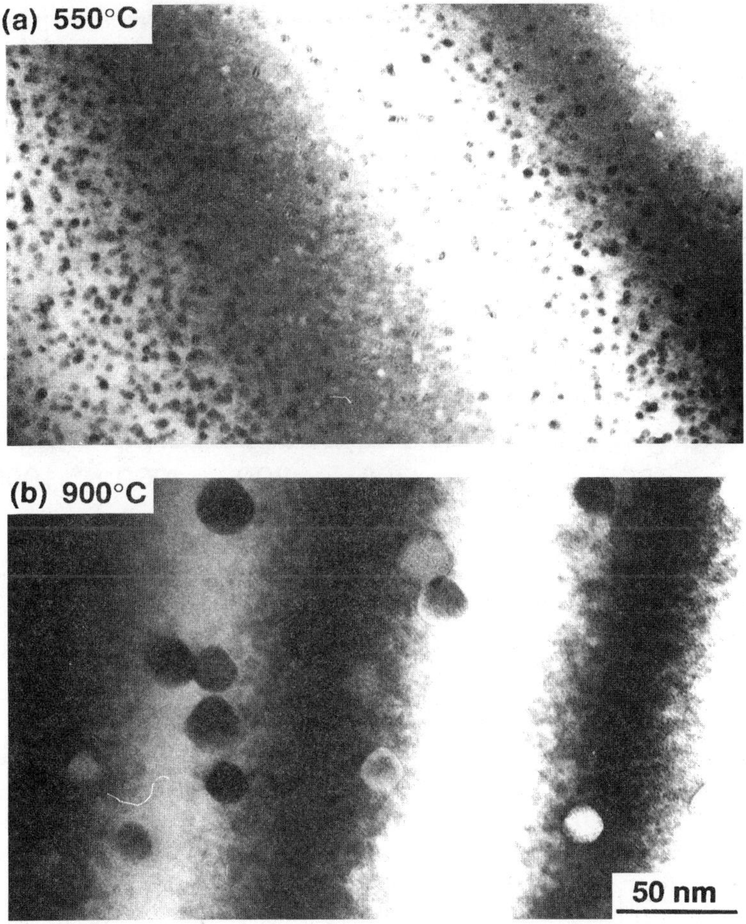

Figure 2 TEM images of two samples from the same LTG-GaAs epilayer that were annealed at (*a*) 550°C for 30 s and (*b*) 900°C for 30 s.

temperature (600°C) MBE, Si is incorporated as a donor for concentrations up to low 10^{18} cm^{-3}. At higher concentrations, a significant amount of the Si is incorporated on the arsenic site and the material becomes compensated. As the substrate temperature is lowered from 600°C, the concentration of Si on donor sites (before compensation occurs) increases (37). When very low substrate temperatures are reached during MBE, there is uncertainty concerning the site for the incorporation of Si. Using localized vibration mode (LVM) spectroscopy, Manasreh et al (38) have seen that below a substrate temperature of 300°C there is a significant decrease in the amount of the incorporated Si that is substitutional. Man-

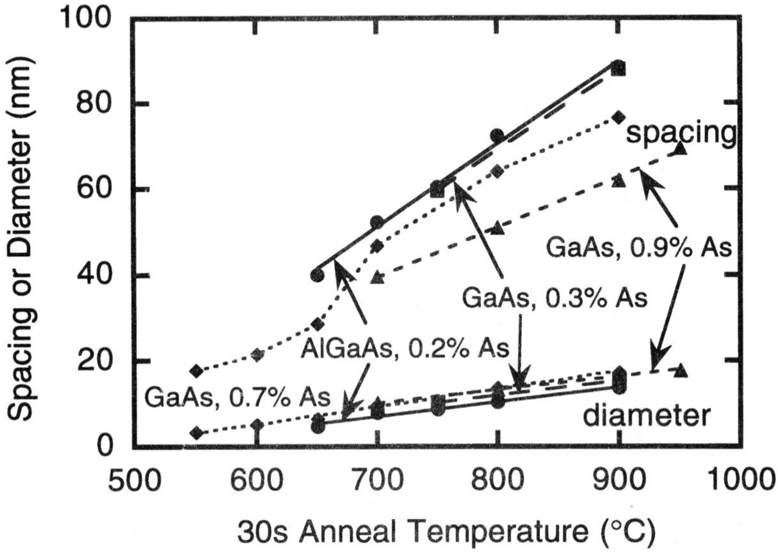

Figure 3 Density and average size of arsenic clusters in four different epilayers as a function of temperature for a 30 s anneal. Three of the epilayers are GaAs, one containing 0.3% excess arsenic, one containing 0.7% excess arsenic, and one with 0.9% excess arsenic. The fourth epilayer is AlGaAs containing 0.2% excess arsenic.

asreh et al (38) report a similar decrease in the incorporation of Be on substitutional sites for substrate temperatures below 300°C. However, McQuaid et al (39) found that almost all Si atoms were on Ga sites for a GaAs layer grown at 200°C with Si concentrations of 1×10^{19} cm^{-3}. The results of Manasreh and McQuaid indicate that the incorporation of Si on Ga sites at low substrate temperatures may be sensitive to the growth condition, i.e. group III-to-IV flux ratio, as has been observed by O'Hagan & Missous (40).

The effect of planes of dopants on the arsenic cluster coarsening was investigated by Melloch et al (14) and is illustrated with the TEM image in Figure 4. The white lines in the image are AlAs markers, the rest of the regions are GaAs. Planes of Si or Be dopants were incorporated into the epilayers at the positions indicated. These dopant planes were accomplished by interrupting growth by closing the shutter in front of the Ga effusion cell while maintaining the arsenic flux. The shutter in front of the Si (or Be) dopant effusion furnace was then opened to deposit the dopant to a sheet concentration of 5×10^{12} cm^{-2}. To make sure there was no effect of the growth interruption on the arsenic precipitation, the lower GaAs region in Figure 4 contained a growth interruption of the same

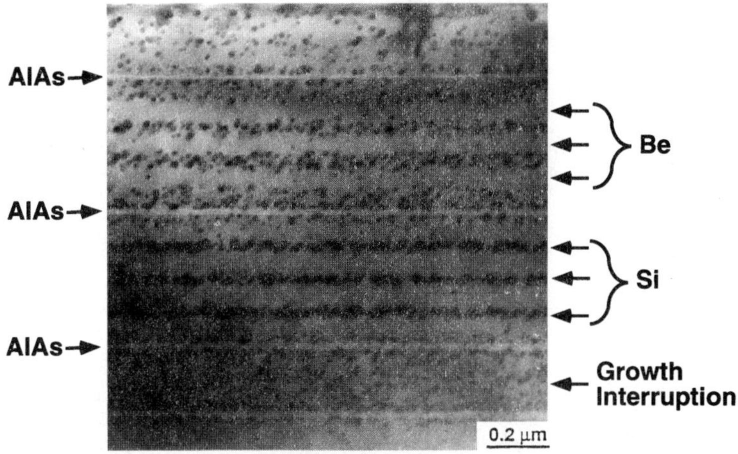

Figure 4 TEM image of a film structure grown at 250°C using As_2 and annealed for 30 s at 800°C. The white lines are 5 nm AlAs layers. Between the lower two AlAs layers is a 200-nm GaAs region that contained a 3-min growth interruption in the center. There is no apparent effect on the arsenic cluster coarsening due to the growth interruption. The next 400-nm region is a GaAs layer that contains three planes of Si dopants. The arsenic clusters are lined up on the planes of Si dopants. The next 400-nm region is a GaAs layer that contains three planes of Be dopants. The arsenic clusters are seen to coarsen preferentially between the planes of Be dopants.

duration (3 min) as used to deposit the planes of dopant atoms. After growth, the epilayer was annealed for 30 s at 800°C. Clearly seen in the figure is a preferential coarsening of the arsenic clusters at the planes of Si atoms, the *n*-type dopant, and away from the planes of Be atoms, the *p*-type dopant. The growth interruption had no effect on the arsenic cluster coarsening process. Cheng et al (41) have used this preferential coarsening of arsenic precipitates at Si delta-doped regions to demonstrate a GaAs/As superlattice.

To further investigate the effect of doping on the arsenic cluster coarsening process, Melloch et al (15) grew a *pn* doping superlattice where each doped region was 200-nm thick and doped to a concentration of 1×10^{18} cm^{-3}. Shown in Figure 5 is the TEM image of the epilayer after annealing to (*a*) 700°C and (*b*) 900°C for 30 s. With increased annealing there is a clear coarsening of the arsenic clusters and a preferential coarsening of the arsenic from the *p*-LTG-GaAs regions to the *n*-LTG-GaAs regions. Similar observations have been reported by Ibbetson et al (16). The behavior of the arsenic cluster–coarsening with doping is believed to be caused by the Schottky nature of the arsenic clusters (15). In *n*-GaAs : As, the arsenic clusters are negatively charged, whereas in *p*-GaAs : As they are positively

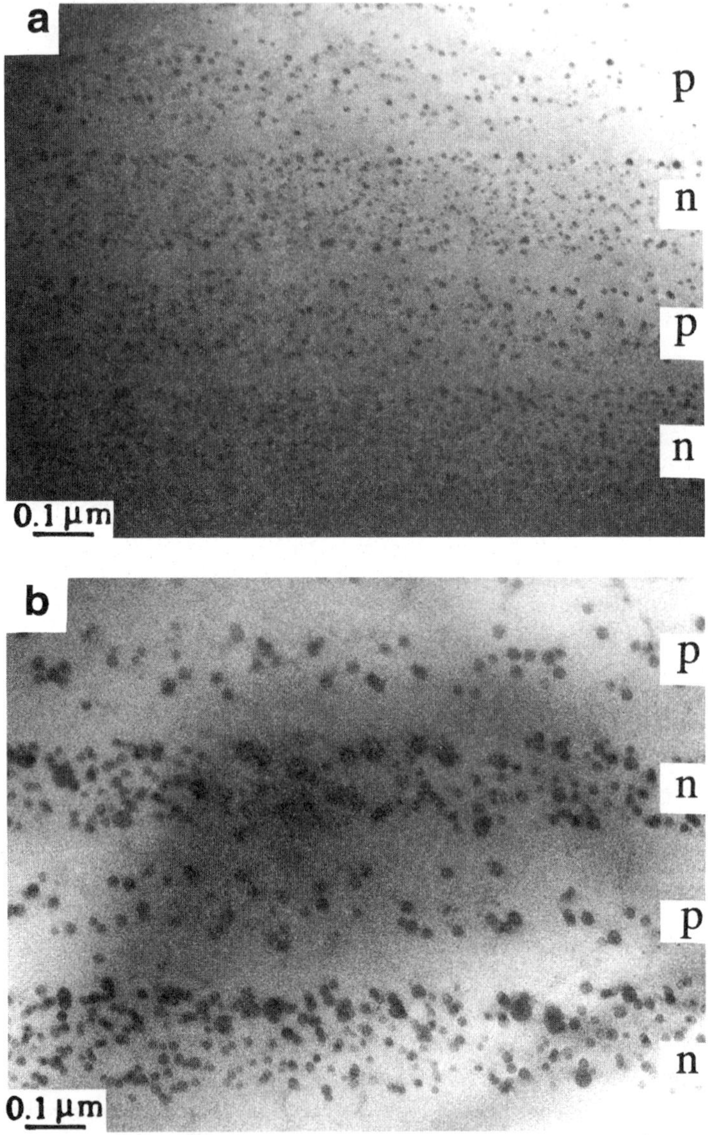

Figure 5 TEM image of a series of 200-nm GaAs regions uniformly doped with Si at 1×10^{18} cm^{-3} followed by a 200-nm GaAs region uniformly doped with Be at 1×10^{18} cm^{-3}, and subsequently annealed at (*a*) 700°C for 30 s and (*b*) 900°C for 30 s.

charged. Therefore, if an arsenic antisite or interstitial is thermally emitted from an arsenic cluster and is ionized, it will be positively charged and attracted to the arsenic cluster in *n*-GaAs : As but repelled from an arsenic cluster in *p*-GaAs : As. Thus the arsenic clusters will be more stable in *n*-GaAs : As than in *p*-GaAs : As and account for the preferential coarsening in *n*-doped regions.

Secondary ion mass spectroscopy (SIMS) of samples annealed at 800°C for 30 s and 950°C for 30 s are shown in Figure 6. There is no detectable diffusion of the dopants even for the 950°C 30 s anneal. In fact, SIMS characterization of a sample annealed for 10 min at 900°C also showed no discernible diffusion of the Be, unlike what is observed for Be in GaAs grown at normal (600°C) substrate temperatures. Apparently the excess arsenic in the film is preventing the Be from being interstitial, where it is a fast diffuser, and keeping it on substitutional sites, where it is a slow diffuser. Also seen in Figure 6 is the excess arsenic in the layers grown at 250°C (top 1.2 μm) compared to the underlying stoichiometric layers grown at 600°C. The variation in the arsenic concentration is clearly seen, with decreased concentrations in the regions corresponding to the *p-n* junction depletion regions, consistent with the TEM observations of Figure 5.

Effect of Heterojunctions on Arsenic Cluster Formation

Mahalingam et al (13, 42) first noticed that if a heterojunction were present, the coarsening with anneal resulted in an accumulation of arsenic precipitates on the GaAs side of the heterojunction and a depletion of arsenic clusters on the AlGaAs side of the heterojunction. The behavior of the arsenic cluster formation and coarsening in a superlattice is illustrated in the TEM image of Figure 7. This is a TEM image of a series of 10-period $Al_{0.3}Ga_{0.7}As$/GaAs regions of progressively wider $Al_{0.3}Ga_{0.7}As$ barriers or GaAs wells, grown at low substrate temperatures and then annealed at 700°C for 30 s. For the wide GaAs wells, the arsenic clusters are completely contained within the GaAs well regions. As the well regions become narrower, the arsenic clusters are centered in the GaAs well regions but extend into the AlGaAs barrier regions. Finally the 10-period superlattice with the narrowest well and barrier widths is almost devoid of arsenic clusters, but the regions surrounding it (a GaAs region below and a superlattice above) contain arsenic clusters.

The behavior of the arsenic precipitate coarsening at heterojunctions and superlattices can be explained based on the difference in interfacial energies between an arsenic precipitate in GaAs and an arsenic precipitate in AlGaAs. Because the Al-As bond is stronger than the Ga-As bond, the interfacial energy will be lower between an arsenic precipitate and GaAs

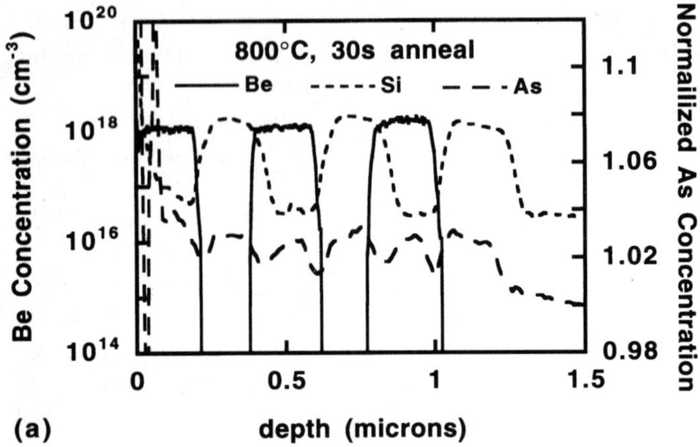

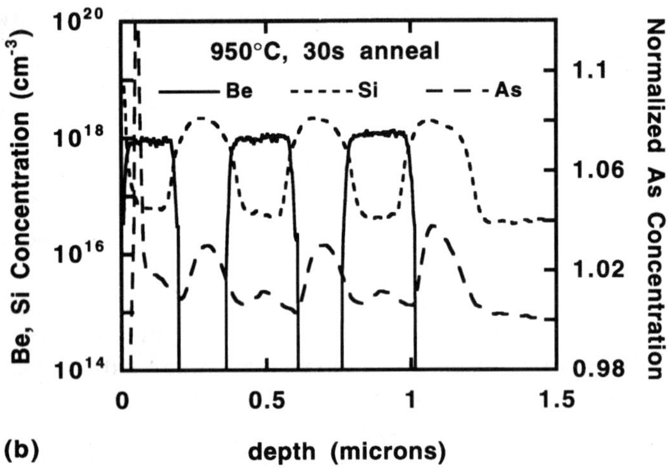

Figure 6 Secondary ion mass spectra of the doping superlattice after (*a*) 800°C 30 s annealing and (*b*) 950°C 30 s annealing. There is virtually no movement of the dopants, but the redistribution of the arsenic is evident. The arsenic concentration was normalized to its value in the film region grown at a substrate temperature of 600°C. Note that the Cs profiling of the Si gives much poorer depth resolution than the O profiling of the Be. This is why the Si profiles in the *n*-regions do not appear flat compared with the Be profiles in the *p*-regions.

than between an arsenic precipitate and AlGaAs. Therefore, the arsenic precipitates will preferentially coarsen from AlGaAs regions to GaAs regions. This is why the arsenic precipitates in the regions in Figure 7 with the thicker quantum wells are all contained in the quantum well regions. However, for the particular anneal condition used for the sample of Figure

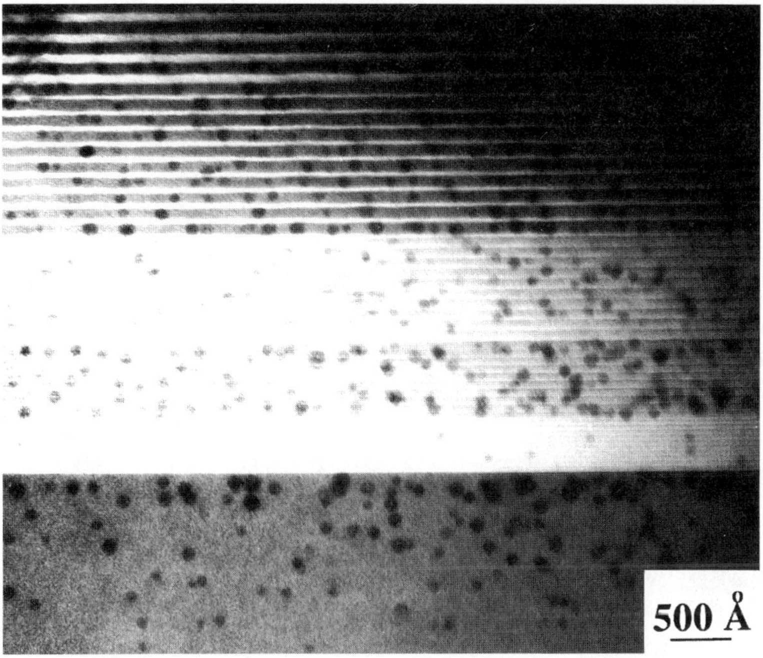

Figure 7 A series of 10-period superlattices of progressively wider $Al_{0.3}Ga_{0.7}As$ barriers or GaAs wells grown at low temperatures and then annealed at 700°C for 30 s.

7, in the thinner quantum well regions the arsenic precipitates are centered in the quantum wells but extend into the AlGaAs barriers because completely enclosing the arsenic precipitates in the quantum wells would require a larger number of smaller precipitates. This larger number of smaller precipitates would result in a larger arsenic precipitate-to-matrix interfacial area ratio and hence larger interfacial energy than fewer precipitates that are centered in the GaAs well regions that extend into the AlGaAs regions. Finally, the 10-period superlattice with the thinnest GaAs barriers is almost devoid of arsenic precipitates because it is thin compared with an arsenic diffusion length, and it is surrounded by regions of lower arsenic precipitate-to-matrix interfacial energies, i.e. the GaAs region below, and the 10-period superlattice above that has thicker GaAs well regions.

The behavior of the arsenic precipitates at AlGaAs/GaAs heterojunctions can be used to form planes of arsenic precipitates with high density. Shown in Figure 8 is the TEM image of a sample with 10-nm GaAs wells that are separated by 100-nm AlGaAs barriers. This structure was grown at low substrate temperatures so that excess arsenic was con-

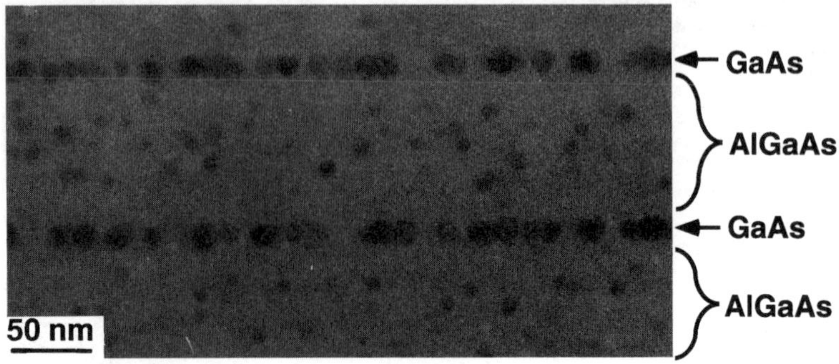

Figure 8 Bright-field image of 10-nm thick GaAs wells separated by 100-nm AlGaAs barriers for samples annealed at 700°C for 5 min.

tained throughout the GaAs and AlGaAs regions. The TEM image in Figure 8 is of a sample annealed at 700°C for 5 min. Even after the 5 min anneal at 700°C there is a clear tendency for the arsenic precipitates to coarsen to the GaAs well regions. After a 1.5 h anneal at 700°C there were very few arsenic clusters remaining in the AlGaAs regions (13). It may be possible to use this behavior of the arsenic precipitates at heterojunctions to get the arsenic precipitates close enough that they will behave as a buried conducting layer. Although the density of arsenic precipitates in Figure 8 appears high, the figure is a projection of a three-dimensional sample onto two dimensions. Therefore, the planes of arsenic precipitates in Figure 8 were not conducting—a higher density of arsenic precipitates is required than has been obtained so far. Similar behavior has been observed by Cheng et al (19) investigating precipitate coarsening in a structure consisting of InGaAs quantum wells and GaAs barriers. Cheng et al (19) observed preferential coarsening of the arsenic precipitates to the InGaAs quantum wells, the region of lower precipitate-to-matrix interfacial energy. Cheng et al (43) have also demonstrated preferential coarsening of arsenic precipitates at In delta-doped regions in GaAs and AlGaAs.

For engineering composite structures containing arsenic clusters, it would be desirable to be able to vary the placement of arsenic clusters between GaAs quantum wells and AlGaAs barrier regions. Melloch et al (44) have demonstrated such a capability. When growing at low substrate temperatures, a technique called migration enhanced epitaxy (MEE) can be used to grow highly stoichiometric epilayers (45). MEE consists of separately supplying about one monolayer of the group III and group V fluxes to the growing surface, which allows the excess arsenic to desorb from the surface. Therefore, by switching between MBE and MEE growth modes within a monolayer, one can change from stoichiometric to non-

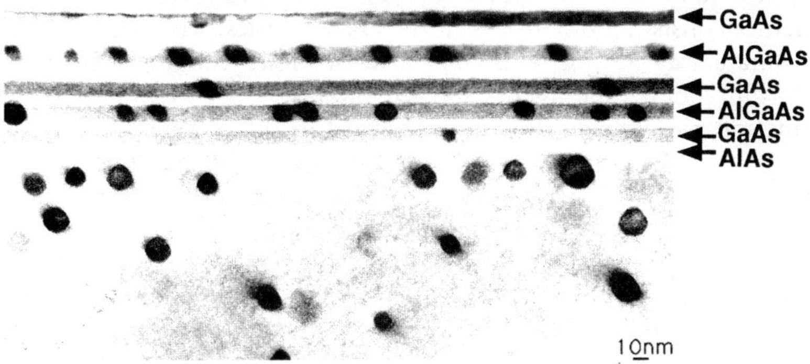

Figure 9 TEM image of a film structure grown at 260°C using As_4. The bright lines are AlAs layers (the lowest one is marked) and are either 6- or 12-nm thick. The region below the marked AlAs layer is the GaAs temperature transition region. The three GaAs wells in the image are indicated with arrows. The AlAs layers and the GaAs regions between the AlAs layers were grown by MEE, and as-grown are highly stoichiometric. The $Al_{0.2}Ga_{0.8}As$ layers were grown by MBE, and hence as-grown have a large excess of arsenic.

stoichiometric epitaxy. This technique then could allow one to grow GaAs wells that are stoichiometric and AlGaAs barriers that contain excess arsenic. However with annealing, the excess arsenic would still coarsen at the GaAs well regions. To prevent this, Melloch et al (44) used thin AlAs layers as arsenic diffusion barriers between the stoichiometric GaAs wells and the nonstoichiometric AlGaAs barriers to retain the excess arsenic and the arsenic precipitates in the AlGaAs barrier regions, as illustrated in Figure 9. The effectiveness of AlAs as a diffusion barrier to the large amount of excess arsenic in LTG-materials was first demonstrated by Yin et al (46).

Composite Formation Using Ion Implantation

Claverie et al (47) have demonstrated the fabrication of composites of arsenic clusters in GaAs by arsenic implantation and thermal annealing. Implantation allows lateral control of the placement of the excess arsenic with possible application for isolation between devices. With the use of implantation, one is not limited to arsenic as the only species introduced into the semiconductor. Chang et al (48) have formed Fe_3GaAs and Ni_3GaAs precipitates in GaAs using implantation of Fe and Ni followed with a precipitation anneal. The GaAs samples were implanted with

1×10^{16} ions/cm^2 of Fe or Ni at an energy of 170 keV and at room temperature. The samples were annealed at 950°C for 30 s or at 600°C for 30 min with a GaAs proximity cap. TEM analysis revealed a composite structure consisting of clusters in the GaAs matrix. The size of a typical precipitate was 35 nm in diameter, with moiré fringes clearly visible. Microdiffraction experiments showed the presence of extra spots in the diffraction pattern, which were attributed to the presence of hexagonal Fe_3GaAs or Ni_3GaAs in a particular orientation with respect to the GaAs matrix. The (004) X-ray rocking curves of the annealed samples showed only a sharp substrate peak and no extra peak as observed before annealing. This indicates that most of the strain in the matrix is relaxed by precipitation. The optical properties of these new composites are yet to be explored. Preliminary Hall-effect measurements on the Fe- and Ni-implanted samples that were annealed at 950°C for 30 s showed they are both *p*-type, with carrier concentrations of 2.5×10^{17} cm^{-3} and 7.5×10^{17} cm^{-3}, with mobilities of 179 and 252 cm^2/V-s, respectively. This large hole concentration is probably due to residual Fe and Ni atoms that act as acceptors. The ability to form these composites with different metals may allow an additional degree of control of the composite properties because of the different Schottky barriers associated with different metal/GaAs interfaces. Precipitates of noble metals such as Cu and Ag in GaAs produce particularly strong absorption and scattering of optical radiation and may represent attractive new composite materials for optical applications (49).

ROLE OF POINT DEFECTS AND ARSENIC CLUSTERS

Electrical Conductivity

In LTG-GaAs much of the excess arsenic, with concentrations of 1 to 2%, can be in the form of the arsenic antisite (As_{Ga}), the arsenic interstitial (As_i), the Ga vacancy (V_{Ga}), and complexes of these defects. The observation of the ionized arsenic antisite ($As_{Ga}+$) with concentrations of up to 1 to 5×10^{18} cm^{-3} in LTG-GaAs was made using electron paramagnetic resonance (EPR) (7). This concentration of $As_{Ga}+$ is equal to the concentration of ionized acceptors, presumably the ionized Ga vacancy ($V_{Ga}-$). The observation of the neutral As_{Ga} has been made using absorption measurements (50, 51). Because of the large concentration of point defects, LTG-GaAs exhibits hopping conductivity with resistivities as low as 10 Ω-cm (52). However, with annealing, the resistivity increases dramatically to as high as 10^6 Ω-cm for annealed LTG-GaAs (52), and resistivities as high as 6×10^{11} Ω-cm have been reported for annealed LTG-$Al_{0.3}Ga_{0.7}As$ (53). For intermediate growth temperature, 350–450°C, as-grown GaAs is semi-insulating (10^7 Ω-cm) due to the incorporated point defects (4, 54).

There has been considerable debate concerning the role of point defects vs the role of arsenic precipitates in explaining the electrical properties displayed after annealing LTG-materials (21, 52, 55, 56). The defect model predicts a decrease in hopping conductivity as the excess arsenic precipitates but assumes that the arsenic precipitates play little role and that the material is compensated by the residual arsenic antisites. The model based on arsenic precipitates as embedded Schottky contacts assumes a depletion region around each arsenic precipitate. If the background doping is low enough, compared with the spacing of the precipitates, the material will be completely depleted, as illustrated in Figure 10*a*, and will exhibit a high resistivity. The degree to which the defects or the embedded Schottky barriers control the material properties depends on the anneal that the material has experienced. The transition from a material dominated by point defects to one dominated by the arsenic precipitates occurs around 600°C—the only anneal temperature or the maximum anneal temperature used in much of the data reported in the literature.

The conductivity as a function of anneal temperature of LTG-GaAs has been reported by Ibbetson et al (23). They found that the room-temperature conductivity of an LTG-GaAs sample annealed for 30 s at 600°C was due to hopping conduction. For higher-temperature anneals, the room-temperature conductivity was not the result of hopping conduction but due to a thermally assisted tunneling process with an activation energy of ~0.6 eV, which Ibbetson et al attribute to the arsenic precipitates. In fact, this activation energy of ~0.6 eV is very close to the reported Schottky barrier height for arsenic deposited on GaAs (57).

Atique et al (58) have studied the structural and electronic properties of doped LTG-GaAs as a function of anneal for three 1 μm-thick epilayers with Si doping concentrations ranging from 5×10^{17} cm^{-3} to 5×10^{18} cm^{-3} and five 1 μm-thick epilayers with Be doping concentrations ranging from 5×10^{17} cm^{-3} to 5×10^{19} cm^{-3}. These epilayers were all grown under similar conditions; the Si-doped epilayers contained 0.9% excess arsenic, and the Be-doped epilayers contained 0.6% excess arsenic. The epilayers were grown on semi-insulating GaAs substrates; after epitaxy, each wafer was cleaved into a series of samples. Samples from the Si-doped epilayers were annealed for 30 s at temperatures ranging from 650 to 1000°C, and samples from the Be-doped epilayer were annealed at temperatures ranging from 650 to 900°C. Van der Pauw structures were fabricated on the samples to ascertain the electrical properties. Shown in Figure 11 are the resistivities of these samples as a function of anneal temperature. The 30 s anneal at 700°C results in a significant increase in resistivity of the samples (limited

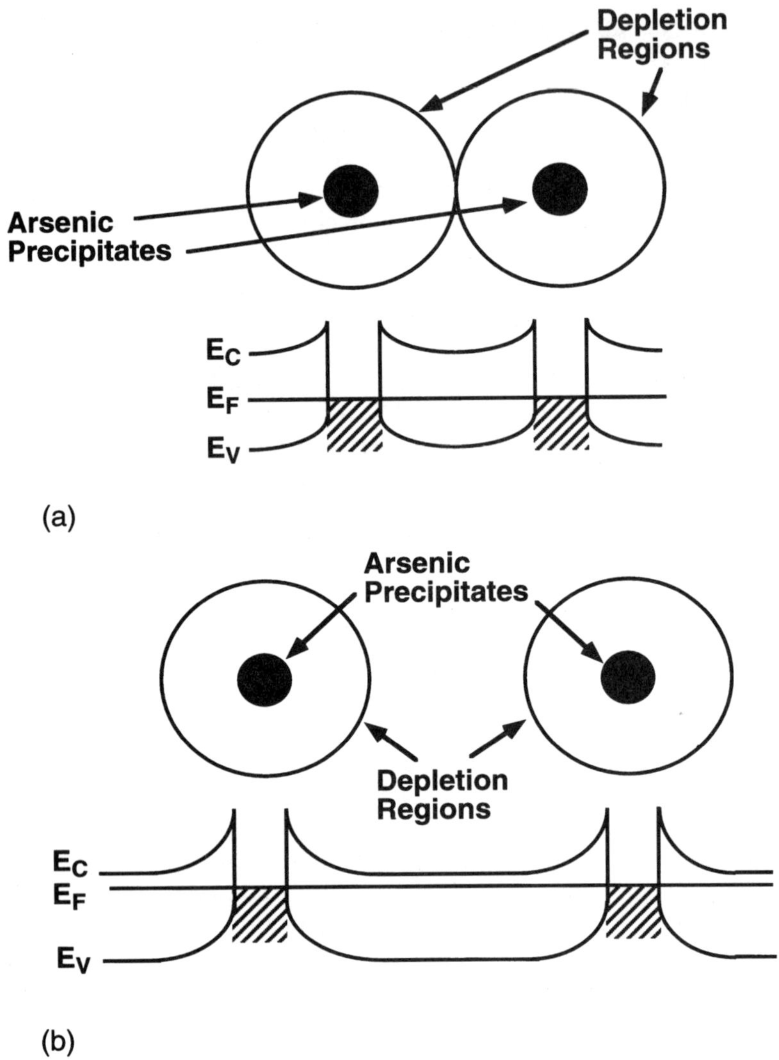

Figure 10 (*a*) Depletion regions around arsenic precipitates such that the depletion regions just overlap. (*b*) Depletion regions around arsenic precipitates with spacing such that there is undepleted material between the precipitates.

by substrate conduction) from the as-grown values, except for the most heavily doped Si sample. The resistivity is seen to decrease as the anneal temperature is raised above 700°C because of arsenic cluster coarsening, which causes the depletion regions around the arsenic clusters to no longer

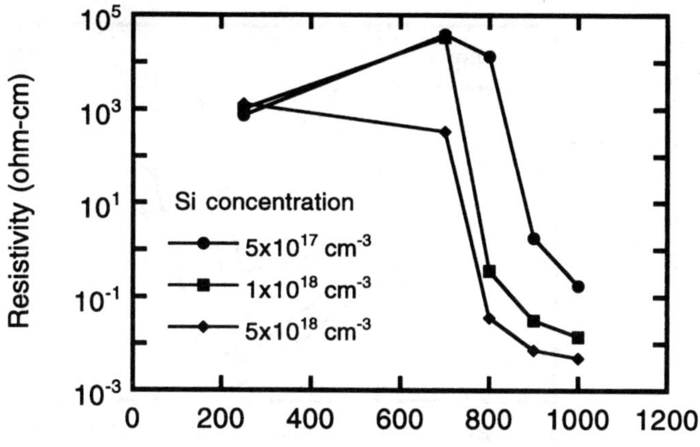

(a)

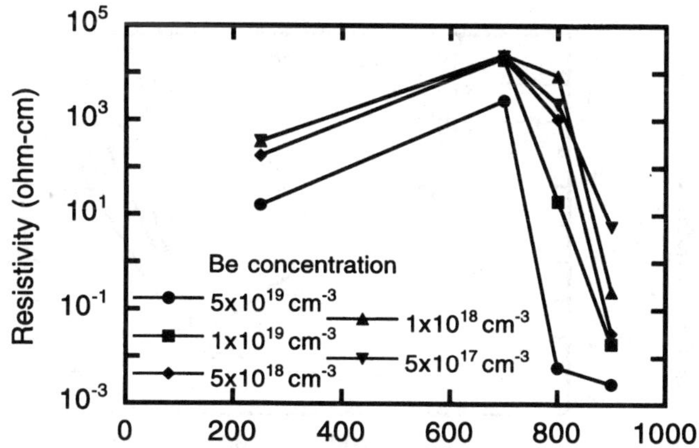

(b)

Figure 11 Resistivity as a function of temperature for 30-s isochronal annealing of GaAs epilayers containing excess arsenic for (*a*) three different *n*-type doping concentrations, and (*b*) five different *p*-type doping concentrations.

overlap, as illustrated in Figure 10*b*. This effect is clearly seen in the carrier concentrations (as determined by Hall measurements) as a function of anneal temperature for these epilayers (Figure 12). Also of note is that some of the *p*-doped epilayers are *n*-type as-grown or when annealed at low temperature, as has been observed by others (40, 59). The mobilities as a function of anneal for the *n*-GaAs : As samples are shown in Figure 13.

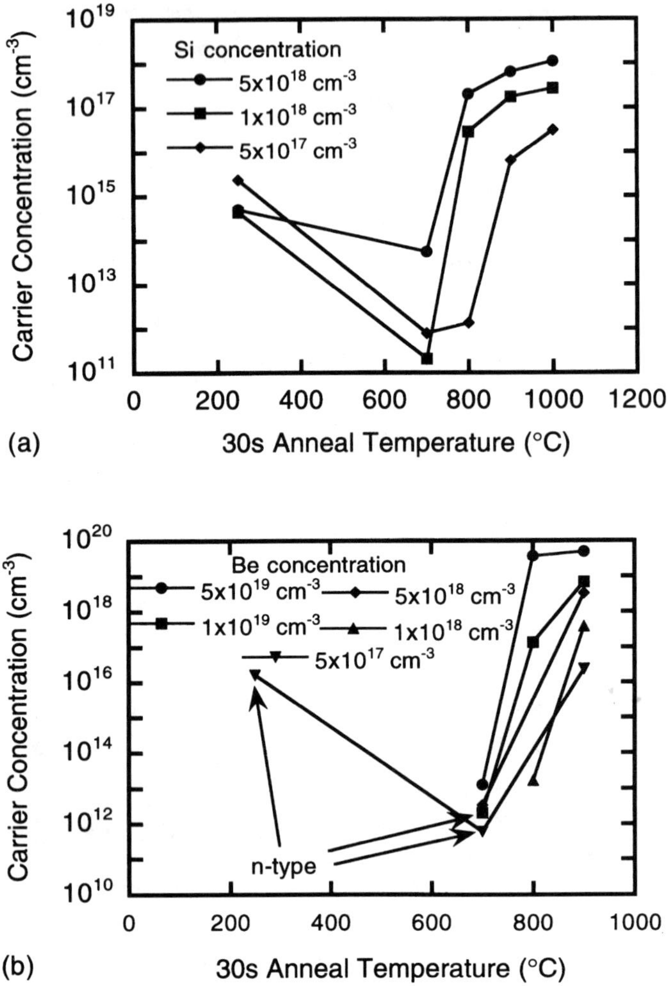

Figure 12 Carrier concentration as a function of temperature for 30-s isochronal annealing of GaAs epilayers containing excess arsenic for (*a*) three different *n*-type doping concentrations and (*b*) five different *p*-type doping concentrations.

Atique et al (58) have investigated the correlation between the measured carrier concentrations and the average arsenic precipitate size and average arsenic precipitate spacing. Figure 14 shows the model and band structure of an arsenic precipitate surrounded by its spherical depletion region. In this model, the arsenic precipitate forms a Schottky barrier with the GaAs matrix analogous to a planar Schottky barrier. By solving Poisson's equa-

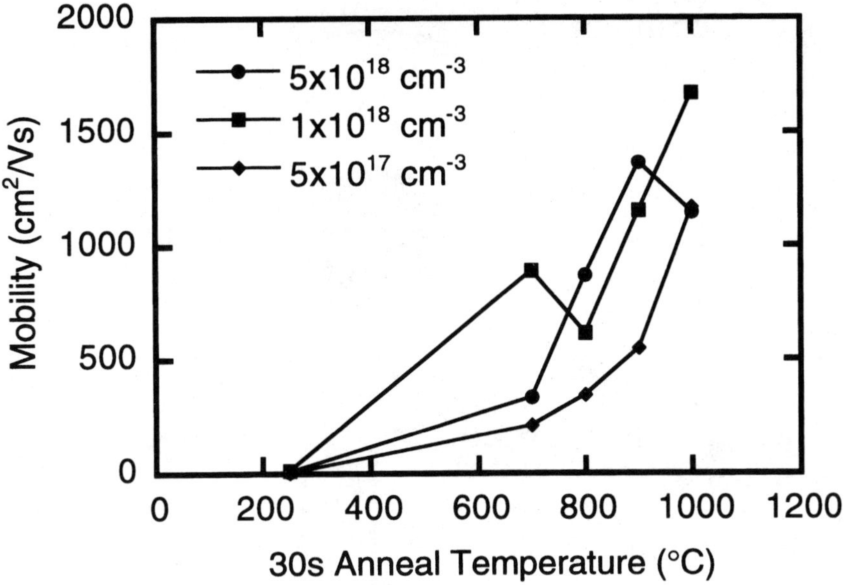

Figure 13 Mobility as a function of anneal temperature for 30-s isochronal anneals for the Si-doped LTG-GaAs epilayers.

tion, the built-in potential V_o in Figure 14—determined from the doping level of the material—is related to the arsenic precipitate radius r_o and the depletion radius r_s by (21)

$$V_o = -(qN_D/6\varepsilon)[(2r_s^3/r_o)+r_o^2-3r_s^2], \qquad 1.$$

where N_D is the doping density, and ε is the permittivity of the semiconductor matrix. This equation allows the calculation of the depletion radius for each of our samples, which enables an estimation of the free carrier concentration from the fraction of material depleted. The expected free carrier concentration N_f for each sample is related to the doping density N_D, the depletion radius r_s, and the distance between precipitates d by

$$N_f = N_D[1-(2r_s/\mathrm{d})^3]. \qquad 2.$$

Atique et al (58) define $2r_s/d$ as the degree of depletion. (A degree of depletion of 1 occurs when the depletion regions just touch, and a degree of depletion of 1/2 is the case when the depletion regions extend to include half the distance between precipitates.) Plotted in Figure 15*a,b* are the free electron and free hole carrier concentrations determined from Hall measurements vs the degree of depletion as determined from TEM

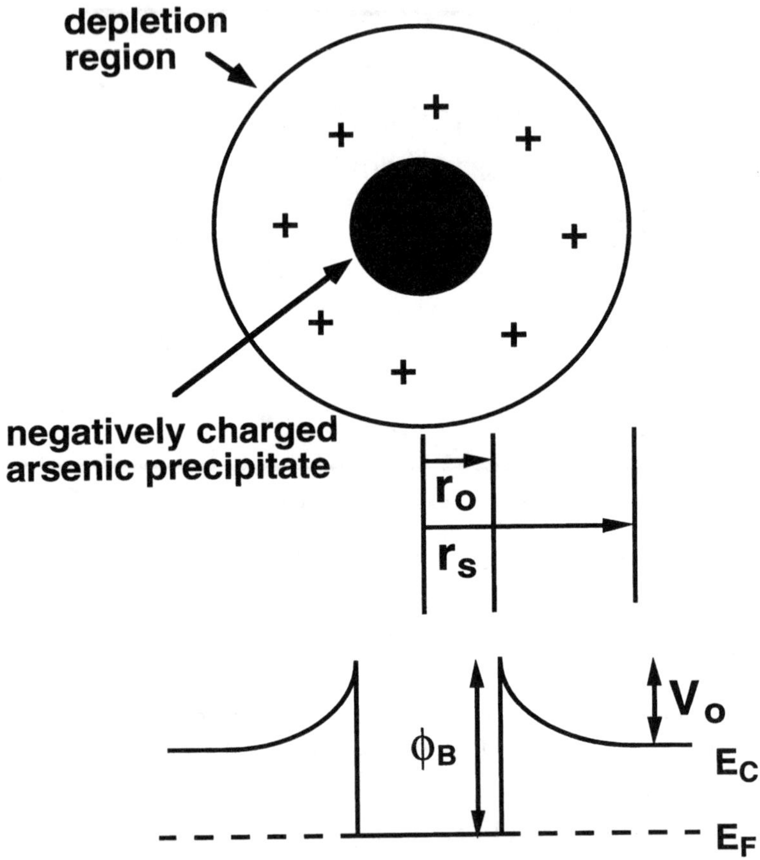

$$V_o = \frac{-qN_D}{3\varepsilon}\left[\frac{r_s^3}{r_o} + \frac{r_o^2}{2} - \frac{3r_s^2}{2}\right]$$

Figure 14 Schottky barrier model for the embedded arsenic precipitates.

measurements. Also plotted in Figure 15 is Equation 2, the expected free carrier concentration based on the arsenic clusters acting as internal Schottky diodes. The measured free electron concentration data correlate well with the model of the arsenic clusters acting as internal Schottky barriers. Note that any residual AsGa defects, As_{Ga} deep donors, would not be able to decrease the free electron densities. In Figure 15*b*, as the Be concentration increases, the drop in the free hole concentration is seen to

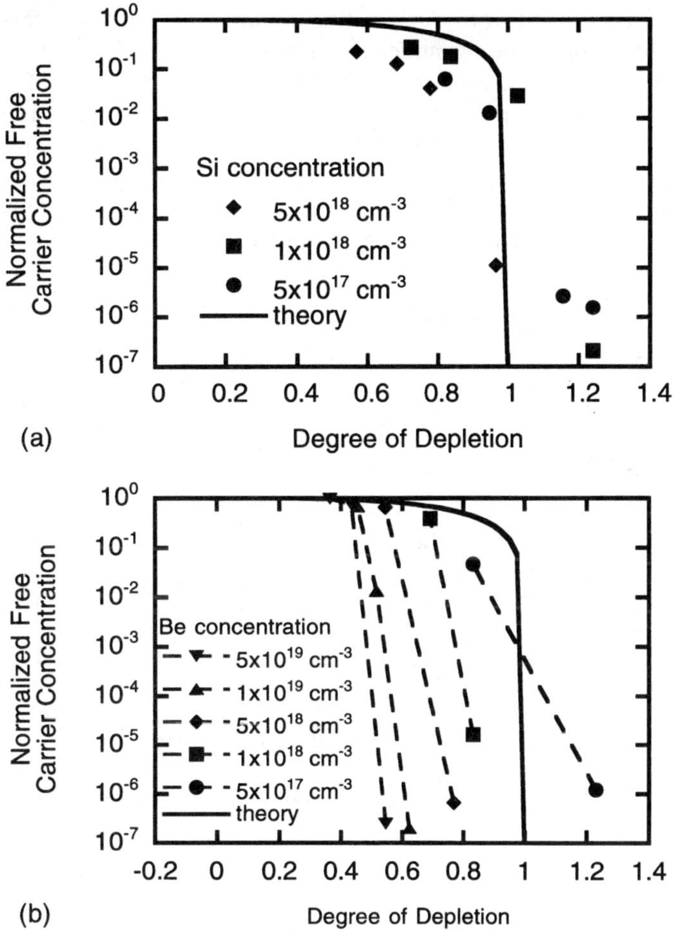

Figure 15 Normalized carrier concentration vs degree of depletion for (*a*) three different Si-doping levels and (*b*) five different Be-doping levels. Degree of depletion is $2\ r_s/d$, where r_s is the depletion sphere radius and d is the distance between precipitates.

occur at lower values of degree of depletion, or arsenic cluster spacing, than expected from the internal Schottky barrier model. This is attributed to As_{Ga} partially compensating the shallow Be acceptors. For a given anneal condition, the concentration of As_{Ga} will be higher in *p*-doped LTG-GaAs than in *n*-doped LTG-GaAs because there is less of a driving force for precipitation because of the locations of the Fermi levels (15). In LTG-*p*-GaAs, the As_{Ga} will compensate the shallow acceptors, thus the Fermi level begins at midgap. In LTG-*n*-GaAs, the As_{Ga} defects do not compensate the shallow Si donors, thus precipitation of the excess arsenic is required to move the Fermi level to midgap.

The average arsenic cluster size and spacing as a function of anneal is shown in Figure 16*a* for the Si-doped epilayers and in Figure 16*b* for the Be-doped epilayers, illustrating the coarsening of the arsenic clusters with annealing. Note that as the doping density increases for a given anneal condition, the average diameter and spacings of the arsenic clusters are smaller. The reason for this trend is that for higher doping densities at a

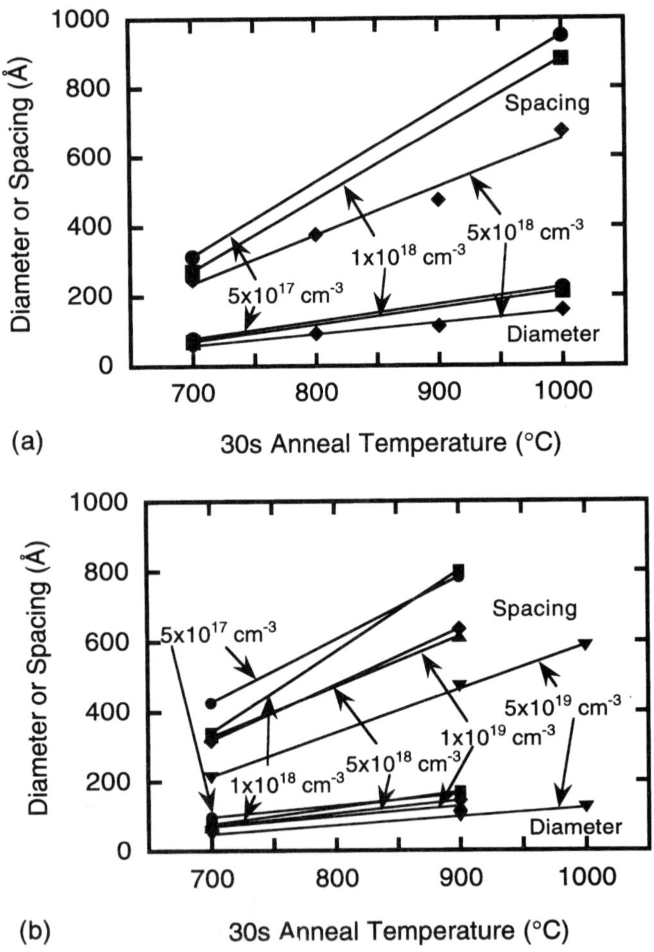

Figure 16 Average cluster sizes and spacings as a function of anneal temperature for 30-s isochronal annealing of LTG-GaAs epilayers for (*a*) three different Si-doping levels and (*b*) five different Be-doping levels.

given anneal condition, a smaller spacing between precipitates is required to move the Fermi level toward midgap—toward a state of lower free energy.

Fermi Level Position

Several groups have made effective use of the contactless electro-modulation methods of photoreflectance (60) and contactless electro-reflectance (60) to study the nature of Fermi level pinning on LTG-GaAs and GaAs:As (56, 61–65) and InGaAs:As (66). Certain features, i.e. Franz-Keldysh oscillations, observed in electromodulation from special structures with large, uniform electric fields have been found to be extremely useful for evaluating Fermi-level positions (60). Although the work of Warren et al (56, 61, 62) was carried out on both *n*- and *p*-type material, the investigations of References 63–65 were performed on only *n*-type samples. In addition, Warren et al (56, 61, 62) used piezoreflectance to further evaluate the nature of the electric fields created by the Fermi-level pinning. The experiments of Warren et al showed that the Fermi level in LTG-*p*-GaAs is pinned 0.4 eV below the conduction band edge, whereas in *p*-GaAs:As the Fermi level is pinned 0.67 eV below the conduction band edge, i.e. midgap. In addition, these experiments of Warren et al showed that the Fermi level is unpinned in LTG-*n*-GaAs and pinned at midgap in *n*-GaAs:As. Thus the controlling mechanism in the GaAs:As material is indeed that of Schottky barrier-controlled pinning on metallic arsenic precipitates, consistent with previously reported investigations for these materials.

To probe the electric fields in and adjacent to the LTG layers, special three-layer heterostructure stacks were grown for the electromodulation measurements. They consisted of a doped/undoped (layer I)/undoped (layer II) configuration, where undoped layer I is either GaAs or InGaAs grown at normal MBE substrate temperatures, and undoped layer II is either LTG-GaAs, GaAs:As, or InGaAs:As. In the work of Warren et al (56, 61, 62), the structures were n^+ (N) and p^+ (P) GaAs buffer layers (doped at 5×10^{18} cm^{-3}) grown on (001) substrates, followed by 50-nm undoped (U) GaAs grown at a substrate temperature of 600°C. The substrate temperature was then dropped to 250°C, and 150 nm of GaAs was deposited for samples referred to as LTUN and LTUP, respectively. In other samples, this recipe was followed by an in situ ramp back up to 600°C for a 30 s anneal in an As_2 flux. These samples are referred to as ALTUN and ALTUP layers. The key assumption in using this method with the three-layer structure is that the Fermi level will be pinned at some characteristic energy in layer II. If so, the electric field will be (*a*) nearly zero in layer II and (*b*) nonzero and nearly constant in undoped layer I. A knowledge of the thickness of undoped layer I and the doping level of the buffer layer makes it possible to evaluate the Fermi level in layer II. Electric fields in the undoped regions of all these samples were then measured from the Franz-Keldysh oscillations observed, using both photoreflectance and contactless electroreflectance. These two techniques gave identical fields; in addition, contactless electroreflectance gave the field sign from the phase of the reflected signal (60). There were no observable photovoltage effects in the ALTUN and ALTUP materials at 300 K. In the LTUN and LTUP samples, measurements were made at elevated temperatures ($\sim$400 K) to eliminate the influence of the photovoltage (60).

It is important to note that recent experiments have demonstrated the need to study both *n*- and *p*-type material to obtain unambiguous results about Fermi-level positions (67). In addition to the above studies, Warren et al also have performed piezoreflectance measurements at 300 K on all samples to determine the nature of the fields (i.e. uniform or nonuniform) responsible for the Franz-Keldysh oscillations. It has been demonstrated that only uniform fields produce Franz-Keldysh oscillations in piezoreflectance spectra (60).

In the LTUN structure, the measured field from the Franz-Keldysh oscillations in the electromodulation traces was 1.1×10^5 V/cm, which corresponds to the built-in field at the U-N interface. For this sample, no Franz-Keldysh oscillations were observed in piezoreflectance, indicating that the field is nonuniform. This value places an upper limit of about 0.5 eV for the Fermi-level distance from the conduction band edge at the LT-U interface. In the LTUP sample, the measured field from electromodulation was 1.8×10^5 V/cm, which corresponds to a Fermi-level position in the LT material $\sim$0.45–0.5 eV below the conduction band. The uniform nature of this field was verified by the Franz-Keldysh oscillations in the piezoreflectance data, which also gave the same value of the field as the electromodulation measurement. These two measurements are exactly what one would expect from a single, deep-donor level 0.4 eV below the conduction band, which corresponds to the EPR and infrared absorption measurement of arsenic-antisite defects (7, 50, 51). Figure 17*a,b* shows our modeling results for the LTUN and LTUP cases, respectively, with such a defect, illustrating the lack of firm pinning for this system. This model also serves to explain the presence of a photovoltaic effect in these samples. These figures are consistent with the piezoreflectance data, i.e. nonuniform field (1.1×10^5 V/cm) in the U region near the U-N interface for the LTUN sample and uniform field (1.8×10^5 V/cm) in the U region for the LTUP material.

Data from the ALTUN and ALTUP samples, however, indicate a completely different system. Shown in Figures 18 and 19 are the contactless

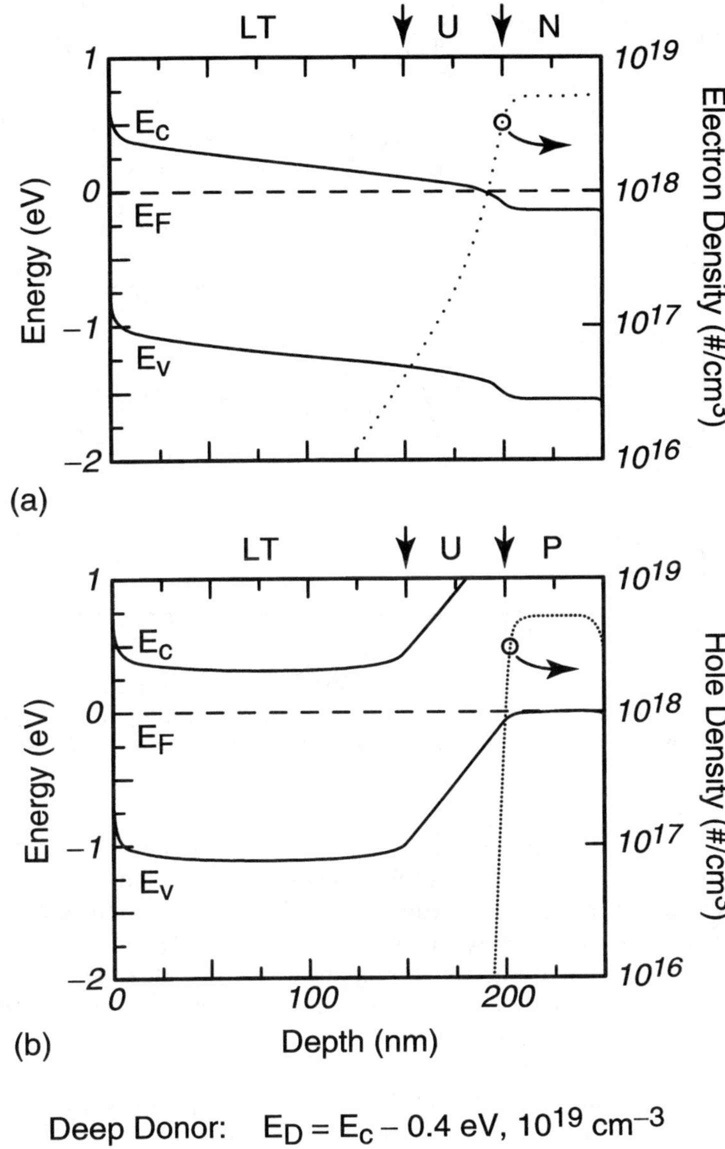

Figure 17 Band profile for the as-grown (un-annealed) (*a*) LTUN and (*b*) LTUP structures. A deep donor is included in the LT material with energy 0.4 eV below the conduction band. Arrows indicate the interface between the LT, U, and N layers.

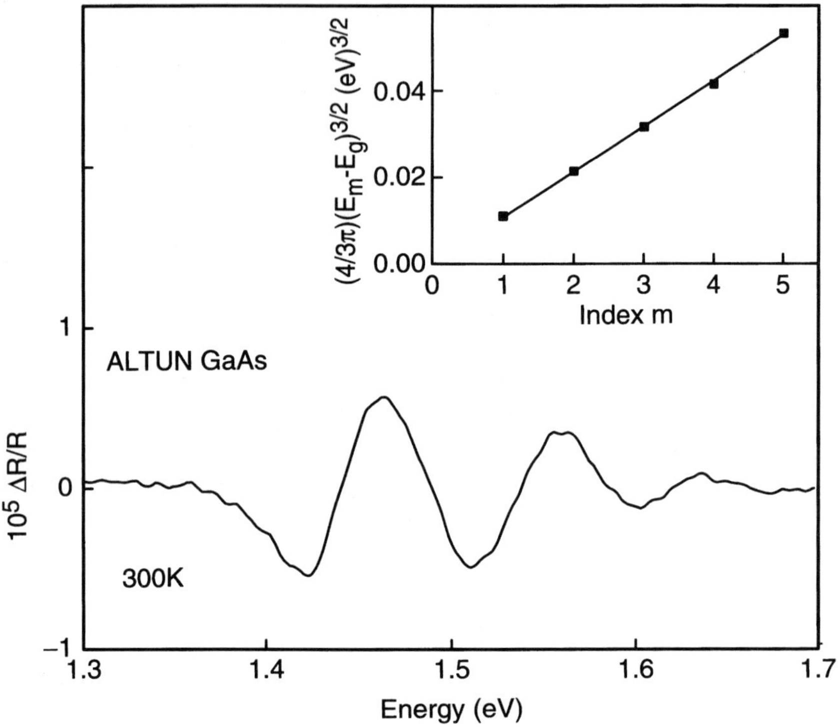

Figure 18 Contactless electroreflectance spectrum from the ALTUN structure at 300 K. The inset shows a plot of $(4/3\pi)(E_m\text{-}E_g)^{3/2}$ as a function of the Franz-Keldysh oscillations extrema index *m*.

electroreflectance spectra at 300 K for these two samples, respectively. Both samples display a number of Franz-Keldysh oscillations, although the signals exhibit opposite phases consistent with *n*- and *p*-type band bending, respectively. Also the piezoreflectance traces from both materials exhibited Franz-Keldysh oscillations, indicating linear electric fields. The positions of the mth extrema in the Franz-Keldysh oscillations are given by (60)

$$m\pi = (4/3)[(2\mu_{\parallel})^{1/2}(E_m - E_g)^{3/2}/q\hbar F] + \chi, \qquad 3.$$

where E_m is the photon energy of the mth extrema, E_g is the band gap, F is the electric field, $\mu_{\parallel}$ is the reduced interband effective mass in the direction of $\vec{F}$ and χ is an arbitrary phase factor. Therefore, if $\mu_{\parallel}$ is known, a plot of $(4/3\pi)(E_m\text{-}E_g)^{3/2}$ as a function of index number *m* yields a straight line from which *F* can be evaluated from Equation 3. In the case of the uniform electric fields in these structures, the relation between F and the barrier height (V_B) created by the Fermi-level pinning can be written as (60)

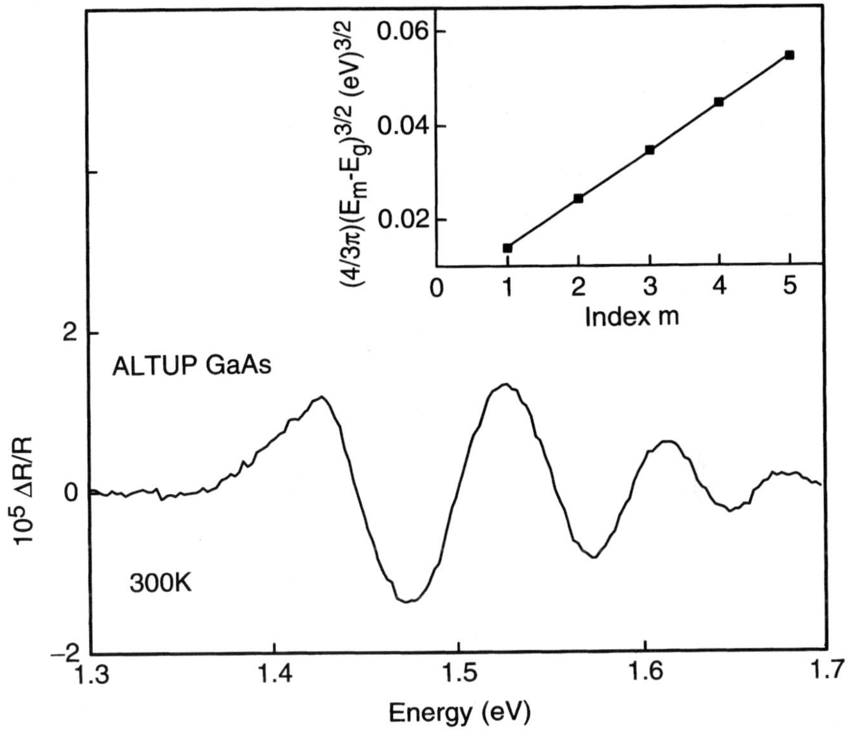

Figure 19 Contactless electroreflectance spectrum from the ALTUP structure at 300 K. The inset shows a plot of $(4/3\pi)(E_m\text{-}E_g)^{3/2}$ as a function of the Franz-Keldysh oscillations extrema index m.

$$V_B = FL + (kT/q) + SCC, \qquad 4.$$

where L is the thickness of undoped layer I and the second and third terms are the Debye length and space charge (doping layer) corrections, respectively.

The insets of Figures 18 and 19 display such plots from which the authors have deduced fields of 1.25×10^5 V/cm for both samples, which indicate a single midgap GaAs : As Fermi-level position for both materials. Modeling for the ALTUN and ALTUP structures are shown in Figure 20*a,b*, respectively. Allowing for doping degeneracy and Debye tailing in the doping substrates, and an 8-nm Debye tail in the GaAs: As layer, their calculations yield fields of 1.3×10^5 V/cm with a pinning of 0.67 eV below the conduction band edge. The extrapolated pinning position is in excellent agreement with a measured Fowler-Nordheim photoresponse threshold of 0.7 eV (25) and is what one would expect from the barrier height of

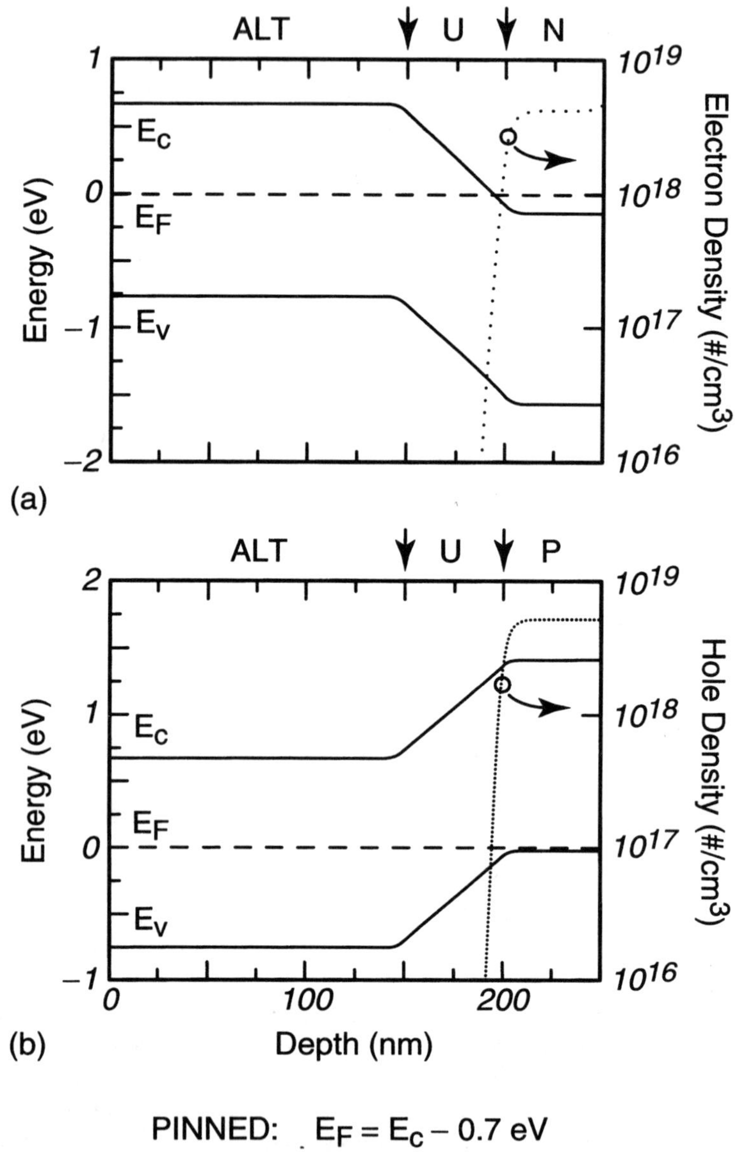

Figure 20 Band profile for the annealed (*a*) ALTUN and (*b*) ALTUP structures. The Fermi level is pinned at 0.67 eV below the conduction band edge. Arrows indicate the interface between the LT, U, and N layers.

metallic arsenic to GaAs (57). The fact that there was no measurable photovoltage in these samples is consistent with metallic Fermi-level pinning by the arsenic precipitates. In addition, it is clear that such pinning could not be produced by a single donor or acceptor defect.

Woodall et al also have reported a similar experiment on two ALTUP (sample A and B) and one UP (sample C) GaInAs/InP (001) samples (66). Samples A and B were three-layer structures—the doped layer was *p*-type GaInAs, the undoped layer I was GaInAs 100-nm thick, and the undoped layer II was GaInAs : As 100-nm thick. Sample C is the same as samples A and B but without the GaInAs : As layer II. For sample C, the Franz-Keldysh oscillations measure the pinning position of the free surface of GaInAs. The photoreflectance spectra at 300 K for these three samples are shown in Figure 21. All three materials exhibit well-pronounced Franz-Keldysh oscillations. For clarity, the spectra are plotted in terms of E-E_g. The values of E_g listed in the figure were obtained from the three-point method. The electric fields were 46 ± 3, 44 ± 3, and 50 ± 3 kV/cm, from which the authors deduced Fermi levels of 0.49 ± 0.03 V, 0.47 ± 0.03 V, and 0.53 ± 0.03 V above the valence band edge for samples A, B, and C, respectively. If the GaInAs : As did not pin the Fermi level, then because

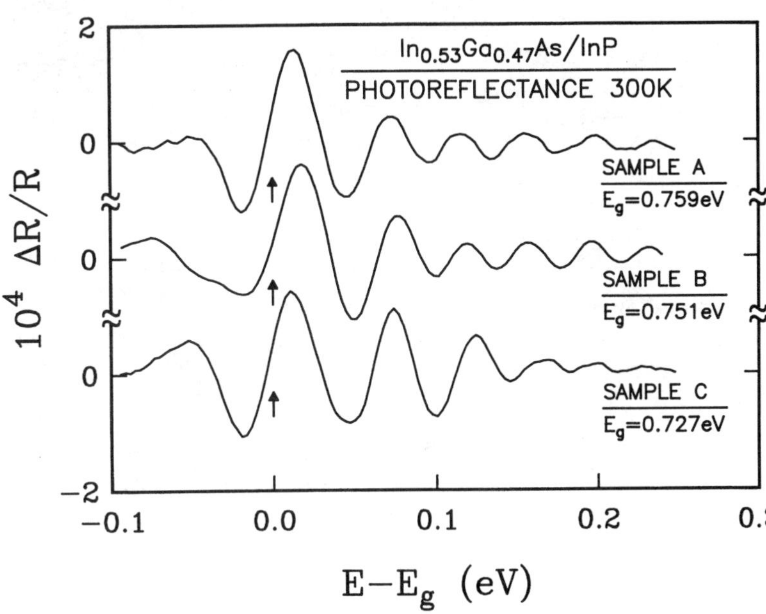

Figure 21 Photoreflectance spectra at 300 K for two ALTUN (samples A and B) and one UP (sample C) GaInAs/In/InP materials plotted as a function of (E-E_g).

of the differences in undoped GaInAs thickness between the samples A, B and sample C, the electric fields in samples A and B would have to be about one half the value of sample C in order for the Fermi levels at the surface to be the same for all three materials. This is clearly not the case. Thus this study demonstrated that the Fermi level is pinned at about the same value for both GaInAs surfaces and GaInAs : As.

Scanning Tunneling Microscopy Studies

Feenstra et al (27, 68–70) have used scanning tunneling microscopy (STM) to investigate LTG-material as it changes with annealing from one with a distribution of point defects to one with a distribution of arsenic clusters. An STM image of as-grown material is shown in Figure 22, revealing a concentration of arsenic antisites of about 1×10^{20} cm^{-3} (68, 69). The characteristic appearance of the antisites in the STM images arises from

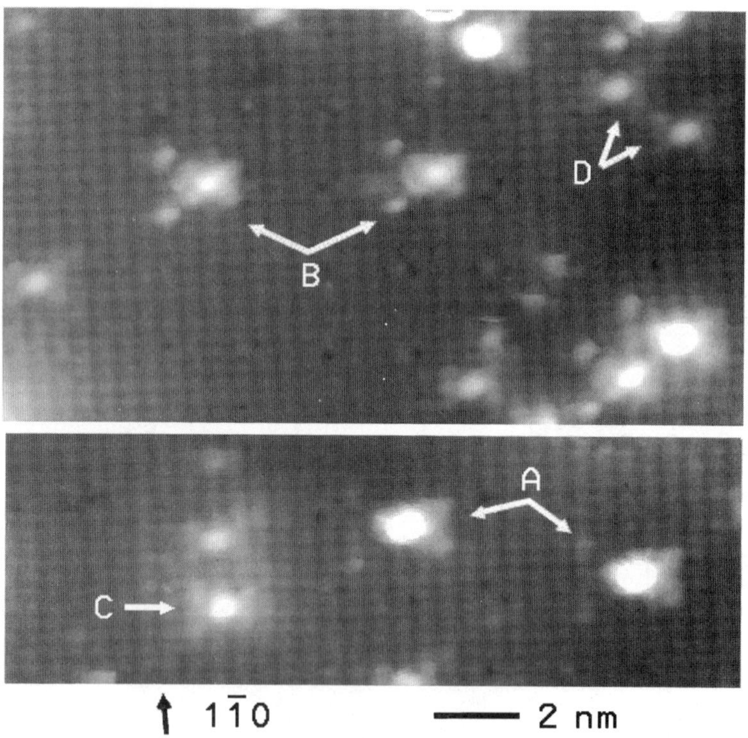

Figure 22 STM images of the (110) cleaved surface of as-grown LTG-GaAs, acquired with 0.1 nA tunnel current and at a sample voltage of −2.0 V. Various point defects can be seen, classified as types A, B, C, and D, as indicated.

the wave function of the defect donor states, and the labels A, B, C, D in Figure 22 refer to antisites located in crystal planes located 0, 1, 2, or 3 planes respectively below the (110) surface. Tunneling spectroscopy directly reveals the spectrum of midgap donor states (0/+ and +/++), as shown in Figure 23. As the background doping of the material changes from *n*-type to *p*-type, the Fermi level moves into the middle of this band

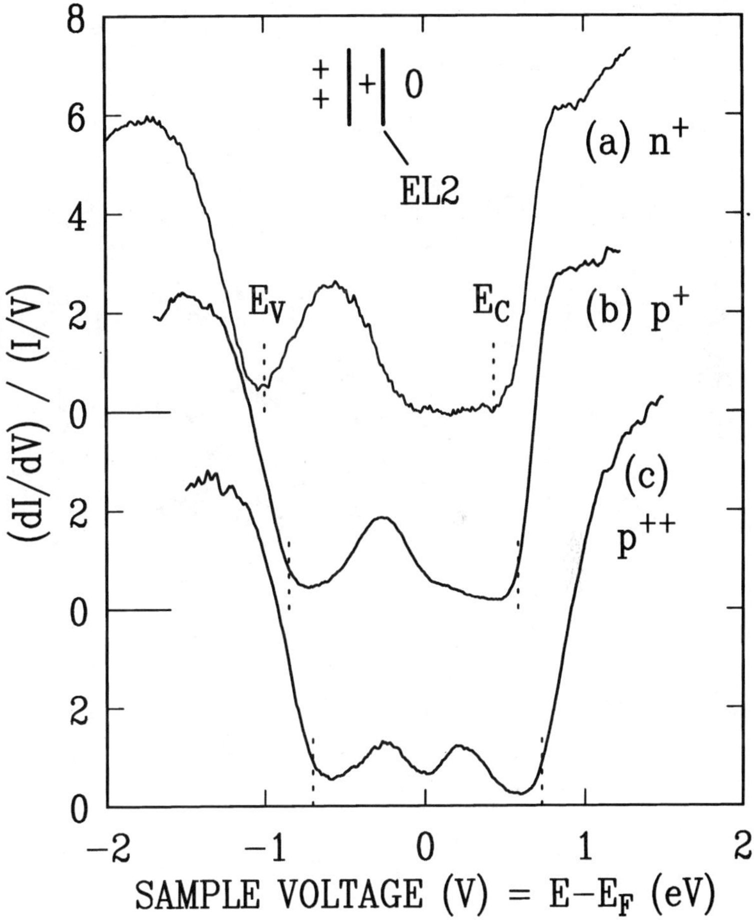

Figure 23 Tunneling spectra acquired from layers of as-grown LTG-GaAs containing varying amount of compensating shallow dopants. The valence-band maximum E_V and conduction-band minimum E_C are indicated by dashed lines in each spectrum. An intense band of states, arising from arsenic-related defects, appears within the band gap. The states of a bulk arsenic antisite defect (EL2) are shown in the upper part of the figure, relative to the band edges of spectrum (*a*).

of donor states. A local minimum in conductance forms at the Fermi level and is attributed to Coulomb interactions (both on-site and off-site) at the defects (68, 69).

STM studies of annealed material are shown in Figures 24–29. In all cases, the structures studied consist of a layer of LTG-GaAs with thickness of 500 to 1500 Å, surrounded on both sides by highly doped GaAs layers grown at temperatures near 600°C. Figure 24 is obtained from n-type LTG-material doped at 1×10^{19} cm^{-3}, with a 30 min anneal at 600°C (27). Arsenic clusters appear there as either protrusions or depressions in the LTG annealed (denoted LTA) layer, with typical diameter of 50 Å. The density of these features is found to be 1.2×10^{17} cm^{-3}, in reasonable

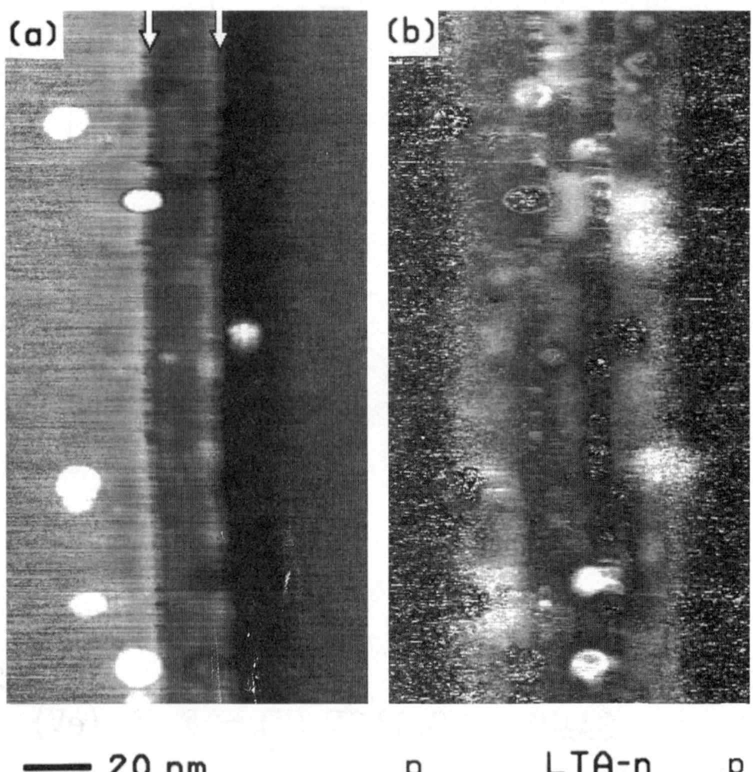

Figure 24 STM image of a cleaved GaAs structure, showing (*a*) topography and (*b*) conductance, acquired at a sample voltage of −2.5 V. The gray-scale range in (*a*) is 8 Å. As indicated in (*b*), the structure contains an LTA n-type layer, surrounded by p-type layers grown at higher temperature. The sample was annealed at 600°C. Two steps occur in the LTA-layer, as marked by arrows in (*a*).

agreement with that found in TEM studies (56). Antisite defects, with their characteristic images, are never seen in this *n*-type annealed material, implying a concentration of less than 10^{16} cm^{-3}. Higher temperature anneals produce coarsening of the arsenic clusters; results for an additional 800°C, 30 s anneal are shown in Figure 25. The cluster density is found to be about 7×10^{15} cm^{-3} in this case, with a typical diameter of 150 Å. A large cluster is observed near the center of Figure 25*a*; line cuts through this precipitate at the location indicated by arrows are shown in Figure 25*c*. The line cuts reveal a characteristic reversal of the topographic contrast with reversed sample-tip bias, as indicated by the arrows. This behavior is a clear indication of local depletion around a charged metal cluster (27, 71). The range of the depletion seen in Figure 25*c* is about 25 Å. Spectroscopic results from the 1×10^{19} cm^{-3} *n*-type material, annealed at 600 and 800°C, are shown in Figure 26 (27). For the 600°C anneal, the Fermi level far from the clusters is located near midgap, $E_V + 0.65$ eV, indicating that the depletion spheres of the precipitates overlap. For the 800°C case, the Fermi level is located at $E_V + 1.05$ eV far from the clusters, demonstrating Fermi-level movement toward flat-band conditions. At the clusters, distributions of states extending out from the valence and conduction bands are observed in both spectra. The smaller precipitates (Figure 26*a*) are certainly not good metals, but the density of midgap states is easily sufficient to provide a charge reservoir for midgap pinning of the Fermi level.

We recently performed STM studies on *n*- and *p*-type LTG-GaAs with doping of 1×10^{18} cm^{-3}, annealed at 600°C for 30 min plus 700°C for 30 s. The *n*-type material revealed a cluster density of 6×10^{16} cm^{-3}, with typical diameter of 80 Å. A close-up view of one precipitate is shown in Figure 27. A depletion sphere around the precipitate, appearing dark in this filled-state conductance image of *n*-type material, is clearly seen in Figure 27*b*. The radius of the depletion sphere is about 75 Å [including the white fringe around the dark circle (72)] for this 35-Å radius precipitate. Spectroscopic results, acquired from a point in the LTA-layer far from a cluster are shown in Figure 27*c*. The characteristic band-gap region of GaAs is seen, but the observed gap is about 2.0 eV, compared with the correct value of 1.4 eV. Thus some tip-induced band bending is clearly occurring in the spectrum. It is well known that tip-induced band bending can occur in STM spectroscopy of low-doped semiconductors (72–74), although the 0.6 eV band bending seen in Figure 27*c* is significantly larger than that generally observed for 1×10^{18} cm^{-3} doped material. It is likely that the composite nature of this LTG material, containing a collection of depletion spheres, will lead to increased tip-induced band bending. For this 700°C annealed material, carrier activation studies (56, 58) indicate

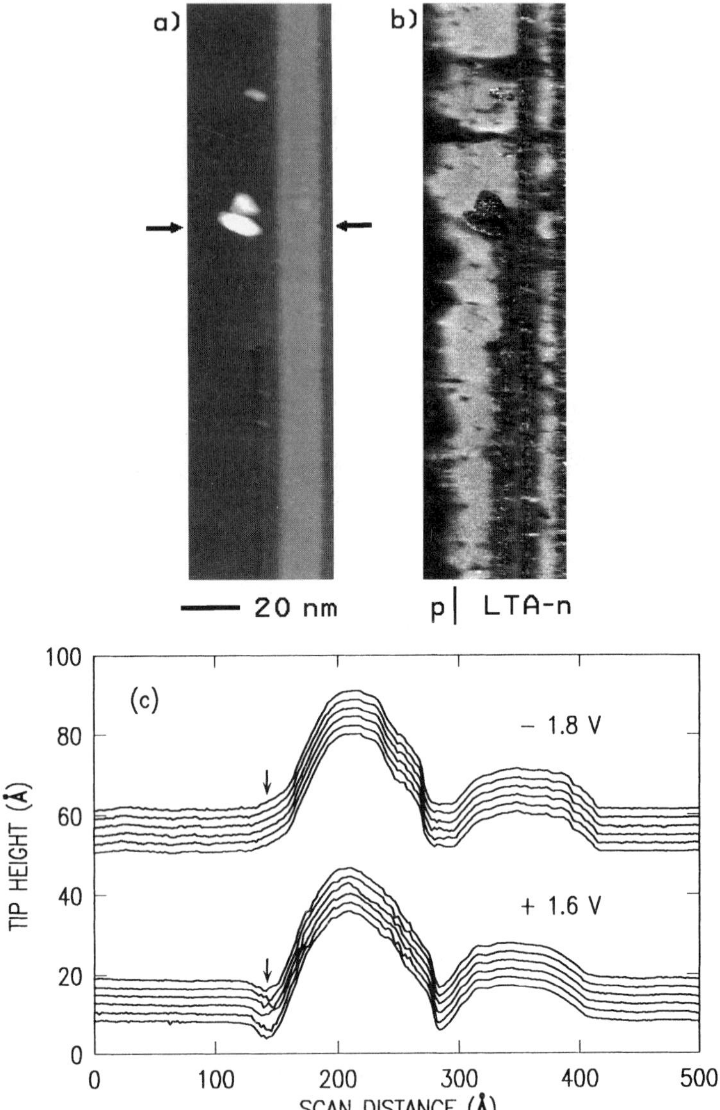

Figure 25 STM image of a cleaved GaAs structure, showing (*a*) topography and (*b*) conductance, acquired at a sample voltage of −2.3 V. The gray-scale range in (*a*) is 33 Å. As indicated in (*b*), the structure contains an LTA *n*-type layer, bordered on the left-hand side by a *p*-type layer grown at higher temperature. The sample was annealed at 800°C. Cross sectional cuts along the line indicated by arrows in (*a*) are displayed in panel (*c*), acquired at sample voltages of −1.8 and +1.6 V.

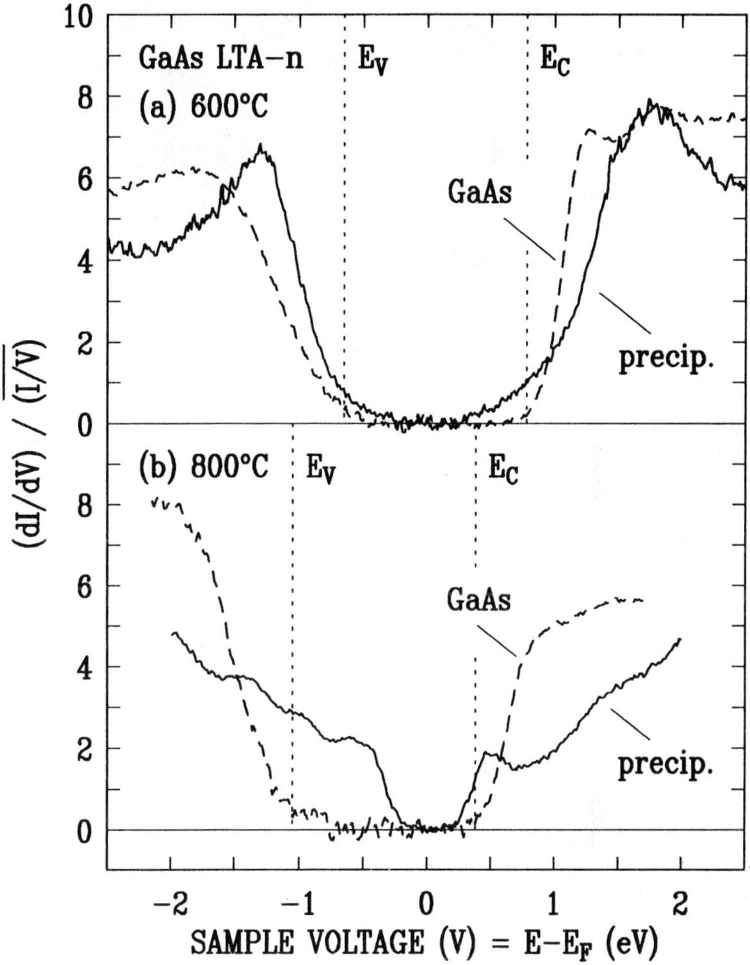

Figure 26 STM spectra acquired from LTA *n*-type layers, annealed at (*a*) 600°C and (*b*) 800°C. Spectra are shown acquired on the arsenic precipitates (*solid line*) and on regions of bare LTA GaAs in between the precipitates (*dashed line*). The valence and conduction band edges are marked by E_V and E_C respectively.

that the depletion spheres are nearly overlapping so that tip-induced band bending will be significant.

According to the buried Schottky barrier model of Figure 14, midgap pinning by the precipitates is expected to lead to depletion regions around each precipitate. The barrier height (or built-in potential) of the depletion regions is related to the depletion radius r_s and cluster radius r_o (Equation

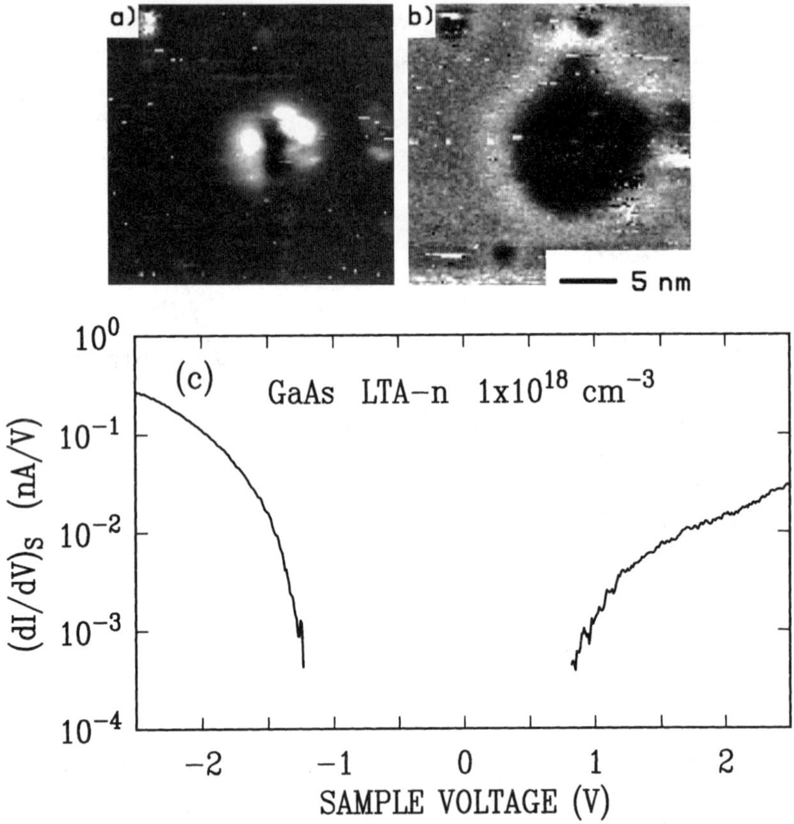

Figure 27 STM image of LTA *n*-type layer, showing (*a*) topography and (*b*) conductance, acquired at a sample voltage of −2.5 V. The gray-scale range in (*a*) is 2.0 Å. The sample was annealed at 700°C. A tunneling spectrum (conductance at constant tip-sample separation) is shown in (*c*), acquired on a flat region of LTA-GaAs between precipitates.

1). With a barrier height of 0.8 eV for *n*-type material (21), this formula leads to depletion widths r_s-r_o of 90 and 165 Å, respectively for the cases pictured in Figures 25*c* and 27*b*, which are significantly larger than the observed depletion widths of 25 and 40 Å, respectively. One reason for this discrepancy is simply that the nonzero size of the STM probe tip will tend to mask the edges of the precipitates, especially for large clusters such as those seen in Figure 25. In addition, the above-mentioned effects of tip-induced band bending will act to reduce the observed depletion regions. An analogous effect occurs in *pn*-junctions, where for doping levels less than 10^{18} cm^{-3}, observed depletion widths can be significantly smaller than the actual widths (75) (e.g. in figure 5 of Reference 72 the observed junction width is two times smaller than expected for the $n = 2 \times 10^{17}$ cm^{-3} doping).

Figures 28 and 29 display STM results obtained from p-type material with doping concentration of 1×10^{18} cm^{-3}, annealed at 600°C for 30 min, and at 700°C for 30 s. In the topographic image of Figure 28*a*, clusters appear as the protrusion seen in the center of the image and the depressions seen at the upper and lower edges of the image. The density of clusters in this material is found to be 2×10^{16} cm^{-3}, with typical diameter of 80 Å. Faint white fringes appear around the clusters in the conductance image of Figure 28*b*, indicative of local band bending in this filled-state image of p-type material. In addition, small point defects, appearing as white points, are seen in the conductance image. Figure 29 shows an expanded view of one point defect. Their appearance is identical to that of the antisite defects seen in the as-grown material, and thus we identify them as being antisite defects. The number of these defects is estimated from the images to be 1.6×10^{18} cm^{-3}, i.e. close to the nominal dopant concentration. Thus within a factor of two, there are just the right number of charged antisites to compensate the bulk acceptors, and the material becomes semi-insulating. Spectroscopic results from this p-type material are shown in the lower part of Figure 29. Tip-induced band bending is very large here, with observed band gaps of about 4 eV, indicating a carrier density many orders of magnitude below the nominal acceptor concentration of 1×10^{18} cm^{-3}.

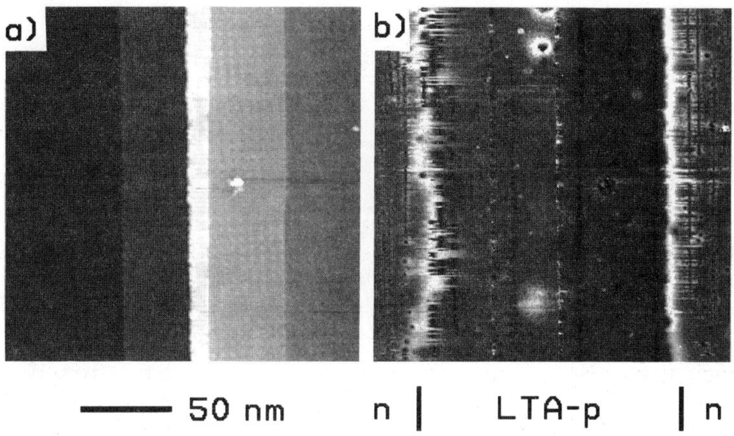

Figure 28 STM image of a cleaved GaAs structure, showing (*a*) topography and (*b*) conductance, acquired at a sample voltage of −2.2 V. The gray-scale range in (*a*) is 22 Å. As indicated in (*b*), the structure contains an LTA p-type layer, surrounded by n-type layers grown at higher temperature. The sample was annealed at 700°C. Several steps occur in the LTA-layer, as seen in (*a*).

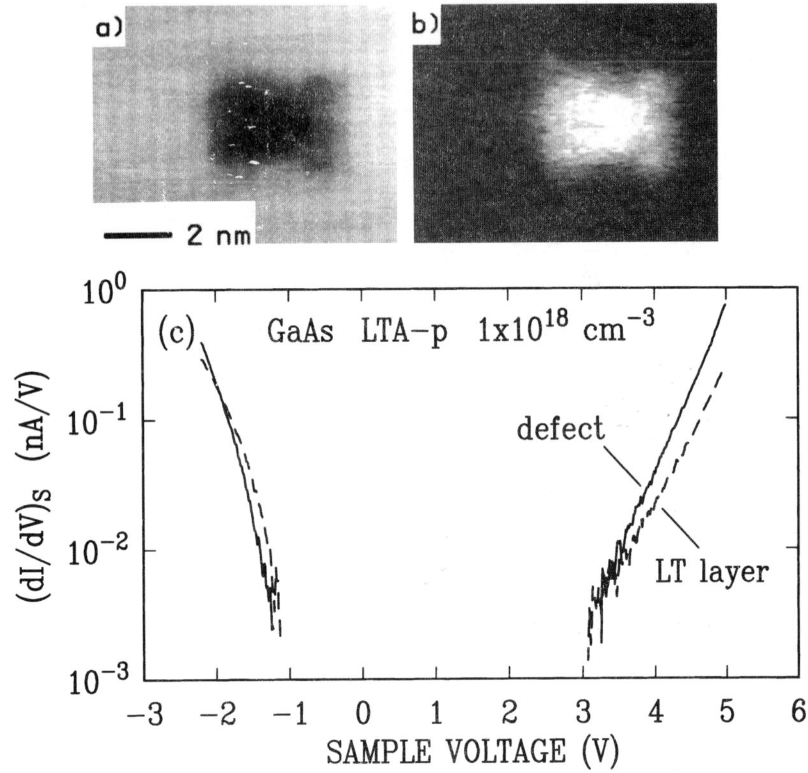

Figure 29 STM image of same sample as in Figure 28, displaying an expanded view of a point defect (arsenic antisite). Image (*a*) shows topography and (*b*) conductance, acquired at a sample voltage of -2.2 V. The gray-scale range in (*a*) is 1.1 Å. Tunneling spectra are shown in (*c*), acquired on the point defect and on a region of flat LTA-GaAs between defects.

Spectra acquired on top of the antisites show a small shift in the band edge positions, precisely as expected for a charged defect (76). Spectra acquired on top of the clusters shown in both Figures 27 and 28 display characteristics similar to those acquired from locations far from the clusters, except that glitches in the spectra are observed on the clusters, indicating charging effects. Apparently, in these cases, the clusters have insufficient electrical contact to the conductive layers surrounding the LTG-layer to permit direct observation of midgap states. Nevertheless, the observed depletion regions around the clusters definitively establish their effectiveness in Fermi-level pinning.

Summarizing the STM results for annealed LTG-GaAs, one finds that in *n*-type material, precipitates are observed with no accompanying antisite defects, whereas in *p*-type material, the two species are found to coexist.

This result is consistent with the formation dynamics of clusters proposed by Melloch et al (15), in which the precipitates form (*a*) to drive the Fermi level to a midgap position, thereby minimizing the chemical potential, and (*b*) to minimize the interfacial energy between incorporated arsenic and the GaAs matrix. For *p*-type material, the midgap pinning position is most effectively obtained by the donor antisite defects, so that in annealed material, a concentration of the antisites equal to the acceptor concentration will remain, and the rest of the arsenic is incorporated in precipitates. In *n*-type material, the antisite defects, being donors, cannot compensate the shallow donors, so they all form clusters. These precipitates produce midgap pinning, and thus depletion spheres are observed around them. For the case of undoped material, STM observations are not possible, but it is expected that the antisite defect concentration will be low because few if any of the defects are needed for midgap pinning. The relative effect on material properties of residual defects and arsenic clusters depends on their relative concentrations.

Carrier Lifetime

The lifetime of photogenerated carriers in LTG-GaAs is of interest for photoconductive applications of the material (77–81). Harmon et al (82) have used a pump-probe measurement of the differential transmission to determine the lifetime in LTG-GaAs as a function of anneal. A Coherent Mira 900f titanium : sapphire laser was used to produce $\sim$125 fs pulses tuned to a wavelength of 866 nm to measure the GaAs epilayers and tuned to a wavelength of 720 nm to measure the $Al_{0.25}Ga_{0.75}As$ epilayer. These wavelengths were chosen to provide photons of energies about 10 meV above the band gaps to probe the states near the conduction and valence band edges. The pulse from the titanium: sapphire laser, which was running at 73 MHz, was split into a pump and probe beam. The pump beam had an average power of 100 to 200 mW and was focused to a spot size of approximately 50 μm. The probe beam had an average power of 4 to 5 mW and was focused to a slightly smaller spot. The pump beam fills conduction band states and causes a reduction in the absorption. As the electrons recombine, the absorption increases back to its equilibrium value. This transient in the absorption is measured as a function of time, using the probe beam, which is delayed relative to the pump beam by using an Aerotech linear translation stage to vary the optical path length.

The transient responses—plotted as normalized differential transmissions—are shown in Figure 30 for a GaAs epilayer containing 0.9% excess arsenic for different anneal conditions. The carrier lifetime is determined by fitting the transient in Figure 30 with a single exponential. To ensure that the fit is dominated by carrier recombination rather than by

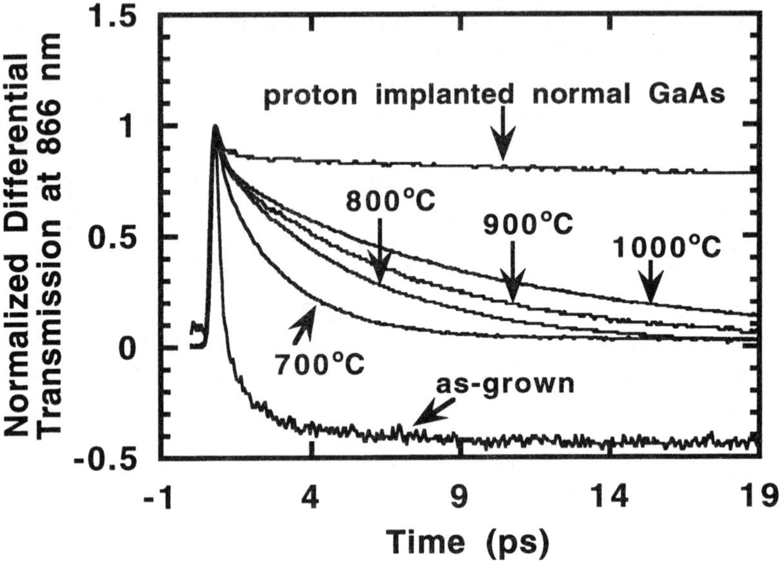

Figure 30 Normalized differential transmission measurements of low-temperature-grown GaAs showing the dependence of the decay of the differential transmission on the arsenic cluster coarsening anneal. The anneals were of duration 30 s at the temperatures indicated. The normal GaAs film is 1-μm MBE material implanted with 1×10^{12} cm^{-2} protons in order to assure recovery to equilibrium between laser pulses.

carrier cooling or system noise, the fit begins 1.5 ps after the peak and extends to the point where the transient reaches 10% of the final value. The results of carrier lifetime vs precipitate spacing are shown in Figure 31 for a GaAs epilayer containing 0.9% excess arsenic, a GaAs epilayer containing 0.7% excess arsenic, a GaAs epilayer containing 0.3% excess arsenic, and an $Al_{0.25}Ga_{0.75}As$ epilayer containing 0.2% excess arsenic. These results indicate an increase in carrier lifetime with an increase in precipitate spacing.

To interpret the data in Figure 31, a single recombination center is assumed. The carrier lifetime can be described by

$$\tau = \frac{1}{\sigma N v_t} \qquad 5.$$

where σ is the capture cross section of the trap, N is the number of empty traps, and v_t is the electron thermal velocity. If we suppose the trap is the arsenic cluster, then we would expect the capture cross section to be proportional to the area of the precipitate. This model does not accurately describe the observed trends because the GaAs epilayer with 0.3% excess

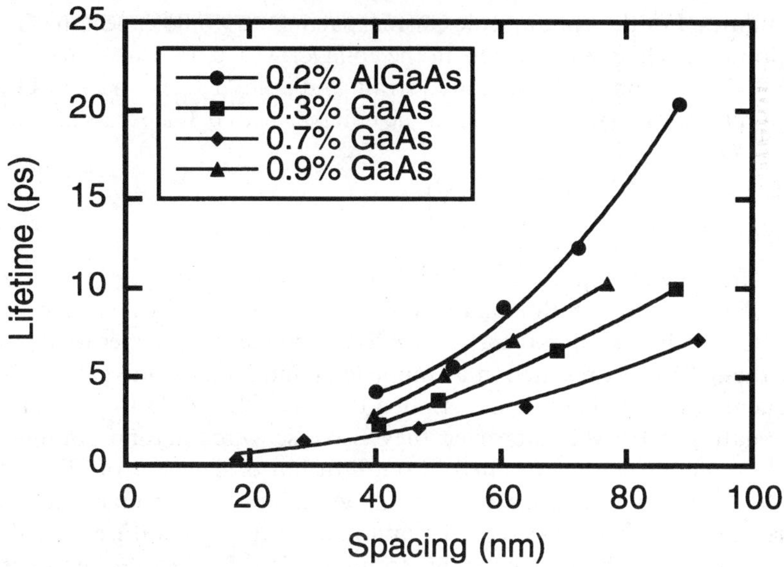

Figure 31 Carrier lifetime vs cluster spacing, fit by a square dependence on cluster spacing for four epilayers. Three of the epilayers are GaAs, one containing 0.3% excess arsenic, one containing 0.7% excess arsenic, and one with 0.9% excess arsenic. The fourth epilayer is AlGaAs containing 0.2% excess arsenic.

arsenic has a shorter carrier lifetime than the GaAs epilayer with 0.9% excess arsenic when the cluster spacings are the same. For the same arsenic cluster spacing, one would expect the GaAs epilayer with 0.9% excess arsenic to have the shorter carrier lifetime because the arsenic clusters are larger—hence the capture cross section is larger. In addition to the arsenic clusters, it is possible that other point defects affect the carrier lifetimes. However, Equation 5 describes a system of traps capable of absorbing only one electron at a time. For a system containing traps that can be multiply ionized, there would be terms of the form of Equation 5 for each ionization level. The arsenic precipitates, indeed, can accept many electrons and are perhaps better thought of as sinks for electrons rather than discrete, single-electron traps. In this view, electron removal from the GaAs matrix is limited by diffusive transport to the precipitates rather than random thermal impacting on the precipitates. This model predicts a square dependence of measured lifetime to precipitate spacing, which is in better agreement with the results observed. A square dependence of the carrier lifetime with the spacing between clusters is observed in the experimental data illustrated in Figure 31—the variation of lifetime in the three films for a given arsenic cluster spacing is the result of differences in mobilities. For the same arsenic cluster spacing, we would expect the GaAs region between arsenic clusters in the epilayer with 0.9% excess arsenic to have a lower mobility (more initial strain and point defects) than the GaAs region between arsenic clusters in the epilayer with 0.3% excess arsenic.

DEVICE APPLICATIONS

Buffer Layers

The first device application of LTG-GaAs was as a buffer layer under MESFETs (4, 5). Smith et al (5) found that such a buffer layer grown at 200°C, which was subsequently annealed when the top epilayer was grown for the MESFET, resulted in a complete elimination of side-gating. Subsequently, Lin et al (83) demonstrated the use of LTG-GaAs to eliminate side-gating in HEMTs; however, they also discovered that non-optimized LTG-buffer structures resulted in outdiffusion of defects from the LTG buffer to the active regions, resulting in degradation of the high-frequency performance. Lin et al (83) used a multi-substrate temperature procedure during MBE of the buffer regions to minimize this effect. Solomon et al (84) also have used a LTG-GaAs layer to reduce back-gating in GaAs semiconductor-insulator-semiconductor FETs (SISFETs).

Another application of LTG-GaAs for MESFETs was the placement of the LTG-GaAs region on top of the MESFET channel to improve the breakdown voltage (46). Lin et al (46) discovered that they needed a 20-nm AlAs layer between the LTG-GaAs and the *n*-GaAs channel to prevent arsenic and Ga-vacancy diffusion from the LTG-GaAs layer into the *n*-GaAs channel, which would compensate the channel. Substantial enhancement of the breakdown voltage was achieved without degradation of the drive capability of the MESFET.

The semi-insulating properties have also been used for isolation in optical devices. Subramanian et al (85) have demonstrated the utility of GaAs : As for isolation between tandem solar cells. There have also been many demonstrations of the applications of GaAs : As as a current-blocking layer in diode lasers (86–88).

Photoconductors

Because these composites are semi-insulating, exhibit reasonable mobilities, and in some cases subpicosecond lifetimes (77–79), they are attractive materials for applications where one needs a high-speed photoconductor. GaAs : As has been used as the photoconducting switch to launch freely propagating electromagnetic pulses (79–81). The freely propagating electromagnetic pulses have applications for radar systems. Rahman et al (81) have used LTG-GaAs to photoconductively switch coplanar-strip antennas to generate picosecond bursts of freely propagating electromagnetic energy, which was used for measuring target resonances.

Because the arsenic precipitates form embedded Schottky barriers, it is possible to detect light with photon energies below the band gap of the semiconductor matrix, relying on internal photoemission from the arsenic precipitates. Using this concept, GaAs-based *p-i-n* photodiodes have been demonstrated with reasonable room-temperature responsivities out to 1.7 μm by incorporating a composite as the *i*-layer (89, 90). The below-band-gap responsivity is the result of internal photoemission from the arsenic precipitates in the *i*-layer. This below-band-gap response is illustrated in Figure 32 for a GaAs (25) and an AlGaAs (91) *p-i-n* diode. The *i*-layers in these two diodes were grown at 250°C to incorporate excess arsenic. Upon raising the substrate temperature to 600°C to grow the top *p*-layers, the excess arsenic precipitates, forming the composite *i*-layer. Linear extrapolation of the curves in Figure 32 results in internal barrier heights of 0.7 eV for GaAs : As and 0.94 eV for $Al_{0.25}Ga_{0.75}As$: As.

Below band-gap, metal-semiconductor-metal photodetectors (MSM-PDs) on GaAs : As have also been investigated (92). The MSM-PD was buried within a ground-signal-ground coplanar stripline, with the Ti/Pt/Au

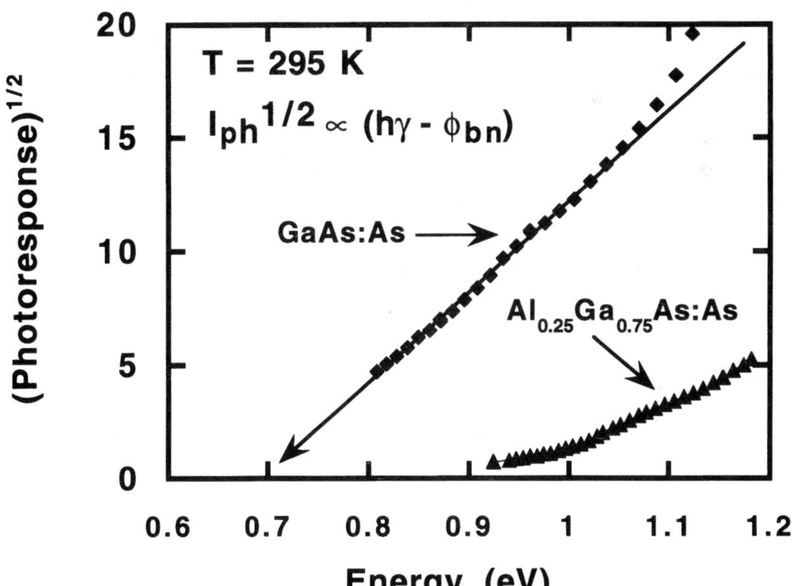

Figure 32 Photoresponse vs photon energy for two PiN diodes. One *p-i-n* diode has an *i*-layer of GaAs containing arsenic clusters, and the other has an *i*-layer of $Al_{0.25}Ga_{0.75}As$ containing arsenic clusters. The *p*- and *n*-regions are normal highly stoichiometric material.

MSM-PD fingers and striplines defined in one mask step. The detector area was $75 \times 75\ \mu m^2$, with finger spacing of 3 μm. The optical excitation used was 1.3-μm pulses of width 700 to 900 fs. The generated electrical pulse was coupled into a 40 GHz Tektronix sampling head. The total system rise time was 13 ps. The measured pulse response was at the system limit, with 10 to 90% rise and fall times of about 14 ps, independent of bias from 2–35 V. Integration of the pulse produced an estimated responsivity of 0.25 mA/W. By deconvolving the response of the probe and the scope from the temporal measurement, a detector bandwidth well beyond 50 GHz was estimated.

Although light below the band gap of the semiconductor matrix can be detected by internal photoemission from the arsenic clusters, the responsivity is low because of the low absorption of the incident photons. Recently, devices have been exhibited that use internal gain to increase the responsivity. One was a HEMT version, where the HEMT channel was grown at low temperatures and thus contained arsenic clusters (93). The other was a photoconductor with an *n*-GaAs:As channel and ohmic contacts separated by 1 μm, as illustrated in Figure 33 (94). The device in Figure 33 was annealed so that the precipitates coarsen to a point where the channel is just depleted. This device exhibits gain for both above- and below-band-gap illumination. The gain is a consequence of an increase in carrier lifetime due to the band bending present because of the *n*-type doping. For above-band-gap illumination, the holes and electron are separated by the band bending, with the holes going to the precipitates and the electrons staying in the conducting channels between the precipitates. Similarly for below-band-gap illumination, the electrons emitted from the precipitates are contained in the regions between the precipitates due to

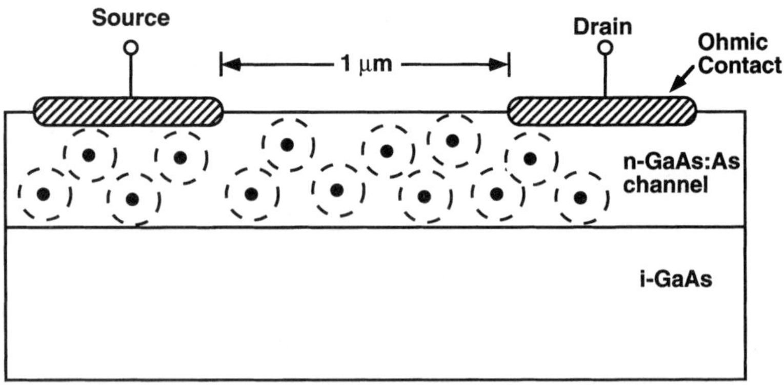

Figure 33 Cross section of metal-semiconductor-metal photoconductor showing depletion regions surrounding the arsenic precipitates in the channel.

the band bending. This effectively increases the lifetime of photogenerated electrons because the electrons are prevented from recombining at the precipitates by internal electric fields, thus resulting in a photoconductive gain effect. The greater the density of photogenerated electrons, the smaller the effective increase in electron lifetimes because the higher hole concentration on the precipitates reduces the band bending. Therefore, one is trading bandwidth for gain (responsivity) with these devices. Shown in Figure 34 is the photocurrent as a function of bias, with illumination power as a parameter, for 850-nm illumination (Figure 34*a*) and for 950-nm illumination (Figure 34*b*). The saturation in the photocurrent is caused by velocity saturation. The responsivity as a function of wavelength is shown in Figure 35 with optical power as a parameter. The responsivity varies considerably with input power, which indicates that the gain of the device varies with input power, consistent with the argument above for the decrease in carrier lifetime with increase in illumination intensity.

Nonlinear Optics

LTG technology has opened the door to a new class of nonlinear-optical applications, called photorefractive nanostructures, that require semi-insulating quantum well structures. Photorefractive quantum wells (95–98) are photo-active semiconductor thin films that can be used to record time-varying holograms. Such dynamic holographic media will form the core of future imaging applications that range from parts inspection on assembly lines, to error removal in high-data-rate fiber telecommunications, to information processing in optical computers.

LTG materials have moved into several key functional roles in the designs of photorefractive nanostructures. The value of the LTG material for holography lies in their high defect densities and their ultrafast carrier lifetimes. One of the most important issues during hologram recording is the density of optical information storage. This data density is limited primarily by carrier diffusion, with the minimum data spacing given by $L = 2\pi L_D$, where L_D is the diffusion length. The ultrafast lifetimes in LTG layers (82) produce ultrashort diffusion lengths down to 0.1 μm, yielding two-dimensional data densities of 3×10^8 bits/cm^2. These high data densities were demonstrated experimentally in LTG GaAs thin films (99).

In photorefractive experiments, the LTG layer performed as the short-transport-length material as well as an electro-optic layer. Photorefractive effects need electro-optic effects such as electric field–induced changes in the exciton absorption. But the excitonic transitions in LTG materials with large excess arsenic concentrations are severely broadened, making them appear less useful for electro-optic applications (100). However, it was discovered that the excitonic transitions in strongly annealed LTG-

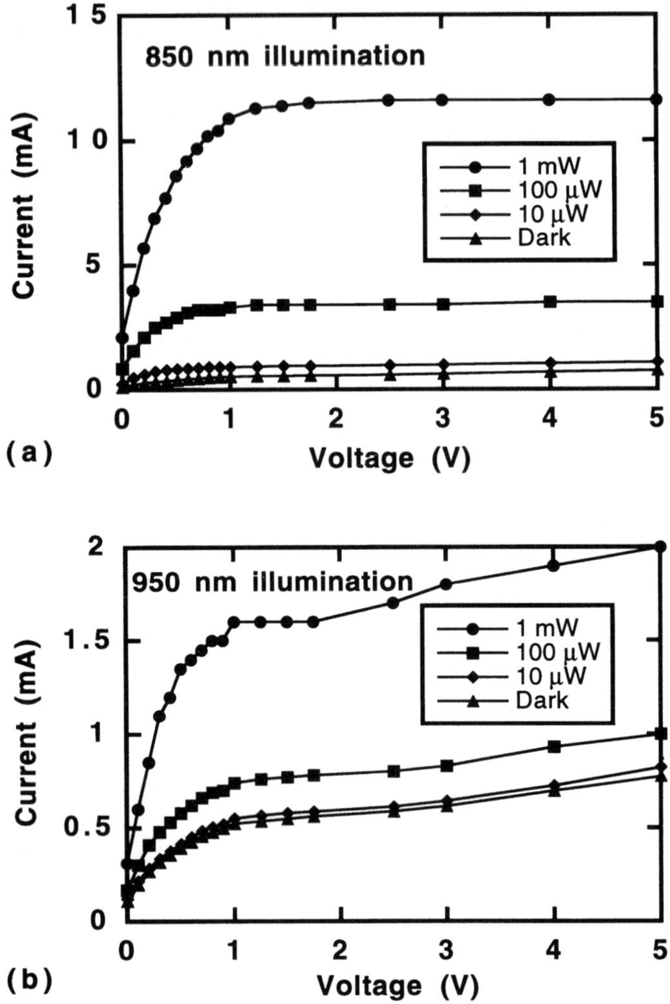

Figure 34 Current-voltage characteristics of the Figure 23 photodetector with 1-μm finger spacing under (*a*) 850- and (*b*) 950-nm illumination.

AlGaAs were particularly sensitive to applied electric fields, producing large electro-optic effects at low field strengths that exceeded the quantum-confined Stark effect at comparable fields, as shown in Figure 36 (101, 102). The excitons in the LTG material responded to applied electric fields as if a much larger field had been applied. It has been suggested (92, 103) that the arsenic precipitates become highly charged during current transport, leading to internal microfields that can significantly exceed

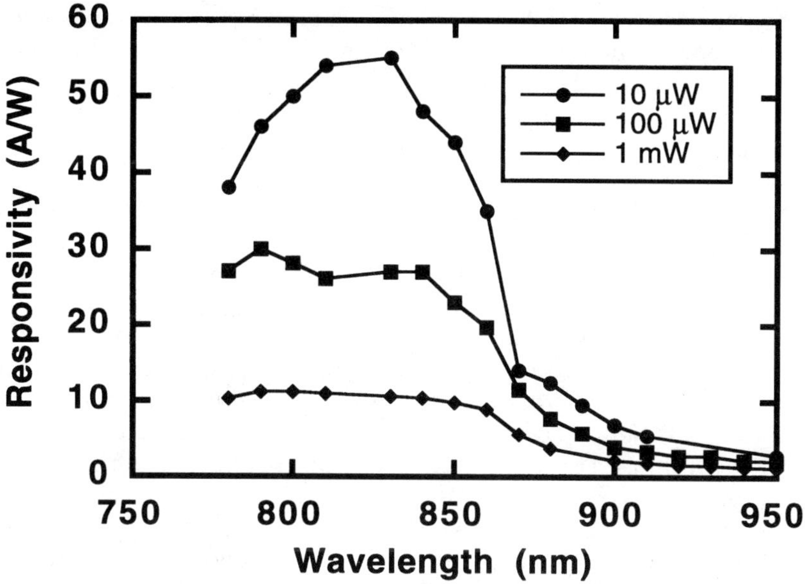

Figure 35 Responsivity vs wavelength of the Figure 23 photodetector with optical power as a parameter.

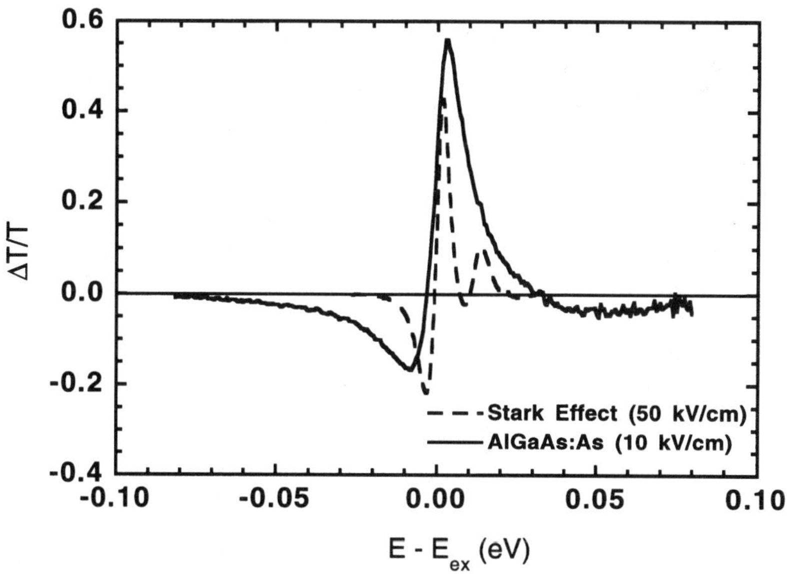

Figure 36 Comparison of the quantum-confined Stark effect in a high-quality, highly stoichiometric multiple-quantum well structure and in an AlGaAs epilayer containing arsenic clusters.

macroscopic fields. This electro-optic mechanism represents a fundamental departure from traditional electro-optic mechanisms that rely entirely on macroscopic fields. Furthermore, the broadened excitons in the LTG materials lead to broad-band electro-optic modulation, which can be important for femtosecond pulse applications (104), or for other broad-band applications such as wavelength-division multiplexing.

More complicated photorefractive quantum wells structures have recently incorporated LTG layers as ultrafast trapping layers to store space charge (105, 106). The LTG layers in these structures were not used for electro-optic properties. Traditional high-temperature grown quantum wells provided quantum-confined excitons with large quantum-confined Stark effects. These quantum wells required proton implantation to make them semi-insulating, which complicates the device processing. Recent studies of LTG multiple quantum wells grown at moderate temperatures (107) have exhibited extremely sharp excitonic transitions, with large quantum-confined Stark shifts that produce absorption changes up to $\Delta\alpha = 9000\ \text{cm}^{-1}$. This ability to grow high-quality semi-insulating multiple quantum well structures with sharp excitonic features had been an elusive goal. The discovery of sharp spectra in LTG quantum wells makes this material an important new option for many electro-optic applications.

Non-Alloyed Ohmic Contacts

The dominant contact technology for GaAs-based devices includes an alloying step, which makes these contacts unsuitable as optical mirrors. Yamamoto et al (108) have shown that ex situ non-alloyed contacts can be formed on LTG-GaAs with a specific resistivity of $1.5 \times 10^{-3}\ \Omega\text{-cm}^2$. Patkar et al (109) have further developed non-alloyed contact technology based on LTG epilayers for both *p*- and *n*-GaAs, demonstrating a three-order-of-magnitude reduction in specific contact resistivity to *n*-GaAs as compared with results from Yamamoto et al (108). The fact that these contacts are non-alloyed opens the possibility of their use as optical mirrors in applications such as solar cells, light-emitting diodes (LEDs), and vertical-cavity surface-emitting lasers (VCSELs).

Low-temperature growth has been used in the past to improve the Be-doping efficiency in GaAs (110) and InGaAs devices (111, 112). When doping GaAs *p*-type with Be during MBE, hole concentrations greater than $10^{20}\ \text{cm}^{-3}$ can be achieved with substrate temperatures in the range of 300 to 400°C due to a slight arsenic excess in the epilayer. If this LTG *p*-GaAs is annealed subsequently, with an arsenic over-pressure, the excess arsenic in the bulk ensures that any interstitial Be atoms are driven onto acceptor sites, whereas the excess arsenic should go onto regular arsenic sites in the lattice. A non-alloyed ohmic contact is readily formed when metal is deposited on such a heavily doped layer due to holes tunneling through the narrow depletion region. Patkar et al (109) have demonstrated tunneling contact resistances of $<10^{-6}$ Ω-cm^2 to such LTG-*p*-GaAs epilayers.

Achieving high *n*-type doping concentrations is more difficult than achieving high *p*-type concentrations in bulk GaAs. Silicon is the typical *n*-type dopant used in MBE of GaAs. In a depletion region, Si atoms on Ga sites (the donor site) can approach 10^{20} cm^{-3}, which is the case for the near-surface during MBE (113). However, the electron concentration in the bulk of the GaAs—below the surface depletion region—reverts to only 5×10^{18} cm^{-3} because of a large fraction of the Si atoms switching to arsenic sites. Upon removal from the MBE system and exposure of the *n*-GaAs surface to the atmosphere, the near-surface, heavily *n*-doped region oxidizes—unlike a heavily *p*-doped GaAs surface that is resistant to oxidation. A metal deposited on this oxidized, *n*-GaAs surface will not form an ohmic contact. Patkar et al (109) have found that a thin, 1- to 2-nm undoped GaAs layer that contains about 1% excess arsenic can be used to protect the $>10^{20}$ cm^{-3} donor density GaAs region. After air exposure of these surfaces, deposition of Ti, Ag, or Au results in ohmic contacts with a specific contact resistance of $<10^{-6}$ Ω-cm^2.

SUMMARY

In summary, GaAs, AlGaAs, and InGaAs epilayers containing excess arsenic are interesting new materials. The nature of the material can be dramatically changed by annealing from a material with large concentrations of point defects to a two-phase system of metallic arsenic clusters in an arsenide semiconductor matrix. The two-phase form shows promise for many electronic and optoelectronic applications.

ACKNOWLEDGMENTS

This work was partially supported by the United States Air Force Office of Scientific Research under Grants No. F49620–93–1–0031 and F49620–93–1–0388 and the MRSEC Program of the National Science Foundation under award No. DMR-9400415. FHP acknowledges the support of United States Army Research Office contract No. DAAL03–92–G-0189, National Science Foundation grant No. DMR-9120363, and the New York State Science and Technology Foundation through its Centers for Advanced Technology Program.

Literature Cited

1. Cho AY, Arthur JR. 1975. *Prog. Solid State Chem.* 10: 157–91
2. Cho AY. 1970. *J. Appl. Phys.* 41: 2780–86
3. Foxon CT, Joyce BA. 1977. *Surf. Sci.* 64: 293–305
4. Murotani T, Shimanoe T, Mitsui S. 1978. *J. Cryst. Growth* 45: 302–8
5. Smith FW, Calawa AR, Chen C-L, Mantra MJ, Mahoney LJ. 1988. *IEEE Electron Device Lett.* 9: 77–80
6. Melloch MR, Miller DC, Das B. 1989. *Appl. Phys. Lett.* 54: 943–45
7. Kaminska M, Weber ER, Liliental-Weber Z, Leon R, Rek ZU. 1989. *J. Vac. Sci. Technol. B* 7: 710–13
8. Melloch MR, Otsuka N, Woodall JM, Warren AC, Freeouf JL. 1990. *Appl. Phys. Lett.* 57: 1531–33
9. Kaminska M, Liliental-Weber Z, Weber ER, George T, Kortright JB, et al. 1989. *Appl. Phys. Lett.* 54: 1881–83
10. Melloch MR, Mahalingam K, Otsuka N, Woodall JM, Warren AC. 1991. *J. Cryst. Growth* 111: 39–42
11. Liliental-Weber Z, Cooper G, Mariella R Jr, Kocot C. 1991. *J. Vac. Sci. Technol. B* 9: 2323–27
12. Mahalingam K, Otsuka N, Melloch MR, Woodall JM, Warren AC. 1991. *J. Vac. Sci. Technol. B* 9: 2328–32
13. Mahalingam K, Otsuka N, Melloch MR, Woodall JM. 1992. *Appl. Phys. Lett.* 60: 3253–55
14. Melloch MR, Otsuka N, Mahalingam K, Chang CL, Kirchner PD, et al. 1992. *Appl. Phys. Lett.* 61: 177–79
15. Melloch MR, Otsuka N, Mahalingam K, Chang CL, Woodall JM, et al. 1992. *J. Appl. Phys.* 72: 3509–13
16. Ibbetson JP, Speck JS, Gossard AC, Mishra UK. 1993. *Appl. Phys. Lett.* 62: 169–71
17. Ibbetson JP, Speck JS, Gossard AC, Mishra UK. 1993. *Appl. Phys. Lett.* 62: 2209–11
18. Melloch MR, Woodall JM, Otsuka N, Mahalingam K, Chang CL, et al. 1993. *Mater. Sci. Eng. B* 22: 31–36
19. Cheng TM, Chin A, Chang CY, Huang MF, Hsieh KY, Huang JH. 1994. *Appl. Phys. Lett.* 64: 1546–48
20. Baranowski JM, Liliental-Weber Z, Yau W-F, Weber ER. 1991. *Phys. Rev. Lett.* 66: 3079–3082
21. Warren AC, Woodall JM, Freeouf JL, Grischkowsky D, McInturff DT, et al. 1990. *Appl. Phys. Lett.* 57: 1331–33
22. Verma AK, Tu J, Smith JS, Fujioka H, Weber E. 1993. *J. Electron. Mater.* 22: 1417–20
23. Ibbetson JP, Speck JS, Nguyen NX, Gossard AC, Mishra UK. 1993. *J. Electron. Mater.* 22: 1421–24
24. Fang Z-Q, Look DC. 1993. *J. Electron. Mater.* 22: 1429–32
25. McInturff DT, Woodall JM, Warren AC, Braslau N, Pettit GD, et al. 1992. *Appl. Phys. Lett.* 60: 448–50
26. Jager ND, Verma AK, Dreszer P, Newman N, Liliental-Weber Z, et al. 1993. *J. Electron. Mater.* 22: 1499–1502
27. Feenstra RM, Vaterlaus A, Woodall JM, Pettit GD. 1993. *Appl. Phys. Lett.* 63: 2528–30
28. Cho AY. 1969. *Surf. Sci.* 17: 494–503
29. Eaglesham DJ, Pfeiffer LN, West KW, Dykaar DR. 1991. *Appl. Phys. Lett.* 58: 65–67
30. Liliental-Weber Z, Swider W, Yu KM, Kortright J, Smith FW, Calawa AR. 1991. *Appl. Phys. Lett.* 58: 2153–55
31. Ibbetson JP, Mirin RP, Mishra UK, Gossard AC. 1994. *J. Vac. Sci. Technol. B* 12: 1050–52
32. Wie CR, Xie K, Look DC, Evans KR, Stutz CE. 1991. *Mater. Res. Soc. Symp. Proc.* 198: 178
33. Matyi RJ, Melloch MR, Woodall JM. 1992. *Appl. Phys. Lett.* 60: 2642–44
34. Liliental-Weber Z, Claverie A, Washburn J, Smith FW, Calawa AR. 1991. *Appl. Phys. A* 53: 141–46
35. Matyi RJ, Melloch MR, Woodall JM. 1993. *J. Cryst. Growth* 129: 719–27
36. Voorhees PW. 1985. *J. Stat. Phys.* 38: 231–52
37. Köhler K, Ganser P, Maier M. 1993. *J. Cryst. Growth* 127: 720–23
38. Manasreh MO, Stutz CE, Solomon JG, Mier MG, Kaspi R, Evans KR. 1993. In *Semi-Insulating III-V Materials*, ed. GJ Miner, W Ford, ER Weber, Ixtapa, Mexico, 1992. Bristol: Inst. Phys.

39. McQuaid SA, Newman RC, Missous M, O'Hagan S. 1992. *Appl. Phys. Lett.* 61: 3008–10
40. O'Hagan S, Missous M. 1994. *J. Appl. Phys.* 75: 7835–41
41. Cheng TM, Chang CY, Chang TC, Huang JH, Huang MF. 1994. *Appl. Phys. Lett.* 64: 3626–28
42. Mahalingam K, Otsuka N, Melloch MR, Woodall JM, Warren AC. 1992. *J. Vac. Sci. Technol. B* 10: 812–14
43. Cheng TM, Chang CY, Chin A, Huang MF, Huang JH. 1994. *Appl. Phys. Lett.* 64: 2517–19
44. Melloch MR, Chang CL, Otsuka N, Mahalingam K, Woodall JM, Kirchner PD. 1993. *J. Cryst. Growth* 127: 499–502
45. Horikoshi Y, Kawashima M, Yamaguchi H. 1986. *Jpn. J. Appl. Phys.* 25: L868–70
46. Yin L-W, Hwang Y, Lee JH, Kolbas RM, Trew RJ, Mishra UK. 1990. *IEEE Electron Device Lett.* 11: 561–63
47. Claverie A, Namavar F, Liliental-Weber Z. 1993. *Appl. Phys. Lett.* 62: 1271–73
48. Chang JCP, Otsuka N, Harmon ES, Melloch MR, Woodall JM. 1994. *Appl. Phys. Lett.* 65: 2801–3
49. Nolte DD. 1994. *J. Appl. Phys.* 76: 3740–45
50. Manasreh MO, Look DC, Evans KR, Stutz CE. 1990. *Phys. Rev. B* 41: 10272–75
51. Look DC, Walters DC, Mier M, Stutz CE, Brierley SK. 1992. *Appl. Phys. Lett.* 60: 2900–2
52. Look DC, Walters DC, Manasreh MO, Sizelove JR, Stutz CE, Evans KR. 1990. *Phys. Rev. B* 42: 3578–81
53. Campbell AC, Crook GE, Rogers TJ, Streetman BG. 1990. *J. Vac. Sci. Technol. B* 8: 305–7
54. Look DC, Robinson GD, Sizelove JR, Stutz CE. 1993. *J. Electron. Mater.* 22: 1425–28
55. Look DC. 1991. *J. Appl. Phys.* 70: 3148–51
56. Warren AC, Woodall JM, Kirchner PD, Yin X, Pollak FH, et al. 1992. *Phys. Rev. B* 46: 4617–20
57. Rossi DV, Fossum ER, Pettit GD, Kirchner PD, Woodall JM. 1987. *J. Vac. Sci. Technol. B* 5: 982–84
58. Atique N, Harmon ES, Chang JCP, Woodall JM, Melloch MR. 1995. *J. Appl. Phys.* 77: 1471–76
59. Bliss DE, Walukiewicz W, Ager JW, Haller EE, Chan KT, Tanigawa S. 1992. *J. Appl. Phys.* 71: 1699–1707
60. Pollak FH, Shen H. 1993. *Mater. Sci. Eng. Rep.* 10: 275–374
61. Warren AC, Woodall JM, Kirchner PD, Yin X, Guo X, et al. 1992. *J. Vac. Sci. Technol. B* 10: 1904–7
62. Warren AC, Woodall JM, Borroughes JH, Kirchner PD, Heinrich HK, et al. 1992. *Mater. Res. Soc. Symp. Proc.* 241: 15–25
63. Shen H, Rong FC, Lux R, Pamulapati J, Dutta M, et al. 1992. *Appl. Phys. Lett.* 61: 1585–87
64. Shen H, Rong FC, Lux R, Pamulapati J, Dutta M, et al. 1992. *Proc. Soc. Photo-Opt. Instrum. Eng.* 1678: 211
65. Look DC, Hoelscher JE, Grant JT, Stutz CE, Evans KR, Numan M. 1992. *Mater. Res. Soc. Symp. Proc.* 241: 87–92
66. Woodall JM, Kirchner PD, Freeouf JL, McInturff DT, Melloch MR, Pollak FH. 1993. *Philos. Trans. R. Soc. London Ser. A* 344: 521–32
67. Yin X, Chen H-M, Pollak FH, Cao Y, Montano PA, et al. 1991. *J. Vac. Sci. Technol. B* 9: 2114–17
68. Feenstra RM, Woodall JM, Pettit GD. 1994. *Mater. Sci. Forum* 143–147: 1311–18
69. Feenstra RM, Woodall JM, Pettit GD. 1993. *Phys. Rev. Lett.* 71: 1176–79
70. Feenstra RM. 1994. *Semicond. Sci. Technol.* 9: 2157–68
71. Feenstra RM, Martensson P. 1988. *Proc. 19th Int. Conf. Phys. Semiconductors,* ed. W Zawadzki, pp. 629–36. Warsaw: Polish Inst. Phys.
72. Feenstra RM, Vaterlaus A, Yu ET, Kirchner PD, Lin CL, et al. 1993. *Semiconductor Interfaces at the Sub-Nanometer Scale*, ed. HWM Salemink, MD Pashley, pp. 127–37. Netherlands: Kluwer
73. Feenstra RM, Stroscio JA. 1987. *J. Vac. Sci. Technol. B* 5: 923–29
74. Feenstra RM. 1994. *Phys. Rev. B* 50: 4561–70
75. Chapman R, Kellam M, Goodwin-Johansson S, Russ J, McGuire GE. 1992. *J. Vac. Sci. Technol. B* 10: 502–7
76. Stroscio JA, Feenstra RM. 1988. *J. Vac. Sci. Technol. B* 6: 1472–78
77. Smith FW, Lee HW, Diadiuk V, Hollis MA, Calawa AR, et al. 1989. *Appl. Phys. Lett.* 54: 890–92
78. Gupta S, Battacharya PK, Pamaulapati J, Mourou G. 1990. *Appl. Phys. Lett.* 57: 1543–45
79. Warren AC, Katzenellenbogen N, Grischkowsky D, Woodall JM, Melloch MR, Otsuka N. 1991. *Appl. Phys. Lett.* 58: 1512–14
80. Carin L, Kralij DR, Melloch MR, Woodall JM. 1993. *IEEE Microwave Guided Wave Lett.* 3: 339–41

81. Rahman A, Kralj D, Carin L, Melloch MR, Woodall JM. 1994. *Appl. Phys. Lett.* 64: 2178–80
82. Harmon ES, Melloch MR, Woodall JM, Nolte DD, Otsuka N, Chang CL. 1993. *Appl. Phys. Lett.* 63: 2248–50
83. Lin BJF, Kocot CP, Mars DE, Jaeger R. 1990. *IEEE Trans. Electron Devices* 37: 46–50
84. Solomon PM, Wright SL, Canora FJ. 1991. *IEEE Electron Device Lett.* 12: 117–19
85. Subramanian G, Dodabalapur A, Campbell JC, Streetman BG. 1991. *Appl. Phys. Lett.* 58: 2514–16
86. Sin YK, Horikawa H, Matsuyama I, Kamijoh T. 1992. *Electron. Lett.* 28: 803–4
87. Rogers TJ, Lei C, Streetman BG, Deppe DG. 1993. *J. Vac. Sci. Technol. B* 11: 926–28
88. Takamori T, Sin YK, Watanabe K, Kamijoh T. 1992. *Appl. Phys. Lett.* 61: 2266–68
89. Warren AC, Burroughes JH, Woodall JM, McInturff DT, Hodgson RT, Melloch MR. 1991. *IEEE Electron Device Lett.* 12: 527–29
90. Srinivasan A, Sadra K, Campbell JC, Streetman BG. 1993. *J. Electron. Mater.* 22: 1457–59
91. McInturff DT, Woodall JM, Warren AC, Braslou N, Pettit GD, et al. 1993. *Appl. Phys. Lett.* 62: 2367–68
92. Melloch MR, Otsuka N, Harmon ES, Nolte DD, Woodall JM, McInturff DT. 1993. *J. Jpn. Appl. Phys.* 32: 771–74
93. Subramaniam S, Schulte D, Ungier L, Zhao P, Plant TK, Arthur JR. 1995. *IEEE Electron Device Lett.* 16: 20–22
94. Harmon ES, McInturff DT, Melloch MR, Woodall JM. 1995. *J. Vac. Sci. Technol. B* 13: In press
95. Nolte DD, Melloch MR. 1995. In *Photorefractive Effects Materials*, ed. DD Nolte, 7: 373–451. Dordrecht: Kluwer
96. Nolte DD, Olson DH, Doran GE, Knox WH, Glass AM. 1990. *J. Opt. Soc. Am. B* 7: 2217–25
97. Wang QN, Brubaker RM, Nolte DD, Melloch MR. 1994. *J. Opt. Soc. Am. B* 9: 1626–41
98. Nolte DD, Melloch MR. 1994. *MRS Bull.* 19: 44–49
99. Nolte DD, Melloch MR, Ralph SE, Woodall JM. 1992. *Appl. Phys. Lett.* 61: 3098–3100
100. Knox WH, Doran GE, Asom M, Livescu G, Leibenguth R, Chu SNG. 1991. *Appl. Phys. Lett.* 59: 1491–93
101. Nolte DD, Melloch MR, Woodall JM, Ralph SJ. 1993. *Appl. Phys. Lett.* 62: 1356–58
102. Melloch MR, Nolte DD, Otsuka N, Chang CL, Woodall JM. 1993. *J. Vac. Sci. Technol. B* 10: 795–97
103. Nolte DD, Lahiri I, McKenna J, Steldt FR, Chang JCP, et al. 1994. In *Proc. 22nd Int. Conf. Phys. Semicond. Vancouver, BC, Canada*
104. Weiner AM, Leaird DE, Reitze DH, Paek EG. 1992. *IEEE J. Quantum Electron.* 28: 2251–61
105. Partovi A, Glass AM, Olson DH, Zydzik GJ, O'Bryan HM, et al. 1993. *Appl. Phys. Lett.* 62: 464–66
106. Kyono CS, Ikossi-Anastasiou K, Rabinovich WS, Bowman SR, Katzer DS. 1994. *Appl. Phys. Lett.* 64: 2244–46
107. Lahiri I, Nolte DD, Harmon ES, Melloch MR, Woodall JM. 1995. *Appl. Phys. Lett.* Submitted
108. Yamamoto H, Fand Z-Q, Look DC. 1990. *Appl. Phys. Lett.* 57: 1537–39
109. Patkar MP, Chin TP, Woodall JM, Lundstrom MS, Melloch MR. 1995. *Appl. Phys. Lett.* 66: 1412–14
110. Lievin JL, Dubon-Chevallier C, Alexandre F, Leroux G, Dangla J, Ankri D. 1986. *IEEE Electron Device Lett.* 7: 129–31
111. Hamm RA, Panis MB, Nottenburg RN, Chen YK, Humphrey DA. 1989. *Appl. Phys. Lett.* 54: 2586–88
112. Metzger RA, Hafiji M, Stanchina WE, Liu T, Wilson RG, McCray LG. 1993. *Appl. Phys. Lett.* 63: 1360–62
113. Kirchner PD, Jackson TN, Petitt GD, Woodall JM. 1985. *Appl. Phys. Lett.* 47: 26–28

Annu. Rev. Mater. Sci. 1995. 25: 601–46

LUMINESCENCE AS A DIAGNOSTIC OF WIDE-GAP II–VI COMPOUND SEMICONDUCTOR MATERIALS

B. J. Skromme

Department of Electrical Engineering and Center for Solid State Electronics Research, Arizona State University, Tempe, Arizona 85287–5706

KEY WORDS: photoluminescence, zinc selenide, impurities, strain, acceptors

ABSTRACT

The use of photoluminescence and related optical techniques to characterize the effects of strain and impurity levels in zinc selenide is reviewed. A theoretical discussion of the strain splittings occurring in heteroepitaxial ZnSe grown by molecular beam epitaxy or metalorganic chemical vapor deposition (MOCVD) on GaAs or other III–V substrates is given and compared to extensive experimental data. The binding energies and excited-stage energies of shallow acceptor levels due to N, Li, P, O, and two as-yet unidentified impurities are determined. The shallow acceptor level previously attributed to As is due to Li. The C impurities in MOCVD ZnSe have no observable effect on its opto-electronic properties but can stimulate relaxation of thermal strain at heavy doping levels. We investigate bound excitons involving N, Li, P, and two unidentified acceptor levels and show that Haynes's Rule linear relationship does not apply to acceptors in ZnSe.

INTRODUCTION

Luminescence techniques have formed the cornerstone of characterization methodology for II–VI compound semiconductor materials ever since the

0084–6600/95/0801–0601$05.00

early work on these substances in the form of phosphors, and even more so over the last several decades, as high-quality single crystals and epitaxial layers have been developed. Many of the fundamental luminescence processes now recognized in most semiconductor materials were first identified and studied in II–VI materials. However, the applications of the single-crystal materials were severely hindered until recently by the inability to dope the wide-gap selenides and sulfides *p*-types, and even today the inability to dope ZnTe *n*-type. The recent dramatic successess in *p*-type doping of ZnSe and some related materials using N plasma sources during growth by molecular beam epitaxy (MBE) (1, 2) have opened up many potential applications for these materials, including the first demonstration of a blue laser diode (3), and subsequent extensive developments in this area (4). This success in crystal growth owes a great deal, however, to earlier work by Dean and others using luminescence spectroscopy to show the importance of impurities, as opposed to native defects, in controlling the conductivity of these materials (5), and in demonstrating the desirable characteristics of the N acceptor level in relation to other available dopants (6).

Several prior reviews of the luminescent properties of wide-gap II–VI materials have been given (5, 7–12). We therefore concentrate on a critical review of recent developments in this field and on those issues that have been most controversial and crucial to the development of *p*-type epitaxial ZnSe suitable for laser and LED applications. In particular we emphasize the effects of strain in heteroepitaxial ZnSe on its luminescence properties and the optical characterization of acceptor levels in ZnSe.

EFFECTS OF STRAIN ON THE LUMINESCENCE OF WIDE-GAP II–VI COMPOUNDS

The presence of strain is an unavoidable feature of heteroepitaxial growth. The vast majority of MBE and organometallic chemical vapor deposition (OMCVD) experiments to date have utilized III–V substrate materials, principally GaAs. There main reasons for this have been the difficulty in obtaining high quality, conductive, twin-free II–VI substrates at a reasonable cost, and difficulties in preparing ZnSe substrates for growth. Advantages of GaAs substrates include availability of high quality, large area substrates with a reasonable lattice match (0.27% mismatch at 300 K) to ZnSe, and the possibility of monolithic integration of III–V and II–VI technologies. The disadvantages include the potential problem of autodoping and/or Ga and As out-diffusion into the II–VI material, and the finite lattice and thermal expansion coefficient mismatch. Both the lattice and thermal mismatches may produce strain in the epitaxial layers, and this strain is conveniently analyzed using optical techniques. Conversely, the strain complicates the accurate interpretation of the optical spectra and has sometimes led to erroneous conclusions in the literature regarding impurity and defect behavior in the material. In the following, we discuss these effects for the various features typically observed in luminescence.

Free Excitons

The most important element in analyzing the low-temperature photoluminescence (PL) spectra of heteroepitaxial materials is to identify the intrinsic, free exciton/polariton features correctly. We label these features X, and the shear strain-split light- and heavy-hole components of the exciton as X_{lh} and X_{hh}, respectively. These features are most readily distinguished using absorption-like measurements such as photoluminescence excitation spectroscopy (PLE), simple or modulated reflectance, or just transmission (although the latter usually requires substrate removal, which will generally change the strain). Because of their strong exciton (bound electron-hole pair) binding energies, the wide-gap materials exhibit pronounced excitonic features at low temperature in their simple reflectance spectra.

This fact was utilized by Yao et al in a detailed study of the strain relaxation in MBE ZnSe/GaAs as a function of ZnSe layer thickness (13, and references therein). They showed that layers below the practical critical thickness of around 1700 Å remain pseudomorphic to GaAs and thus exhibit a biaxial compressive strain, which gradually relaxes in thicker layers. The relaxation is mostly complete after around 1 μm. Layers thicker than this show a reversal of the X_{lh}/X_{hh} splitting, because of residual thermal expansion mismatch strain, which is biaxial and tensile in character. The use of reflectance enabled an earlier, incorrect interpretation (14) of the free exciton splitting in thick ZnSe/GaAs layers as being due to upper and lower polariton branches to be corrected. Unfortunately, many subsequent workers have not followed this excellent example of distinguishing extrinsic from intrinsic features, which is experimentally as simple as if not simpler than PL measurements. The more complicated techniques of photo- and piezo-modulated reflectivity have also been applied to characterize the lattice mismatch strain splitting in pseudomorphic ZnSe/GaAs but yielded no fundamentally new information compared to the simple reflectivity measurements (15).

Additional methods can also distinguish the intrinsic features from others in the PL spectra, including excitation and temperature-dependent measurements. The free exciton peaks normally dominate at moderately high excitation, where the extrinsic features saturate due to the finite number of impurity centers. An example of this effect is shown in Figure

1, where the neutral donor-bound excitons (I_{2o}) saturate with respect to the free excitons in an OMCVD sample with a relatively large thermal tensile strain. Note the further confirmation of the free exciton identity from the reflectance spectrum at the top. Earlier work based on PL alone failed to discern the larger than usual strain in this sample and led to erroneous identification of one of the I_{2o} features as a new carbon-related exciton (16, and references therein). Temperature-dependent measurements are also useful because bound excitons generally quench in favor of free excitons (and possibly donor-to-valence band transitions) at high temperature. This effect is illustrated in Figure 2 for the sample of Figure 1. Figure 2 also illustrates thermalization between the strain-split components of the neutral donor-bound excitons, which is discussed below.

The free exciton strain splittings have been varied in a systematic manner in pseudomorphic material by growing thin ZnS_xSe_{1-x} alloy layers on GaAs substrates as a function of S content (17), and by growing thin ZnSe layers on bulk $In_xGa_{1-x}As$ substrates with various In contents (18, 19). The results of the latter experiment are illustrated in Figure 3, where the

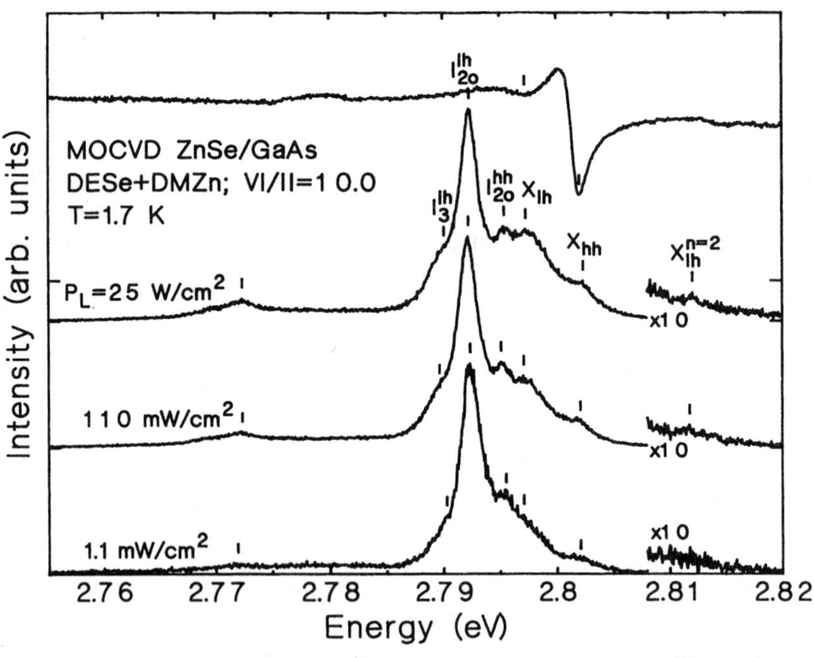

Figure 1 Low-temperature PL (lower three spectra) as a function of excitation intensity for a ZnSe/GaAs layer grown by MOCVD, together with the corresponding reflectance spectrum (at top). Note the increase in X_{lh} and X_{hh} with respect to the I_{2o} peaks at high intensity (P_L).

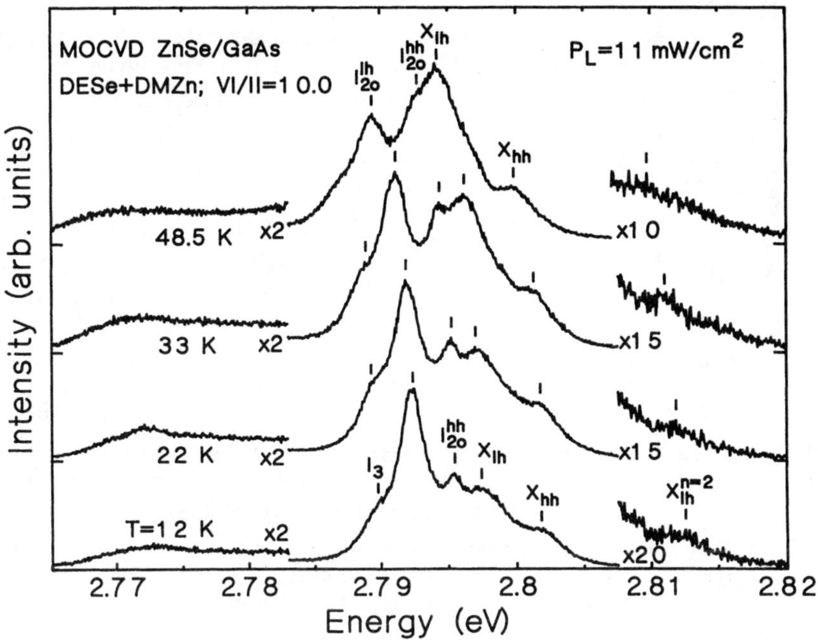

Figure 2 Variable-temperature excitonic PL spectra of the sample of Figure 1, showing the quenching of donor-bound exciton (I_{2o}) peaks with respect to the free excitons and the thermalization between strain-split components of I_{2o} as the temperature is raised.

symbols are experimental data points, and the solid lines represent a fit to conventional deformation potential theory, neglecting the very small interaction between the light hole and spin-orbit split-off bands. (Some of the corresponding spectra are discussed below.) In this theory, for a biaxial stress in the (100) plane, the excitonic peak positions are given by the expressions (20),

$$E(\mathrm{X}_{\mathrm{hh,lh}}) = E_{\mathrm{g}} - E_x + A \pm B, \qquad 1.$$

where

$$A = -2a\left(\frac{c_{11}-c_{12}}{c_{11}}\right)\varepsilon \quad \text{and} \quad B = b\left(\frac{c_{11}+2c_{12}}{c_{11}}\right)\varepsilon,$$

a and b are the hydrostatic and tetragonal shear deformation potentials, c_{11} and c_{12} are elastic stiffness coefficients, E_{g} is the band gap at $T \approx 0$, E_{x} is the free exciton binding energy, and $\varepsilon = -\varepsilon_{\mathrm{xx}} = -\varepsilon_{\mathrm{yy}}$ is the biaxial strain at the measurement temperature ($T \approx 0$) ($\varepsilon_{\mathrm{xx}}$ and $\varepsilon_{\mathrm{yy}}$ are components of the strain tensor, so that $\varepsilon > 0$ for a compressive strain). The strain is

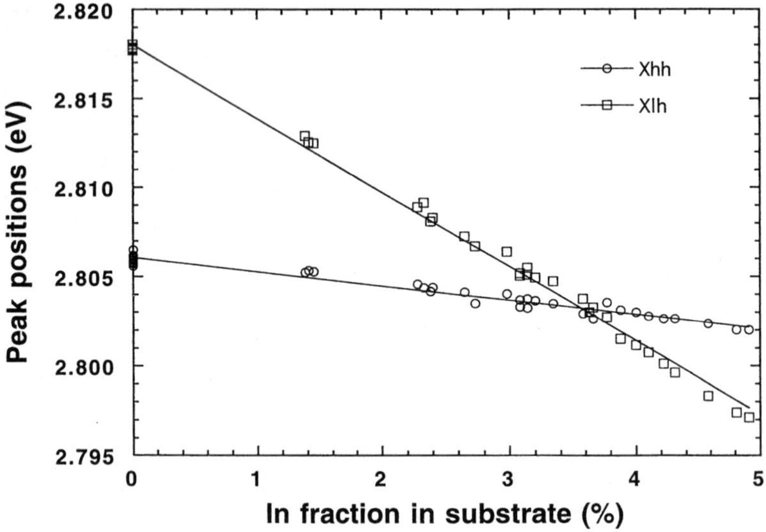

Figure 3 Positions of the heavy- and light-hole free exciton PL peaks (*open symbols*) in pseudomorphic ZnSe layers grown on bulk $In_xGa_{1-x}As$ substrates as a function of In content in the substrate, together with a fit to deformation potential theory (*solid lines*), as described in the text.

calculated from the measured In composition of the substrate and the room-temperature lattice constants of GaAs, InAs, and ZnSe, ignoring the ~0.01% differential thermal contraction between GaAs and ZnSe from 0 to 300 K. The layers are all assumed to be perfectly pseudomorphic, which is confirmed by their narrow (~0.3 meV FWHM) bound exciton linewidths.

The fit is performed using a ZnSe lattice constant of 5.6678 Å, $c_{11} = 88.3$, and $c_{12} = 53.1$ GPa (the last two values are averages of all the data cited in Reference 21, omitting the early values by Berlincourt et al). It yields $a = -4.34 \pm 0.04$, $b = -1.05 \pm 0.01$ eV, and $E_g - E_x = 2.8032 \pm 0.0001$ eV (error bars represent the standard errors in the fit; absolute errors are probably larger). The last value is in good agreement with previous values for bulk ZnSe of 2.8027 (22) and 2.80315 eV (23), and the deformation potentials are in good agreement with the values $a = -4.4$ and $b = -1.0$ eV, determined from a strain splitting study of ZnSSe/GaAs (17). The deformation potentials are in only fair agreement with earlier values of $a = -5.4$ and $b = -1.2$ eV from a reflectance study under uniaxial stress (24). However, the earlier study depended significantly on an arbitrary assumption that $b = (4/\sqrt{3})d$, where d is the rhombohedral shear deformation potential (i.e. $C_2 = 2C_3$ in the original notation of Reference 24).

This assumption is unlikely to be correct [$d \sim (2.5\text{–}3.0) \times b$ in many III–V materials], and renders these values suspect. Kudlek et al (23) found agreement with these earlier larger values, but their strain determinations were based on a calculation using the thermal expansion coefficients of GaAs and ZnSe, and thus depended on the values of thermal expansion coefficients used, for which there is considerable variation in the literature. Their calculation also depends on the assumption that relaxation of thermal strain at the growth temperature is complete. We have found in the case of OMCVD material, for example, that the relaxation is often incomplete and may depend strongly on growth parameters (16).

The strain dependence of transitions involving the $E_o + \Delta_o$ spin-orbit split-off band has also been studied in reflectivity by Stoehr et al and found to follow the hydrostatic shift of the band gap as expected in the simplest theories (25). In other words, no strain dependence of the spin-orbit splitting was found. The Zeeman splitting of the strain-split heavy- and light-hole excitons in a magnetic field has been studied by Coquillat et al to determine the effective g-values of these states (26). Rockwell et al have shown that sufficient externally applied hydrostatic pressure induces a reversal in the thermal strain splitting of the free exciton peaks detected in the photoreflectance spectra of thick ZnSe/GaAs layers, by virtue of the differing compressibilities of the epilayer and substrate (27).

Neutral Donor-Bound Excitons

The neutral donor-bound exciton peaks (usually denoted I_2 in ZnSe) are also shifted and split by strain as indicated, e.g. in Figures 1, 2 and 4, although there has been considerable controversy on this point. Dean et al (28, 29) and some later workers (30–32) attributed the two observed peaks to strain splitting. However, subsequent work attributed the strain-shifted light-hole component of I_{2o} (the lowest energy component of I_2) to native donors different from the chemical species that had been identified in bulk material; the peak was referred to as I_x, the x denoting the uncertainty associated with this peak (14, 33, 34). A key experiment demonstrating that the two observed peaks are in fact derived from the same complex was the thermalization described in Reference 35 and also visible in Figure 2; the higher energy peak increases in intensity at the expense of the lower energy peak as the temperature is raised. The thermalization behavior is also apparent from reported results using excitation spectroscopy (28, 29), although the assignment of heavy- and light-hole components was inverted, and the strain was incorrectly attributed to the effects of lattice (rather than thermal) mismatch in this earlier work. Following the presentation of the results in Reference 35, other workers repeated these thermalization experiments and obtained identical results (36, 37). The

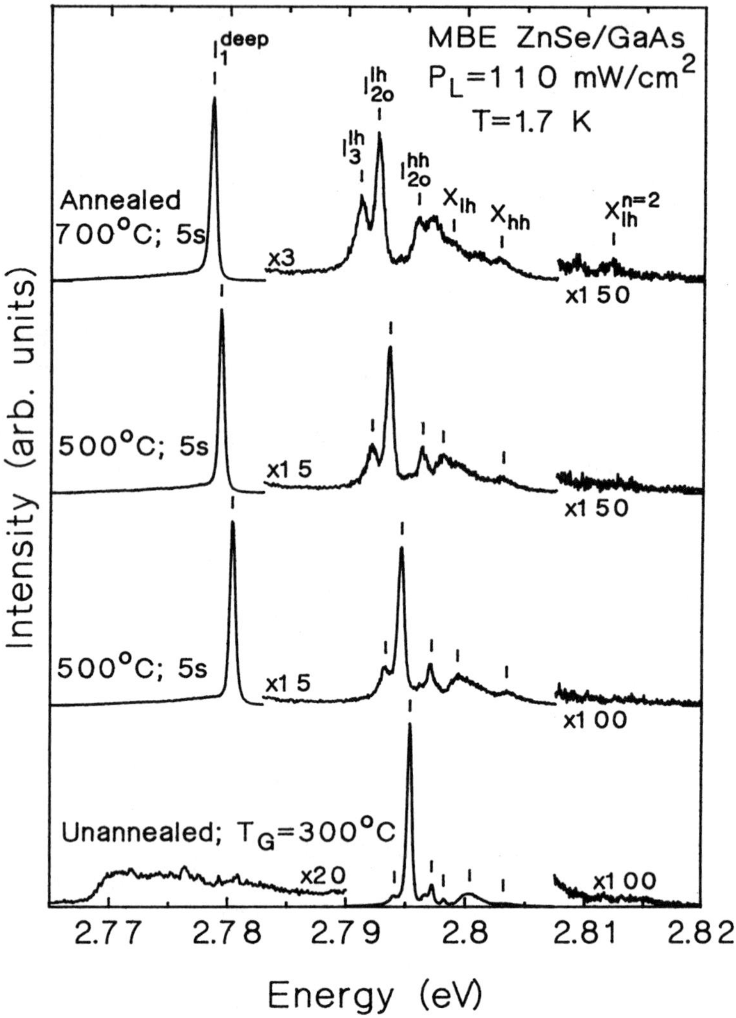

Figure 4 Low-temperature excitonic PL spectra for an as-grown 8.4 μm-thick ZnSe/GaAs layer (*bottom spectrum*) and for the same sample after rapid thermal annealing at the indicated temperatures and times. The appearance of the Cu-bound exciton (I_1^{deep}) is due to inadvertent Cu contamination during the SiN_x cap deposition and during the annealing process. The two samples annealed at 500°C differed only in the system used to deposit the cap prior to annealing but experienced different degrees of thermal strain relaxation.

analysis of Ohkawa et al in particular showed that the intensities of the two strain-split components are perfectly correlated at a given temperature and that the thermal activation energy is equal to the energy splitting (36).

Observations that using purer Se source material during MBE growth yielded reduced intensities for both peaks (30, 38), as well as lower carrier concentrations (30, 35, 38), provided additional support to rule out native donors, as well as to rule out Ga out-diffusion from the GaAs substrates as the principal source of these donors. The intensities of both strain-split I_2 peaks involving residual Cl donors in OMCVD material were also shown to be simultaneously reduced by raising the growth temperature (39). Further support for the strain-splitting model was provided by the observation that intentional doping with Ga (40, 41), In (42), or Cl (43) increases the intensities and/or linewidths of both I_2 peaks. The observation of a two-electron replica of the I_{2o}^{lh} peak gave additional evidence (16, 28, 41, 44). An increase in the splitting of the I_{2o} peaks with increasing layer thickness, and therefore increasing net strain, was also reported (37). Observation of increased shifts in both I_2 peaks after rapid thermal annealing (RTA) of ZnSe/GaAs, as well as an increased separation between the peaks due to relaxation of the larger thermal mismatch strain at the annealing temperature, provided more strong evidence for the strain-splitting model (45–48). The elimination of the splitting in homoepitaxial material is also consistent with this model (49, 50). In addition, very recent confirmation comes from a circular polarization study of the luminescence under circularly polarized excitation (51).

Although the increased strain splitting of the I_2 peaks after a RTA was unmistakable (45–48), an even better-resolved set of spectra as a function of annealing temperature is shown in Figure 4 for a high purity MBE ZnSe/GaAs layer (16). As the I_{2o} peaks shift down in energy because of increased thermal strain introduced during the annealing process, the splitting between the light- and heavy-hole components clearly increases, as does that between the corresponding components of the free exciton. The ionized donor-bound exciton peak (I_3) also shifts down at the same rate as I_{2o}^{lh}, showing that it involves a light hole (see below). A similar shift is observed in the deep (probably Cu) acceptor-bound exciton peak introduced during the capping and annealing process, I_1^{deep}.

A similar dependence of the I_2 splitting on substrate temperature should be evident in as-grown epitaxial material. However, this effect has not been clearly discerned in all prior studies. Yao claimed that the splitting was temperature independent over the 250 to 400°C range, even though the peaks shifted somewhat lower in energy; this observation was interpreted as evidence for a new native donor (33). However, the shifts observed in this paper were only on the order of 1 meV, much less than the up to $\sim$3 meV shifts visible in Figure 4, making the conclusion rather uncertain. Cheng et al clearly observed a reduced splitting in I_{2o} as the MBE growth teperature was lowered from 350 to 150° (52). Substrate temperature-dependent splitting was also observed by Kudlek et al in material grown by hot-wall epitaxy (HWE) from 375 to 450°C (53), and similar strain-dependent effects have been observed in OMCVD material, although the degree of relaxation of thermal strain in this case depends on the degree of C incorporation at the interface (16). Thus there is ample evidence for the expected temperature dependence of the strain splitting.

The existence of any strain splittings in I_2 was strongly contested by Kudlek et al based on a magneto-optical study of the splitting of the lowest energy I_2 PL peak in strained MBE material (54, 55). They claimed to observe five of the six components that would be expected from this peak if it were unsplit by strain and argued that the absence of strain splitting might be due to the effect of averaging I_2 emission from different depths within the layer, subject to different degrees of compressive or tensile strain. This model is untenable, however, because such a spread of strain values would lead to strongly inhomogeneously broadened exciton peaks, in which sharp magnetic field-split subcomponents would not be observable at all. There is no physically reasonable model that can explain these results, and it appears that the authors might have confused PL peaks originating from different donor species or other origins and thus assigned "too many" subcomponents to the original I_{2o} peak. In more recent work, the same authors assume the existence of strain splittings and present a new model (23, 53, 56, 57), which we discuss below.

The splitting of the I_{2o} peak in Figure 4 is clearly less than that of the free exciton for any given degree of strain, however. Averaging over data on 128 samples with various degrees of strain, we find that the I_{2o}^{lh}/I_{2o}^{hh} splitting is about constant at $57 \pm 9\%$ of the free exciton splitting. The significant difference in the magnitude of these splittings has not always been recognized. Dean (29) and Shahzad et al (37) indicated that the splitting of the two systems was comparable, as would be expected for a hole that is very weakly bound in both the free exciton and in the I_2 complex. However, the difference in splittings has frequently been evident in other published data and was emphasized particularly by Kudlek et al (12, 23, 54, 56, 57). These authors proposed that the difference in the splittings was due to the depth dependence of the strain distribution in heteroepitaxial layers, combined with an assumption that in all cases most of the donors in the epilayer lie close to the substrate interface. Their argument is that the free exciton emission orginates closer to the sample surface and therefore reflects a larger tensile strain, whereas the I_2 emission originates deeper in the sample, due to the lower absorption coefficient at this energy, and therefore might reflect more of the compressive lattice mismatch strain, yielding a smaller net strain.

Some support for this idea comes from resonant electronic Raman scattering (RERS) (which may actually be selectively excited luminescence because the two processes are very difficult to distinguish without time-resolved measurements), where a dye laser is used to select emission preferentially from regions with specific strains. The existence of satellites that track the laser energy while it is tuned over an ~1 meV range in an MBE ZnSe/GaAs sample shows that there is some strain inhomogeneity in the material (23, 56). However, the claim that one can deduce the depth distributions of donors from these measurements (23) is unfounded, unless one has independently characterized the strain profile in the films by another method. Moreover, lateral and depth-wise inhomogeneities in strain cannot be distinguished this way. In any case, the notion that most residual donors lie near the substrate interface (and therefore presumably originate from out-diffusion from the substrate) is unsupported by any direct evidence, such as C-V profiling data, and is inconsistent with strong evidence discussed above that most residual donors originate from the source materials, rather than from the substrate.

The above model is based on the assumption that the strain splittings of the free and neutral donor-bound excitons are both linear with strain, and equal in uniformly strained material. In this case, the shift of the average of the I_{2o}^{lh} and I_{2o}^{hh} peak positions should be reduced compared to the shift in the average of the X_{lh} and X_{hh} peak positions from the unstrained values for each peak, in exactly the same proportion as the observed reduction (~57%) in the splittings of the two sets of components. This conclusion follows directly from Equation 1, which shows that the average of the positions of the two components reflects the hydrostatic strain only, which must be reduced proportionately to the shear strain if the strain is biaxial. The hydrostatic deformation potentials should be identical for any states derived from band-edge Bloch functions (58). Using values from Table I of Reference (23), the average of the I_2 components shifts down by 1.59 meV, while the average of the X components shifts by 1.85 meV; the ratio is 0.86, whereas the ratio of the splittings of the two systems is 0.58 (the last value is in good agreement with our data mentioned above). The poor correspondence between the two ratios seems to rule out this model. Moreover, in this model the shift of I_{2o}^{lh} should be less than that of X_{lh} with strain. Experimentally, the separation of the two peaks is virtually independent of strain (see below).

We propose a more reasonable alternative model to explain the reduced strain splitting. Most existing data suggest that the splitting of I_{2o}^{lh} and I_{2o}^{hh} is apparently linear in strain, for strains up to about 1.5×10^{-3} (23). However, we suggest that linear splitting should not necessarily be expected and that non-linearity may appear clearly only for larger tensile strain magnitudes than have been studied to date. Specifically, we assume that I_{2o}^{lh} shifts linearly with strain because it is experimentally almost fixed in separation from X_{lh}, but we assume that I_{2o}^{hh} shifts non-linearly with strain. The reason for the non-linearity is the existence of several excited states of the I_2 complex, which lie up to ~4 meV above the ground state in unstrained material (59). These states are believed to involve orbital excitations of the exciton about the neutral donor, in non-rigid rotator molecular-like states. Kudlek et al have modeled the energies of these states by adapting the model originally developed by Rühle & Klingenstein for the case of InP (60). Quantitative agreement with most of the experimental peak positions was claimed in strained ZnSe/GaAs, using the light-hole mass (23, 56). However, the major approximations necessary in such a model, including the neglect of valence band degeneracy, all inter-particle correlations, and the anisotropy of the valence band, imply that the agreement may be fortuitous. In particular, inclusion of correlation effects is essential to model the binding of neutral impurity-bound exciton complexes in a reliable way (61). In addition, the use of an isotropic light-hole bulk mass is inappropriate because the light-hole mass in strained material is strongly anisotropic.

The I_2 complex in unstrained material contains two identical electrons paired into a $j = 0$ singlet state and a $j = 3/2$ hole. The intrinsic angular momentum of this hole couples with the angular momentum of the rotator state in question to produce a single $S_{3/2}$ ground state and $P_{1/2}$, $P_{3/2}$, and $P_{5/2}$ states derived from the $l = 1$ (first excited rotator state. (Many additional states arise from higher excited rotator states). The $P_{5/2}$ state may split further into Γ_7 and Γ_8 components in a crystal field of T_d symmetry. The oscillator strengths of the excited states should be weak compared with that of the ground state due to parity considerations. A uniaxial strain field will act to mix and split most of these states because they are admixtures of $|m_j| = 3/2$ and $|m_j| = 1/2$ components, which split apart in this case. It will be possible to characterize many of the resulting mixed states as mainly heavy- or light-hole states only when the strain splitting substantially exceeds the zero-strain energy differences of the various ground and excited states of the I_2 complex. This is not the case in any sample studied to date, where the free exciton strain splitting is at most 7 to 8 meV, compared to zero-strain splittings up to 4 to 5 meV.

Some of the states involved in this mixing process may not even be detectable optically because their recombination may be strongly forbidden. Indeed, Finkman et al (62) could interpret the radiative lifetime of the neutral donor-bound exciton in GaAs only by postulating a large number of nearly degenerate but non-observable low-lying excited states of the complex. In this interpretation, the I_{2o}^{hh} peak is actually of mixed HH/LH character, which explains why it does not follow either component of the free exciton exactly; it appears to shift linearly with strain because only a limited range of strains has been studied. At higher strains, it should eventually adopt the slope of the X_{hh} component, which is an experimentally testable prediction of this model. Likewise, all other excited states should eventually follow either X_{lh} or X_{hh}; the states following the latter will cross the X_{lh} position at some point and become lifetime broadened and unobservable. Because of the strong interactions between the ground and excited states, it is now evident that it is not possible to characterize the strain behavior of the I_2 peaks by simply adopting a reduced shear deformation potential of the bound hole, as is possible, e.g. for the case of a hydrogenically bound hole (63). (A simple reduced deformation potential was erroneously assumed in Reference 16, based on more limited data.) As mentioned above, modeling this behavior empirically in terms of strain-dependent hole masses (56) neglects effects due to the mass anisotropy. Accurate calculations of the behavior of this very complicated multiparticle system are likely to be exceedingly difficult given the necessity to include correlation effects and a proper Luttinger Hamiltonian for the hole. Moreover, it is not at all clear how one could explain the larger strain splitting of the free exciton peaks by invoking polariton effects, as suggested by Vaziri et al (40).

Experimentally, we have studied the strain splitting of various bound exciton complexes in luminescence in over two hundred samples, including as-grown MBE and OMCVD ZnSe/GaAs layers grown in seven different laboratories under various conditions; thermally annealed layers; pseudomorphic and relaxed ZnSe layers grown on bulk InGaAs substrates of various In contents or on AlAs or InGaAs buffer layers; and free standing layers removed from their substrates by chemical etching. The strains in pseudomorphic layers grown on bulk InGaAs or GaAs are determined as described above. The strains in relaxed layers or those released from their substrates are determined from the observed I_{2o}^{lh} peak positions, when observable, or otherwise from the X_{lh} peak positions using the deformation potentials determined in our study of ZnSe/bulk InGaAs. The bound excitons are preferred for determining strain because they are sharper peaks. To calibrate the I_{2o}^{lh} position vs strain relation, we first perform a linear fit to the I_{2o}^{lh} positions as a function of X_{lh} positions, which yields $E(I_{2o}^{lh}) = (1.0591\ E(X_{lh}) - 0.17015)$ eV. This formula is used to calculate hypothetical X_{lh} positions that are converted to strain values using Equation 1 and the deformation potentials discussed there, i.e. $E(X_{lh}) = (2.80318 + 5.78874\varepsilon)$ eV, where $\varepsilon = -\varepsilon_{xx} = -\varepsilon_{yy}$ is the in-plane strain ($\varepsilon > 0$ for biaxial compression). This procedure is subject to possible small errors from variations in the dominant residual donor species but still yields less scatter than using measured X_{lh} positions directly.

The shallow donor and acceptor-bound exciton peaks in most of the samples under biaxial tension are shown as a function of strain in Figure 5*a*. The slope of the I_{2o}^{lh} position as a function of strain is identical to that of X_{lh} within the accuracy of the fit, as expressed in the equation given above with a slope only 6% larger than unity. This result suggests that this state does not have a significant admixture of heavy-hole character, which would give it a smaller and possibly non-linear slope. The I_{2o}^{hh} position vs ε relation is approximately linear over the studied range, with a slope of 2.977 eV, compared to 1.129 eV for X_{hh}. The fit lines to two I_{2o} components intersect for a strain very close to zero at a value of 2.7986 eV, which is slightly higher in energy than the position reported for I_{2o}^{Al} in bulk ZnSe, 2.79754 eV (59). This result suggests that we may be observing a donor slightly shallower than Al, possibly B. Further work involving a study of the two-electron transitions and RERS from neutral donors in a magnetic field will be necessary to determine the donor identity with certainty. The other excited I_2 states have various slopes, mainly similar to that of X_{lh}. The I_{2a}^{A} and I_{2b}^{A} peaks extrapolate to positions about 0.40 and 2.37 meV above the intersection of the I_{2o} components, strongly suggesting that these peaks are derived from the I_{2a} and I_{2b} peaks known in bulk material, respectively (59). These PL results are largely in agreement with the results of Kudlek et al using excitation spectroscopy (23), except regarding the interpretation of the data and the values of the deformation potentials.

In thin, pseudomorphic ZnSe layers grown on bulk InGaAs substrates, we have observed I_{2o} peaks for two different donor species, denoted A and B, as shown in Figure 6 (19). The relative intensities of the two peaks vary between crystal growth runs (but are fairly constant for different substrates employed during the same run), corresponding to variations in the background doping (the samples were grown at various points in the history of the growth chamber). Corresponding I_3 peaks are observed for each donor (see below), and the intensities of the I_2 and I_3 peaks for each donor show an excellent correlation with each other, although the intensities of the I_2 and I_3 peaks relative to each other for a given donor vary from sample to sample. The identities of the donor species are not yet certain, but seem to correspond most closely to Al, and In or F, respectively.

The effects of releasing the thermal mismatch strain in MBE ZnSe/GaAs by completely removing the substrate by jet etching in a H_2SO_4:H_2O_2:H_2O solution and then placing the free-standing film on a piece of bulk ZnSe are illustrated in Figure 7. The bound exciton peaks all shift to higher energy, and the I_{2o}^{lh}/I_{2o}^{hh} splitting is no longer resolved. Remarkably, the spacings of the excited states of I_2 from the ground state now match those observed in unstrained bulk ZnSe almost exactly (59), even though

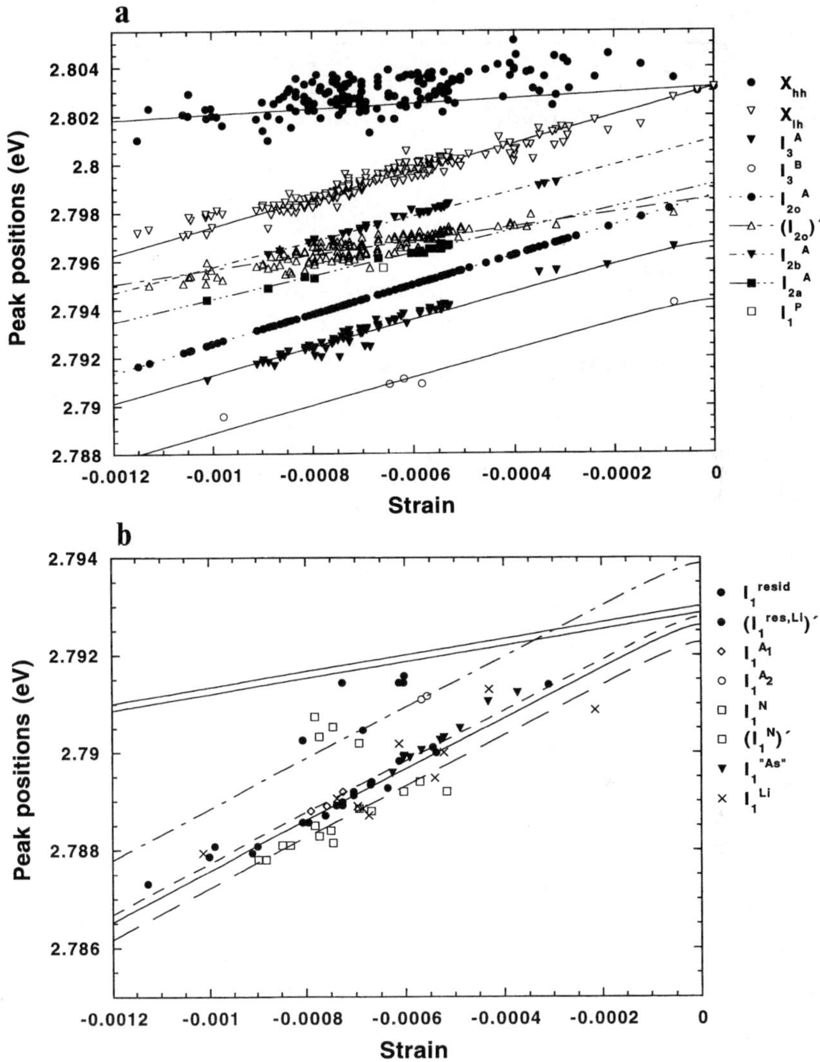

Figure 5 (*a*) Positions of various donor-bound exciton peaks and the neutral P acceptor-bound exciton peak as a function of strain in MBE and OMCVD ZnSe epilayers, where the strain is determined as discussed in the text. The solid lines through the X_{lh}, X_{hh}, I_3^A, and I_3^B peaks are plots of Equations 1 and 3 (the I_3^B line is the same as I_3^A, shifted down to match the data for this deeper donor), whereas the dashed and dotted lines through the I_2 components are least-squares linear fits to the data. The superscripts A and B denote two residual donor species (as illustrated, for example, in Figure 6). The scatter in X_{hh} positions is due to the usually weak and broad nature of this peak. (*b*) Positions of neutral acceptor-bound exciton peaks in the same samples. The solid lines are a plot of Equations 5 for Li/"As"/residual acceptors (all believed to be due to Li), whereas the dash-dot, short-dashed, and long-dashed lines are corresponding plots of the lowest energy component (only), shifted to match the data for A_2, A_1, and N acceptors, respectively.

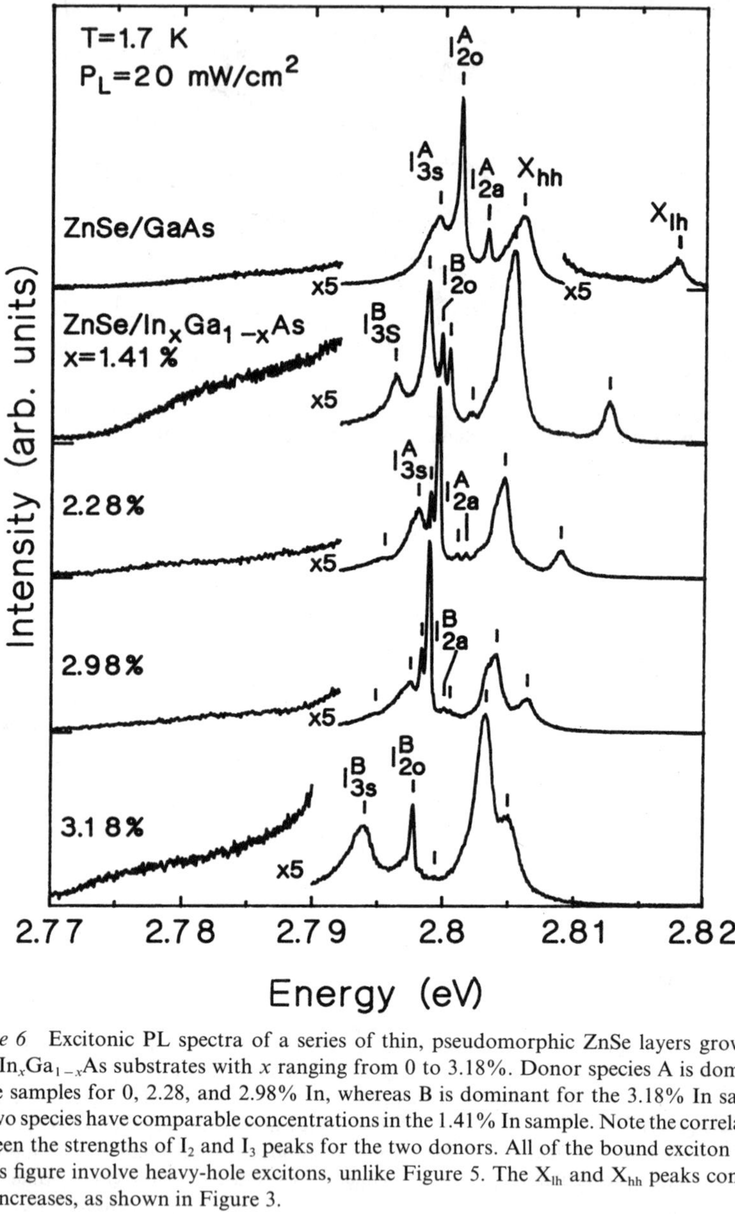

Figure 6 Excitonic PL spectra of a series of thin, pseudomorphic ZnSe layers grown on bulk $In_xGa_{1-x}As$ substrates with x ranging from 0 to 3.18%. Donor species A is dominant in the samples for 0, 2.28, and 2.98% In, whereas B is dominant for the 3.18% In sample; the two species have comparable concentrations in the 1.41% In sample. Note the correlations between the strengths of I_2 and I_3 peaks for the two donors. All of the bound exciton peaks in this figure involve heavy-hole excitons, unlike Figure 5. The X_{lh} and X_{hh} peaks converge as x increases, as shown in Figure 3.

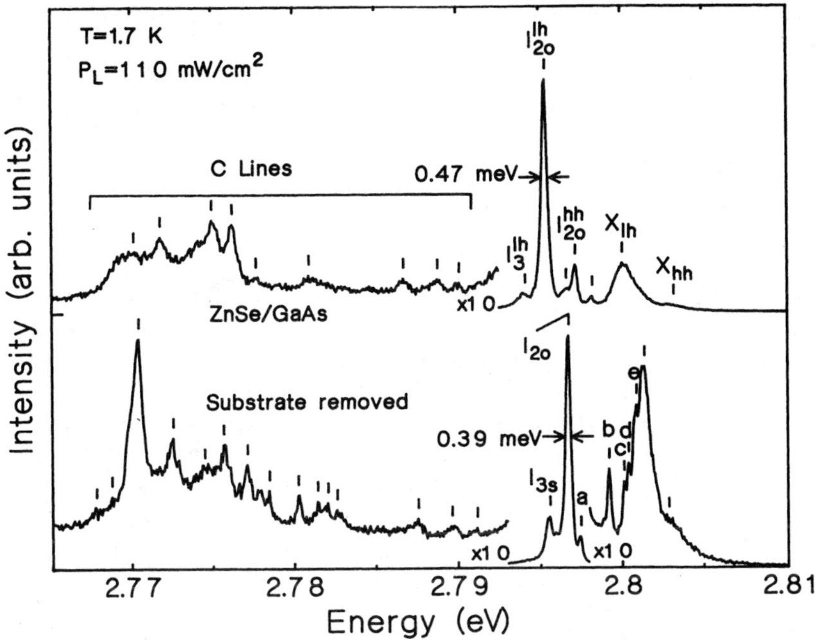

Figure 7 Excitonic PL spectra of a thick MBE ZnSe/GaAs layer both before (*upper spectrum*) and after (*lower spectrum*) releasing it from its substrate. The letters a–e denote the excited states I_{2a}–I_{2e} in the notation of Merz et al (59).

residual peak shifts suggest that a small strain is still present. The linewidths of the bound excitons are not broadened by the substrate removal process, as was the case in earlier studies (57); in fact, we observe a slight narrowing of the linewidths as indicated on the figure.

The weak peaks labeled C-lines in the range 2.77 to 2.79 eV in Figure 7 are not two-electron transitions, as one might be tempted to conclude from their positions. They are, in fact, principal bound exciton peaks of a new (as-yet-unidentified) independent system, which is relatively strongly phonon-coupled with a Huang-Rhys parameter of $S \sim 0.23$. This conclusion is supported by excitation below the band gap, which makes the LO phonon satellites of these peaks visible, and by studies of other samples in which they are much more intense (too intense relative to I_{2o} to be two-electron satellites!). We have observed these peaks not infrequently in MBE samples from three different laboratories, which suggests that they are relatively common. In some samples they are absent, and the normal two-electron peaks (59) dominate this region. Further work is needed to establish their origin, but for now it is important to recognize their existence so that residual donors are not misidentified on the basis of these peaks.

Finally, the linewidths of donor-bound (and other bound) excitons are useful in assessing the degree of inhomogeneous strain that is present due to extended interfacial defects in the material. Examples of this type of study and correlations with structures and growth conditions are given elsewhere (18, 35, 45, 64).

Ionized Donor-Bound Excitons

A definitive and strongly supported study of ionized donor-bound exciton (I_3) peaks was performed in bulk ZnSe by Merz et al (59), who observed strong ($j = 1$) and weak ($j = 2$) components of these peaks for five different donor species and correlated them with corresponding I_2 peaks. This interpretation was further supported by Dean et al. (65), although it was questioned by Isshiki et al (66). Their objections seem to be based on arguments involving the excitation intensity dependence and excitation spectra of the I_2 and I_3 peaks; in particular, that excitation below E_g but above the free exciton energy cannot photo-neutralize ionized donors. However, this assertion is unfounded because the electron in a free exciton can readily be captured at an ionized donor, releasing the hole into the valence band in an exothermic reaction. In any case, Isshiki et al offer no viable alternative model for these peaks, and the convincing arguments of Merz et al and Dean et al have been widely accepted.

The I_3 peaks have been much less commonly reported in ZnSe/GaAs, presumably because inhomogeneous strain broadening of the bound exciton peaks caused by interfacial defects, combined with Stark effect broadening in material with high residual impurity content, tends to cause these peaks to merge with I_{2o}^{lh}. They were apparently first observed and identified in purified, relaxed ZnSe/GaAs layers under tensile strain by Yoneda et al, who concluded that they are shifted from their positions in bulk material by strain (30). They were subsequently also identified in a variety of compressively strained pseudomorphic ZnSe layers on GaAs (with or without various buffer layers) (18, 35, 45, 64), and in pseudomorphic ZnSe on bulk InGaAs substrates and therefore subject to various degrees of strain (18, 19). The existence of I_3 peaks in relaxed ZnSe/GaAs was confirmed in a comparison of heteroepitaxial to homoepitaxial MBE ZnSe by Park et al (49) and later by Ohkawa et al (67). These peaks have also been observed in relaxed material by several other workers (19, 39, 44, 68–72).

To our knowledge, no quantitative analysis of the strain behavior of I_3 in ZnSe has been given. The theory of this complex is formally identical to that of an isoelectronic-bound exciton, however, as discussed for the case of ZnTe:O in Reference 73. The spherical electron-hole exchange interaction in the bound exciton can be modeled as $\Delta_1 \mathbf{J_h} \cdot \sigma$, where $\mathbf{J_h}$ is the intrinsic ($j_h = 3/2$) angular momentum of the hole, σ is the $s = 1/2$ electron spin, and the splitting is $4\Delta_1$. In the case of I_3, the experimental observation that the allowed $j = 1$ (strong) peak lies below the forbidden $j = 2$ (weak) peak(s) (59, 65) implies that Δ_1 is positive, in opposition to the more normal ZnTe:O case. A small cubic crystal field splitting of the $j = 2$ state is possible and has been observed for Cl donors (65) but is ignored here for simplicity. The total effective Hamiltonian in the presence of a (100) biaxial strain is then

$$H_{\text{eff}} = \Delta_1 \mathbf{J_h} \cdot \sigma + A + B(J_{hz}^2 - \mathbf{J_h}^2/3), \qquad 2.$$

where J_{hz} is the z-component of $\mathbf{J_h}$, A and B are the same as defined in Equation 1, and higher order terms such as strain dependence of the exchange splitting have been neglected. We assume that the shear deformation potential of the very weakly bound hole in this complex is the same as in the free exciton. Diagonalizing the above yields a total five states for finite strain (73), but recombination of only two of these is dipole-allowed in σ-polarization, the normal configuration for PL measurements. The energies of these σ-allowed states are

$$E(I_{3w,3s}) = -\Delta_1/2 + A \pm (B^2 - 2\Delta_1 B + 4\Delta_1^2)^{1/2} \qquad 3.$$

measured from the hypothetical I_3 position in the absence of exchange splitting, which is $5\Delta_1/2$ above the position of I_{3s} in unstrained material. Mixing of the $j = 1$ and $j = 2$ states by the strain redistributes the oscillator strengths until I_{3w} attains three times the oscillator strength of I_{3s} in the limit of large strain ($|B| \gg \Delta_1$), where the I_{3w} and I_{3s} states are essentially heavy and light hole in character, respectively (73). Numerically, $\Delta_1 = +0.1$ meV for all known donors in ZnSe (59), so that this limit is reached even for small strains. In this limit, we have

$$E(I_{3s})(\varepsilon) \approx E(I_{3s})(\varepsilon = 0) + A - B + 3\Delta_1,$$

$$E(I_{3w})(\varepsilon) \approx E(I_{3w})(\varepsilon = 0) + A + B - 3\Delta_1, \qquad 4.$$

where we have shifted the energy origins to the strain-free position of each respective peak. Experimentally, thermalization and interference from the I_2 peaks ensure that only the lowest energy peak (I_{3s} for biaxial tension and I_{3w} for biaxial compression, respectively) will be observable in most cases.

Experimentally, the data in Figure 5*a* show through I_3^A peak positions that the shift of I_3 for tensile strain is indeed linear in strain. The solid line is a plot of $E(I_{3s})$ as given by Equation 3, where the zero strain position is adjusted to 2.7972 eV to achieve the best fit to the data. This value is slightly larger than the values for known donors in bulk material, similar to our finding for I_{2o}^{lh}. A least-squares linear fit to the data is virtually indistinguishable from this curve (except for strains smaller than that of the data point for smallest strain); this observation confirms that the I_{3s} (i.e. I_3^{lh}) peak is a fixed distance below X_{lh} and has almost exactly the same slope. The I_{3s} and I_{2o}^{lh} lines converge slightly (by about 8% over the range of the plot), due to the slightly larger slope of I_{2o}^{lh} compared with X_{lh}, but the latter difference is probably within experimental error.

For pseudomorphic layers grown on bulk InGaAs substrates, I_{3w} (i.e. I_3^{hh}) peaks for two different donors (denoted *A* and *B*) are apparent in the data of Figure 6*a*,*b*, corresponding to the I_{2o}^{hh} peaks discussed above. These peaks show an anomalous broadening in some of the samples, which is not understood. However, the I_3 nature of these peaks is clearly confirmed by exciting just above E_g, where they are strongly favored over the corresponding I_2 peaks as observed in unstrained material by Dean et al (65) (see Figure 8). Excitation just below the X_{lh} position (possibly at an excited state of I_2), also shown in Figure 8, favors the I_2 peaks instead (65). Finally, the I_3^{lh} peak is apparently transformed into a normal I_{3s} peak (with I_{3w} visible just above it) after removing the substrate of the sample in Figure 7, further confirming our model.

Neutral Acceptor-Bound Excitons

Acceptor-bound exciton (I_1) peaks were apparently first reported in ZnSe/GaAs in N-doped OMCVD material (6, 74) and in undoped OMCVD material, where the residual acceptor was evidently Li (75). Intentionally Li-doped OMCVD material was also subsequently shown to exhibit I_1 peaks (76, 77). All of these peaks were observed to be shifted down in energy compared to bulk material, due to strain. Similarly shifted I_1 peaks were observed first in undoped (14) and then in N- (78, 79) and Li-doped (80) ZnSe/GaAs grown by MBE, as well as in N-implanted MBE ZnSe/GaAs (46). A doublet nature ($\sim$2 meV separation) for I_1 in material under biaxial tension was first noted for I_1^{Li} by Cheng et al (80), and for I_1^N by Ohkawa et al (36). (In retrospect, the I_3 peak observed in Reference 77 was probably also the higher energy component of an I_1 doublet.) Ohkawa et al showed the existence of thermalization between the two components of the I_1^N doublet, as was later shown for I_1^{Li} by Haase et al (81) [this thermalization was also apparent in the data of Shahzad (37), who did not, however, discuss it]. Ohkawa et al modeled the I_1 doublet in terms of strain splitting into light- and heavy-hole exciton states but ignored exchange and crystal field interactions and assumed a shear deformation potential for the hole unchanged from that in the free exciton

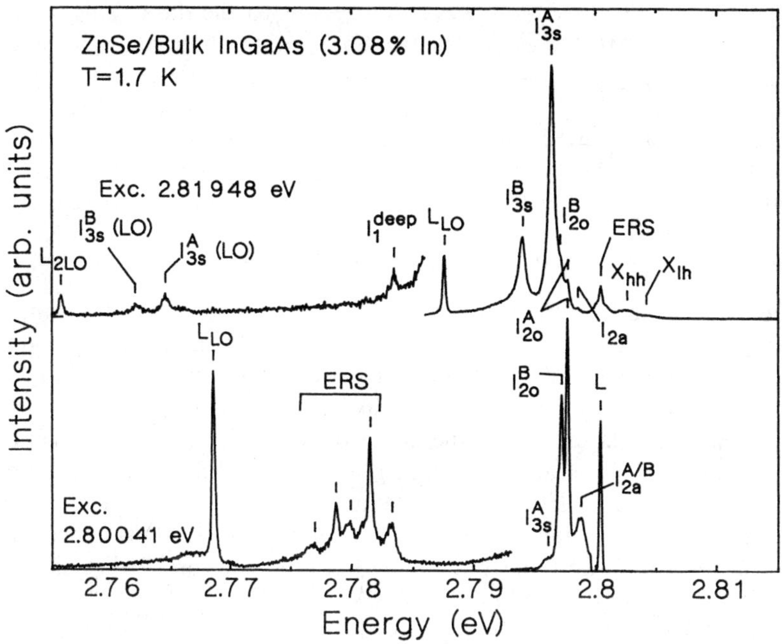

Figure 8 Excitonic PL spectra of a pseudomorphic ZnSe layer on a bulk $In_xGa_{1-x}As$ substrate with $x = 0.0308$, for excitation at the indicated laser energies just above the energy gap (*upper spectrum*) and just below the heavy-hole free exciton peak, possibly in resonance with an excited state of I_2 (*lower spectrum*). The I_3 and I_2 peaks respectively are selectively enhanced in these two cases.

(36). In addition, the hydrostatic deformation potential they used was incorrectly extracted from Reference 24, whose data in any case are not very accurate (see above). As a result of these assumptions and errors, the quantitative agreement with theory was poor.

A more sophisticated and accurate model of the strain splitting of the I_1 system was given by Zhang et al (82), who included the hole-hole exchange interaction, crystal field-splitting, and reduction in the shear deformation potential of the holes due to the binding (63). Good quantitative agreement with experiment was obtained, although a unique set of model parameters could not be determined due to limited available information on the closely-spaced triplet structure observed for I_1^{Li} in unstrained material (11). The reader is referred to References 82 and 73 for a detailed discussion of the model, where it is shown that the energies of the four possible transitions are given by

$$E_1 = E_o + A + \tfrac{3}{4}\gamma + \tfrac{3}{10}\beta \mp B' + \sqrt{\tfrac{1}{4}(\gamma - \tfrac{3}{5}\beta)^2 + 4B'^2},$$

$$E_2 = E_o + A + \tfrac{1}{4}\gamma - \tfrac{2}{5}\beta \mp B',$$

$$E_3 = E_o + A + \tfrac{1}{4}\gamma + \tfrac{3}{5}\beta \mp B',$$

and

$$E_4 = E_o + A + \tfrac{3}{4}\gamma + \tfrac{3}{10}\beta \mp B' - \sqrt{\tfrac{1}{4}(\gamma - \tfrac{3}{5}\beta)^2 + 4B'^2}, \qquad 5.$$

where E_o is the hypothetical peak position in the absence of strain, exchange, and crystal field interactions, and the upper and lower signs apply under compressive and tensile biaxial stress, respectively. The quantities A and B' are the same as A and B in Equation 1, respectively, with b replaced in the latter by b′, the corresponding value for the acceptor-bound holes. The parameters γ and β are the splittings due to the hole-hole exchange and crystal field interactions, respectively. The highest energy state (E_1 for tensile and E_4 for compressive stress) is expected to be unobservable, and the middle two components are believed to be experimentally unresolved in most cases.

Subsequent direct confirmation of the role of strain in producing the doublet structure was provided by an experiment involving removal of the GaAs substrate, which eliminated the strain splitting (70). These results also proved that the alternative model advanced by Kudlek et al (12, 53, 54, 56), who claimed that the splitting was entirely the result of the hole-hole exchange interaction and that the strain splitting was somehow totally quenched, is incorrect. This group later adopted the model of Zhang et al and performed thermalization measurement of the I_1^{Li} triplet in bulk ZnSe (83) to show that the best choice of parameters in that model is $\gamma = -0.30$ meV and $\beta = +0.14$ meV, in the original notation of Reference 82 (which was modified in Reference 83).

The model of Zhang et al is plotted in Figure 5*b* using the parameters given above, where it is seen to give a good representation of the position of the lower energy I_1^{Li} peak as a function of strain, as well as of the I_1^{As} and residual I_1 peaks, both of which are now believed to be due to Li as well (see discussion below under As acceptors). The higher energy component does not seem to fit very accurately with these parameters, although this could be due to the scatter in the data. The model is also plotted for the lowest energy component of I_1^N, $I_1^{A_1}$, and $I_1^{A_2}$ (see below for a discussion of the latter two acceptors), with E_o treated as a fitting parameter; it does not seem to predict the position of the higher energy peak of I_1^N accurately using the Li parameters (not shown). It is to be noted that the I_1 peaks for a given acceptor species do not have a fixed separation from the I_{2o}^{lh} or I_3^{lh} peaks as a function of strain because of the reduced shear deformation potentials of the acceptor-bound holes. A residual I_1 singlet has also been observed in compressively strained pseudomorphic ZnSe layers grown on AlAs (35, 64) and InGaAs (18, 45) buffer layers.

Donor-to-Acceptor and Band-to-Acceptor Transitions

No clear dependence on biaxial tensile strain has been observed in our data for donor-to-acceptor (D°-A°) transitions involving various acceptors in ZnSe/GaAs probably because of the considerable shifts in (D°-A°) peak positions that naturally result from changes in dopant concentration and nonradiative recombination rates (84). A high energy shift has been observed for residual acceptors in pseudomorphic (compressively strained) MBE ZnSe on AlAs (35) or InGaAs (45) buffer layers, and in ZnSe grown directly on GaAs by vapor phase epitaxy (VPE) (85), implying that the heavy-hole acceptor level is the one involved in the recombination in this case.

Conduction band-to-acceptor transitions have recently been shown not only to shift to lower energy but also to exhibit a resolvable strain splitting into distinct heavy- and light-hole components, provided that the tensile strain is sufficiently large (86). The splitting enabled a direct determination of the shear deformation potential of the Li acceptor-bound hole as $b' = -0.56 \pm 0.11$ eV, which is smaller than the value of -0.81 eV calculated on the basis of perturbation theory (63) using the Luttinger parameters measured by Venghaus (22). An example of such a resolved splitting is shown in Figure 9, for a OMCVD ZnSe/GaAs sample grown at 480°C. It is seen that (similarly to the data in Reference 86) the higher energy heavy-hole component dominates for temperatures above 4.5 K because of its three times larger oscillator strength. For this reason, the dominant (e-A°) peak in material under smaller tensile strains is essentially always the heavy-hole component, even though the splitting is not resolved in most cases. This fact contradicts expectations based only on thermalization and is important to the determination of acceptor binding energies from (e-A°) peak positions, as described below. To our knowledge, this is the first example for any semiconductor material where strain splittings of acceptor levels have been directly resolved.

OPTICAL SPECTROSCOPY OF ACCEPTOR LEVELS IN ZnSe

Determination of Acceptor-Binding Energies

Accurate determination of shallow acceptor-binding energies is important to distinguish clearly the levels produced by different dopants in the effort

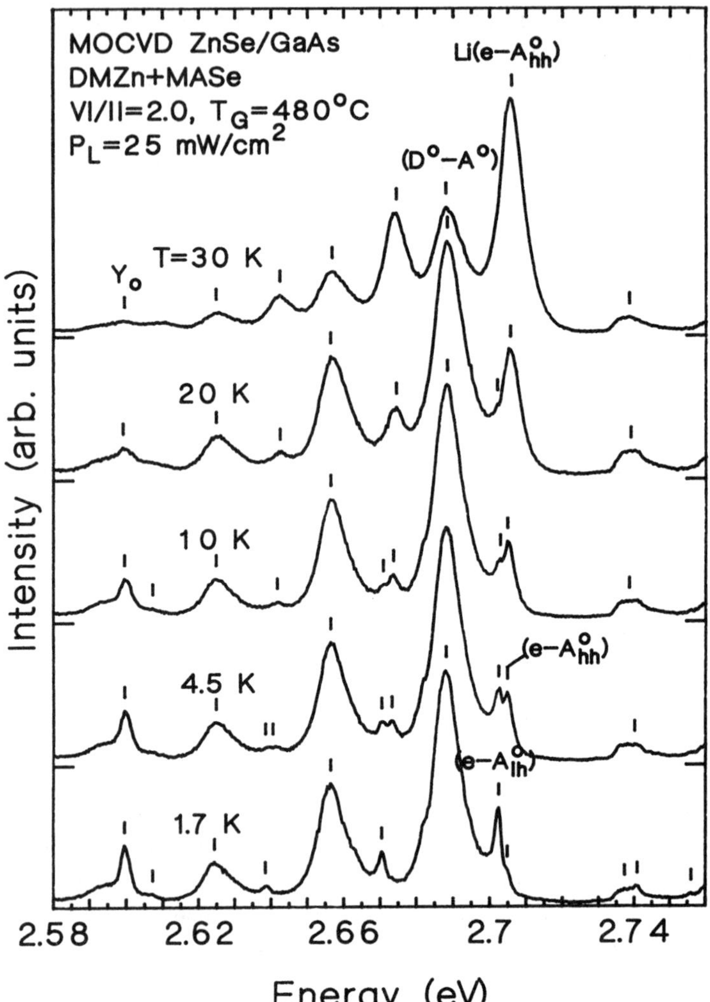

Figure 9 Temperature-dependent (D°-A°) and (e-A°) peaks in an unintentionally Li-doped OMCVD ZnSe/GaAs layer grown at the University of Minnesota. Due to the relatively large thermal strain, separate light- and heavy-hole (e-A°) peaks are resolved.

to dope the material *p*-type. Several methods have been used for this purpose. Acceptor-binding energies are sometimes estimated from distant (D°-A°) peak positions under nonresonant excitation, using arbitrary assumption on the magnitude of the Coulomb term (see typical peak position in Table 1). However, the latter can vary significantly depending

Table 1 Measured energies of (D°-A°) and (e-A°) PL peaks in unstrained ZnSe[a]

	(D°-A°) peak position at 1.7 K[b] (eV)				(e-A°) peak position at 0 K[c] (eV)			
Acceptor	Average	Number of samples	Standard deviation	Range	Average	Number of samples	Standard deviation	Range
O	2.7256[d]	20	0.0029	2.7204–2.7330	2.7380	1	—	—
P	2.7188	3	0.0008	2.7179–2.7194	2.7339	1	—	—
A_2	2.7126	2	(0.0004)[e]	2.7122–2.7129	2.7270	1	—	—
A_1	2.7047	4	0.0011	2.7035–2.7061	2.7162	1	—	—
N	2.6991	11	0.0019	2.6963–2.7021	2.7108	1	—	—
Li	2.6933	10	0.0011	2.6915–2.6954	2.7077	6	0.0003	2.7073–2.7081
As[f]	2.6962	18	0.0014	2.6949–2.6995	2.7078	4	0.0010	2.7068–2.7086
Na	2.6825	2	(0.0003)[e]	2.6823–2.6828	2.6938	1	—	—

[a] Extrapolated from values measured in our laboratory (in strained material). For corresponding values in strained material, add (2.2327 eV)ε, where ε is the biaxial strain magnitude ($\varepsilon > 0$ for biaxial compression). E.g. in typical relaxed (>1–2 μm thick) MBE ZnSe/GaAs layers grown at ~270°C, $\varepsilon \approx -5 \times 10^{-4}$, thus peaks will be ~0.0011 eV lower than values shown above. Slope is based on heavy-hole acceptor level for Li, using experimentally determined value of b' for the bound hole. The acceptor binding energies determined from the $T = 0$ (e-A°) peak positions in this table, using $E_g(T = 0) = 2.8218$ eV, differ slightly from the values in Table 2, due to differences in the method used to estimate the strain. The differences are all within the quoted error limits on E_A. Strains are usually determined from I_{2o}^{lh} peak positions in deriving these values.

[b] Measured at a fixed excitation condition, using an unfocused 8 mW beam from an Ar^+ laser (3.53 eV). Position varies as a function of donor species, excitation level, doping, and nonradiative recombination rate.

[c] Hypothetical peak positions, extrapolated from linear fit to (e-A°) peak positions measured as a function of temperature using ~1 mW unfocused excitation from an Ar^+ laser at 3.53 eV, and correcting for $\Delta E_g(T)$. Samples with limited number of data points or poor fits excluded.

[d] Includes samples with residual acceptor believed to be O. Excluding these values gives an average of 2.7274 eV (4 points), $\sigma = 0.0007$ eV, and range = 2.7267 to 2.7284 eV.

[e] Insufficient number of data points to be meaningful.

[f] Acceptor now believed to be residual Li contamination.

on the doping level, excitation intensity and nonradiative recombination rate; the chemical identity of the donor (often unknown) introduces additional variations. The nonradiative rate is important because it competes more effectively with distant pairs (long lifetimes) than with close ones (short lifetimes), so that it shifts the (D°-A°) peak to higher energy (84, 87). Thus this method is accurate only within several meV, which can be comparable to the differences between acceptor species. (See Table 1, where it can be deduced that for an assumed fixed donor binding energy of 25.6 meV, the average value of the Coulomb term varies from 10.5 to 14.2 meV among data on different acceptor species. Comparing data on individual samples, we find an average Coulomb term of 12.3 meV, a standard deviation of 1.7 meV, and a range from 9.5 to 16.0 meV, under our fixed measurement conditions.

Based on a very early paper by Halsted & Aven (88), many authors have used a Haynes's-Rule linear relationship, to estimate acceptor binding energies from I_1 exciton localization energies. However, Dean et al have clearly pointed out that no linear relationship between exciton localization energy and acceptor-binding energy actually exists in ZnSe, or for that matter in any other direct gap semiconductor material (6, 89). To emphasize this point, we show a plot of this relationship for known acceptor levels in ZnSe in Figure 10, where it is obvious that no linear relationship of any kind exists, even if the somewhat anomalous case of the P shallow acceptor level is excluded.

Other methods include fits to the energies of discrete (D°-A°) line spectra, which can be a reasonable method but requires an accurate knowledge of the dielectric constant and a well-resolved, high-quality spectrum. Discrete pair lines have been resolved only a few times in heteroepitaxial material (35, 47, 82), and the spectra are rarely of sufficient quality for this method. Calculations of binding energies can be made using effective mass theory from excited state energies determined from selectively excited (D°-A°) pair (SPL) measurements but depend on the accuracy of the theory, knowledge of the valence band parameters, and the use of an appropriate method to incorporate central cell corrections into the theory.

A method that is free of most of the above limitations is the use of (e-A°) peak positions, typically measured as a function of temperature, both as a consistency check on the theory and as an averaging method to increase the accuracy. The results depend on only the value used for the band gap at 0 K, but differences in binding energies among acceptors are independent of this value. The method was first applied in the case of ZnSe : Li/GaAs, where the (e-A°) peaks were found to be visible even down to 1.7 K in *p*-type samples (82). After correcting for the temperature dependence of the band gap, the peak positions were found to fall on a

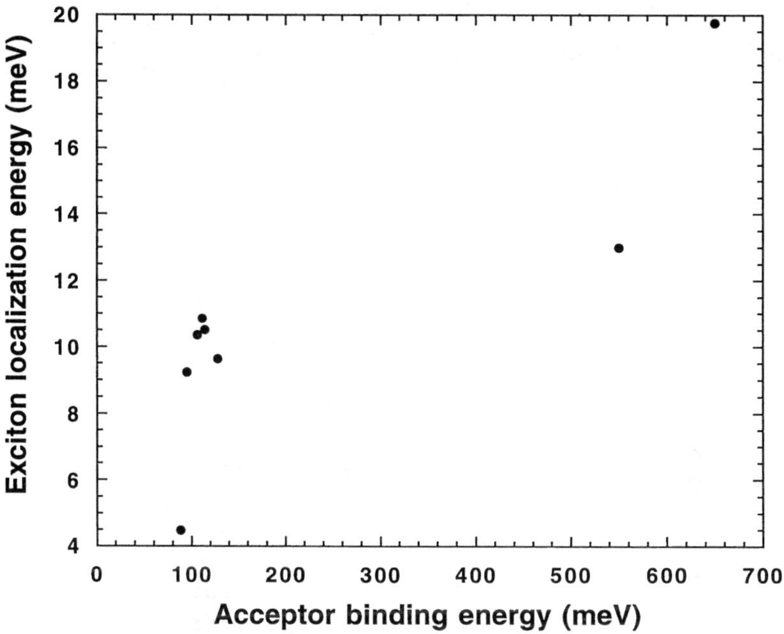

Figure 10 A plot of exciton localization energy at neutral acceptors as a function of the corresponding acceptor-binding energy for P, A_2, A_1, N, Li, Na, Au, and Cu acceptor levels (in order of increasing binding energy) in ZnSe. Based on the data of Figure 5*b* and Table 2, and values from References 65, 89. Note that the relationship between these two quantities is not linear nor even monotonic.

straight line with a slope of approximately $k_B/2$, where k_B is Boltzmann's constant, in agreement with the theory of (e-A°) lineshapes from Eagles (90). The intercept is corrected for strain effects as described in Reference 82, using the assumption proven above that the dominant peak involves the heavy-hole acceptor level. The resulting value yields the binding energy in unstrained material, assuming $E_g(T = 0) = 2.8218$ eV in that case (65). In recent work (86), we use the experimentally determined deformation potential discussed above for the bound hole, rather than the theoretical value used in Reference 82. Application of this method to eight different shallow acceptor levels in ZnSe has been described elsewhere (69, 70, 82, 86, 91, 92), and is summarized in the column labeled E_A of Table 2.

This method requires an accurate knowledge of the temperature variation of the band gap, which can be determined from the temperature-dependent shifts in bound exciton PL peak positions, with corrections when necessary for changes in the dominant strain-split sub-component as a function of temperature resulting from thermalization. Data measured

Table 2 PL peak positions, binding energies, and ground state-to-excited state energy separations for acceptors in unstrained ZnSe[a]

	References (E_A)	References (SPL data)	E_A ($1s_{3/2}$)	Energy separations (meV)[b] $1s_{3/2}$–$2p_{3/2}$	$1s_{3/2}$–$2s_{3/2}$	$1s_{3/2}$–$2p_{5/2}$ (Γ_8)	$1s_{3/2}$–$2p_{5/2}$ (Γ_7)	$1s_{3/2}$–$3s_{3/2}$	$1s_{3/2}$–$2p_{1/2}$	$1s_{3/2}$–$4s_{3/2}$
S	(99)	(99)	~56	37.4	40.2	—	—	—	—	—
O	Present work	Present work	84±2	47.8	56.7	—	63.9	67.3	—	—
P	Present work	Present work	88.4±0.5	—	69.3[c]	72.6[c]	73.9[c]	75.6[c]	—	—
A_2	Present work	—	95.0±0.4	—	—	—	—	—	—	—
A_1	Present work	Present work	106.1±0.6	—	75.4	79.4	84.2	—	—	—
N	Present work	(6)	111.3±0.5	67.8	79.9	82.6	89.8	92.8	98.5	102.2
Li	Present work	(98)	114.2±0.3	72.9	82.6	85.8	93.0	97.8	100.0	102.5
As[d]	Present work	Present work	114.3±0.5	73.6	83.9	—	93.5	98.4	—	—
Na	Present work	(98)	127.9±0.5	83.1	97.6	100.4	106.8	110.5	113.0	114.7
Ag	(89)	(89)	~431	—	393.5	—	—	412.5	—	419.3
Au	(89)	—	~550	—	—	—	—	—	—	—

[a] Based on fits to (e-A°) peak positions measured in strained ZnSe/GaAs as a function of temperature, corrected for temperature dependence of E_g. Strain corrections made using value of b′ measured for Li, assumed to be the same for all other acceptors (except P, where value measured for that acceptor was used). Strains determined from measured X_{lh} PL peak positions, and from $X_{lh}-X_{hh}$ PL peak splittings (values usually averaged), using deformation potentials from ZnSe/bulk InGaAs study.

[b] Typical accuracies are ±0.1 to 0.2 meV.

[c] Assignment of these levels is arbitrary because they do not conform to effective mass theory.

[d] Now believed to be due to accidental Li contamination; note similarities in values for As and Li acceptors.

in our laboratory on a total of 23 heteroepitaxial ZnSe/GaAs samples are shown in Figure 11, together with several fits to semi-empirical formulas. Not surprisingly, two-parameter functions like those given by Varshni (93) or O'Donnell & Chen (94) [the former being the type quoted in previous studies of E_g (T) in ZnSe (95, 96)] cannot provide sufficiently accurate fits over the entire range of the data (we include a band gap of 2.67 eV at 300

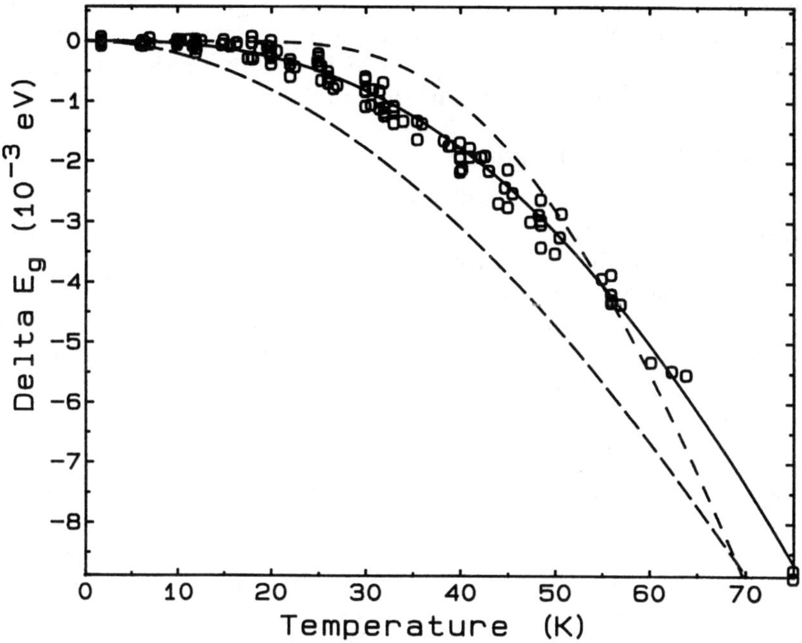

Figure 11 Temperature-dependent shifts in bound exciton (or occasionally free exciton) PL peak positions for a total of 23 ZnSe/GaAs samples (*open symbols*), together with plot of a two-parameter Varshni function (*long dashes*), $\Delta E_g(T) = -\alpha T^2/(T+\beta)$, using parameters taken from Reference 95 ($\alpha = 0.859$ meV/K and $\beta = 405$ K). Using parameters from Reference 96 gives a similar plot (not shown). The short-dashed and solid lines, respectively, are least-squares fits of our data to a two-parameter O'Donnell-Chen function, $\Delta E_g(T) = -S\langle\hbar\omega\rangle \times [\coth(\langle\hbar\omega\rangle/2k_BT) - 1]$, with $S = 4.142$ and $\hbar\omega = 16.95$ meV, and to the three parameter function of Equation 6, with $\alpha = 1.182$ meV/K, $\beta = 304.6$ K, and $\gamma = 28921$ K^2. In the two least-squares fits, we include a data point of $E_g = 2.67$ eV at $T = 300$ K (not shown); eliminating this point substantially improves the fit from 0 to 75 K for the O'Donnell-Chen case at the cost of predicting an unrealistic E_g of 2.76 eV at 300 K, but does not significantly improve the fit for the function of Equation 6. The improved O'Donnell-Chen fit (not shown) is still inferior to the one using Equation 6. We do not fit our own data to a two-parameter Varshni function because the limited range of the data causes the fit to converge to a parabola (i.e. large values of α and β), which is not good fit.

K in the fits, although this is not shown on the plot). A three-parameter function of the form

$$\Delta E_g(T) = \frac{-\alpha T^3}{T^2 + \beta T + \gamma} \qquad 6.$$

does, however, yield a satisfactory fit over the range 0–77 K (and to the point at 300 K) with the parameters indicated in the figure caption. We adopted this formula to compute the results in Table 2, based on data such as that illustrated for various acceptors in Figure 12. (However, we caution that our parameters may not yield accurate results between 77 and 300 K, because we included no data in this range in the fits.)

Excited States of Acceptors

Excited state energies are useful to provide a more precise distinction between different levels and to determine if a given acceptor behaves according to effective mass theory. They are frequently determined using SPL measurements, according to the technique discussed in References 97, 98. The results are summarized in Table 2, based on data from measurements in our laboratory (69, 86, 91, 92) and elsewhere (47, 89, 98, 99). Specific cases are discussed below.

Properties of Particular Impurities

NITROGEN We discuss the N acceptor first because it has achieved the highest doping levels in MBE material [$\sim 1 \times 10^{18}$ cm^{-3} (2)] and is the only *p*-type dopant to be used widely in device applications to date. The properties of the shallow N acceptor level have been investigated extensively by Dean et al (6) and subsequently by Shahzad et al (47). They found binding energies of 111 and 110.0 meV, respectively, in excellent agreement with the value we determined from (e-A$^\circ$) positions (Table 2). They also found excited state energies that clearly confirm the effective-mass-like nature of the level (6, 47). The only challenge to the view that the N acceptor is substitutional and well-behaved has been by Schetzina et al, who argued on the basis of variable temperature Hall effect data (100), linear polarization of the luminescence (101), and the temperature dependence of the (e-A$^\circ$) peak position at temperatures in the $\sim$77 to 260 K range (96, 101) that the acceptor is non-effective mass-like and possibly involves twofold coordinated interstitial N atoms. However, the polarization was a measurement artifact (JF Schetzina, private communication), whereas the alleged strong dependence of binding energy on temperature depends on the nature of the dominant band-edge emission processes at high temperature, which is highly controversial, and the resulting values

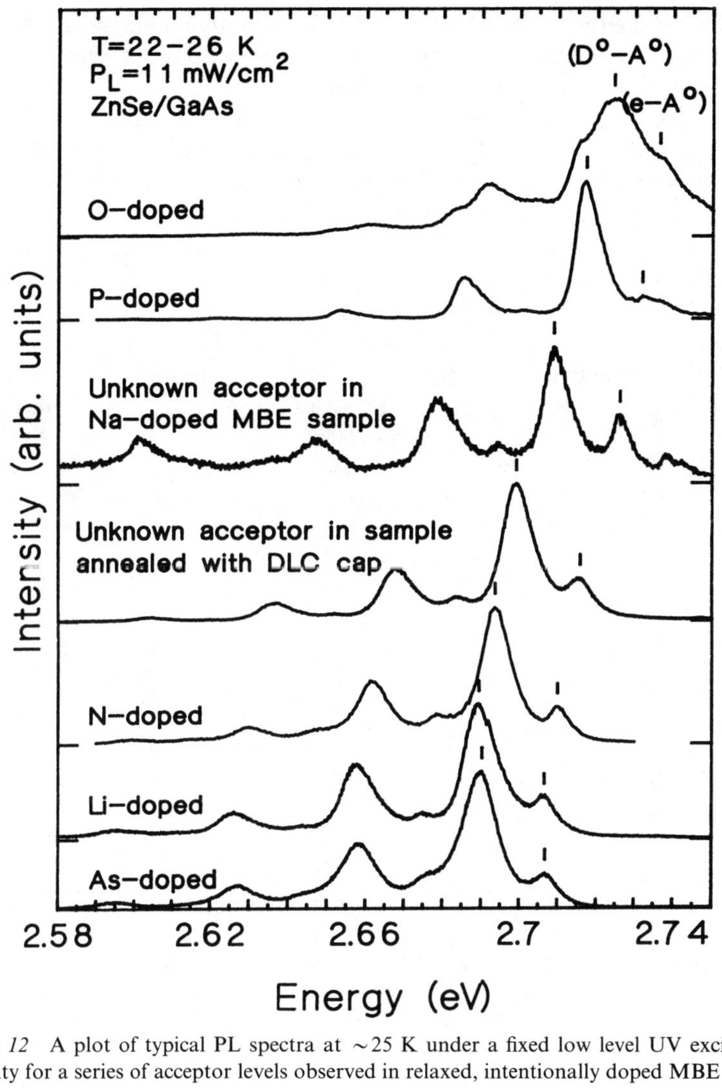

Figure 12 A plot of typical PL spectra at ~25 K under a fixed low level UV excitation intensity for a series of acceptor levels observed in relaxed, intentionally doped MBE ZnSe/GaAs layers and in undoped ZnSe/GaAs subject to an anneal with DLC caps, as discussed in the text. Both (D°-A°) and (e-A°) peaks and their LO phonon replicas are visible in each case. The shallow acceptor level in the As-doped sample is now believed to be from accidental Li contamination.

derived for $E_g(T)$. The reduced electrical activation energy of the acceptor has been interpreted more conventionally by Fan et al as the usual effect of screening and banding of the excited states, which effectively lowers the valence band edge (102). Moreover, the ion channelling measurements of Yao et al show no detectable interstitial N atoms (103) in heavily N-doped ZnSe, and this model has found little acceptance.

The I_1 peak due to N acceptors is reported to lie about 0.3 meV lower in energy than that for I_1^{Li} (6). The data in Figure 5*b* show that the strain dependence of I_1^N accurately parallels that of I_1^{Li}, whereas the separation between the two is 0.35 meV, in good agreement with the previous results. The excited states of I_1^N have not been studied in unstrained material, thus a precise model can not be given. A very sharp I_1^N peak is visible in Figure 13 in the spectrum of a lightly N-doped MBE layer. In more heavily doped material, as shown in the lower spectrum of Figure 13, the, I_1^N peak frequently exhibits a low energy wing known as an undulation structure (91, 104), which has also been observed in ZnSe for I_1^{deep} (46) and I_1^P (70), and is common to acceptor-bound excitons in many semiconductors (105, 106). We assign this structure to excitons bound to pairs of N acceptors at various separations within the lattice. It dominates the spectrum in

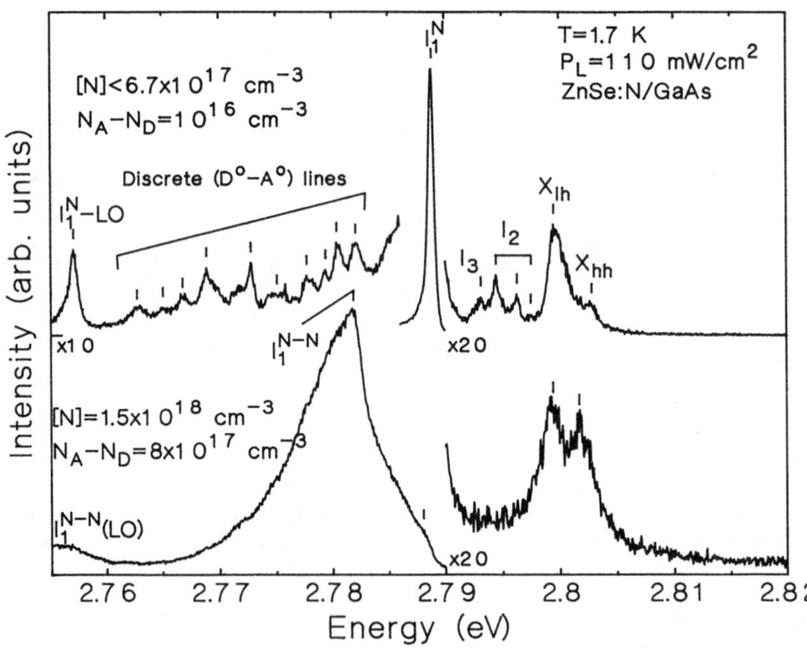

Figure 13 Low-temperature excitonic PL spectra for lightly (*upper spectrum*) and heavily (*lower spectrum*) N-doped ZnSe MBE layers. The broad structure observed below the normal I_1^N position in the heavily doped layer is assigned to an undulation structure involving excitons bound to N-N pairs; it quenches with respect to a peak that appears at the normal position as the temperature is raised to ~20–30 K (not shown).

some cases at low temperatures ($\sim$1.7 K), as shown e.g. in Figure 13, because of the tendency of excitons to tunnel to the lowest available states. As the temperature is raised above 20–30 K, this feature decreases rapidly with respect to a normal I_1^N peak that becomes visible at $\sim$2.789 eV, as the excitons begin to percolate to the more common distant pairs.

A determination of the binding energy of the shallow N acceptor is illustrated in Figures 14 and 15, where we display variable-temperature data for the (e–A°) transition involving the heavy-hole acceptor level in a lightly N-doped layer. The excellent linear fit to the corrected peak positions (illustrated in Figure 15) allows us to deduce an accurate binding energy (after correction for strain effects, see above) of 111.3 ± 0.5 meV, as listed in Table 1. This value is in good agreement with previous determinations by other methods (6, 47).

An important consideration in N doping is the factor that limits the maximum achievable doping level. In the case of OMCVD material, it is now well established that H passivation plays a major role in limiting the activation of N (107), as has recently been confirmed independently by two different groups by detection of N H bonds in infrared and Raman spectroscopy of OMCVD ZnSe : N (108, 109) and by correlations in the secondary ion mass spectrometry (SIMS) profiles of N and H (109). In the MBE case, studies of deep-hole traps by deep-level transient spectroscopy (DLTS) have shown concentrations at most a few percent of the dopant concentration for net doping levels up to $\sim 3 \times 10^{17}$ cm^{-3} (110, 111), giving no proof of significant compensation. Deep-donor levels have been detected in photoconductivity and photo-Hall effect measurements on MOCVD ZnSe : N (112), although this material probably also contained H. Trigonal complexes believed to involve V_{Se} paired with an acceptor other than N have recently been detected in optically detected magnetic resonance (ODMR) studies of MBE ZnSe : N (113), but the concentration of the acceptor involved in the complex is stated to be less than 10^{15} cm^{-3}, and this complex therefore does not appear to be a significant factor in compensating material with a net dopant concentration of 4×10^{17} cm^{-3}.

Photoluminescence measurements have given evidence for an $\sim$44-meV deep compensating donor level in heavily doped ZnSe : N by virtue of a second (D°-A°) pair band at $\sim$2.68 eV (114), deeper than the normal N (D°-A°) band observed at $\sim$2.695 eV in lightly doped material, involving shallow substitutional donor impurities. Haukkson et al proposed that the deep donor might be a N_{Se}-Zn-V_{Se} complex (114). Recent SPL measurements have shown conclusively that both (D°-A°) pair bands involve the same $\sim$111 meV shallow N acceptor level and must therefore involve different donors (91, 104). This conclusion is also supported by variable-temperature PL measurements showing that only a single (e-A°) band is

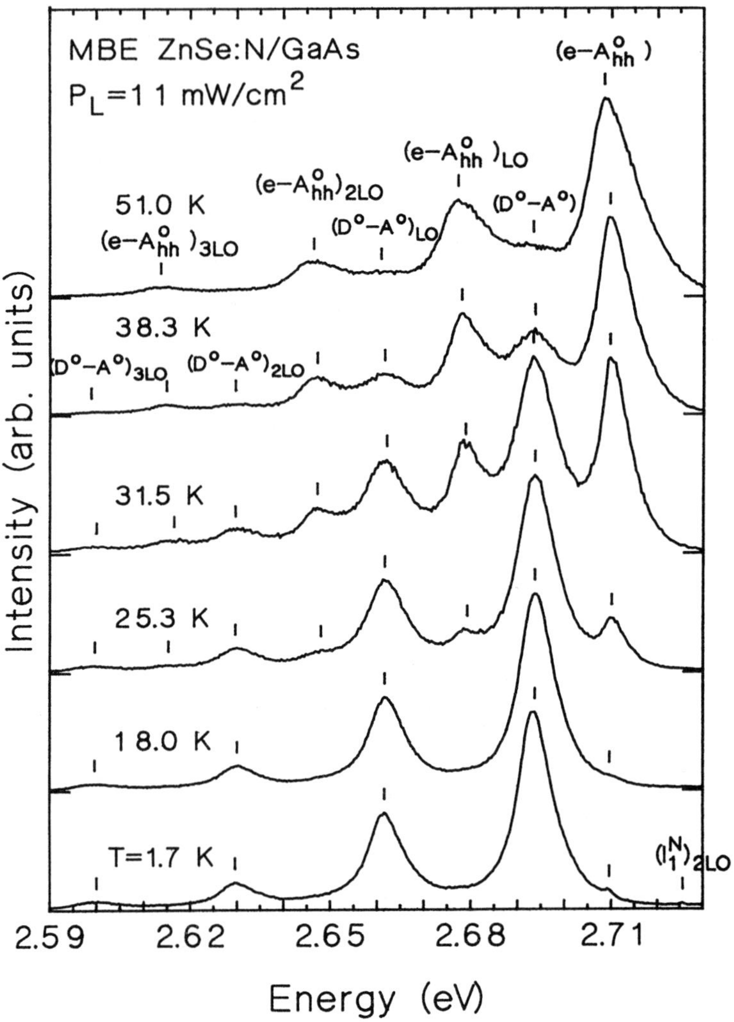

Figure 14 N-related (D°-A°) and (e-A°) PL peaks as a function of temperature for the lightly doped sample of Figure 13. The (e-A°) peak initially quenches with increasing temperature (due to increasing nonradiative recombination rates) and then increases as the shallow donors start to thermally ionize.

observed for both (D°-A°) bands at high temperature (104). The existence of a deep donor is further supported by ODMR measurements on heavily doped ZnSe : N, which showed a new anisotropic deep-donor resonance with $g \sim 1.38$ (115). More recent spin-flip Raman scattering measurements also showed this resonance (116). High-pressure PL measurements of the

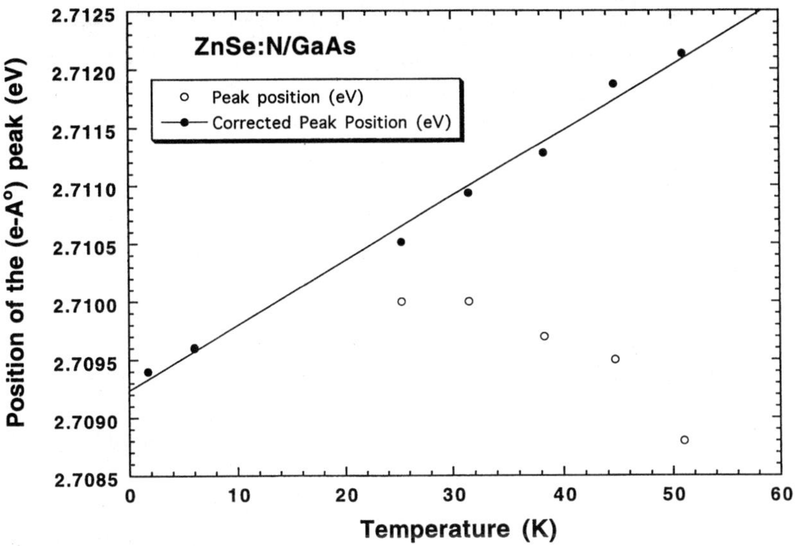

Figure 15 Positions of the (e-A°) peaks in Figure 14 as a function of temperature (*open circles*), together with corresponding values for the temperature variation of the band gap (*closed circles*), and a linear fit to the latter values (*solid line*). The slope of the fit is 0.0558 meV/K (compared to a theoretical value of $k_B/2 = 0.0431$ meV/K) and the intercept is 2.7092 eV.

two (D°-A°) pair bands have shown that the deep (D°-A°) pair band has a smaller pressure coefficient than the shallow (D°-A°) band, which follows the band gap (117). This result suggests that the deep donors involve localized states with pressure coefficients more characteristic of the whole conduction band than of the Γ point. One theoretical possibility for these donors is a ⟨100⟩ split interstitial configuration involving N (118).

A new, relatively broad bound exciton peak at ~2.765 eV has also recently been observed in heavily doped ZnSe : N and tentatively assigned to excitons bound to the new deep-donor level (91, 104). Unfortunately, the width of this peak makes it difficult to use perturbation spectroscopy to determine the symmetry of the related center. Although there is considerable evidence for the existence of the ~44 meV donor centers, a quantitative assessment of their importance in compensating the material is difficult using luminescence techniques and has not been performed. The question remains open, therefore, whether they are the factor limiting the ultimate doping level with N or whether simple solubility considerations for the N acceptor itself are governing (119, 120). The formation of N-N pairs may be one limitation at high doping (121).

LITHIUM While Li acceptors have produced *p*-type doping of ZnSe up to about 2×10^{17} cm^{-3} (122), the extremely high interstitial diffusivity of this dopant and its tendency to electromigration make it of little interest in practical applications. For example, Akopyan et al found evidence for diffusion of Li into ZnSe from LiOH solutions at room temperature (123)! The spectroscopic properties of Li acceptors are well established, and have been treated in detail for ZnSe: Li/GaAs by Zhang et al (70, 82). A particularly interesting aspect was the observation of a (e-A°) peak (at 2.706 eV) even at 1.7 K. This peak decreased in intensity relative to the (D°-A°) peak as the temperature was raised from 1.7 to 15 K because of temperature-dependent nonradiative recombination rates (82). At higher temperatures, where the donors start to thermally ionize into the conduction band, the (e-A°) peak increases normally. Although the (e-A°) peak happens to be coincident with the R_o peak in bulk ZnSe (124), it has very different properties and is not related to the latter.

Unintentional Li contamination may occur more frequently in as-grown MBE ZnSe/GaAs and annealed material than previously realized because it has frequently been misidentified as As (see below).

ARSENIC Early studies of bulk As-doped ZnSe showed only deep levels believed to involve Jahn-Teller distortions of isolated As_{Se} into C_{3v} symmetry (125, 126). Investigations in As-doped epitaxial material have, however, given some evidence of a shallow acceptor level (127–130), although the reported properties have varied somewhat. Moreover, a shallow acceptor observed in as-grown or thermally annealed ZnSe/GaAs has been widely attributed to As acceptors originating from the GaAs substrate (14, 46, 48, 71, 131, 132). A detailed spectroscopic analysis of the most commonly observed level believed to be As-related was reported in Reference 69, where it was found that this level is spectroscopically indistinguishable from the well-known Li acceptor level in virtually all respects.

More recently, we have found (91) that in some As-doped MBE layers exhibiting shallow acceptor peaks, this level is in fact due to accidental Li contamination, presumably from the Zn_3As_2 source material used to dope the material with As. This determination is based on resonant dye laser excitation of the I_1 peak appearing in these samples, which produces sharp local phonon mode replicas known to correspond uniquely to substitutional Li isotopes (82, 89). Successful observation of the Li local modes requires that the I_1 peaks be reasonably sharp and strong compared with the (D°-A°) peaks in the sample; conditions are not always favorable for its observation, even in intentionally Li-doped material. However, the fact that we have observed this feature in several As-doped MBE layers and in several undoped MBE samples showing an ~ 114 meV acceptor level after annealing, combined with the low intensity of the shallow acceptor peaks in most As-doped samples and the general lack of repeatability and success with As doping in the literature, strongly suggests that As may in fact not exhibit any shallow level at all in ZnSe. If this is correct, As produces only deep levels as originally reported in bulk material. This result implies that accidental Li contamination is more common than previously recognized and is responsible for most of the shallow acceptor-related peaks previously attributed to As acceptors in as-grown and annealed material.

PHOSPHORUS Like As, P was originally reported to produce only deep level PL peaks, which were attributed to Jahn-Teller distorted isolated substitional P_{Se} with C_{3v} symmetry (125, 126). Subsequent ODMR measurements (133, 134) confirmed the deep acceptor nature of this level and the results of the original electron paramagnetic resonance (EPR) measurements (126). A center involving P_4 clusters has also been detected by EPR (135) and ODMR (136). Recent Raman scattering measurements detected local phonon modes in ZnSe : P, which were assigned to undistorted, ionized (nonmagnetic) P_{Se}^{-} centers having T_d symmetry, tending to confirm the model that the deep acceptors involve isolated P_{Se} (137). However, there have been persistent reports of a shallow, P-related acceptor level appearing in the low temperature PL spectrum of epitaxial ZnSe : P (5, 38, 70, 130, 138–140), as evidenced by a (D°-A°) band whose reported position ranges from ~2.716 to 2.728 eV, although it ranges from 2.716 to 2.721 eV in most cases. [The 2.728 eV value was for unstrained homoepitaxial material grown by liquid phase epitaxy (LPE), which had very broad peaks possibly indicating heavy doping (130, 138); most of the other samples were relaxed ZnSe/GaAs layers with narrower peaks.] There is even a recent, preliminary report claiming to observe $p \sim 10^{15}$ cm^{-3} in OMCVD ZnSe : P (71). However, the shallow peak has not been observed in all P-doped material, some of which has shown only deep acceptor-related peaks (74, 125, 141).

The shallow level is studied in Reference 70, where temperature-dependent measurements revealed a (e-A°) doublet with a splitting of 2.4 meV; the origin of this splitting is not entirely clear (it may not be entirely due to strain). The (e-A°) peak decreases initially with respect to the corresponding (D°-A°) peak from 1.7 to 20 K similar to the Li case, presumably because of temperature-dependent nonradiative recombination rates, but increases normally above ~25 K. Our most recent calculations of the binding energy in unstrained material, using the temperature-dependent (e-A°) peak positions and assuming that the splitting is due to strain, yield the 88.4 ± 0.5 meV binding energy listed in Table 2

(86). Measurements of the excited states of this level using SPL (Table 2) do not fit to effective mass theory, suggesting that the level may be nonhydrogenic in nature.

The existence of an I_1 peak due to P was first recognized by Bhargava, who showed that it lies just above the normal positions of I_{2o} lines in ZnSe (142); it is distinguished from the latter by its larger LO phonon coupling strength ($S \sim 0.03$). In earlier work, this peak was denoted I_α^o (11, 65), although it was not identified there beyond showing that it is not an I_2 peak. We have confirmed the existence of this peak (70), as shown in Figure 16, and also identified a low-energy undulation wing that is visible in the figure at ~2.787 to 2.791 eV, below the I_1^P peak at 2.7957 eV. This structure is similar to that described for N acceptor above. We believe that the rather broad features observed at 2.789 and 2.791 eV in MBE ZnSe : P/GaAs by Yao & Okada (140) also form part of this undulation structure in their sample. They found an increase in the strength of an I_2 peak at 2.796 eV with P-doping, which they attributed to Ga donors

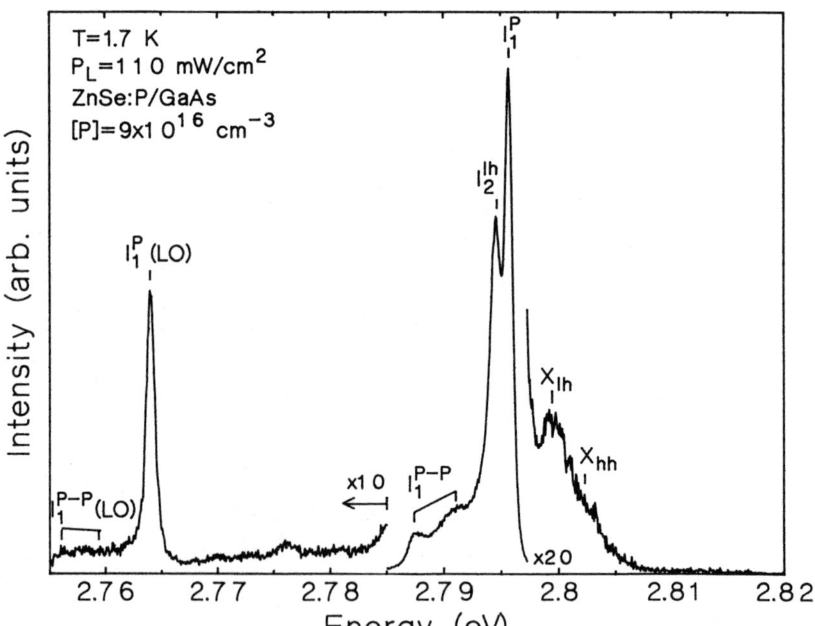

Figure 16 Excitonic PL spectrum of a lightly doped ZnSe : P layer grown by MBE on GaAs. The structure just below I_{2o}^{lh} is assigned to an undulation wing involving excitons bound to pairs of P-related shallow acceptors (note its presence in the LO phonon replica, and the absence of I_{2o}^{lh} in the phonon replica because the latter is less strongly phonon-coupled than I_1^P).

diffused from the substrate, with respect to that of an I_x peak at 2.795 eV assigned to native donors, and they attributed the 2.789 and 2.791 eV peaks to I_1 peaks. However, the I_x peak is now known to be I_{2o}^{lh} (see above), so that the increase of the I_2 peak relative to I_x cannot be understood if it were due to Ga donors. Instead, the I_2 peak should be identified as I_1^P, and the broad lower energy structure (which is broader than normal I_1 peaks) is better understood as an undulation wing involving excitons bound to pairs of the shallow P-related acceptors at various spacings within the lattice.

Similarly, DePuydt et al (38) and Kawakyu et al (141) reported increased intensities of the I_2 peaks in their samples with P doping, which we suggest was actually from introduction of I_1^P. This feature thus seems to be very common in ZnSe doped with P. Careful examination of the LO phonon replicas can help to distinguish this peak from I_2 peaks. The unusually shallow binding of this I_1 complex, the lack of the usual doublet splitting in strained material, and the apparently non-effective mass-like nature of this level provide some support for the interpretation that the shallow acceptor level involves a complex, although further experiments are needed to establish this result firmly. It cannot be the result of the P_4 clusters detected by EPR because the corresponding level is reported to lie about 1.2 eV below the conduction band (136).

OXYGEN Oxygen would not normally be expected to be an acceptor in ZnSe because it is a column VI element like Se. However, Akimoto et al showed that O does in fact produce acceptor activity in ZnSe doped during MBE growth with ZnO, and in undoped material implanted with O (143). Net acceptor-doping levels up to 10^{16} cm^{-3} were determined from C-V profiles of a *p-n* junction (144), and O acceptor behavior was also observed in CdTe, CdS, and ZnS (145). Photoluminescence features similar to those observed by Akimoto et al have also been detected in O-doped material grown by other workers (146–148). A recent spectroscopic study by Chen et al provided a thorough characterization of this shallow level, based on observations of the corresponding (D°-A°) and (e-A°) peaks and SPL measurements on the former peak (92). However, the I_1 peak originally reported by Akimoto et al was found to be I_{2o}^{lh}, by use of reflectance measurements to identify X_{lh} and X_{hh}, due to the previously unrecognized effects of strain. No true I_1 peak involving the shallow acceptor level has yet been identified. Also, our analysis suggests that the "O isoelectronic bound excitons" reported by Shahzad et al in O-implanted material (149) may actually be I_1 peaks resulting from accidental contamination during the annealing process and may not be related to O (92). The results of this study (92) are summarized in Table 2; the excited state energies were fit successfully to effective mass theory, proving that the acceptor level is indeed effective mass-like. We also found that O occurs as a residual acceptor in undoped, Ga-, and As-doped MBE ZnSe.

One important result of our recent work is that shallow donors were introduced when doping with ZnO, probably by donor impurities in the source material (92). Improved doping methods (such as the plasma-doping technique used successfully for N) should be investigated to determine the maximum doping levels that can be achieved with O when avoiding unintentional donor contamination. This dopant might be particularly interesting for wider gap materials such as ZnMgSSe, where N is now suspected to undergo a shallow-deep transition (150). A recent theoretical model suggests that the acceptor might involve interstitial O bonded to Zn atoms (121), but further experimental work is necessary to establish the true nature of this acceptor.

CARBON AND NEW LEVELS Carbon contamination is generally a significant concern in MOCVD growth. In MOCVD ZnSe, Giapis et al reported observing a bound exciton at ~ 2.792 eV, denoted I^c, which they originally attributed to excitons bound to C-related complexes (44, 151, 152). However, later studies revealed that this peak was really I_{2o}^{lh}, which was shifted lower in energy by strain in heavily C-contaminated material (16) (see Figure 5*a*). The increased strain apparently resulted from stimulation of the relaxation of the thermal mismatch strain by high interfacial C levels, which might have formed precipitates. No direct effect of C on the optoelectronic properties of ZnSe was found; in particular, it does not seem to enhance nonradiative recombination (16). This study shows the importance of including strain effects in the analysis of PL data on heteroepitaxial material.

Doping of ZnSe with Na in low-temperature MBE growth (38, 153–155) has consistently resulted in the observation of a (D°-A°) peak around 2.710 to 2.716 eV, much higher in energy than the P_o peak at ~ 2.68 eV normally attributed to Na acceptors in bulk material (156). An I_1 peak was also observed around 2.7909 eV (154). This level could be that observed in bulk ZnSe : Na (157) and speculated there to involve a Na_{Zn}-V_{Se} complex. The (D°-A°) and (e-A°) peaks for this acceptor were studied as a function of temperature in Reference 86, where the binding energy of the acceptor (denoted A_2) was found to be 95.0 ± 0.4 meV, as shown in Table 2. Further work is needed to determine if this level involves Na, an impurity in the Na source material such as K, or some other type of complex. The position of the I_1 peak for this acceptor is shown as a function of strain in Figure 5*b*, where it is seen to lie about 1.3 meV higher in energy than I_1^{Li} for the same degree of strain (about -5.6×10^{-4}).

Finally, we have also discovered a new level with a binding energy of about 106 meV in undoped MBE ZnSe/GaAs annealed using diamond-like C (DLC) caps (86); this level is denoted A_1 (see Table 2). The corresponding I_1 peak seems to be about 0.16 meV shallower than that of Li (Figure 5*b*). The excited states of this level fit well to effective mass theory, confirming its effective mass-like nature. This level shows promise as a new *p*-type dopant if it can be chemically identified.

SUMMARY AND CONCLUSIONS

We have discussed and modeled the important effects of strain on excitonic and other features in the low-temperature luminescence spectrum of heteroepitaxial ZnSe and have shown in particular how failure to recognize strain effects can lead to serious misinterpretations of the spectra. In several cases, examples of the latter include misidentification of the light-hole neutral donor-bound exciton peak as a different type of transition believed to involve excitons bound to native donors, oxygen acceptors, or C-related complexes. A careful analysis of the spectra, however, supported by reflectance or other measurements to identify the free exciton features with certainty, can yield a wealth of useful information about the material to aid the crystal grower in improving its properties. Analysis of strain effects is also crucial in other heteroeptiaxial II–VI materials such as ZnTe (158) that are not discussed here.

We have also reviewed the spectroscopic properties of various acceptor levels in ZnSe, to aid in assessing the prospects for novel doping methods and improvements in the now standard N doping processes used in MBE growth. The numerical results are summarized in Table 2 and in Figure 12, which shows PL spectra for a series of dopants under nearly identical experimental conditions. The energies of the (D°-A°) and (e-A°) peaks are also summarized in Table 1. The (D°-A°) positions there are based on data recorded under a fixed set of conditions ($T = 1.7$ K, using an unfocused 8 mW, 3.53 eV laser beam) but still show substantially more scatter than the values for the (e-A°) peaks. The latter peak positions are actually found by linear extrapolations of full temperature-dependent data down to 1.7 K, for a fixed excitation level (unfocused $\sim$1 mW, 3.53 eV laser beam). The (e-A°)-type of transition is therefore much more reliable for the identification of acceptor levels, even though it has been little used because most workers record data at a fixed, low temperature where (e-A°) peaks are not always visible. The advantage of using these peaks is apparent when comparing the standard deviations in Table 2 for (D°-A°) and (e-A°) peak positions; for cases with four or more data points, they average 1.7 and 0.7 meV, respectively. The larger scatter for the (D°-A°)

case is due to the doping level and non-radiative recombination rate dependence of these peaks, as well as the smaller amount of averaging because peak positions are measured under only one condition. The scatter for the (D°-A°) peaks would be larger if the excitation intensity had varied.

Qualitatively, we have shown that it is important to determine the nature of the deep-donor level occurring in heavily N-doped ZnSe, as this level may be causing significant compensation. Also, quantitative evaluation of the concentration of this level is needed to assess its importance, although no method has yet been devised to do so. Among other acceptors, we have shown that As does not, in fact, produce a shallow-acceptor level because the level previously observed in As-doped material is actually due to accidental Li contamination. On the other hand, a shallow level does occur for P, but its nature is not yet understood. The unusual acceptor O deserves to be investigated more carefully and might be useful especially in wide-gap quaternary alloy materials; another as-yet-unidentified acceptor also appears to have potentially useful properties. To support the active device development now taking place, optical spectroscopy in various forms will continue to be of crucial importance in expanding our knowledge.

Acknowledgments

The author is supported by the National Science Foundation under Grant Nos. DMR-9106359 and ECS-9202664. He is very grateful to the many crystal growers who supplied samples for the investigations described here, including K Akimoto, BC Cavenett, H Cheng, KP Giapis, KF Jensen, RM Park, SM Shibli, S Sivananthan, MC Tamargo, and FS Turco-Sandroff, and to Y Zhang, W Liu, and J Chen for their assistance in the acquisition and interpretation of much of the data.

Literature Cited

1. Park RM, Trotter MB, Rouleau CM, DePuydt JM, Haase MA. 1990. *Appl. Phys. Lett.* 57: 2127–29
2. Qiu J, DePuydt JM, Cheng H, Haase MA. 1991. *Appl. Phys. Lett.* 59: 2992–94
3. Haase MA, Qiu J, DePuydt JM, Cheng H. 1991. *Appl. Phys. Lett.* 59: 1272–74
4. For example, see *Proc. Int. Conf. on II-VI Compounds, 6th, Newport, RI,* 1994. *J. Cryst. Growth* 138:
5. Dean PJ. 1980. *Czech. J. Phys. B* 30: 272–87
6. Dean PJ, Stutius W, Neumark GF, Fitzpatrick BJ, Bhargava RN. 1983. *Phys. Rev. B* 27: 2419–28

6a. Aven M, Prener JS, eds. 1967. *Physics and Chemistry of II-VI Compounds.* Amsterdam: North Holland

7. Halsted RE. 1967. See Ref. 6a, pp. 383–431
8. Hopfield JJ. 1967. In *Proc. Int. Conf.*

on II-VI Semiconducting Compounds, Providence, RI, ed. DG Thomas, pp. 786–804. New York: Benjamin

9. Curie D, Prener JS. 1967. See Ref. 6a, pp. 435–85
10. Hartmann H, Mach R, Selle B. 1982. In *Current Topics in Materials Science*, ed. E Kaldis, pp. 1–414. Amsterdam: North Holland
11. Bhargava RN. 1982. *J. Cryst. Growth* 59: 15–26
12. Gutowski J, Presser N, Kudlek G. 1990. *Phys. Status Solidi* (*a*) 120: 11–59
13. Yao T, Okada Y, Matsui S, Ishida K, Fujimoto I. 1987. *J. Cryst. Growth* 81: 518–23. The equations for $E(X_{hh})$ ("E1") and $E(X_{lh})$ ("E2") in this paper contain several errors (compare our equations), and the wavelength to energy conversions are incorrect (the refractive index of air was apparently omitted).
14. Yao T, Ogura M, Matsuoka S, Morishita T. 1983. *Jpn. J. Appl. Phys.* 22: L144–46
15. Lee YR, Ramdas AK, Kolodziejski LA, Gunshor RL. 1988. *Phys. Rev. B* 38: 13143–49
16. Skromme BJ, Liu W, Jensen KF, Giapis KP. 1994. *J. Cryst. Growth* 138: 338–45
17. Matsumura N, Tsubokura M, Saraie J. 1989. *J. Cryst. Growth* 95: 525–28
18. Skromme BJ, Tamargo MC, Turco FS, Shibli SM, Nahory RE, Bonner WA. 1989. In *Heteroepitaxial Approaches in Semiconductors: Lattice Mismatch and Its Consequences*, ed. AT Macrander, TJ Drummond, pp. 335–46. Pennington, NJ: Electrochem. Soc.
19. Skromme BJ, Zhang Y, Tamargo MC, Bonner WA. 1991. *Strain effects in heteroepitaxial ZnSe grown on bulk InGaAs by MBE.* Presented at Int. Conf. II-VI Compounds, 5th, Tamano, Japan. #Wela-17
20. Asai H, Oe K. 1983. *J. Appl. Phys.* 54: 2052–56
21. Lee S, Hillebrands B, Stegeman GI, Cheng H, Potts JE, Nizzoli F. 1988. *J. Appl. Phys.* 63: 1914–16
22. Venghaus H. 1979. *Phys. Rev. B* 19: 3071–82
23. Kudlek G, Presser N, Pohl UW, Gutowski J, Lilja J, et al. 1992. *J. Cryst. Growth* 117: 309–15
24. Langer DW, Euwema RN, Era K, Koda T. 1970. *Phys. Rev. B* 2: 4005–22
25. Stoehr M, Hamdani F, Lascaray JP, Maurin M. 1991. *Phys. Rev. B* 44: 8912–17
26. Coquillat D, Hamdani F, Lascaray JP, Briot O, Briot N, Aulombard RL. 1993. *Phys. Rev. B* 47: 10489–96
27. Rockwell B, Chandrasekhar HR, Chandrasekhar M, Ramdas AK, Kobayashi M, Gunshor RL. 1991. *Phys. Rev. B* 44: 11307–14
28. Dean PJ, Pitt AD, Wright PJ, Young ML, Cockayne B. 1983. *Phys. B* 116: 508–13
29. Dean PJ. 1984. *Phys. Status Solidi* (*a*) 81: 625–46
30. Yoneda K, Hishida Y, Toda T, Ishii H, Niina T. 1984. *Appl. Phys. Lett.* 45: 1300–2
31. Mar HA, Park RM. 1986. *J. Appl. Phys.* 60: 1229–32
32. Menda K, Takayasu I, Minato T, Kawashima M. 1988. *J. Cryst. Growth* 86: 342–47
33. Yao T. 1985. *J. Cryst. Growth* 72: 31–40
34. Potts JE, Smith TL, Cheng H. 1987. In *Modern Optical Characterization Techniques for Semiconductors and Semiconductor Devices*, ed. OJ Glembocki, FH Pollak, JJ Song, pp. 27–33. Bellingham, WA: Soc. Photo-opt. Instr. Eng.
35. Skromme BJ, Tamargo MC, de Miguel JL, Nahory RE. 1988. In *Epitaxy of Semiconductor Layered Structures*, ed. RT Tung, LR Dawson, RL Gunshor, pp. 577–82. Pittsburgh: Mater. Res. Soc.
36. Ohkawa K, Mitsuyu T, Yamazaki O. 1988. *Phys. Rev. B* 38: 12465–69
37. Shahzad K. 1988. *Phys. Rev. B* 38: 8309–12
38. DePuydt JM, Smith TL, Potts JE, Cheng H, Mohapatra SK. 1988. *J. Cryst. Growth* 86: 318–23
39. Giapis KP, Lu D-C, Jensen KF, Potts JE. 1990. *J. Cryst. Growth* 104: 291–96
40. Vaziri M, Reifenberger R, Gunshor RL, Kolodziejski LA, Venkatesan S, Pierret RF. 1989. *J. Vac. Sci. Technol. B* 7: 253–58
41. Skromme BJ, Shibli SM, de Miguel JL, Tamargo MC. 1989. *J. Appl. Phys.* 65: 3999–4005
42. Matsumoto T, Iijima T, Goto H. 1990. *J. Cryst. Growth* 99: 427–31
43. Ohkawa K, Mitsuyu T, Yamazaki O. 1987. *J. Appl. Phys.* 62: 3216–21
44. Giapis KP, Jensen KF, Potts JE, Pachuta SJ. 1990. *J. Electron. Mater.* 19: 453–62
45. Skromme BJ, Tamargo MC, Turco FS, Shibli SM, Bonner WA, Nahory RE. 1989. In *GaAs and Related Compounds*, ed. JS Harris, pp. 205–10. Bristol, England: Inst. of Phys.
46. Skromme BJ, Stoffel NG, Gozdz AS,

Tamargo MC, Shibli SM. 1989. In *Advances in Materials, Processing, and Devices for III-V Compound Semiconductors*, ed. DK Sadana, LE Eastman, R Dupuis, pp. 391–96. Pittsburgh: Mater. Res. Soc.

47. Shahzad K, Khan BA, Olego DJ, Cammack DA. 1990. *Phys. Rev. B* 42: 11240–44
48. Yodo T, Yamashita K. 1988. *J. Cryst. Growth* 93: 656–61
49. Park RM, Rouleau CM, Trotter MB, Koyama T, Yodo T. 1990. *J. Mater. Res.* 5: 475–77
50. Ohkawa K, Mitsuyu T. 1991. *J. Appl. Phys.* 70: 439–42
51. Chou WC, Twardowski A, Chern-Yu K, Chen FR, Hua CR, et al. 1994. *J. Appl. Phys.* 75: 2936–40
52. Cheng H, DePuydt JM, Haase M, Potts JE. 1990. *Appl. Phys. Lett.* 56: 848–50
53. Kudlek G, Presser N, Gutowski J, Hingerl K, Sitter H, et al. 1990. *J. Appl. Phys.* 68: 5630–35
54. Kudlek G, Presser N, Gutowski J, Menke DR, Kobayashi M, Gunshor RL. 1990. In *20th Int. Conf. Phys. Semiconductors*, ed. EM Anastassakis, JD Joannopoulos, pp. 1381–84. Singapore: World Scientific
55. Kudlek G, Presser N, Gutowski J, Durbin S, Menke D, et al. 1990. *J. Cryst. Growth* 101: 667–72
56. Kudlek G, Presser N, Gutowski J. 1991. *Semicond. Sci. Technol.* 6: A83–89
57. Kudlek G, Gutowski J. 1992. *J. Lumin.* 52: 55–69
58. Bir GL, Butikov EI, Pikus GE. 1963. *J. Phys. Chem. Solids* 24: 1467–74
59. Merz JL, Kukimoto H, Nassau K, Shiever JW. 1972. *Phys. Rev. B* 6: 545–56
60. Rühle W, Klingenstein W. 1978. *Phys. Rev. B* 18: 7011–21
61. Stébé B, Munschy G. 1980. *Solid State Commun.* 35: 557–60
62. Finkman E, Sturge MD, Bhat R. 1986. *J. Lumin.* 35: 235–38
63. Schmidt M. 1977. *Phys. Status Solidi (b)* 79: 533–38
64. Skromme BJ, Tamargo MC, de Miguel JL, Nahory RE. 1988. *Appl. Phys. Lett.* 53: 2217–19
65. Dean PJ, Herbert DC, Werkhoven CJ, Fitzpatrick BJ, Bhargava RN. 1981. *Phys. Rev. B* 23: 4888–901
66. Isshiki M, Kyotani T, Masumoto K, Uchida W, Suto S. 1987. *Phys. Rev. B* 36: 2568–77
67. Ohkawa K, Ueno A, Mitsuyu T. 1992. *J. Cryst. Growth* 117: 375–84
68. Shahzad K, Cammack DA. 1990. *Appl. Phys. Lett.* 56: 180–82
69. Zhang Y, Skromme BJ, Shibli SM, Tamargo MC. 1993. *Phys. Rev. B* 48: 10885–92
70. Zhang Y, Skromme BJ, Cheng H. 1993. In *Semiconductor Heterostructures for Photonic and Electronic Applications*, ed. DC Houghton, CW Tu, RT Tung, pp. 567–72. Pittsburgh: Mater. Res. Soc.
71. Kuhn WS, Driad R, Stanzl H, Lusson A, Wolf K, et al. 1994. *J. Cryst. Growth* 138: 448–54
72. Skromme BJ, Zhang Y, Liu W, Parameshwaran B, Smith DJ, Sivananthan S. 1994. In *Growth, Processing, and Characterization of Semiconductor Heterostructures*, ed. G Gumbs, B Weiss, S Luryi, GW Wicks, pp. 15–20. Pittsburgh: Mater. Res. Soc.
73. Zhang Y, Skromme BJ, Turco FS. 1990. In *Electronic, Optical, and Device Properties of Layered Structures*, ed. J Hayes, MS Hybertsen, ER Weber, pp. 133–36. Pittsburgh: Mater. Res. Soc.
74. Stutius W. 1982. *Appl. Phys. Lett.* 40: 246–48
75. Dean PJ, Pitt AD, Skolnick MS, Wright PJ, Cockayne B. 1982. *J. Cryst. Growth* 59: 301–6
76. Yasuda T, Mitsuishi I, Kukimoto H. 1988. *Appl. Phys. Lett.* 52: 57–59
77. Yoshikawa A, Muto S, Yamaga S, Kasai H. 1988. *J. Cryst. Growth* 93: 697–702
78. Park RM, Mar HA, Salansky NM. 1985. *J. Appl. Phys.* 58: 1047–49
79. Mitsuyu T, Ohkawa K, Yamazaki O. 1986. *Appl. Phys. Lett.* 49: 1348–50
80. Cheng H, DePuydt JM, Potts JE, Smith TL. 1988. *Appl. Phys. Lett.* 52: 147–49
81. Haase MA, Cheng H, DePuydt JM, Potts JE. 1990. *J. Appl. Phys.* 67: 448–52
82. Zhang Y, Skromme BJ, Cheng H. 1993. *Phys. Rev. B* 47: 2107–21
83. Pohl UW, Kudlek GH, Klimakow A, Hoffmann A. 1994. *J. Cryst. Growth* 138: 385–90
84. Kamiya T, Wagner E. 1977. *J. Appl. Phys.* 48: 1928–34
85. Matsumoto T, Iijima T, Ishida T. 1988. *Jpn. J. Appl. Phys.* 27: L1942–44
86. Zhang Y, Liu W, Skromme BJ, Cheng H, Shibli SM, Tamargo MC. 1994. *J. Cryst. Growth* 138: 310–17
87. Skromme BJ, Stillman GE. 1984. *Phys. Rev. B* 29: 1982–92
88. Halsted RE, Aven M. 1965. *Phys. Rev. Lett.* 14: 64–65
89. Dean PJ, Fitzpatrick BJ, Bhargava RN. 1982. *Phys. Rev. B* 26: 2016–35

90. Eagles DM. 1960. *J. Phys. Chem. Solids* 16: 76–83
91. Skromme BJ. 1995. *J. Kor. Phys. Soc.* Submitted
92. Chen J, Zhang Y, Skromme BJ, Akimoto K. 1995. *J. Appl. Phys.* In press
93. Varshni YP. 1967. *Physica* 34: 149–54
94. O'Donnell KP, Chen X. 1991. *Appl. Phys. Lett.* 58: 2924–26
95. Shirakawa Y, Kukimoto H. 1980. *J. Appl. Phys.* 51: 2014–19
96. Bowers KA, Yu Z, Gossett KJ, Cook JW Jr, Schetzina JF. 1994. *J. Electron. Mater.* 23: 251–54
97. Tews H, Venghaus H. 1979. *Solid State Commun.* 30: 219–21
98. Tews H, Venghaus H, Dean PJ. 1979. *Phys. Rev. B* 19: 5178–84
99. Blanconnier P, Hogrel JF, Jean-Louis AM, Sermage B. 1981. *J. Appl. Phys.* 52: 6895–900
100. Yang Z, Bowers KA, Ren J, Lansari Y, Cook JW Jr, Schetzina JF. 1992. *Appl. Phys. Lett.* 61: 2671–73
101. Gossett KJ, Yu Z, Bowers KA, Ren J, Schetzina CA, Schetzina JF. 1994. *Optical properties of p-type nitrogen-plasma-doped ZnSe and ZnSSe thin films.* Presented at Int. Conf. on II-VI Compounds and Related Mater., 6th, Newport, RI. Pap. PB-34
102. Fan Y, Han J, He L, Saraie J, Gunshor RL, et al. 1993. *Appl. Phys. Lett.* 63: 1812–14
103. Yao T, Matsumoto T, Sasaki S, Chung CK, Zhu Z, Nishiyama F. 1994. *J. Cryst. Growth* 138: 290–94
104. Hauksson IS, Wang SY, Simpson J, Prior KA, Cavenett BC, et al. 1995. *Phys. Rev. B.* Submitted
105. Molva E, Magnea N. 1980. *Phys. Status Solidi (b)* 102: 475–86
106. Dean PJ, White AM. 1978. *Solid State Electron.* 21: 1351–55
107. Taskar NR, Khan BA, Dorman DR, Shahzad K. 1993. *Appl. Phys. Lett.* 62: 270–72
108. Wolk JA, Ager JW III, Duxstad KJ, Haller EE, Taskar NR, et al. 1993. *Appl. Phys. Lett.* 63: 2756–58
109. Kamata A, Mitsuhashi H, Fujita H. 1993. *Appl. Phys. Lett.* 63: 3353–54
110. Ando K, Kawaguchi Y, Ohno T, Ohki A, Zembutsu S. 1993. *Appl. Phys. Lett.* 63: 191–93
111. Hu B, Karczewski G, Luo H, Samarth N, Furdyna JK. 1993. *Appl. Phys. Lett.* 63: 358–60
112. Ohki A, Kawaguchi Y, Ando K, Katsui A. 1991. *Appl. Phys. Lett.* 59: 671–73
113. Kennedy TA, Glaser ER, Murdin BN, Pidgeon CR, Prior KA, Cavenett BC. 1994. *Appl. Phys. Lett.* 65: 1112–14
114. Hauksson IS, Simpson J, Wang SY, Prior KA, Cavenett BC. 1992. *Appl. Phys. Lett.* 61: 2208–10
115. Murdin BN, Cavenett BC, Pidgeon CR, Simpson J, Hauksson I, Prior KA. 1993. *Appl. Phys. Lett.* 63: 2411–13
116. Boyce PJ, Davies JJ, Wolverson D, Ohkawa K, Mitsuyu T. 1994. *Appl. Phys. Lett.* 65: 2063–65
117. Chen AL, Walukiewicz W, Haller EE. 1994. *Appl. Phys. Lett.* 65: 1006–8
118. Chadi DJ, Troullier N. 1993. *Physica B* 185: 128–31
119. Van de Walle CG, Laks DB, Neumark GF, Pantelides ST. 1993. *Phys. Rev. B* 47: 9425–34
120. Fan Y, Han J, He L, Gunshor RL, Brandt MS, et al. 1994. *Appl. Phys. Lett.* 65: 1001–3
121. Chadi DJ. 1994. *J. Cryst. Growth* 138: 295–300
122. DePuydt JM, Haase MA, Cheng H, Potts JE. 1989. *Appl. Phys. Lett.* 55: 1103–5
123. Akopyan IC, Novikov BV, Grigorev RV, Grigorev SR, Röppischer H, Stein N. 1990. *Phys. Status Solidi (a)* 119: K167–71
124. Dean PJ, Merz JL. 1969. *Phys. Rev.* 178: 1310–18
125. Reinberg AR, Holton WC, de Wit M, Watts RK. 1971. *Phys. Rev. B* 3: 410–16
126. Watts RK, Holton WC, de Wit M. 1971. *Phys. Rev. B* 3: 404–9
127. Hingerl K, Jantsch W, Juza P, Lang M, Sitter H, et al. 1992. *J. Cryst. Growth* 117: 341–47
128. Shibli SM, Tamargo MC, Skromme BJ, Schwarz SA, Schwartz CL, et al. 1990. *J. Vac. Sci. Technol. B* 8: 187–91
129. Okajima M, Kawachi M, Sato T, Hirahara K, Kamata A, Beppu T. 1986. In *Extended Abstracts of the 18th Conf. on Solid State Dev. and Mater.*, pp. 647–50. Tokyo: Business Center for Academic Society Japan
130. Fitzpatrick BJ, Werkhoven CJ, McGee TF III, Harnack PM, Herko SP, et al. 1981. *IEEE Trans. Electron. Dev.* ED-28: 440–44
131. Williams JO. 1991. In *Spectroscopy of Advanced Materials*, ed. RJH Clark, RE Hester, pp. 215–49. New York: Wiley
132. Yodo T, Yamashita K. 1988. *Appl. Phys. Lett.* 53: 2403–5
133. Davies JJ, Nicholls JE. 1979. *J. Lumin.* 18/19: 322–26
134. Nicholls JE, Davies JJ. 1979. *J. Phys. C* 12: 1917–26

135. Watts RK, Holton WC. 1970. *Phys. Rev. B* 2: 4882–4
136. Verity D, Davies JJ, Nicholls JE. 1981. *J. Phys. C* 14: 485–96
137. Nakano K, Boyce PJ, Davies JJ, Wolverson D. 1992. *J. Cryst. Growth* 117: 331–35
138. Kosai K, Fitzpatrick BJ, Grimmeiss HG, Bhargava RN, Neumark GF. 1979. *Appl. Phys. Lett.* 35: 194–96
139. Aoki M, Washiyama M, Nakamura H, Sakamoto K. 1982. *Jpn. J. Appl. Phys.* (Suppl. 21–1)21: 11–17
140. Yao T, Okada Y. 1986. *Jpn. J. Appl. Phys.* 25: 821–27
141. Kawakyu Y, Sasaki M, Uemoto T, Beepu T. 1987. In *Extended Abstracts of the 19th Conf. on Solid State Dev. and Mater.*, pp. 259–62. Tokyo: Business Center for Academic Society Japan
142. Bhargava RN. 1988. *J. Cryst. Growth* 86: 873–79
143. Akimoto K, Miyajima T, Mori Y. 1989. *Phys. Rev. B* 39: 3138–44
144. Akimoto K, Miyajima T, Mori Y. 1989. *Jpn. J. Appl. Phys.* 28: L531–34
145. Akimoto K, Okuyama H, Ikeda M, Mori Y. 1992. *Appl. Phys. Lett.* 60: 91–93
146. Ohki A, Kawaguchi Y, Ando K, Zembutsu S. 1992. *J. Cryst. Growth* 117: 102–6
147. Kulakov MP, Negriy VD. 1990. *J. Cryst. Growth* 101: 525–27
148. Ren J, Bowers KA, Hwang S, Cook JW Jr, Schetzina JF. 1991. *J. Cryst. Growth* 111: 772–75
149. Shahzad K, Jones KS, Lowen PD, Park RM. 1991. *Phys. Rev. B* 43: 9247–50
150. Han J, Ringle MD, Fan Y, Gunshor RL, Nurmikko AV. 1994. *Appl. Phys. Lett.* 65: 3230–32
151. Giapis KP, Jensen KF. 1990. *J. Cryst. Growth* 101: 111–17
152. Giapis KP, Jensen KF. 1989. *Appl. Phys. Lett.* 55: 463–65
153. Potts JE, Cheng H, DePuydt JM, Haase MA. 1990. *J. Cryst. Growth* 101: 425–29
154. Cheng H, DePuydt JM, Potts JE, Haase MA. 1989. *J. Cryst. Growth* 95: 512–16
155. Yao T, Taguchi T. 1985. In *Proc. 13th Int. Conf. Defects in Semiconductors*, ed. LC Kimerling, JM Parsey Jr, pp. 1221–27. Warrendale, PA: Metallurg. Soc. of AIME
156. Bhargava RN, Seymour RJ, Fitzpatrick BJ, Herko SP. 1979. *Phys. Rev. B* 20: 2407–19
157. Swaminathan V, Greene LC. 1976. *Phys. Rev. B* 14: 5351–62
158. Zhang Y, Skromme BJ, Turco-Sandroff FS. 1992. *Phys. Rev. B* 46: 3872–85

Annu. Rev. Mater. Sci. 1995. 25: 647–78

METAL-OXIDE HETEROSTRUCTURES

R. Ramesh[1] *and V. G. Keramidas*

Bellcore, Red Bank, New Jersey 07701

KEY WORDS: high-temperature superconductivity, pulsed laser deposition, ferroelectric heterostructures

---

ABSTRACT

This review provides an introduction to the idea of metal-oxide heterostructures. The potential application of metal-oxide heterostructures is demonstrated with the case study of ferroelectric nonvolatile memories, in which the ferroelectric stack contains a multicomponent metal oxide. The advantages of using metallic oxides as contact electrodes compared with conventional metal electrodes are demonstrated through results of reliability tests. Some novel ideas for the integration of multicomponent metallic oxides with conventional semiconductor electronics are also presented. Finally, a short discussion of the evolution of the general field of metal-oxide heterostructures is provided.

INTRODUCTION

The inherent complexity of the crystal structure and microstructure of modern high technology materials and the interdependence of the properties on microstructure and processing mean that detailed characterization of these inter-relationships is crucial. Thus in order to improve the performance, ensure reliable operation, and explore the limits of the technologies involved, it is imperative to establish these structure- and property-processing inter-relationships iteratively. This philosophy is germane for the emerging field of electronic oxide thin films and heterostructures.

[1] Current address: Department of Materials and Nuclear Engineering, and Center for Superconductivity Research, University of Maryland, College Park, Maryland 20742

0084–6600/95/0801–0647$05.00

The surface of the earth is almost completely comprised of oxides, which we use in various forms and levels of technological sophistication. The discovery of high-temperature superconductivity in cuprate materials has in some sense rejuvenated scientific and technological interest in oxides, especially those useful in high technology, value-added applications. Consistent with the trend towards miniaturization and the fabrication of artificially engineered heterostructures in semiconductor materials, the broad range of physical properties available in oxides is now beginning to be exploited in thin-film form.

The past few years have witnessed some dramatic advances in metal-oxide thin-film technologies, spurred by the discovery of superconductivity at or above liquid nitrogen temperatures in perovskite copper oxides (1). A variety of technologies are being employed to deposit the thin-film materials. The intrinsic properties of the high-temperature cuprate superconductors require a high degree of crystalline quality and stoichiometry in the films; consequently, epitaxial growth has been preferred. As illustrated in Figure 1, the inherent similarity in crystal structure, chemistry, and lattice dimensions of the cuprate superconductors with other perovskite or perovskite-type materials has also led to the possibility for the growth of metal-oxide heterostructures incorporating one or more of these oxides. They are the structural analogues of compound semiconductor heterostructures and are built upon the face-centred cubic oxygen sublattice. For example, heterostructures and superlattices consisting of the cuprate superconductor such as Y-Ba-Cu-O, and chemically substituted derivatives such as Pr-Ba-Cu-O, have been grown as possible starting materials for S-N-S- or S-I-S-type junction applications (2). The metallic properties of the cuprate superconductors have also been used to fabricate metal-ferroelectric-metal capacitor heterostructures for nonvolatile memory applications (3a). These heterostructures can also be used for integrated microwave applications because both the ferroelectric materials and the cuprate superconductors possess properties useful in such applications (4). The degree of structural perfection and chemical homogeneity that is required of these cuprate superconductors (with relatively short coherence lengths) has led to the rapid development of thin-film deposition technologies such as pulsed laser deposition (PLD), off-axis sputtering, etc that are capable of providing high-quality films. Consequently, there has been rapid progress in most of these areas in understanding the key issues involved in the growth, testing, and processing of the relevant device heterostructure.

Developments in the growth of high-quality, cuprate superconductor oxide thin films have influenced, to a large extent, the development of other microelectronic and related high technology areas that involve metal-

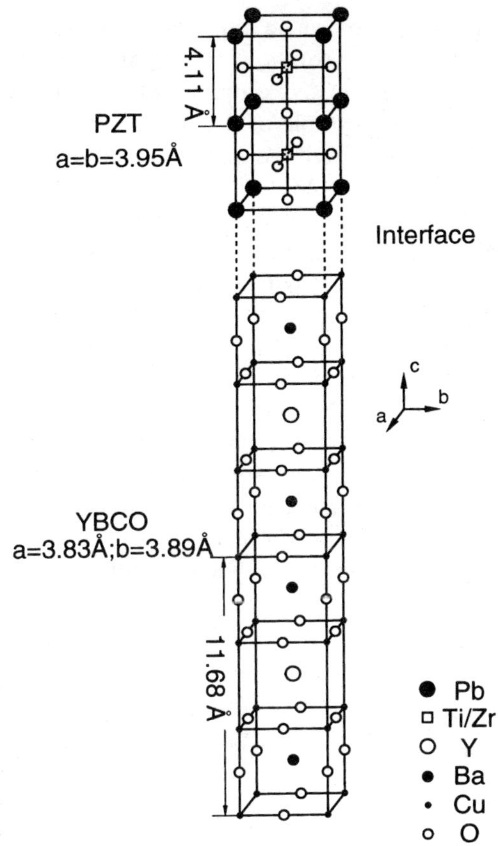

Figure 1 A schematic illustration of the intrinsic similarity in crystal structure and chemistry between two types of perovskite oxides, namely the cuprate superconductors and the ferroelectric oxides. Similar examples can be found for metal oxides with the spinel, garnet, or layered crystal structures.

oxide thin films and heterostructures as the active medium. The most important feature is the wide range of physical properties available in metal oxides, ranging from dielectrics to superconductors—diamagnetics to ferromagnetic materials (5). An excellent example of these properties is found in ferroelectric perovskite thin films, which are actively being considered as the storage medium in nonvolatile, solid-state memories (6a–c, 7a,b). Depending upon the type of ferroelectric material, i.e. switchable material such as lead zirconate titanate (PZT) and cationically substituted derivatives (PLZT), or a perovskite that is used in the paraelectric state such as barium strontium titanate (BST), these thin films can be used either for nonvolatile storage [ferroelectric nonvolatile random access memories (FRAM)], using PZT as the storage element, or for volatile storage in a conventional dynamic random access memory (DRAM), using BST as the storage element. Another example of the use of high-quality metal-oxide thin-film heterostructures is in cuprate superconductor/ferroelectric heterostructures in microwave circuits (4). Similarly, the magnetic, magneto-optic, and microwave properties of garnets (a typical example is yttrium iron garnet $Y_3Fe_5O_{12}$) and cationically substituted derivatives can be used to create thin-film heterostructures (Figure 2) for magnetic and magneto-optic memories, microwave filters, isolators, etc. Indeed, the wide range of physical properties available in metal oxides suggests that new and novel devices and applications are only limited by our imagination and our ability to fabricate these thin films with the desired quality. In order to maintain focus within this review, we have picked the ferroelectric metal oxide as a prototypical example. This choice was prompted by several reasons. Firstly, ferroelectric thin films are very likely to play an important role in future generation memories, both volatile and nonvolatile, and there is considerable worldwide research and development effort

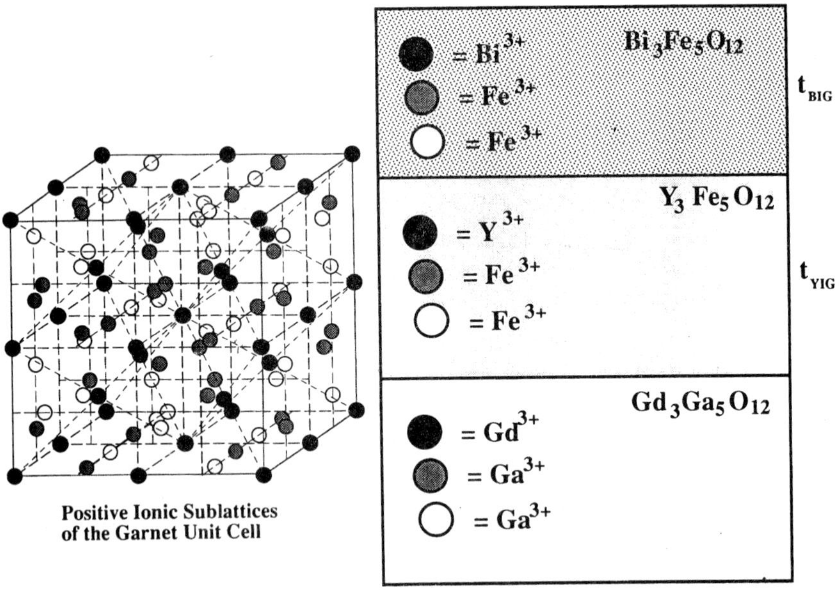

Figure 2 A schematic illustration of the possibility of growing oxide heterostructures within the magnetic and magneto-optic garnet family using yttrium iron garnet (YIG) and bismuth iron garnet (BIG) as the prototypical examples and gadolinium gallium garnet (GGG) as the single-crystal substrate.

surrounding these heterostructures. Secondly, this is a case study of a problem in which integration with conventional semiconductor electronics is of paramount importance; consequently, methods to grow these complicated multicomponent metal oxides on a substrate such as Si have to be discovered. This is also an example in which metal-oxide heterostructures have dramatically improved ferroelectric properties compared with capacitors grown on the conventional Pt metal. Finally, ferroelectric thin films comprise an area that the authors are actively engaged in. Consequently, we are acutely aware of the problems and pitfalls involving the integration of metal oxides on semiconductors.

Ferroelectric Metal-Oxide Heterostructures for Nonvolatile Memories

A strong research and development effort is currently directed at producing a commercially viable, solid-state, nonvolatile (FRAM) technology. Many laboratories are focused on integrating submicron thin ferroelectric capacitors of, for example, (PZT), with the mature silicon-based transistor technology to yield capacitor-transistor memory architectures (6a–c, 7a, 8–11a, 12), as schematically illustrated in Figure 3. This figure also shows the

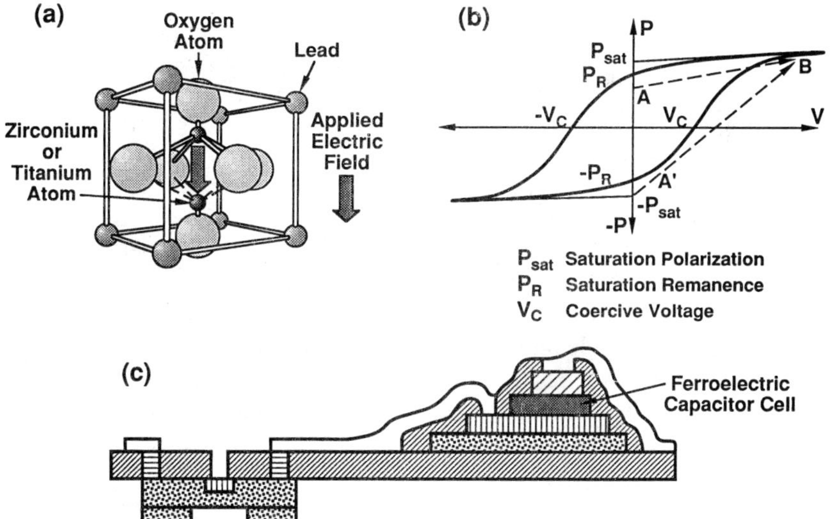

Figure 3 A schematic view of a ferroelectric capacitor integrated on a transistorized silicon wafer to create a 1-pass gate transistor, 1 capacitor nonvolatile memory element. This figure also shows (*a*) the unit cell of the perovskite PZT ferroelectric in which the displacement of the body-centered cation leads to the two logic states (*b*) and the different types of interfaces that can arise during the growth and fabrication of such integrated structures (*c*).

perovskite unit cell of the ferroelectric PZT and the variety of structural, chemical, electronic, and ionic interfaces that arise during the fabrication and integration of these metal-oxide heterostructures on a Si substrate. Realization of a commercially viable FRAM technology has been hampered by one or more problems related either to the reliable performance of the PZT ferroelectric capacitor or to the growth and processing of capacitors that translate to high-density memory elements. The electrode-ferroelectric interface is crucial in determining the reliability characteristics, such as fatigue resistance (which is the stability of switchable polarization when subjected to repeated bipolar pulses) of the ferroelectric capacitor. Similarly, internal interfaces in the ferroelectric, such as domain walls and grain boundaries, and microstructural defects, such as oxygen vacancies, influence the ferroelectric properties. In another approach, a hithertofore undisclosed ferroelectric material termed as Y1 has also yielded capacitor structures with desirable ferroelectric properties. Because the exact nature of this material remains undisclosed, we do not discuss or compare the properties of the ferroelectric heterostructures presented in this review with those of the Y1 material.

In order to integrate these ferroelectric and other metal-oxide heterostructures into a commercially viable memory technology, several aspects have to be considered. Firstly, a scalable deposition process (i.e. ability to grow on 6-inch wafers) such as chemical vapor deposition (CVD) or sputtering that also provides the ability to grow the thin films on nonplanar surfaces (i.e. conformal growth) is essential. Secondly, a reliable etching technology has to be available to fabricate the memory elements of the desired size. Finally, the memory elements need to function in a reliable fashion over conventionally accepted performance conditions of write and read voltages, usage temperature, etc. Ultimately all these requirements have to be met in a timely fashion (i.e. with respect to other competing technologies) and also in a cost-effective and competitive manner.

We begin this review by addressing two fundamental issues: (*a*) the use of alternative electrode materials to contact the ferroelectric layer, e.g. the use of metallic oxide electrodes in place of Pt; and (*b*) the question of integration of these multicomponent metal oxides with Si-CMOS in general, with progression to the aspects of growth, device fabrication, and finally reliability testing of the prototype capacitor structures. Direct growth of the lead-based compounds on Si (or SiO_2/Si) at high temperatures leads to the formation of stable binary silicates such as lead silicate. This is generally true if the film contains any of the highly reactive Group I or II cations (such as Ba, Ca, K, etc). Consequently, integration of such metal oxides on Si requires novel approaches to provide chemical isolation of the overlayers from the substrate and also to yield the desired crystallographic phase and orientation in the film. In our research program, we address these two issues, related intimately to the engineering of interfaces in the capacitor stack and to the Si wafer, through two novel approaches. We use perovskite metallic oxides as top and bottom electrodes instead of Pt; this, under optimized processing conditions, yields capacitors that show very little bipolar fatigue at least up to 10^{12} cycles. Two different types of oxide electrodes have been discussed, namely, the perovskites such as Y-Ba-Cu-O (YBCO) (3) and La-Sr-Co-O (LSCO) (13a, 14, 15) and the metallic oxides with the rutile structure such as RuO_2 (16, 17a). These perovskite oxides have room-temperature resistivities of the order of a few hundred $\mu\Omega$ cm (for example, epitaxial LSCO has a resistivity of 90 $\mu\Omega$ cm at room temperature). In comparison, Pt has a resistivity in the range of 5 to 20 $\mu\Omega$ cm at room temperature. In an attempt to lower the effective sheet resistance of the bottom electrode while maintaining the good fatigue characteristics of the LSCO-based capacitors, the growth of the LSCO/PLZT/LSCO capacitor stack on epitaxial or highly [001]-oriented Pt (18) or even polycrystalline Pt (19) has been used. The integration of highly oriented multicomponent oxides on Si can be accomplished using an approach that involves the growth of a thin layer or an anisotropic perovskite template layer such as bismuth titanate (BTO) on the thermally oxidized Si wafer prior to the deposition of the Pt/LSCO/PLZT/LSCO capacitor stack (20). This BTO template layer grows with a preferred c-axis texture, even on an amorphous surface such as SiO_2/Si. It not only provides a good chemical barrier between the overlayers and the Si wafer but also provides a perovskite-like surface (because it is c-axis oriented, i.e. the a-b plane is exposed) for the subsequent growth of the cubic perovskite layers. The fact that this template layer itself has a perovskite structure and crystal chemistry is exploited to nucleate the [001] orientation of the perovskite phase preferentially over competing orientations of the perovskite phase and over competing phases such as the nonferroelectric pyrochlore phase. Figure 4 schematically illustrates both these concepts.

Notwithstanding the earlier caveat that a scalable deposition technology is required for commercial implementation of these metal-oxide capacitors, these preliminary experiments are valuable to the rapid exploration of the viability of materials combinations, the consequences of the complicated interface chemistry in these heterostructures, and the effect of various process and microstructural variables on the ferroelectric's performance. Such experiments can be carried out with relative ease by pulsed-laser deposition (PLD). Having said this, it is imperative to outline the attributes of this deposition process and also to explore its capabilities and limitations. Figure 5 schematically depicts the PLD process for the growth

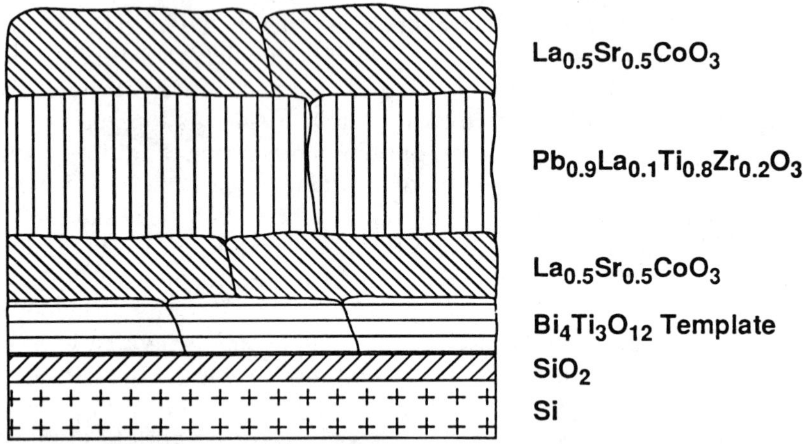

Figure 4 A schematic description of the use of layered perovskites such as bismuth titanate as a structural and chemical template and the use of oriented cubic metallic oxides such as lanthanum strontium cobalt oxide as top and bottom electrodes for the ferroelectric capacitor heterostructure.

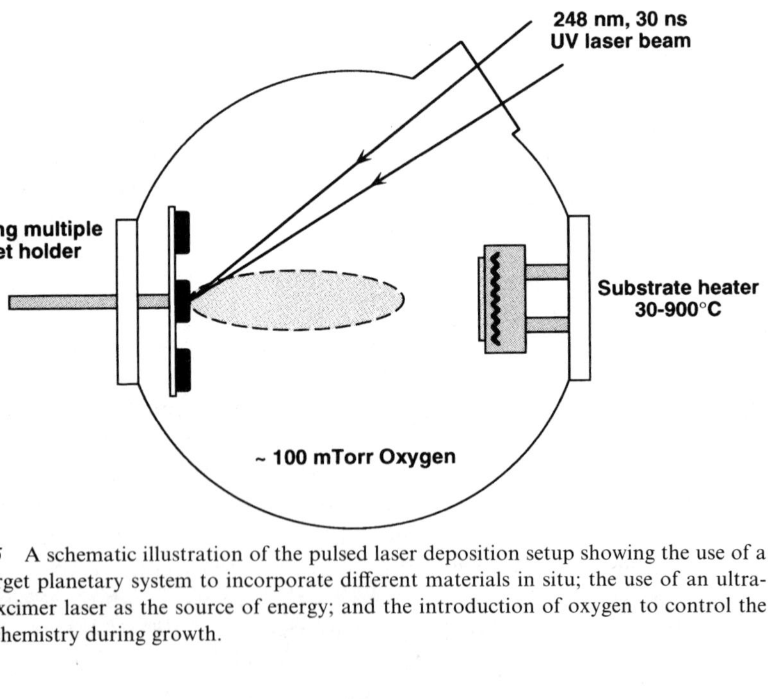

Figure 5 A schematic illustration of the pulsed laser deposition setup showing the use of a multitarget planetary system to incorporate different materials in situ; the use of an ultraviolet excimer laser as the source of energy; and the introduction of oxygen to control the plume chemistry during growth.

of the metal-oxide heterostructures. PLD has achieved vast popularity, primarily in the oxide thin-film deposition community, through the demonstration of the growth of very high-quality thin films and heterostructures (21, 22). Thin films of the desired material and composition are deposited by the ablation of material from this target's surface by a short laser pulse. Typically, an ultra-violet (e.g. 248 nm) excimer laser with a pulse width of a few nanoseconds is used for this purpose, although other types of lasers (for example, a frequency tripled YAG laser) have also been used. The instantaneous interaction of the pulsed-laser beam with the target produces a plume of material that is forward directed and is transported towards the heated substrate that is placed directly in the line of this plume. The contents of this plume are complicated and are not described in this review. Under appropriate deposition conditions, the composition of the thin film is the same as that of the target. Herein lies the intrinsic advantage of PLD, in comparison with other techniques such as sputtering, which require considerable enrichment in the target of the more volatile species such as Pb. This compositional congruency is dependent on the substrate temperature, the interface chemistry between the substrate surface and the film, and the oxygen partial pressure during deposition. Another advantage of PLD is the fact that multiple targets can be loaded inside the chamber before deposition using a planetary target loading system, thereby enabling the growth of heterostructures and superlattices with relatively clean interfaces, i.e. in situ growth is available. The heterostructures can be cooled in a variety of ambients after the deposition is completed. Finally, other advantages include a rapid throughput of films that is useful in process optimization, and the ability to include in situ processing, such as etching, through the use of a load-lock and etching chamber in conjunction with the deposition chamber.

It is also important to point out the deficiencies of PLD, or aspects that have not been demonstrated by PLD. The first is that uniform coating of large-area wafers (for example, 4–6-inch diameter) has not been routinely demonstrated. To achieve this uniformity will require not only the uniform heating of such large wafers, but also the deposition of material with uniform thickness and composition over the whole wafer. The second issue that remains to be demonstrated is the ability to carry out conformal growth. Finally, films with complete absence of particulates have yet to be demonstrated, although the particulate density and size distribution can be reduced considerably through the use of high-density targets.

With this general introduction, we now describe some of the key advances made in this field through the use of the above approaches, namely the use of anisotropic perovskite templates and metallic-oxide electrodes to influence the phase stability and orientation distribution as well as the ferroelectric properties of the test capacitors. We illustrate with examples primarily from the research currently in progress at Bellcore, although most of the conclusions are quite relevant to those obtained from other research laboratories. Wherever appropriate, results from other research groups are also discussed. Details of the laser deposition conditions are available in earlier papers (22). In this review, we present only general results, conclusions, discussions, and prognostications for future research and development work. For the sake of completeness, we present results on one specific heterostructure system, namely the LSCO/PLZT/LSCO capacitor stack, because this system is not only interesting from the fundamental oxide heterostructure growth point of view, but also has a strong chance of being used in a commercial memory technology in the future. The PLZT compositions typically being focused on in our program include $Pb_9La_1Zr_2Ti_8O_3$ (10/20/80), $Pb_{1.0}Z_{r.2}Ti_{.8}O_3$ (0/20/80), and $PbZr_{.5}Ti_{.5}O_3$ (0/50/50), although the whole range of compositions in the PLZT phase diagram can be and have been grown by PLD.

STRUCTURE AND MICROSTRUCTURE

Both LSCO and PLZT are cubic at the growth temperature, which is typically in the range of 550 to 670°C, depending upon the substrate. PLZT undergoes a cubic-tetragonal transformation upon cooling to room temperature. Most of our preliminary experiments have been carried out using [100]-oriented single-crystal $LaAlO_3$ substrates because they are closely lattice matched to both LSCO and PLZT (within a few percent). Very high-quality heterostructures have been grown on these substrates, as evidenced by X-ray diffraction pole figure scans (discussed below) and electron microscopy studies. X-ray Bragg scans of the LSCO/PLZT show that the PLZT layer is typically c-axis oriented. The degree of alignment of the c-axis depends on the PLZT composition and the postdeposition cooling conditions. PLZT(10/20/80) compositions yield only c-axis-oriented films, whereas the 0/20/80 composition has a stronger propensity to yield a mixture of c-axis- and a-axis- (i.e. c-axis in the substrate plane) oriented films. The volume fraction of the film with the c-axis normal to the substrate increases as the cooling rate from deposition temperature is faster.

Transmission electron microscopy (TEM) studies of the films grown on $LaAlO_3$ substrates confirm the X-ray diffraction results with respect to their crystalline quality (23). Planar sections reveal the presence of only low angle grain boundaries and dislocations. Ninety degree Domain walls have been detected in 0/20/80 films, both by X-ray diffraction (i.e. the presence of a-axis-oriented regions) and by diffraction contrast electron microscopy. Figure 6 shows one such example of 90°C domains observed in a planar section micrograph. These domains appear to nucleate at the interface between the electrode and the ferroelectric film. The growth crystallography of these films on $LaAlO_3$ is the cause for the 45° angle these domain walls make with the substrate normal. On the other hand, 10/20/80 films have few, if any, 90° domains. This observation is consistent with X-ray diffraction data that show a completely c-axis-oriented film. This difference can be understood in terms of two facts: (*a*) The c-axis/a-axis lattice parameter ratio for the 10/20/80 films is much smaller, and the a-axis lattice parameter is much better matched to the LSCO layer; (*b*) the saturation polarization value is also much smaller (and consequently the depolarizing field is also much smaller). In films with the 0/20/80 composition, the tendency to form 90° domains is primarily driven by the necessity to lower the depolarizing field, which can happen if the polarization direction is in-plane; however, this is opposed by the larger mismatch of the c-axis lattice parameter with that of LSCO or the $LaAlO_3$ substrate.

Growth on either YSZ-buffered Si or SiO_2/Si is accomplished through the use of a BTO template layer. The efficacy of this layer in controlling the phase stability and PLZT layer orientation can be established from X-

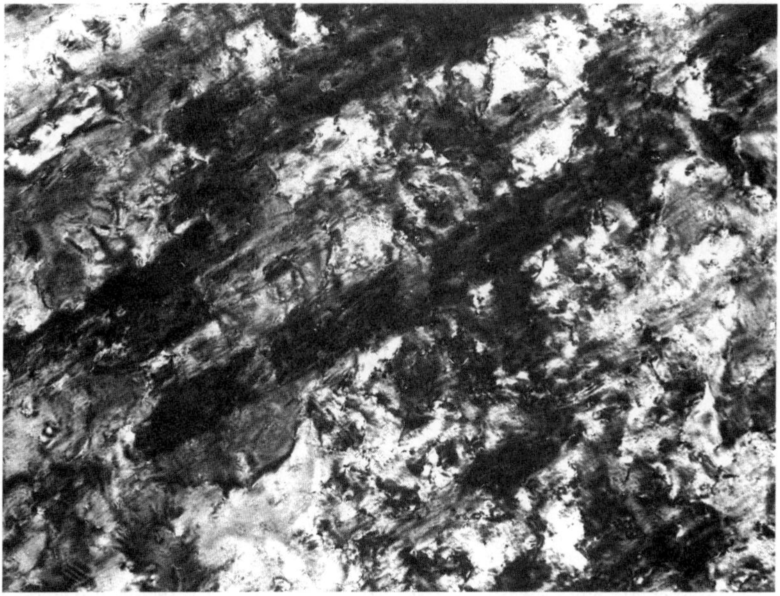

Figure 6 A planar section transmission electron micrograph that shows the formation of 90°C domains in a PZT film denoted by the composition 0/20/80.

ray diffraction patterns. Figure 7 shows the effect of using the BTO template layer on the crystalline quality of the ferroelectric stack grown on [100] YSZ/Si substrates. In Figure 7*a*, the X-ray diffraction patterns show

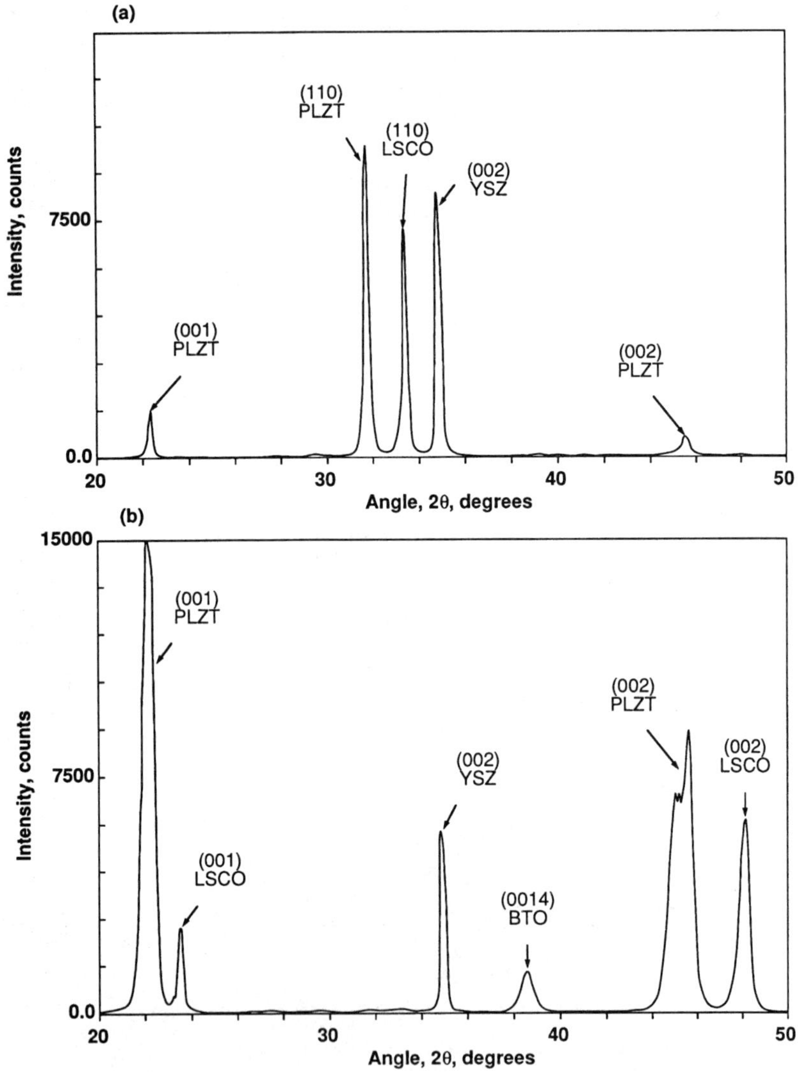

Figure 7 X-ray Bragg scans demonstrating the effect of the bismuth titanate template layer on the film orientation in the case of a LSCO/PLZT/LSCO heterostructure grown on YSZ-buffered [100] Si. Without the template layer (*a*), the film is polycrystalline with a preferred [110] texture, with a thin template layer (*b*), it is very strongly [001] oriented.

the existence of predominant [110] PLZT and LSCO peaks along with the [002] YSZ peaks, although one might expect to see the [001] peaks instead. The reason for this is the intrinsic mismatch between the LSCO (a = 3.83 Å) and YSZ ($a/\sqrt{2}$ = 3.64 Å; mismatch of 4.7°) lattice parameters. This mismatch, in conjunction with the inherent lack of structural anisotropy in the cubic perovskites, leads to a situation where the growth of the LSCO layer is predicted by the minimization of the growth surface energy. In general, this is accomplished by the stacking of the planes with the highest atomic density (i.e. the close-packed planes). In the case of the perovskites, these are the (110)-type planes, and consequently the films have a predominant [110] surface normal texture, although weak [001] and [002] diffraction peaks are also observed in this scan. The BTO template layer dramatically alters this because of its intrinsic structural anisotropy. When BTO is grown at the appropriate substrate temperatures, there is a strong preference to grow with the *001+ direction normal to the substrate, i.e. the a-b plane is the lowest energy surface. A thin layer, sufficient to cover the YSZ substrate surface, can provide a perovskite-like surface for the subsequent growth of the cubic layers. The distinct effect of such a template layer on the orientation of the LSCO/PLZT/LSCO ferroelectric heterostructure is demonstrated in the X-ray scan in Figure 7*b*. The complete absence of the [110] diffraction peaks is to be noted. This is not unique to BTO; indeed, earlier research on high-temperature superconductors (such as YBCO) had already demonstrated the growth of high-quality c-axis-oriented YBCO thin films on the same YSZ/Si substrate. Although the difference in structural quality may be of fundamental interest for researchers investigating growth mechanisms, the films, with and without the template layer, show a dramatic difference in ferroelectric properties (see below).

TEM studies show that the LSCO and the PLZT layers in the heterostructures grown on $LaAlO_3$ and YSZ/Si substrates are highly c-axis-oriented with very little, if any, large-angle grain boundaries (24). Figure 8*a*, a micrograph of all the layers in this complicated heterostructure, shows that the BTO layer is c-axis oriented and has a high density of stacking defects, a phenomenon well-known in these layered perovskites. High-resolution images of the LSCO/PLZT interface, Figure 8*b*, show no indication of possible interfacial reaction. The LSCO/PLZT interface is highly defective, primarily due to interface dislocations as a consequence of the lattice mismatch. Furthermore, the possibility for localized cationic interdiffusion cannot be ruled out and is a topic that requires further attention.

A better demonstration of the effect of the crystallographic anisotropy in BTO is the growth on SiO_2/Si (20). The thermal oxide on Si is important

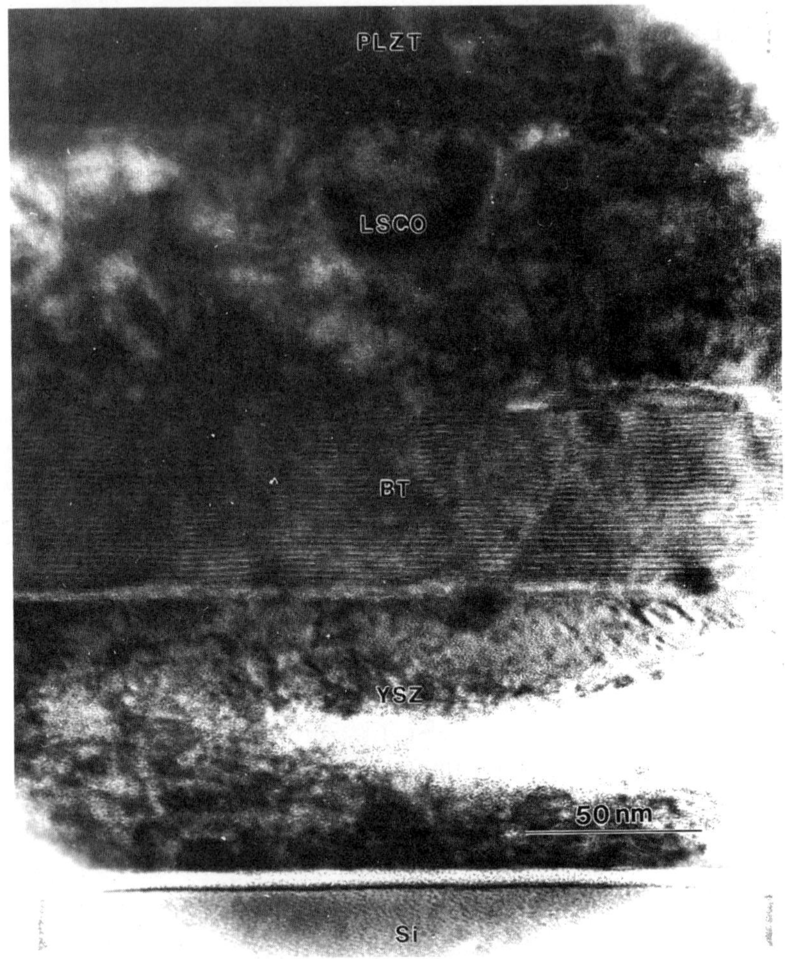

Figure 8 A transmission electron micrograph of the LSCO/ZPLZT/LSCO/BTO/YSZ/Si heterostructure showing that all the layers are single crystalline, although very defective.

because it forms the basis for fabricating the pass-gate transistors in the Si-CMOS wafer. At first glance, one imagines that BTO would be completely polycrystalline when grown on the amorphous surface; however, this is not the case. At the appropriate growth temperatures, it grows completely c-axis-oriented, although the film has very little crystallographic long-range correlation in the plane. However, for subsequent growth of the LSCO/PLZT/LSCO stack, this layer provides a sufficient perovskite template and promotes the growth of this phase in the [001] orientation, as illustrated in the X-ray diffraction pattern in Figure 9. Rocking measurements about the [002] PLZT peak show a width of about 1.0–1.2°, which is larger than that obtained on $LaAlO_3$ and YSZ/Si. The efficacy of this BTO template layer to control the orientation of the overlayer is further demonstrated by the fact that even metals such as Pt (fcc; a = 3.92 Å) can be grown with a specific orientation. Figure 10 shows a Bragg scan from a heterostructure that is similar to that schematically illustrated in Figure 4 but has about 800 Å of Pt grown on the BTO template layer before the growth of the LSCO/PLZT/LSCO ferroelectric stack. Note that only the [002] Pt peak is observed with a very strong intensity. The inset to this figure compares the Bragg scan of a Pt layer grown without the BTO template layer, indicating that it is weakly [111] oriented. X-ray pole figure scans of the LSCO/PLZT/LSCO heterostructures grown on the three types of substrates discussed here are shown in Figure 11. In this experiment, the pole figure is obtained about the [101] PLZT diffraction peak. In the case of films grown on $LaAlO_3$ and YSZ/Si, we observe the strong fourfold intensity distribution, indicating that there is a significant degree of orientational locking of the in-plane [010] and [100] directions on these substrates. On the other hand, the films grown on SiO_2/Si appear to be completely random in-plane. The effect of these

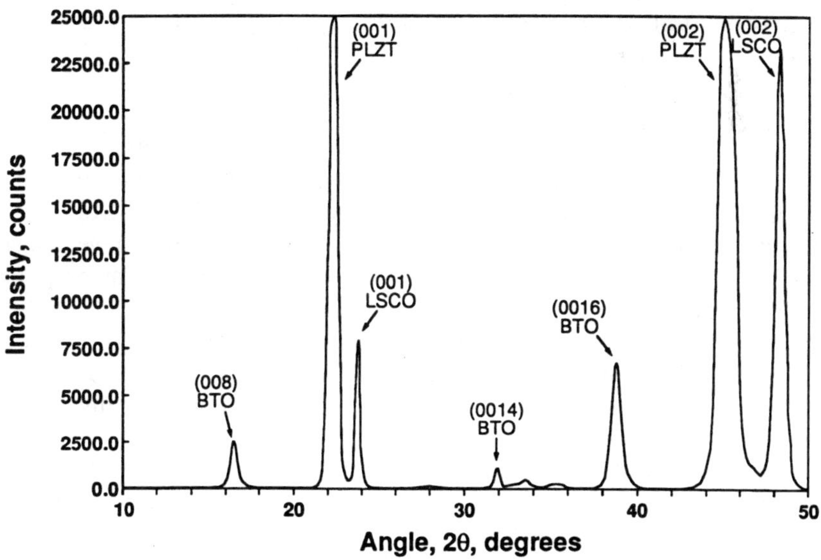

Figure 9 An X-ray Bragg scan of a LSCO/PLZT/LSCO heterostructure grown on a SiO_2/Si substrate with the bismuth titanate template layer, showing the efficacy of this template layer to induce the [001] orientation in the ferroelectric capacitor stack.

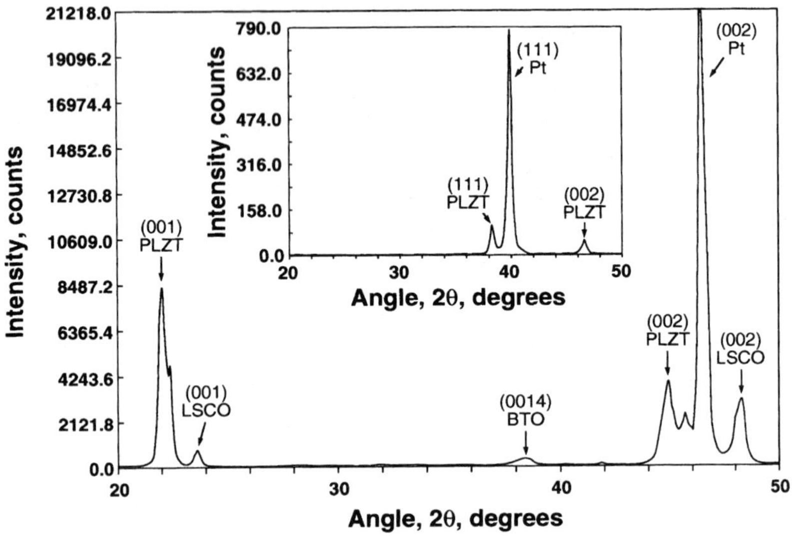

Figure 10 An X-ray Bragg scan from a heterostructure that is similar to that shown in Figure 4, but having an additional Pt layer grown on top of the bismuth titanate template layer. The Pt layer is used to reduce the overall sheet resistance of the bottom electrode. The template layer also influences the orientation of the Pt layer, which is strongly [002] oriented.

microstructural differences on the ferroelectric properties is discussed below.

FERROELECTRIC PROPERTIES

Notwithstanding the ability to grow high-quality metal-oxide thin films, the main reason there is considerable interest in these heterostructures is that they have highly desirable ferroelectric properties. Herein we discuss only the salient features and some aspects idiosyncratic to films prepared by PLD. Through a combination of photolithographic lift-off and either wet etching or dry ion-beam milling, we have been able to successfully define capacitors in the range of 600 to 2 μm in diameter. Ferroelectric measurements in these capacitors are primarily by pulsed testing so that the results are reasonably relevant to the memory operation. Figure 12 schematically illustrates the standard set of pulses that are used to measure the pulsed polarization values. The first pulse in this set of five pulses is applied to the test capacitor to polarize it into the negatively poled state, i.e. $-P_r$ state. The second pulse, which is in the positive direction, is termed a positive-switching pulse and measures the polarization obtained from the capacitor upon switching it from the $-P_r$ state to the positively poled

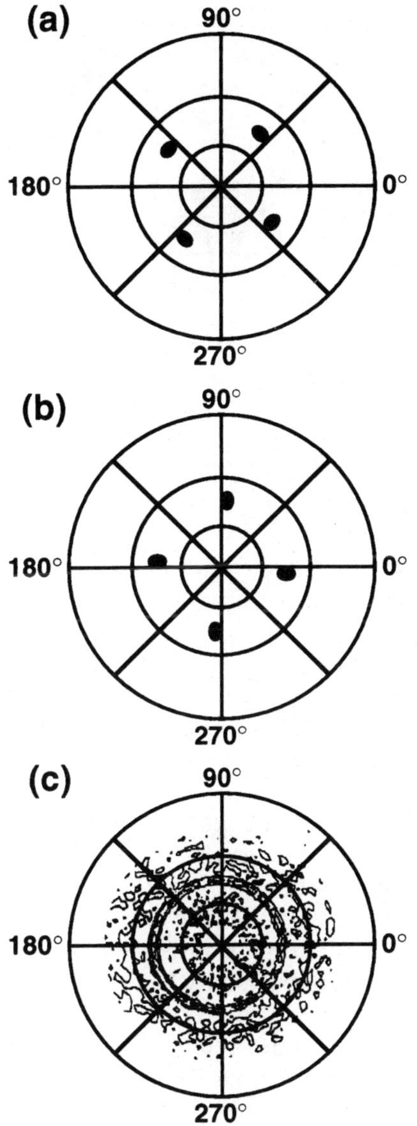

Figure 11 X-ray pole figure scans from LSCO/PLZT/LSCO heterostructures grown on three types of substrates: (*a*) on [001] $LaAlO_3$; (*b*) on BTO/YSZ/Si; (*c*) on BTO/SiO_2/Si. Note that whereas the films show a strong degree of in-plane orientational locking on the first two types of substrates, on the thermally oxidized Si substrate there is no in-plane structural coherence.

state; the polarization value is termed as P*. The second positive pulse measures the nonswitched polarization; this is a feature that is observed only during pulsed testing of capacitors and is highly relevant for a real memory operation. This polarization value is termed P^. The difference between P* and P^ is the actual switched polarization and is twice the

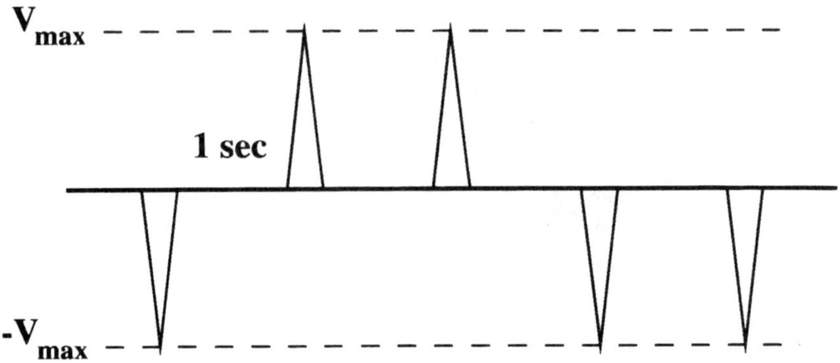

Figure 12 A schematic illustration of the standard set of pulses that are used to test the ferroelectric properties of the capacitor heterostructures.

remanent polarization that is obtained from a conventional Sawyer-Tower hysteresis loop. The next two pulses in the negative direction yield $-P^*$ and $-P^\wedge$ values. This set of pulses provides all the information about the hysteresis loop that is relevant for a memory operation.

Figure 13 shows a family of pulsed hysteresis loops from an LSCO/PLZT/LSCO heterostructure grown on SiO_2/Si with the BTO template and a Pt under-layer. In the PLZT-oriented films with the 10/20/80 composition, we have been able to obtain pulsed remanent polarization values in the range of 15 to 20 μC cm^2 at room temperature, with an applied field of 5 V. The coercive field, which is also an important memory element design variable, is typically in the range of 0.7 to 1.5 V in these heterostructures. The film resistivity is typically in the range of 5×10^9 to 5×10^{10} Ω cm. Ferroelectric hysteresis properties for films grown on YSZ/Si are similar to properties of films grown on $LaAlO_3$. We note that these values are more than sufficient for nonvolatile memory operation if they satisfy the reliability requirements. The film crystalline quality has an important effect on the hysteresis properties, as illustrated in Figure 14. The three hysteresis loops shown in this figure correspond to the three pole figure scans in Figure 11. The important correlation that arises from this figure is that as the film becomes more and more polycrystalline (i.e. the degree of in-plane registry becomes poorer), the polarization value is correspondingly lower. As is illustrated below, this decrease in polarization is not detrimental to the reliable performance of the capacitor. In addition, the ∂P values obtained from capacitors grown on SiO_2/Si are more than sufficient for nonvolatile memory operation, which typically require about 2 μC cm^2 of switched charge under the whole range of operating conditions.

Some interesting correlations of the symmetry and nature of the hys-

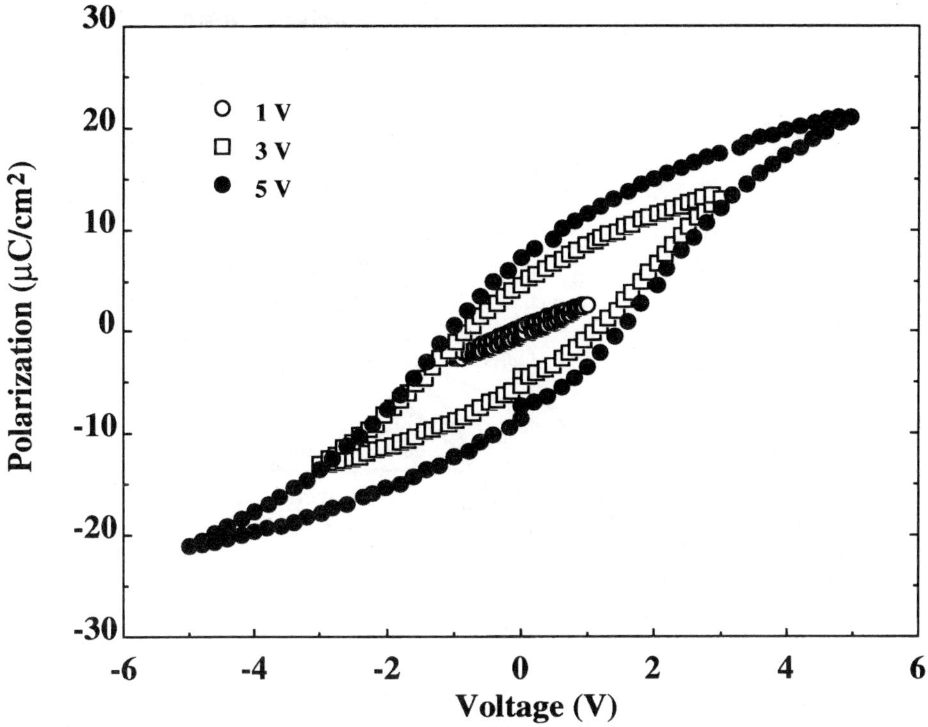

Figure 13 A family of pulsed hysteresis loops obtained from a LSCO/PLZT/LSCO capacitor grown on a Pt/BTO/SiO_2/Si substrate. The hysteresis loops were obtained at different applied fields.

teresis loops as a function of the deposition conditions have been observed. For example, in an earlier publication (25), we demonstrated that the postdeposition cooling rate impacts the polarization values. More recently, we have observed that the symmetry of the hysteresis loop is sensitive to the ambient in which the heterostructure is cooled to room temperature. This is especially obvious for LSCO/PLZT/LSCO heterostructures grown on single-crystal $LaAlO_3$ substrates. Figure 15 shows hysteresis loops from three films that were grown under identical conditions, i.e. at a substrate temperature of 640°C and in an oxygen ambient of 100 mtorr. However, they were cooled in oxygen ambients ranging from 10^{-5} to 760 torr. The capacitors cooled in 760 torr show symmetric hysteresis loops, and this is our optimum processing condition. As the cooling ambient reduces, the loops progressively shift to the right, indicating the development of an internal field. As illustrated in the inset to Figure 15, this internal field is directed from the bottom electrode toward the top electrode (25). The

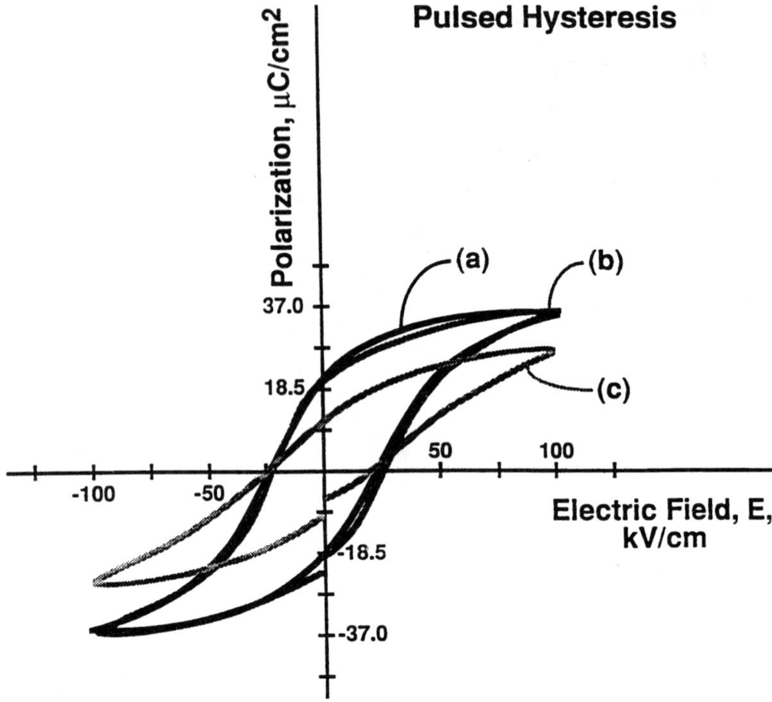

Figure 14 Pulsed hysteresis loops from test capacitors grown on (*a*) $LaAlO_3$; (*b*) BTO/YSZ/Si; and (*c*) BTO/SiO_2/Si showing that the polarization properties scale with the crystalline quality of the film (see Figure 11).

asymmetry of the hysteresis loop is important in understanding the failure mechanisms and reliability of the capacitors and is discussed below.

An important area of application of these ferroelectric capacitors is in high-density nonvolatile memories. Such memories have storage densities above 1 Mbit, which translates to a capacitor of lateral dimension in the range of 2 to 4 μm (26). These capacitors cannot be fabricated by conventional lift-off or shadow-masking processes and require full-wafer lithographic processing, etching, and the use of interlevel dielectrics to isolate the top and bottom electrodes. Consequently, several additional steps are introduced in the fabrication process, and each step may produce additional complications that lead to degradation in the ferroelectric properties. Only very limited information has been published on this topic. In our research, we have demonstrated that a simple ion-beam etching process can be used to delineate capacitor structures down to 2 μm in diameter, without serious loss of reliability. Figure 16 illustrates this point

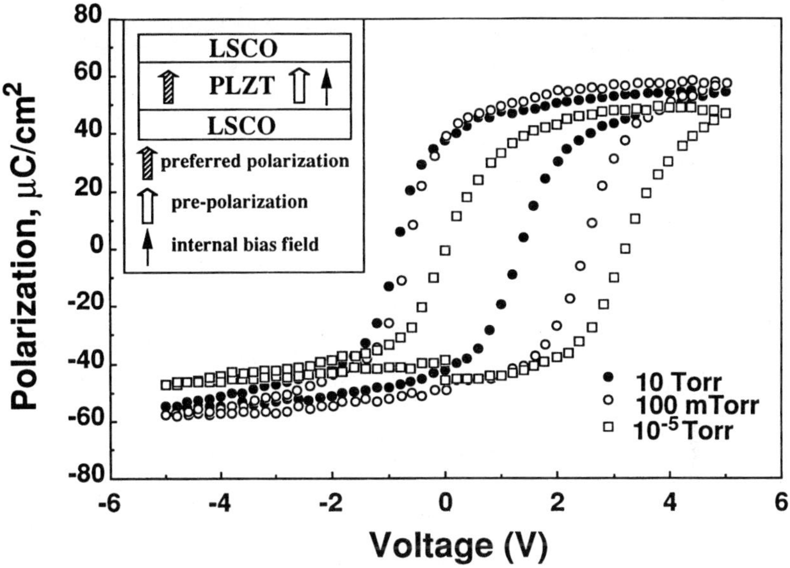

Figure 15 A set of pulsed hysteresis loops illustrating the effect of one deposition parameter, namely the oxygen ambient during cool down to room temperature, on the symmetry of the hysteresis loop. As the film is cooled under more reducing ambients, the hysteresis loop becomes more asymmetric, i.e. an internal field directed from the bottom electrode to the top electrode is developed.

in terms of the remanent polarization (i.e. $P^* - P^\wedge$) as a function of the area of the capacitor, ranging from 100 μm diameter discrete capacitors produced by shadow masking to 2 μm capacitors produced by ion-beam etching. For each size, at least 20 to 30 capacitors were measured, leading to the spread in the data. The main conclusion from this figure is that the polarization characteristics scale reasonably well over this size range. Further studies of the fatigue, aging, and retention characteristics have shown that these properties are also commensurate with those obtained from large capacitors. Notwithstanding these encouraging preliminary results, considerable work is necessary to elucidate all the details involved in the scaling of the ferroelectric properties with the capacitor size.

Reliability Testing of Ferroelectric Heterostructures

Our discussion thus far has focused on the basic properties of the test capacitors fabricated from the metal-oxide heterostructures. However, in commercial applications, these capacitors have to perform extremely reliably under a broad range of ambient conditions. For example, in order to be a viable nonvolatile memory technology, these ferroelectric capacitor

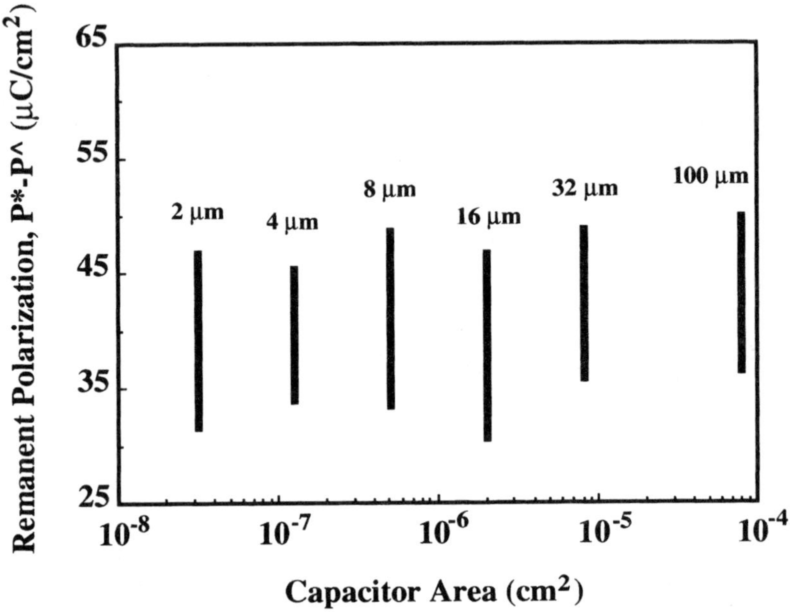

Figure 16 A plot of remanent polarization (P* − P^) as a function of capacitor size and area demonstrating that within this size range the basic ferroelectric properties are relatively well behaved.

heterostructures should be capable of switching from one polarization state to another at least 10^{12} times and preferably 10^{15} times, without any significant loss of the remanent polarization. This is so because these capacitors will be used in applications involving repeated access and switching from one logic state to another, a process that could happen at least 10^{12} times over the lifetime of the memory. This property, termed fatigue resistance, is one of the prime reliability features expected of the capacitors. Similarly, the capacitor (and the corresponding memory element) should maintain a polarization state (i.e. a logic state) and clearly discriminate it from the opposite state almost indefinitely, i.e. it should have good retention characteristics. Another feature that is desirable in the ferroelectric capacitor is the ability to be switched from one logic state to another, a property termed aging. An important reliability feature identified more recently is imprint, which in simple terms is the preference of the capacitor to remain in a particular logic state (i.e. polarization state), especially when it is maintained in that state for a long time under thermal or electrical stress. It should be noted that these requirements have to be met not only for room temperature operations but also for temperature requirements as high as 155°C (for military applications).

Many research groups are carrying out detailed pulsed testing of these capacitors to understand the effect of process variables and microstructure on the key reliability parameters. It has been found that using metallic oxide electrodes instead of the conventional Pt electrodes makes a dramatic difference in the fatigue characteristics. This difference is illustrated in Figure 17, for the case of a LSCO/PLZT/LSCO capacitor grown on $LaAlO_3$. In capacitors grown under the optimum conditions, no significant fatigue is observed even after 10^{12} cycles (13a, 14, 15). These results are promising, considering the fact that capacitors with Pt electrodes typically fatigue and lose more than 50° of their initial polarization after just 10^8 cycles, as shown in Figure 17. Ferroelectric capacitor heterostructures grown on BTO/YSZ/Si, BTO/SiO_2/Si, and Pt/BTO/SiO_2/Si, where BTO is the template layer, show a similar trend in fatigue behavior, as illustrated in Figure 18. As seen in the case of the hysteresis loops, the important difference between the epitaxial heterostructures grown on $LaAlO_3$ and the c-axis-oriented films grown on SiO_2/Si substrates is the switched polar-

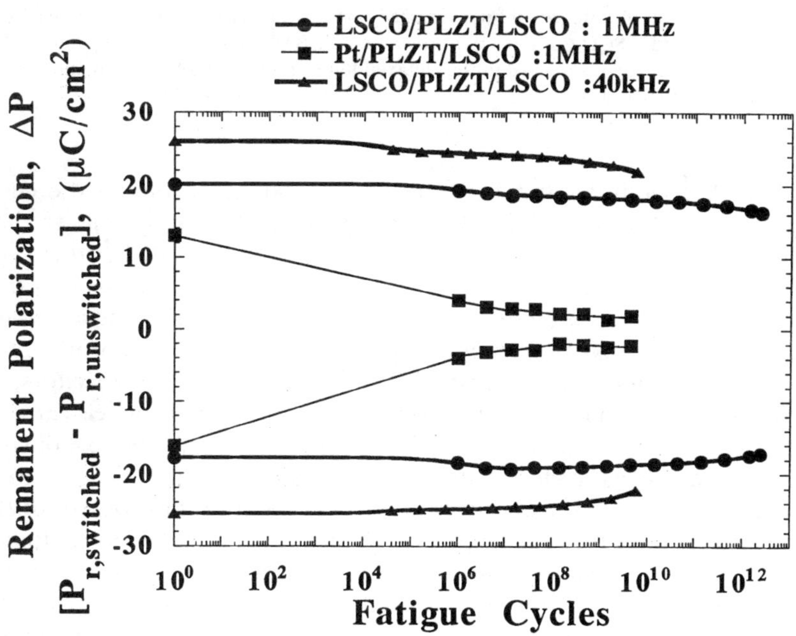

Figure 17 A plot of remanent polarization as a function of bipolar fatigue cycles at a test frequency of 1 MHz and amplitude of 5 V comparing the performance of capacitors with Pt electrodes with that of capacitors with LSCO electrodes. Whereas the capacitors with LSCO top and bottom electrodes do not show any fatigue, the capacitor with a Pt top electrode shows significant fatigue.

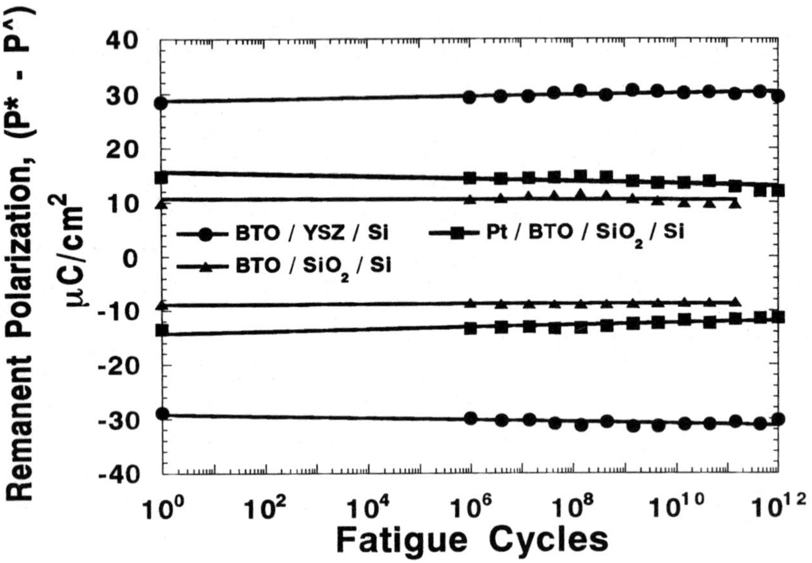

Figure 18 A plot of switched and nonswitched polarization as a function of fatigue cycles for a LSCO/PLZT/LSCO test capacitor grown on a Pt/BTO/SiO_2/Si substrate showing that the fatigue-free behavior obtained in the epitaxial films grown on $LaAlO_3$ is obtained in the oriented capacitors as well.

ization. This desirable trend in fatigue characteristics is observed even at a testing temperature of 100°C, although the ∂P value is smaller than that observed at room temperature. These results suggest that there is a fundamental difference in fatigue characteristics between capacitors that are fabricated directly on Pt and those that are fabricated using a metallic oxide electrode. Several phenomenological models have been proposed to explain the fatigue behaviour of PZT-based ferroelectric capacitors; to date there has been no rigorous proof for or against these models (27–30). The difference between Pt and LSCO electrodes suggests the role of oxygen-related defects, such as charged vacancies.

Although fatigue is extensively studied as a reliability property of the ferroelectric capacitors, in reality many other properties have to be studied concurrently and one needs to ensure that all are within the range needed for the specific application. Among them, logic state retention is an important parameter. A retention experiment should actually be carried out on a memory element because it is a measure of the element's ability to maintain a programmed logic state and also discriminate between the two states. However, retention tests on discrete capacitors have significant value, especially in evaluating the effect of various process parameters and in studying fundamental mechanisms of loss in retention, which is ultimately related to the ferroelectric material properties, the domain structure in the capacitor, etc. The retention test in a discrete capacitor is carried out by the following set of pulsed testing experiments. The retention for a logic 1 is measured by writing and reading the capacitor with pulses of opposite polarities (thus enabling a large amount of charge to be switched), whereas that for logic 0 is measured by writing and reading with pulses of the same polarity. Measurements typically have to be carried out over several decades of time and over a range of testing conditions, including pulse amplitudes, widths and test temperature. Figure 19 shows one typical set of data for a LSCO/PLZT/LSCO capacitor grown on a Pt/BTO/SiO_2/Si substrate. There is very little change in the logic 0 polarization value as a function of retention time, whereas the logic 1 polarization, in general, shows a weak, log-time decay. The failure criterion is determined by the ability of the amplifier in the memory circuit to discriminate between the voltage signal for the logic 1 and that for logic 0. This differential is typically 100 mV, which translates into a difference in polarization between logic 1 and 0 of ~ 1–$2\ \mu C\ cm^2$ at the end of the test period. This failure criterion is valid for all the reliability tests that need to be carried out on the capacitor. Extrapolation of the data in Figure 19

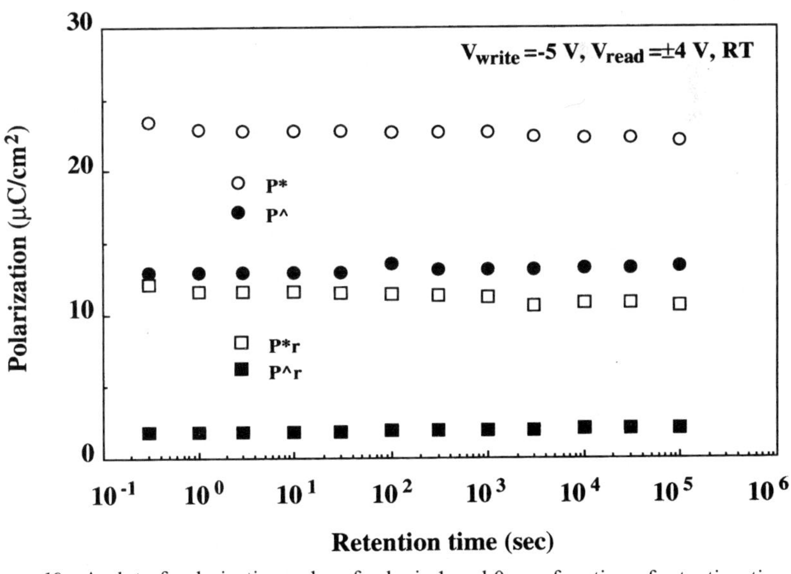

Figure 19 A plot of polarization values for logic 1 and 0 as a function of retention time in that state. This behavior is representative for both room temperature and 100°C testing conditions, although the difference between logic 1 and 0 is smaller at 100°C.

to this criterion yields lifetimes of better than 10^9 s. Similar measurements at temperatures up to 100°C have shown that such behavior is consistent over this temperature range, although the difference in polarization between logic 1 and 0 is smaller at higher temperatures.

An aging test provides a measure of the ability to switch the capacitor from one polarization state to the opposite state after being maintained in the previous state for a period of time. Thus, aging, which is a measure of the ferroelectric material property, is related to retention but measures the switchability of the capacitor from one state to another rather than the retentivity of a given state. Figure 20 shows a typical plot of the aging characteristics of the same family of capacitors for which retention data are shown in Figure 19. Imprint is the preference of the capacitor (or the memory element) to remain in a particular logic state or to revert to that state after having been written into the opposite state. Testing for imprint, in general, and specifically for imprint in LSCO/PLZT/LSCO capacitor structures, is still in its infancy (31–33). The tendency of the capacitors to imprint was studied using a modified form of the retention test, as illustrated in Figure 21. (In the above section, symmetric hysteresis loops obtained from the test capacitors were shown in Figure 14.) The corresponding pulse polarization measurements show symmetric behavior, i.e. $|P^*| \approx |-P^*|$ and $|P^\wedge| \approx |-P^\wedge|$, as is expected from the symmetric

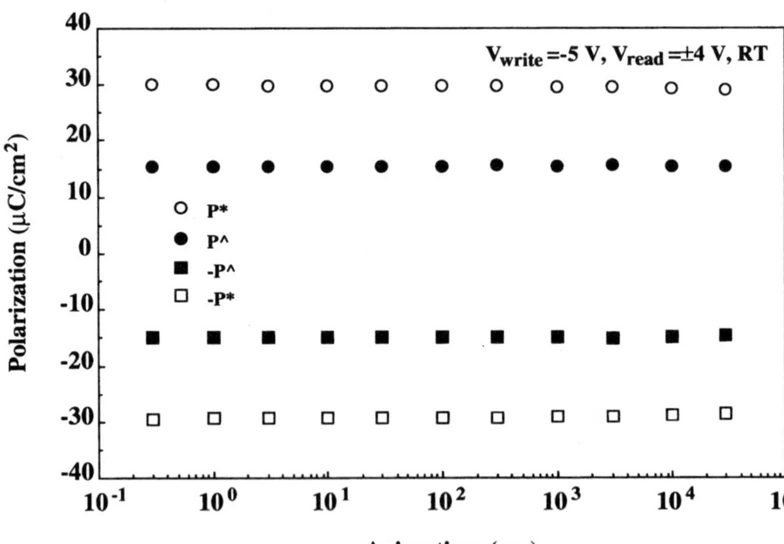

Figure 20 A plot of switched and nonswitched polarization values as a function of aging time for the same family of capacitors used in the retention test in Figure 19.

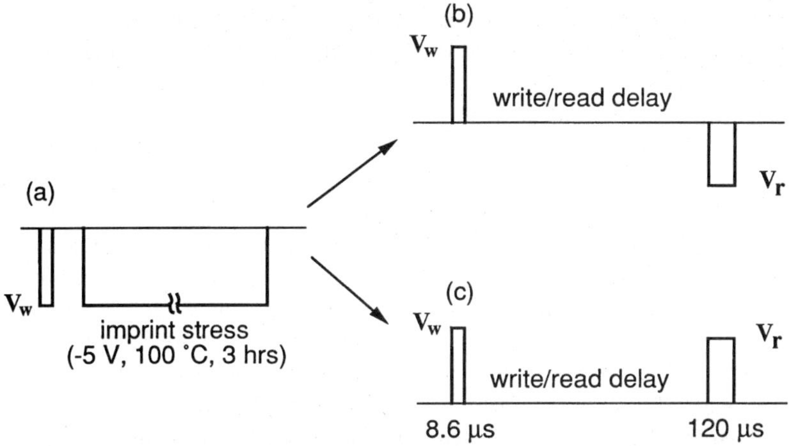

Figure 21 A schematic illustration of the testing sequence used to measure the tendency of the capacitors to imprint, i.e. to prefer one polarization state to another.

hysteresis loops. However, if a ferroelectric capacitor prefers one polarization state over the other, it will show asymmetric pulse polarization values as well as asymmetric hysteresis loops. Because the preferred polarization state, for example the logic state 1, is more stable than the other (logic state 0), polarization in the logic state 0 switches back to the logic state 1 after writing to the logic state 1. The level of the polarization returning to the logic state 0 depends on the imprint stress (such as temperature and dc bias) and operating conditions (mainly temperature). If the preferred polarization state, for example, is a negatively poled state, the polarization written in a positively poled state returns to the negatively poled state after a write pulse ($-$Vw), resulting in an apparently larger nonswitched polarization value in the positively poled state (P^) than in the negatively poled state ($-$P^). In fact, this nonswitched polarization value (P^) contains switched polarization as well as nonswitched polarization. Therefore, nonswitched polarizations, P^ and $-$P^, reflect the degree of negative and positive preferred states, respectively. More importantly, the time evolution of P^ and $-$P^ (depending upon the imprinted state) reveals important information regarding the tendency of the memory element to imprint. Among the many test variables, three are likely to be important. The first is the method of imprint, i.e. dc bias or single-sided fatigue, both of which simulate the repeated read operation of that logic state. The second is the thermal activation that accelerates the degradation process leading to imprint. The third, the delay between the write pulse (Figure 21) and the read pulse, is an important variable because it indicates the ability of the capacitor to maintain the switched state. Consequently, the test consists of a write-read delay experiment after the imprint stress.

In this imprint test, the test capacitor is exposed to an imprint stress consisting of a dc bias (e.g. −5 V) at 100°C for 3 h. The capacitor is subsequently cooled to room temperature with the contacts electrically closed (to prevent pyroelectric effects from interfering with the measurement). The capacitor is then written into the opposite state (i.e. with a +5 V-pulse) and read with either a +4 V-pulse or a −4 V-pulse. The important parameter here is the delay between the write and the read pulse. If the capacitor were to imprint, i.e. prefer to remain in the negatively poled state, then it would try to revert to this state after having been poled into the positive state by the +5 V pulse. Consequently, the +4 V-read pulse, which should ideally measure only the nonswitched polarization, will progressively yield larger polarization values. Similarly, a −4 V-read pulse should show a continuously decreasing value of polarization. Figure 22 shows results of this experiment for a dc bias of −5 V, while Figure 23 shows similar data for a dc bias of +5 V.

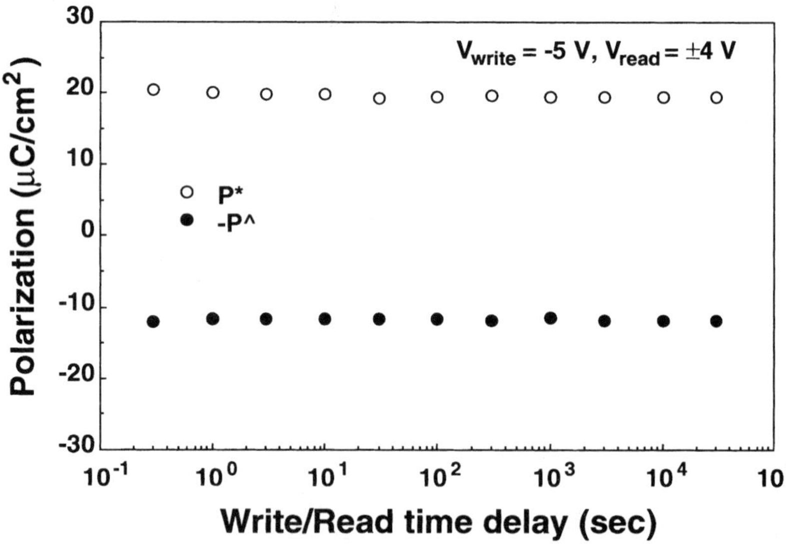

Figure 22 Results of the write-read delay experiment carried out to detect a tendency to imprint in the test capacitors. This plot shows results for a capacitor that was imprinted at −5 V for 3 h and a temperature of 100°C. The relevant values are the positive nonswitched polarization P^ and the negative switched polarization −P*. A tendency to imprint would be manifested by a monotonic increase in P^ and a corresponding decrease in −P* (although the decrease in −P* could also be due to retention losses).

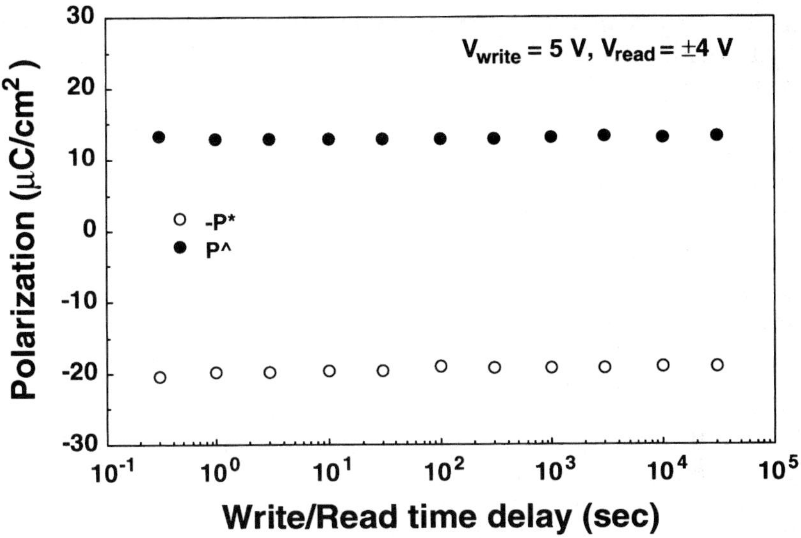

Figure 23 Results of an imprint test similar to that shown in Figure 22, except that the imprint stress was +5 V. Note that the trend in the two values, namely $-P^\wedge$ and P*, is similar to that observed in Figure 22.

DISCUSSION

We hope this review conveys perspectives to the reader from two points of view, namely that it (*a*) specifically addresses issues related to ferroelectric metal-oxide heterostructures for nonvolatile memory applications and (*b*) addresses the general area of metal-oxide heterostructures. The field of ferroelectric thin films for nonvolatile memory applications is clearly at the crossroads. Research and development effort has been ongoing for at least the past eight years, and there have been several breakthroughs that have led to a better understanding of these complicated systems and consequently have brought us closer to the goal of attaining a reliable high-density solid-state nonvolatile memory technology. However, many questions still remain. The use of metallic-oxide electrodes such as RuO_2 and LSCO has clearly led to a new way of fabricating the ferroelectric capacitor. The fact that these oxides heterostructures can be integrated on a Si wafer with a Pt contact means that their integration with the transistor drive electronics is feasible. Recent studies of the growth of the LSCO/PLZT/LSCO heterostructure on a polycrystalline Pt surface have revealed that reliable capacitors can be fabricated (19). Consequently, the role of the template approach has to be examined in greater detail. An important area of research and development is the scaling of the ferroelectric capacitor size and growth on wafers large enough for commercial viability. The issue of scaling has not been addressed by many groups. For example, a 1 Mbit ferroelectric memory technology requires capacitors with lateral dimensions in the 2 to 4 μm range. There is only limited information available on capacitors within this size range. Hence, this must be a high priority for research effort in the immediate future. This leads directly to issues related to etching of the heterostructure stack to delineate the capacitors and test them, which requires sensitive electrical testing capability. When the dimension of the test capacitor becomes smaller than the microstructural dimension such as the grain size (in polycrystalline materials), then the lateral uniformity may begin to vary as the capacitor size is further reduced. This may lead to a need for epitaxial heterostructures. Another important issue that is specifically related to the manufacturability of these oxide heterostructures is the ability to grow them on nonplanar substrate surfaces that are at least 6 inches in diameter. The need for nonplanarity arises from the fact that the capacitor has to be integrated with the pass-gate transistor, a process that involves the use of interlevel dielectrics and metallization lines. Consequently, conformal growth techniques, such as chemical vapor deposition, need to be explored with greater urgency. In very high-density memory architectures, it is likely that the ferroelectric stack will be grown directly on top of the source (in the DRAM-type architecture), i.e. without the luxury of the SiO_2 layer or any other barrier layer between the metal oxides and Si. Therefore, the bottom electrode in the ferroelectric stack has to make direct chemical contact with the Si surface, a very challenging problem. Finally, there is a need to understand the fundamentals of the reliability and transport properties of these novel heterostructures.

CONCLUSION

The mushrooming interest in the area of metal-oxide thin films and heterostructures is most likely to increase (1) because of the wide range of physical properties available in multicomponent metal oxides. Research into prototypical systems such as the perovskites (in which there is already considerable activity), spinels (e.g. the magnetic ferrites), garnets, and layered compounds provides good systems to test fundamental aspects of oxide heteroepitaxy related to accommodation of lattice mismatch, interface chemistry, and relative phase stability. These systems are analogous to heteroepitaxial compound semiconductors and from an application point of view, their physical properties can be exploited. Magnetic spinel ferrites and garnet thin-film heterostructures are already being grown by a variety of techniques for applications that include magnetic and magneto-optic recording media, integrated thin-film magnetic recording heads, and integrated microwave components (especially when grown on GaAs substrates). The nonlinear optical properties of ferroelectric perovskites are being explored for spatial light modulation, holographic storage, etc. The intrinsically large dielectric constants of the ferroelectrics in the paraelectric state will be used in future, ultra-high density dynamic random access memories. Recently, large values of magnetoresistance have been observed in perovskite oxides such as La-Ca-Mn-O, which are very similar structurally and chemically to the ferroelectric perovskites (34). All these developments point to exciting research on metal-oxide thin film heterostructures.

ACKNOWLEDGMENTS

The work on ferroelectric metal oxides at Bellcore is supported by the Advanced Research Projects Agency under contract No. N00014-93-C-0127. We have benefited from many discussions with numerous colleagues around the world and we thank them all. Specifically we wish to acknowledge the ongoing interactions with JT Evans, Jr. (Radiant Technologies), WE Warren, D Dimos, and G Pike (Sandia), O Auciello and A Kingon (MCNC/NCSU), and S Ghonge and E Goo (University of Southern California).

Literature Cited

1. Bednorz JG, Muller KA. 1986. *Z. Phys. B* 64: 189–91
2. Venkatesan T, Inam A, Dutta B, Ramesh R, Hegde MS, et al. 1990. *Appl. Phys. Lett.* 56: 391–93
3. Ramesh R, Inam A, Chan WK, Wilkens B, Myers K, et al. 1991. *Science* 252: 944–46
3a. Ramesh R, Chan WK, Wilkens B, Gilchrist H, Sands T, et al. 1992. *Appl. Phys. Lett.* 61: 1537–39
4. Carroll KR, Pond JM, Chrisey DB, Horwitz JS, Leuchtner RE, Grabowski KS. 1993. *Appl. Phys. Lett.* 62: 1845–47
5. Galasso FS. 1970. *Structure and Properties of Inorganic Solids*, New York: Pergamon
6. Kingon A, Myers ER, Tuttle BA, eds. 1991. *Proc. Mater. Res. Soc. Symp. Ferroelectric Thin Films II, Boston*, Pittsburgh: Mater. Res. Soc. 136 pp.
6a. Myers ER, Tuttle BA, Desu SB, Larsen PK, eds. 1993. *Proc. Mater. Res. Soc., San Francisco*, Pittsburgh: Mater. Res. Soc. 499 pp.
6b. Paz de Araujo CA, ed. 1992. *Proc. 4th Int. Symp. Integrated Ferroelectrics, Monterey, CA*, Colorado Springs, CO: Univ. Colorado Press. 683 pp.
6c. Paz de Araujo CA, ed. 1993. *Intgr. Ferroelect.* 4: 107–291
7. Scott JF, Paz de Araujo CA. 1989. *Science* 246: 1400–5
7a. Sayer M, Sreenivas K. 1990. *Science* 247: 1056
7b. Haertling GH. 1991. *J. Vac. Sci. Technol.* 9: 414

8. Evans JT, Womack RD. 1988. *IEEE J. Solid State Circuits* 23: 1171–74
9. Sinharoy S, Buhay H, Lampe DR, Francombe MH. 1992. *J. Vac. Sci. Technol.* A10: 1554
10. Dey SK, Zuleeg R. 1990. *Ferroelectrics* 108: 37–45
11. Abt NE, Misic P, Zehngut D, Regan E. 1992. *Proc. 4th Int. Symp. on Integrated Ferroelectrics, Monterey, CA*, ed. CA Paz de Araujo, pp. 533–40. Colorado Springs, CO: Univ. Colorado Press
11a. Desu SB, Yoo IK. 1992. *Proc. 4th Int. Symp. on Integrated Ferroelectrics, Monterey, CA*, ed. CA Paz de Araujo, pp. 640–56. Colorado Springs, CO: Univ. Colorado Press
12. Mihara T, Watanabe H, Paz de Araujo CA, Cuchiaro J, Scott M, McMillan LD. 1992. *Proc. 4th Int. Symp. Integrated Ferroelectrics, Monterey, CA*, ed. CA Paz de Araujo, p. 137. Colorado Springs, CO: Univ. Colorado Press
13. Wolf RM. 1991. *Proc. Mater. Res. Soc. Symp. Ferroelectric Thin Films II*, ed. A Kingon, ER Myers, BA Tuttle, Pittsburgh: Mater. Res. Soc.
13a. Cheung JT, Neurgaonkar RR. 1992. *Proc. 4th lnt. Symp. on Integrated Ferroelectrics, Monterey, CA*, ed. CA Paz de Araujo, pp. 158–70. Colorado Springs, CO: Univ. Colorado Press
14. Ramesh R, Gilchrist H, Sands T, Keramidas VG, Haakenaasen R, Fork DK. 1993. *Appl. Phys. Lett.* 63: 3592
15. Lichtenwalner DJ, Dat R, Auciello O, Kingon AI. 1994. *Ferroelectrics* 152: 97
16. Abt N. 1990. In *Proc. Mater. Res. Soc. Symp. Ferroelectric Thin Films*, ed ER Myers, AI Kingon, 200: 302–12. Pittsburgh: Mater. Res. Soc.
17. Kwok CK, Vijay DP, Desu SB, Parikh NR, Hill EA. 1992. *Proc. 4th Int. Symp. Integrated Ferroelectrics*, ed. CA Paz de Araujo, p. 412. Colorado Springs, CO: Univ. Colorado Press
17a. Al-Shareef HN, Bellur KR, Auciello O, Kingon AI. 1994. *Ferroelectrics* 152: 85
18. Ramesh R, Lee J, Sands T, Keramidas VG, Auciello O. 1994. *Appl. Phys. Lett.* 64: 2511–13
19. Dat R, Lichtenwalner DJ, Auciello O, Kingon AI. 1994. *Appl. Phys. Lett.* 64: 2673–75
20. Ramesh R, Sands T, Keramidas VG. 1994. *J. Electronic Mater.* 23: 19–26
21. Chrisey DB, Hubler GK, eds. 1994. *Pulsed Laser Deposition of Thin Films*, New York: Wiley & Sons. 613 pp.
22. Auciello O, Kingon AI, Krauss AR, Lichtenwalner DJ. 1993. In *Proc. of NATO Advanced Study Institute*, ed. O Auciello, J Engemann, 234: 1–633. Dordrecht: Kluwer Academic
23. Ghonge SG, Goo E, Ramesh R, Sands T, Keramidas VG. 1993. *Appl. Phys. Lett.* 63: 1628–30
24. Ghonge SG, Goo E, Ramesh R, Haakenaasen R, Fork DK. 1994. *Appl. Phys. Lett.* 64: 3407–9
25. Lee J, Ramesh R, Keramidas VG, Warren WL, Pike GE, Evans JT Jr. 1995. *Appl. Phys. Lett.* In press
26. Ramesh R, Dutta B, Ravi TS, Lee J, Sands T, Keramidas VG. 1994. *Appl. Phys. Lett.* 64: 1588–90
27. Duiker HM, Beale PD, Scott JF, Paz de Araujo CA, Melnick BM, et al. 1990. *J. Appl. Phys.* 68: 5783
28. Miller SL, Nasby RD, Schwank JR, Rodgers MS, Dressendorfer PV. 1990. *J. Appl. Phys.* 68: 6463–71
29. Scott JF, Paz de Araujo CA, Melnick BM, McMillan LD, Zuleeg R. 1991. *J. Appl. Phys.* 70: 382
30. Yoo IK, Desu SB, Xing J. 1993. In *Proc. Mater. Res. Soc. Symp. Ferroelectric Thin Films III*, ed. ER Myers, BA Tuttle, SB Desu, PK Larsen, p. 165. Pittsburgh: Mater. Res. Soc.
31. Mihara T, Watanabe H, Paz de Araujo CA. 1993. *Jpn. J. Appl. Phys.* 32: 4168
32. Dat R, Lichtenwalner DJ, Auciello O, Kingon AI. 1995. *Integr. Ferroelect.* In press
33. Lee J, Ramesh R, Keramidas VG. 1995. *Integr. Ferroelect.* Submitted
34. Jin S, Tiefel TH, McCormack M, Fastnacht RA, Ramesh R, Chen LH. 1994. *Science* 264: 413

Annu. Rev. Mater. Sci. 1995. 25: 679–709

HIGH-TEMPERATURE SUPERCONDUCTING MULTILAYERS AND HETEROSTRUCTURES GROWN BY ATOMIC LAYER-BY-LAYER MOLECULAR BEAM EPITAXY

J. N. Eckstein and I. Bozovic

E. L. Ginzton Research Center, Varian Associates, Palo Alto, California 94304

KEY WORDS: high-temperature superconductor, Josephson junction, molecular beam epitaxy, superlattice, multilayers, proximity effect, tunneling, interface

ABSTRACT

Atomic layer-by-layer molecular beam epitaxy (ALL-MBE) of high-temperature superconductors (HTSC) and other complex oxides has been developed. Thin films with atomically flat surfaces and abrupt interfaces can be produced an atomic layer at a time. Samples are engineered by stacking molecular layers of different compounds, by adding or omitting atomic monolayers, and by doping within specified monolayers. Novel artificial HTSC compounds such as $Bi_2Sr_2Ca_7Cu_8O_x$ (2278), as well as various heterostructures, have been synthesized in this way. This unique synthetic capability has allowed several fundamental physics issues, such as the dimensionality of the HTSC state, long-range proximity effects, resonant tunneling with a specified number of hops, etc, to be addressed. Trilayer Josephson junctions with uniform and reproducible properties have been fabricated; phase-locked operation of two junctions has also been demonstrated. Finally, titanate slabs only 4-Å thick have been grown without pin-holes and shown to provide tunneling barriers for *c*-axis transport.

0084–6600/95/0801–0679$05.00

Introduction

Oxides—which exhibit most of the interesting phenomena known to occur in solid-state systems—may be richer in phenomenology than any other comparable class of materials. Oxides can be insulators, semiconductors, or metals. The copper-oxide-based compounds that we have studied are superconductors with the highest critical temperatures. In some oxides, electrons manifest simple single-particle transport properties, with a high mobility; in other oxides, electrons show strongly correlated behavior that can result in a Mott-Hubbard transition, localization, or charge- or spin-density waves. In some oxides electron-phonon coupling leads to polaronic transport. Other oxides show collective states such as magnetism, or have large local magnetic moments that can couple to form ferromagnetic or antiferromagnetic phases, both of which exist up to high temperatures. Yet other oxides have large nonlinear dielectric and optical properties (1, 2).

Why do so many different phenomena occur in oxides? It is probably impossible to give the complete answer, but any response would have to acknowledge the diversity of the electronic states that exists in the metal atoms that are oxidized and arranged with the oxygen atoms to form a crystal. These states range from the simple one-electron-like states of the alkali metals through the *d* and *f* states of electrons near the Fermi energy contained in transition and rare earth metals. Some metal atoms participate in highly covalent bonding to oxygen that leads to relatively broad bands and delocalized states. Depending on the charge transfer that occurs, these bands can be partially or completely occupied. In other crystals the metal atom and oxygen states do not hybridize much at all; the bonds are predominantly ionic and the bands narrow. Sometimes both types of bonding occur within the same crystal. Furthermore, the symmetry of the crystalline structure and the occupation of states in metal atoms determine local magnetic moments that underlie magnetic order.

With such a wide variety of physical states, oxides provide a flexible system to study interactions, structures, and transport at heterointerfaces. Many of the phenomena of interest, however, have very small characteristic length scales, so-called coherence lengths. Thus heterojunctions made to study such phenomena must be fabricated, studied, and understood on an atomic level, and the interfaces need to be nearly crystallographically perfect. In general, lattice mismatch between different materials makes this difficult. Fortunately, many oxides belong to broad structural classes that have similar atomic spacing. In such classes the lattice constant is predominantly determined by the size of the oxygen ion, with a smaller influence due to the particular metal atoms involved. In these cases the lattice spacing can be thought of as being governed by large touching oxygen anions, with the smaller metal cations fitting in the space left over. The existence of many different oxides with dramatically different physical properties but similar atomic spacing provides opportunities to investigate a wide spectrum of interface phenomena.

A class of oxides that exhibits wide-ranging properties but has a limited range of lattice spacing are the so-called perovskites (1). Simple perovskites have a chemical formula ABO_3 and form a cubic (or nearly cubic) structure with the A ion in a body-centered position, the B ion at the corners, and the O ions on the edges. Typically, the states near the Fermi energy are derived from the atomic B and O states, while the states derived from the A atom are farther away in energy. In particular, the A atoms can be thought of as electron donors to the BO_x states. Often the A atoms are alkali metal, alkaline earth, or rare earth atoms, while the B atoms are transition metals with rich *d*-state properties.

Besides the simple perovskites, there are related compounds that are partially perovskite. Such materials—such as some of the high-critical-temperature (high-T_c) superconductors, e.g. ed$Bi_2Sr_2CaCu_2O_{8+x}$ (often referred to as BSCCO-2212, where the letters denote the metal atoms plus oxygen and the numbers the metal atom stoichiometry)—have complex, unit cell structures in which some atoms form an oxygen-deficient perovskite-like structure while others appear to be in a rock-salt structure. The diverse physical properties of materials in the perovskites and related compounds make this family a good choice for assembling heterostructures and for studying and exploiting interfacial phenomena.

In this review we describe a technique we have developed for fabricating atomically precise heterojunctions and metastable materials, as well as some recent experiments we have performed to explore charge transport across and in the neighborhood of such interfaces. By accurately sequencing the supply of reactant atoms, we are able to assemble complex oxide heterostructures, depositing one metal-oxide monolayer at a time. We call this technique atomic layer-by-layer molecular beam epitaxy, ALL-MBE (3). We have used this to grow superlattices and other multilayers to study fundamental aspects of superconductivity in cuprates (4). We have used multilayer films with thin barrier layers—such as single molecular layers of the metastable compound $Bi_2Sr_2Ca_7Ca_8O_{20}$ (BSCCO-2278) (5)—inserted between cuprate slabs to make trilayer Josephson junctions, and we have shown that by fine-tuning the conductivity of such barriers, their working characteristics such as the critical current can be engineered (6). More recently, we have found that certain sandwich structures can easily be stacked into series arrays of Josephson junctions (7–9). In such arrays we have observed coherent phase locking of the stacked junctions, as we describe below. Besides cuprate barriers, we have also studied trilayer structures containing very thin titanate layers sandwiched between cuprate layers (7, 10). Complete titanate layers as thin as 4 Å can be grown. These provide barriers to *c*-axis transport, and we have shown that for thin (4–20 Å) titanate barriers the transport across the barrier is via tunneling. We have also found, however, that the properties of the adjacent CuO_2 planes are affected by the neighboring titanate layers across the heterojunction, leading to localization, and this localization determines the density of states between which the tunneling process occurs. Thus understanding and controlling the electronic structure on the scale of individual atomic oxide layers are necessary in order to use various oxide interfaces for practical applications.

Atomic Layer-by-Layer Molecular Beam Epitaxy

The hallmark of ALL-MBE is very accurate sequencing of atoms supplied to the growing crystalline surface in such a way as to control kinetically the evolution of the chemical reaction coordinate for each molecular layer. As we have described elsewhere, evidence points to the conclusion that under the conditions in which we operate, only the top molecular layer is involved in the growth chemistry, so that precise sequences of molecular layers can be assembled into heterostructures such as superlattices, etc, with little or no intermixing (3).

Key features of the ALL-MBE system include individual Knudsen cell (thermal) sources for each cation component of the film to be grown, computer-controlled shuttering of each source, oxidation using ozone, monitoring of film surface structure by in situ reflection high-energy electron diffraction (RHEED), and pseudo-double-beam atomic absorption measurement of source fluxes. A schematic diagram of the ALL-MBE system is shown in Figure 1.

The thermal evaporation sources provide atomic beam fluxes that are stable over periods of hours to within a few percent. Currently, eight sources are installed in the system: Bi, Sr, Ca, Cu, Dy, Ba, Ti, and Ag. Computer-controlled beam shutters allow accurate sequencing of the atoms supplied to the surface and control the absolute stoichiometry of each atomic layer grown. This configuration permits the growth of both Dy-Ba-Cu-O compounds and compounds from the Bi-Sr-Ca-Cu-O family, as well as several titanates. The Ag source is used for in situ deposition of electrical contacts.

The atomic flux from each source is monitored to an absolute accuracy of better than one percent using a technique we have developed based on pseudo-double-beam atomic absorption spectroscopy (11). All of the atomic absorption equipment resides outside the high vacuum chamber.

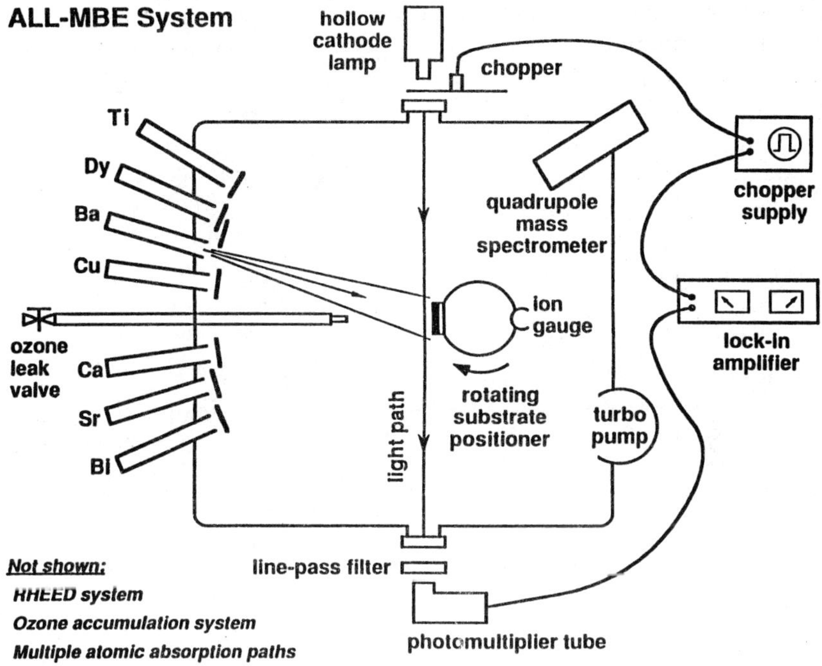

Figure 1 Schematic diagram of ALL-MBE growth system showing in situ atomic absorption apparatus for accurate beam flux measurement.

In this technique, a mechanically chopped beam of light from a hollow cathode lamp passes through the atomic beam path just in front of the substrate position and is detected using a photomultiplier tube and a lock-in amplifier. Each atomic source requires its own lamp and optical path. By opening and closing the shutter during the measurement, a pseudo-double-beam effect is achieved, which makes the measurement immune to drifts in reflection and absorption by the viewport windows, in gain settings on amplifiers, and in lamp emission intensity. The technique is sufficiently fast and accurate to detect and correct for changes in beam flux of less than one percent during growth. In order to relate the atomic absorption signal from each source to the actual beam flux, calibration films are grown under thermodynamic conditions identical to those used for real growths; the resulting surface densities of atoms in the films are obtained by Rutherford back scattering (RBS) analysis. Beam flux measurement and control by atomic absorption spectroscopy have proved critical to atomic layer-by-layer growth of high-quality, high-temperature superconductor (HTS) thin films. In particular, we have found that films that are atomically flat and free of second-phase defects can be made only if the growth accurately targets the correct elemental-layer stoichiometries to within one to three percent, depending on the specific phase and element in question. The actual layer composition that gives defect-free material is not exactly stoichiometric, e.g. 2212. This may in part be due to small, systematic but reproducible errors in the RBS calibration analysis but also may reflect the influence of vacancies and antisite substitutions in determining the actual chemical composition of the phase.

Oxidation of the metal atoms in a growing oxide film is achieved using a controlled flux of ozone molecules (12). In the case of cuprate superconductors, growth usually takes place between 600 and 750°C; sufficient oxidation power is obtained to oxidize the film fully even with a relatively low flux of ozone to the sample. The typical oxygen background pressure is between 5.0×10^{-7} and 2.0×10^{-5} Torr, which permits line-of-sight beam deposition and abrupt beam flux modulation using shutters. We estimate these pressures to correspond to fluxes of molecular ozone of between 1.8×10^{14} and 7×10^{15} molecules/cm^{2}s. Such a pressure is also low enough to utilize electron diffraction during growth and not to reduce seriously the lifetime of hot filaments used in heaters and electron guns.

During growth, the surface structure is monitored on an atomic scale using RHEED. This utilizes a thin, monoenergetic electron beam (10 KeV), which is incident on the crystal surface at a glancing angle of incidence, θ_i. The electrons, with a de Broglie wavelength of about 0.15 Å, diffract from whatever periodic regularity is present on the surface, and in the far field form a k-space representation of the surface periodicity. From a perfectly flat crystalline surface, the need to conserve energy and all the components of momentum (which include allowed momentum transfer vectors from the periodic surface structure) simultaneously in the diffraction process causes the diffracted beam to form a pattern of spots in the far field. Real surfaces are not perfect, however, and have disorder. This causes the spots to have a finite width, which in the transverse direction is related simply to the size of the area of coherent diffraction via an uncertainty relation, $\Delta x \Delta k \approx 1$. In the vertical (longitudinal) direction, the same k-space linewidth leads to a stretching of the spots by a factor of $1/\theta_i$. Because θ_i is a very small angle, typically less than 0.02 radians, each spot is stretched into a streak, and the diffraction pattern seen during molecular beam epitaxial growth looks similar to the streaky patterns shown in Figure 2, in which the separation of the streaks in the transverse direction contains information about the transverse periodicity of the surface. In contrast, electrons incident on a surface that consists of oriented three-dimensional microcrystals can pass through thin slabs of crystalline material, diffracting in both directions and giving a spotty far field pattern

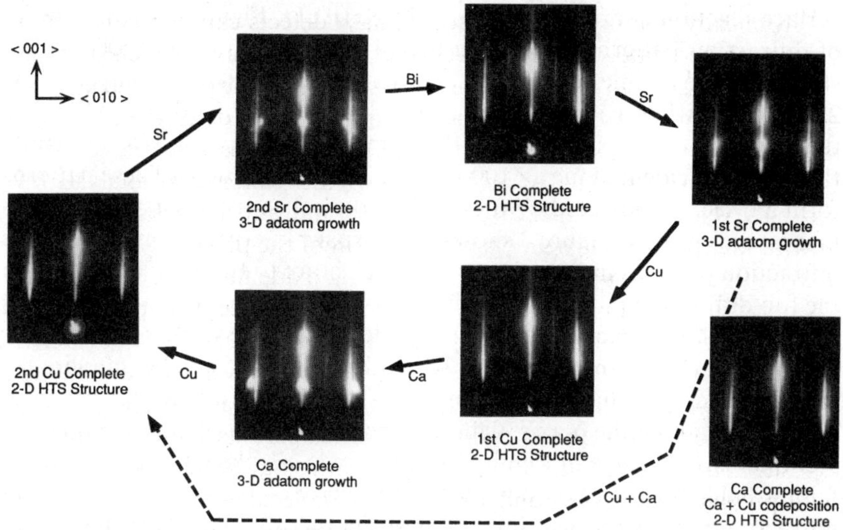

Figure 2 The cycle of RHEED diffraction patterns observed during growth of a single molecular layer of 2212 at times indicated. The patterns showing vertical streaks are flat on an atomic scale, while the spotty patterns show evidence for three-dimensional accumulation. The spots are large and diffuse, indicating that the microstructure is small, about 10 to 15 Å in size.

similar to that seen in transmission electron microscopy (TEM). Thus the diffraction pattern displayed provides a k-space view of the crystallographic microstructure of the surface (specifically, the lattice constants and the symmetry), of the relation of the surface to the underlying film, and of the surface flatness, roughness, and dimensionality on an atomic scale.

When the sequence and number of atoms deposited on the growing surface are chosen appropriately, the nature of chemical reactions at the surface during the crystal growth—specifically, the sequence of transient chemical intermediate states characterizing the reactions that occur during the growth of each unit cell—can be studied, identified, and controlled. By studying the crystal structure of the surface with RHEED during layered growth, we obtain a real-time microscopic window on the chemical processes occurring during the synthesis of each molecular layer. We have used this to study intermediate chemical states that form during the growth of various members of the Bi-Sr-Ca-Cu-O family of superconducting compounds, in both single-phase and heterostructure films.

We illustrate this approach by studying the growth of a molecular monolayer of the BSCCO-2212 phase, as indicated by the evolution of the surface reaction coordinate revealed in RHEED. Figure 2 shows a series of diffraction patterns seen along the ⟨110⟩ azimuth of the $SrTiO_3$ substrate at various times during the growth of a molecular monolayer of the 2212 phase, when each atomic layer is sequentially deposited. (Because the unit cell of 2212 is rotated by 45° from that of the perovskite substrate, the beam is incident along a ⟨100⟩-like azimuth of the film.) These patterns form a cycle, as indicated by the solid arrows, that repeats as each molecular layer is grown. Figure 3 shows a real-space interpretation of what the diffraction patterns indicate the surface structure to be along this cycle. In the top diffraction pattern in Figure 2, taken after the growth of the Bi_2O_2 bilayer, the large spacing of intense bundles of satellite streaks corresponds to an in-plane distance of ~2.7 Å, while the much smaller satellite-to-satellite spacing is characteristic of the ~25 Å incommensurate structural modulation along the *b*-axis. (This can be seen in parts of the TEM cross-sectional lattice image in Figure 4.) This film is *a-b* plane twinned, and the fainter half-order streaks indicate a periodicity of ~5.4, which is seen in diffraction patterns measuring *a*-axis structure (13). The streaky pattern, without any spots, indicates a surface that is locally two dimensional, without three-dimensional, second-phase inclusions. Observations along other azimuths are consistent with these data.

Following the Bi_2O_2 bilayer, a monolayer of Sr is deposited (by this we mean the number of Sr atoms required for the SrO layer in 2212) in the presence of ozone, resulting in the diffraction pattern shown in the next image. The diffuse spots superimposed on the streaky pattern indicate the growth of three-dimensional polycrystalline domains about 10 to 15 Å in size. This pattern shows a textured orientation that can be indexed to the

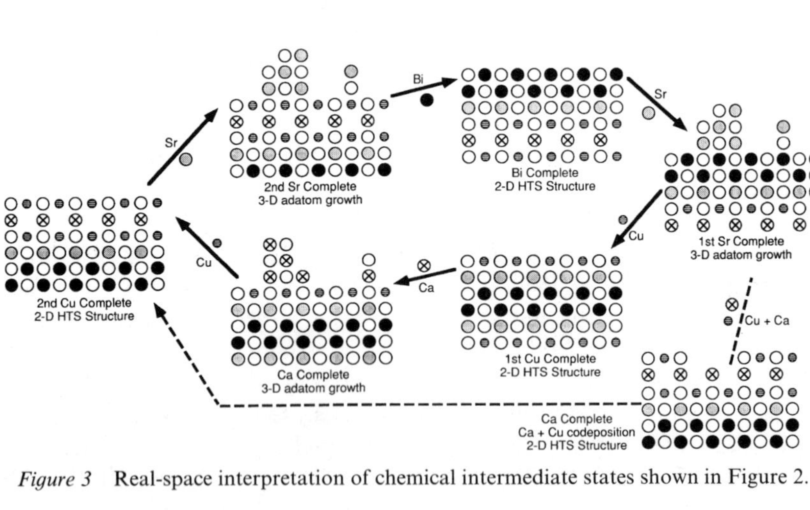

Figure 3 Real-space interpretation of chemical intermediate states shown in Figure 2.

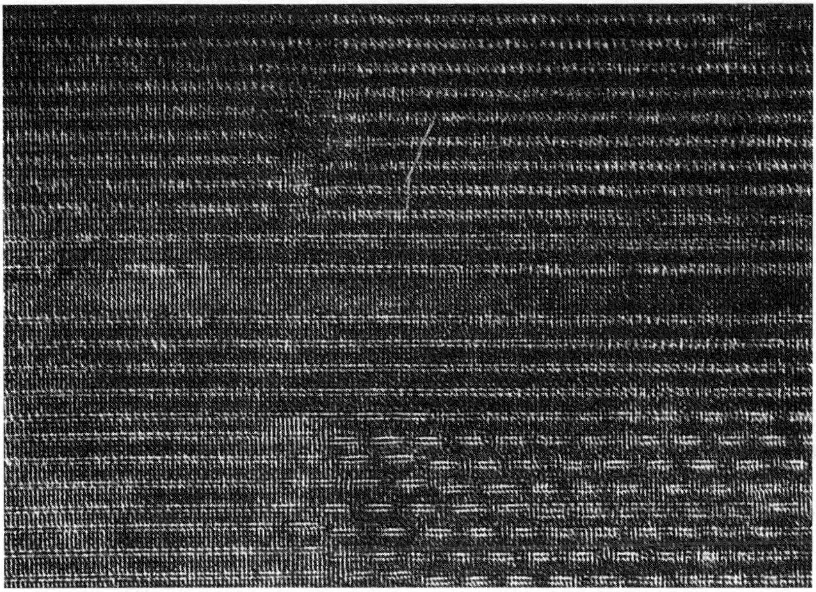

Figure 4 Cross-sectional lattice image transmission electron micrograph showing metastable single $Bi_2Sr_2Ca_7Ca_8O_{20}$ (2278) layer inserted in a film of $Bi_2Sr_2CaCu_2O_8$ (2212). TEM courtesy of Mark Wall, Lawrence Livermore National Laboratory.

crystal structure of SrO. Next, a monolayer of Cu is deposited and oxidized, which results in the next pattern in the cycle. During deposition of the copper layer, the diffuse spots characteristic of three-dimensional domains disappear; by the time the copper deposition is complete the surface is once again two dimensional with an in-plane lattice constant and symmetry identical to that seen after the Bi_2O_2 bilayer.

An effect similar to that seen after SrO growth appears in the next pattern, which shows surface diffraction along the same azimuth, after deposition of the Ca monolayer. The spotty RHEED pattern can be indexed to small three-dimensional domains of CaO. When the growth of the subsequent monolayer of oxidized copper is completed, the surface once again displays only the streaky structure characteristic of the 2212 superconducting phase itself, with somewhat less evidence for the incommensurate structural modulation. A formula unit sequence continues with the growth of the final SrO layer, with dynamics similar to those observed during the growth of the first SrO layer. Finally, as the Bi_2O_2 bilayer is grown, the streaky pattern shown in the first diffraction pattern re-emerges, indicating a flat surface once again. This sequence is repeated for every 2212 molecular formula unit deposited. Similar sequences of diffraction patterns are also seen during the growth of metastable phases, such as $Bi_2Sr_2Ca_{n-1}Cu_nO_{2n+4}$ for large *n*, and when certain types of substitutions are used. Together with postgrowth structural analysis, this information quantifies the stoichiometric and thermodynamic conditions that allow for the synthesis of single-phase metastable compounds.

These pattern sequences can be interpreted in terms of a kinetically driven reaction coordinate of the top molecular layer. In particular, the data indicate the formation of stable subphases after the growth of Cu or Bi layers, which have the same crystallographic structure as the superconductor phase. This can be attributed to two related facts. First, strong electric fields exist between the positively charged $Bi_2O_2^{2+}$ layers and the negatively charged CuO_2^{2-} and $(CuO_2\text{-}Ca\text{-}CuO_2)^{2-}$ layers, which lead to a tightly bonded laminated structure. Second, in order to form the tightly bonded perovskite structure with touching O^{2-} ions, both the relatively larger Sr^{2+} or Ca^{2+} ions and the smaller Cu^{2+} ions are required. When the Cu layers are complete, both of these criteria are satisfied. On the other hand, surfaces with stoichiometries that do not allow such closely packed structures to form do the next best thing—using the available atoms, they form intermediate chemical states of other stable binary or ternary oxide phases on top of the underlying stable crystal structure. These phases are not present in postgrowth X-ray diffraction (XRD) analysis, but their transient existence is evidenced by the transient diffuse spots seen in the RHEED patterns.

The formation of the intermediate state shown in Figure 2 after deposition of Ca can be avoided by depositing both Cu and Ca at the same time. In that case, no evidence for three-dimensional CaO growth is seen; instead a RHEED pattern similar to that shown along the dotted path is seen throughout the growth of the $CaCu_2O_4^{2-}$ layer. The absence of spots indicates that no three-dimensional structure occurs as an intermediate reaction state; only streaks characteristic of the superconductor structure are seen. For larger n-layer compounds, streaky patterns are also seen when $CaCuO_2$ layers are codeposited, indicating that this is the basic two-dimensional building block for the oxygen-deficient perovskite subcell. In this way, the evolution of the reaction coordinate can be directed by manipulating the sequence of atoms available for surface reaction. In particular, this shows that the crystallographic order of the superconducting phase is better maintained when the dotted reaction path is followed, employing partial codeposition of layers that like to grow two dimensionally.

Under microscopic examination the surfaces of such films exhibit flat, featureless, single-crystal field regions that are sometimes interrupted with isolated second-phase defects. On films grown sufficiently near the correct phase stoichiometry, second-phase defects are absent, leaving only the single-crystal field region. Atomic force microscopy (AFM) and scanning tunneling microscopy (STM) characterization indicates surfaces that are smooth to $< \pm 10$ Å over several micron-sized areas, which is consistent with the expected smoothness of the substrate (14). The surface topography is of a terraced structure, with atomically flat regions interrupted by formula unit–sized steps, mainly 15 Å, but sometimes 12 or 18 Å high. The density and orientation of such steps are consistent with the typically small degree of unintentional vicinality the substrate has. For well-oriented substrates the terraces are several thousand angstroms wide. Thus such films are well suited for electronic device applications involving layered heterostructures.

We have used RHEED monitoring during ALL-MBE growth to study intermediate chemical states that form during the growth of various cuprate superconductors, titanates, and rock-salt oxides in both single-phase and heterostructure films. Figure 4 illustrates the type of material engineering that ALL-MBE makes possible. This lattice-image TEM picture shows a film of BSCCO-2212 into which one molecular layer of the metastable 2278 phase was inserted to act as a thin barrier to *c*-axis supercurrent transport. This was done simply by adding six extra $CaCuO_2$ planes to the computer-controlled growth recipe in the perovskite part of that particular BSCCO unit cell. In this way, novel metastable materials can be layered and their properties studied. Moreover, artificial stacking of different kinds of molecular layers to form specific heterostructures can be accomplished.

The transport and morphology of films of the stable-phase BSCCO-2212 are good benchmarks to use to judge the quality of materials grown by ALL-MBE. Figure 5 shows a typical resistive transition measured in a 500-Å thick film of 2212 phase material. The T_c here is >85 K. The residual resistivity, calculated by linear extrapolation to 0 K from higher temperatures, is actually negative, which indicates very little disorder-related scattering. The normal-state resistivity just above the onset of the superconducting transition is $<100\ \mu\Omega$ cm. Such films have also been patterned with small constrictions for transport critical-current (J_c) measurements. These show $J_c > 2.0 \times 10^7$ A/cm^2 at 4.2 K. These transport data indicate that such single-crystal films—although grown at much lower temperatures than bulk single crystals and far from thermodynamic equilibrium—have properties good enough to use in engineered heterostructures designed to study fundamental aspects of superconductivity.

Superconducting Superlattices

One of the principal advantages of the ALL-MBE technique is that it is ideally suited for synthesis of atomically precise metastable multilayers

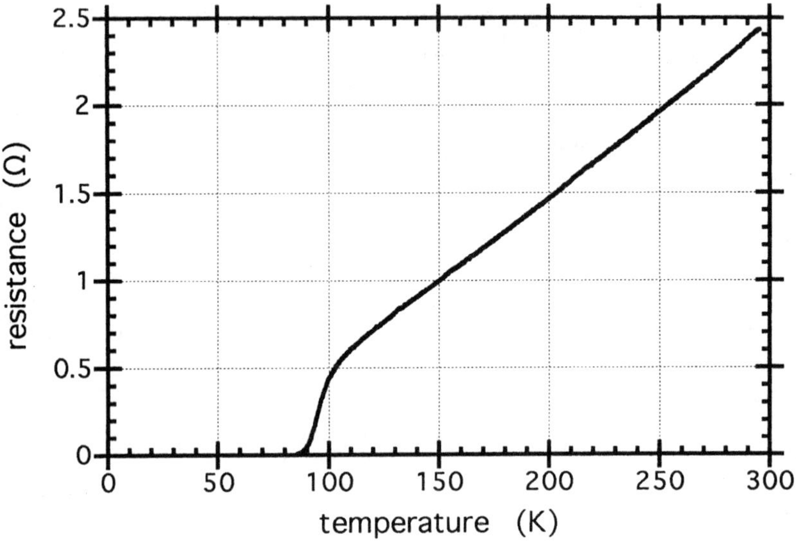

Figure 5 Resistive transition of 500-Å thick film of 2212 phase grown by ALL-MBE.

and superlattices. We have synthesized a number of superlattice samples consisting of alternating layers of various members of the Bi-Sr-Ca-Cu-O family of compounds. In one particularly interesting case (15), molecular monolayers of the 2201 phase ($\sim$12.5 Å) were interleaved with molecular monolayers of the 2223 phase ($\sim$18.5 Å); this directly tests the ability to layer metastable structures atomically. Had the atoms mixed from one molecular layer to another, some fraction of the film would have converted to the 2212 phase, and in this case the 001 XRD reflection, characteristic of about 31 Å spacing, would have been reduced or even absent. Instead, what is observed experimentally is a strong 001 reflection near 31 Å as shown in the top curve of Figure 6, along with the 002 reflection. The second curve in Figure 6 is a numerical simulation (15) obtained from the known unit cell structures of the 2201 and 2223 compounds. The good match of relative intensities of the 001 and 002 reflections is clear, and indicates that under the growth conditions we use ($T_{sub} \approx 730°C$), there is no significant mixing between adjacent molecular layers during growth. The relatively large contribution of the strong laminating electric fields in the *c*-axis direction in cuprates to the total cohesive energy should tend to quench bulk diffusion. Apparently, once an electrostatically neutral unit cell is complete, it is quite inert to participation in subsequent chemistry involved in the growth of the next unit cell.

High-T_c superlattices have also been used to address a question of prime

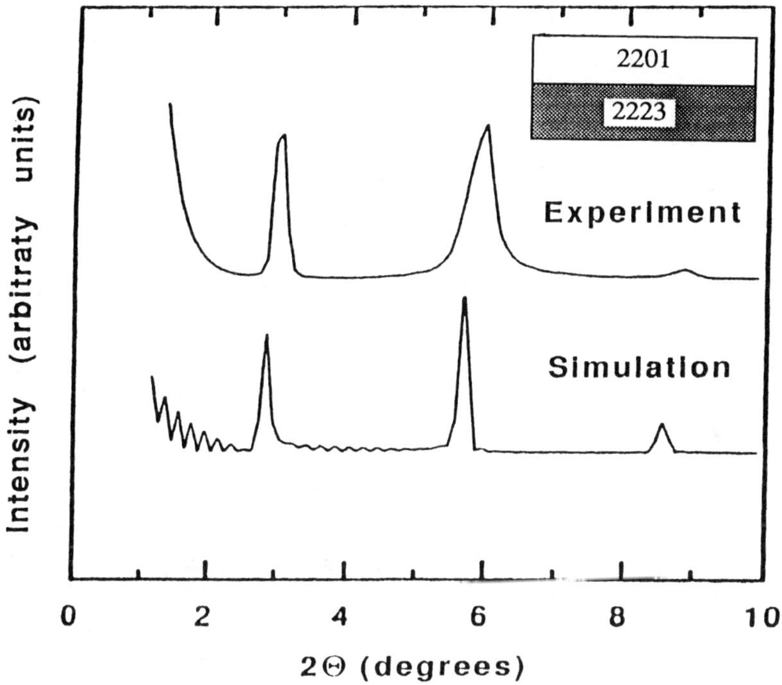

Figure 6 X-ray diffraction measurement of 2201/2223 superlattice. In this film, single molecular layers of the 2201 phase were spaced apart by single molecular layers of the 2223 phase. The top curve shows the measured X-ray diffraction pattern, while the bottom curve shows the results of a numerical simulation. If the molecular layers had mixed to form layers of the 2212 phase, the 001 peak would be reduced.

theoretical interest: Is the superconductivity in HTS materials essentially two dimensional? To test this, superlattices were grown in which the extended super-unit-cell consisted of a single molecular monolayer of the high-T_c 2212 phase ($\sim$15.5 Å) followed by n layers of the low-T_c 2201 phase, effectively isolating the high-T_c molecular monolayers from each other without causing carrier depletion from the high-T_c layers into the adjacent material (16). (For 2201, the superconducting transition occurs at about 10 to 15 K, and the carrier concentration is similar to that of 2212.) Samples with $n = 1, 2, 3, 4, 5$, and 10 were grown in order to study the coupling between unit cells required for superconductivity to occur.

The superlattice samples were characterized by XRD. A clear signature of the desired superlattice structure is shown in the XRD pattern in Figure 7. Here all the peaks can be indexed to a single extended super-unit-cell lattice parameter. They are as sharp as the peaks in single-phase 2201 or

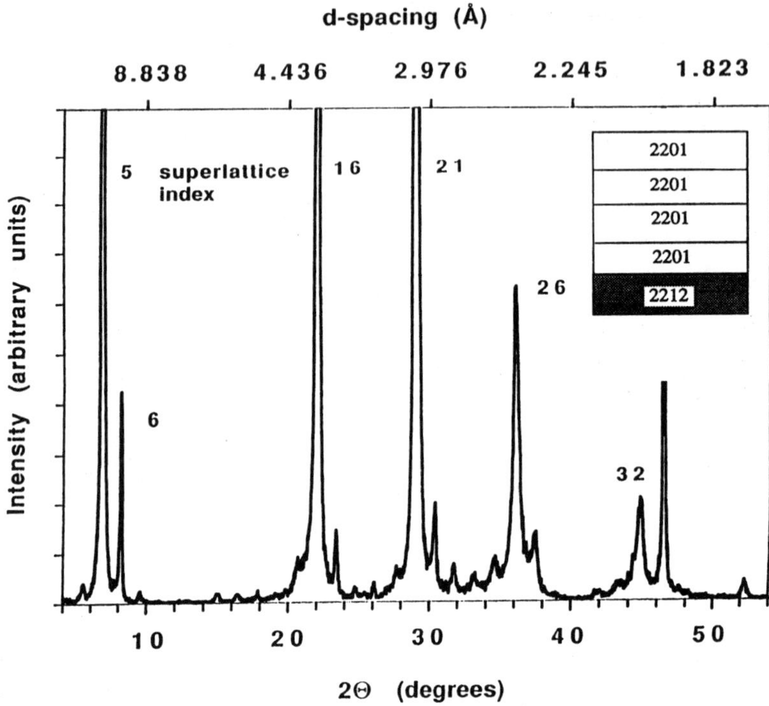

Figure 7 X-ray diffraction measurement of $n \times 2201/2212$ superlattice. In this film, several molecular layers, $n = 4$, of the 2201 phase separated single molecular layers of the 2212 phase.

2212 samples. The *a-b* plane transport properties of the films were then studied. Figure 8 shows the temperature dependence of the resistance for three of the samples with $n = 1$, 2, and 5, as well as transitions measured for single-phase films of the 2201 and 2212 phases. The data clearly indicate that the superconducting transitions in the dilute superlattices were almost identical to those seen in the bulk 2212 film (16). This implies that very little coupling between adjacent high-T_c unit cells is necessary for the formation of the superconducting condensate and shows that the pairing interaction responsible for the high-T_c superconductivity is essentially two dimensional.

These transport results are in contrast to conclusions reported by several other groups that have grown and studied dilute superlattices of the HTS compound $YBa_2Cu_3O_7$ (YBCO) in nonsuperconducting $PrBa_2Cu_3O_7$ (PBCO) (17–19). They reported significantly reduced superconductive transition temperatures when single molecular layers of YBCO were

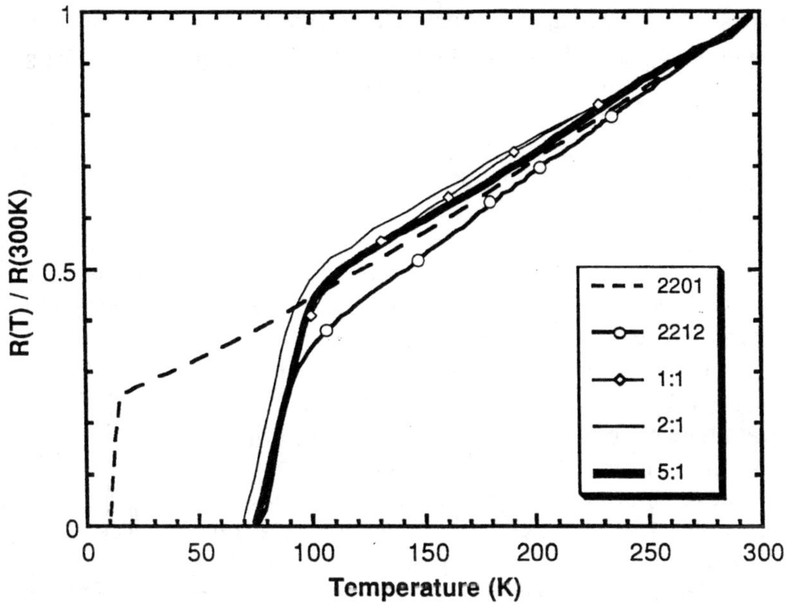

Figure 8 Temperature dependence of the resistance in five films. Results for single-phase films are indicated as well as results for dilute superlattices in which single molecular layers of the 2212 phase were spaced by one, two, and five molecular layers of the lower T_c 2201 phase. Essentially the same resistive transition is observed for the superlattice films as is seen in the bulk 2212 film.

spaced apart with layers of PBCO. Such disagreement may be due to intrinsic differences between superconductivity in YBCO and its derivatives compared with Bi-family superconductors. More likely, the layer architecture chosen for the BSCCO work may be more suitable for investigating the fundamental dimensionality of the high-T_c condensate because its structure does not suffer from significant parasitic effects that may influence the YBCO/PBCO superlattices. In particular, because the carrier concentration of the superconductor, 2212, is similar to that of the spacer layer, 2201, in the case of Bi-family superlattices, charge depletion in the single, separated layers of 2212 would not be expected to cause a parasitic reduction of T_c. This result also shows that each of the molecular layers of the higher-T_c phase grown in such superlattice samples individually exhibits near-ideal superconducting properties.

High-T_c Trilayer Josephson Junctions

Another useful application of heterostructures grown by ALL-MBE is artificially synthesizing accurate trilayer structures for Josephson junctions

(20). In such devices, superconducting transport between base and counter-electrode layers of HTS compounds—2212 in our case—is impeded by a barrier layer. By engineering the properties of barrier layers, device properties have been systematically altered in much the same way as bandgap engineering is used in layered semiconductor structures (21). Although transport occurs in the difficult *c*-axis direction, this is apparently more than compensated by the high degree of control ALL-MBE has for *c*-axis growth and the near perfect heterointerfaces that result between superconducting and barrier layers. In a first set of experiments, single molecular layers of such metastable phases as BSCCO-2278 were used as barrier layers, with the Ca site doped with trivalent cations (Dy or Bi) to suppress free carriers. Such a film architecture is shown in the TEM image in Figure 4. Four-contact devices to test vertical transport were fabricated by wet chemical etching for device isolation, followed by ion milling, to expose the bottom 2212 base electrode and to define a 30 μm-by-30 μm active mesa. Sputtered SiO_2 served as an isolation layer for metallization; evaporated silver was used for ohmic contacts, interconnect lines, and probe pads. In an attempt at barrier engineering, several barrier designs were investigated as listed in Table 1 (21). Specifically, by doping similar metastable single molecular layers with Dy, significantly more resistive barriers were obtained. Two values of Dy doping were studied. In one, 0.5 of a Ca atom was substituted by Dy per formula unit, which is expected to produce a composition near the metal-to-insulator transition (22). In the other, 1.85 Dy atoms were substituted for Ca per formula unit, which is expected to dope well beyond the metal-to-insulator transition. Vertical-transport, trilayer Josephson junctions were fabricated using the same process described above. Figure 9 summarizes the results obtained. Devices doped more lightly and with Bi had higher critical currents and lower normal-state resistance; devices doped more heavily and with Dy had

Table 1 Barrier layer structure and device characteristics for a series of nine films[a]

Film	Barrier layer structure	Barrier dopant	Critical current (μA)
A–D	$Bi_2Sr_2(Ca, Sr, Bi)_7Cu_8O_{20}$ (34 Å)	Bi = 0.50	100–400
E		Bi = 0.84	10–300
F, G	$Bi_2Sr_2(Ca, Dy)_{10}Cu_{11}O_{26}$ (43 Å)	Dy = 0.50	2–10
H	$Bi_2Sr_2(Ca, Dy)_4Cu_5O_{14}$ (25 Å)	Dy = 1.85	0.1–10
I	$Bi_2Sr_2(Ca, Dy)_9Cu_{10}O_{24}$ (40 Å)		

[a] The amount of Bi or Dy doping in the barriers is given in terms of atomic monolayers of dopant per molecular layer of barrier material; for cuprate superconductors, this is equivalent to the formula unit concentration, e.g. Bi = 2.0 in $Bi_2Sr_2CaCu_2O_8$.

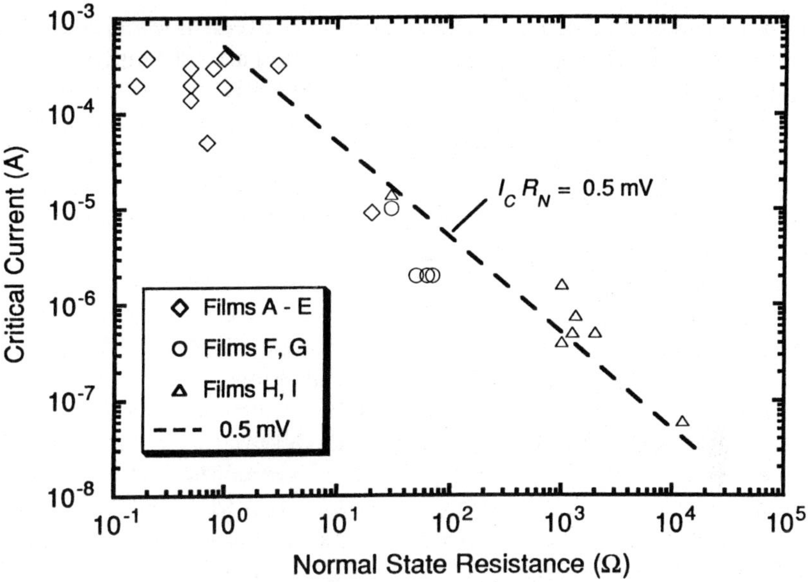

Figure 9 Critical current vs normal-state resistance for devices from nine growths. Films A–E incorporated single molecular barriers in which the Ca sites were doped with 0.85 and 0.5 Bi atoms per formula unit. Films F and G had similar barriers, but with 0.5 Dy atom doping per formula unit. In films H and I the Dy doping was increased to 1.85 Dy atoms per formula unit. The results indicate that such doping changes critical current and normal-state resistance by three orders of magnitude while maintaining nearly constant $I_c R_n$ voltage.

lower critical currents and much higher normal-state resistance. Scaling of the junction critical current and normal-state resistance over three orders of magnitude was obtained, while the $I_c R_n$ product, measured at 4.2 K, remained approximately constant at 0.5 mV.

Stacked Series Arrays of Superconductor-Normal Metal-Superconductor (SNS) Josephson Junctions

Although it is certainly possible to grow heterostructures using the thick barrier layers described above and shown in Figure 4, the growth of BSCCO-2212 layers on top of such a barrier can suffer from a large number of crystallographic point and line defects such as stacking faults and antiphase boundaries, especially on significantly vicinal substrates. This problem is evidenced in RHEED growth as a change in the typical sequence and timing of intermediate-state features in the first 10 or so layers after a large unit cell like 2278. In TEM such features have been directly observed. The use of a single molecular layer of doped 2212 as a barrier layer for making SNS-like junctions appears to offer a convenient way to avoid this problem and to stack many Josephson junctions close together in a multilayer structure. The advantage of 2212-based barriers is that their crystal structures are virtually identical to those of all other superconducting 2212 layers. During growth of these structures, the transient surface chemistry dynamics as seen in RHEED (see Figure 2) is very similar for doped and undoped unit cells. Among other advantages, this makes it easy to stack more than one closely spaced barrier layer, and to explore mutual interactions among junctions.

The layer architecture for stacked junctions using doped 2212 barrier layers is shown in Figure 10. The base and counter-electrode layers are several hundreds of angstroms thick, and five 2212 layers (75 Å) are used for the middle electrode that separates the upper and lower barriers. Dysprosium is used as a dopant because it readily occupies the Ca site and compared with Ca, donates one more electron to the cuprate states at the Fermi energy. Thus it acts to reduce the local hole concentration. This has been studied in bulk samples, where it was found that an increasing Dy concentration led to increased normal-state resistivity and a lower superconducting transition temperature (22). In that work, for 50% substitution no sign of superconductivity was seen, and for higher concentrations higher-resistivity insulating behavior was observed. To

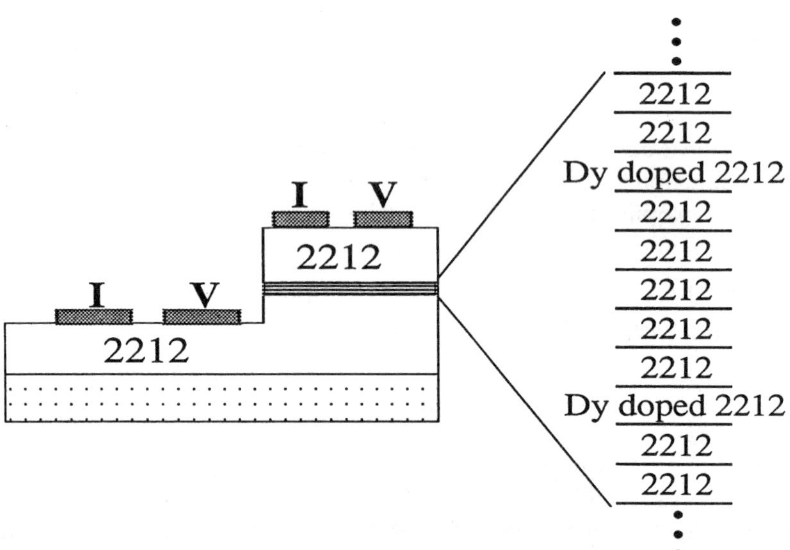

Figure 10 Schematic of layer architecture and device structure for stacked series arrays of Josephson junctions formed by inserting single Dy-doped molecular layers of BSCCO-2212 as barriers. The center superconducting electrode layer consists of five molecular layers of 2212 and is 75-Å thick.

characterize our material, single-phase, thin-film samples of $Bi_2Sr_2(Ca_{.65}Dy_{.35})Cu_2O_8$ were grown by ALL-MBE to characterize the nature of the barrier material by itself, not subject to any proximity effect from adjacent superconducting 2212. The resistance vs temperature measured is shown in Figure 11. The material is clearly not a superconductor in and of itself at temperatures above 20 K, and its resistivity is comparable to what has been observed in bulk samples with similar chemical composition.

Multilayer samples using single 2212 formula unit layers substituted with Dy as barriers were grown and fabricated into trilayer transport devices. The devices described here were either square 30 μm × 30 μm mesas or smaller circular mesas with diameters of 9 or 7 μm, and true four-point top- and bottom-layer contacts were defined using a five-step microfabrication process. The first structure grown replaced all of the Ca^{2+} ions in one formula unit layer with Dy^{3+} ions to make a barrier. In bulk samples such a substitution is known to make the material highly insulating. The junctions made out of this material showed nonlinear *I-V* curves but no critical current. At zero bias the resistance was about 0.1 ohm, while at large bias it looked like a >3 ohm resistance. This doping was too great for Josephson coupling, as no *c*-axis critical current was seen. In a second try, 35% of the Ca was replaced with Dy. This is the same concentration that has by itself the transition shown in Figure 11. The *c*-axis transport-resistive transition that was obtained is shown in

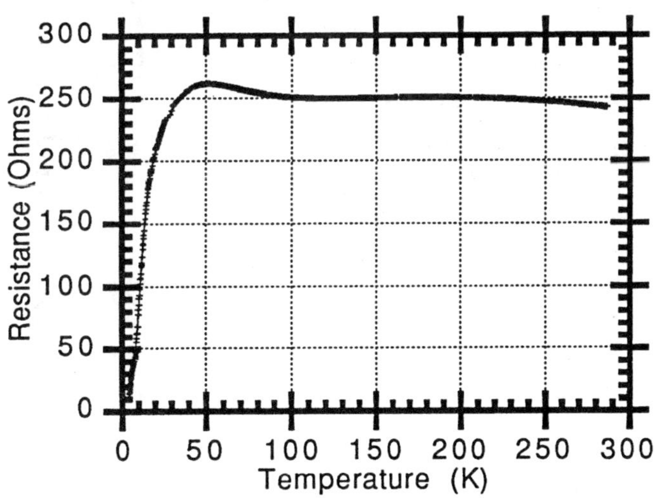

Figure 11 Basal plane resistive transition of a 2000 Å thick film of $Bi_2Sr_2(Ca_{.65}Dy_{.35})Cu_2O_8$. Single molecular layers of this material are used as barriers in the devices reported below.

Figure 12. This transport geometry is somewhat more complicated than that used to obtain the data in Figure 11, and it is necessary to separate out the in-plane resistive transition of the electrodes from the c-axis transition through the barrier. The electrodes become superconducting at about 80 K. The c-axis transition through the barrier is seen, then, at temperatures below this and looks like a "foot" added to the film transition, typical of SNS junctions with a proximity-effect transition. Clearly, there is a qualitative difference between the temperature at which the single-phase material begins to go superconducting (<20 K) compared to what is seen when an isolated molecular layer of this material is placed near superconducting 2212. By placing the Dy-substituted 2212 molecular layer next to two superconducting 2212 slabs, the transition through the barrier is raised to more than 60 K. When 50% Dy-substituted barriers are used, similar transitions are obtained, but with the proximity-effect transition complete at 45 K. This is also substantially different from what is seen in single-phase samples of this composition and shows that proximity coupling of superconductivity into nominally nonsuperconducting cuprate layers readily occurs across such interfaces between doped and undoped layers, at least for doped layers with sufficient carrier concentrations. It suggests that when the carrier concentration is too low for intrinsic superconductivity to arise at some temperature, it may still be high enough to

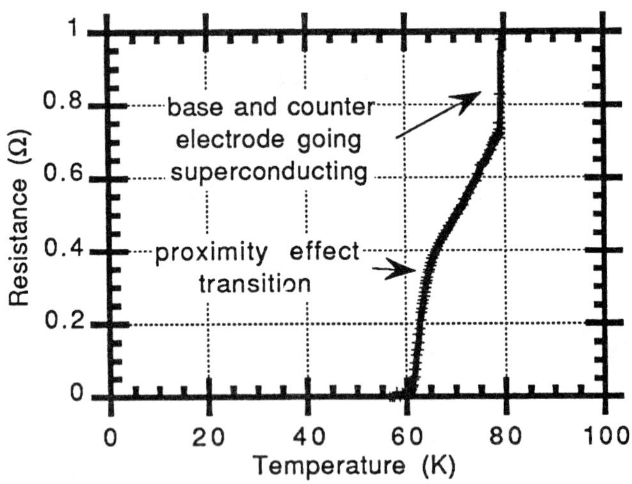

Figure 12 Resistive transition of series-stacked Josephson junction array. The device has two barriers spaced apart by five molecular layers of the superconducting BSCCO 2212 phase. Each barrier consists of one molecular layer of Dy-doped 2212. The Dy concentration used is the same as in the sample of Figure 11.

sustain proximity-coupled superconductivity from a strongly superconducting neighboring layer.

The trilayer transport devices made with these wafers were all good Josephson junctions. For a Dy substitution of 50% they worked best at 25 to 45 K, while the 35%-substituted barriers worked best from 50 to 65 K. The *I-V* curve for a typical junction with a 50% Dy-substituted barrier measured at 43 K is shown in Figure 13. Below this temperature range, the critical current became larger and the *I-V* curves became distorted and appeared to exhibit flux-flow behavior as expected from a long junction. The *I-V* curves showed only one voltage-state transition, suggesting that the stacked junctions are nominally identical, that is, they have similar I_c and R. When microwave power was applied, steps were seen only at twice the voltage expected for a single junction. This is shown in Figure 14 for the same stacked junction device as in Figure 13. The mechanism of phase locking in these samples is not yet understood. It is not possible to characterize the junctions in each stack independently, but the fact that the *I-V* curves for the stack with and without microwave power show no independent junction features suggests the possibility of mutual phase locking between the stacked junctions. Unlike in earlier studies of phase locking, this two-junction stack has very few pairs in the superconducting 2212 layer between the two barriers because the middle layer is so thin and cuprates have a relatively low carrier concentration. Because the equations describing each Josephson junction are nonlinear and are coupled by the magnitude of the order parameter in the middle layer,

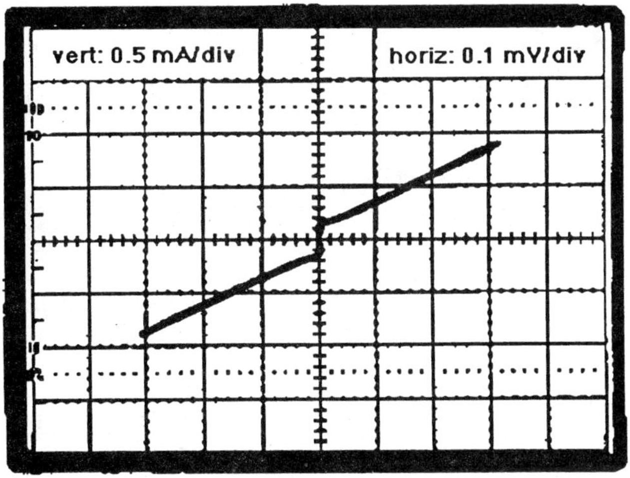

Figure 13 *I-V* characteristics of series-stacked Josephson junction array measured at 43 K.

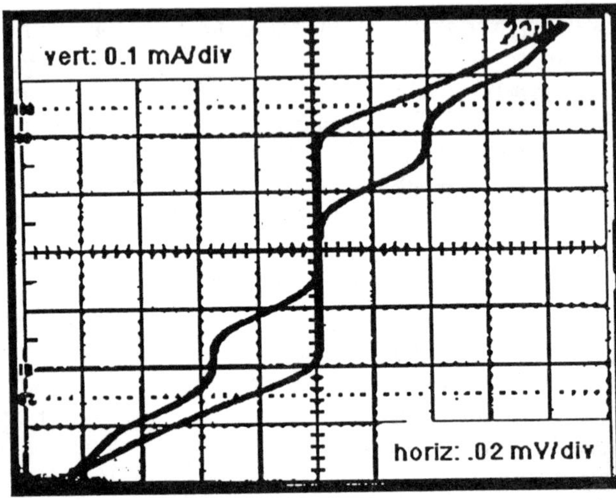

Figure 14 Double exposure showing a close-up *I-V* curve of the device of Figure 4 with and without microwaves at 40 K. The current steps happen at twice the voltage they would for a single junction, indicating phase coupling of the series array.

the phase locking observed may occur through direct coupling of order parameters. Because the ac Josephson currents are more easily able to modulate the magnitude of the order parameter when the number of pairs in the middle layer is small, this coupling may be quite strong. This can be studied by making samples with variable spacing and intentionally mismatched critical currents within a single stack.

The limited temperature range over which the large junctions showed resistively shunted junction (RSJ)-like curves was extended by using a self-aligned process that allowed the fabrication of smaller-area devices. Wafers with similar epitaxial structures were processed into circular mesas with diameters of 9 and 7 μm, thus reducing the area by as much as a factor of 23. This allowed devices to be tested over a much wider temperature range without large-junction effects distorting the *I-V* characteristics. Figure 15 shows the *I-V* curves recorded from one wafer at 4.0 K. These devices showed a remarkable uniformity in critical current density over a large part of the wafer area. Five of the devices had values of J_c within $\pm 5\%$. The one device with a larger critical current density was patterned in an area near second-phase defects, which arose from the intrinsic stoichiometry spread—determined by source position—that exists over the wafer. This trend has been observed on other wafers as well. Variations in layer stoichiometry can be correlated with variations in the transport critical current of the layered devices. By the same token,

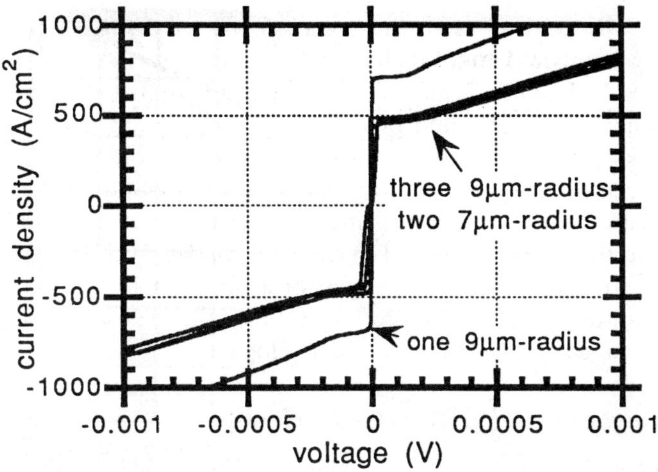

Figure 15 Critical current density of devices measured on one wafer at 4.0 K. Five devices have current densities that are the same within $\pm 5\%$. The device with the larger critical current was patterned near a region of second-phase defects.

control over layer stoichiometry leads to more uniform device properties. The response of the smaller devices to higher-frequency microwaves was also measured. Figure 16 shows microwave steps observed when a 9 μm radius device was illuminated with 75 GHz power. Here, sharp current

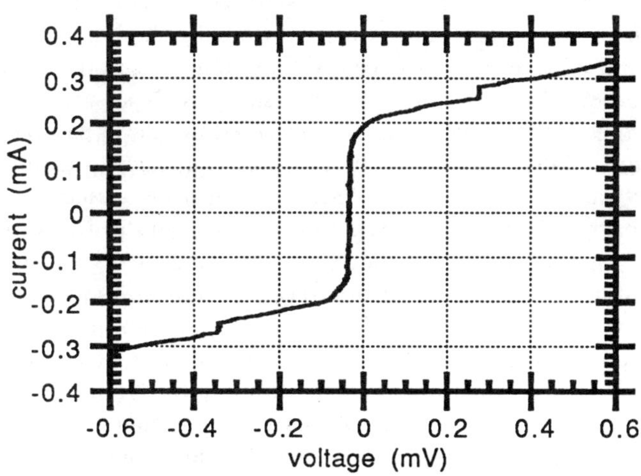

Figure 16 *I-V* characteristics of a 9μ diameter device with 75 GHz microwave illumination at 4.0 K. The current steps occur at voltages twice what a single junction gives. The curve is offset in voltage by approximately 50 μV.

steps are observed at twice the voltage expected for a single junction, indicating phase locking of the two series junctions.

Such stacked junctions may be technologically useful. If many junctions can be stacked together and phase-coherent operation maintained, then such stacks may be utilized to make very precise voltage standards like those currently made by the National Institute of Standards and Technology using low-temperature superconducting series arrays (23). Also, stacks like these may be useful in digital applications. There, the higher voltage that a phase-coherent stack of Josephson junctions has in the switched state could help in interfacing low-voltage Josephson logic with higher-voltage cryogenic semiconductor logic.

Tunneling Through Titanate Barriers

Heterostructures of cuprate slabs and very thin insulating titanate layers designed to make tunneling structures have also been grown by ALL-MBE, and trilayer transport through them was studied in the way described above for SNS trilayers. Such structures are interesting because they can in principle be grown perfectly lattice matched, e.g. the lattice constants of simple perovskites such as $CaTiO_3$ can be within 0.3% that of BSCCO 2212. Such structures can thus bring together superconducting layers and very thin insulating layers at atomically precise heterojunctions in order to look for tunneling phenomena. A variety of such structures were grown in which the insulating titanate layers were incorporated in different ways, and these were processed into *c*-axis trilayer transport devices. Titanate layers were inserted within BSCCO unit cells in some samples and were inserted between BSCCO unit cells at the natural cleavage plane between BiO layers in other samples. No *c*-axis supercurrent through the titanate barriers was observed in these samples, but an exponential relationship between zero bias resistance and titanate layer thickness was observed for thin barriers, indicating tunneling transport. Thicker barriers showed this relationship to saturate. More than seven orders of magnitude of resistance was spanned in this set of samples. The resistance vs temperature observed, as well as the current voltage characteristics, followed power-law behavior, suggestive of one-phonon–assisted tunneling between localized states. By analyzing the nature of transport in the cuprate layers immediately adjacent to the titanate layers, it was discovered that these CuO_2 planes were depleted of carriers, implying that the interface to the titanate layer modified the nature of the neighboring cuprate layers significantly. In particular, along with reduced hole density, carrier localization occurs in these interface layers, and we have found that the device *c*-axis current-voltage characteristics can be explained in terms of phonon-assisted hopping over the titanate barrier between localized interface cuprate states on either side of the barrier.

The first barriers we studied consisted of unit cells of $SrTiO_3$ grown in the middle of thick cuprate slabs, such as BSCCO-2278, like that shown in Figure 4. An example of such a structure is indicated schematically in Figure 17. Because these c-axis transport barriers are inserted within the structure of a thick BSCCO unit cell, we describe these as "intracell" barriers. Samples were grown in which one, three, five, and seven unit cells of $SrTiO_3$ were inserted as shown. These gave barriers between 4 and 28 Å thick. Below and above this barrier layer, films of 2212 were grown to form the base and counter-electrode layers. Figure 18 shows a log-log plot of the resistive transitions observed in this series. In these samples the measured four-point resistance consists of the base and counter-electrode spreading resistances in series with c-axis transport through the barrier. For the thinnest barrier, one unit cell of $SrTiO_3$ (4 Å thick), the superconducting transition in the base and counter-electrode can clearly be seen. It is also noticeable in the film with three unit cells of $SrTiO_3$. For thicker barriers, however, resistance through the barrier dominates the measured resistance at all temperatures.

Below T_c these samples have a power-law resistance vs temperature

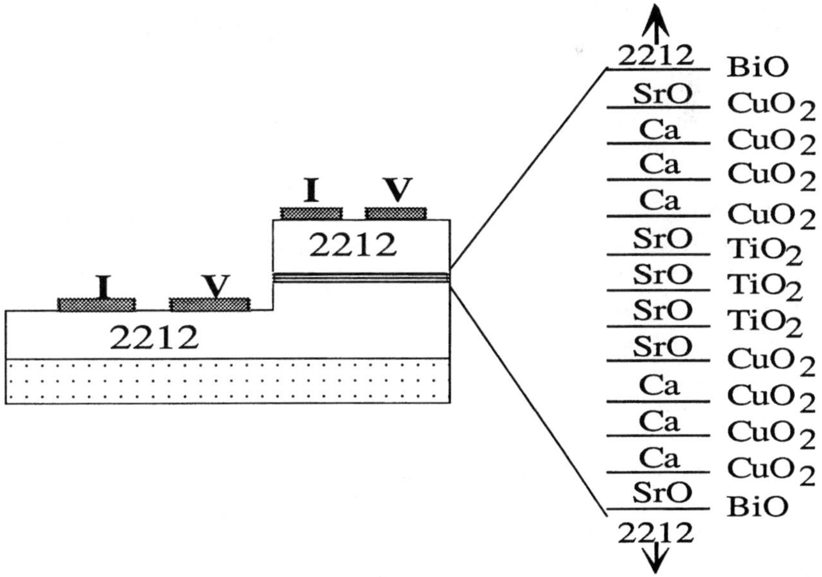

Figure 17 Schematic of atomic oxide layering scheme within an intracell tunnel barrier. This shows 3.5 unit cells of $SrTiO_3$ layered within a molecular layer of BSCCO-2278.

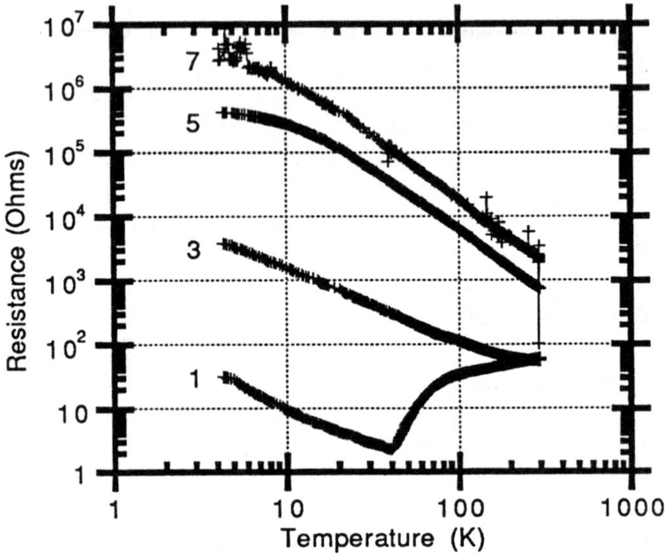

Figure 18 Resistance vs temperature for samples with one, three, five, and seven unit cells of $SrTiO_3$ inserted into BSCCO 2278 to act as a tunnel barrier.

relationship. The resistance was measured at low bias currents where the sample conductance was constant. The deviation of power-law behavior for the five-layer data at low temperatures is not real, but rather due to a bias current that was too large for such a large resistance to remain in the constant-conductance voltage range. Below about 50 K, the power-law $R(T)$ for the one and three $SrTiO_3$ unit cell barriers was about $T^{1.2}$. For the five- and seven-layer barriers it was somewhat higher, about $T^{1.6}$.

Below 50 K the zero bias resistance is clearly exponentially related to the barrier thickness. This implies that tunneling is controlling the device transport. Increasing the barrier thickness by one layer of $SrTiO_3$ increases the zero bias resistance by about a factor of 10. This is true even for the thinnest barrier used, i.e. one unit cell layer of $SrTiO_3$. This says that the exponential behavior holds even down to the thinnest barrier possible, which implies that the barrier/cuprate microstructure is quite abrupt and not dominated by diffusion. Picking the data at 10 K, a plot of zero bias resistance vs barrier thickness is shown in Figure 19. An exponential relationship holds for more than four orders of magnitude change in zero bias resistance. Expressing the zero bias resistance as $R(d) = R_o \exp(2\gamma d)$, where γ is the magnitude of the imaginary wavevector characteristic of the wave function while tunneling through the barrier of thickness d, the slope of the line in Figure 19 implies that $\gamma = 3.3 \times 10^9\ m^{-1}$. This in turn can be

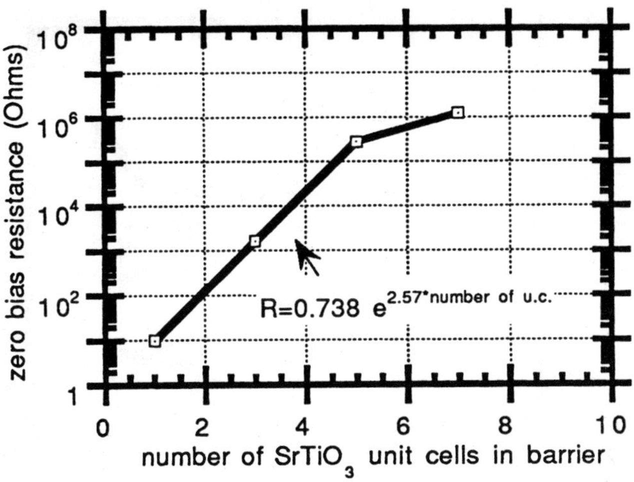

Figure 19 Exponential dependence of zero bias resistance upon the thickness of the titanate layer demonstrates tunneling.

related to a barrier height through which the tunneling occurs, $\Delta V = 340$ mV, if the effective mass in the barrier is equal to one electron mass. Such a value for a potential difference between the Fermi energy in the cuprate and in the edge of the titanate oxygen-derived band through which the current tunnels is not unreasonable.

Interestingly, the *I-V* curves also show predominantly power-law dependence of current on voltage. Figure 20 shows a plot of the differential conductance, dI/dV, vs voltage for a sample with a barrier made of three titanate unit cells (12 Å thick) measured at three different temperatures. The conductance is constant at low bias voltages. At a bias of about $2\,kT/e$, it transitions to being somewhat stronger than linearly dependent on voltage. This suggests that a temperature-controlled population of a density of states (DOS), either electronic or phononic, is involved. In particular, it is clear that the *G-V* data do not reveal a conventional superconductive DOS with a sharp gap as is seen in low T_c superconductor-insulator-superconductor (SIS) tunnel junctions.

At this point, it is useful to recall that tunnel junctions are made up of two parts, the barrier and the electrodes. Because the DOS measured by tunneling is dominated by the DOS of the interface layers, it is necessary to study the nature of the electronic states and carrier density in CuO_2 planes immediately adjacent to the titanate planes. In any heterojunction, band line-up and carrier depletion can change the position of the Fermi energy at the interface from its bulk equilibrium value. In our case, this

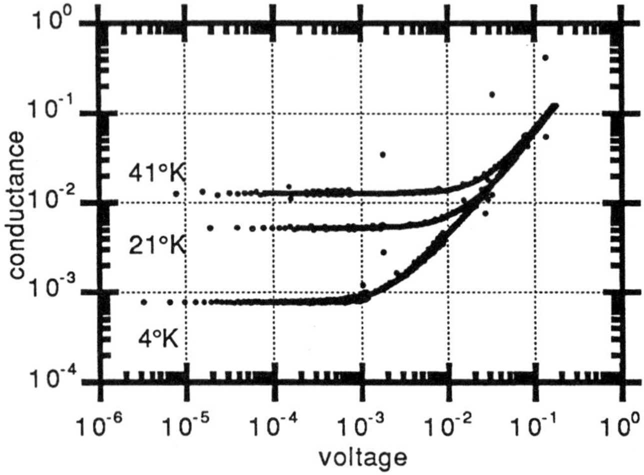

Figure 20 Temperature dependence of differential conductance as a function of bias voltage. For voltages less than about kT/e the conductance is constant. For higher voltages it asymptotically approaches a universal power law. No gap features are seen.

was tested by growing films consisting entirely of barrier unit cells such as the structure shown schematically in Figure 17. It is the DOS of these cuprate planes in the tunnel junction structure that determines the *I-V* characteristics, and studying such samples is an independent measure of the nature of the DOS that acts as the source and sink of carrier transport across the barrier. Ideally, in order to hope to see superconductive DOS in these planes, the transport in such test films should be superconducting. When several such samples were fabricated, however, another picture emerged. Figure 21 shows the basal *a-b*-plane (not *c*-axis) resistive transition measured. Here we plot the logarithm of the resistance vs $1/T^{1/3}$; this is the temperature dependence of two-dimensional variable-range hopping, 2D-VRH (24). Fitting with other transport laws was attempted, but none of them showed as good a fit to the low-temperature data as two-dimensional variable-range hopping. In particular, the transport is clearly not simply thermally activated nor is it three-dimensional variable-range hopping. This transport result implies that the CuO_2 planes nearest the titanate layer are, in fact, not superconducting. Furthermore, these cuprate layers have a reduced density of carriers and enough disorder to lead to Anderson localization. Thus the proper view of the tunneling structures is not SIS, but rather S-L-I-L-S, where L stands for the thin nonsuperconducting layer of localized states at the interface.

Trilayer transport in the structures we made should then be closely

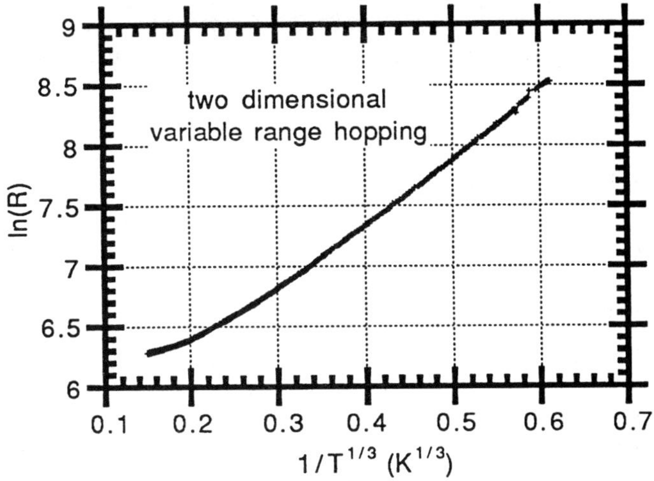

Figure 21 Resistance vs temperature of a superlattice film consisting entirely of cuprate/titanate interface layers. The curve shows that the cuprate layers next to the titanate layers have a reduced carrier concentration and exhibit carrier localization. Thus the transport through the tunnel junction is dominated by localization at the interface.

related to tunneling through semiconductor layers with localized DOS in metal-semiconductor-metal samples as analyzed by Glazman & Matveev (25). They showed, for thin barrier layers with localized states that are smeared out in energy, that the tunneling process can be broken down into a series of parallel conductance channels. In order to hop from one localized state to another, a phonon makes the process more likely by opening up channels that involve hopping between states of different energy. In our superlattice samples, the two-dimensional-VRH occurs for temperatures less than about 50 K. This, together with the slope of the 2D-VRH curve, implies a smearing out of the band-edge states over an energy interval of about 5 to 15 meV. Thus even for relatively high temperatures, the one-phonon–assisted tunneling probability is enhanced. While the geometry and density of states in our structures are not exactly what was analyzed by Glazman & Matveev, it is interesting to note that their model predicts nearly the same power-law behavior for the dependence of resistance on temperature and of critical current on voltage that we observe. In their model, the one-phonon–assisted tunneling process leads to a resistance that is proportional to $T^{1.33}$ and a current for voltages larger than kT/e that is proportional to $V^{2.33}$. These exponents are close to what we have measured in our samples, and we conclude that the transport in our S-L-I-L-S junctions is dominated by one-phonon–assisted tunneling between localized states on either side of the barrier layer and that the role of the superconducting electrodes is simply to supply current in a lossless manner to the interface region.

The question remains, What causes the CuO_2 planes next to the titanate layers to have a low carrier density and to be characterized by a localized DOS? If the titanate grows with a deficiency of oxygen, then extra electrons from the titanium atoms that would normally be transferred to oxygen ions in the titanate lattice would be available to fill states locally that are normally occupied by holes at the heterointerface. In addition, if there is any irregularity in the lattice structure, such as vacancies or intersite substitutions, then Anderson localization is likely at a low enough carrier concentration, similar to what is seen in semiconductor systems. Whatever the cause of the carrier depletion/localization observed at these interfaces, it remains a problem that must be solved in order to obtain true high-T_c SIS tunnel junctions that have transport determined by the superconducting states DOS, i.e. supercurrent and quasi-particle DOS-derived I-V characteristics.

Conclusion

Oxide heterojunctions offer the possibility of bringing a wide range of physical phenomena together at nearly lattice-matched interfaces. Because of the large number of oxides and the possibility of substitutional doping, independent control over many material properties in individual molecular layers is possible. We have studied heterojunction samples including high-T_c superconductors as well as thin nonsuperconducting layers that act as barriers to c-axis transport. Cuprate layers can be doped with trivalent cations to make them inherently nonsuperconducting. Such layers when placed as barriers in trilayer samples act as proximity-coupled normal layers and make Josephson junctions that work well up to 65 K. Moreover, when these barriers are made up of the underlying superconductor phase, BSCCO-2212 in our case, they can easily be stacked to make series arrays. These exhibit coherent phase locking and may be technologically useful. We have also studied heterojunctions between titanate and cuprate atomic oxide layers. We have found that these layers grow well together: $SrTiO_3$ layers as thin as one unit cell completely and uniformly cover cuprate layers. Tunnel junctions made from such structures show tunneling through $SrTiO_3$ layers ranging from one to seven unit cells thick. At the heterojunction, the adjacent cuprate layers are depleted of carriers by the interface, and a smeared-out density of states at the Fermi energy immediately adjacent to the barrier layer results. This leads to a suppression of superconductivity at the interface and causes the current-voltage characteristics to not be derived from the superconductive density of states.

ACKNOWLEDGMENTS

The authors would like to thank the Naval Research Laboratory and the Office of Naval Research for supporting this work.

Literature Cited

1. Goodenough J. 1971. In *Progress in Solid State Chemistry*, ed. H Reiss, 5: 145. Oxford: Pergamon
2. Tsuda N, Nasu K, Yanase A, Siratori K. 1991. *Electronic Conduction in Oxides*. Berlin: Springer
3. Eckstein JN, Bozovic I, Klausmeier-Brown ME, Virshup GF, Ralls KS. 1992. *Mater. Res. Soc. Bull.* 17: 27
4. Bozovic I, Eckstein JN, Klausmeier-Brown ME, Virshup GF. 1992. *J. Supercond.* 5: 19
5. Virshup GF, Klausmeier-Brown ME, Bozovic I, Eckstein JN. 1992. *Appl. Phys. Lett.* 60: 2288
6. Klausmeier-Brown ME, Virshup GF, Bozovic I, Eckstein JN. 1992. *Appl. Phys. Lett.* 61: 2806
7. Eckstein JN, Bozovic I, Virshup GF. 1994. SPIE Proc. 2157: *Superconducting Superlattices and Multilayers*. Bellingham: SPIE
8. Eckstein JN, Bozovic I, Virshup GF, Ono RH, Benz SP. 1995. *IEEE Trans. Appl. Supercond.* In press
9. Benz SP, Reintsema CD, Ono RH, Eckstein JN, Bozovic I, Virshup GF. 1995. *IEEE Trans. Appl. Supercond.* In press
10. Eckstein JN, Bozovic I, Virshup GF. 1995. *IEEE Trans. Appl. Supercond.* In press
11. Klausmeier-Brown ME, Eckstein JN, Bozovic I, Virshup GF. 1992. *Appl. Phys. Lett.* 60: 657
12. Berkley DD, Johnson BR, Anand N, Beauchamp KM, Conroy LE, Goldman AM. 1988. *Appl. Phys. Lett.* 53: 1973
13. Eckstein JN, Bozovic I, Schlom DG, Harris JS Jr. 1990. *Appl. Phys. Lett.* 57: 1049
14. Bozovic I, Eckstein JN, Schlom DG, Harris JS Jr. 1990. In *Science and Technology of Thin Film Superconductors II*, ed. R McConnell, S Wolf, p. 267. New York: Plenum
15. Schlom DG, Marshall AF, Harris JS Jr, Bozovic I, Eckstein JN. 1991. In *Advances in Superconductivity III*, ed. K Kajimura, H Hayakawa, p. 1011. Tokyo: Springer
16. Bozovic I, Eckstein JN, Klausmeier-Brown ME, Virshup GF. 1992. *J. Supercond.* 5: 19
17. Triscone JM, Fischer O, Bruner O, Antognaza L, Kent AD, Karkut MG. 1990. *Phys. Rev. Lett.* 64: 804
18. Lowndes DH, Norton DP, Budai JD. 1990. *Phys. Rev. Lett.* 65: 1160
19. Li Q, Xi XX, Wu XD, Inam A, Vadlamannati S, et al. 1990. *Phys. Rev. Lett.* 64: 3086
20. Virshup GF, Klausmeier-Brown ME, Bozovic I, Eckstein JN. 1995. *Appl. Phys. Lett.* In press
21. Klausmeier-Brown ME, Virshup GF, Bozovic I, Eckstein JN. 1995. *Appl. Phys. Lett.* In press
22. Tamegai T, Koga K, Suzuki K, Ichihara M, Sakai F, Iye Y, et al. 1989. *Jpn. J. Appl. Phys. Lett.* 28: 112
23. Hamilton CA, Burroughs CJ, Kautz RL. 1995. *IEEE Trans. Instrum. Meas.* In press
24. Shklovski BI, Efros AL. 1984. *Electronic Properties of Doped Semiconductors*. Berlin: Springer
25. Glazman LI, Matveev KA. 1988. *Sov. Phys. JETP* 67: 1276

Annu. Rev. Mater. Sci. 1995. 25: 711–53

WIDE-BANDGAP II-VI HETEROSTRUCTURES FOR BLUE/GREEN OPTICAL SOURCES: Key Materials Issues

Leslie A. Kolodziejski

Department of Electrical Engineering and Computer Science, Massachusetts Institute of Technology, Cambridge, Massachusetts 02139

Robert L. Gunshor

School of Electrical Engineering, Purdue University, West Lafayette, Indiana 47907

Arto V. Nurmikko

Division of Engineering and Physics Department, Brown University, Providence, Rhode Island 02912

KEY WORDS: blue/green lasers, ZnMgSSe, graded ZnSeTe contact, ZnSe : N *p*-type doping, II-VI/III-V heterovalent interfaces

ABSTRACT

Substantial advances have been realized in the quest to achieve blue/green light-emitting devices and injection lasers based on wide-bandgap II-VI semiconductor materials. The lasers operate continuously at room temperature and emit a significant amount of power with wavelengths ranging from 463 to 514 nm (depending on the actual structure). The various laser structures are composed of (Zn,Mg)(Se,S) and Zn(Se,S) cladding layers with (Zn,Cd)Se quantum wells and possess a graded ohmic contact consisting of Au metal on a pseudo-alloy of Zn(Se,Te). In addition to describing the advantages and disadvantages offered by various epitaxial growth

0084–6600/95/0801–0711$05.00

techniques, issues related to *n*- and *p*-type doping of the ZnSe-based materials are addressed, and various methods to form an ohmic contact to *p*-type ZnSe are described. The graded gap contact enables electrical measurements of the *p*-type layers, which suggest the presence of an AX center as the band gap widens in the (Zn,Mg)(Se,S). The II-VI/III-V heterovalent interface of ZnSe/GaAs affects not only the electrical injection of carriers into the active region of the laser but also plays a role in the laser lifetime due to defect generation, including stacking faults and dark line defects.

INTRODUCTION

The 1990s have been a period of rapid advance and development for the field of wide-bandgap ZnSe–based II-VI semiconductors and heterostructures due to (*a*) the attainment of *p*-type conductivity, (*b*) the discovery of appropriate heterostructure combinations for spatial manipulation of charge carriers and optical confinement, and (*c*) the ability to inject holes by formation of an ohmic contact to *p*-type Zn-chalcogenide materials. These accomplishments seemed to have occurred virtually simultaneously over the last four years, injecting excitement into the field and providing the optoelectronics community with a wide variety of II-VI–based heterostructure devices operating at short wavelength, including blue/green light-emitting diodes (LEDs) and *pn* injection diode lasers. The race was officially started in 1991 by the early reports of green lasers, and the II-VI community has been exploding with advances ever since. Continuous wave blue/green *pn* injection laser diodes operating at room temperature, and having lifetimes of one hour, have been fabricated, and most likely this accomplishment will soon be replaced with another. Therefore, the intent of this chapter is to provide an overview of the race and to highlight the significant advances, specifically the history surrounding the key technologic barriers, the accomplishments, and the future difficulties or directions. The many contributions to the field cannot be described adequately in the space provided, but various references have attempted to compile the numerous contributions (1–5). The period prior to the achievement of *p*-type conductivity is summarized in reference (6) and provides a historical perspective of the research accomplishments of the 1980s.

Device and System Applications

The driving force behind the remarkable effort put forth by the II-VI community revolves around the fact that the II-VI ZnSe–based family of binaries, ternaries, and quaternaries possesses direct energy band gaps that cover a wide spectral region, particularly in the blue and green. Figure 1 shows the energy band gap (in eV) vs the lattice parameter for a number of technologically important compound semiconductors. The shaded region highlights the ZnSe-based II-VI materials system that is the focus of this chapter, but also shown for comparison is the major competing materials system based on GaN. From a commercial viewpoint, the ability to achieve short wavelength optical sources will enable a substantial impact to be made in optical recording technology. The decrease in wavelength (from red to blue) offers a factor of four increase in the density of recorded information, but improvements in the system design are also enabled, which could result in a factor of ten increase in the density of optically recorded information. Optical storage techniques today represent one of the fastest growing segments in the multibillion dollar data storage industry. A compact, short wavelength blue laser is crucial for the next generation of optical storage systems, whether based on CD-ROM-type equipment or magneto-optical systems. Other short-wavelength optical source applications involve LEDs to complete the spectrum for full-color moving panel displays, as well as for bright emissive full-color flat panel displays. In printing applications, the role for blue lasers includes color xerography

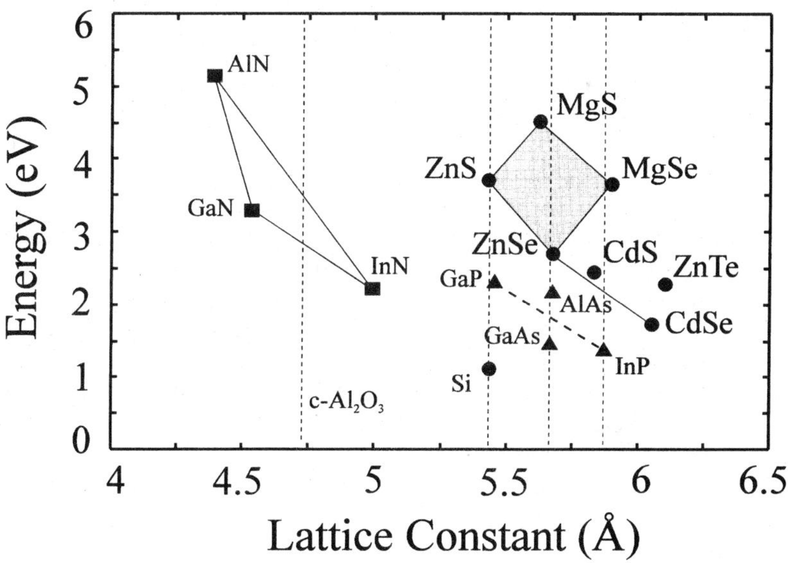

Figure 1 Energy bandgap (eV) vs lattice constant for a variety of technologically important compound semiconductors. Solid lines connect heterostructure systems currently under investigation for short-wavelength sources. Dashed lines indicate lattice constants for key substrates that have been investigated.

to replace the necessity of using photo-sensitive chemicals in color laser printers while utilizing a direct write capability with the short wavelength source. Medical diagnostic and imaging applications, where a short wavelength laser is required to activate fluorescent tracers, present another broad range of applications. Other smaller, commerical niche markets will undoubtedly arise as long-lived, reliable, short wavelength sources become available.

Competing Material Systems

As seen in Figure 1, alternative compound semiconductor systems exist for achievement of short wavelength optical sources (for a current review see reference 7). Specifically, SiC and the (Al,Ga,In)N-based material family also cover the spectral range, which includes the blue/green wavelengths. The use of SiC for optical devices is complicated because the material possesses an indirect energy bandgap that ultimately limits the intensity of light emitted from a SiC diode and also eliminates the ability to form a short wavelength coherent source. Nevertheless, SiC blue LEDs emitting approximately 18 μW of power (25 mA, 3 V) at a wavelength of 470 nm are commercially available. The SiC LED, however, is rapidly being replaced by the GaN-based materials system because of the wide spectral range accessible by the wide-bandgap III-Vs, the ability to form heterostructures composed of direct energy bandgap alloys, the ability to produce both *p*- and *n*-type materials, and the robustness of these high-temperature semiconductors. In particular, the introduction of a high-brightness blue LED by the Nichia Chemical Company in 1994 was a major technological advance in the history of blue LEDs. Although this chapter is devoted to the II-VI light emitters, we draw occasional comparisons with the III-V nitrides. However, major drawbacks currently plaguing the wide-bandgap III-Vs are the unavailability of lattice-matched high-quality substrates and the tendency for the materials to crystallize in both the zinc blende and hexagonal crystal configurations. Similar difficulties, as faced by the II-VIs in achieving *n*- and *p*-type conductivity, had to be surmounted; these difficulties involved the very high temperatures (800–1200°C) necessary for the growth and the resulting high density of vacancies that occur at these temperatures. In many respects, the materials difficulties presented to the wide-bandgap II-VI and III-Vs are similar, involving non-equilibrium growth techniques and doping, lattice-mismatched substrates, the need to form lattice-matched heterostructures, and device reliability.

Historical Material Limitations

Intense investigations of wide-bandgap II-VI bulk materials originated in the 1960s (8, 9), but achievement of optical sources was never realized.

The difficulty stemmed from the inability to incorporate acceptors (or donors) into the bulk material and to achieve *p*-type (or *n*-type) conductivity. The very high temperatures ($\sim$800–1200°C) necessary for the bulk growth resulted in a situation where the impurities were compensated by the formation of native defects, and the resulting material exhibited very high resistivities. The application of non-equilibrium growth techniques for the growth of epilayers, and ultimately heterostructures, of ZnSe-based materials alleviated the problem of controlling impurity incorporation and defect generation.

COMPARISON OF NON-EQUILIBRIUM EPITAXIAL GROWTH TECHNIQUES

The research movement focusing on the development of short-wavelength optical sources based on wide-bandgap II-VI materials was started by the application of non-equilibrium growth techniques, which were kinetic in nature, and thus typically occurred at significantly lower growth temperatures than those used for bulk growth. Some of the key differences between current non-equilibrium growth techniques are summarized below, but the primary focus of the chapter considers growth of the ZnSe-based materials by molecular beam epitaxy (MBE). The use of MBE has thus far proved to be the most successful. The difficulties surrounding the epitaxial growth of Zn-chalcogenides using metalorganic chemical vapor deposition (MOCVD), or alternatively metalorganic vapor phase epitaxy (MOVPE), and gaseous source ultrahigh vacuum epitaxy technologies, such as metalorganic MBE (MOMBE) or gas source MBE (GSMBE), are briefly highlighted. A potential advantage offered by growth techniques that capitalize on some degree of chemistry occurring at the surface involves the use of either photon or electron beams to modify the surface chemistry and thus the growth process.

MBE vs MOCVD vs Gaseous Source UHV Epitaxy: MOMBE, GSMBE

MBE of the ZnSe-based wide-bandgap II-VI materials system is performed primarily with elemental sources of the constituent species, although in some cases compound sources are employed to facilitate control of the flux. Typically, for growth of the (Zn,Mg)(S,Se) quaternary alloy, effusion ovens containing elements of Zn, Se, and Mg are employed, with ZnS used as a compound source to alleviate the difficulties associated with using elemental S. For the ternary material (Zn,Cd)Se, an effusion oven containing Cd is used. The dopant species consist of primarily Cl, originating from a compound source of $ZnCl_2$ for *n*-type species, and the nitrogen radio frequency (rf) plasma cell as a source of *p*-type acceptor species (discussed below). The vapor pressure of all of the constituent species is very high, similar to that for arsenic or phosphorus, and the sticking coefficient for each species is typically not unity. Thus the growth by MBE is extremely difficult and requires long equilibration periods to stabilize the flux from each of the effusion ovens. To ensure the ability to control the composition, and ultimately the quality of the epilayer, water-cooled quartz crystal monitors are engaged to measure the various fluxes at the growth front. For the growth of any of the wide-bandgap II-VI compounds, the growth does not occur in an overpressure of one particular species, as is typically the case for the growth of GaAs. The stoichiometry at the growth front can be assessed by monitoring the reconstruction pattern observed by reflection high-energy electron diffraction. Optimum growth conditions are typically used with approximately one-to-one flux ratios, but for *p*-type doping experiments, a slightly anion-rich surface has been found to result in greater incorporation levels of acceptor species. The growth temperature is typically maintained between 245 and 320°C and is calibrated daily to ensure accurate reproducibility. Calibration techniques include the use of low-temperature optical pyrometers, and the observation of a eutectic phase-change occurring at 356°C for 500 Å of Au deposited onto Ge and mounted next to the GaAs substrate. In addition, because the sticking coefficients are temperature sensitive, care is required not only to monitor the substrate temperature, but to compensate for variations in surface emissivity during growth. Using calibrated optical pyrometers, theoretical predictions of pyrometric interference profiles (due to the presence of the ZnSe/GaAs heterostructure) were matched to experimental data obtained during film growth in order to maintain the proper substrate temperature (10). Characterization of the material had previously involved X-ray diffraction and optical techniques; today, with the availability of ohmic contacts, electrical transport measurements are used. The issue of point defects, arising from the relatively low growth temperature, has not been adequately assessed. At present, the lifetime of optical sources is quite short and is attributed to problems including the presence of a high density of point defects present within the active layer and in the adjacent cladding layers. Minimization of point defects during the MBE growth, by carefully exploring the growth parameter space, has only recently been addressed and is clearly necessary to advance II-VI optical sources to long-lived commercial short-wavelength sources.

From the standpoint of growth mechanisms occurring during non-equilibrium epitaxial growth, MBE is by far the simplest growth technique. The use of gaseous metalorganic or hydride sources, as typically used in MOCVD (MOVPE), MOMBE, or GSMBE, complicates growth by increasing the occurrence of gas phase reactions but also adds various degrees of freedom to the growth by capitalizing on the reactions occurring at the growth front. The key advantage offered by these gaseous source epitaxy techniques is the use of precision mass flow controllers to control each flux, thus eliminating the need for thermally controlled low-temperature effusion cells. The selection and chemical stability of particular source precursors dictate the pressure and temperature at which the growth process will proceed. The primary issues associated with utilizing these other growth approaches is the purity of the source gas and the required growth temperature. Higher growth temperatures will encourage defect generation and will affect the incorporation of nitrogen (the *p*-type acceptor), which has a very low sticking coefficient. To dope the ZnSe-based compounds *p*-type with nitrogen, a source of atomic nitrogen is required, and thus the issue of growth pressure becomes critical because the mean free path will decrease as pressure increases. In addition, as discussed below, the presence of hydrogen at the growth front may have dire consequences on the electrical activity of donors and acceptors. An issue that is presently being addressed involves the use of metalorganic and hydride gas sources that have hydrogen atoms as part of the molecular chain. Source purity is of particular concern when selecting a gaseous source and is reminiscent of the source purity problem plaguing the MBE community in the early 1980s when elemental Zn or Se was used. Multiple vacuum distillations of the elements were necessary to guarantee low unintentional impurity incorporation.

Beam-Assisted Growth Techniques

An additional largely unexplored flexibility for the growth of II-VI materials is the use of the beam-assisted growth approach, i.e. illumination with either a photon or an electron beam, in conjunction with either MBE or gaseous source epitaxy technologies. Employing MOVPE or MOMBE with beam-assistance provides opportunities to explore a variety of lateral patterning techniques, in addition to the more standard composition modulation in the growth direction; the key difference is the fact that in situ lateral placement of atoms cannot be achieved in conventional MBE.

The effects of photons on growth are dependent on the energy and power density of the illuminating photons and on the type of molecule (or precursor) selected for the constituent species (which is directly related to the type of growth technology employed). Photo-assisted epitaxy has been reported for the growth of narrow- (11–14) and wide-bandgap II-VI materials (15–23), as well as for a variety of III-V semiconductors (24–34). The physical mechanisms that have been reported or verified experimentally to explain the effect of photon illumination include (*a*) selective desorption of adsorbed surface species; (*b*) pyrolysis, or thermal decomposition, of source precursors at the surface due to local increases in the substrate temperature; (*c*) direct photo-decomposition of source precursors that are present in the gas phase; and (*d*) photo-catalysis of adsorbed molecular species through the creation of free charge carriers present at the surface. Similar effects on the epitaxial growth process have been reported when the surface is illuminated with an electron beam (35, 36). In this case, the physical mechanisms that have been postulated include kinetically induced decomposition of adsorbed precursor molecules, and/or the presence of charge carriers that interact with surface species.

In the area of II-VI materials, photo-assisted epitaxy has been primarily utilized (*a*) for photo-decomposition of the source materials (such as in MOVPE); (*b*) to modify the desorption rate of various species to improve dopant incorporation and activation (as in MBE); and (*c*) for selective area epitaxy via significant growth rate enhancements observed in MOVPE and MOMBE. Narrow bandgap II-VI materials such as CdTe, HgTe, and the alloy (Hg,Cd)Te have been grown using MOVPE at extremely low temperatures ($\sim 150°C$) by the photo-decomposition of the metalorganic molecules with ultraviolet radiation (11–14). The quality of the layers appeared similar to layers grown at higher temperatures. Early work in photo-assisted MBE addressed the difficulty in amphoteric doping of the II-VI compounds by modifying the surface stoichiometry via the selective desorption of a particular species and, hence, enhancing the incorporation of the dopant. By stimulating the desorption of the group VI species, more sites were available on the anion sublattice, and thus the level of dopant incorporation dramatically increased. Films grown without photon illumination were found to be insulating (15, 16, 18). An additional photo-induced effect, which has been postulated, involves an increase in the surface mobility of adatoms on a metal-rich surface, which will occur by the preferential desorption of the chalcogenide. A photo-catalytic effect is believed to occur in the MOVPE (17, 19) and MOMBE (23, 36) growth of ZnSe and is selective to specific metalorganic precursors. Use of above-bandgap photon illumination was found to substantially increase the growth rate (~ 5 to 15 times); these particular photo-assisted effects are believed to occur via the presence of surface charge carriers, which interact with adsorbed surface species. The physical phenomena associated with the beam-assisted growth approach of ZnSe are substantial and may provide a multitude of new ideas for manipulating lasers and electron beams during the growth, to enhance understanding, but to also engineer new laterally defined semiconductor device structures.

Homoepitaxy vs Heteroepitaxy

For the heteroepitaxial growth of ZnSe on bulk material or epilayers, GaAs has overwhelmingly been the substrate of choice for the II-VI community due to the relatively close lattice-match (0.27%), ease of substrate preparation for epitaxy, and availability of large-area, high-quality, affordable GaAs material. A disadvantage is the small, but finite, thermal expansion mismatch between the ZnSe overlayer and the GaAs substrate. In addition, formation of the heterovalent II-VI/III-V interface appears to affect the quality of the ZnSe epitaxy either by the formation of extended defects at (or near) the interface, or from the presence of a large density of interface states (that ultimately affect the electronic characteristics of optical devices). Perhaps the optimum choice for epitaxy is to use ZnSe bulk substrates, but such alternatives have, until recently, been encumbered by the extremely high cost, poor material quality, and the inability to form a surface suitable for subsequent epitaxy. From the point of view of optical devices, available ZnSe bulk substrates dictate that only lateral injection device configurations are possible due to the lack of *p*-type and *n*-type conductivities for bulk ZnSe material. For application to light-emitting devices, the use of a transparent substrate is highly desirable to facilitate the maximum amount of light emission. The first few attempts of homoepitaxy of ZnSe achieved uneven success because of the varying quality of the bulk substrate and the starting surface preparation. At the present time, the use of bulk ZnSe substrates has been revived because of the much higher quality substrate material obtained from Eagle-Picher Laboratory and Sumitomo Electric Industries Ltd.; the availability is very limited and the cost still relatively high, but the preparation of the substrate surface appears to be better understood.

The early use of melt-grown ZnSe substrates was investigated by Park et al (37) in 1985, and later by several groups (38–43), but was relatively unsuccessful because the bulk material typically contained a significant concentration of impurities. Substrate surface preparation was primarily performed using chemical polishing and etching with bromine-methanol solutions, followed by in-vacuum thermal annealing prior to epitaxy. Surface morphology was found to be very dependent on the surface pretreatment (38). Perfecting the bulk substrate growth using a seeded physical vapor transport method (44, 45) eliminated the difficulties previously encountered with low-angle grain boundaries and unintentional impurity incorporation; X-ray rocking curve measurements indicated full-width at half-maximum values of 14 arc s to confirm the high degree of microstructural quality that was achieved. Additional advancements (46) were made by Sumitomo by fabricating a polycrystalline ingot; with proper annealing and the application of stress, high-quality bulk material was obtained. The high-quality epitaxial growth of ZnSe on these new bulk substrates was further advanced by use of a hydrogen atom–cleaning technique to prepare the surface for epitaxy. Jeon and co-workers (46) implemented a hydrogen plasma source (the same as that used for the nitrogen doping experiments discussed in the section on P-Type Conductivity: the Bottleneck) to remove the native oxide present on the ZnSe surface following chemical treatment. In the presence of an atomic hydrogen beam, the oxide is removed at temperatures between 350 and 380°C; the reflection high-energy electron diffraction (RHEED) pattern exhibits a 1x1 unreconstructed ZnSe surface following oxide removal. Upon nucleation of the ZnSe epilayer, no noticable change in the RHEED pattern is observed, except for the appearance of the slightly Se-stabilized reconstructed surface occurring from the growth conditions employed.

Early work in the heteroepitaxy of ZnSe onto GaAs primarily focused on the optimization of the ZnSe growth parameters, whereas the nucleation condition of the GaAs substrate was not an issue. Indications that the condition of the GaAs substrate was, in fact, critical originated with the observation that the growth proceeded differently depending on how the starting surface of GaAs was prepared. In retrospect, one would expect the nucleating conditions to be extremely critical when forming the II-VI/III-V heterovalent interface which, from an electronic standpoint, is unique compared with the more typical homovalent interface. The first suggestion that the growth of the overlayer of ZnSe was affected by the substrate occurred as Gunshor and co-workers (47) compared the epitaxy using thermally cleaned bulk substrates with the epitaxy using epitaxial layers of GaAs. The MBE growth of ZnSe on bulk substrates, which were thermally cleaned (~580°C) to remove the native GaAs oxide, exhibited a three-dimensional growth behavior, whereas nucleation on an epitaxial surface, which was thermally prepared by desorbing an amorphous layer of As (~270°C), exhibited two-dimensional growth characteristics. Subsequent experiments designed to modify the II-VI/III-V heterovalent interface by post-growth annealing indicated that the interface state density could be modified significantly and depended critically on the stoichiometry of the heterointerface (48).

To assess the effect of interface stoichiometry, a series of experiments (49) was performed in which *p*-type GaAs epilayers were grown on *p*-type (100) GaAs substrates, after which the epilayer was transferred under ultrahigh vacuum to a separate growth chamber for the subsequent growth of ZnSe and formation of the heterovalent interface. Prior to the nucleation of ZnSe, the GaAs epilayers were heated in the ZnSe growth chamber to temperatures in the range of 320 to 535°C to reduce the surface As content in stages; this resulted in four different GaAs RHEED reconstruction patterns: c(4x4), (2x4), (4x6), and (4x3) (50). In each case, after forming a particular GaAs surface, the substrate temperature was reduced to 320°C for the nucleation of ZnSe. Using conventional capacitance-voltage (C-V) measurements to determine the interface state densities, workers determined that the electrical properties improved as the GaAs surface became increasingly As deficient.

To examine the consequence of a Ga-rich (4x3) surface, the ZnSe/GaAs heterovalent interfaces were formed by two methods: The first method involved the nucleation of ZnSe directly on the Ga-rich (4x3) GaAs surface. The second involved the intentional reaction of Se with the (4x3) GaAs surface prior to the nucleation of ZnSe. The Se was reacted with the GaAs epilayer by first heating the GaAs epilayer until the (4x3) surface was formed and then exposing the surface to a Se flux while at the elevated temperature. C-V profiles of the sample having this Se-reacted interface were quite similar to those of the sample having the Ga-rich (4x3) surface.

Early workers speculated that the ZnSe/GaAs interfaces exhibiting the lower interface state densities were associated with the presence of an interfacial layer of Ga_2Se_3 (48, 49, 51). Transmission electron microscopy (TEM) cross-sectional dark-field and high-resolution images of heterojunctions formed on (4x3) reconstructed GaAs surfaces indicated the presence of a uniform band, which was interpreted as an interfacial compound having an average thickness of two monolayers. Image simulations strongly supported the identification of zinc blende Ga_2Se_3 as the interfacial layer formed on As-deficient GaAs epilayers (52). Further insight into the bonding at the heterovalent interface was provided by in situ X-ray photoelectron spectroscopy (XPS) studies. A major component of the XPS study was the comparison of interfacial features with the spectra obtained from a Ga_2Se_3 epilayer, which had been specifically grown as a calibration source for the experiment. The comparison of the Se_{3d} core level features from the ZnSe epilayer surface, from a Se-reacted As-deficient GaAs surface, and from the separately grown Ga_2Se_3 epilayers supported the identification of zinc blende Ga_2Se_3 as the interfacial layer.

SUBSTITUTIONAL INCORPORATION OF IMPURITIES

The overwhelming research concentration on the ZnSe-based compound semiconductors has been devoted to the incorporation of acceptors to achieve *p*-type conductivity. Prior to the early 1980s, undoped ZnSe grown by MBE, using commercially available source materials having 0.999999 purity, exhibited *n*-type conductivity with low resistivity (53). The low resistivity of the ZnSe material implied that the ZnSe was stoichiometric because a relatively small deviation from a unity Zn-to-Se flux ratio, toward either Zn-rich or Se-rich growth conditions, was found to result in high-resistivity material (54). The defects generated during growth under non-stoichiometric conditions appeared to compensate the unintentionally incorporated impurities. However, by enhancement of the purity of the source materials, either by zone-refining or vacuum distillation, ZnSe, deposited with stoichiometric growth conditions, typically exhibits high resistivity ($\sim 10^4\ \Omega$cm) and undoped carrier concentrations in the high 10^{14} to low $10^{15}\ cm^{-3}$ (55). With regard to the wide-bandgap quaternary alloys of (Zn,Mg)(S,Se), the high reactivity of Mg dictates that vacuum distillation is required, followed by extended periods of source out-gassing, to achieve low undoped carrier concentrations.

Intentional incorporation of donors to obtain *n*-type conductivity is relatively straightforward and has been achieved using elemental Ga, In, and a compound source of $ZnCl_2$ for Cl. The highest level of donor incorporation has been achieved with Cl having concentrations approaching $10^{20}\ cm^{-3}$ through techniques such as MBE (56), planar doping with MBE (57), and by GSMBE (58). The donor level is relatively shallow ($\sim$28 meV), the crystal quality remains very high for significant donor incorporation, and hydrogen does not appear to passivate the Cl atoms and affect electrical conductivity. Most optical devices reported thus far utilize Cl for the *n*-type dopant species in ZnSe-based binaries, ternaries, and quaternaries.

P-Type Conductivity: the Bottleneck

A major difficulty preventing the use of wide-bandgap II-VI compounds for formation of light-emitting devices was the inability to incorporate acceptors while simultaneously controlling the *p*-type conductivity. As mentioned above, *n*-type conductivity was relatively easy to control once source purity and minimization of unintentional impurity incorporation were achieved. The large degree of ionicity found in the covalent bonding of wide-bandgap II-VI materials results in a situation where the acceptor is predicted to have a relatively high probability to induce local lattice relaxation, which could result in the creation of a deep, electrically inactive acceptor level within the band gap. The maximum incorporation level of acceptors will also be affected by solid solubility limits for various impurities. Impurities residing in Column V of the periodic table are candidates for acceptors in II-VI Zn-chalcogenides, for example ZnSe: column I elements such as Li (59, 60) are also potential candidates for acceptors, but the relatively small size of the atom allows substantial interstitial incorporation and significant lattice diffusion, particularly at room temperature and during device operation, and thus Li is precluded as a technologically viable *p*-type acceptor. Using MBE, workers investigated the incorporation of both arsenic (61, 62) and phosphorus (61, 63) as *p*-type dopant species. In the case of As, very high levels ($\sim 10^{17}$–10^{21} cm^{-3}) of impurity incorporation were achieved using Zn_3As_2 or an As cracker for a source of As (62). Secondary ion mass spectrometry (SIMS) indicated that As was easily incorporated into the lattice, with photoluminescence spectra indicating that shallow acceptor levels were present. In all cases, however, the ZnSe : As layers were found to be highly resistive. Similar results were reported (61, 63) for the use of phosphorus as a *p*-type acceptor; photoluminescence suggested the presence of both a shallow acceptor state and a deep center. From these early experimental data, Chadi (64) formulated a theory using first-principle pseudopotential calculations, which suggested that the presence of either P or As within the ZnSe lattice actually forms two distinct atomic states. When located substitutionally on a Se site having fourfold coordination, the column V element was suggested to give rise to a shallow effective-mass state. However, a deep state was also calculated, which resulted from a large lattice relaxation associated with the breaking of a Zn-impurity bond. Chadi (64) proposed that self-compensation was not the dominant cause of the high resistivity of the ZnSe : column V layers, but in fact was due to the presence of the shallow-deep acceptor states appearing during the use of As or P as acceptor species. Chadi also postulated nitrogen to be the optimum acceptor for ZnSe because the calculations indicated that whether in the fourfold coordinated state or in the large lattice-relaxed state, nitrogen would behave as a shallow acceptor.

Utilizing nitrogen as an efficient *p*-type acceptor species represented a significant technological challenge. From this perspective, the issue revolved around the very low sticking coefficient of nitrogen (65, 66) during the MBE of ZnSe using N_2 gas as a dopant source. Initial results by Park and associates (65) illuminated the difficulty associated with the low sticking coefficient because the use of N_2 gas at relatively high flow levels resulted in incorporated nitrogen atoms in the 10^{15}–10^{16} cm^{-3} range. Utilizing nitrogen species in the form of a low-energy N^+ or N_2^+ ion beam did not alleviate the doping difficulties; the resulting ZnSe : N layers still exhibited high resistivity (67). The major technological breakthrough (68, 69) was realized by implementing a rf plasma cell in conjunction with ultrapure nitrogen gas during the MBE of ZnSe. The plasma source was built by Oxford Applied Research (model MPD21), operated at 13.5 MHz under conditions of molecular flow using nitrogen gas, and provided a source of atomic nitrogen (70) (in addition to a flux of nondissociated N_2 molecules) to the ZnSe growth front. By comparing the photoluminescence

(PL) spectra of films grown with either N_2 or atomic $N+N_2$, Park and associates were able to discern clearly the incorporation of nitrogen acceptors by the presence of the donor-to-acceptor (D-to-A) pair transitions dominating the PL spectrum for the layers grown with the rf plasma source activated. To assess the net acceptor incorporation, capacitance-voltage (C-V) measurements were utilized because an ohmic contact to *p*-type ZnSe could not be formed. (The metal-to-semiconductor barrier for Au on ZnSe is 1.6 eV.) The C-V measurements, indeed, confirmed the *p*-type conductivity and indicated net acceptor carrier concentrations around 3×10^{17} cm^{-3} range. An electron cyclotron resonance (ECR) plasma cell was also found to give rise to *p*-type ZnSe:N (71, 72). Following the announcement by Park et al (68) and Ohkawa et al (69) of the successful *p*-type doping of ZnSe utilizing the nitrogen plasma cell, numerous groups world-wide (2) initiated investigations into the epitaxial growth and characterization of ZnSe:N layers in a quest for achieving blue/green light-emitting devices.

The most successful growth technique for achieving *p*-type conductivity to date is MBE using the nitrogen plasma cell for ZnSe-based materials. By controlling the proper Se-to-Zn flux ratio and flux of atomic nitrogen, calibrated substrate temperature, and extended defect density for the heteroepitaxy of ZnSe on GaAs, relatively uncompensated ($\sim$5–15% compensated) ZnSe:N was routinely achieved. As the amount of nitrogen incorporated into the lattice increased, either by reducing the substrate temperature or increasing the nitrogen flux, self-compensation was also seen to increase (73). As a more typical number, atomic incorporation levels of 1×10^{18} cm^{-3} can be obtained for ZnSe:N, but net acceptor concentrations remain near 5×10^{17} cm^{-3}. Theoretical models have been proposed that incorporate the presence of various types of point defects to explain the occurrence of self-compensation (74, 75). Other experimental results (76) suggest that as more and more nitrogen atoms are incorporated, a deep donor level appears and effectively compensates the net acceptor concentration. The energy of the dominant D-to-A pair transition measured in PL is speculated to be related to the presence of a deep donor (76, 77); a shallow acceptor level is characterized by the D-to-A pair transition having a higher energy.

In 1991, Chadi proposed that the ideal *p*-type impurity species in wide-bandgap ZnSe would be nitrogen, based primarily on pseudopotential calculations (64). However, because of the reported passivation of dopants in silicon and III-V-based compound semiconductors by hydrogen, Chadi speculated that the use of nitrogen-containing gas species and the presence of hydrogen during growth may also give rise to a similar acceptor passivation phenomenon for the II-VI materials. A preliminary experiment by Pong et al (78) indicated that hydrogen may, in fact, affect the electrical activation of donor or acceptor species in wide-bandgap ZnSe or ZnS. In their work, post-growth hydrogenation through a hydrogen-permeable cap of SiO_2 was employed to incorporate deuterium. The decrease in intensity of the D-to-A pair transitions measured by PL was speculated to be the result of neutralization of one of the dopant impurities contributing to the D-to-A pair transition; the particular dopant that was affected was not identified.

Utilization of gaseous source epitaxial growth techniques, in which hydrogen typically plays a major role either as a carrier gas, as part of the dopant molecule (for example NH_3), or as a decomposition by-product, has provided further information regarding hydrogen and the passivation of dopants in wide-bandgap II-VI semiconductors. Recent Fourier-transform infrared absorption studies (79) of ZnSe : N grown by MOVPE (using NH_3 as the dopant source and H_2 as the carrier gas) indicate the existence of N-H bonds in ZnSe : N films; in this case, the ZnSe : N layers were highly resistive, and these results are in agreement with the work of Kamata and associates (80). In the case of MOCVD, the incorporation mechanism of hydrogen is extremely difficult to isolate and most likely occurs through a variety of mechanisms. The growth chemistry is complex and typically occurs near atmospheric pressure where the mean free path of an atomic nitrogen species is very short. In addition, the presence of hydrogen as a carrier gas and as a major decomposition by-product, either during decomposition of the dopant source or as the constituent source species, provides a multitude of pathways for incorporation of hydrogen atoms. The use of ultrahigh-vacuum gaseous source epitaxy techniques offers the ability to isolate the degree of hydrogen incorporation and thus investigate the electrical passivation of various II-VI dopants (81). Figure 2 shows SIMS data obtained from a ZnSe : N film grown by gas source molecular beam epitaxy (GSMBE); here the only source of hydrogen originates from the decomposition of the H_2Se gas, either as H_2 or HSe because the group II source is elemental zinc. The film consists of a uniformly doped ZnSe : N layer having a nitrogen concentration near 10^{18} cm^{-3} and containing two relatively thin undoped layers of ZnSe. The data clearly show the consistent tracking of the hydrogen and nitrogen signals as the nitrogen is intentionally varied. In Figure 3, the SIMS profile is shown for the MBE growth of a layer of ZnSe during impingement of a H_2 flux and during normal growth conditions. The presence of H_2 molecules at the growth front is found to increase the incorporation of the hydrogen above the background level in the SIMS profile. Films grown with a nitrogen plasma cell, in conjunction with GSMBE, are typically found to be highly resistive, indicating the potential passivation of acceptors by hydrogen (81). In contrast,

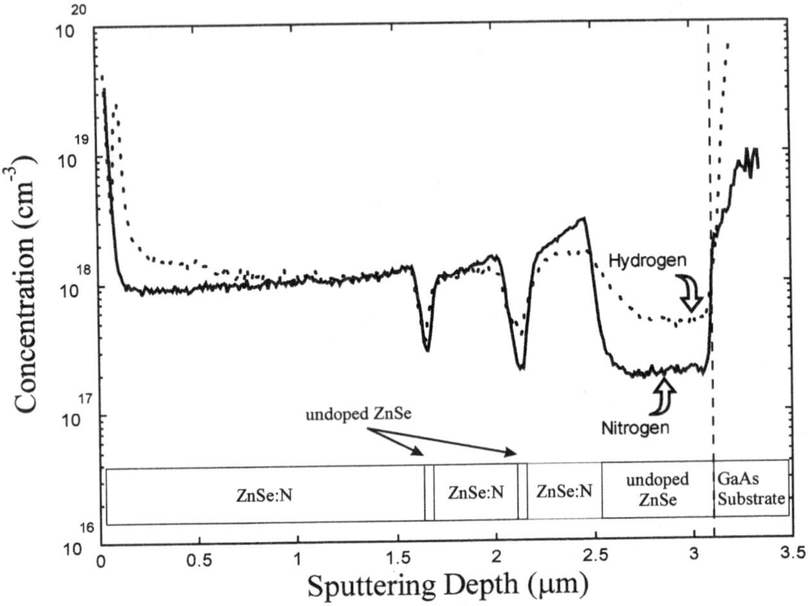

Figure 2 SIMS profile of a ZnSe: N structure grown by GSMBE having an initial undoped layer near the ZnSe/GaAs interface, followed by two thin, undoped regions in an otherwise uniformly nitrogen-doped film, as schematically indicated in the figure. The undoped layer next to the GaAs interface enables the background concentration of nitrogen and hydrogen to be determined. The sudden rises in the hydrogen and nitrogen at the ZnSe/GaAs interface are artifacts of the data normalization process.

however, ZnSe grown by GSMBE and doped with donors of Cl does not indicate any deleterious effects from the presence of hydrogen at the growth front and does not appear to be electrically passivated (58). The severe difficulties encountered for achieving *p*-type conductivity in ZnSe during growth using MOCVD (or MOVPE) and GSMBE suggest that hydrogen is effectively incorporated into the lattice when ZnSe is doped with nitrogen. Alternate doping sources are required to offer the potential to eliminate the hydrogen species prior to incorporation. The passivation of impurities by hydrogen has been observed for both Si and III-V based semiconductors, although it is only severe for growth occuring at temperatures typically below ~500°C. In the case of the II-VIs, however, growth is limited in all cases to low temperatures near 300°C because of the desorption of the constituent species at growth temperatures above 400°C. The passivation of acceptors, specifically Mg, by hydrogen has also been observed for the GaN-based wide-bandgap III-V materials. The remedy that can be employed for Si or III-V materials involves post-

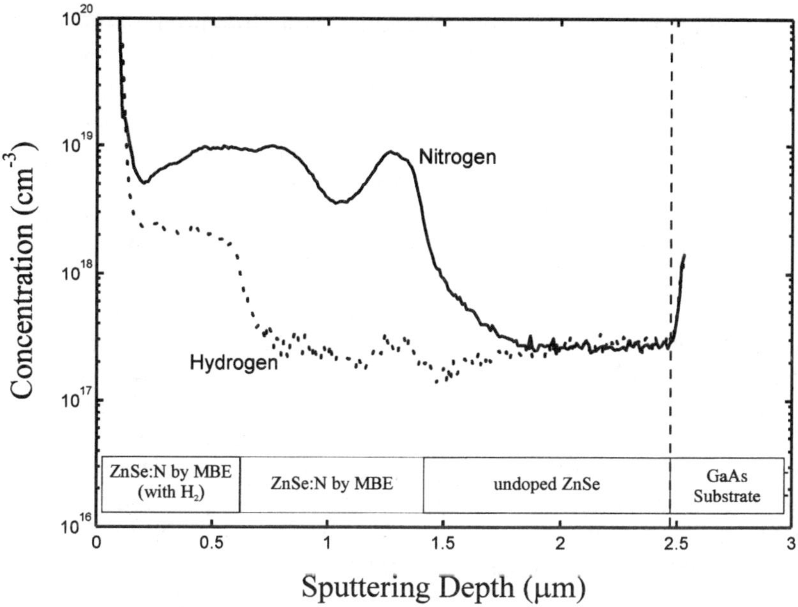

Figure 3 SIMS profile of a ZnSe : N structure (as indicated schematically in the figure) grown by MBE both with and without intentional injection of hydrogen through the H_2Se cracker held at 300°C. The sudden rises in the hydrogen and nitrogen at the ZnSe/GaAs interface are artifacts of the data normalization process.

growth, high-temperature thermal annealing, usually performed at temperatures above 400°C, or exposure to a low energy electron beam. In the case of the wide-bandgap II-VIs, however, thermal annealing above the growth temperature may result in defect generation, although it may not affect the hydrogen incorporation. Preliminary thermal-annealing (both rapid and long term) experiments (82) of ZnSe : N have not resulted in a degradation of the PL characteristics but also have not indicated an observable change in the degree of hydrogen passivation.

Formation of Ohmic Contacts

Once the barrier to achieve useful *p*-type carrier concentrations had been surmounted, a new one quickly became evident and involved the electronic barrier associated with injection of holes from a metal into *p*-type ZnSe to form an ohmic contact. An ohmic contact can easily be formed on lightly doped *n*-type ZnSe by using indium alloyed with mercury, whereas pure indium and even indium tin oxide can form a relatively low-resistance contact to highly doped *n*-type layers such as ZnSe : Ga or ZnSe : Cl. However, a metal having a sufficiently high work function, which is necessary to eliminate the barrier for injection of holes into *p*-type ZnSe, has not been found. Metals that have been investigated include gold, palladium, platinum, and various multilayer combinations. For many prototype devices initially explored, *n*-type ZnSe was grown on *p*-type ZnSe (grown on a *p*-type GaAs substrate) to facilitate injection of holes from the *p*-type GaAs into the *p*-type ZnSe; the barrier was still large but reduced to around 1 eV. To overcome this new barrier, scientists employed a variety of techniques including bandgap engineering. Thus far, four approaches to forming an ohmic contact to *p*-type ZnSe have been reported: (*a*) the use of a ZnTe/ZnSe graded contact scheme (83); (*b*) the use of a semimetal of HgSe grown directly on the top of the ZnSe heterostructure (84, 85); (*c*) the growth of a highly nitrogen-doped, highly defective, low-temperature layer of ZnSe (86); and (*d*) injection of holes from a graded III-V layer on *p*-type GaAs designed such that the valence band offset is reduced (87). Two additional proposals for alleviating the difficulty in forming an ohmic contact to *p*-type ZnSe have been put forth: (*a*) the use of a metal/Si-As interlayer/BeTe/ZnSe multilayer structure (88); and (*b*) the use of BeZnTeSe-graded bandgap layers (89).

ZnTe/ZnSe GRADED CONTACTS The concept for utilizing bandgap engineering and applying it to the ZnTe/ZnSe heterostructure resulted from an investigation (90) of the *p*-type doping of ZnTe using the nitrogen plasma cell (typically used for the *p*-type doping of ZnSe). Han and co-workers (90) found that under the apparently identical growth conditions used for ZnSe:N, a substantially higher level of nitrogen incorporation was obtained during the MBE-growth of ZnTe, as compared with that achieved with ZnSe. In fact, levels in excess of 1×10^{19} cm^{-3} were routinely achieved for ZnTe grown on highly lattice-mismatched GaAs. In addition, contacts formed by evaporating Au (without annealing) were found to readily form an ohmic contact to ZnTe:N; standard transmission line model measurements of the specific contact resistivity indicated values of 1×10^{-3} Ωcm^2. Epitaxial growth of ZnTe:N onto *p*-type ZnSe was not a solution, however, as can be seen in Figure 4*a*. The expected energy band lineup at the p-ZnTe/p-ZnSe interface has a valence band offset of approximately 1 eV and thus still provides a barrier to hole injection from the p-ZnTe into the p-ZnSe. The energy spike in the valence band was removed by forming a graded bandgap consisting of a pseudo-alloy of Zn(Se,Te). (A schematic diagram of the resulting energy band lineup is seen in Figure 4*b*.) In order to avoid the difficulty in controlling the alloy fraction during the MBE growth of the Zn(Se,Te) alloy (91), a multilayer structure was fabricated to simulate a continuously graded layer; the difficulty in the growth of the alloy originates in the competition for incorporation between Te and Se

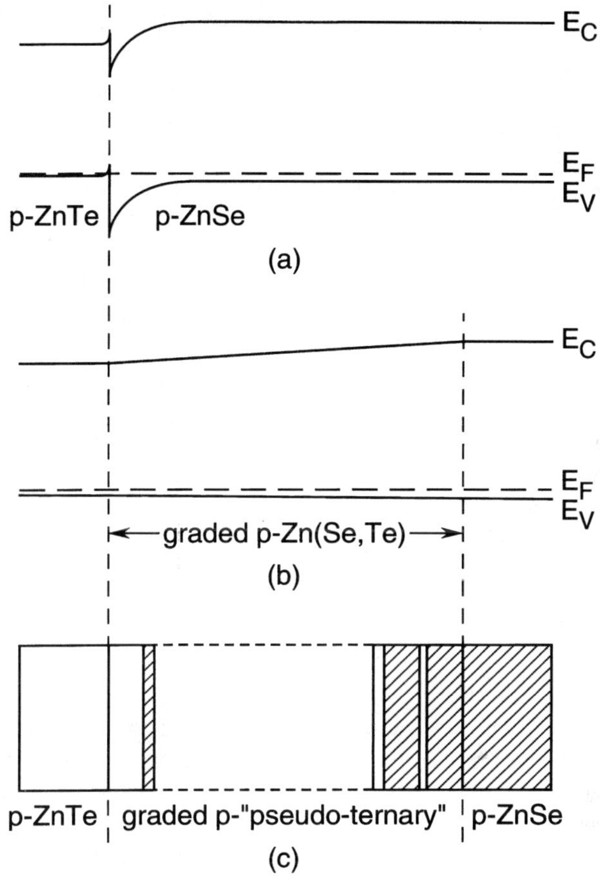

Figure 4 Schematic diagram of the energy band lineup at the (*a*) p-ZnTe/p-ZnSe interface and the (*b*) p-ZnTe/p-Zn(Se,Te) graded layer/p-ZnSe. (*c*) Schematic illustration of the multi-layered structure of the pseudo-graded Zn(Se,Te) contact layer.

species at the growth front. Figure 4*c* illustrates the pseudo-alloy Zn(Se,Te) multilayered structure consisting of a variety of cells. The total thickness of each cell is 20 Å, and there is a total of 17 cells; the thicknesses of the ZnTe and ZnSe layers are varied, as shown in Figure 4*c*, to simulate a continuously graded Zn(Se,Te) alloy layer. The use of the ZnTe/ZnSe-graded contact scheme, with a Pd/Au metal contact on the *p*-type ZnTe topmost layer, enabled a series of transmission line model measurements with specific contact resistivities of 1×10^{-4} Ωcm^2. Figure 5 shows the room temperature current (I) vs voltage (V) for the various metal-to-semiconductor contact schemes described above. The specific contact resis-

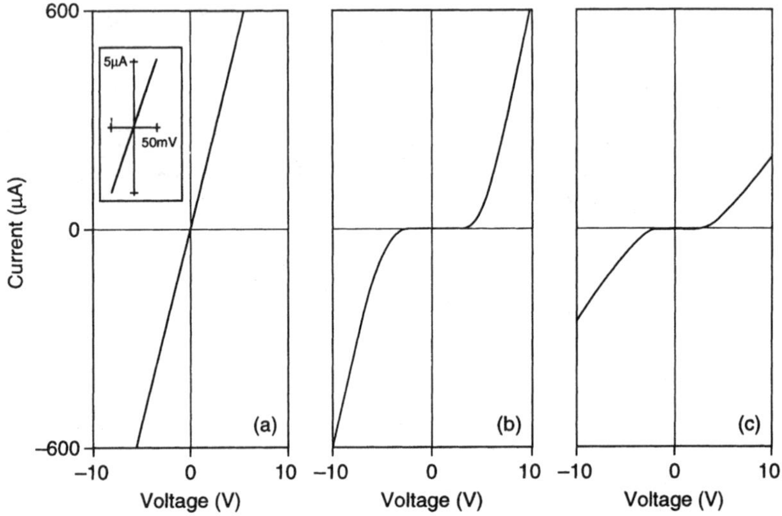

Figure 5 I-V characteristics of different contact schemes measured by an HP-4145 parameter analyzer at room temperature: (*a*) Au/Zn(Se,Te) graded layer/p-ZnSe, with the inset showing the I-V characteristic where the voltage range has a 50 mV scale maximum; (*b*) Au/p-ZnSe; and (*c*) Au/p-ZnTe/p-ZnSe.

tivities are approximately one order of magnitude larger than what would be considered optimal for laser diode operation (92). Ohmic behavior between ZnSe and the Au/Zn(Se,Te) graded contact was maintained over the temperature range of 77 to 300 K, even though the sample resistivity was seen to vary by three orders of magnitude due to carrier freeze-out (93). Optimization of the growth scheme for forming the graded Zn(Se,Te) contact is still required in order to further reduce the contact resistance. In addition, a significant reduction in the contact resistance was achieved by the use of a multilayered metal contact to the p-ZnTe and consisted of Au(Pt)Pd, as opposed to pure Au (94).

The graded Zn(Se,Te) contact structures have been incorporated into a variety of Zn(Se,S)- and (Zn,Mg)(Se,S)-based light-emitting diodes and *pn* injection devices (see section on Blue/Green Light-Emitting Devices). The improvement in reducing the forward bias voltage necessary to reach lasing conditions has been dramatic; in 1991, using Au contacts, the threshold voltage neared 30 V, whereas today, the threshold voltage is typically near 5 V for room temperature operation of laser diodes. In the initial period of studying optical sources based on II-VI wide-bandgap semiconductors, the primary device failure originated at the Au/p-ZnSe contact because of excessive heating. By employing the graded contact, the *pn* diode-injection laser operation progressed from a liquid nitrogen temperature, pulsed operation to a room temperature, continuous-wave operation.

The lack of a suitable metal to form an ohmic contact to wide-bandgap Zn-chalcogenides restricted researchers from obtaining substantial information about the transport properties of p-ZnSe and related materials. Specifically, the variation of mobility and free-hole concentration as a function of temperature and the minimization of self-compensation via optimization of growth parameters were unavailable. The ohmic behavior of the ZnTe/ZnSe contact enabled temperatures to be varied from 300 to 77 K. In addition, an apparent limitation in achieving high *p*-type doping levels had surfaced. Contrary to the results observed in *p*-type ZnSe, incorporation of Cl as an *n*-type dopant in ZnSe was achieved by MBE (56, 57) and GSMBE (58) at levels exceeding 1×10^{19} cm^{-3}. In comparison with ZnSe, ZnTe was found to be easily doped *p*-type but was not easily doped *n*-type. For *p*-type doping ZnTe experiments (95, 96) using MBE, the dopant species flux levels in many cases approached those of the host constituent species, with an associated decrease in the crystalline quality. However, use of the nitrogen rf plasma cell for experiments designed to investigate the growth of *p*-type ZnTe illuminated a key difference between ZnSe and ZnTe. When the ZnTe epilayers were grown nominally at the same temperature (245°C) and growth rate, and with the same nitrogen plasma operating conditions as those employed for the growth of ZnSe : N, free-hole concentrations exceeding 1×10^{19} cm^{-3} in ZnTe : N were easily obtained; whereas in ZnSe : N, free-hole concentrations of 10^{17} to 10^{18} cm^{-3} were measured. The large difference in the level of *p*-type doping for ZnTe, as compared with ZnSe, provided significant theoretical speculation. Two proposals exist suggesting that self-compensation [by either native defects (97) or dopant lattice relaxation (64)] is dominating the observed behavior, or that a solubility limit has been reached in the case of ZnSe : N (60, 98). In the work by Fan and co-workers (93), compensation is not the dominant mechanism limiting the level of nitrogen incorporation because temperature-dependent Hall effect measurements indicate very low compensation ratios of 6–11%. In addition, for ZnTe : N, Han et al (90) found that essentially 100% of the nitrogen atoms incorporated into ZnTe are situated at Te lattice sites and serve as acceptors. In contrast, careful studies (99) correlating the transport measurements with SIMS indicate that the solid solubility limit may have been reached in the case of ZnSe : N. A model for computing the solid solubility of dopant atoms in various host semiconductors has been proposed (60, 98). The calculations predict that the solubility limit for nitrogen in ZnTe is significantly larger than that for nitrogen in ZnSe. Although equilibrium theory is not expected to predict the behavior of solid solubility under the conditions of non-equilibrium MBE growth, the trends that are suggested may help to explain the difficulty surrounding the *p*-type doping of ZnSe, but not plaguing the *p*-type doping of ZnTe.

Although free-hole concentrations exceeding the mid 10^{17} cm^{-3} were difficult to achieve in ZnSe : N; as the band gap widened for ZnSe-based ternary and quaternary compounds, the activation energy for acceptors appeared to increase (100). The ionization energy for acceptors in Zn(Se,S) was found to be greater than that for ZnSe ($E_A \sim 92$ meV), and the ionization energy for acceptors in (Zn,Mg)(Se,S) was still greater ($E_A \sim 177$ meV). The increased ionization energy of acceptors, occurring as the band gap widens, appeared to be associated with a lattice relaxation phenomenon and is reminiscent of the behavior of DX centers observed in many *n*-type semiconductors (101). Figure 6 shows the transport behavior obtained by measuring the resistivity of a (Zn,Mg)(S,Se) bar as a function of the inverse temperature (with illumination and in the dark). Persistent

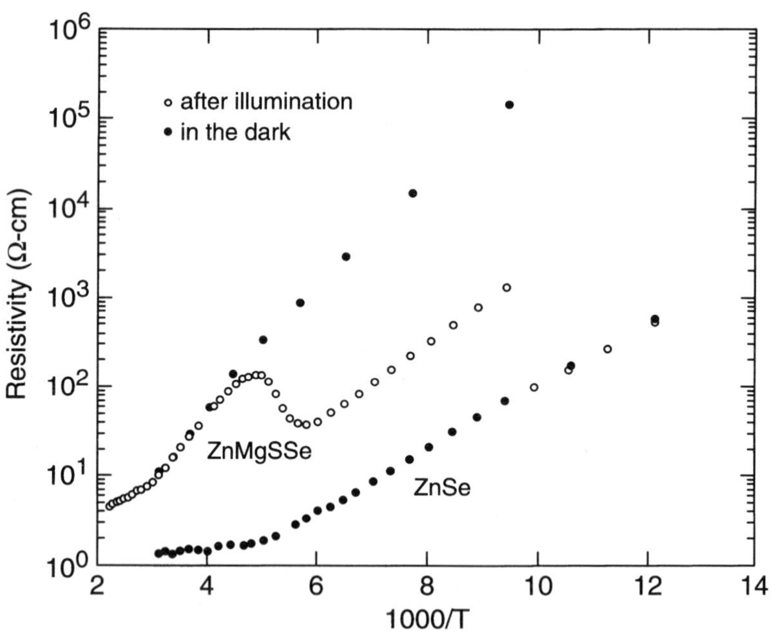

Figure 6 Resistivity of a rectangular bar vs inverse temperature. The persistent photoconductivity in the (Zn,Mg)(S,Se) quaternary alloy is associated with a drop in resistivity of more than two orders of magnitude. Under similar illumination, the ZnSe epilayer shows about a 10% reduction in resistivity at low temperature.

photoconductivity, with a resistivity decrease of more than two orders of magnitude, was observed at low temperatures. As a comparison with the quaternary alloy, an identical experiment performed on nitrogen-doped ZnSe shows approximately a 10% reduction in the sample resistivity at low temperatures. The observed behavior of the resistivity and comparisons with other III-V compounds suggest that the nitrogen acceptors in *p*-type (Zn,Mg)(S,Se) and Zn(S,Se) possess a DX nature, although in keeping with the *p*-type conductivity, they may be more appropriately described as an A(acceptor)X center. The deepening of the acceptor level in the wide-bandgap II-VI materials system most likely reflects the position of the Γ-point valence band edge (which is sensitive to alloy composition) and the highly localized deep state (which is less sensitive to the alloy composition). The addition of Mg and S to ZnSe had the effect of pushing the valence band edge (and the hydrogenic acceptor level) downward and caused the AX level (now lying within the band gap) to be energetically more favorable. Interestingly, prior to the study just described, a DX-like behavior had not been reported for acceptors in any semiconductor. Of course, the appearance of a deep level for nitrogen as the band gap increases presents a problem for the development of wider bandgap devices. Thus the motivation remains to find suitable acceptors for the wide-bandgap II-VI compounds, or a means to neutralize the tendency for nitrogen to form an AX center in wide-bandgap alloys of ZnSe.

ALTERNATIVE APPROACHES TO OHMIC CONTACTS The large Schottky barrier height (>1 eV) present between most metals and *p*-type ZnSe, and the large valence band offset (~ 1 eV) present between the ZnSe and the GaAs substrate, are the two obstacles that prohibit formation of an ohmic contact. As a method to replace the use of metals, the epitaxial growth of either a semimetal of HgSe or an alloy of ZnHgSe effectively reduces the metal/p-ZnSe barrier to a barrier consisting of HgSe/p-ZnSe. The deposition of a binary mercury-containing layer has been reported (84, 102), using interconnected MBE growth reactors, where the HgSe deposition is carried out in a dedicated chamber. Separate reactors are required because of the high vapor pressure of the liquid Hg source. An alternative method proposed by P Holloway (private communication) for depositing the HgSe contact layer consists of completing the MBE growth of the top *p*-type ZnSe and reducing the substrate temperature in the presence of a Se flux to allow the deposition of ~ 100 Å of amorphous Se. Following removal from the ultrahigh vacuum reactor, the sample is exposed to Hg vapor and confined to a quartz ampule held at ~ 200 °C to allow the formation of the HgSe top contact layer. Theoretical calculations (85) indicate that the valence band offset may be reduced to approximately 0.5 eV for doping concentrations of around 1×10^{18} cm^{-3}. By forming an alloy of ZnHgSe with a barrier height of 0.3 eV, a doping concentration of 1×10^{17} cm^{-3} is postulated to be required to make an ohmic contact (85). One potential disadvantage of the use of the HgSe contact layer is a degree of lattice mismatch with ZnSe. Another is a significant optical absorption over a very wide spectral region covering visible wavelengths. These difficulties impose limitations on the thickness of the HgSe layer, which functions as a transparent contact.

A second approach to injection of holes into the active region of the *pn* junction of the optical device incorporates the GaAs substrate as part of the contacting scheme; in this case, of course, the III-V substrate is *p*-type. The valence band discontinuity is around 1 eV, but it can be reduced substantially by the use of a buffer layer having a graded III-V layer initially composed of GaAs and graded to the wide-bandgap (In,Al)P, where the expected valence band offset is probably around 0.34 eV. The growth on the p-GaAs substrate would initially consist of an (In,Ga)P layer graded to (In,Ga,Al)P, and by eliminating Ga, would be graded to (In,Al)P; careful selection of the alloy composition enables the entire buffer layer stucture to be lattice-matched to the GaAs substrate. [The voltage drop across the (In,Ga)P/GaAs interface becomes insignificant by appropriate *p*-type doping, whereas the valence band discontinuity between (In,Ga,Al)P/GaAs is minimized by the appropriate selection of Al composition (103).] A preliminary investigation by Onomura and associates (87) found that the voltage drop across the II-VI/III-V interface is reduced by the presence of an (In,Ga)P buffer layer on p-GaAs. In this case, the (In,Ga)P layer is grown on GaAs by MOCVD and transferred ex situ into a II-VI MBE reactor for ZnSe-based epitaxy. Heteroepitaxy of ZnSe onto epitaxial layers of (In,Ga)P, (In,Ga,Al)P, and (In,Al)P, grown by GSMBE in an interconnected MBE system, has been investigated (104); preliminary results suggest that the ZnSe/III-V interface is formed via a layer-by-layer growth mechanism as seen by reflection high energy diffraction intensity oscillations. The epitaxial layer of ZnSe exhibits a dominant free-exciton transition in photoluminescence and the absence of a deep-level, defect-related luminescence. Through use of a III-V buffer layer for injection of holes from the back-side *p*-type substrate, laser structures having an *n*-type top ZnSe-based layer grown on the *p*-type ZnSe-based layer, all fabricated on the *p*-type graded III-V buffer layer, can be grown. The ohmic contact to the topmost ZnSe-based layer would be greatly simplified and most likely would consist of elemental In.

BLUE/GREEN LIGHT-EMITTING DEVICES

Short Wavelength Light-Emitting Diodes

The wide, direct energy bandgap of the ZnSe-based compound semiconductor family enables the prospect of achieving light-emitting diodes at short visible wavelengths. Significant progress has been made to improve the quantum efficiency of such light emitters, as well as to increase their lifetime.

The wide-bandgap II-VI optical devices take direct advantage of quantum well heterostructure configurations to provide, both in principle and increasingly in practice, highly efficient LEDs. The internal radiative efficiency at cryogenic temperatures ($\sim$77 K) of a (Zn,Cd)Se quantum well is very close to 100% but is found to decrease significantly as the sample temperature is raised toward room temperature. From a materials point of view, the density of point defects, which contribute nonradiatively to the recombination rate of electrons and holes, dominates the observed decay in radiative efficiency. Inherently, however, the ZnSe-based quantum well emitters possess strong electron-hole correlations (exciton effects) that offer a more robust radiative recombination probability. Recent experiments by Ding et al (105) support a model containing exciton recombination; the measured radiative lifetime at room temperature was 0.9 ns, whereas an estimate of lifetime for an electron-hole density in a free electron-hole pair model would be considerably longer. Furthermore, the LED spectra matches both the photoluminescence and absorption spectra for the quantum well structures, confirming the fact that the emission originates at the (Zn,Cd)Se quantum well. At 77 K, where the spectral widths are very narrow ($\sim$10 meV), the LED emission is clearly seen to be excitonic in origin, but the electron-hole pairwise Coulomb correlations appear to persist as the temperature is raised to room temperature. By way of comparison with the (In,Ga)N LEDs, owing to the much more localized nature of the impurity levels that are involved in the light emission mechanism, radiative recombination is a much slower process, with carrier lifetimes on the order of 400 ns at room temperature.

As an illustration of the gradual improvements in controlling the number of nonradiative centers, Figure 7 shows the photoluminescence spectra, under low levels of excitation obtained from a ZnCdSe/ZnSSe/ZnMgSSe separate-confinement heterostructure as a function of temperature. The PL efficiency from the (Zn,Cd)Se quantum well decreases by a factor of five from 77 K to room temperature. These results are extremely promising for any wide-bandgap semiconductor because the data suggest a moderate density of point defects or other defects acting as nonradiative recom-

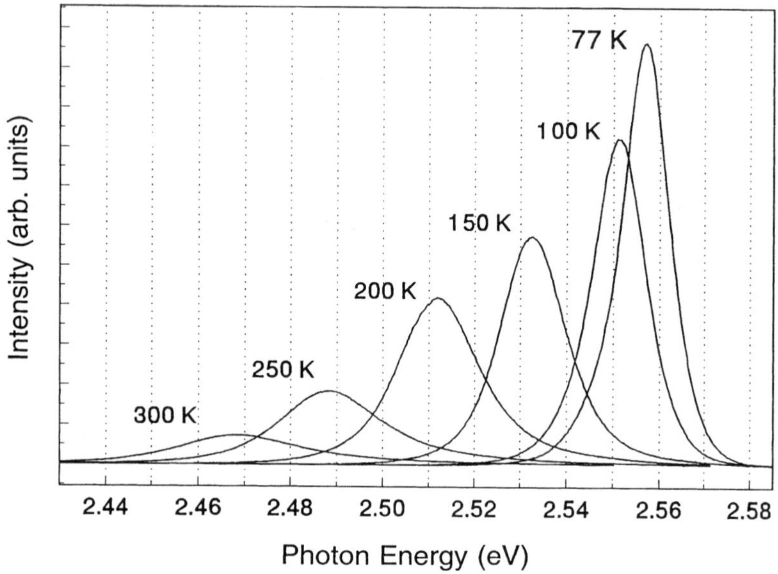

Figure 7 Electron-hole pair lifetime from $T = 10$ K to room temperature of a ZnCdSe/ZnSSe/ZnMgSSe diode laser structure at a carrier density corresponding to diode laser threshold (105).

bination centers. The typical room temperature emission wavelength of Cd-containing LEDs is in the range of 480 to 490 nm, although effort is underway to decrease the wavelength by compositional changes in the well layer by incorporating other ternaries and quaternaries. Figure 8 shows a room temperature numeric device emitting at 490 nm. By employing a ZnSe binary quantum well, a blue LED operating at 457 nm has been obtained (106). The use of the low-resistance ZnTe/ZnSe graded contact (see section on Formation of Ohmic Contacts) enables several hundred μW of output power to be obtained with diode voltages of 3 to 3.5 V, and with current densities in the range of several tens of A cm^{-2}. With the use of transparent substrates (ZnSe and its alloys) and transparent contact electrodes (107), the actual design of the blue LEDs can be improved considerably from the viewpoint of electrical/optical efficiency. A crucial element in terms of technological viability for the II-VI-based blue LEDs concerns device lifetime, which at present is a key challenge. This topic is discussed below, but for both LEDs and lasers, the degradation process is highly superlinear, with respect to the current density. For typical values of current density, J $\sim$ 20 A cm^{-2}, LED lifetimes approaching 1000 h have been reached for the ZnSe-based quantum well emitters.

Additional device opportunities for visible II-VI LEDs arise from the physical phenomenon of exciton self-trapping, which is inherent in Zn(Se,Te)-based II-VI materials. Earlier optical studies (91, 108) revealed that the incorporation of small concentrations of tellurium in ZnSe epilayers and quantum well structures has the dramatic effect of significantly modifying the emission wavelength of these materials to lower energies. In the case of tellurium-doped II-VI LEDs, excitonic binding with strong lattice relaxation is responsible for the large spectral shift in emission wavelength. The cooperative effect of two short-range interactions—the free-exciton-phonon interaction and the Te impurity-exciton binding interaction—contribute to binding the exciton and thus reducing the energy of light emission. The usefulness of deeply trapped, localized, radiatively efficient centers for a light emitter is that the electronic excitations are sheltered from extrinsic defects that compete with radiative recombination. Incorporation of Te within the quantum well of ZnSe-based LEDs enabled Hagerott and co-workers (109) to demonstrate both green and yellow LEDs. The heterostructure design that was selected for the investigation was virtually identical to that used for normal blue/green (Zn,Cd)Se/Zn(Se,S) quantum well LEDs with the exception that one or two monolayers of ZnTe were placed within the center of the (Zn,Cd)Se quantum well, and thus provided the isoelectronic Te center. As mentioned above, the controlled homogeneous incorporation of Te to form an alloy of Zn(Se,Te) is difficult because of the competition for incorporation of Te and Se species (91). Magneto-optical studies of similar quantum well structures containing (Zn,Mn)Se barriers determined that the hole wavefunction is strongly confined to the center of the well, whereas the electron wavefunction is much less affected (110). Therefore, the exciton is microscopically described as a self-trapped hole, with the electron coulombically bound to it. Any shallow donor sites in the vicinity of the Te centers will also be beneficial. Figure 9 is a photograph of LEDs fabricated with a (Zn,Cd)Se quantum well, with the incorporation of one monolayer of ZnTe to achieve green emission and two monolayers of ZnTe to achieve yellow-green emission. The dramatic difference in color that can be achieved by planar control of isoelectronic centers is clearly seen (109). Additional uses of Te-containing II-VI layered structures and their application to LEDs have been demonstrated (111–113); it is important to note that the LEDs take advantage of the close lattice-matching conditions and transparency of ZnSe bulk substrates. The device characteristics are impressive, with an output power of 792 μW (10 mA, 4 V) at a peak wavelength of 510 nm in the green spectral region; these performance characteristics correspond to an external efficiency of 2%, illustrating the advantages gained by use of a transparent substrate. Further developments include variations on the heterostructure configuration to vary the output wavelength, while focusing attention on defect generation and device lifetime (113).

Recently, other materials systems employing wide-bandgap II-VI heterostructures have been demonstrated as possible candidates for green LEDs. Faschinger et al (114) introduced MBE-grown (Zn,Mg)(Te,Se) as the active material, suggesting that with even a relatively small amount of Te incorporation, both the *p*-type doping and formation of ohmic contacts are facilitated. A very different approach to green LEDs has been proposed and reported (115, 116) wherein the design consists of a *n*-type CdSe/*p*-type ZnTe heterojunction, separated by a graded $Cd_{1-x}Mg_xSe$ layer. The graded $Cd_{1-x}Mg_xSe$ layer enables the injection of electrons, at low voltage, into the heavily doped *p*-type ZnTe layer to participate in radiative recombination events (115). The approach circumvents the difficulty of forming a ZnTe *pn* junction and utilizes bandgap engineering techniques to achieve the same goal; these approaches may be particularly useful for application to LEDs, which are not as constrained as laser diodes for device design.

Laser Device Design and Fabrication: Materials Issues

It is interesting to trace the evolution in the succession of structures that have been employed from the birth of II-VI laser diodes to the present. The first demonstration (117, 118) of a *pn* injection diode employed a configuration as shown in Figure 10*a*. The Zn(S,Se) ternary layers were introduced to serve as cladding layers for the optical waveguide region with the ZnSe layers and thus provide the electronic barriers for the (Zn,Cd)Se quantum well(s). One significant disadvantage of this structure involves the lattice-constant mismatch between the Zn(S,Se) and ZnSe layers. The Zn(S,Se) alloy composition was selected to provide a lattice-match to GaAs; unfortunately, the lattice-constant alloy composition mismatch between GaAs and ZnSe is translated to the interface between the cladding and waveguide layers. In a typical device, the ZnSe waveguide layer thickness will exceed the critical thickness (1500 Å) for the formation of misfit dislocations; hence such a structure is inherently non-pseudomorphic and will contain arrays of misfit dislocations. In an effort to prevent the formation of dislocations originating from lattice mismatch, and instead preserve a pseudomorphic structure, the laser diode structures shown in Figure 10*b*,*c* were fabricated (119). These structures incorporated several (typically six) quantum wells for the confinement of electrons and holes as well as photons; the slightly higher refractive index of the quantum well region provided some limited optical waveguiding. The structure of Figure 10*b* is essentially all ZnSe except for the six (Zn,Cd)Se quantum wells; the In mol fraction in the (In,Ga)As buffer layer on GaAs was

Figure 8 A numeric display device fabricated by patterning a (Zn,Cd)Se QW structure and including an indium tin oxide transparent top electrode (107).

Figure 9 Tellurium isoelectronic centers at work: The 75- Å- thick (Zn,Cd)Se QW region contains nominally one monolayer of ZnTe to produce green electroluminescence, whereas the introduction of two monolayers of ZnTe results in yellow-green electroluminescence (109).

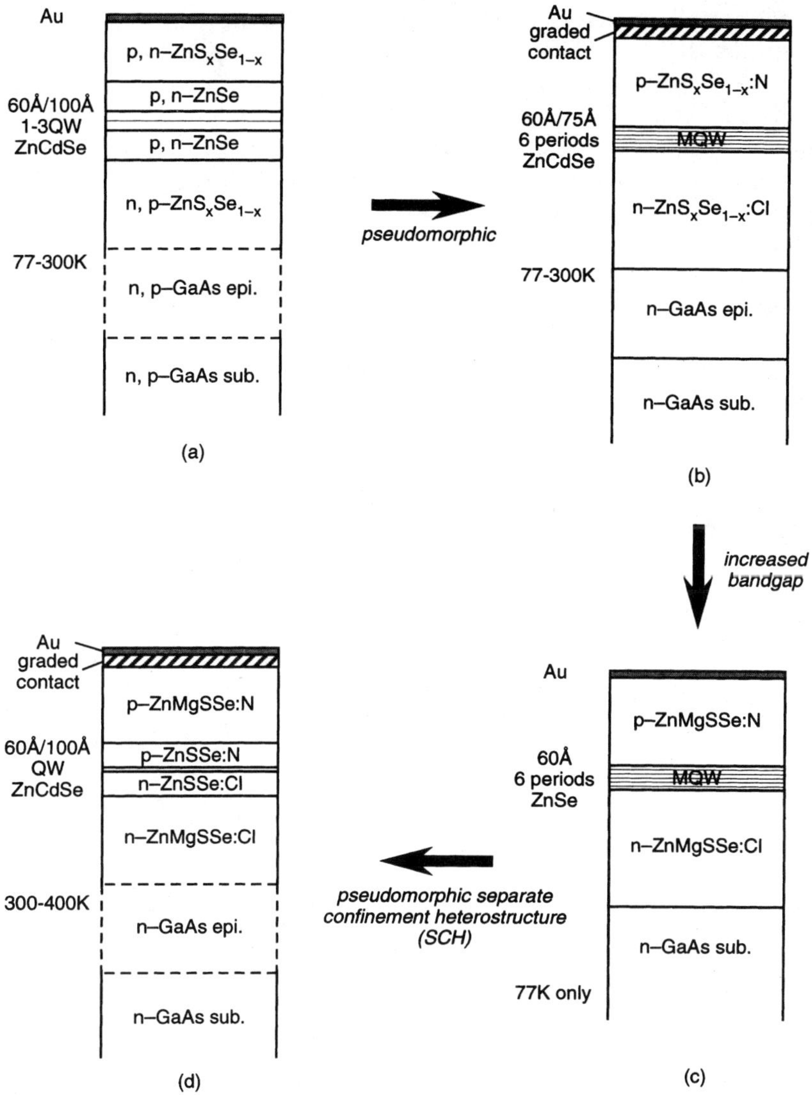

Figure 10 Schematic diagrams of key blue/green laser diode configurations and their evolution from the initial laser design to the current laser design. See text for description.

designed such that the 4-μm buffer layer relaxes to the lattice constant of ZnSe. The structure of Figure 10*c* replaces the ZnSe with the wider bandgap Zn(S,Se) ternary. Okuyama et al (120) reported a laser diode operating at 77 K that was similar in design to the structure in Figure 10*c*, except that the constituent layers were replaced by wider bandgap alloys, specifically the first use of the quaternary (Zn,Mg)(S,Se). The structure shown in Figure 10*d* includes the incorporation of Mg, which is advantageous because the increased energy bandgap results in photon emission at shorter wavelengths. Preservation of the pseudomorphic nature of the II-VI epilayer to the GaAs was enabled by selection of the different alloy compositions. The successful growth of the quaternary alloy encouraged several groups (121–123) to use the wide-bandgap quaternary material in the implementation of a separate confinement heterostructure design (SCH), as shown in Figure 11. Here, the optical waveguiding is enhanced by the wide-bandgap cladding layer; simultaneously the pseudomorphic structure is maintained. An additional feature of the structures under consideration is the quality of the heterovalent GaAs/Zn(S,Se) (or (In,Ga)As/Zn(S,Se)) interface through which hole (or electron) injection takes place. Earlier work (49, 51) indicated that the growth of a III-V epilayer on the GaAs substrate provided a considerable degree of control in the electrical and microstructural characteristics of the II-VI/III-V interface. Currently, the aggressive development of a useful laser device is at the focus of much research, and therefore small improvements in the structural quality of the II-VI structure when nucleated on a stoichiometrically engineered III-V buffer layer is a crucial factor.

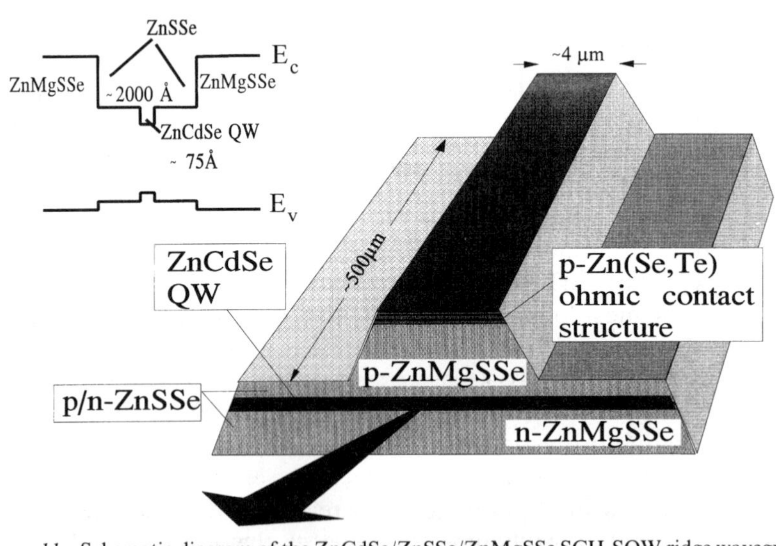

Figure 11 Schematic diagram of the ZnCdSe/ZnSSe/ZnMgSSe SCH-SQW ridge waveguide laser structure.

Pseudomorphic, Separate-Confinement Heterostructure Lasers

The focus of this section is the current performance features of pseudomorphic, separate confinement heterostructures, shown schematically in Figure 11. This particular structure was used to demonstrate room temperature, continuous-wave operation for a laser operating at 505–510 nm (124–126) with a device lifetime of several tens of seconds. Very recently, the lifetime has been extended to 1 h for the same structures (127). For all of these laser structures, the importance of a GaAs homoepitaxially grown buffer layer cannot be over-emphasized as a method to reduce crystalline defects at the II-VI/III-V interface. On the other hand, the issue of utilizing ZnSe substrates remains clearly advantageous, although unique fabrication difficulties remain to be overcome. An important element in the present continuous-wave (cw) devices is the application of the ZnTe/ZnSe-graded contact.

Early work by Okuyama et al (120) on the MBE growth of the quaternary (Zn,Mg)(Se,S) alloy indicated that these materials would prove crucial in improving the performance of ZnSe-based lasers. By utilizing the quaternary alloy in a ZnCdSe/ZnSSe/ZnMgSSe separate-confinement heterostructure (SCH), single quantum well (SQW) laser (121), the diode lasers were able to operate beyond room temperature under conditions of low duty cycle and very short pulse injection ($\sim$50 ns). In these gain-guided lasers, the room temperature injection current density was measured at $\sim$500 A cm^{-2} for devices of about 1 mm in length, which is an excellent figure of merit. Differential quantum efficiency of about 17% was also obtained. Although reliable values for the index of refraction for (Zn,Mg)(Se,S) are being measured, a plausible value for the electronic/optical confinement factor is about $\Gamma \sim 0.03$ for typical quantum well thicknesses of 75 Å and Zn(S,Se) optical guiding layers with thicknesses of $\sim$2000 Å.

Further improvement in the SCH-SQW laser device performance was realized by utilizing index-guided ridge waveguide (mesa) structures, which also featured the ZnTe/ZnSe graded contact. The use of a self-aligned process is the most logical fabrication approach, whereas the mesa itself is presently produced by various dry-etching techniques. Generally, the device processing of the II-VI lasers presents many challenges because the initial low-temperature ($\sim$300°C) MBE growth correspondingly requires low-temperature processing techniques. Futhermore, great care must be taken to minimize the introduction of mechanical defects during the handling of the devices. Figure 12 shows an example of the performance characteristics for three nominally identical laser devices, under room temperature cw conditions, configured as a ridge waveguide device ($\sim$4.2 μm

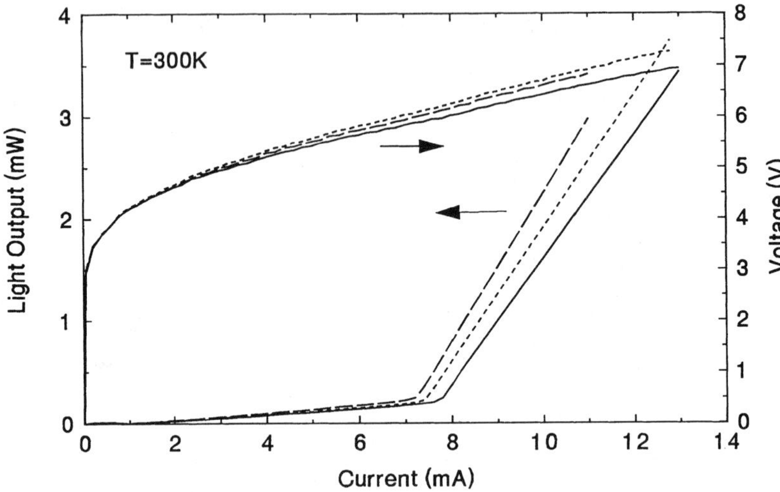

Figure 12 Optical output power vs current, and the voltage vs current traces of three room temperature (Zn,Cd)Se SQW ridge waveguide lasers under continuous-wave conditions. Reflective coatings were applied to reduce threshold current (126).

wide), and incorporating the ZnTe/ZnSe graded contact. With reflective facet coatings in these ~500 μm-long devices, laser threshold currents about 5 to 7 mA and 5.8 V have been measured. Single transverse mode characteristics have been obtained in this geometry for both single and multiple quantum well devices with average output powers of up to 10 mW. Differential quantum efficiencies of about 50% have also been reached. Excellent device characteristics have been reported (123) on pulsed SCH-SQW diode lasers in very narrow ($\leqslant 2$ μm wide) ridge waveguide structures using p^+ ZnSe contacts (86); well-defined transverse mode operation and low threshold currents of 20 mA are demonstrated. By adding high-reflectivity facet coatings, threshold current values as low as 2.5 mV for 300 μm-long devices are the lowest reported to date for a II-VI laser. However, at room temperature, the duty cycle was only several tens of percent, and the longevity of the devices was very short. The figures of merit for SCH-SQW diode laser performance are expected to develop rapidly, especially with regard to threshold voltage and device lifetime under continuous-wave operation, as the issue of defect generation is brought under control.

There have also been a number of incremental advances in shortening the wavelength of operation of the wide-bandgap II-VI diode lasers. The challenges for reaching the deep blue (wavelengths <450 nm) region are twofold within the sphere of the currently employed ZnSe-based semiconductors (see Figure 1). On the one hand, the demand for strong electronic and optical confinement in a pseudomorphic heterostructure necessitates the growth of the (Zn,Mg)(S,Se) quaternary with large Mg and S content (up to several tens of percent). This alloy range is largely unexplored thus far, and it is important to note that the binary endpoints consisting of MgSe and MgS do not crystallize in the zinc blende form as a bulk material. Alternative means of composing wide-bandgap ternaries and quaternaries from ZnSe, ZnTe, and ZnS also require pursuit. Second, as already shown above, the *p*-type doping of (Zn,Mg)(S,Se) becomes rapidly more difficult with increasing band gap due to the strongly nonhydrogenic deepening of the acceptor level. For these very wide-bandgap quaternary materials, electrical injection into a *pn* junction device requiring high current densities may not be feasible. Band structure engineering solutions may provide some needed flexibility in this instance.

Diode laser devices that operate in the blue wavelength range have been demonstrated (106; A Ishibashi, unpublished data). In particular, by employing a ZnSe quantum well with two (Zn,Mg)(S,Se) layers for electronic and optical confinement, pulsed room temperature operation of a separate-confinement heterostructure diode laser at $\lambda \sim 463$ nm has been demonstrated (106), with average powers in excess of 1 mW. In these devices, the Mg concentration was nominally 10%, whereas the sulfur concentration was 14%; in the outer optical cladding layer, the Mg concentration was 24%, with a sulfur concentration of 28%.

Although the lifetime of the cw SCH diode lasers is still very short, the devices are robust enough for room temperature characterization measurements. For example, measurements of the quantum well gain for the II-VI lasers have been practically nonexistent, especially in terms of quantitative spectroscopic precision. Recently results have appeared that employ the Hakki-Paoli method for the measurement of the gain spectra in the SCH diode lasers. Figure 13 shows the modal gain/loss spectra (α vs λ) determined for a range of different injection current levels (128). The threshold current for the particular device was 28 mA. The peak modal gain reaches the value of $\alpha = -38$ cm^{-1} at laser threshold. From ellipsometric measurements of the index of refraction for the constituent layers in the SCH device, a value for the optical confinement factor $\lambda \sim 0.02$–0.03 is calculated. Hence, the peak quantum well gain at threshold approaches the value $g \sim 1500$ cm^{-1} and beyond. The transparency condition ($\alpha = 0$) is reached at a current of approximately 5 mA. A current independent loss ($\gamma_{sc} \sim 6$ cm^{-1}) is attributed to scattering losses in the ridge waveguide structure.

In probing for the optical constants of a diode laser at room temperature over a broader spectral range, and specifically covering the $n = 1$ heavy-

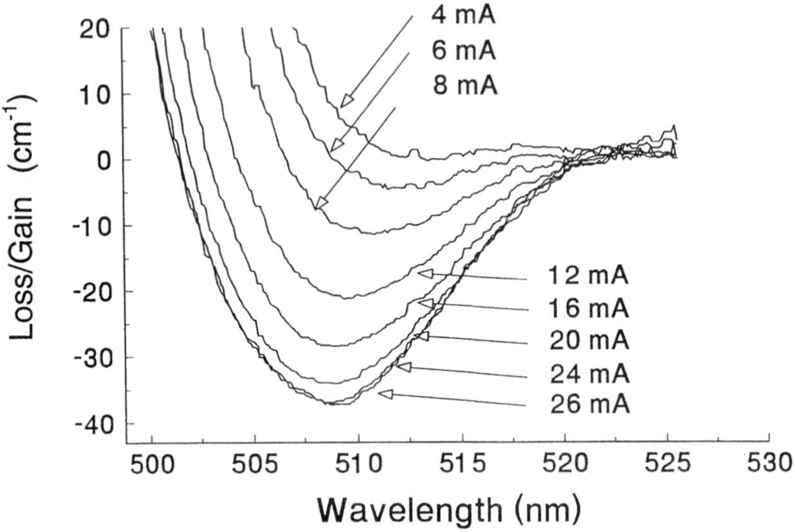

Figure 13 Gain spectra obtained for the SCH-SQW diode laser at 300 K for different levels of current injection by the Hakki-Paoli method (128).

hole quantum well exciton resonance, work has been performed recently that employs the fundamental and general relationships between spontaneous emission and gain/absorption spectra. The method was first introduced by Henry et al (129) for GaAs diode lasers. Figure 14 shows the gain/absorption spectra near the $n = 1$ interband transition of the active 75-Å thick (Zn,Cd)Se quantum well at 300 K (130). Under low current injection, both the $n = 1$ heavy-hole (HH) and light-hole (LH) exciton absorption peaks are well defined. With increasing injection, both absorption peaks begin to saturate due to many-body effects. At laser threshold (I $\sim$ 300 mA in gain-guided devices), the LH exciton peak remains well defined, and even the HH exciton resonance remains discernible, while supplying gain to its low energy side. Hence, it appears that a form of electron-hole pairwise coulomb correlation remains important in the diode laser under operating conditions. Such a correlation would result in an enhancement in the electron-hole wavefunction overlap when compared with a one-electron picture. The corresponding increase in the optical matrix element would decrease the electron-hole pair density (injection current) necessary for reaching lasing conditions. The presence of excitonic effects in the gain for strongly confined II-VI quantum wells has been demonstrated earlier in optical pumping experiments at cryogenic temperatures (130, 131); such pump-probe experiments continue to demon-

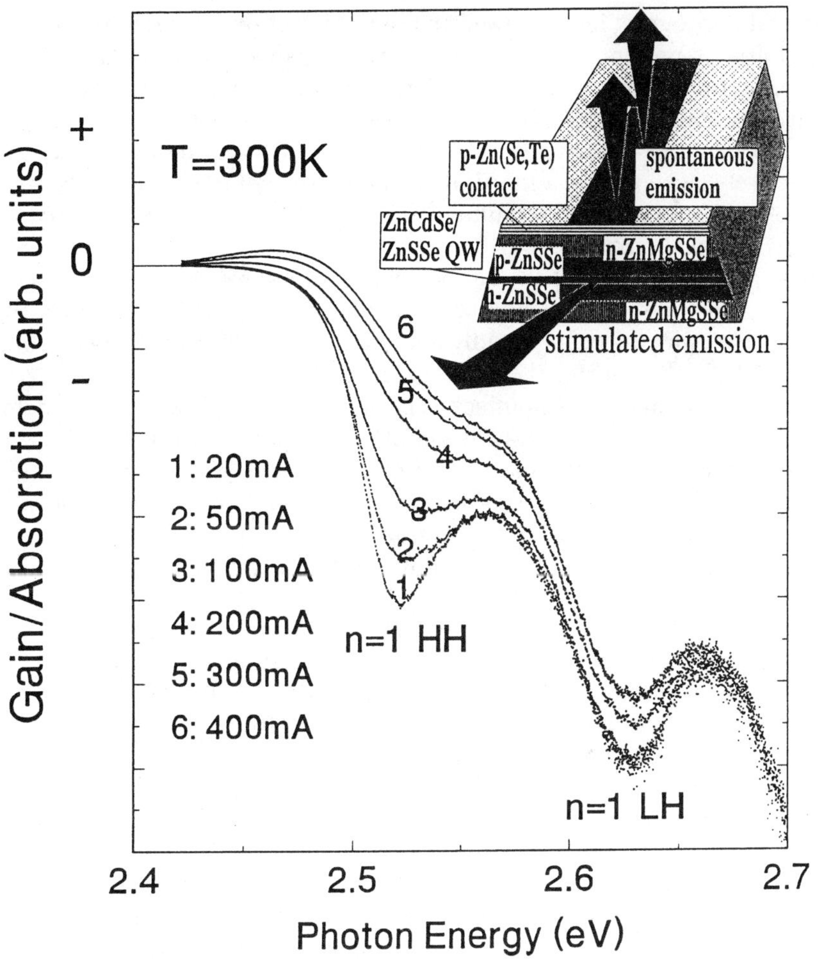

Figure 14 A wider spectral view of the gain/absorption spectrum of the same SCH device showing the effects of increasing injection on the HH and LH exciton states from below to above laser threshold. The inset displays the experimental geometry (105).

strate the importance of the coulomb correlations in active (Zn,Cd)Se quantum wells (132).

THE PROMISE AND FUTURE DIRECTIONS

Device Failure and Lifetime

The main barrier currently blocking the commercial exploitation of the wide-bandgap II-VI light emitters is their limited lifetime. Although considerable progress has followed the initial injection laser demonstrations of 1991, at present the longest-lived room temperature cw laser diodes have lifetimes ranging from 1 to 60 min. Room temperature cw threshold currents as low as 250 A cm^{-2} have been obtained, and threshold voltages can be slightly below 5 V. The devices are usually index-guided with reactively etched mesas of typically 4 μm in width; flip-chip heat sinking is employed, and SiO_2/TiO_2 multilayers are used for reflective facet coatings (126). Typical output power for room temperature cw operation is a few mW; pulsed operation provides powers of several hundred mW. It should be emphasized that these numbers are not particularly significant because new developments occur rapidly and continually. At present, two groups (133–136) have reported the development of dark-line defects using electroluminescence and photoluminescence spectroscopy. The resulting defect microstructure of the degraded samples was examined using TEM, with both cross-sectional and plan view imaging. The degradation phenomena bear a strong resemblance to phenomena that occurred during the early development of (Al,Ga)As injection laser devices. The presence of extended defects such as stacking faults and threading dislocations associated with stacking faults combine with point defects to produce a growth of dislocation networks within the active region. These dislocation networks act to reduce the optical gain by substituting nonradiative recombination and ultimately stop the lasing process. Therefore, a primary concern is the reduction of the density of point defects as well as the number of extended defects within the device. In dealing with these problems, one naturally considers the extent to which the defects are intrinsic to the material and/or the growth method. The thrust of much of the current research deals with this question.

The dislocation networks appear to develop within the quantum well(s), in the gain/injection current region, and nucleate at points where threading dislocations or stacking faults pass through the quantum well. The origin of the threading dislocations is not from lattice misfit because the structures are grown to be pseudomorphic, but instead is associated with stacking faults that arise at or near the II-VI/III-V heterovalent interface. It is significant that nucleation of the stacking faults is rarely observed in the regions above the quantum wells and that the stacking faults are most numerous (1×10^5 to 1×10^6 cm^{-2}) near to the II-VI/III-V interface. The tentative conclusion is that the generation of stacking faults, at least at these densities, is not intrinsic to the growth of wide-bandgap II-VI compounds, but is instead associated with some heteronucleation event that causes a phase uncertainty for stacking during the early stages of growth.

A natural question arises with regard to the point defects. Are the low temperatures used for the growth of the wide-bandgap II-VI compounds, typically ranging from 240 to 350°C, too far below the melting point of the compounds so that the elements impinging on the substrate are unable to incorporate at appropriate lattice sites without formation of vacancies, interstitials, and antisite defects? One insight regarding the density of point defects is provided by the temperature dependence of photoluminescence. It has not been unusual for the quantum efficiency to decrease by more than two orders of magnitude, often approaching 10^{-3}, as a sample is heated from cryogenic to room temperature. This decrease in quantum efficiency is usually attributed to the presence of thermally activated, nonradiative recombination centers. Such centers are usually associated with point defects or point defect complexes. The presence of point defect-related nonradiative recombination centers has implications both for internal quantum efficiency and for degradation. The TEM evidence suggests that the dislocation networks are developed by a mechanism of climb in which point defects play an important role. At the wide energy bandgaps inherent in visible light emitters, a significant amount of energy is released during a nonradiative recombination event (where the photon absorption is accompanied by a local multiphonon relaxation, which produces considerable local heating). Such nonradiative recombination events provide the driving force for motion of point defects and, hence, dislocation climb. The nonradiative event also can produce additional point defects.

In an effort to enhance the device lifetime, we devised a study in which the growth parameters such as substrate temperature and cation/anion flux ratio were varied in order to ascertain the optimum growth conditions needed to minimize the density of nonradiative recombination centers. The objective was to improve room temperature quantum efficiency while reducing the point defects available for dislocation climb within the active region of the laser. Definite improvements in quantum efficiency and reduction in point defects were realized. For example, the photoluminescence emission from a (Zn,Cd)Se quantum well decreases by only a factor of five from cryogenic to room temperature by the choice of appropriate growth conditions (see Figure 7). Because this decrease in quantum efficiency includes the (essentially unknown) effects of carrier confinement in the quantum well at room temperature, the implications are that the point defect density is significantly reduced by optimization of the growth parameters, and high density is not necessarily to be expected at the low growth temperatures.

Solution to the Problems?

When considering the actual technological prospects for the wide-bandgap II-VI semiconductor light emitters, it is important to underline the brief span during which device research has brought the pulsed, low-temperature ZnSe-based quantum well diode laser in 1991 to the demonstration of a room temperature cw laser in late 1993. Only about a year ago had the II-VI diode laser as a device progressed to the point that fundamentally important questions could be raised about its technological prospects, specifically in terms of lifetime, degradation, and reliability. Meanwhile, the lifetime of the cw (Zn,Cd)Se SCH diode has increased from a few seconds up to one hour, a pace of evolution that is quite rapid when compared with the earlier development of the GaAs diode laser. On the other hand, device operation on the scale of thousands of hours is required even for limited technological applications, such as in some optical storage and display devices. Parallel progress has also taken place in the blue/green LED area.

Any semiconductor laser is a demanding, active microelectronic device, both from the standpoint of required material quality, and device design and processing. Typically, current densities on the order of 1 kA cm^{-2} run through the multilayer structure of a volume on the order of 1000 μm^3. Crystalline defects of almost any kind generally act as sources of degradation, which limit the device lifetime and lead to gradual or precipitous failure. Wide-bandgap semiconductors are fundamentally susceptible to crystalline defects; e.g. point defects that typically contribute to nonradiative recombination and reduction of electrical activity due to the strong coupling of electronic (defect) states with the lattice. As discussed above, two parts of the device structure have been identified through recent research as points of weakness in the as-grown material. First, point defects in the strained (Zn,Cd)Se active quantum well region provide undesired nonradiative recombination and release highly local electronic energy into the lattice. Second, the II-VI/III-V heterovalent heterointerface can generate a specific form of a stacking fault-like defect, bounded by threading dislocations. The density of these defects in better quality II-VI structures grown on GaAs is presently about 10^5 cm^{-2}, which ensures the presence of at least a few such defects within the area of a diode laser device. Major current research is focusing on the control of these defects, which are viewed as extrinsic during epitaxial growth. For example, it is presently unclear what microscopic chemical forces drive the formation of the stacking faults in the growth of ZnSe on GaAs. Other III-V buffer layers are being researched, including phosphide-based compounds (104), and work is actively under way to improve the quality and (*n*-type) doping of bulk ZnSe for homoepitaxy. If the quality of ZnSe substrates can reach a quality comparable to that of GaAs, a range of new possibilities for diode lasers and LEDs will emerge.

Although a considerable number of II-VI heterostructures have been studied so far (chiefly for their optical properties), many opportunities exist in fundamental materials science related to new heterostructures for light-emitting devices. The requirements of electronic and optical confinement in a pseudomorphic heterostructure will continue to demand the presence of at least one quaternary semiconductor. Presently (Zn,Mg)(S,Se) and (Zn,Mg)(Se,Te) are the only such alloys introduced by epitaxial growth techniques. Useful alternatives, e.g. for Mg, among other alkaline earth elements may find their place in future devices.

Device design and processing are other important elements on the road toward technologically viable II-VI visible lasers. At this stage, device processing techniques have been developed largely by empirical means, and little systematic work exists to analyze the role, e.g. of fabrication-induced stresses, on the defect landscape. However, because the propagation of dislocations, slip, and motion of related mechanical defects occurs more readily in ZnSe than in GaAs, it is clear that any future fabrication and manufacturing schemes must address these issues. There remains much room for device innovation with the II-VI diode lasers. Although most attention has been directed to the edge-emitting configuration, other laser geometries offer viable future alternatives.

Among laser device configurations that are currently pursued, we mention here briefly those in which the optical resonator arrangement departs from the conventional edge emitter. By optical pumping methods, both the so-called whispering gallery mode (137) and vertical cavity surface-emitting laser (VCSEL) operation (138) have been demonstrated in ZnSe-based quantum well heterostructures. Although both configurations have been pursued in III-V semiconductor lasers for some time, there are particularly good reasons to develop such blue/green-emitting devices. For example, the optical confinement (overlap with the active quantum well medium) is relatively modest in the edge-emitting double heterostructures but can be significantly increased in the whispering gallery and VCSEL geometries. The gain coefficient is large in the II-VI lasers, so that with a high-Q resonator, one may reach threshold current densities on the order of $\sim$J 100 A cm^{-2} (and much lower in the so-called quantum microcavity limits). Attaining a low current density is, at least in the short run, highly desirable given the rapid increase in device degradation with current density. Yet another future direction in blue/green diode lasers involves quantum-confined heterostructures, such as quantum wires, where the singularities in the conduction-valence band states coupled with enhanced electron-hole coulomb interaction may yet provide the ultimate II-VI laser.

Literature Cited

1. Gunshor RL, Kukimoto H, Ruth RP, guest eds. 1993. Special issue. Wide-Bandgap II-VI Semiconductor Materials. *J. Electron. Mater.* 22: 429–550
2. *Proc. 6th Int. Conf. II-VI Compounds and Related Optoelectronic Materials, Newport, RI.* 1993. ed. RN Bhargava, RP Ruth, T Yao AV Nurmikko, 1-4: 1–1081. Amsterdam/London/New York/Tokyo: Elsevier
3. Ruda HE, ed. 1992. *Wide Gap II-VI Compounds for Optoelectronic Applications*, New York: Chapman & Hall. 415 pp.
4. Kolodziejski LA, guest ed. 1994. *J. Electron. Mater.* 23: 243–318
5. Kolodziejski LA, guest ed. 1995. *J. Electron. Mater.* 24: In press
6. Kolodziejski LA, Gunshor RL, Nurmikko AV, Otsuka N. 1995. In *Molecular Beam Epitaxy: Applications to Key Materials*, ed. RFC Farrow, pp. 344–452. Park Ridge, NJ: Noyes
7. Morkoc H, Strite S, Gao GB, Lin ME, Sverdlov B, Burns M. 1994. *J. Appl. Phys.* 76: 1363–98
8. Aven M, Prener JS, eds. 1967. *Physics and Chemistry of II-VI Compounds*, Amsterdam: North-Holland. 844 pp.
9. Hartmann H, Mach R, Selle B. 1982. In *Current Topics in Materials Science*, ed. E Kaldis, 9: 1–414. Amsterdam: North-Holland
10. Ringle MD, Grillo DC, Han J, Gunshor RL, Hua GC, et al. 1994. *Compositional control of (Zn,Mg)(ZnMg) (S,Se) epilayers grown by MBE for II-VI blue/green laser diodes*. Presented at 21st Int. Symp. Compound Semiconductors, San Diego
11. Morris BJ. 1986. *Appl. Phys. Lett.* 48: 865–67
12. Fujii S, Fujita Y, Iuchi T. 1988. *J. Cryst. Growth* 93: 750–54
13. Zinck JJ, Brewer PD, Jensen JE, Olson GL, Tutt LW. 1988. *Appl. Phys. Lett.* 52: 1434–36
14. Fujita Y, Terada T, Fujii S, Iuchi T. 1991. *J. Cryst. Growth* 107: 621–25
15. Bicknell RN, Giles NC, Schetzina JF. 1986. *Appl. Phys. Lett.* 49: 1095–97
16. Bicknell RN, Giles NC, Schetzina JF. 1986. *Appl. Phys. Lett.* 49: 1735–37
17. Fujita Sz, Tanabe A, Sakamoto T, Isemura M, Fujita Sg. 1988. *J. Cryst. Growth* 93: 259–64
18. Harper RL, Hwang S, Giles NC, Schetzina JF, Dreifus DL, Myers TH. 1989. *Appl. Phys. Lett.* 54: 170–72
19. Yoshikawa A, Okamota T, Fujimoto T, Onoue K, Yamaga S, Kasai H. 1990. *Jpn. J. Appl. Phys.* 29: L225–28
20. Gunshor RL, Kolodziejski LA, Nurmikko AV, Otsuka N. 1990. In *Semiconductors and SemiMetals*, ed. TP Pearsall, 33: 337–400. Boston: Academic
21. Simpson J, Adams SJA, Wallace JM, Prior KA, Cavenett BC. 1992. *Semicond. Sci. Technol.* 7: 460–63
22. Matsumura N, Fukada T, Saraie J. 1990. *J. Cryst. Growth* 101: 61–66
23. Coronado CA, Ho E, Kolodziejski LA, Huber CA. 1992. *Appl. Phys. Lett.* 61: 534–36
24. Nishizawa J, Abe H, Kurabayashi T, Sakurai N. 1986. *J. Vac. Sci. Technol.* A4: 706–10
25. Aoyagi Y, Kanazawa M, Doi A, Iwai S, Namba S. 1986. *J. Appl. Phys.* 60: 3131–35
26. Kukimoto H, Ban Y, Komatsu H, Takechi M, Ishizaki M. 1986. *J. Cryst. Growth* 77: 223–28
27. Bedair SM, Whisnant JK, Karam NH, Griffis D, El-Masry NA, Stadelmaier HH. 1986. *J. Cryst. Growth* 77: 229–34
28. Tu CW, Donnelly VM, Beggy JC, Baiocchi FA, McCrary VR, et al. 1988. *Appl. Phys. Lett.* 52: 968–70
29. Donnelly VM, Tu CW, Beggy JC, McCrary VR, Lamont MG, et al. 1988. *Appl. Phys. Lett.* 52: 1067–69
30. DenBaars SP, Dapkus PD. 1989. *J. Cryst. Growth* 98: 195–208
31. Donnelly VM, McCaulley JA. 1989. *Appl. Phys. Lett.* 54: 2458–60
32. Sugiura H, Yamada T, Iga R. 1990. *Jpn. J. Appl. Phys.* 29: L1–3
33. Nagata K, Iimura Y, Aoyagi Y, Namba S. 1990. *J. Cryst. Growth* 105: 52–56
34. Yamada T, Iga R, Sugiura H. 1992. *Appl. Phys. Lett.* 61: 2449–51

35. Takahashi T, Arakawa Y, Nishioka M, Ikoma T. 1992. *Appl. Phys. Lett.* 60: 68–70
36. Ho E, Coronado CA, Kolodziejski LA. 1993. *J. Electron. Mater.* 22: 473–78
37. Park RM, Mar HA, Salansky. 1985. *J. Vac. Sci. Technol.* B3: 1637–40
38. Yodo T, Koyama T, Yamashita K. 1988. *J. Cryst. Growth* 86: 273–78
39. Ohishi M, Ohmori K, Fujii Y, Saito H. 1988. *J. Cryst. Growth* 86: 324–28
40. Menda K, Takayasu I, Minato T, Kawashima M. 1988. *J. Cryst. Growth* 86: 342–47
41. Koyama T, Yodo T, Oka H, Yamashita K, Yamasaki T. 1988. *J. Cryst. Growth* 91: 639–46
42. Yodo T, Ueda K, Yamashita K. 1990. *J. Cryst. Growth* 99: 403–7
43. Ohkawa K, Ueno A, Mitsuyu T. 1992. *J. Cryst. Growth* 117: 375–84
44. Cantwell G, Harsch WC, Cotal HL, Markey BG, McKeever SWS, Thomas JE. 1992. *J. Appl. Phys.* 71: 2931–36
45. Cotal HL, Markey BG, McKeever SWS, Cantwell G, Harsch WC. 1993. *Phys. B* 185: 103–8
46. Jeon MH, Calhoun LC, Park RM. 1995. *J. Electron. Mater.* 24: In press
47. Gunshor RL, Kolodziejski LA, Melloch MR, Vaziri M, Choi C, Otsuka N. 1987. *Appl. Phys. Lett.* 50: 200–2
48. Qian QD, Qiu J, Melloch MR, Cooper JA, Kolodziejski LA, et al. 1989. *Appl. Phys. Lett.* 54: 1359–61
49. Qiu J, Menke DR, Kobayashi M, Gunshor RL, Qian QD, et al. 1990. *Appl. Phys. Lett.* 56: 1272–74
50. Kobayashi N. 1988. *Jpn. J. Appl. Phys.* 27: L1597–99
51. Qian QD, Qiu J, Kobayashi M, Gunshor RL, Melloch MR, Cooper JA. 1989. *J. Vac. Sci. Technol.* B7: 793–98
52. Li D, Gonsalves JM, Otsuka N, Qiu J, Kobayashi M, Gunshor RL. 1990. *Appl. Phys. Lett.* 57: 449–51
53. Yao T, Makita Y, Maekawa S. 1979. *Appl. Phys. Lett.* 35: 97–98
54. Yao T. 1985. In *The Technology and Physics of Molecular Beam Epitaxy*, ed. EHC Parker. pp. 313–43. New York: Plenum
55. Yoneda K, Yuji H, Toda T, Ishii H, Niina T. 1984. *Appl. Phys. Lett.* 45: 1300–2
56. Ohkawa K, Mitsuyu T, Yamazaki O. 1987. *J. Appl. Phys.* 62: 3216–21
57. Zhu Z, Takebayashi K, Yao T. 1993. *Jpn. J. Appl. Phys.* 32: 654–56
58. Fisher PA, Ho E, House JL, Petrich GS, Kolodziejski LA, et al. 1994. *Mater. Res. Soc. Symp. Proc.* 340: 451–56
59. Cheng H, DePuydt JM, Potts JE, Smith TL. 1988. *Appl. Phys. Lett.* 52: 147–49
60. Van de Walle CG, Laks DB, Neumark GF, Pantelides ST. 1993. *Phys. Rev. B* 47: 9425–34
61. Yao T. 1985. *J. Cryst. Growth* 72: 31–40
62. Shibli SM, Tamargo MC, Skromme BJ, Schwarz SA, Schwartz CL, et al. 1990. *J. Vac. Sci. Technol. B* 8: 187–91
63. Yao T, Okada Y. 1986. *Jpn. J. Appl. Phys.* 25: 821–27
64. Chadi DJ. 1991. *Appl. Phys. Lett.* 59: 3589–91
65. Park RM, Mar HA, Salanski NM. 1985. *J. Appl. Phys.* 58: 1047–49
66. Mitsuyu T, Ohkawa K, Yamazaki O. 1986. *Appl. Phys. Lett.* 49: 1348–50
67. Ohkawa K, Mitsuyu T, Yamazaki O. 1988. *J. Cryst. Growth* 86: 329–34
68. Park RM, Troffer MB, Rouleau CM, DePuydt JM, Haase MA. 1990. *Appl. Phys. Lett.* 57: 2127–29
69. Ohkawa K, Karasawa T, Mitsuyu T. 1991. *J. Cryst. Growth* 111: 797–801
70. Vaudo RP, Yu Z, Cook JW, Schetzina JF. 1993. *Opt. Lett.* 18: 1843–45
71. Ito S, Ikeda M, Akimoto K. 1992. *Jpn. J. Appl. Phys.* 31: L1316–18
72. Vaudo RP, Cook JW, Schetzina JF. 1994. *J. Vac. Sci. Technol. B* 12: 1232–35
73. Qiu J, DePuydt JM, Cheng H, Haase MA. 1991. *Appl. Phys. Lett.* 59: 2992–94
74. Chadi DJ, Troullier N. 1993. *Phys. B* 185: 128–31
75. Kwak KW, King-Smith RD, Vanderbilt D. 1993. *Phys. B* 185: 154–58
76. Hauksson IS, Simpson J, Wang SY, Prior KA, Cavenett BC. 1992. *Appl. Phys. Lett.* 61: 2208–10
77. Zhu A, Takebayashi K, Tanaka K, Ebisutani T, Kawamata J, Yao T. 1994. *Appl. Phys. Lett.* 64: 91–93
78. Pong C, Johnson NM, Street RA, Walker J, Feigelson RS, De Mattei RC. 1992. *Appl. Phys. Lett.* 61: 3026–28
79. Wolk JA, Ager JW, Duxstad KJ, Haller EE, Taskar NR, et al. 1993. *Appl. Phys. Lett.* 63: 2756–58
80. Kamata A, Mitsuhashi H, Fujita H. 1993. *Appl. Phys. Lett.* 63: 3353–55
81. Ho E, Fisher PA, House JL, Petrich GS, Kolodziejski LA, et al. 1995. *Appl. Phys. Lett.* 66: 1062–64
82. Fisher PA, Ho E, House JL, Petrich GS, Kolodziejski LA, et al. 1995. *J. Cryst. Growth*. In press
83. Fan Y, Han J, He L, Saraie J, Gunshor RL, et al. 1992. *Appl. Phys. Lett.* 61: 3160–62
84. Lansari Y, Ren J, Sneed B, Bowers KA,

Cook JW, Schetzina JF. 1992. *Appl. Phys. Lett.* 61: 2554–56
85. Yang Z, Schetzina JF. 1994. *J. Electron. Mater.* 23: 1071–74
86. Walker CT, DePuydt J, Haase M, Qiu J, Cheng H. 1993. *Phys. B* 185: 27–35
87. Onomura M, Ishikawa M, Nishikawa Y, Saito S, Parbrook PJ, et al. 1993. *Electron. Lett.* 29: 2114–15
88. Dandrea RG, Duke CB. 1994. *Appl. Phys. Lett.* 64: 2145–47
89. Mensz PM. 1994. *Appl. Phys. Lett.* 64: 2148–50
90. Han J, Stavrinides TS, Kobayashi M, Gunshor RL, Hagerott MM, Nurmikko AV. 1993. *Appl. Phys. Lett.* 62: 840–42
91. Kolodziejski LA, Gunshor RL, Fu Q, Lee D, Nurmikko AV, et al. 1988. *Appl. Phys. Lett.* 52: 1080–82
92. Robinson G. 1985. In *Physics and Chemistry of III-V Compound Semiconductor Interfaces*, ed. CW Wilmsen, p. 129. New York: Plenum
93. Fan Y, Han J, He L, Saraie J, Gunshor RL, et al. 1993. *Appl. Phys. Lett.* 63: 1812–14
94. Ozawa M, Hiei F, Takasu M, Ishibashi A, Akimoto K. 1994. *Appl. Phys. Lett.* 64: 1120–22
95. Kitagawa F, Mishima T, Takahashi K. 1981. *J. Electrochem. Soc.* 127: 937–43
96. Hishida Y, Ishii H, Toda T, Niina T. 1989. *J. Cryst. Growth* 95: 517–21
97. Jansen RW, Sankey OF. 1989. *Phys. Rev. B* 39: 3192–206
98. Laks DB, Van de Walle CG, Neumark GF, Pantelides ST. 1993. *Appl. Phys. Lett.* 63: 1375–77
99. Fan Y, Han J, He L, Gunshor RL, Brandt MS, et al. 1994. *Appl. Phys. Lett.* 65: 1001–3
100. Han J, Ringle MD, Fan Y, Gunshor RL, Nurmikko AV. 1994. *Appl. Phys. Lett.* 65: 3230–32
101. Malloy KJ, Khachaturyan K. 1993. In *Semiconductors and SemiMetals*, ed. RK Willardson, AC Beer, ER Weber, p. 235. Boston: Academic
102. Yang Z, Bowers KA, Ren J, Lansari Y, Cook JW, Schetzina JF. 1992. *Appl. Phys. Lett.* 61: 2671–73
103. Itaya K, Ishikawa M, Hatakoshi G. 1993. *Jpn. J. Appl. Phys.* 32: 1919–22
104. Ho E, House JL, Petrich GS, Kolodziejski LA. 1994. *The doping of ZnSe using gas source molecular beam epitaxy*. Presented at SPIE Meet. on Blue/Green Laser Diodes, Boston
105. Ding J, Hagerott M, Kelkar P, Nurmikko AV, Grillo DC, et al. 1994. *Phys. Rev. B* 50: 5787–90
106. Grillo DC, Han J, Ringle M, Hua G, Gunshor RL, et al. 1994. *Electron. Lett.* 25: 2131–32
107. Hagerott M, Jeon H, Nurmikko AV, Grillo DC, Kobayashi M, Gunshor RL. 1992. *Appl. Phys. Lett.* 60: 2825–27
108. Lee D, Mysyrowicz A, Nurmikko AV, Fitzpatrick BF. 1987. *Phys. Rev. Lett.* 58: 1475–78
109. Hagerott M, Ding J, Jeon H, Nurmikko AV, Fan Y, et al. 1993. *Appl. Phys. Lett.* 62: 2108–10
110. Fu Q, Lee D, Nurmikko AV, Gunshor RL, Kolodziejski LA. 1989. *Phys. Rev. B* 39: 3173–77
111. Ren J, Eason DB, Yu Z, Sneed B, Cook JW, Schetzina JF, et al. 1994. *J. Vac. Sci. Technol.* B12: 1262–65
112. Eason DB, Hughes WC, Ren J, Riegner M, Yu Z, et al. 1994. *Electron. Lett.* 30: 1178–80
113. Eason D, Ren J, Yu Z, Hughes C, El-Masry NA, et al. 1995. *J. Cryst. Growth.* In press
114. Faschinger W, Krump R, Brunthaler G, Fereira S, Sitter H. 1994. *Appl. Phys. Lett.* 65: 3215–17
115. Phillips MC, Wang MW, Swenberg JF, McCaldin JO, McGill TC. 1992. *Appl. Phys. Lett.* 61: 1962–64
116. Yu ET, Phillips MC, McCaldin JO, McGill TC. 1992. *Appl. Phys. Lett.* 61: 1962–64
117. Haase MA, Qiu J, DePuydt JM, Cheng H. 1991. *Appl. Phys. Lett.* 59: 1272–74
118. Jeon H, Ding J, Patterson W, Nurmikko AV, Xie W, et al. 1991. *Appl. Phys. Lett.* 59: 3619–21
119. Jeon H, Ding J, Nurmikko AV, Xie W, Grillo DC, et al. 1992. *Appl. Phys. Lett.* 60: 2045–47
120. Okuyama H, Miyajima T, Morinaga Y, Hiei F, Ozawa M, Akimoto K. 1992. *Electron. Lett.* 28: 1798–99
121. Gaines JM, Drenten RR, Haberern KW, Marshall T, Mensz P, Petruzzello J. 1993. *Appl. Phys. Lett.* 62: 2462–64
122. Grillo DC, Fan Y, Han J, He L, Gunshor RL, et al. 1993. *Appl. Phys. Lett.* 63: 2723–25
123. Haase MA, Baude JF, Hagedore MS, Qiu J, DePuydt J, et al. 1993. *Appl. Phys. Lett.* 63: 2315–17
124. Nakayama N, Itoh S, Nakano K, Okuyama H, Ozawa M, et al. 1993. *Electron. Lett.* 29: 1488–89
125. Nakayama N, Itoh S, Nakano K, Okuyama H, Ozawa M, et al. 1993. *Electron. Lett.* 29: 2194–95
126. Salokatve A, Jeon H, Ding J, Hovinen M, Nurmikko AV, et al. 1993. *Electron. Lett.* 29: 2192–93
127. Ishibashi A. 1994. *Blue/green laser diodes based on ZnMgSSe*. Presented at

SPIE Meet. on Blue/Green Laser Diodes, Boston

128. Kozlov V, Salokatve A, Nurmikko AV, Grillo DC, He L, et al. 1994. *Appl. Phys. Lett.* 65: 1863–65
129. Henry CH, Logan RA, Merritt FR. 1980. *J. Appl. Phys.* 51: 3042–50
130. Ding J, Ishihara T, Hagerott M, Jeon H, Nurmikko AV, et al. 1992. *Phys. Rev. Lett.* 60: 1707–10
131. Ding J, Ishihara T, Hagerott M, Jeon H, Nurmikko AV. 1993. *Phys. Rev. B* 47: 10528–37
132. Cingolani R, Calcagnile L, Rinaldi R, Prete P, Vanzetti L, et al. 1994. *Excitons and electron-hole plasma in ZnCdSe/ZnSe quantum wells.* Presented at SPIE Meet. on Blue/Green Laser Diodes, Boston
133. Guha S, DePuydt JM, Haase MA, Qiu J, Cheng H. 1993. *Appl. Phys. Lett.* 63: 3107–9
134. Gua GC, Otsuka N, Grillo DC, Fan Y, Han J, et al. 1994. *Appl. Phys. Lett.* 65: 1331–33
135. Guha S, Cheng H, Haase MA, DePuydt JM, Qiu J, et al. 1994. *Appl. Phys. Lett.* 65: 801–3
136. Hovinen M, Ding J, Salokatve A, Nurmikko AV, Hua GC, et al. 1995. *J. Appl. Phys.* In press
137. Hovinen M, Ding J, Nurmikko, AV, Grillo DC, Fan Y, et al. 1993. *Appl. Phys. Lett.* 63: 3128–30
138. Jeon H, Kozlov V, Kelkar P, Nurmikko AV, Grillo DC, et al. 1995. *Electron. Lett.* 25: In press

SUBJECT INDEX

CUMULATIVE INDEXES

CONTRIBUTING AUTHORS, VOLUMES 21–25

CHAPTER TITLES, VOLUMES 21–25

ANNUAL REVIEWS SERIES *Volumes not listed are no longer in print*		**Prices, postpaid, per volume. USA/other countries**	Regular Order Please send Volume(s):	Standing Order Begin with Volume:
❑ *Annual Review of*	**CELL AND DEVELOPMENTAL BIOLOGY**	(new title beginning with volume 11)		
Vols. 1-7	(1985-91)	$41 / $46		
Vols. 8-9	(1992-93)	$46 / $51		
Vol. 10-11	(1994 and Nov. 1995)	$49 / $54	Vol(s). ______	Vol. ______
❑ *Annual Review of*	**COMPUTER SCIENCE** (Series suspended)			
Vols. 1-2	(1986-87)	$41 / $46		
Vols. 3-4	(1988-89/90)	$47 / $52	Vol(s). ______	
Special package price for				
Vols. 1-4	(if ordered together)	$100 / $115	❑ Send all four volumes.	
❑ *Annual Review of*	**EARTH AND PLANETARY SCIENCES**			
Vols. 1-6, 8-19	(1973-78, 80-91)	$55 / $60		
Vols. 20-21	(1992-93)	$59 / $64		
Vol. 22-23	(1994 and May 1995)	$62 / $67	Vol(s). ______	Vol. ______
❑ *Annual Review of*	**ECOLOGY AND SYSTEMATICS**			
Vols. 2-12, 14-17, 19-22	(1971-81, 83-86, 88-91)	$40 / $45		
Vols. 23-24	(1992-93)	$44 / $49		
Vol. 25-26	(1994 and Nov. 1995)	$47 / $52	Vol(s). ______	Vol. ______
❑ *Annual Review of*	**ENERGY AND THE ENVIRONMENT**			
Vols. 1-16	(1976-91)	$64 / $69		
Vols. 17-18	(1992-93)	$68 / $73		
Vol. 19-20	(1994 and Oct. 1995)	$71 / $76	Vol(s). ______	Vol. ______
❑ *Annual Review of*	**ENTOMOLOGY**			
Vols. 10-16, 18, 20-36	(1965-71, 73, 75-91)	$40 / $45		
Vols. 37-38	(1992-93)	$44 / $49		
Vol. 39-40	(1994 and Jan. 1995)	$47 / $52	Vol(s). ______	Vol. ______
❑ *Annual Review of*	**FLUID MECHANICS**			
Vols. 2-4, 7	(1970-72, 75)			
9-11, 16-23	(1977-79, 84-91)	$40 / $45		
Vols. 24-25	(1992-93)	$44 / $49		
Vol. 26-27	(1994 and Jan. 1995)	$47 / $52	Vol(s). ______	Vol. ______
❑ *Annual Review of*	**GENETICS**			
Vols. 1-12, 14-25	(1967-78, 80-91)	$40 / $45		
Vols. 26-27	(1992-93)	$44 / $49		
Vol. 28-29	(1994 and Dec. 1995)	$47 / $52	Vol(s). ______	Vol. ______
❑ *Annual Review of*	**IMMUNOLOGY**			
Vols. 1-9	(1983-91)	$41 / $46		
Vols. 10-11	(1992-93)	$45 / $50		
Vol. 12-13	(1994 and April 1995)	$48 / $53	Vol(s). ______	Vol. ______
❑ *Annual Review of*	**MATERIALS SCIENCE**			
Vols. 1, 3-19	(1971, 73-89)	$68 / $73		
Vols. 20-23	(1990-93)	$72 / $77		
Vol. 24-25	(1994 and Aug. 1995)	$75 / $80	Vol(s). ______	Vol. ______
❑ *Annual Review of*	**MEDICINE: Selected Topics in the Clinical Sciences**			
Vols. 9, 11-15, 17-42	(1958, 60-64, 66-42)	$40 / $45		
Vols. 43-44	(1992-93)	$44 / $49		
Vol. 45-46	(1994 and April 1995)	$47 / $52	Vol(s). ______	Vol. ______